Igor A. Karnovsky · Olga Lebed

Advanced Methods of Structural Analysis

Igor A. Karnovsky
811 Northview Pl.
Coquitlam BC V3J 3R4
Canada

Olga Lebed
Condor Rebar Consultants, Inc.
300-1128 Hornby St.
Vancouver BC V6Z 2L4
Canada

ISBN 978-1-4899-8335-0 ISBN 978-1-4419-1047-9 (eBook)
DOI 10.1007/978-1-4419-1047-9
Springer New York Dordrecht Heidelberg London

Printed on acid-free paper

Springer is part of Springer Science+Business Media (www.springer.com)

Dedicated to
Tamara L'vovna Gorodetsky

Preface

Theory of the engineering structures is a fundamental science. Statements and methods of this science are widely used in different fields of engineering. Among them are the civil engineering, ship-building, aircraft, robotics, space structures, as well as numerous structures of special types and purposes – bridges, towers, etc. In recent years, even micromechanical devices become objects of structural analysis.

Theory of the engineering structures is alive and is a very vigorous science. This theory offers an engineer-designer a vast collection of classical methods of analysis of various types of structures. These methods contain in-depth fundamental ideas and, at the present time, they are developed with sufficient completeness and commonness, aligned in a well-composed system of conceptions, procedures, and algorithms, use modern mathematical techniques and are brought to elegant simplicity and perfection.

We now live in a computerized world. A role and influence of modern engineering software for analysis of structures cannot be overestimated. The modern computer programs allow providing different types of analysis for any sophisticated structure. As this takes place, what is the role of classical theory of structures with its in-depth ideas, prominent conceptions, methods, theorems, and principles? Knowing classical methods of Structural Analysis is necessary for any practical engineer. An engineer cannot rely only on the results provided by a computer. Computer is a great help in modeling different situations and speeding up the process of calculations, but it is the sole responsibility of an engineer to check the results obtained by a computer. If users of computer engineering software do not have sufficient knowledge of fundamentals of structural analysis and of understanding of physical theories and principal properties of structures, then he/she cannot check obtained numerical results and their correspondence to an adopted design diagram, as well as explain results obtained by a computer. Computer programs "... can make a good engineer better, but it can make a poor engineer more dangerous" (Cook R.D, Malkus D.S, Plesha M.E (1989) Concepts and applications of finite element analysis, 3rd edn. Wiley, New York). Only the knowledge of fundamental theory of structures allows to estimate and analyze numerical data obtained from a computer; predict the behavior of a structure as a result of changing a design diagram and parameters; design a structure which satisfies certain requirements; perform serious scientific analysis; and make valid theoretical generalizations. No matter how sophisticated the structural model is, no matter how effective the numerical algorithms are, no matter how powerful the computers are that implement these algorithms, it is the engineer who analyzes the end result produced from these algorithms. Only an individual who has a deep knowledge and understanding of the structural model and analysis techniques can produce a qualitative analysis.

In 1970, one of the authors of this book was a professor at a structural engineering university in Ukraine. At that time computers were started to be implemented in all fields of science, structural analysis being one of them. We, the professors and instructors, were facing a serious methodical dilemma: given the new technologies, how to properly teach the students? Would we first give students a strong basis in classical structural analysis and then introduce them to the related software, or would we directly dive into the software after giving the student a relatively insignificant introduction to classical analysis. We did not know the optimal way for solving this problem. On this subject we have conducted seminars and discussions on a regular basis. We have used these two main teaching models, and many different variations of them. The result was somewhat surprising. The students who were first given a strong foundation in structural analysis quickly learned how to use the computer software, and were able to give a good qualitative analysis of the results. The students who were given a brief introduction to structural analysis and a strong emphasis on the computer software, at the end were not able to provide qualitative results of the analysis. The interesting thing is that the students themselves were criticizing the later teaching strategy.

Therefore, our vision of teaching structural analysis is as follows: on the first step, it is necessary to learn analytical methods, perform detailed analysis of different structures *by hand* in order to feel the behavior of structures, and correlate their behavior with obtained results; the second step is a computer application of engineering software.

Authors wrote the book on the basis of their many years of experience of teaching the Structural Analysis at the universities for graduate and postgraduate students as well as on the basis of their experience in consulting companies.

This book is written for students of universities and colleges pursuing Civil or Structural Engineering Programs, instructors of Structural Analysis, and engineers and designers of different structures of modern engineering.

The objective of the book is to help a reader to develop an understanding of the ideas and methods of structural analysis and to teach a reader to estimate and explain numerical results obtained by hand; this is a fundamental stone for preparation of reader for numerical analysis of structures and for use of engineering software with full understanding.

The textbook offers the reader the fundamental theoretical concepts of Structural Analysis, classical analytical methods, algorithms of their application, comparison of different methods, and a vast collection of distinctive problems with their detailed solution, explanation, analysis, and discussion of results; many of the problems have a complex character. Considered examples demonstrate features of structures, their behavior, and peculiarities of applied methods. Solution of all the problems is brought to final formula or number.

Analyses of the following structures are considered: statically determinate and indeterminate multispan beams, arches, trusses, and frames. These structures are subjected to fixed and moving loads, changes of temperature, settlement of supports, and errors of fabrication. Also the cables are considered in detail.

In many cases, same structure under different external actions is analyzed. It allows the reader to be concentrated on one design diagram and perform complex analysis of behavior of a structure.

In many cases, same structure is analyzed by different methods or by one method in different forms (for example, Displacement method in canonical, and matrix forms). It allows to perform comparison analysis of applied methods and see advantages and disadvantages of different methods.

Distribution of Material in the Book

This book contains introduction, three parts (14 chapters), and appendix.

Introduction provides the subject and purposes of Structural Analysis, principal concepts, assumptions, and fundamental approaches.

Part 1 (Chaps. 1–6) is devoted to analysis of statically determinate structures. Among them are multispan beams, arches, trusses, cables, and frames. Construction of influence lines and their application are discussed with great details. Also this part contains analytical methods of computation of displacement of deformable structures, subjected to different actions. Among them are variety loads, change of temperature, and settlements of supports.

Part 2 (Chaps. 7–11) is focused on analysis of statically indeterminate structures using the fundamental methods. Among them are the force and displacement methods (both methods are presented in canonical form), as well as the mixed method. Also the influence line method (on the basis of force and displacement methods) is presented. Analysis of continuous beams, arches, trusses, and frames is considered in detail.

Chapter 11 is devoted to matrix stiffness method which is realized in the modern engineering software. Usually, the physical meaning of all matrix procedures presents serious difficulties for students. Comparison of numerical procedures obtained by canonical equations and their matrix presentations, which are applied to the same structure, allows trace and understands meaning of each stage of matrix analysis. This method is applied for fixed loads, settlement of supports, temperature changes, and construction of influence lines.

Part 3 (Chaps. 12–14) contains three important topics of structural analysis. They are plastic behavior of structures, stability of elastic structures with finite and infinite number of degrees of freedom, including analysis of structures on the basis of the deformable design diagram (P–Δ analysis), and the free vibration analysis.

Each chapter contains problems for self-study. Answers are presented to all problems.

Appendix contains the fundamental tabulated data.

Authors will appreciate comments and suggestions to improve the current edition. All constructive criticism will be accepted with gratitude.

Coquitlam, Canada *Igor A. Karnovsky*
Vancouver, Canada *Olga I. Lebed*

Acknowledgments

We would like to express our gratitude to everyone who shared with us their thoughts and ideas that contributed toward the development of our book.

We thank the members of the Springer team: specifically Steven Elliot (Senior Editor) thanks to whom this book became a reality, as well as, Andrew Leigh (Editorial Assistant), Kaliyan Prema (Project Manager), and many other associates who took part in preparing and publishing our book.

We wish to express our great appreciation to the following professors for their help and support during the process of writing our book:

Isaac Elishakoff (Florida Atlantic University, USA)

Luis A. Godoy (University of Puerto Rico at Mayaguez)

Igor V. Andrianov (Institute of General Mechanics, Germany)

Petros Komodromos (University of Cyprus, Greece)

One of the authors (I.A.K) is grateful to Dr. Vladimir D. Shaykevich (Civil Engineering University, Ukraine) for very useful discussions of several topics in structural mechanics.

We are especially grateful to Dr. Gregory Hutchinson (Project Manager, Condor Rebar Consultants, Vancouver) for the exceptionally useful commentary regarding the presentation and marketing of the material.

We would like to thank Dr. Terje Haukaas (University of British Columbia, Canada) and Lev Bulkovshtein, P.Eng. (Toronto, Canada) for providing crucial remarks regarding different sections of our book.

We thank the management of Condor Rebar Consultants (Canada) Murray Lount, Dick Birley, Greg Birley and Shaun de Villiers for the valuable discussions related to the construction of the special structures.

We greatly appreciate the team of SOFTEK S-Frame Corporation (Vancouver, Canada) for allowing us to use their extremely effective S-Frame Software that helped us validate many calculations. Special thanks go to George Casoli (President) and John Ng (Vice-President) for their consistent attention to our work.

Our special thanks go to David Anderson (Genify.com Corporation) for useful discussions related to the use of computer software in the teaching of fundamental subjects.

We would like to express our gratitude to Evgeniy Lebed (University of British Columbia, Canada) for assisting us with many numerical calculations and validations of results.

Particular appreciation goes to Sergey Nartovich (Condor Rebar Consultants, Vancouver), whose frequent controversial statements raised hell and initiated spirited discussions.

We are very grateful to Kristina Lebed and Paul Babstock, for their assistance with the proofreading of our book.

Our special gratitude goes to all members of our families for all their encouragement, patience, and support of our work.

Contents

Introduction xxi

Part I Statically Determinate Structures

1 Kinematical Analysis of Structures 3
1.1 Classification of Structures by Kinematical Viewpoint 3
1.2 Generation of Geometrically Unchangeable Structures 5
1.3 Analytical Criteria of the Instantaneously Changeable Structures 7
1.4 Degrees of Freedom 11
Problems 13

2 General Theory of Influence Lines 15
2.1 Analytical Method for Construction of Influence Lines 15
2.1.1 Influence Lines for Reactions 16
2.1.2 Influence Lines for Internal Forces 20
2.2 Application of Influence Lines for Fixed and Moving Loads 27
2.2.1 Fixed Loads 27
2.2.2 Moving Loads 30
2.3 Indirect Load Application 33
2.4 Combining of Fixed and Moving Load Approaches 35
2.5 Properties of Influence Lines 36
Problems 37

3 Multispan Beams and Trusses 39
3.1 Multispan Statically Determinate Beams 39
3.1.1 Generation of Multispan Statically Determinate Hinged Beams 39
3.1.2 Interaction Schemes and Load Path 40

3.1.3 Influence Lines for Multispan Hinged Beams 42
3.1.4 Summary 45
3.2 The Generation of Statically Determinate Trusses 47
3.2.1 Simple Trusses 47
3.2.2 Compound Trusses 48
3.2.3 Complex Trusses 49
3.3 Simple Trusses 49
3.4 Trusses with Subdivided Panels 54
3.4.1 Main and Auxiliary Trusses and Load Path 55
3.4.2 Baltimore and Subdivided Warren Trusses 57
3.5 Special Types of Trusses 61
3.5.1 Three-Hinged Trusses 61
3.5.2 Trusses with a Hinged Chain 64
3.5.3 Complex Trusses 68
3.5.4 Summary 70
Problems 72

4 Three-Hinged Arches 77
4.1 Preliminary Remarks 77
4.1.1 Design Diagram of Three-Hinged Arch 77
4.1.2 Peculiarities of the Arches 78
4.1.3 Geometric Parameters of Circular and Parabolic Arches 79
4.2 Internal Forces 80
4.3 Influence Lines for Reactions and Internal Forces 86
4.3.1 Influence Lines for Reactions 88
4.3.2 Influence Lines for Internal Forces 88
4.3.3 Application of Influence Lines 92
4.4 Nil Point Method for Construction of Influence Lines 94
4.4.1 Bending Moment 94
4.4.2 Shear Force 95
4.4.3 Axial Force 96
4.5 Special Types of Arches 97
4.5.1 Askew Arch 97
4.5.2 Parabolic Arch with Complex Tie 100
Problems 103

5 Cables 109
5.1 Preliminary Remarks 109
5.1.1 Direct and Inverse Problems 110
5.1.2 Fundamental Relationships 111
5.2 Cable with Neglected Self-Weight 113
5.2.1 Cables Subjected to Concentrated Load 113

5.2.2 Cable Subjected to Uniformly Distributed Load 116
5.3 Effect of Arbitrary Load on the Thrust and Sag 122
5.4 Cable with Self-Weight 125
5.4.1 Fundamental Relationships 125
5.4.2 Cable with Supports Located at the Same Level 127
5.4.3 Cable with Supports Located on the Different Elevations 130
5.5 Comparison of Parabolic and Catenary Cables 135
5.6 Effect of Axial Stiffness 137
5.6.1 Elastic Cable with Concentrated Load 137
5.6.2 Elastic Cable with Uniformly Distributed Load 139
Problems 140

6 Deflections of Elastic Structures 145
6.1 Introduction 145
6.2 Initial Parameters Method 147
6.3 Maxwell–Mohr Method 159
6.3.1 Deflections Due to Fixed Loads 159
6.3.2 Deflections Due to Change of Temperature 165
6.3.3 Summary 170
6.4 Displacement Due to Settlement of Supports and Errors of Fabrication 170
6.5 Graph Multiplication Method 176
6.6 Elastic Loads Method 185
6.7 Reciprocal Theorems 189
6.7.1 Theorem of Reciprocal Works (Betti Theorem) 189
6.7.2 Theorem of Reciprocal Unit Displacements (Maxwell Theorem) 190
6.7.3 Theorem of Reciprocal Unit Reactions (Rayleigh First Theorem) 192
6.7.4 Theorem of Reciprocal Unit Displacements and Reactions (Rayleigh Second Theorem) 193
6.7.5 Summary 193
Problems 195

Part II Statically Indeterminate Structures

7 The Force Method 211
7.1 Fundamental Idea of the Force Method 211
7.1.1 Degree of Redundancy, Primary Unknowns and Primary System 211
7.1.2 Compatibility Equation in Simplest Case 214

7.2 Canonical Equations of Force Method 217
7.2.1 The Concept of Unit Displacements 217
7.2.2 Calculation of Coefficients and Free Terms of Canonical Equations 219
7.3 Analysis of Statically Indeterminate Structures 222
7.3.1 Continuous Beams 222
7.3.2 Analysis of Statically Indeterminate Frames 224
7.3.3 Analysis of Statically Indeterminate Trusses 233
7.3.4 Analysis of Statically Indeterminate Arches 237
7.4 Computation of Deflections of Redundant Structures 243
7.5 Settlements of Supports 246
7.6 Temperature Changes 251
Problems 259

8 The Displacement Method 271
8.1 Fundamental Idea of the Displacement Method 271
8.1.1 Kinematical Indeterminacy 272
8.1.2 Primary System and Primary Unknowns 274
8.1.3 Compatibility Equation. Concept of Unit Reaction 275
8.2 Canonical Equations of Displacement Method 276
8.2.1 Compatibility Equations in General Case 276
8.2.2 Calculation of Unit Reactions 277
8.2.3 Properties of Unit Reactions 279
8.2.4 Procedure for Analysis 280
8.3 Comparison of the Force and Displacement Methods 291
8.3.1 Properties of Canonical Equations 292
8.4 Sidesway Frames with Absolutely Rigid Crossbars 294
8.5 Special Types of Exposures 296
8.5.1 Settlements of Supports 296
8.5.2 Errors of Fabrication 300
8.6 Analysis of Symmetrical Structures 302
8.6.1 Symmetrical and Antisymmetrical Loading 302
8.6.2 Concept of Half-Structure 303
Problems 305

9 Mixed Method 313
9.1 Fundamental Idea of the Mixed Method 313
9.1.1 Mixed Indeterminacy and Primary Unknowns 313
9.1.2 Primary System 314
9.2 Canonical Equations of the Mixed Method 316
9.2.1 The Matter of Unit Coefficients and Canonical Equations 316

9.2.2 Calculation of Coefficients and Free Terms 317
9.2.3 Computation of Internal Forces 318
Problems 319

10 Influence Lines Method 323
10.1 Construction of Influence Lines by the Force Method 323
10.1.1 Continuous Beams 325
10.1.2 Hingeless Nonuniform Arches 331
10.1.3 Statically Indeterminate Trusses 339
10.2 Construction of Influence Lines by the Displacement Method 344
10.2.1 Continuous Beams 346
10.2.2 Redundant Frames 353
10.3 Comparison of the Force and Displacements Methods 355
10.4 Kinematical Method for Construction of Influence Lines 358
Problems 364

11 Matrix Stiffness Method 369
11.1 Basic Idea and Concepts 369
11.1.1 Finite Elements 370
11.1.2 Global and Local Coordinate Systems 370
11.1.3 Displacements of Joints and Degrees of Freedom 371
11.2 Ancillary Diagrams 372
11.2.1 Joint-Load (J-L) Diagram 372
11.2.2 Displacement-Load (Z-P) Diagram 376
11.2.3 Internal Forces-Deformation (S-e) Diagram 377
11.3 Initial Matrices 379
11.3.1 Vector of External Joint Loads 379
11.3.2 Vector of Internal Unknown Forces 380
11.4 Resolving Equations 381
11.4.1 Static Equations and Static Matrix 381
11.4.2 Geometrical Equations and Deformation Matrix 386
11.4.3 Physical Equations and Stiffness Matrix in Local Coordinates 387
11.5 Set of Formulas and Procedure for Analysis 390
11.5.1 Stiffness Matrix in Global Coordinates 390
11.5.2 Unknown Displacements and Internal Forces 391
11.5.3 Matrix Procedures 392

11.6 Analysis of Continuous Beams 393
11.7 Analysis of Redundant Frames 404
11.8 Analysis of Statically Indeterminate Trusses 410
11.9 Summary 414
Problems 415

Part III Special Topics

12 Plastic Behavior of Structures 423
12.1 Idealized Stress–Strain Diagrams 423
12.2 Direct Method of Plastic Analysis 427
12.3 Fundamental Methods of Plastic Analysis 430
12.3.1 Kinematical Method 430
12.3.2 Static Method 430
12.4 Limit Plastic Analysis of Continuous Beams 432
12.4.1 Static Method 433
12.4.2 Kinematical Method 435
12.5 Limit Plastic Analysis of Frames 441
12.5.1 Beam Failure 442
12.5.2 Sidesway Failure 444
12.5.3 Combined Failure 444
12.5.4 Limit Combination Diagram 444
Problems 445

13 Stability of Elastic Systems 449
13.1 Fundamental Concepts 449
13.2 Stability of Structures with Finite Number Degrees of Freedom 453
13.2.1 Structures with One Degree of Freedom 453
13.2.2 Structures with Two or More Degrees of Freedom 458
13.3 Stability of Columns with Rigid and Elastic Supports 461
13.3.1 The Double Integration Method 461
13.3.2 Initial Parameters Method 466
13.4 Stability of Continuous Beams and Frames 471
13.4.1 Unit Reactions of the Beam-Columns 471
13.4.2 Displacement Method 473
13.4.3 Modified Approach of the Displacement Method 481
13.5 Stability of Arches 483
13.5.1 Circular Arches Under Hydrostatic Load 484
13.5.2 Complex Arched Structure: Arch with Elastic Supports 490

13.6 Compressed Rods with Lateral Loading 491
13.6.1 Double Integration Method 492
13.6.2 Initial Parameters Method 494
13.6.3 P-Delta Analysis of the Frames 499
13.6.4 Graph Multiplication Method for Beam-Columns 502
Problems 504

14 Dynamics of Elastic Systems 513
14.1 Fundamental Concepts 513
14.1.1 Kinematics of Vibrating Processes 513
14.1.2 Forces Which Arise at Vibrations 513
14.1.3 Degrees of Freedom 515
14.1.4 Purpose of Structural Dynamics 519
14.1.5 Assumptions 519
14.2 Free Vibrations of Systems with Finite Number Degrees of Freedom: Force Method 520
14.2.1 Differential Equations of Free Vibration in Displacements 520
14.2.2 Frequency Equation 521
14.2.3 Mode Shapes Vibration and Modal Matrix 522
14.3 Free Vibrations of Systems with Finite Number Degrees of Freedom: Displacement Method 530
14.3.1 Differential Equations of Free Vibration in Reactions 530
14.3.2 Frequency Equation 532
14.3.3 Mode Shape Vibrations and Modal Matrix 532
14.3.4 Comparison of the Force and Displacement Methods 538
14.4 Free Vibrations of One-Span Beams with Uniformly Distributed Mass 538
14.4.1 Differential Equation of Transversal Vibration of the Beam 540
14.4.2 Fourier Method 541
14.4.3 Krylov–Duncan Method 543
Problems 546

Appendix 551

Bibliography 587

Index 589

13.6 Comparison of [illegible] with [illegible] Methods [illegible]
13.6.1 [illegible] Method [illegible]
13.6.2 [illegible] Parameter Method [illegible]
13.6.3 [illegible]
13.6.4 [illegible] Method [illegible]
[illegible] Rigid [illegible]
Problems [illegible]

14 Dynamics of [illegible] [illegible]
14.1 [illegible]
14.1.1 [illegible]
14.1.2 [illegible]
14.1.3 [illegible]
14.1.4 [illegible]
14.1.5 [illegible]
14.2 [illegible] Degree-of-Freedom [illegible] Method [illegible]
14.2.1 [illegible]
14.2.2 [illegible]
14.2.3 [illegible]
14.3 [illegible] with [illegible]
Degree-of-Freedom [illegible]
14.3.1 [illegible]
14.3.2 [illegible]
14.3.3 [illegible]
14.3.4 [illegible]
14.4 [illegible]
14.4.1 [illegible]
14.4.2 [illegible]
14.4.3 [illegible]
Problems [illegible]

Appendix [illegible]

Bibliography [illegible]

Index [illegible]

Introduction

The subject and purposes of the Theory of Structures in the broad sense is the branch of applied engineering that deals with the methods of analysis of structures of different types and purpose subjected to arbitrary types of external exposures. Analysis of a structure implies its investigation from the viewpoint of its strength, stiffness, stability, and vibration.

The purpose of analysis of a structure from a viewpoint of its *strength* is determining internal forces, which arise in all members of a structure as a result of external exposures. These internal forces produce stresses; the *strength* of each member of a structure will be provided if their stresses are less than or equal to permissible ones.

The purpose of analysis of a structure from a viewpoint of its *stiffness* is determination of the displacements of specified points of a structure as a result of external exposures. The *stiffness* of a structure will be provided if its *displacements* are less than or equal to permissible ones.

The purpose of analysis of *stability* of a structure is to determine the *loads* on a structure, which leads to the appearance of *new forms of equilibrium*. These forms of equilibrium usually lead to collapse of a structure and corresponding loads are referred as *critical* ones. The stability of a structure will be provided if acting loads are less than the critical ones.

The purpose of analysis of a structure from a viewpoint of its vibration is to determine the frequencies and corresponding shapes of the vibration. These data are necessary for analysis of the forced vibration caused by arbitrary loads.

The Theory of Structures is fundamental science and presents the rigorous treatment for each group of analysis. In special cases, all results may be obtained in the close analytical form. In other cases, the required results may be obtained only numerically. However, in all cases algorithms for analysis are well defined.

The part of the Theory of Structures which allows obtaining the analytical results is called the classical Structural Analysis. In the narrow sense, the purpose of the classical Structural Analysis is to establish relationships between external exposures and corresponding internal forces and displacements.

Types of Analysis

Analysis of any structure may be performed based on some assumptions. These assumptions reflect the purpose and features of the structure, type of loads and operating conditions, properties of materials, etc. In whole, structural analysis may be divided into three large principal groups. They are static analysis, stability, and vibration analysis.

Static analysis presumes that the loads act without any dynamical effects. Moving loads imply that only the position of the load is variable. Static analysis combines the analysis of a structure from a viewpoint of its strength and stiffness.

Static linear analysis (SLA). The purpose of this analysis is to determine the internal forces and displacements due to time-independent loading conditions. This analysis is based on following conditions:

1. Material of a structure obeys Hook's law.
2. Displacements of a structure are small.
3. All constraints are two-sided – it means that if constraint prevents displacement in some direction then this constraint prevents displacement in the opposite direction as well.
4. Parameters of a structure do not change under loading.

Nonlinear static analysis. The purpose of this analysis is to determine the displacements and internal forces due to time-independent loading conditions, as if a structure is nonlinear. There are different types of nonlinearities. They are physical (material of a structure does not obey Hook's law), geometrical (displacements of a structure are large), structural (structure with gap or constraints are one-sided, etc.), and mixed nonlinearity.

Stability analysis deals with structures which are subjected to compressed time-independent forces.

Buckling analysis. The purpose of this analysis is to determine the critical load (or critical loads factor) and corresponding buckling mode shapes.

P-delta analysis. For tall and flexible structures, the transversal displacements may become significant. Therefore we should take into account the additional bending moments due by axial compressed loads P on the displacements caused by the lateral loads. In this case, we say that a structural analysis is performed on the basis of the *deformed* design diagram.

Dynamical analysis means that the structures are subjected to time-dependent loads, the shock and seismic loads, as well as moving loads with taking into account the dynamical effects.

Free-vibration analysis (FVA). The purpose of this analysis is to determine the natural frequencies (eigenvalues) and corresponding mode shapes (eigenfunctions) of vibration. This information is necessary for dynamical analysis of any structure subjected to arbitrary dynamic load, especially for seismic analysis. FVA may be considered for linear and nonlinear structures.

Stressed free-vibration analysis. The purpose of this analysis is to determine the eigenvalues and corresponding eigenfunctions of a structure, which is subjected to additional axial time-independent forces.

Time-history analysis. The purpose of this analysis is to determine the response of a structure, which is subjected to arbitrarily time-varying loads.

In this book, the primary emphasis will be done upon the static linear analysis of plane structures. Also the reader will be familiar with problems of stability in structural analysis and free-vibration analysis as well as some special cases of analysis will be briefly discussed.

Fundamental Assumptions of Structural Analysis

Analysis of structures that is based on the following assumptions is called the elastic analysis.

1. Material of the structure is continuous and absolutely elastic.
2. Relationship between stress and strain is linear.
3. Deformations of a structure, caused by applied loads, are small and do not change original design diagram.
4. Superposition principle is applicable.

Superposition principle means that any factor, such as reaction, displacement, etc., caused by different loads which act simultaneously, are equal to the algebraic or geometrical sum of this factor due to each load separately. For example, reaction of a movable support under any loads has one fixed direction. So the reaction of this support due to different loads equals to the *algebraic* sum of reactions due to action of each load separately.Vector of total reaction for a pinned support in case of any loads has different directions, so the reaction of pinned support due to different loads equals to the *geometrical* sum of reactions, due to action of each load separately.

Fundamental Approaches of Structural Analysis

There are two fundamental approaches to the analysis of any structure. The first approach is related to analysis of a structure subjected to given fixed loads and is called the fixed loads approach. The results of this analysis are diagrams, which show a distribution of internal forces (bending moment, shear, and axial forces) and deflection for the entire structure due to the given fixed loads. These diagrams indicate the most unfavorable point (or member) of a structure under the given fixed loads. The reader should be familiar with this approach from the course of mechanics of material.

The second approach assumes that a structure is subjected to unit concentrated moving load only. This load is not a real one but imaginary. The results of the second approach are graphs called the influence lines. Influence lines are plotted for reactions, internal forces, etc. Internal forces diagrams and influence lines have a fundamental difference. Each influence line shows distribution of internal forces in the *one specified section* of a structure due to location of imaginary unit moving load only. These influence lines indicate the point of a structure where a load should be placed in order to reach a maximum (or minimum) value of the function under consideration at the specified section. It is very important that the influence lines may be also used for analysis of structure subjected to any fixed loads. Moreover, in many cases they turn out to be a very effective tool of analysis.

Influence lines method presents the higher level of analysis of a structure, than the fixed load approach. Good knowledge of influence lines approaches an immeasurable increase in understanding of behavior of structure. Analyst, who combines both approaches for analysis of a structure in engineering practice, is capable to perform a complex analysis of its behavior.

Both approaches do not exclude each other. In contrast, in practical analysis both approaches complement each other. Therefore, learning these approaches to the analysis of a structure will be provided in parallel way. This textbook presents sufficiently full consideration of influence lines for different types of statically determinate and indeterminate structures, such as beams, arches, frames, and trusses.

Part I
Statically Determinate Structures

Chapter 1
Kinematical Analysis of Structures

Kinematical analysis of a structure is necessary for evaluation of ability of the structure to resist external load. Kinematical analysis is based on the concept of *rigid disc*, which is an unchangeable (or rigid) part of a structure. Rigid discs may be separate members of a structure, such as straight members, curvilinear, polygonal (Fig. 1.1), as well as their special combination.

Any structure consists of separate rigid discs. Two rigid discs may be connected by means of link, hinge, and fixed joint. These types of connections and their static and kinematical characteristics are presented in Table 1.1.

The members of a structure may be connected together by a hinge in various ways. Types of connection are chosen and justified by an engineer as follows:

1. *Simple hinge*. One hinge connects *two* elements in the joint.
2. *Multiple hinge*. One hinge connects *three or more* elements in the joint. The multiple hinge is equivalent to n–1 simple hinges, where n is a number of members connected in the joint. Hinged joints can transmit axial and shear forces from one part of the structure to the other; the bending moment at the hinge joint is zero.

1.1 Classification of Structures by Kinematical Viewpoint

All structures may be classified as follows:

- *Geometrically unchangeable structure*. For this type of structure, any distortion of the structure occurs only with deformation of its members. It means that this type of structure with absolutely rigid members cannot change its form. The simplest geometrically unchangeable structure is triangle, which contains the pin-joined members (Fig. 1.2a).
- *Geometrically changeable structure*. For this type of structure, any finite distortion of the structure occurs without deformation of its members. The simplest geometrically changeable system is formed as hinged four-bar linkage (Fig. 1.2b, c). In Fig. 1.2c, the fourth bar is presented as ground. In both cases, even if the system would be made with absolutely rigid members, it still can change its form.

I.A. Karnovsky and O. Lebed, *Advanced Methods of Structural Analysis*,
DOI 10.1007/978-1-4419-1047-9_1, © Springer Science+Business Media, LLC 2010

Fig. 1.1 Types of the rigid discs

Table 1.1 Types of connections of rigid discs and their characteristics

Type of connection	Link	Hinge	Fixed joint
Presentation and description of connection	D_1 D_2 Rigid discs D_1 and D_2 are connected by link (rod with hinges at the ends)	D_1 D_2 Rigid discs D_1 and D_2 are connected by hinge	D_1 D_2 Rigid discs D_1 and D_2 are connected by fixed joint
Kinematical characteristics	Mutual displacement of both discs *along the link* is zero	Mutual displacements of both discs in *both* horizontal and vertical directions are zeros	*All* mutual displacements of both discs (in horizontal, vertical, and angular directions) are zeros
Static characteristics	Connection transmits one force, which prevents mutual displacement along the link	Connection transmits two forces, which prevent mutual displacements in vertical and horizontal directions	Connection transmits two forces, which prevent mutual displacements in vertical and horizontal directions, and moment, which prevents mutual angular displacement

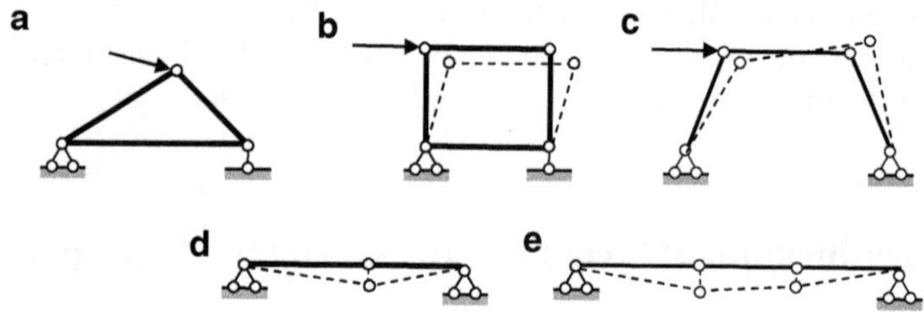

Fig. 1.2 Types of structure by kinematical viewpoint

It is pertinent to do the following important remark related to terminology. Sometimes terms "stable" and "unstable" are applied for the above-mentioned types of structures. However, the commonly accepted term "stable/unstable" in classical theory of deformable systems is related to concept of critical load, while term "geometrically unchangeable/changeable" is related to way of connection of rigid disks. There is the fundamental difference between kinematical analysis of a structure on the one hand, and analysis of stability of a structure subjected to compressed load, on the other hand. Thus, in kinematical analysis of structures, we will use the term "unchangeable structure" instead of "stable structure," and term "changeable structure" instead of "unstable structure." Stability analysis is considered in Chap. 13.

- *Instantaneously changeable structure*. This system allows infinitesimal relative displacements of its members without their deformation and after that the structure becomes geometrically unchangeable (Fig. 1.2d).
- *Instantaneously rigid structure*. This system allows infinitesimal relative displacements of its members without their deformation and after that the structure becomes geometrically changeable (Fig. 1.2e). The term "instantaneously" is related to initial condition of the structure only.

In structural engineering only geometrically unchangeable structures may be accepted.

1.2 Generation of Geometrically Unchangeable Structures

In order to produce a rigid structure in a whole, the rigid discs should be connected in specific way. Let us consider general rules for formation of geometrically unchangeable structures from two and three rigid discs.

If a structure is formed from *two discs*, then their connections may be as follows:

1. Connection by fixed joint (Fig. 1.3a)
2. Connection by hinge C and rod AB, if axis of AB does not pass through the hinge C (Fig. 1.3b)
3. Connection by three nonparallel rods. Point of intersection of any two rods presents a fictitious hinge C'. In this case, the other rod and fictitious hinge corresponds to Case 2 (Fig. 1.3c)

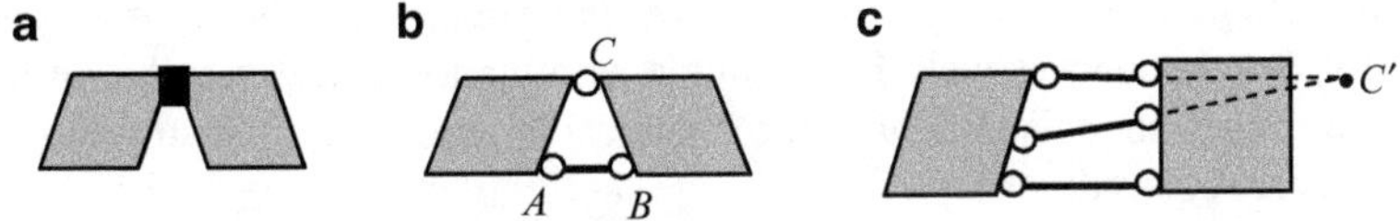

Fig. 1.3 Geometrically unchangeable structures formed from two discs

If a structure is formed from *three discs*, then their connections may be as follows:

1. Connection in pairs by three hinges A, B, and C, which do not belong to one line (Fig. 1.4a).
2. Connection in pairs by two (or more) concurrent links, if points of their intersections A, B, and C do not belong to one line (Fig. 1.4b).

Case 1.4a may be presented as shown in Fig. 1.4c: additional joint A is attached to rigid disc D by two links 1 and 2. This case leads to a rigid triangle (Fig. 1.2a), which is a simplest geometrically unchangeable structure. This is the main principle of formation of simplest trusses.

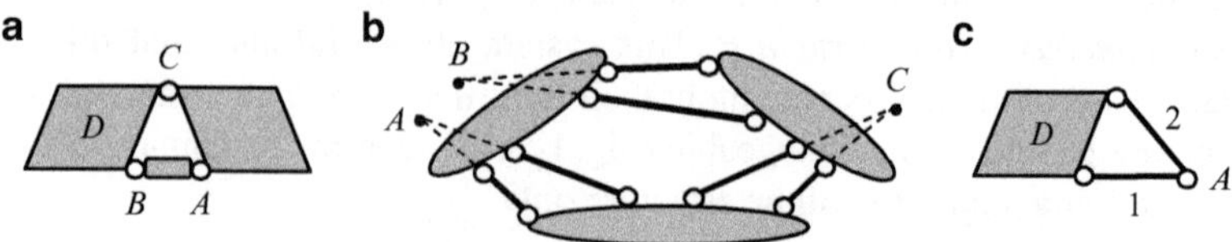

Fig. 1.4 Geometrically unchangeable structures formed from three discs

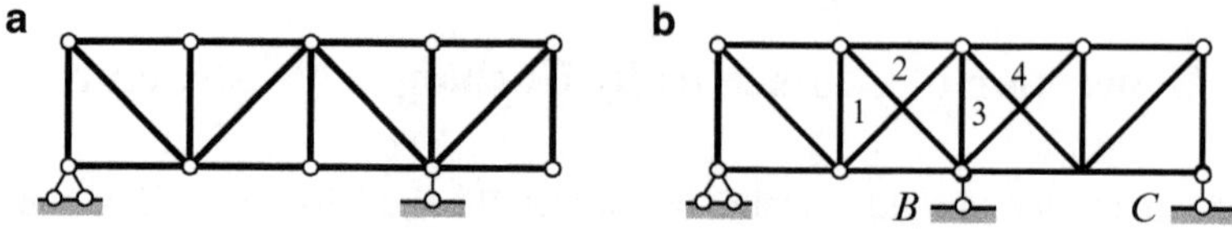

Fig. 1.5 (**a**) Structure with required constraints; (**b**) structure with redundant constraints

Required and Redundant Constraints

All constraints of any structure are divided into two groups.

Required constraint of a structure is such a constraint, eliminations of which change a kinematical characteristic of the structure. It means that the entire unchangeable structure transforms into changeable or instantaneously changeable one, instantaneously changeable transforms into changeable, and changeable transforms into changeable with mobility more by unity. Note that *constraint* assumes not only supports, but elements as well. For example, elimination of *any* member of truss in Fig. 1.5a transforms this structure into changeable one, so for this structure all the elements are required.

Redundant constraint of a structure is such a constraint, eliminations of which do not change a kinematical characteristic of the structure. It means that the entire unchangeable structure remains the unchangeable (Fig. 1.5b), changeable structure remains changeable one, and instantaneously changeable remains instantaneously changeable structure. The structure in Fig. 1.5a has no redundant constraints. For structure in Fig. 1.5b, the following constraints may be considered as redundant: 1 or 2, and 3 or 4, and one of the supports – B or C, so the total number of redundant constraints is three.

Constraint replacing. In case of unchangeable structure, the constraints may be replaced. It means that the *required* constraint may be eliminated and instead of that *another* required constraint should be introduced, or the *redundant* constraint is replaced by *another* redundant constraint. This procedure allows from a given structure to create a lot of other structures.

1.3 Analytical Criteria of the Instantaneously Changeable Structures

Kinematical analysis of a structure can also be done using the static equations. The following criteria for instantaneously changeable systems may be used:

(a) Load has finite quantity but the internal forces have infinite values
(b) Load is absent and the internal forces are uncertain (type of 0/0)

Let us discuss these criteria for structure shown in Fig. 1.6a.

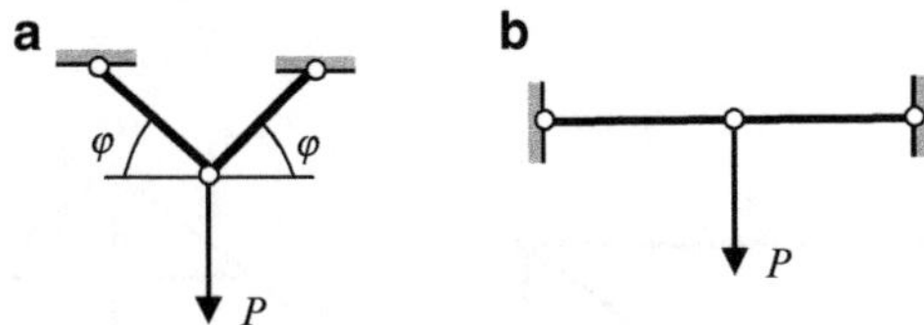

Fig. 1.6 Kinematical analysis using the static equations

Internal forces in members of the structure are $N = P/(2\sin\varphi)$. If $\varphi = 0$ (Fig. 1.6b), then $N = \infty$. Thus, external load P of finite quantity leads to the internal forces of infinite values. It happens because the system is instantaneously changeable. Indeed, two rigid discs are connected using three hinges located on the one line.

Figure 1.7 presents the design diagram of the truss. This structure is generated from simplest rigid triangle; each next joint is attached to previous rigid disc by two end-hinged links. The structure contains three support constraints, which are necessary minimum for plane structure. However, location of these supports may be wrong. Thorough kinematical analysis of this structure may be performed by static equations.

Reaction R of support may be calculated using equilibrium condition

$$\sum M_A = 0 \rightarrow R \times a - P \times b = 0 \rightarrow R = \frac{Pb}{a}.$$

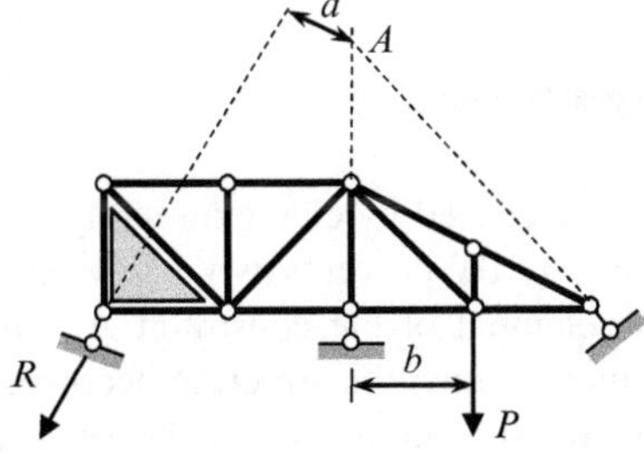

Fig. 1.7 Kinematical analysis of the truss

1. If $a = 0$, then for any external load P the reaction of the left support is infinitely large ($R = Pb/0$)
2. If $a = 0$ and $P = 0$, then reaction R is uncertain ($R = 0/0$). Thus, if all lines of support constraints are concurrent at one point ($a = 0$), then this case leads to instantaneously changeable system.

Instantaneously changeable systems may occur if two rigid discs of a structure join inappropriately. Two such connections of rigid discs are shown in Fig. 1.8. If a system may be separated into two rigid discs (shown by solid color) by a section cutting three elements, which are parallel (Fig. 1.8a, elements 1, 2, 3), or concurrent in one point (Fig. 1.8b, point A, elements a, b, c), then the system is instantaneously changeable one.

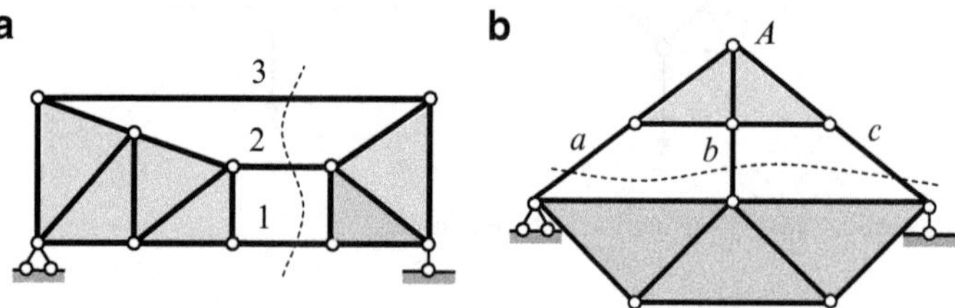

Fig. 1.8 Instantaneously changeable systems

However, in practice, connection of two rigid discs by two (or more) parallel members may be used in special condition of loading. Figure 1.9 presents the rigid beam (disc D_1), which is supported by vertical hinged-end rods 1–3 (disc D_2 is a support part). The system may be used if the axial forces in members 1–3 are tensile (Fig. 1.9a). However, the system cannot be used if the axial forces in members 1–3 are compressive (Fig. 1.9b).

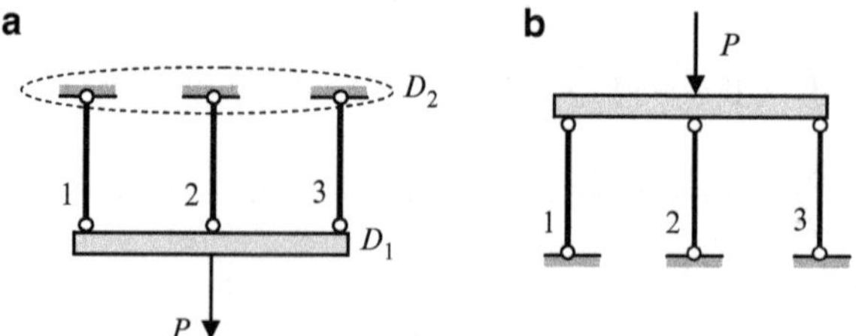

Fig. 1.9 Geometrically changeable systems

Evolution of the structure caused by changing the type of supports is shown in Fig. 1.10. Constraint A (Fig. 1.10a) prevents two displacements, in vertical and horizontal directions. If one element of the constraint A, which prevents horizontal displacement, will be removed, then the structure becomes *geometrically changeable* (Fig. 1.10b), so the removed constraint is the *required* one. In case of any horizontal displacement of the structure, all support constraints A, B, and C remain parallel to each other.

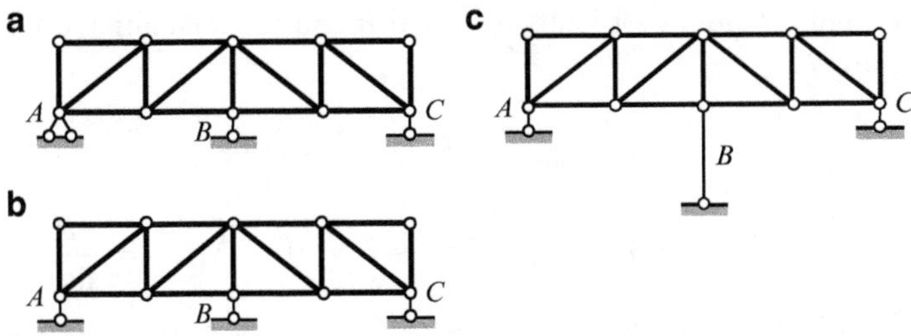

Fig. 1.10 (**a**) Geometrically unchangeable structure; (**b**) geometrically changeable structure; (**c**) instantaneously changeable system

The next evolution is presented in Fig. 1.10c. If any support, for example supporting element B will be longer than other supports, then structure becomes *instantaneously changeable* system. Indeed, in case of any horizontal displacement of the structure, the support constraints will not be parallel any more.

It is worth to mention one more static criterion for instantaneously changeable and geometrically changeable structures: internal forces in *some element* obtained by two different ways are *different*, or in another words, analysis of a structure leads to *contradictory* results. This is shown in the example below.

Design diagram of the truss is presented in Fig. 1.11a: the constraint support B is directed along the element BC. The system has the necessary minimum number of elements and constraints to be geometrically unchangeable structure. Let us provide more detailed analysis of this structure. Free body diagrams and equilibrium conditions for joints A, 1, and B are shown in Fig. 1.11b.

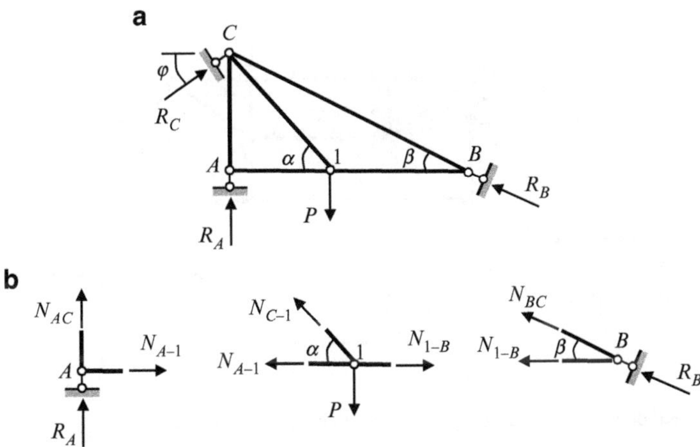

Fig. 1.11 Kinematical analysis of instantaneously changeable system

Equilibrium conditions for joints A, 1, and B and corresponding results are presented below.

Joint A. $$\sum X = 0 \rightarrow N_{A-1} = 0,$$

$$\sum Y = 0: \quad N_{C-1}\sin\alpha - P = 0 \rightarrow N_{C-1} = \frac{P}{\sin\alpha}$$

Joint 1. $\sum X = 0: \quad N_{1-B} - N_{C-1}\cos\alpha = 0 \rightarrow N_{1-B} = N_{C-1}\cos\alpha = P\cot\alpha$

$$\sum Y = 0: \quad N_{BC}\sin\beta + R_B\sin\beta = 0 \rightarrow N_{BC} = -R_B$$

Joint B. $\sum X = 0: \quad -N_{1-B} - N_{BC}\cos\beta - R_B\cos\beta = 0 \rightarrow N_{1-B} = 0$

Two different results for internal force N_{1-B} have been obtained, i.e., $P\cot\alpha$ and zero. This indicates that the system is defective. From mathematical point of view, this happens because the set of equilibrium equations for different parts of the structure is incompatible. From physical point of view, this happens because three support constraints are concurrent in one point C for any angle φ. Any variation of this angle φ remains the system as instantaneously changeable. If constraint at support C will be removed to point A, or angle of inclination of support B will be different, then system becomes geometrically unchangeable structure.

Let us show this criterion for system presented in Fig. 1.12. Note that hinges D and E are multiple similarly to hinges C and F.

Reaction

$$R_A \rightarrow \sum M_B = 0 \rightarrow R_A = \frac{Pb}{a+b}.$$

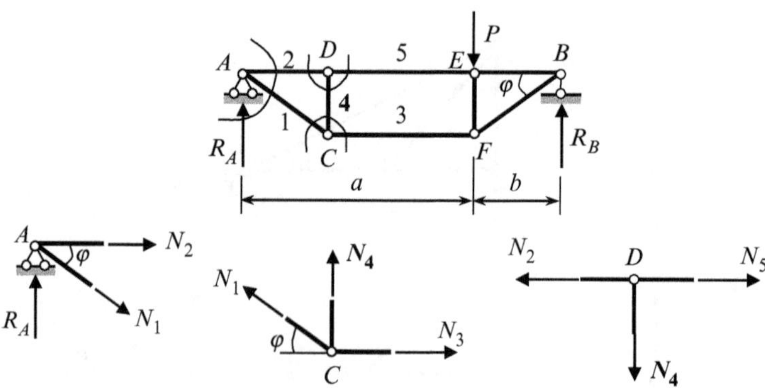

Fig. 1.12 Kinematical analysis of geometrically changeable system

Equilibrium conditions leads to the following results:

$$\text{Joint } A.\ \sum Y = 0: \quad R_A - N_1 \sin\varphi = 0 \to N_1 = \frac{R_A}{\sin\varphi} = \frac{Pb}{(a+b)\sin\varphi},$$

$$\text{Joint } C.\ \sum Y = 0: \quad N_4 + N_1 \sin\varphi = 0 \to N_4 = -N_1 \sin\varphi = -\frac{Pb}{a+b},$$

$$\text{Joint } D.\ \sum Y = 0: \quad N_4 = 0.$$

We received for internal force N_4 two contradictory (or inconsistent) results. Why this happens? To answer on this question let us consider the very important concept "degrees of freedom."

1.4 Degrees of Freedom

A number of independent parameters, which define configuration of a system without deformation of its members is called the degree of freedom. The number of degrees of freedom for any structure may be calculated by the Chebushev' formula

$$W = 3D - 2H - S_0, \tag{1.1}$$

where D, H, and S_0 are the number of rigid discs, simple hinges, and constraints of supports, respectively.

For trusses the degrees of freedom may be calculated by formula

$$W = 2J - S - S_0, \tag{1.2}$$

where J and S are the number of joints and members of the truss, respectively.

Special cases. There are three special cases possible.

1. $W > 0$. The system is geometrically changeable and cannot be used in engineering practice.
2. $W = 0$. The system has the necessary number of elements and constraints to be geometrically unchangeable structure. However, the system still can be inappropriate for engineering structure. Therefore, this case requires additional structural analysis to check if the formation of the structure and arrangement of elements and constraints is correct. This must be done according to rules, which are considered above. For example, let us consider systems, which are presented in Fig. 1.13a–c.

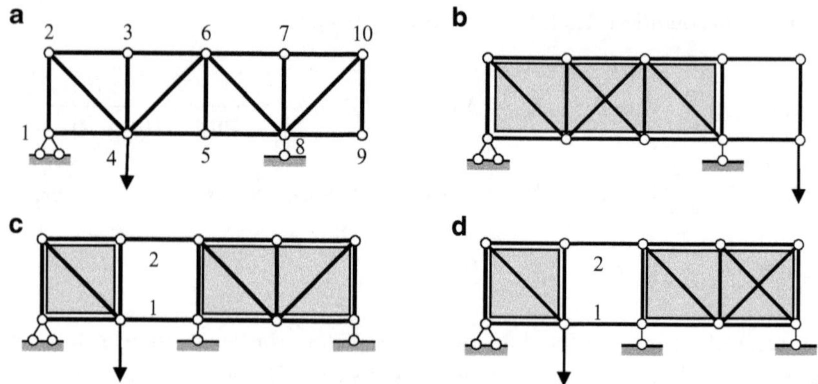

Fig. 1.13 Different ways of truss formation. For cases (**a**), (**b**), and (**c**) degrees of freedom $W = 0$, for case (**d**) $W = -1$. Only case (**a**) may be adopted as engineering structure

Degrees of freedom W for these cases according to formula (1.2) equal to zeros. Indeed:

$$\text{Case a}: J = 10, \quad S = 17, \quad S_0 = 3, \quad \text{so} \quad W = 2 \times 10 - 17 - 3 = 0,$$
$$\text{Case b}: J = 10, \quad S = 17, \quad S_0 = 3, \quad \text{so} \quad W = 2 \times 10 - 17 - 3 = 0,$$
$$\text{Case c}: J = 10, \quad S = 16, \quad S_0 = 4, \quad \text{so} \quad W = 2 \times 10 - 16 - 4 = 0.$$

The degrees of freedom for trusses may be calculated also by the Chebushev formula. The truss in Fig. 1.13a contains 17 bars, which are considered as the rigid discs, and 3 support constraints. The number of the *equivalent simple hinges H* can be calculated as follows: only two hinges are simple (joints 1 and 9) and all other hinges are multiple. The each multiple hinge at joints 2, 3, 5, 7, 10 are equivalent to two simple hinges; the each multiple hinge 4, 6, 8 are equivalent to four simple hinges. Thus, the total number of equivalent simple hinges is $H = 2 \times 1 + 5 \times 2 + 3 \times 4 = 24$. The Chebushev formula leads to the following result $W = 3D - 2H - S_0 = 3 \times 17 - 2 \times 24 - 3 = 0$. The same results may be obtained for Cases (b) and (c).

Even if $W = 0$ for systems in Fig. 1.13a–c, only system (a) may be used as engineering structure. System (b) has a rigid disc (shown as solid) and geometrically changeable right part. System (c) consists of two rigid discs, which are connected by *two* members 1 and 2 (while for generation of geometrically unchangeable structures two rigid discs must be connected by *three* nonconcurrent members), and therefore it is geometrically changeable system.

3. $W < 0$. The system has redundant constraints. However, existence of redundant constraints still does not mean that the structure can resist load, because the structure can be generated incorrectly. The system in Fig. 1.13d contains one redundant constraint; indeed, the degrees of freedom is $W = 2 \times 10 - 17 - 4 = -1$.

However, the assembly of the elements is wrong and therefore, this system cannot be considered as engineering structure. Indeed, the left and right rigid disks (shown by solid) are connected by *two* elements 1 and 2. Therefore, this structure is geometrically changeable.

Let us return to Fig. 1.12. The system contains members with hinges at ends. The number of joints is $J = 6$, the number of members is $S = 8$, and the number of support constraints is $S_0 = 3$. Degrees of freedom of the system is calculated as for truss, i.e., $W = 2\times6-8-3 = 1$. (The Chebushev formula leads to the same result, i.e., $W = 3\times8-2\times10-3 = 1$, where 10 is a total number of *simple* hinges). Therefore, the system does not have a required minimum of elements in order to provide its geometrical unchangeability, and cannot be used as engineering structure. This system is geometrically changeable, because two rigid discs *ACD* and *BEF* are connected incorrectly, i.e., by *two* members 3 and 5. In other words, this system contains the hinged four-bar linkage *CDEF*. Therefore, if *one more* rigid element would be introduced *correctly*, then system would become geometrically unchangeable. Such additional element may connect joints C and E or joints D and F.

Problems

1.1. Perform the kinematical analysis of the following design diagrams:

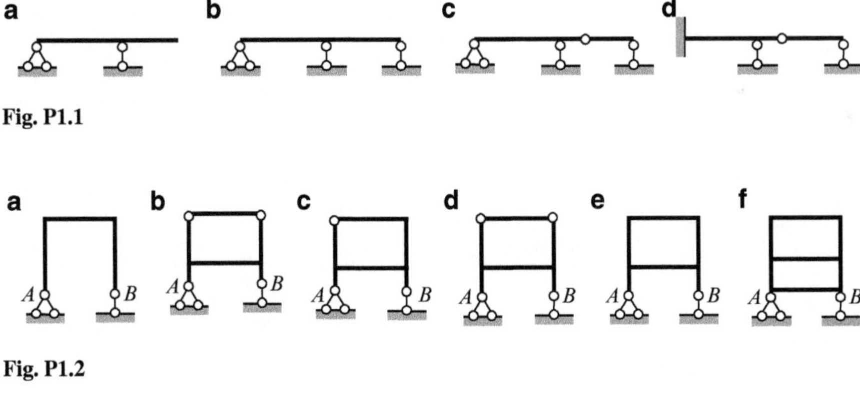

Fig. P1.1

Fig. P1.2

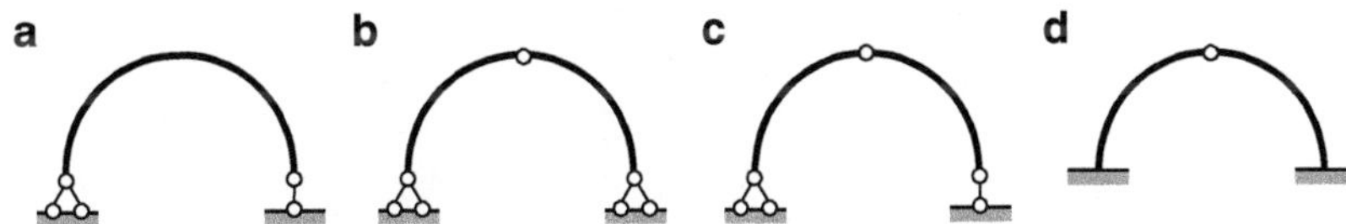

Fig. P1.3

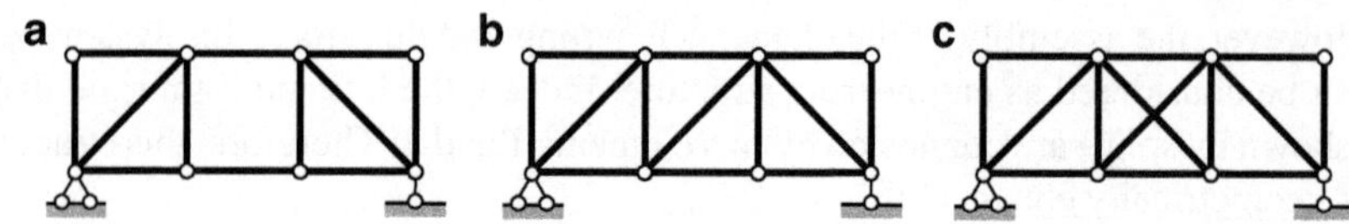

Fig. P1.4

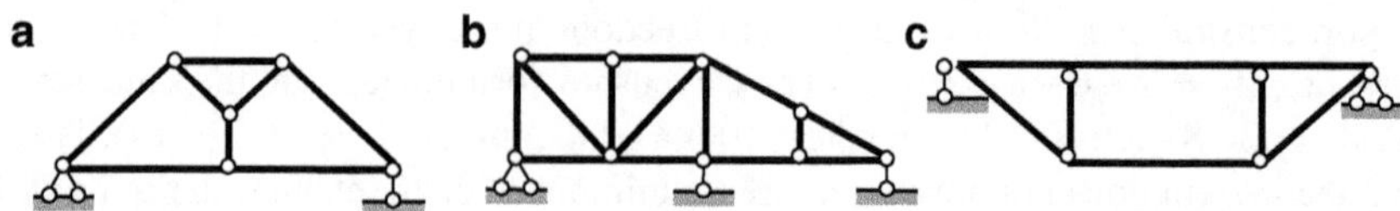

Fig. P1.5

Chapter 2
General Theory of Influence Lines

Construction of influence lines for one span simply supported and cantilevered beams in case of direct and indirect load applications are considered. All influence lines are constructed using analytical expressions for required factor. Applications of influence lines for fixed and moving loads are discussed. This chapter forms the set of concepts which creates a framework for comprehensive analysis of different statically determinate structures.

2.1 Analytical Method for Construction of Influence Lines

The engineering structures are often subjected to moving loads. Typical examples of moving loads on a structure are traveling cars, trains, bridge cranes, etc. In classical structural analysis, the term "moving load" requires one additional comment: this concept means that only the load *position* on the structure may be *arbitrary*.

It is obvious that internal forces and displacements in any section of a beam depend on the position of a moving load. An important problem in analysis of structures is the determination of maximum internal forces in a structure caused by a given moving load and the corresponding most unfavorable position of this load. This problem may be solved using influence lines. Influence line is a fundamental and very profitable concept of structural analysis. Their application allows perform a deep and manifold analysis of different types of structures subjected to *any* type of fixed and moving loads. Influence lines method becomes especially effective tool analysis if a structure is subjected to *different groups* of loads.

Definition: Influence line is a graph, which shows variation of some particular function Z (reaction, shear, bending moment, etc.) in the *fixed* cross section of a structure in terms of *position* of unit concentrated dimensionless load $P = 1$ on the structure.

Each ordinate of influence line means the value of the function, for which influence line is constructed, if the unit load is located on the structure above this ordinate. Therefore, the unit load P, which may have different positions on the structure, is called a moving (or traveling) load. The term "moving load" implies

I.A. Karnovsky and O. Lebed, *Advanced Methods of Structural Analysis*,
DOI 10.1007/978-1-4419-1047-9_2,

only that position of the load is arbitrary, i.e., this is a *static* load, which may have different positions along the beam. The time, velocity of the moving load, and any dynamic effects are not taken into account. Thus, for convenience, from now on, we will use notion of "moving" or "traveling" load for static load, which may have *different* position along the beam.

Influence line for any function Z at a specified section of a structure may be constructed using its definition as follows: the required function should be calculated for different position of unit load within the loaded portion. These values are plotted at the points, which correspond to the position of the load. After that all ordinates should be connected. However, such procedure is extremely annoying and cumbersome (especially for statically indeterminate structures). Because of repetitive procedure of construction of influence lines, their advantages are questionable.

In this book, the construction of influence lines is performed on the bases of a different approach, i.e., deriving an *equation* of influence line for required function Z; this equation relates values of Z and position x of the unit load P. Thus the required factor Z is presented as *analytical function of the position of the load*. Such way of construction of influence lines is called static method. Application of this method for construction of influence lines for reactions is presented below.

2.1.1 *Influence Lines for Reactions*

The following types of beams are considered: simply supported beam, beam with overhang, and cantilever beam.

2.1.1.1 Simply Supported Beam (Fig. 2.1)

The beam *AB* is loaded by moving load $P = 1$. The moving load is shown by circle and the dotted line indicates the loaded contour for possible positions of the load on the structure The distance from the left-hand support to the load is x.

Influence Line for R_A

Equilibrium equation in form of moments of all external forces about the support B allows determine the reaction R_A in terms of x:

$$R_A \to \sum M_B = 0: \quad -R_A \cdot l + P(l-x) = 0 \to R_A = \frac{P(l-x)}{l}.$$

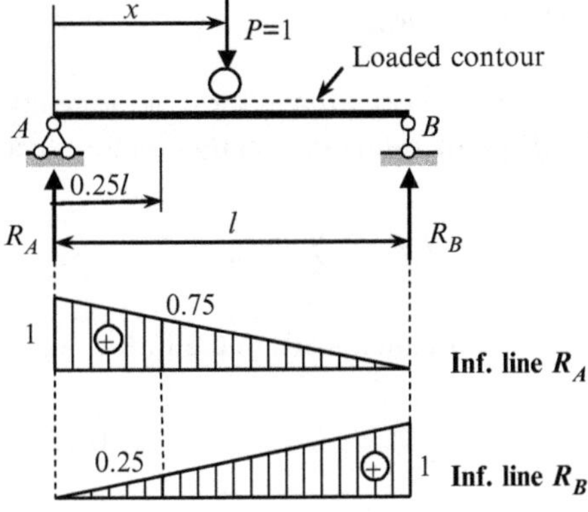

Fig. 2.1 Simply supported beam. Influence lines for reactions

Reaction R_A may be calculated for any location x of the load P. Therefore, last equation should be considered as a *function* $R_A(x)$. This function is called the influence line for R_A and denoted as $\text{IL}(R_A)$. Since $P = 1$, then equation of influence line for reaction R_A becomes

$$\text{IL}(R_A) = \frac{l - x}{l}. \tag{2.1}$$

If $x = 0$ (at support A), then ordinate of influence line $\text{IL}(R_A) = 1$. If $x = l$ (at support B), then ordinate of influence line $\text{IL}(R_A) = 0$. These two points are connected by straight line, since function (2.1) is linear. Influence line for R_A is presented in Fig. 2.1.

We can see that units of influence line ordinates for reaction is dimensionless. In general, units of influence line ordinates for any factor Z are defined as quotient of two units, mainly, unit of the factor Z and unit of the load P (kN). Thus, unit of influence line for reactions and shear is dimensionless because kN/kN; for bending moment: kN× m/kN = m; for linear deflection m/kN; for angular deflection rad/kN.

Influence line $\text{IL}(R_A)$ can be used for analysis of reaction R_A only. Positive ordinates mean that reaction of R_A is directed upward for any position of the concentrated load. If load $P = 1$ is located above point A, then reaction of R_A is equal to 1; it means that load P completely transmits on the support A. If load $P = 1$ is located above point B, then reaction of R_A is equal to zero. If load $P = 1$ has, for example, coordinate $x = 0.25l$, then reaction of R_A is equal to 0.75.

Analytical presentation of equation of influence line allows to avoid many times repeated computation of function Z for different location of the force P; this is a huge advantage of analytical approach for construction of influence lines. Note a following fundamental property: influence lines for reactions and internal forces of any statically determined structures are always presented by straight lines.

Influence Line for R_B

Equilibrium equation in form of moments of all the external forces about support A leads to the following expression for reaction R_B in terms of position x

$$R_B \to \sum M_A = 0: \quad R_B \cdot l - P \cdot x = 0 \to R_B = \frac{P \cdot x}{l}.$$

The last equation leads to the following equation of influence line:

$$\mathrm{IL}(R_B) = \frac{x}{l}. \tag{2.2}$$

If $x = 0$ (at support A), then ordinate of influence line $\mathrm{IL}(R_B) = 0$. If $x = l$ (at support B), then ordinate of influence line $\mathrm{IL}(R_B) = 1$. Influence line for R_B is presented in Fig. 2.1. This graph can be used for analysis of reaction R_B only. If load $P = 1$ is located above point A, then reaction of R_B is equal to zero. It means that load P does not get transmitted on to the support B, when the load P is situated directly over the left-hand support. If load $P = 1$ is located above point B, then reaction of R_B is equal to 1. If load $P = 1$ has, for example, coordinate $x = 0.25l$, then reaction of R_B is equal to 0.25.

2.1.1.2 Simply Supported Beam with Overhang (Fig. 2.2)

The equilibrium equations and corresponding equations for influence lines of reactions are

$$R_A \to \sum M_B = 0:$$

$$-R_A l - P(x - l) = 0 \to R_A = -P\frac{x - l}{l} \to \mathrm{IL}(R_A) = -\frac{x - l}{l},$$

$$R_B \to \sum M_A = 0: \; R_B l - Px = 0 \to R_B = P\frac{x}{l} \to \mathrm{IL}(R_B) = \frac{x}{l}. \tag{2.3}$$

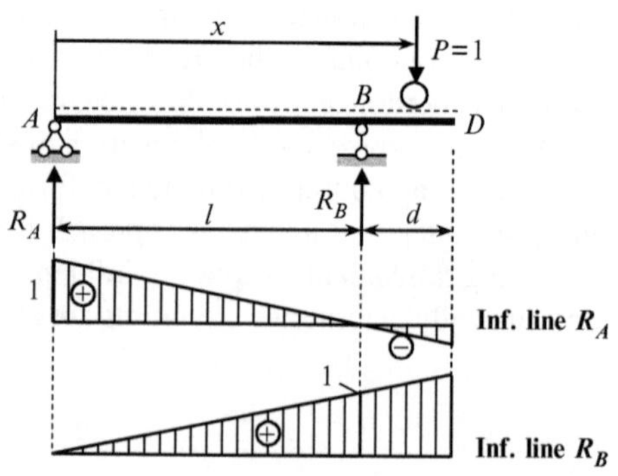

Fig. 2.2 Simply supported beam with overhang. Influence lines for reactions

Influence lines of reactions are shown in Fig. 2.2. If load $P = 1$ is situated at point D $(x = l+d)$, then reaction $R_A = -d/l$. The negative sign means that the reaction R_A is directed downward. The maximum positive reaction R_A occurs if load $P = 1$ stands at point A, the maximum negative reaction R_A occurs if load $P = 1$ stands at point D.

If load $P = 1$ is situated at point D, then $R_B = (l + d)/l$. This means, that reaction $R_B > P = 1$ and is directed upward. The maximum positive reaction R_B occurs if load $P = 1$ stands at point D; the negative reaction R_B does not be arise.

Equations (2.1)–(2.3) for influence lines of reactions show that overhang does not change the equations of influence lines; therefore an influence line within the overhang is an extension of influence line within the span. This is a common property of influence lines for any function (reaction, bending moment, and shear). Thus, in order to construct the influence lines for reaction of a simply supported beam with overhang, the influence lines for reaction between supports should be extended underneath the overhang.

2.1.1.3 Cantilevered Beam (Fig. 2.3)

At the fixed support A, the following reactions arise: vertical and horizontal forces R_A and H_A, and moment M_0; for the given design diagram the horizontal reaction $H_A = 0$. Positive reactions R_A and M_0 are shown in Fig. 2.3.

The Vertical Reaction R_A

This reaction may be calculated considering the equilibrium equation in form of the projections of the all external forces on the vertical axis

$$R_A \to \sum Y = 0: \quad R_A - P = 0 \to R_A = P.$$

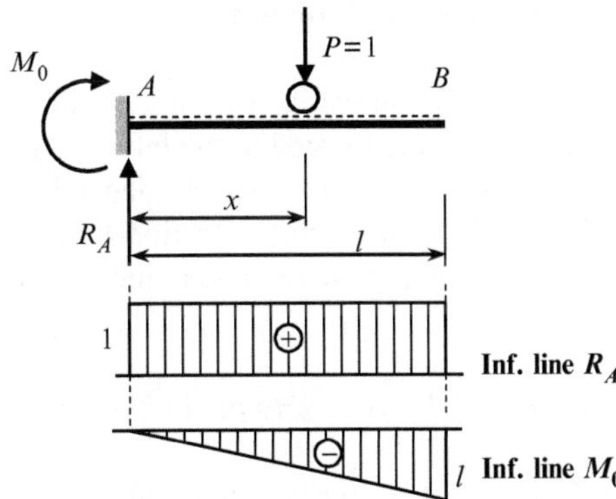

Fig. 2.3 Cantilevered beam. Design diagram and influence lines for reactions

Since $P = 1$, the equation of influence line becomes

$$\mathrm{IL}(R_A) = 1. \tag{2.4}$$

It means that reaction R_A equals to 1 for any position of concentrated load $P = 1$.

The Moment M_0 at Support A

This moment may be calculated considering the equilibrium equation in form of moment of all the external forces with respect to point A

$$M_0 \to \sum M_A = 0: \quad -M_0 - Px = 0 \to M_0 = -Px.$$

Since load $P = 1$, then equation of influence line is

$$\mathrm{IL}(M_0) = -x. \tag{2.5}$$

It means that moment varies according to linear law. If the load $P = 1$ is located at $x = 0$ (point A), then the moment M_0 at the fixed support does not arise. Maximum moment at support A corresponds to position of the load $P = 1$ at point B; this moment equals to $-1l$. The units of the ordinates of influence line for M_0 are meters.

2.1.2 Influence Lines for Internal Forces

Simply supported beam subjected to moving unit load P is presented in Fig. 2.4. Construction of influence lines for bending moment and shear force induced at section k are shown below.

2.1.2.1 Bending Moment M_k

The bending moment in section k is equal to the algebraic sum of moments of all forces, which are located to the left (or right) of section k, about point k. Since the expression for bending moment depends on whether the load P is located to the left or to the right from the section k, then two positions of the load P need to be considered, i.e., to the left and to the right of section k.

Load $P = 1$ Is Located to the Left of Section k

In this case, it is convenient to calculate the bending moment M_k using the right forces (Fig. 2.4). The only reaction R_B is located to the right of point k, so the bending moment is

$$M_k \to \sum M_k^{\text{right}} = 0: \quad M_k = R_B \cdot b.$$

If position of the load P is fixed, then reaction R_B is a *number* and the bending moment is a *number* as well. However, if load $P = 1$ changes its position along the left portion of the beam, then reaction R_B becomes a *function* of position of the load P and, thus, the bending moment is a *function* too. Thus, the *expression* for bending moment is transformed to the *equation* of influence line for bending moment

$$\text{IL}(M_k) = b \cdot \text{IL}(R_B). \tag{2.6}$$

So for the construction of influence line for bending moment we need to construct the influence line for reaction R_B, after that multiply all ordinates by parameter b, and, as the last step, show the operating range of influence line. Since load P is located to the *left* of section k, then the operating range is *left-hand portion* of influence line, i.e., the above equation of influence line is true when the load P is changing its position on the *left* portion of the beam. Hatching the corresponding part of the influence line reflects this fact.

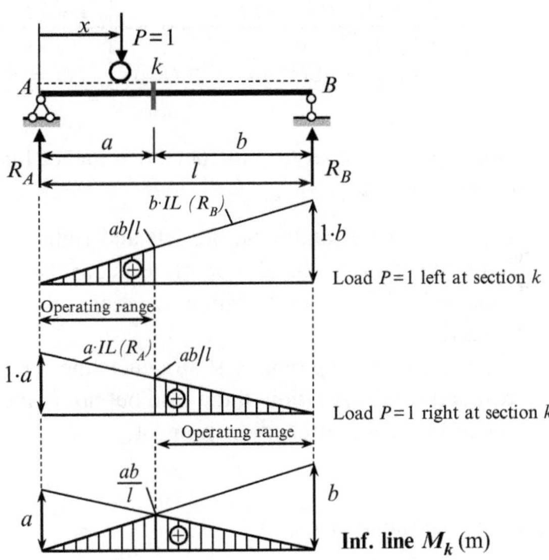

Fig. 2.4 Simply supported beam. Construction of influence line for bending moment at section k

Load $P = 1$ Is Located to the Right of Section k

In this case, it is convenient to calculate the bending moment M_k using the left forces. The only reaction R_A is located to the left of point k, so the bending moment is

$$M_k \to \sum M_k^{\text{left}} = 0: \quad M_k = R_A \cdot a.$$

As above, the expression for bending moment is transformed to the equation of influence line for bending moment

$$\mathrm{IL}(M_k) = a \cdot \mathrm{IL}(R_A). \tag{2.7}$$

For construction of this influence line, we need to construct the influence line for reaction R_A, then multiply all ordinates by parameter a, and finally, to show the operating range of influence line. Since load P is located to the *right* of section k, we obtain *right-hand portion* of influence line as the operating range. Corresponding influence line M_k is presented in Fig. 2.4.

Since units of ordinates $\mathrm{IL}(R_A)$ and $\mathrm{IL}(R_B)$ are dimensionless, then units of ordinates $\mathrm{IL}(M_k)$ are units of length (for example, meter). Ordinate of influence line at section k is (ab/l). Sign of influence line for bending moment M_k for simply supported beam is positive, which means that extended fibers at section k are located below longitudinal axis for any position of the load P.

Henceforward, all mathematical treatment concerning to construction of influence lines will be presented in tabulated form. The above-mentioned discussions are presented in the following table.

Load $P = 1$ left at section k	Load $P = 1$ right at section k
$M_k \to \sum M_k^{\mathrm{right}} = 0,$ $M_k = R_B \cdot b \to \mathrm{IL}(M_k) = b \cdot \mathrm{IL}(R_B)$	$M_k \to \sum M_k^{\mathrm{left}} = 0,$ $M_k = R_A \cdot a \to \mathrm{IL}(M_k) = a \cdot \mathrm{IL}(R_A)$

To summarize, in order to construct the influence line for bending moment at section k it is necessary to:

1. Plot ordinates a and b on the left and right vertical lines passing through the left-hand and right-hand support, respectively.
2. Join each of these points with base point at the other support; both lines intersect at section k.
3. Show the operating ranges of influence line. The hatching (operating range) corresponds to the *position* of the load but not to the *part* of the beam which is used for computation of bending moment.

2.1.2.2 Influence Line Q_k

Since the expression for shear depends on whether the load P is located to the left or to the right from the section k, then two positions of the load P need to be considered. The procedure of construction of influence line for shear is presented in tabulated form. In the table we show the *position* of the load (left or right at section k – this is the first line of the table) and *part* of the beam, the equilibrium of which is considered.

Load $P = 1$ left at section k	Load $P = 1$ right at section k
$Q_k \to \sum Y^{\text{right}} = 0,$ $Q_k = -R_B \to \text{IL}(Q_k) = -\text{IL}(R_B)$	$Q_k \to \sum Y^{\text{left}} = 0,$ $Q_k = R_A \to \text{IL}(Q_k) = \text{IL}(R_A)$

Using these expressions, we can trace the *left-hand* portion of the influence line for shear as the influence line for reaction R_B with negative sign and *right-hand* portion of the influence line for shear as influence line for reaction R_A (Fig. 2.5). Units of ordinate $\text{IL}(Q_k)$ are dimensionless.

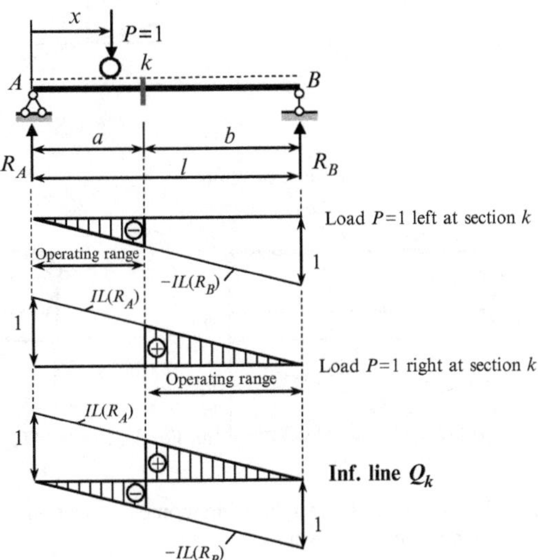

Fig. 2.5 Simply supported beam. Construction of influence line for shear at section k

In order to construct the influence line for shear at section k the following procedure should be applied:

1. Plot ordinate $+1$ (upward) and -1 (downward) along the vertical lines passing through the left-hand and right-hand support, respectively
2. Join each of these points with base point at the other support
3. Connect both portions at section k
4. Show the operating range of influence line: operating range of left-hand portion is negative and operating range of right-hand portion is positive

The negative sign of the left-hand portion and jump at section k may be explained as follows: If the load $P = 1$ is located on the left part of the beam, then shear $Q_k = R_A - P < 0$. When load P is infinitely close to the section k to the left, then shear $Q_k = -R_B$. As soon as the load $P = 1$ moves over section k, then shear $Q_k = R_A$.

It is obvious, that the influence line for shear for the section which is infinitely close to support A coincides with influence line for reaction R_A, i.e., $\text{IL}(Q) = \text{IL}(R_A)$. If section is infinitely close to support B, then the influence line for shear coincides with influence line for reaction R_B with negative sign, i.e., $\text{IL}(Q) = -\text{IL}(R_B)$.

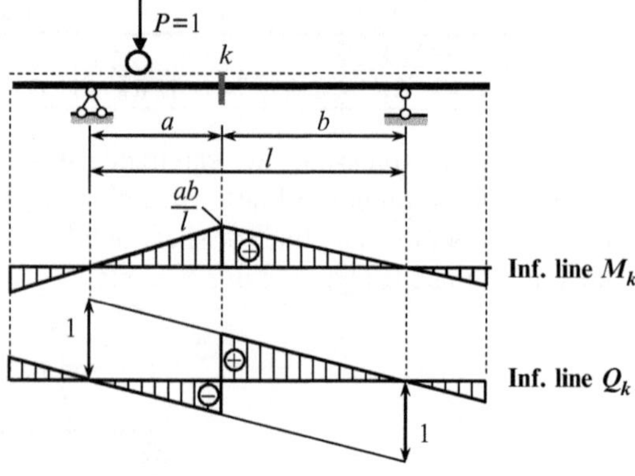

Fig. 2.6 Beam with overhangs. Construction of influence lines for bending moment and shear at section k

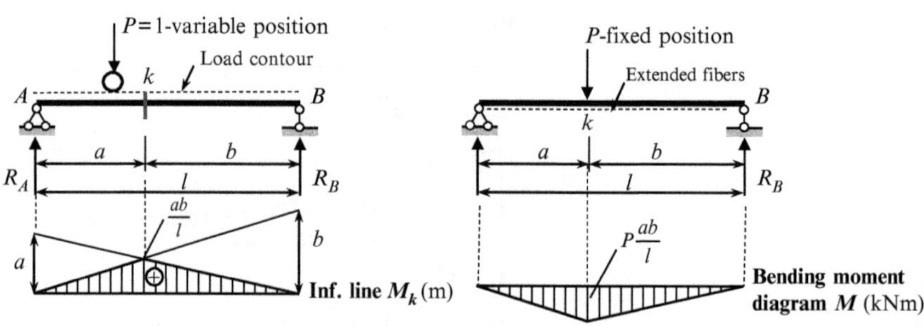

Fig. 2.7 Influence line for bending moment at section k and bending moment diagram due to load P, which is located at point k

Now let us consider a simply supported beam with overhangs (Fig. 2.6). Assume that section k is located between supports. In order to construct influence line for bending moment and shear at section k, it is necessary to:

1. Ignore overhang and construct corresponding influence line for simply supported beam
2. Extent of the latter until its intersection with the vertical passing through the end of overhang

2.1.2.3 Discussion

At this point, emphasis must once again be placed on differentiating between influence line for bending moment at point k and bending moment diagram due to load P, which is located at point k (Fig. 2.7). This diagram is constructed on the tensile fibers.

Even though both diagrams look similar and have same sign (ordinates of bending moment diagram are positive) and same ordinate at section k (if $P = 1$), they have completely different meanings and should not be confused. The difference between them is presented in Table 2.1

Table 2.1 Comparison of influence line for bending moment and bending moment diagram

Influence line M_k	Bending moment diagram M
This graph shows variations of bending moment at the section k *only*; load $P = 1$ is *moving* and has *different* locations along the beam	This graph shows distribution of bending moment in *all* sections of the beam; load P is *fixed* and acts at section k *only*

The following example generalizes all cases considered above for construction of influence lines of internal forces for different sections in case of simply supported beam with overhangs (Fig. 2.8).

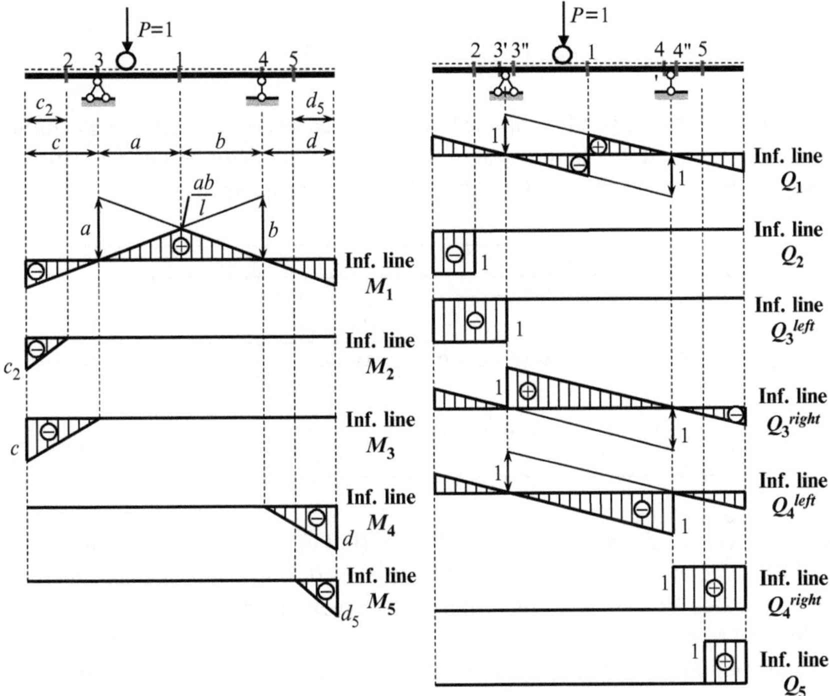

Fig. 2.8 Beam with overhangs. Influence lines for bending moments and shear forces for different sections

Bending Moment M_1 and Shear Force Q_1

The influence line of bending moment at the section 1 for simply supported beam without overhangs presents a triangle with maximum ordinate ab/l, where $l = a + b$. The influence line is extended within the overhang. If load P is located within the span, then bending moment in section 1 is positive, i.e., the extended fibers are located below the longitudinal axis. If load P is located outside of the span, then bending moment in section 1 is negative.

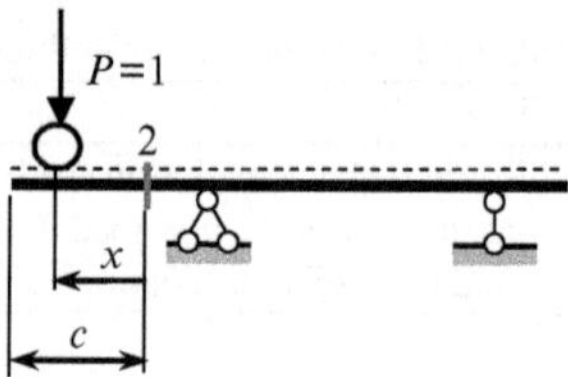

Fig. 2.9 Construction of influence lines for section 2 on the overhang

The influence line for shear in section 1 for the simply supported beam should be extended through support points to the end of overhangs.

Shear Force Q_2 and Bending Moment M_2 (Fig. 2.9)

Since expressions for bending moment and shear at section 2 depend on position of the load P (to the left or to the right of the section 2), then for deriving equations of influence lines two positions of the load should be considered.

$P = 1$ left at section 2	$P = 1$ right at section 2
$Q_2 \to \sum Y^{\text{left}} = 0 \to Q_2 = -P$ $\text{IL}(Q_2) = -1$	$Q_2 \to \sum Y^{\text{left}} = 0 \to Q_2 = 0$ $\text{IL}(Q_2) = 0$
$M_2 \to \sum M_2^{\text{left}} = 0 \to M_2 = -Px$ $\text{IL}(M_2) = -x$ At $x = 0:$ $\text{IL}(M_2) = 0$ At $x = c:$ $\text{IL}(M_2) = -c$	$M_2 \to \sum M_2^{\text{left}} = 0 \to M_2 = 0$ $\text{IL}(M_2) = 0$

Note that for any position of the load $P = 1$ (left or right at section 2) we use the equilibrium equations $\sum Y^{\text{left}} = 0$ and $\sum M_2^{\text{left}} = 0$, which take into account forces that are located *left* at the section. Similarly, for section 5 we will use the forces that are located *right* at the section 5.

Construction of influence lines for bending moments and shear at sections 3, 4, and 5 are performed in the same manner.

Pay attention that influence lines of shear for sections 3^{left} and 3^{right}, that are infinitesimally close to support point, (as well as for sections 4^{left} and 4^{right}) are different. These sections are shown as $3'$ and $3''$, $4'$ and $4''$.

Influence lines for bending moments and shear forces at all pattern sections 1–5 are summarized in Fig. 2.8. These influence lines are good reference source for practical analysis of one-span beams subjected to *any* type of loads. Moreover, these influence lines will be used for construction of influence lines for multispan hinged statically determinate beams.

2.2 Application of Influence Lines for Fixed and Moving Loads

Influence lines, which describe variation of any function Z (reaction, bending moment, shear, etc.) in the fixed section due to moving concentrated unit load $P = 1$ may be effectively used for calculation of this function Z due to *arbitrary fixed* and *moving* loads.

2.2.1 Fixed Loads

Three types of fixed loads will be considered: concentrated loads, uniformly distributed loads, and couples as shown in Fig. 2.10. Let us consider effect of each load separately.

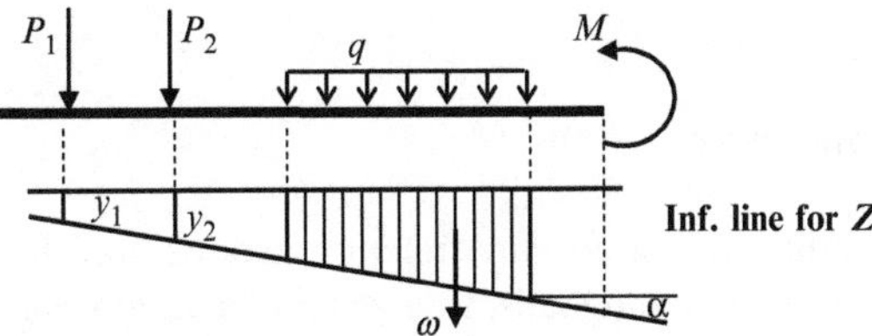

Fig. 2.10 Application of influence line for fixed loads

Concentrated Loads

If a structure is loaded by a load $P (P \neq 1)$, then function Z due to this load is $Z = \pm Py$, with y the ordinate of influence line for function Z at the point where load P is applied. The sign of Z depends on the sign of ordinate y of influence line. If a structure is loaded by several loads P_i, then according to superposition principle

$$Z = \pm \sum_i P_i y_i. \tag{2.8}$$

In order to compute the value of any function (reaction, internal forces in any section of the beam, frame, or any member of the truss, etc.) arising under the action of several concentrated loads P_i, the corresponding influence line must be constructed, each load should be multiplied by the ordinate of the influence line measured at the load point, and the obtained products must be summed up.

It is easy to check that units of ordinate of influence lines, which have been discussed early, lead to the required units for function Z.

Uniformly Distributed Load

Value of any function Z due to action of uniformly distributed load q is determined by formula $Z = \pm q\omega$, with ω the area of influence line graph for function Z within the portion where load q is applied. If the influence line within the load limits has different signs then the areas must be taken with appropriate signs. The sign of the area coincides with sign of ordinates of influence line.

Couple

If a structure is loaded by couple M, then function Z, due to this moment is $Z = \pm M \tan\alpha$, where α is the angle between the base line and the portion of influence line for function Z within which M is applied. If couple tends to rotate influence line toward base line through an angle less than 90 then sign is positive; if angle is greater then 90 then sign is negative.

Summary

Influence line for any function may be used for calculation of this function due to *arbitrary* fixed loads. In a general case, any function Z as a result of application of a several concentrated loads P_i, uniform loads intensity q_j, and couples M_k should be calculated as follows:

$$Z = \sum P_i y_i + \sum q_j \omega_j + \sum M_k \tan\alpha_k, \tag{2.9}$$

where y_i is the ordinates of corresponding influence line, these ordinates are measured at all the load points; ω_j the area bounded by corresponding influence line, the x-axis, and vertical lines passing through the load limits; and α_k is the angle of inclination of corresponding influence line to the x-axis.

The formula (2.9) reflects the superposition principle and may be applied for any type of statically determinate and indeterminate structures. Illustration of this formula is shown below. Figure 2.11 presents a design diagram of a simply supported beam and influence line for reaction R_A.

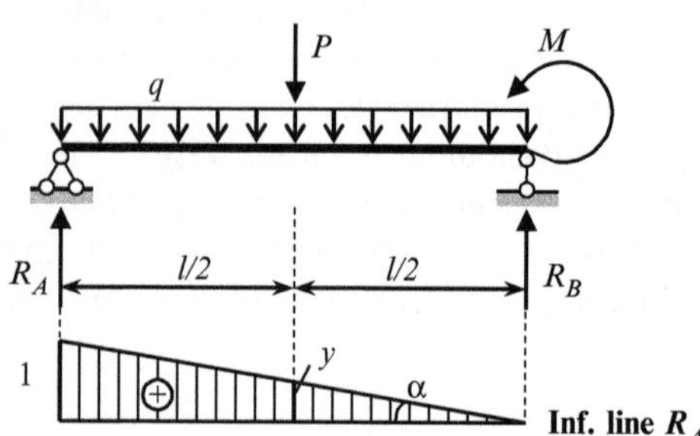

Fig. 2.11 Design diagram of the beam

The value of reaction R_A due to given fixed loads P, q, and M equals

$$R_A = Py + q\omega + M\tan\alpha = P \cdot \underbrace{\frac{1}{2}}_{y} + q \cdot \underbrace{\frac{1}{2} \cdot 1 \cdot l}_{\omega} + M \cdot \underbrace{\frac{1}{l}}_{\tan\alpha} = \frac{P}{2} + \frac{ql}{2} + \frac{M}{l}.$$

The change of values of loads and/or their position does not change the procedure of calculation of any factor using corresponding influence line. From this example we observe an advantage of influence line: once constructed influence line may be used for calculation of relevant function due to arbitrary loads.

Example 2.1. A design diagram of the simply supported beam with overhang is presented in Fig. 2.12. Calculate shear at section n using influence line. The loads are $P_1 = 12\,\text{kN}$; $P_2 = 8\,\text{kN}$; $q_1 = 3\,\text{kN/m}$; $q_2 = 2\,\text{kN/m}$.

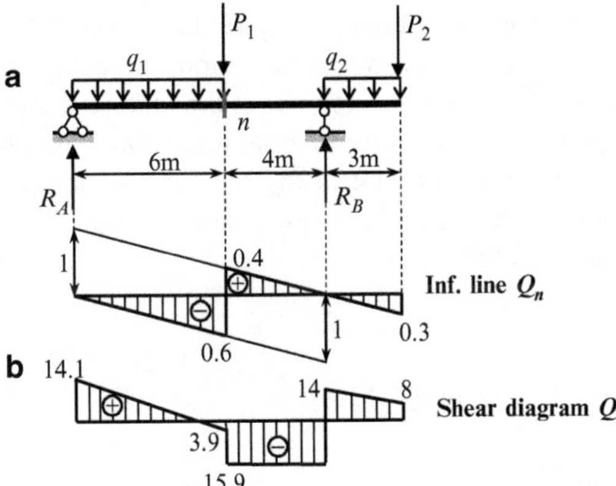

Fig. 2.12 Design diagram of the beam. (**a**) Influence line for shear Q_n. (**b**) Shear force diagram

Solution. First of all, the influence line for required shear at section n should be constructed (Fig. 2.12a).

The shear force at section n caused by the fixed loads is $Q_n = \sum P_i y_i + \sum q_i \omega_i$. If load P_1 is located infinitely close to the left of the section n, then ordinate $y_1 = -0.6$. If load P_1 is located infinitely close to the right of section n, then ordinate $y_1 = 0.4$. Ordinate $y_2 = -0.3$. Areas of the influence line within the distribution loads q_1 and q_2 are

$$\omega_1 = -\frac{1}{2} \times 6 \times 0.6 = -1.8\,\text{m}; \quad \omega_2 = -\frac{1}{2} \times 3 \times 0.3 = -0.45\,\text{m}.$$

The peculiarity of this problem is that force P_1 is located at the section where it is required to find shear. Therefore, we have to consider two cases when this load is located infinitely close from the left and right sides of the section n.

If load P_1 is located to the *left* of section n, then

$$Q_n = 12 \times (-0.6) + 8 \times (-0.3) + 3 \times (-1.8) + 2 \times (-0.45) = -15.9\,\text{kN}.$$

If load P_1 is located to the *right* at section n, then

$$Q_n = 12 \times 0.4 + 8 \times (-0.3) + 3 \times (-1.8) + 2 \times (-0.45) = -3.9\,\text{kN}.$$

In order to understand the obtained results, we need to calculate the shear Q_n for section n and show shear force diagram. Reaction of the support A is $R_A = 14.1$ kN.

If the load P_1 is located to the *left* of section n, then the force P_1 *must* be taken into account

$$Q_n \to \sum Y^{\text{left}} = 0 \to Q_n = R_A - q_1 \cdot 6 - P_1 = 14.1 - 3 \times 6 - 12 = -15.9\,\text{kN}.$$

If the load P_1 is located to the *right* of section n, then the force P_1 *must not* be taken into account

$$Q_n = R_A - q_1 \cdot 6 = 14.1 - 3 \times 6 = -3.9\,\text{kN}.$$

Shear force diagram Q is presented in Fig. 2.12b.

This example shows the serious advantage of influence lines: if we change the type of loading, the amount of the load, and its locations on the beam, then a corresponding reaction (or internal forces) may be calculated immediately.

Since the amount of the load and its location does not affect on the influence lines, then influence lines should be considered as fundamental characteristics of any structure.

2.2.2 *Moving Loads*

Influence line for any function Z allows us to calculate Z for *any position* of a moving load, and that is very important, the most unfavorable position of the moving loads and corresponding value of the relevant function. Unfavorable (or dangerous) position of a moving load is such position, which leads to the maximum (positive or negative) value of the function Z. The following types of moving loads will be considered: one concentrated load, a set of loads, and a distributed load.

The set of connected moving loads may be considered as a model of moving truck. Specifications for truck loading may be found in various references, for example in the American Association of State and Highway Transportation Officials (AASHTO). This code presents the size of the standard truck and the distribution of its weight on each axle. The moving distributed load may be considered as a model of a set of containers which may be placed along the beam at arbitrary position.

The most unfavorable position of a single concentrated load is its position at a section with maximum ordinate of influence line. If influence line has positive and negative signs, then it is necessary to calculate corresponding maximum of the function Z using the largest positive and negative ordinates of influence lines.

In case of set of concentrated moving loads, we assume that some of loads may be connected. This case may be applicable for moving cars, bridge cranes, etc. We will consider different forms of influence line.

Influence Line Forms a Triangle

A dangerous position occurs when one of the loads is located *over the vertex* of an influence line; this load is called a critical load. (The term "critical load" for problems of elastic stability, Chap. 13, has a different meaning.) The problem is to determine *which* load among the group of moving loads is critical. After a critical load is known, all other loads are located according to the given distances between them.

The critical load may be easy defined by a graphical approach (Fig. 2.13a). Let the moving load be a model of two cars, with loads P_i on the each axle. All distance between forces are given.

Step 1. Trace the influence line for function Z. Plot all forces P_1, P_2, P_3, P_4 in order using arbitrary scale from the left-most point A of influence line; the last point is denoted as C.

Step 2. Connect the right most point B with point C.

Step 3. On the base line show the point D, which corresponds to the vertex of influence line and from this point draw a line, which is parallel to the line CB until it intersection with the vertical line AC.

Step 4. The intersected force (in our case P_2) presents a critical load; unfavorable location of moving cars presented in Fig. 2.13a.

Step 5. Maximum (or minimum) value of relevant function is $Z = \sum P_i \cdot y_i$.

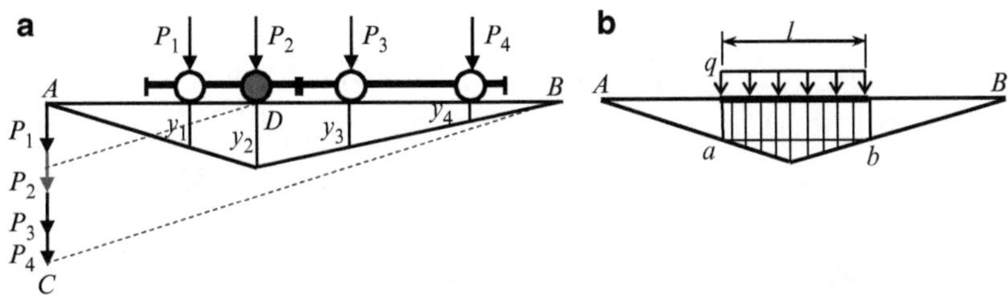

Fig. 2.13 Graphical definition of the unfavorable position of load for triangular influence line. (**a**) Set of concentrated load. (**b**) Uniformly distributed load of fixed length l

Influence Line Forms a Polygon

A dangerous position of the set of moving concentrated loads occurs when one or more loads stand over vertex of the influence line. Both, the load and the apex of the influence line over which this load must stand to induce a maximum (or minimum) of the function under consideration, are called critical. The critical apex of the influence line must be convex.

In case of Uniformly distributed moving load, the maximum value of the function Z corresponds to the location of a distributed load q, which covers maximum one-sign area of influence line. The negative and positive portions of influence line must be considered in order to obtain minimum and maximum of function Z.

The special case of uniformly distributed moving load happens, if load is distributed within the *fixed length l*. In case of triangular influence line, the most unfavorable location of such load occurs when the portion $ab = l$ and base AB will be parallel (Fig. 2.13b).

Example 2.2. Simply supported beam with two overhangs is presented in Fig. 2.14. Determine the most unfavorable position of load, which leads to maximum (positive and negative) values of the bending moment and shear at section k. Calculate corresponding values of these functions. Consider the following loads: uniformly distributed load q and two connected loads P_1 and P_2 (a twin-axle cart with different wheel loads).

Solution. Influence lines for required functions Z are presented in Fig. 2.14.

Action of a uniformly distributed load $q = 1.6\,kN/m$. Distributed load leads to maximum value of the function if the area of influence lines within the distributed load is maximum. For example, the positive shear at the section k is peaked if load q covers all portions of influence line with positive ordinates; for minimum shear in the same section the load q must be applied within portions with negative ordinates.

$$Q_{k(\max+)} = 1.6 \times \frac{1}{2}(0.3 \times 3 + 0.4 \times 4) = 2\,\text{kN};$$

$$Q_{k(\max-)} = -1.6 \times \frac{1}{2}(0.6 \times 6 + 0.3 \times 3) = -3.6\,\text{kN},$$

$$M_{k(\max+)} = 1.6 \times \frac{1}{2} \times 10 \times 2.4 = 19.2\,\text{kN m};$$

$$M_{k(\max-)} = -1.6 \times \frac{1}{2}(1.2 \times 3 + 1.8 \times 3) = -7.2\,\text{kN m}.$$

Positive value of $M_{k\,\max}$ means that if load is located between AB, the tensile fibers of the beam at section k are located below longitudinal axis of the beam. If load is located within the overhangs, then bending moment at section k is negative, i.e., the tensile fibers at section k are located above the longitudinal axis of the beam.

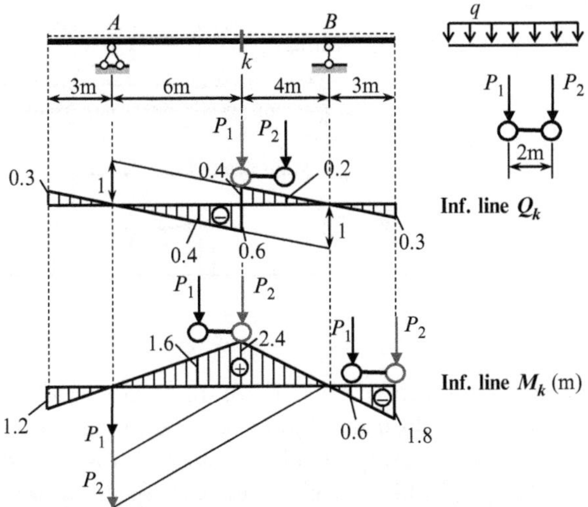

Fig. 2.14 Design diagram of the beam, influence lines, and most unfavorable positions of two connected loads

Action of the set of loads $P_1 = 5\,kN$ *and* $P_2 = 8\,kN$. Unfavorable locations of two connected loads are shown in Fig. 2.14. Critical load for bending moment at section k (triangular influence line) is defined by graphical method; the load P_2 is a critical one and it should be placed over the vertex of influence line.

$$Q_{k(\max +)} = 5 \times 0.4 + 8 \times 0.2 = 3.6\,\text{kN};$$
$$Q_{k(\max -)} = -(5 \times 0.4 + 8 \times 0.6) = -6.8\,\text{kN},$$
$$M_{k(\max +)} = 5 \times 1.6 + 8 \times 2.4 = 27.2\,\text{kNm};$$
$$M_{k(\max -)} = -(5 \times 0.6 + 8 \times 1.8) = -17.4\,\text{kNm}.$$

If a set of loads P_1 and P_2 modeling a crane bridge then the *order* of loads is fixed and cannot be changed. If a set loads P_1 and P_2 is a model of a moving car then we need to consider case when a car moves in opposite direction. In this case the order of forces from left to right becomes P_2 and P_1.

2.3 Indirect Load Application

So far, we have been considering cases when external loads were applied directly to the beams. In practice, however, loads are often applied to secondary beams (or stringers) and then are transmitted through them to the main beam (or girder) as shown in Fig. 2.15. Stringers are simply supported beams. Each stringer's span is called a panel d and each point where the stringer transmits its load onto the main beam is called a panel point or a joint. The load is transmitted from the secondary beams onto the main beam only at panel points.

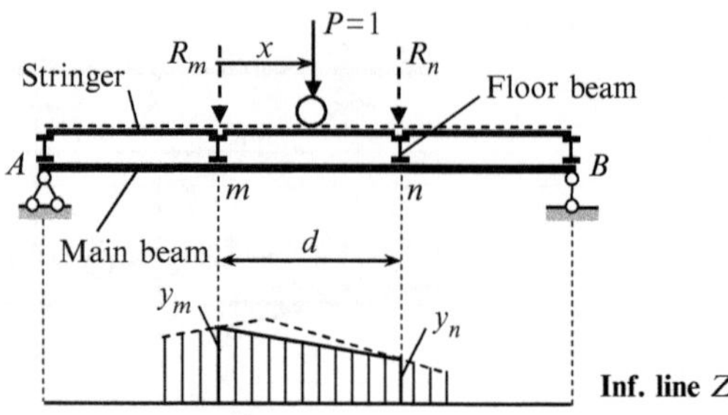

Fig. 2.15 Indirect load application

Influence line for any function Z in case of direct load application is presented in Fig. 2.15 by dotted line. Ordinate of influence line for parameter Z at the panel points m and n are y_m and y_n, respectively. If load $P = 1$ is located at point m or n in case of indirect load application, then parameter Z equals y_m or y_n. If load $P = 1$ is located at any distance x between points m and n, then reactions of stringers R_m and R_n are transmitted on the main beam at points m and n. In this case function Z may be calculated as

$$Z = R_m y_m + R_n y_n.$$

Since

$$R_m = \frac{P(d-x)}{d} = \frac{d-x}{d}, \quad R_n = \frac{Px}{d} = \frac{x}{d},$$

then the required parameter Z becomes

$$Z = \frac{d-x}{d} y_m + \frac{x}{d} y_n = y_m + \frac{1}{d}(y_n - y_m)x. \tag{2.10}$$

Thus, the influence line for any function Z between two closest panel points m and n is presented by a *straight* line. This is the fundamental property of influence lines in case of indirect load application.

Influence lines for any function Z should be constructed in the following sequence:

1. Construct the influence line for a given function Z as if the moving load would be applied directly to the main beam.
2. Transfer the panel points on the influence line and obtained nearest points connect by straight line, which is called as the *connecting line.*

This procedure will be widely used for construction of influence lines for arches and trusses.

Procedure for construction of influence lines of bending moment and shear at section k for simply supported beam in case of indirect load application is shown in Fig. 2.16. First of all we need to construct the influence line for these functions if load P would be applied directly to the main beam AB. Influence line for bending moment presents triangle with vertex ab/l at the section k; influence line for shear is bounded by two parallel lines with the jump at the section k. Then we need to indicate the panel points m and n, which are nearest to the given section k, and draw vertical lines passing through these points. These lines intersect the influence lines at the points m' and n'. At last, these points should be connected by a straight line.

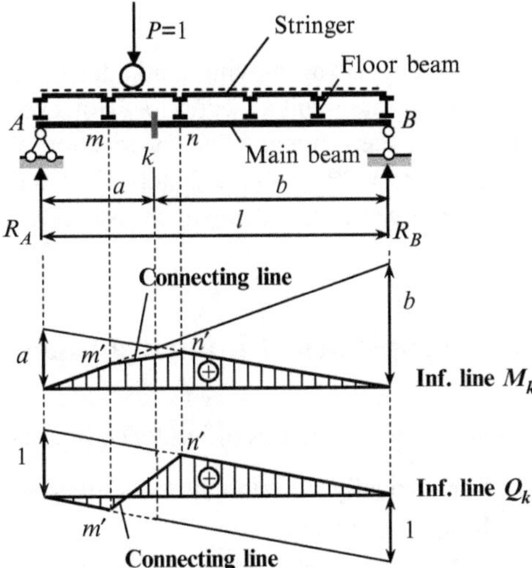

Fig. 2.16 Influence lines in case of indirect load application

Pay attention, that if a floor beam m will be removed, then the influence line for Q_k becomes the *positive one-sign* function instead of a two-sign function, as presented in Fig. 2.16. If the floor beam n and all following ones (except floor beam at the support B) will be removed, then the influence line for Q_k becomes the *one-sign* function too, but a negative one.

2.4 Combining of Fixed and Moving Load Approaches

So far we showed an application of influence lines Z for analysis of this particular function Z. However, in structural analysis, the application of influence lines is of great utility and we can use influence line for Z_1 for calculation of another function Z_2. For example, design diagram of the beam and influence line for reaction R_A is shown in Fig. 2.17. How we can calculate the bending moment at section k?

Using fixed loads approach we need to calculate the reaction, which arises in right stringer; transmit this reaction through floor beam to panel joint onto main beam; determine reactions R_A and R_B and, after that, calculate M_K by definition using

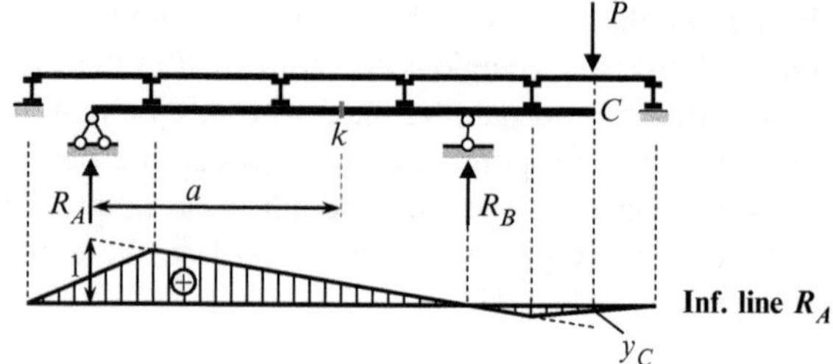

Fig. 2.17 Calculation of bending moment M_k using influence line for R_A

R_A (or R_B). Using moving load approach, we need to construct the influence line for bending moment at section k. However, we can combine both approaches. For this, show the influence line for reaction R_A. According to this influence line, the reaction $R_A = Py_C$. Immediately, the bending moment at section K equals $M_K = R_A a = Py_C a$. This idea will be used later for analysis of structures.

2.5 Properties of Influence Lines

Fundamental properties of influence lines are the following:

1. For statically determinate structures influence lines for reactions and internal forces are linear
2. In case of a simply supported beam with overhangs, the influence lines for reactions and internal forces for section between supports should be constructed as for beam without overhangs and extended along the overhangs
3. In case of indirect load application influence line between two joint points is linear
4. Units of ordinate of any influence line of factor Z are unit of Z divided by unit of load P (kN, lb)

Influence lines describe fundamental properties of a structure, which are independent from the load; therefore, they may be considered as a fundamental and reference data for the structure. Influence lines allow finding internal forces and reactions in case of fixed and moving loads. Influences lines allow performing quick analysis of a structure when it is modified, in particularly in case of indirect load application.

Influence line for some factor is used for determination of this factor only; and after this factor may be used for determination of some other parameters. For example, we can determine reactions due to any fixed loads using influence line. Then this reaction can be used for determination of internal forces by definition. Such combination of two approaches is a very effective way of analysis of complex statically determinate structures and especially of statically indeterminate structures.

Problems

2.1. The simply supported beam AB is subjected to the fixed loads as shown in Fig. P2.1. Analyze the structure using the fixed and moving load approach.

Fixed load approach Determine the reaction at the supports and construct the internal force diagrams, caused by given loads.

Moving load approach

(a) Construct the influence lines for reaction at the supports A and B
(b) Construct the influence lines for bending moment and shear force for section k
(c) Construct the influence lines for bending moment and shear force for sections infinitely close to the left and right the support B
(d) Calculate reaction of the supports, bending moment and shear force for all above mentioned sections using corresponding influence lines
(e) Compare the results obtained by both approaches

Ans. $R_A = 3.9\,\text{kN};\ R_B = 14.1\,\text{kN};\ M_k = 23.4\,\text{kN m};\ Q_k^{\text{left}} = 3.9\,\text{kN};$
$Q_k^{\text{right}} = -8.1\,\text{kN}.$

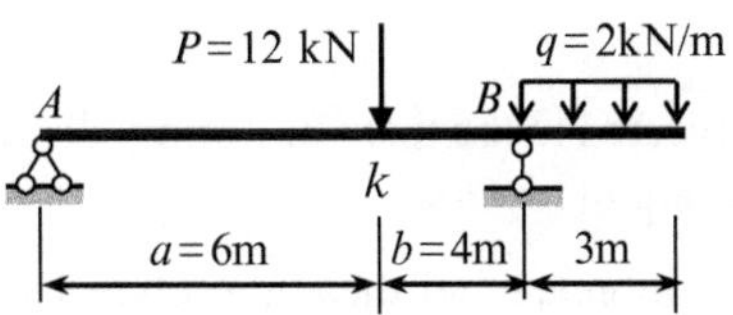

Fig. P2.1

2.2. Simply supported beam is subjected to linearly varying load shown in Fig. P2.2. Determine the reactions of supports and construct the bending moment and shear diagrams. Compute the reaction of the left support using influence line.

Ans. $R_A = \dfrac{ql}{6};\ R_B = \dfrac{ql}{3};\ M_{\max} = \dfrac{\sqrt{3}}{27}ql^2.$

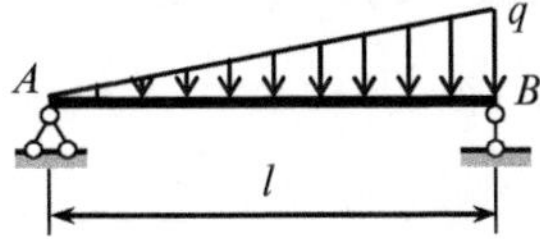

Fig. P2.2

2.3. Cantilever beam AB is subjected to the fixed loads as shown in Fig. P2.3. Analyze the structure using the fixed and moving load approach.

Fixed load approach Determine the reaction of support B; Construct the internal force diagrams.

Moving load approach

(a) Construct the influence lines for vertical reaction R_B and moment M_B at the clamped support
(b) Construct the influence lines for bending moment and shear for section k
(c) Calculate reactions at the support B, bending moment and shear for section k using corresponding influence lines
(d) Compare the results obtained by both approaches

Ans. $R_B = 14.0\,\text{kN}(\uparrow)$; $M_B = 57\,\text{kNm}$.

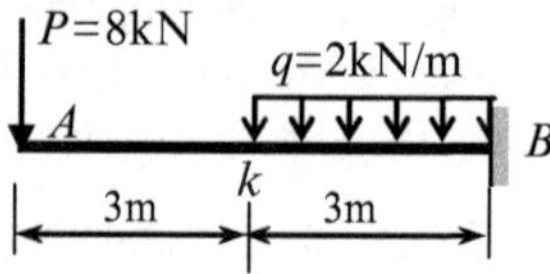

Fig. P2.3

2.4. The beam AB is subjected to the fixed uniformly distributed loads as shown in Fig. P2.4. All dimensions are in meters.

(a) Calculate the bending moment and shear at section k caused by the given fixed load using influence lines for M_K and Q_K, respectively
(b) Calculate the bending moment and shear force at section k caused by the given fixed load, using influence lines for *reaction* R_A
(c) Compare the results

Ans. $M_k = 9\,\text{kN m}$; $Q_k = 1.8\,\text{kN}$.

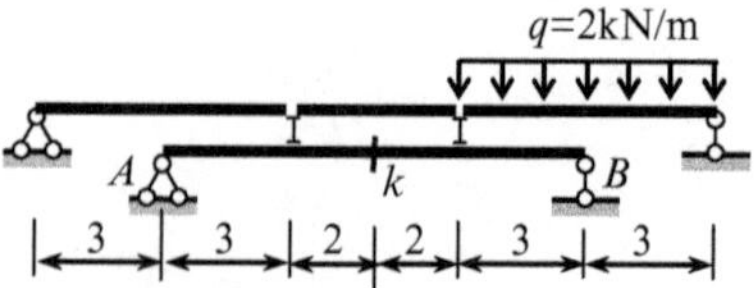

Fig. P2.4

2.5. Explain the meaning of the graph, which is obtained from any influence line by its differentiation with respect to axial coordinate x.

Chapter 3
Multispan Beams and Trusses

This chapter is devoted to the analysis of statically determinate multispan beams and trusses, subjected to moving loads. Methods for the generation of beam and trusses, and the construction of influence lines are discussed. Different types of trusses are considered; among them are trusses with subdivided panels and some special types of trusses.

3.1 Multispan Statically Determinate Beams

Multispan hinged beams (Gerber–Semikolenov beams) are geometrically unchangeable and statically determinate structures consisting of a series of one-span beams with or without overhangs connected together by means of hinges. The simplest Gerber–Semikolenov beams are presented in Fig. 3.1.

Fig. 3.1 Simplest Gerber–Semikolenov beams

3.1.1 Generation of Multispan Statically Determinate Hinged Beams

The following rules of distribution of hinges in beams, which have no clamped ends, provide their unchangeableness and statical determinacy:

1. Each span may contain no more than two hinges.
2. Spans with two hinges must alternate with spans without hinges.
3. Spans with one hinge may follow each other, providing that first (or last) span has no hinges.
4. One of support has to prevent movement in the horizontal direction.

I.A. Karnovsky and O. Lebed, *Advanced Methods of Structural Analysis*,
DOI 10.1007/978-1-4419-1047-9_3,

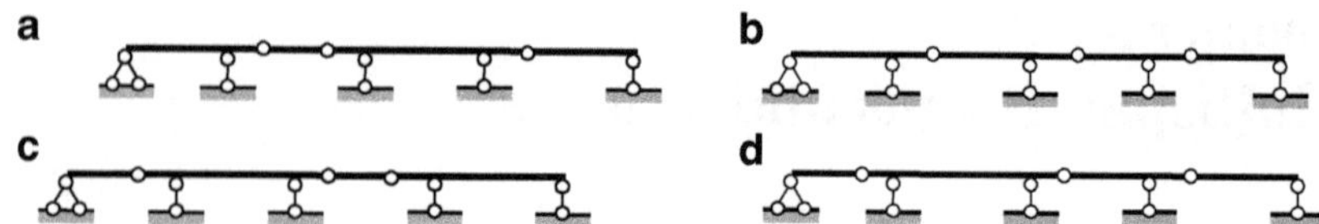

Fig. 3.2 Distribution of hinges in beams with simply supported ends

Figure 3.2 shows some examples of the hinges distribution in beams without fixed ends.

The distinctive properties of multispan statically determinate beams:

1. Structure with intermediate hinges has less stiffness than structure without intermediate hinges. This leads to substantial reduction of bending moments as compared with continuous beams spanning the same opening.
2. Possibility of control stresses by variation of hinges locations.

Advantages of multispan hinged beams are as follows:

1. Change of temperature, settlements of supports, imperfect of assembly does not create stresses.
2. Failure of one of the support may not destroy the entire system.
3. Relatively short members of multispan beams are well suited for prefabrication, transportation, and installation using standard equipment.
4. Multispan hinged beams are usually more economical than a series of disconnected simply supported beams spanning the same opening.

3.1.2 Interaction Schemes and Load Path

Gerber–Semikolenov beams may be schematically presented in the form, which shows the interaction of separate parts and transmission of forces from one part of the beam to another. Gerber–Semikolenov beams consist of two types of beams, namely a main (or primary) and suspended (or secondary) beam.

A main beam is designed to carry a load, which is applied to this beam as well as to maintain a suspended beam. Therefore, the main beam carries a load, which is applied to this beam, as well as a load, which is transmitted on the main beam as a reaction of the suspended beam. Interaction schemes for simplest Gerber–Semikolenov beams are presented in Fig. 3.3.

In these cases, beams 1 are primary and beams 2 are secondary ones. Obviously, that failure of the beam 1 causes the collapse of the entire structure while failure of the beam 2 does not cause failure of the beam 1. There is only one way of presenting multispan beams using an interaction diagram.

Interaction schemes allow to clearly indicate the load pass from one part of a structure to another. Also they are helpful for construction of internal force diagrams as well as for more advanced analysis.

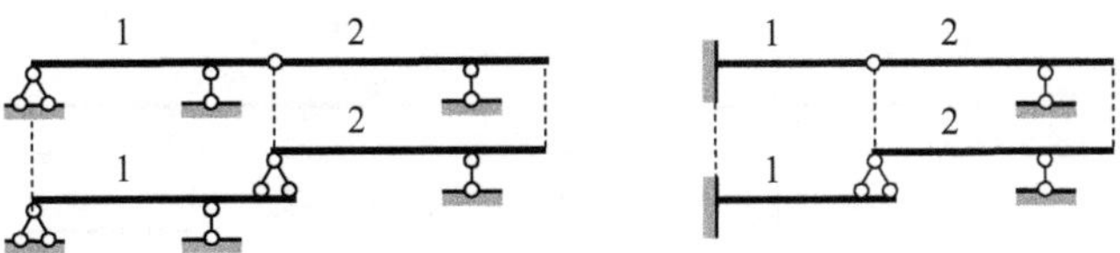

Fig. 3.3 Interaction schemes for simplest Gerber–Semikolenov beams

Five-span beam and its interaction diagram is presented in Fig. 3.4. Since the structure in a whole is restricted against the horizontal displacement owing to support A, then each part of the structure has no displacement in horizontal direction as well. That is why the rolled supports C and E on the interaction diagram are replaced by pinned supports. Restrictions for suspended beams also prevent horizontal displacements. This replacement is conventional.

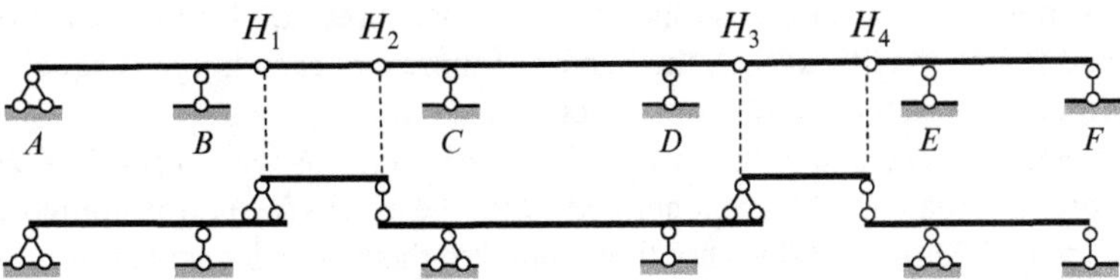

Fig. 3.4 Interaction diagram of Gerber–Semikolenov beam

Beams ABH_1 and H_1H_2 present the main and suspended beams, respectively. It means that collapse of the main beam ABH_1 leads to the collapse of suspended beam, while collapse of suspended beam does not affect on the main beam. Collapse of the beam H_4EF leads to the collapse of the beam H_3H_4 only, and does not affect on other parts of the structure.

The entire structure contains five discs, four simple hinges, and seven constraints of supports. Degree of freedom of the structure is

$$W = 3D - 2H - C_0 = 3 \times 5 - 2 \times 4 - 7 = 0,$$

so entire system is statically determinate and geometrically unchangeable structure. On the other hand, in general case of loads there are seven reactions of supports arise in the structure. For their calculation we can use three equilibrium equations and four additional conditions, i.e., bending moment at each hinge is equal to zero. Therefore, the structure is statically determinate one.

Another four-span beam and its interaction diagram is presented in Fig. 3.5.

The entire structure contains four discs, three simple hinges, and six constraints of supports. Degree of freedom of the structure is

$$W = 3D - 2H - C_0 = 3 \times 4 - 2 \times 3 - 6 = 0,$$

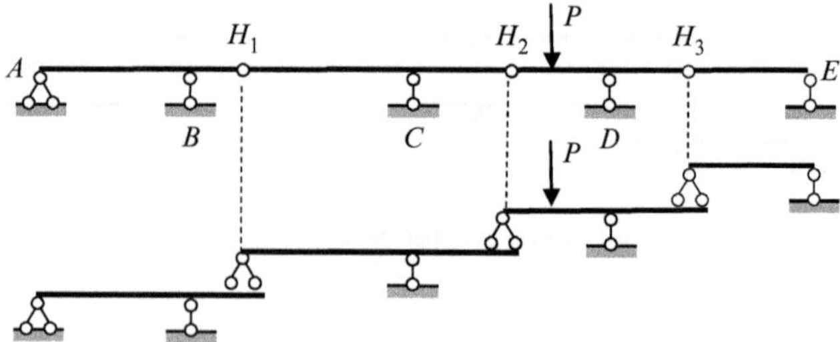

Fig. 3.5 Interaction diagram of Gerber–Semikolenov beam

so entire system is statically determinate and geometrically unchangeable structure. On the other hand, in general case of loads there are six reactions of supports arise in the structure; they are five vertical reactions at supports and one horizontal reaction at support A. For their calculation we can use three equilibrium equations and three additional conditions, i.e., the bending moment at each hinge is equal to zero. Therefore, the structure is statically determinate one.

The beam H_2H_3 is both main for H_3E beam and suspended for H_1CH_2 one. Therefore, the load P, which is applied to the beam H_2H_3 is transmitted only to beams located below and does not transmitted to the beams located above this one.

It is not difficult to form the rules of distribution of hinges which transforms the continuous beam with one (or two) clamped ends into Gerber–Semikolenov beam.

3.1.3 Influence Lines for Multispan Hinged Beams

For construction of influence lines of reactions and internal forces for statically determinate multispan beams the following steps are recommended:

Step 1. The entire multispan hinged beam should be presented in the interaction diagram form. This helps to classify each element of the structure as primary or secondary beams, and visualize the load path from the secondary beam on the primary one.

Step 2. Consider structural element of a multispan beam, which contains a support or section for which the required influence line should be constructed. This element is the primary or secondary simply supported beam with/without overhangs or cantilevered beam. Then we need to construct an influence line for the required function for this beam only.

Step 3. Take into account the influence of moving load which is located on the *adjacent suspended* beam and distribute the influence line which is constructed in step 2 along this secondary beam. For this it is necessary to connect an ordinate of influence line at the end point (hinge) with a zero ordinate at the second support of the suspended beam. This procedure is called as distribution of influence lines within the suspended beam.

Step 4. The beam distribution procedure should be applied for the following suspended beam.

To illustrate this procedure for statically determinate multispan beams we will consider a structure as in Fig 3.6. It is necessary to construct the influence lines for reactions, bending moment, and shear for indicated sections. We are starting from kinematical analysis of a structure: this beam is statically determinate and geometrically unchangeable structure.

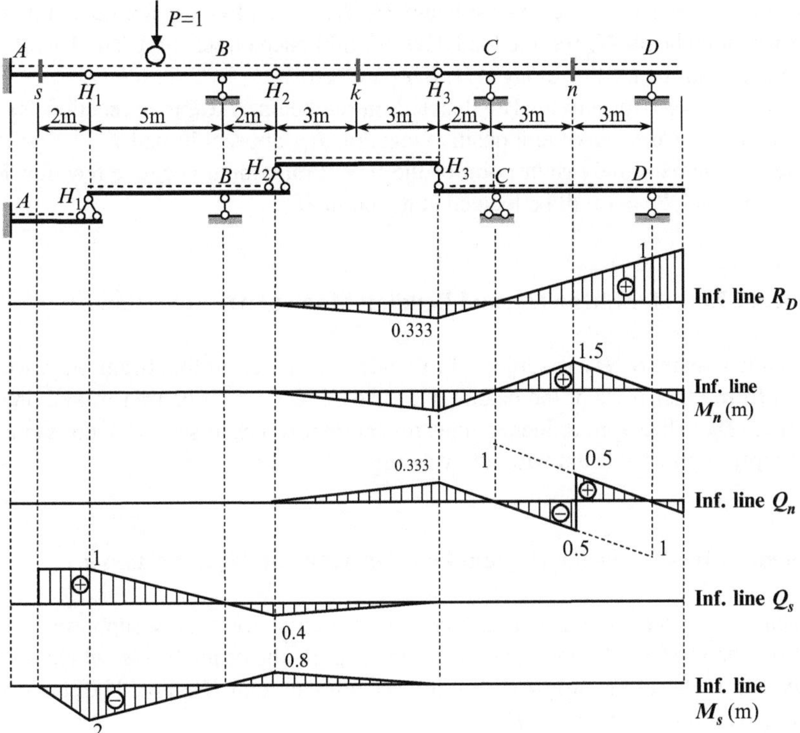

Fig. 3.6 Multispan statically determinate beam. Influence lines for reactions, bending moment and shears at some sections

Influence Line for R_D

The structural element is beam H_3CD. Influence line for reaction R_D within the span CD is a straight line with ordinate 1 at support D, zero at support C, and extended within left-hand and right-hand overhangs.

If load $P = 1$ is located on the suspended beam H_2H_3, then fraction of this force is transmitted to the primary beam H_3CD. If load $P = 1$ located at point H_3 of suspended beam H_2H_3, then this force is completely transmitted to the primary beam H_3CD. Thus, influence line has no discontinuity at point H_3.

If load $P = 1$ is located at point H_2 on the suspended beam H_2H_3, then this force is completely transmitted to the primary beam H_1BH_2 and thus, has no influence on the reaction R_D. Therefore, influence line of R_D has zero ordinate at point H_2. If unit load is located within H_2H_3, then the pressure transmitted from the secondary beam H_2H_3 to the primary beam H_3CD at point H_3 varies proportionally to the distance of the unit load from point H_2. Therefore, ordinate of influence line at point H_3 should be connected with zero ordinate at point H_2.

If load $P = 1$ is located on the beam H_1BH_2, then force is not transmitted to the suspended beam H_2H_3; the load does not influence on reaction R_D. Therefore, ordinates of influence line along AH_1BH_2 are zeros.

If load $P = 1$ (or any load) is located on the beam H_2C, then reaction R_D is directed downward. Maximum positive reaction R_D appears if load $P = 1$ will be located at the extremely right point of the beam; maximum negative reaction R_D appears if load $P = 1$ will be located at the point H_3.

Influence Lines for Shear Q_k, and Bending Moment M_k at Section k

Suspended beam H_2H_3 is subjected to loads, which act on this beam only, while a load from other parts of the beam (AH_2 and H_3D) cannot be transmitted on the beam H_2H_3. Therefore influence lines for internal forces at section k are same as for simply supported beam without overhangs.

Influence Lines for Shear Q_n, and Bending Moment M_n at Section n

Influence lines for section n should be constructed as for simply supported beam with overhang H_3C. At point H_3 this beam supports the beam H_2H_3, which is suspended one. Therefore, ordinates of influence lines at point H_3 should be connected with zeros ordinates at point H_2.

Influence Lines for Shear Q_s, and Bending Moment M_s at Section s

When the load travels along portion AH_1, then construction of the influence lines for section s is exactly the same as for a cantilevered beam.

If load $P = 1$ is located on the suspended beam H_1H_2, then fraction of this force is transmitted to the primary beam AH_1. If load $P = 1$ is located at point H_1 of suspended beam H_1BH_2, then this force is completely transmitted to the primary beam AH_1. Therefore, influence line has no discontinuity at point H_1.

If load $P = 1$ is located at point B on the suspended beam H_1BH_2, then this force is completely transmitted to the support B and no part of this force is taken by the support H_1 and thus, no force is transmitted to the main beam AH_1. Therefore, influence lines for shear and moment at section s has zero ordinates at point B. If unit load is located within the H_1BH_2, then the pressure from the secondary beam on the primary beam at point H_1 varies proportionally to the distance of the unit load from point B.

Similar discussions should be used, when unit load travels along the suspended beam H_2H_3.

If load travels along the beam H_3CD, then no part of this load is transmitted to the beam AH_1. Therefore, ordinates on influence lines for Q_s and M_s along the part H_3CD are zeros.

If load is located within the part s-B, then bending moment at section s is negative. It means, that the extended fibers at section s are located above the neutral line. If any load will be distributed within BH_3, then extended fibers at section s will be located below the neutral line.

3.1.4 Summary

For construction of influence lines for multispan statically determinate hinged beam it is necessary to show interaction diagram, and to show the part of the entire beam, which contains the support or section under consideration; this part is clamped-free or simply supported beam with or without overhangs. Next we need to construct the required influence lines considering the pointed portion of a beam and then distribute influence line along the all beams which are *suspended* with respect to pointed one. Thus construction of influence lines for multispan hinged beams is based on the influence lines for three types of simple beams and does not requires any analytical procedures.

Influence lines of reactions and internal forces for Gerber–Semikolenov beams, as for any statically determinate structure, are linear.

Example 3.1. Load is applied to stringers and transmitted to the Gerber–Semikolenov beam *ABHCD* by floor beams at points m, n, s, and t (Fig. 3.7). Construct the influence lines for reaction at A and for bending moment at section k.

Solution. First of all, show the interaction scheme for entire multispan hinged beam.

Influence line for R_A:

1. Influence line for R_A without floor beams and stringers is presented by polygon *ahd* by dotted line.
2. Draw vertical lines from floor beams' points m, n, s, and t to the intersection with the influence line.
3. Draw vertical lines from extreme left and right points E and F of the stringers to the intersection with the base line. Corresponding points of intersections have the notation e and f.

4. Connect nearest intersection points by connecting lines.
5. Influence line as a result of indirect load application is bounded by broken line *emnstf*.

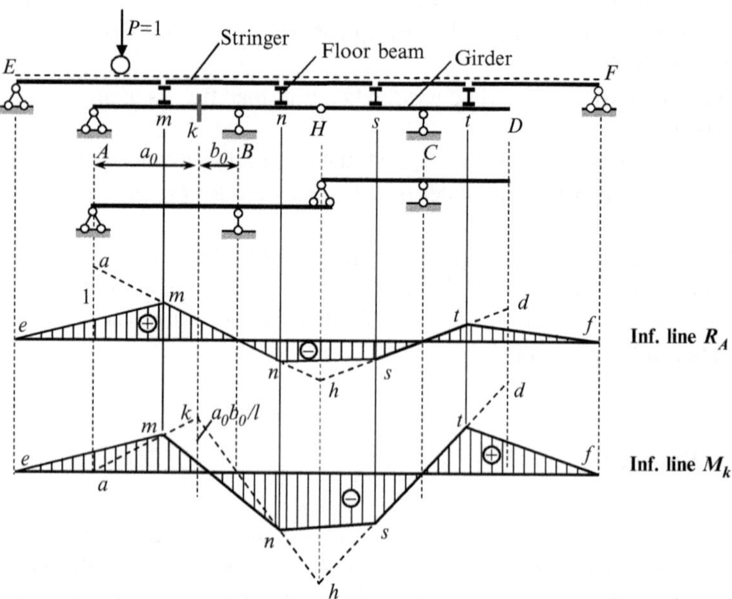

Fig. 3.7 Gerber–Semikolenov beam with indirect load application, its interaction diagram and influence lines

Influence line for M_k:

If the load is applied to the beam *AHD* directly, then the influence line for M_k is bounded by *akhd* points. In case of indirect load application, the influence line is bounded by *emnstf* points.

Discussion

1. Because of the indirect load application, the reactions and internal stresses arise in the beam AD even if the load is applied outside the beam AD (close to supports E or F).
2. Influence lines are convenient to use for analysis of structures, with modified design diagram. For example, for the given structure the bending moment at section k is positive, if any load is located within the stringer beam *Em*. It means that the extended fibers in the section k are located under the neutral axis of the beam. However, if a floor beam m will be removed and two stringers *Em* and *mn* will be replaced by one stringer *En*, then the points e and n on influence line for M_k should be connected. In this case bending moment at section k becomes negative, e.g., the extended fibers in the section k will be located above the neutral line.

3.2 The Generation of Statically Determinate Trusses

The trusses according to their generation are subdivided into simple, compound, and complex ones.

3.2.1 Simple Trusses

Simple trusses are formed on the basis of a hinge-connected triangle with any additional joint attached by means of two additional members. The relationship between the number of bars (S) and joints (J) for simple statically determined and geometrically unchangeable trusses is expressed as

$$S = 2J - 3. \tag{3.1}$$

This formula can be derived using two principally different approaches.

1. *The unchangeability condition of a truss.* A simple truss has an initial triangle disc and each new joint is attached to the previous rigid disk using two bars (Fig. 1.4c). The total number of bars is $S = 3 + 2(J - 3)$. In this formula, the first number represents the three bars used for the initial rigid disc. The 3 within the brackets arises from the three joints of the initial triangle, so $J - 3$ represents the number of joints attached to the initial triangle. The coefficient 2 stems from the number of bars associated with each additional joint. Therefore, the total number of bars is given by (3.1).
2. *The statically determinacy condition of a truss.* For each joint, two equilibrium equations must be satisfied: $\sum X = 0$, $\sum Y = 0$. The number of unknown internal forces is equal to the total number of members S. The total number of unknowns, including the reactions of the supports, is $S + 3$. The number of equilibrium equations equals $2J$. Therefore, a truss is statically determinate if $S = 2J - 3$.

Thus, the two different approaches both lead to the same (3.1) for the kinematical analysis of a simple truss. Therefore, if a *simple* truss is statically determinate, its geometry must be unchangeable and vice versa. Note, however, that (3.1), while necessary, is an insufficient condition on its own to determine that a truss is geometrically unchangeable. It is indeed possible to devise a structure which satisfies (3.1) yet connects the bars in such a way that the structure is geometrically changeable. For more examples see Chap. 1.

If the number of bars $S > 2J - 3$, then the system is statically indeterminate; this case will be considered later in Part 2. If the number of bars $S < 2J - 3$, then the system is geometrically changeable and cannot be used as an engineering structure.

3.2.2 Compound Trusses

A compound truss consists of simple trusses connected to one another. There are different ways to connect simple trusses in order to construct a compound truss.

First approach uses three hinges, not arranged in line. Such an interconnection can be considered as a simple triangle on the basis of its three rigid discs. In the examples below, two of the trusses are represented by disks D_1 and D_2 (Fig. 3.8). Two immovable (pinned) supports may be treated as rigid disk D_3 (the earth).

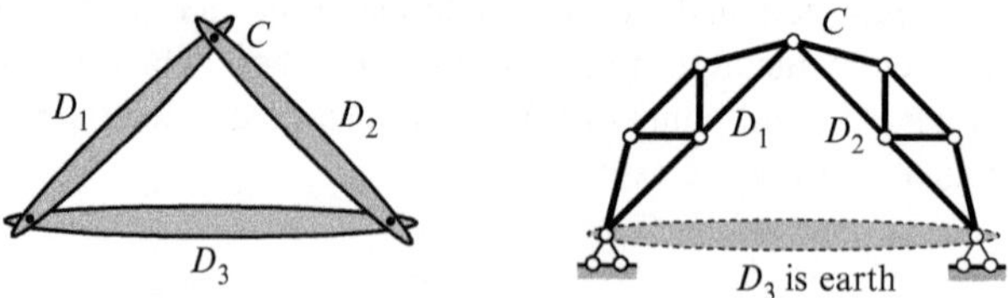

Fig. 3.8 The generation of a rigid disk by means of three hinges

Another approach uses hinge C and a rod R, which does not pass through the hinge (Fig. 3.9).

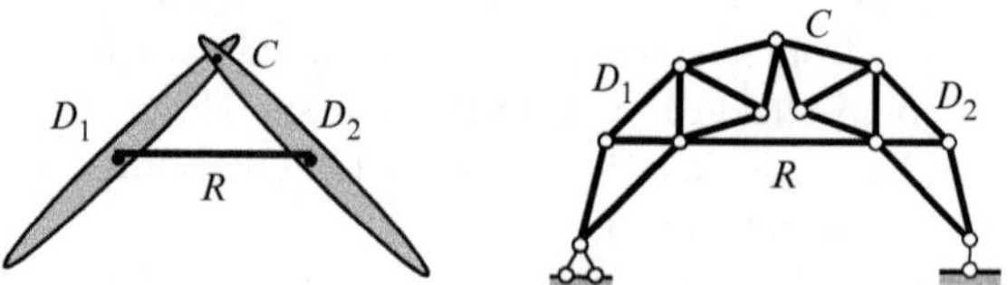

Fig. 3.9 The generation of a rigid disk by means of a hinge and a rod

A third method uses three rods, which are not parallel and do not all intersect at the same single point (Fig. 3.10).

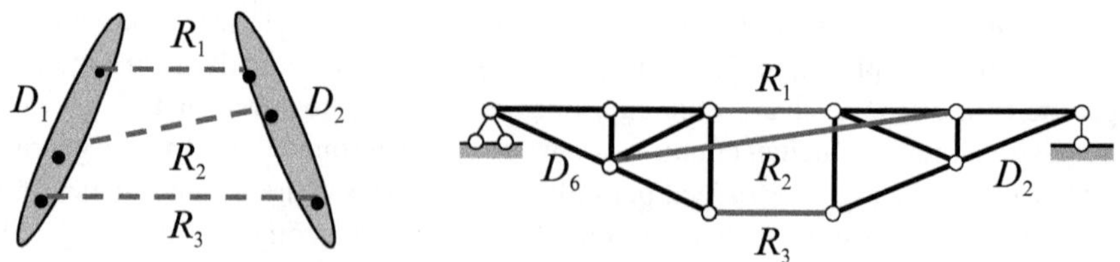

Fig. 3.10 The generation of a rigid disk by means of three rods

In addition, compound trusses can be composed of members that are themselves also trusses (Fig. 3.11). Such trusses suit structures carrying considerable loads, in which the truss designs would require heavy solid members. In these cases it is often more efficient to replace solid members with truss members.

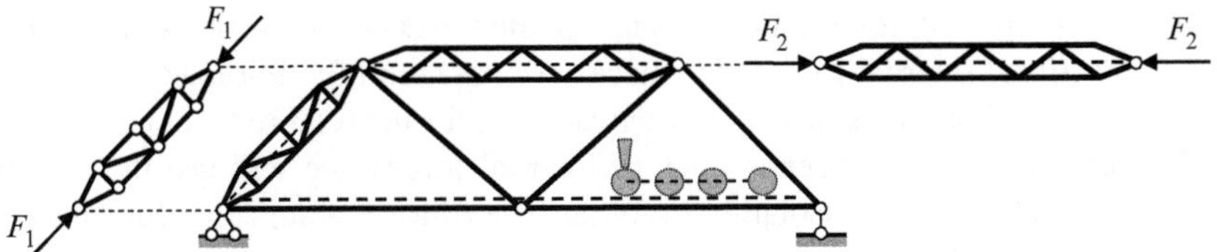

Fig. 3.11 A truss with compound members

The analysis of this truss should be performed as if all the compounded members are solid rods. The obtained internal forces should be applied to the compound member and then that member is considered as a truss. Since each compound member is subjected to two balanced forces, the reactions of the supports are each zero. Therefore, the supports of each compound member can be disregarded.

3.2.3 *Complex Trusses*

Trusses that are complex use other ways to connect two (or three) rigid discs. Figure 3.12a presents a connection of two rigid discs using hinge H, two bars 1 and 2, and an additional support C. This connection arrangement produces a Wichert truss. Figure 3.12b presents a connection of three rigid discs using hinge C and three hinged end bars 1, 2, and 3.

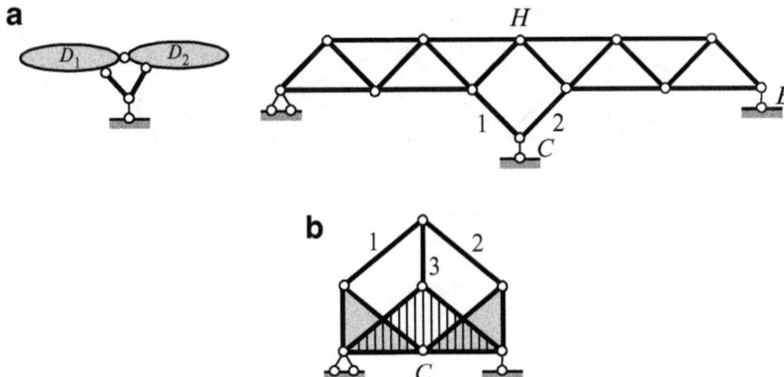

Fig. 3.12 Complex trusses

3.3 Simple Trusses

Construction of influence line for internal forces in trusses is based on analytical methods used for computing internal forces induced by fixed loads. However, construction of influence line for internal forces in a truss has specific features, which will be considered while using this method. We direct the reader's attention to the next important and fundamental point: as in the case of beams, the construction of influence lines for trusses will be based not on the repeated calculation of a required factor (reactions, internal forces) for successive position of a unit load as it moves across the span, but on the *deriving of the function* for the required factor.

For construction of influence lines of internal forces, we will use the method of sections and method of isolation of joints. The following are some fundamental features of the joints and cuts methods.

Using the joint isolation method, *three* types of position of a unit moving load on a load chord should be considered:

1. A moving load at the considered joint
2. A moving load anywhere joint except the considered one
3. A moving load within the dissected panels of a load chord

Using the cuts method, *three* types of position of a unit moving load on load chord should be considered:

1. A moving load on a left-hand part of dissected panel of load chord
2. A moving load on a right-hand part of dissected panel of load chord
3. A moving load within the dissected panel of load chord

Design diagram of the simple truss is shown in Fig. 3.13. Influence lines for reactions R_A and R_B for truss are constructed in the same manner as for reactions for one-span simply supported beam.

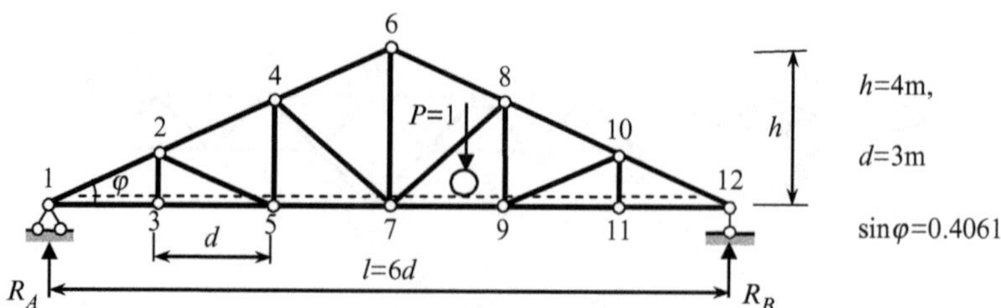

Fig. 3.13 Design diagram of the triangle truss

The following notation for internal forces will be used: U for bottom chord; O for top chord; V for vertical elements; and D for diagonal elements.

Influence Line for Force $O_{4\text{-}6}$ (Section 1-1, Ritter's Point 7, Fig. 3.14a)

The sectioned panel of the loaded chord (SPLC) is panel 5-7. It is necessary to investigate three position of unit load on the loaded chord: outside the panel 5-7 (when load P is located to the right of the joint 7 and to the left of joint 5) and within the panel 5-7. Load-bearing contour (or loaded chord) is denoted by dotted line.

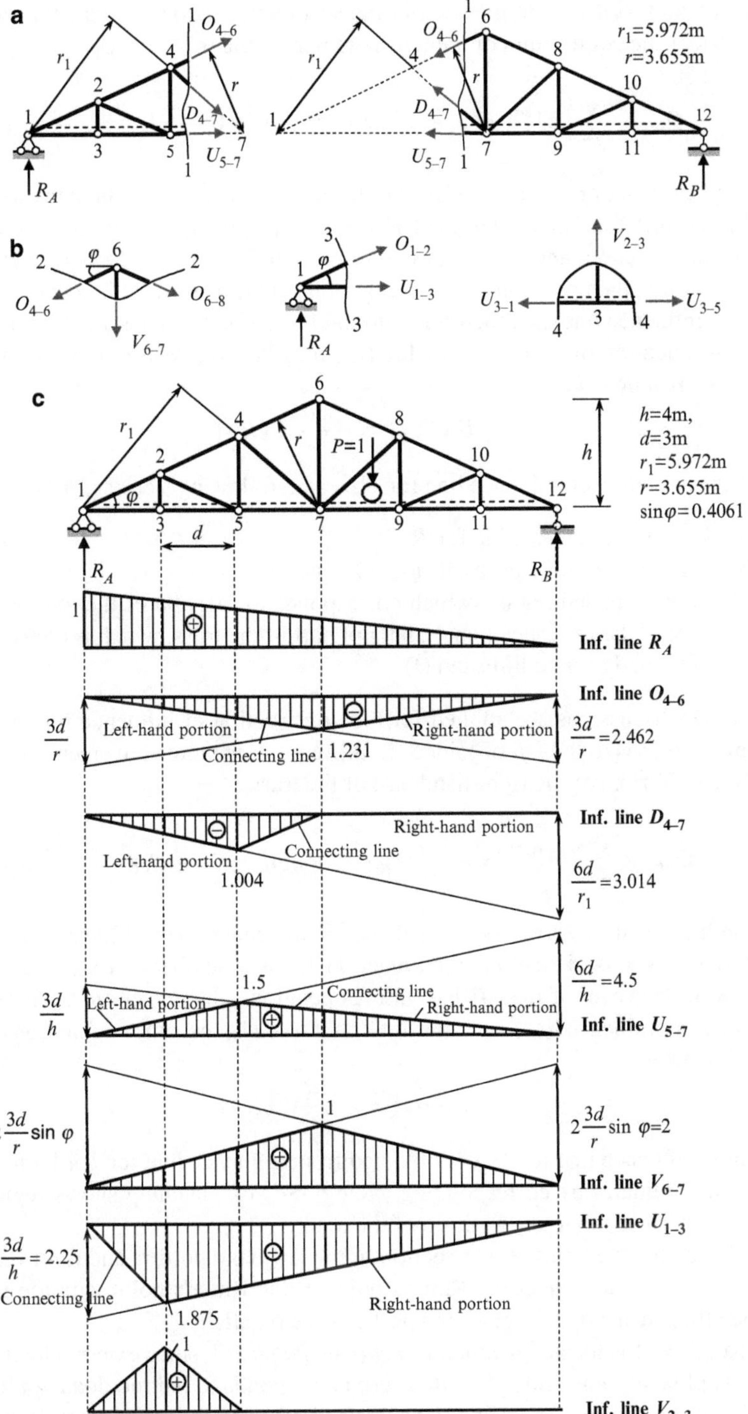

Fig. 3.14 (**a**) Free-body diagrams for left and right parts of the truss. (**b**) Free-body diagrams for specified joints of the truss. (**c**) Influence lines for truss with lower loaded chord

1. When unit load P is to the *right* of the sectioned panel, it is more convenient to consider the equilibrium of the left-hand part of the truss

$$O_{46} \to \sum M_7^{\text{left}} = 0 \to O_{46}r + R_A 3d = 0 \to O_{46} = -\frac{3d}{r} R_A.$$

The last equation for force O_{46} should be transformed into the equation of influence line for O_{46}. When load P has a fixed position, then reaction R_A is a specified number and O_{46} is a number as well. However, if load P is changing its position, then reaction R_A is changing its value as well, so it becomes a *function* (influence line) and therefore internal force O_{46} becomes a function as well. Thus, equation of influence line for O_{46} may be expressed in term of influence line of reaction R_A

$$\text{IL}(O_{46}) = -\frac{3d}{r}\text{IL}(R_A).$$

Thus, to construct influence line for $O_{4\text{-}6}$, the following steps are necessary:

(a) Show the influence line for R_A
(b) Multiply all ordinates by factor $-3d/r$
(c) Draw a shaded region, which corresponds to *location* of the load P (to the right of the sectioned panel), but not part of the truss, which is considered as a free body for equilibrium (!).

2. Similar discussions are applied for the construction of influence line when the unit load P is to the *left* of joint 5. In this case, it is more convenient to consider the equilibrium of the right-hand part of the truss

$$O_{46} \to \sum M_7^{\text{right}} = 0 \to O_{46}r + R_B 3d = 0 \to O_{46} = -\frac{3d}{r} R_B.$$

The last equation should be transformed in the equation of influence line. When load P has a fixed position, then reaction R_B is a number and O_{46} is a number as well. However, if load P has various positions, then reaction R_B becomes a *function* (influence line) and thus, internal force O_{46} becomes a function as well.

Therefore,

$$\text{IL}(O_{46}) = -\frac{3d}{r}\text{IL}(R_B).$$

Thus, influence line for O_{46} is obtained from influence line for R_B by multiplying its ordinates by corresponding value ($-3d/r$). Hatching shows region 1-5, which correspond to the position of the load.

Note, that the point of intersection of the left-hand portion and right-hand one is always located under the Ritter's point. If the Ritter's point is in the infinity, then the left-hand and right-hand portions are parallel.

3. Load $P = 1$ is located *within the sectioned panel* 5-7; since external load should be applied at joints only, then this case corresponds to indirect load application. Therefore, the influence line connects ordinates of influence lines at points 5

and 7; connecting line is always a straight line. The completed influence line is presented in Fig. 3.14c. As can be seen, the element $O_{4\text{-}6}$ is always compressed for any location of the moving load. Maximum internal force $O_{4\text{-}6}$ occurs when concentrated load is located at point 7.

It is very convenient to represent all information related to construction of influence lines in tabulated form as follows.

Force $O_{4\text{-}6}$ (Section 1-1, SPLC Is Panel 5-7, Ritter's Point 7)

$P = 1$ left at SPLC	$P = 1$ right at SPLC
$O_{4\text{-}6} \to \sum M_7^{\text{right}} = 0 \to O_{4\text{-}6}r + R_B 3d = 0,$ $O_{4\text{-}6} = -\frac{3d}{r} R_B \to \text{IL}(O_{4\text{-}6}) = -\frac{3d}{r}\text{IL}(R_B)$	$O_{4\text{-}6} \to \sum M_7^{\text{left}} = 0 \to O_{4\text{-}6}r + R_A 3d = 0,$ $O_{4\text{-}6} = -\frac{3d}{r} R_A \to \text{IL}(O_{4\text{-}6}) = -\frac{3d}{r}\text{IL}(R_A)$

Analytical expressions for influence lines of internal forces for the remaining elements of the truss are obtained in the same manner as above. From now on we will be presenting analytical expressions for construction of influence lines in tabulated form. The tabulated form contains all necessary information for construction of influence line.

Force $D_{4\text{-}7}$ (Section 1-1, SPLC Is Panel 5-7, Ritter's Point 1)

$P = 1$ left at SPLC	$P = 1$ right at SPLC
$D_{4\text{-}7} \to \sum M_1^{\text{right}} = 0 \to D_{4\text{-}7}r_1 + R_B 6d = 0,$ $D_{4\text{-}7} = -\frac{6d}{r_1} R_B \to \text{IL}(D_{4\text{-}7}) = -\frac{6d}{r_1}\text{IL}(R_B)$	$D_{4\text{-}7} \to \sum M_1^{\text{left}} = 0 \to D_{4\text{-}7}r_1 = 0,$ $D_{4\text{-}7} = 0 \to \text{IL}(D_{4\text{-}7}) = 0$

The left-hand portion of influence line for $D_{4\text{-}7}$ is obtained by multiplying all ordinates of influence line for R_B by a constant factor $(-6d/r_1)$. Ordinates of the right-hand portion of influence line for $D_{4\text{-}7}$ are zeros. Connecting line connects the joints 5 and 7. Thus, if any load is located within portion 7-12, then internal force in member $D_{4\text{-}7}$ does not arise. Maximum force $D_{4\text{-}7}$ occurs, if load P is placed at joint 5.

Force $U_{5\text{-}7}$ (Section 1-1, SPLC Is Panel 5-7, Ritter's Point 4)

$P = 1$ left at SPLC	$P = 1$ right at SPLC
$U_{5\text{-}7} \to \sum M_4^{\text{right}} = 0 \to U_{5\text{-}7}\frac{2}{3}h - R_B 4d = 0,$ $U_{5\text{-}7} = \frac{6d}{h} R_B \to \text{IL}(U_{5\text{-}7}) = \frac{6d}{h}\text{IL}(R_B)$	$U_{5\text{-}7} \to \sum M_4^{\text{left}} = 0 \to U_{5\text{-}7}\frac{2}{3}h - R_A 2d = 0,$ $U_{5\text{-}7} = \frac{3d}{h} R_A \to \text{IL}(U_{5\text{-}7}) = \frac{3d}{h}\text{IL}(R_A)$

The left-hand portion of influence line for $U_{5\text{-}7}$ is obtained by multiplying all ordinates of influence line for R_B by a constant factor $6d/h$. The right-hand portion of influence line for $U_{5\text{-}7}$ is obtained by multiplying all ordinates of influence line for R_A by a constant factor $3d/h$. Connecting line runs between points 5 and 7. The force $U_{5\text{-}7}$ is tensile for any location of load on the truss.

Force $V_{6\text{-}7}$ (Section 2-2, Equilibrium of Joint 6, Fig. 3.14b)

$$\begin{aligned}\sum X &= 0 \to -O_{4\text{-}6}\cos\varphi + O_{6\text{-}8}\cos\varphi = 0 \to O_{4\text{-}6} = O_{6\text{-}8},\\ \sum Y &= 0 \to -V_{6\text{-}7} - 2O_{4\text{-}6}\sin\varphi = 0 \to V_{6\text{-}7} = -2O_{4\text{-}6}\sin\varphi \to \mathrm{IL}(V_{6\text{-}7})\\ &= -2\sin\varphi\cdot\mathrm{IL}\,(O_{4\text{-}6})\end{aligned}$$

The influence line for $V_{6\text{-}7}$ is obtained by multiplying all ordinates of influence line for $O_{4\text{-}6}$ by a constant factor $(-2\sin\varphi)$. The maximum ordinate of this influence line equals to $(3d/r)\sin\varphi = 1$. This result may be obtained using the other approach: if load $P = 1$ is located at point 7, then internal forces in all vertical members (except $V_{6\text{-}7}$) and diagonal members are zero; equilibrium equation of joint 7 leads to $V_{6\text{-}7} = 1$.

Force $U_{1\text{-}3}$ (Section 3-3, Equilibrium of Joint 1, Fig. 3.14b)

$P = 1$ applied at joint 1	$P = 1$ applied at joint 3 or to the right of joint 3
$U_{1\text{-}3} \to \sum M_2^{\text{left}} = 0$ $\to U_{1\text{-}3}\frac{h}{3} + 1\cdot d - R_A d = 0,$	$U_{1\text{-}3} \to \sum M_2^{\text{left}} = 0 \to U_{1\text{-}3}\frac{h}{3} - R_A d = 0,$
$R_A = 1 \to U_{1\text{-}3} = 0 \to \mathrm{IL}(U_{1\text{-}3}) = 0$	$U_{1\text{-}3} = \frac{3d}{h}R_A \to \mathrm{IL}(U_{1\text{-}3}) = \frac{3d}{h}\mathrm{IL}(R_A)$

The right-hand portion of influence line for $U_{1\text{-}3}$ is obtained by multiplying all ordinates of influence line for R_A by a constant factor $(3d/h)$. Connecting line connects the joints 1 and 3.

All the above-mentioned influence lines are presented in Fig. 3.14c.

3.4 Trusses with Subdivided Panels

Trusses with subdivided panels present are simple trusses with additional members. Figure 3.15a shows the Pratt truss with two additional members within each panel; for first panel they are 3-*k* and *k*-*n*. Figure 3.15b shows the Parker truss with three additional members within each panel; they are *a*-*b*, *a*-*c* and *b*-*c*. In both examples additional elements divide a panel (vertical member *k*-*n* in first case and inclined members *a*-*c* and *b*-*c* in the second case). In the analysis of such trusses, an immediate application of the joints and through section methods might lead to some difficulties.

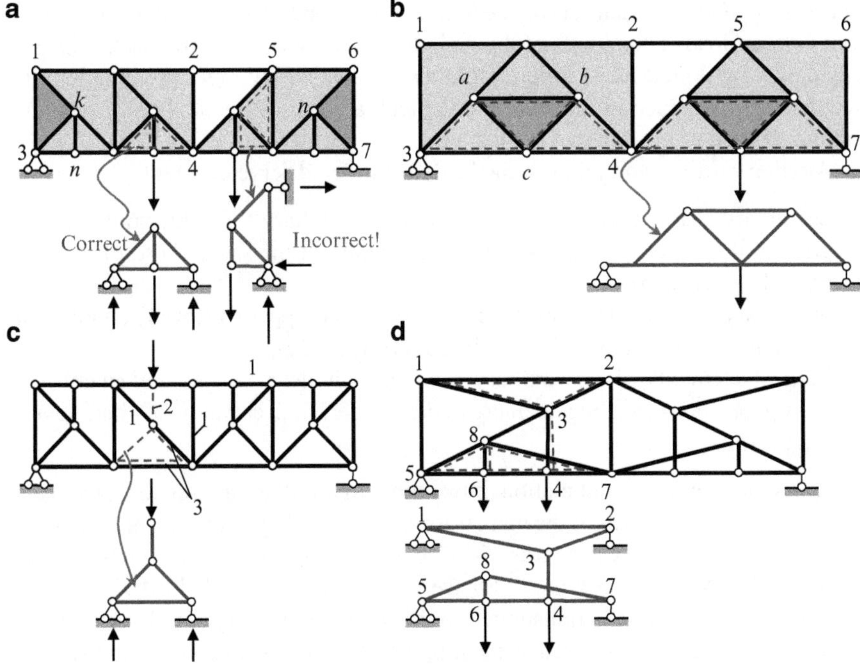

Fig. 3.15 Trusses with subdivided panels with their main and secondary trusses. Note: *Shaded zone* in (**a**) and (**b**) are discussed in the kinematical analysis

3.4.1 Main and Auxiliary Trusses and Load Path

Analysis of trusses with subdivided panels can be performed in an orderly fashion using the concepts of the *main* and *secondary* (*auxiliary*) trusses. The main truss is a simple truss, which is enhanced by additional members. The secondary truss is a *conventional* truss, which contains additional real members (3-k and k-n for truss in Fig. 3.15a) and some imaginary members. These imaginary members coincide with members of the main truss. The secondary trusses are shown by dotted lines. The concepts of the main and secondary trusses allow us to consider trusses with subdivided panels as compound trusses. Combinations of two trusses – a main and sets of secondary trusses are shown in Fig. 3.15. In all cases, the secondary trusses are supported by the joints of the main truss.

The main purpose of a secondary truss is to transmit a load, which is applied *between the joints* of a main truss, *to the joints* of a main truss.

How can we distinguish between the main and the secondary trusses? The rule is that elements of the secondary truss are connected with the main truss in such a way that a *vertical* load, which acts on the joint of the secondary truss, is transmitted only as a *vertical* load to the joints of the main truss. The secondary trusses in Fig. 3.15 are shown by dotted lines. Figure 3.15a also shows an incorrect presentation of a secondary truss, since in this case the vertical, as well as the *horizontal reactions* are transmitted to the joints of the main truss.

Auxiliary trusses add joints to the loaded chord and serve different goals:

1. Secondary elements (i.e., the elements of the auxiliary truss) transmit loads applied to the lower (or upper) chord to the joints of the same chord of the main truss (Fig. 3.15a, b).
2. Secondary elements transmit loads applied to the upper (or lower) chord to the joints of the other chord of the main truss (Fig. 3.15c).
3. Vertical elements of an auxiliary truss can divide a *compressed* member of the upper chord, leading to increased load-carrying capacity of this element (Fig. 3.15c).

It is possible to have a combination of two types of auxiliary trusses so that a load is transmitted to both chords of the main truss as shown in Fig. 3.15d. In this case hinge 6 belongs to bottom auxiliary truss 5-8-7-6, while hinge 4 belongs to top auxiliary truss 1-2-3-4. Members 3-4 and 7-8 have no common points. If load P is located at point 4, then the load is transmitted to the *upper* chord (to joints 1 and 2) of the main truss. If load P is located at point 6, then the load is transmitted to the *lower* chord (to joints 5 and 7) of the main truss. If load P is located between points 4 and 6, then part of the load is transmitted to joint 4 and part to joint 6, and from there it is transmitted from joint 4 to an upper chord and from joint 6 to lower chord.

Kinematical Analysis

For kinematical analysis of the truss in Fig. 3.15a, we show the initial rigid disc 1-3-k. Each subsequent hinge is connected using two bars with hinges at the ends, thus part 1-2-4-3 of the truss presents a rigid disc. The next joint cannot be obtained in this way, so instead consider the truss from the right where the base rigid triangle is 6-7-n. Similarly, each subsequent hinge is connected using two bars with hinges at the ends, so part 5-6-7-4 presents a second rigid disc. Both discs are connected by hinge 4 and member 2-5 according to Fig. 3.9; this compound truss is geometrically unchangeable. The trusses in Fig. 3.15b,c can be analyzed in a similar way.

For all the design diagrams presented in Fig. 3.15, the relationship $S = 2J - S_0$ is satisfied. For example, for structures (a,c) $S = 33$ and $J = 18$; for structures (b,d) $S = 25$, and $J = 14$. All these structures are statically determinate and geometrically unchangeable compound trusses.

The elements of trusses with subdivided panels can be arranged into three groups as follows:

1. *The elements belonging to the main truss only*. The internal forces in these elements are not influenced by the presence of secondary trusses. Such elements are labeled #1 in Fig. 3.15c.
2. *The elements belonging to the secondary system only*. The internal forces in these elements arise if a load is applied to the joints of the secondary system. Such elements are labeled #2.
3. *The elements belonging simultaneously to both the main and the secondary trusses*. The internal forces in such members are obtained by summing the internal forces, which arise in the given element of the main and also of the secondary trusses, calculated separately. Such elements are labeled #3.

3.4.2 Baltimore and Subdivided Warren Trusses

The Baltimore truss is widely used in bridge-building for cases involving particularly long spans and large loads. The classical Baltimore truss is a Pratt truss strengthened by two additional members in each panel (members 2-4 and 3-4 in Fig. 3.16a). The modified Baltimore truss is a Pratt truss strengthened by *three* additional members in each panel (members 3-4, 2-4, and 4-6 in Fig. 3.16b). In both cases, truss 1-2-3-4 should be treated as a secondary truss. This triangular truss is supported by joints 1 and 3 of the main truss. In both cases, the vertical load, which is applied to the loaded chord and acts at joint 2, leads to vertical reactions of the secondary truss. These reactions are transmitted as active loads on the joints of the *same* chord of the main truss.

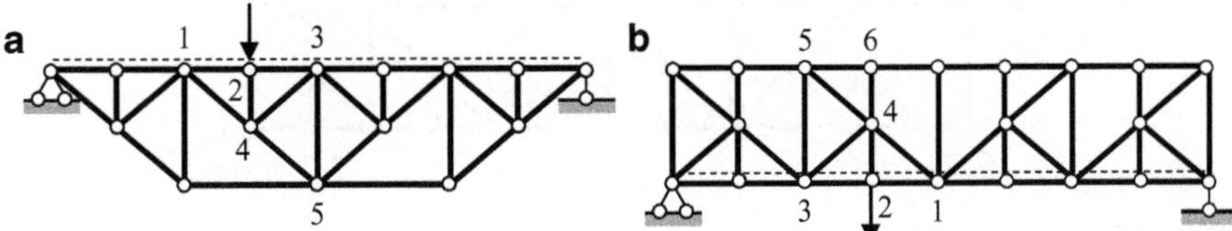

Fig. 3.16 Baltimore truss

In order to reduce the length of the compressed member of the upper chord, member 2-4 is used in case (a) and additional member 4-6 in case (b).

Analysis of such trusses can be performed by considering the entire structure as a combination of a Pratt truss (the main truss) and a set of additional triangular trusses: one of them is additional truss 1-2-3-4.

There are three groups of elements: the ones that belong to the main truss only (bars 3-5 and 4-5), the ones that belong to the secondary truss only (bars 2-4, 3-4, and 4-6); and others that belong to both the main and the secondary trusses (bars 1-4, 1-2, and 2-3).

The internal forces in the elements of the secondary truss should be calculated considering only load, which is applied at the joints of the secondary truss. In other words, in order to find the internal forces in elements 2-4 and 3-4 (due to the load applied at joint 2), we need to consider the secondary truss, so the following notation is used: $V_{2\text{-}4} = V_{2\text{-}4}^{\text{sec}}$ and $D_{3\text{-}4} = D_{3\text{-}4}^{\text{sec}}$. Since joints 1 and 3 represent the supports of the secondary truss, if the load is applied at these joints it has no affect on the elements of the secondary truss. Therefore, this load impacts the elements belonging only to the main truss.

Internal forces in the members of the first group should be calculated considering only the main truss. In this case, the load that acts at joint 2 should be resolved into two forces applied at joints 1 and 3 of the main truss and then $V_{3\text{-}5} = V_{3\text{-}5}^{\text{main}}$; $D_{4\text{-}5} = D_{1\text{-}5}^{\text{main}}$.

Internal forces in the members of the third group should be calculated by considering both the main and the secondary trusses and by summing the corresponding internal forces:

$$D_{1\text{-}4} = D_{1\text{-}5}^{\text{main}} + D_{1\text{-}4}^{\text{sec}};\ O_{1\text{-}2} = O_{1\text{-}3}^{\text{main}} + O_{1\text{-}2}^{\text{sec}}.$$

The same principle will be used for the construction of influence lines.

Another type of this class of trusses is the subdivided Warren truss. This truss is generated from a simple Warren truss using the same generation principle used for a Baltimore truss.

Let us consider a detailed analysis of a subdivided Warren truss presented in Fig. 3.17. The moving load is applied to the lower chord. As usual, we start from a kinematical analysis. This truss is geometrically unchangeable because the rigid discs are connected by means of hinges and bars.

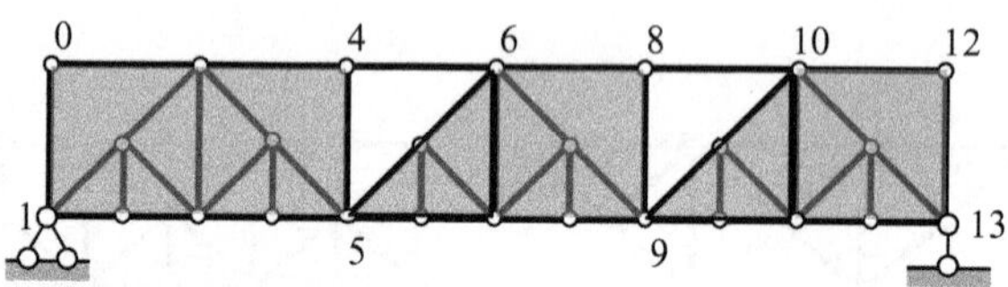

Fig. 3.17 Kinematical analysis of the Warren truss

In order for the structure to be statically determinate the number of bars S and joints J must satisfy the formula $S = 2J - 3$. In our case $S = 49$, $J = 26$, so $S = 2J - 3 = 52 - 3 = 49$. Therefore, this structure is geometrically unchangeable and statically determinate.

Once again the necessity of kinematical analysis must be emphasized, especially for trusses that are not simple and contain many elements. For example, what happens if the entire structure in Fig. 3.18 is modified by removing the following symmetrical set of members: case (1) $4'$-$5'$ and $6'$-$7'$; case (2) $4'$-$5'$, $4'$-7, $6'$-$7'$, and $6'$-$7'$; or case (3) $4'$-7 and $6'$-7? Detailed kinematical analysis, as carried out above, shows that this structure remains geometrically unchangeable for modifications (1) and (2). However, for modification (3) this structure becomes geometrically changeable and cannot be used as an engineered structure. Unfortunately, some modern software will work out a numerical analysis for any structure, even if the structure is changeable. Therefore, in order to avoid structural collapse, each engineer must complete a kinematical analysis as the very first step.

The entire structure in Fig. 3.18 can be represented as a combination of main and secondary trusses. A feature of this subdivided Warren truss is the following: if the load is located on joints 1′, 3′, 5′, etc. of the secondary trusses on the *lower* chord, then the reactions of the secondary trusses are transmitted as active loads to the joints of the *lower* chord of the main truss.

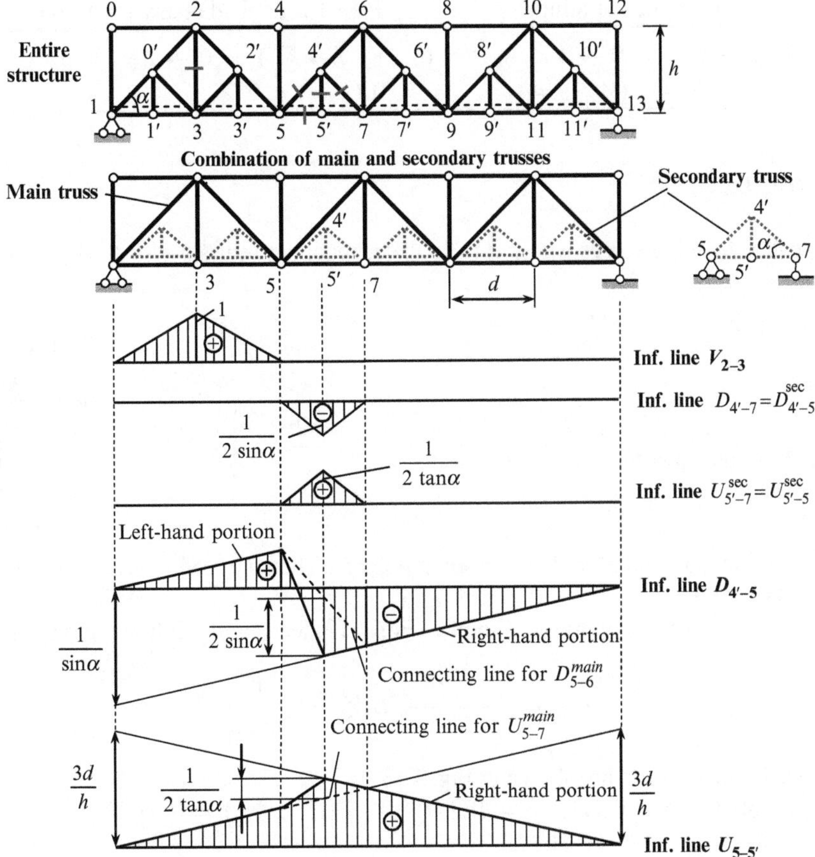

Fig. 3.18 Influence lines for internal forces for subdivided Warren truss

Now we will consider the construction of influence lines for reactions and internal forces of elements belonging to three different groups, i.e., to the main truss only, to a secondary truss only, and to both trusses simultaneously. These elements are indicated in Fig. 3.18.

Influence lines for reactions are constructed in the same manner as in the case of a simple truss. These influence lines are not shown.

The construction of influence lines for internal forces is presented in tabulated form below.

Force $V_{2\text{-}3}$

Vertical member 2-3 belongs to the main truss only; therefore, its influence line is obtained by considering the main truss only. Consider the equilibrium of joint 3 of the main truss:

$P = 1$ is applied at joint 3	$P = 1$ is applied at any joint except 3
$V_{2\text{-}3} \rightarrow \sum Y = 0 \rightarrow V_{2\text{-}3} - P = 0,$ $V_{2\text{-}3} = P \rightarrow \text{IL}(V_{2\text{-}3}) = 1$	$V_{2\text{-}3} \rightarrow \sum Y = 0 \rightarrow V_{2\text{-}3} = 0,$ $\text{IL}(V_{2\text{-}3}) = 0$

Forces $D_{4'\text{-}7}$, $U^{\text{sec}}_{5'\text{-}7}$

To calculate these forces we need to consider secondary truss 5-4′-7-5′ (Fig. 3.19).

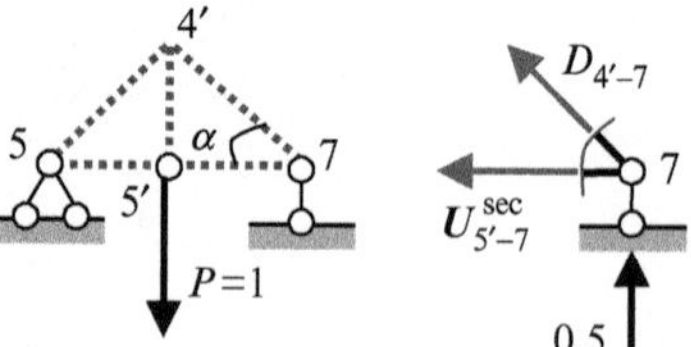

Fig. 3.19 Secondary truss loading

Assume that load $P = 1$ is located at point 5′. Then the reactions at points 5 and 7 will equal 0.5.

The equilibrium condition of joint 7 (Fig. 3.19) leads to the following results:

$$D_{4'\text{-}7} = -\frac{1}{2\sin\alpha}, \quad U^{\text{sec}}_{5'\text{-}7} = \frac{1}{2\tan\alpha}.$$

These influence lines are shown in Fig. 3.18.

It is obvious that $\text{IL}(D_{4'\text{-}7}) = \text{IL}\left(D^{\text{sec}}_{4'\text{-}5}\right)$ and $\text{IL}\left(U^{\text{sec}}_{5'\text{-}7}\right) = \text{IL}\left(U^{\text{sec}}_{5'\text{-}5}\right)$.

Force $D_{4'\text{-}5}$

Diagonal member 4′-5 belongs simultaneously to the main and to a secondary truss. Therefore, an influence line should be constructed considering the main truss and secondary truss 5-4′-7-5′ together:

$$D_{4'\text{-}5} = D^{\text{main}}_{5\text{-}6} + D^{\text{sec}}_{4'\text{-}5} \rightarrow \text{IL}(D_{4'\text{-}5}) = \text{IL}\left(D^{\text{main}}_{5\text{-}6}\right) + \text{IL}\left(D^{\text{sec}}_{4'\text{-}5}\right).$$

Force $D_{5\text{-}6}$, Main Truss: The SPLC Is Panel 5-7; Ritter's Point Is at Infinity

$P=1$ left at SPLC	$P=1$ right at SPLC
$D_{5\text{-}6}^{\text{main}} \to \sum Y^{\text{right}} = 0$ $\to -D_{5\text{-}6}^{\text{main}} \sin\alpha + R_B = 0,$ $D_{5\text{-}6}^{\text{main}} = \dfrac{1}{\sin\alpha} R_B \to \text{IL}\left(D_{5\text{-}6}^{\text{main}}\right) = \dfrac{1}{\sin\alpha}\text{IL}(R_B)$	$D_{5\text{-}6}^{\text{main}} \to \sum Y^{\text{left}} = 0 \to D_{5\text{-}6}^{\text{main}} \sin\alpha + R_A = 0,$ $D_{5\text{-}6}^{\text{main}} = -\dfrac{1}{\sin\alpha} R_A \to \text{IL}\left(D_{5\text{-}6}^{\text{main}}\right)$ $= -\dfrac{1}{\sin\alpha}\text{IL}(R_A)$

This influence line presents two parallel lines with ordinates $1/\sin\alpha$ on the supports and a connecting line within panel 5-7. Ordinate $1/\sin\alpha$ on the right support is not shown.

The influence line for internal force $D_{5\text{-}4'}^{\text{sec}}$ is already known. The required influence line for $D_{4'\text{-}5}$ is presented in Fig. 3.18. We can see that if load P is located at joint $5'$ or within the portion $5'$-7, then the secondary truss within panel 5-7 leads to a significant increase of compressed force $D_{4'\text{-}5}$.

Force $U_{5'\text{-}5}$

Member $5'$-5 belongs simultaneously to the main and to the secondary truss. Therefore, an influence line should be constructed considering the main truss and secondary truss 5-$4'$-7-$5'$:

$$U_{5'\text{-}5} = U_{5\text{-}7}^{\text{main}} + U_{5'\text{-}5}^{\text{sec}} \to \text{IL}\left(U_{5'\text{-}5}\right) = \text{IL}\left(U_{5\text{-}7}^{\text{main}}\right) + \text{IL}\left(U_{5'\text{-}5}^{\text{sec}}\right),$$

where $\text{IL}\left(U_{5'\text{-}5}^{\text{sec}}\right) = \text{IL}\left(U_{5'\text{-}7}^{\text{sec}}\right)$ has already been considered above. The influence line for $U_{5'\text{-}5}$ is presented in Fig. 3.18. We can see that if this truss is subjected to any load within portions 1-5 and/or 7-13, the secondary trusses do not affect internal force $U_{5'\text{-}5}$. The influence of the secondary truss in panel 5-7 on internal force $U_{5'\text{-}5}$ occurs only if any load is located within panel 5-7.

3.5 Special Types of Trusses

The following trusses are considered in this section: Three-hinged trusses, trusses with a hinged chain, and complex trusses. Each of the above-mentioned trusses has some peculiarities.

3.5.1 Three-Hinged Trusses

A three-hinged truss is actually two trusses connected by the hinge C as shown in Fig. 3.20a. Both supports are pinned. The fundamental feature of this structure is that horizontal reactions appear even if the structure is subjected to a vertical load only. These horizontal reactions $H_A = H_B = H$ are called *thrust*. Often such types of structures are called thrusted structures. Some examples of three-hinged trusses are presented in Fig. 3.20. Supports may be located at the same or different levels. Truss (a) contains two pinned supports and the thrust is taken by these supports. A modification of a three-hinged truss is presented in Fig. 3.20b. This structure contains pinned and rolled supports. Instead of one "lost" constraint at support B, we have introduced an additional member that connects both trusses. This member is called a tie; the thrust is taken by the tie.

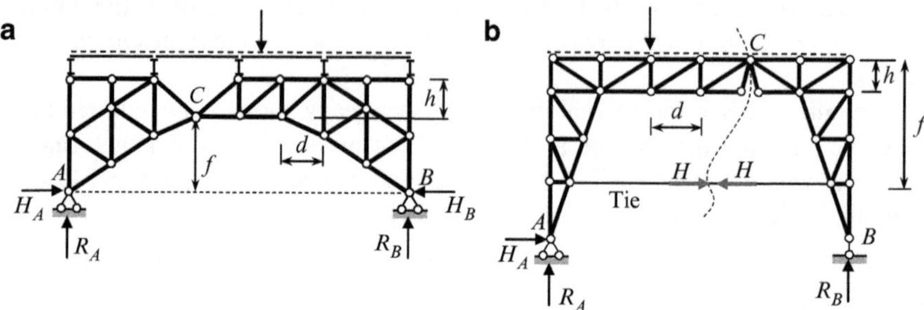

Fig. 3.20 Three-hinged trusses

The structure (a) is geometrically unchangeable. Indeed, two rigid discs, AC and BC are connected to the ground by two hinges, A and B, and line AB does not pass through the intermediate hinge C. Similarly, the kinematical analysis can be carried out for truss (b).

All three-hinged trusses shown in Fig. 3.20 are statically determinate structures. Indeed, the structures in Fig. 3.20a have four unknown reactions, i.e., two vertical reactions, R_A, R_B and two horizontal reactions, H_A, H_B. For scheme (b) we have three unknown reactions (R_A, R_B, and horizontal reaction H_A) as well as internal force H (thrust) in the tie. For their determination, three equilibrium equations can be formulated considering the whole system. Since the bending moment at hinge C is zero, this provides an additional equation of equilibrium. It means that the sum of the moments of all external forces located on the right or on the left part of the structure with respect to hinge C is zero, i.e.,

$$\sum_{\text{left}} M_C = 0 \quad \text{or} \quad \sum_{\text{right}} M_C = 0.$$

These four equations of equilibrium determine all four unknowns.

Three-hinged symmetrical truss is shown in Fig. 3.21; the span $l = 6d$. We need to construct the influence lines for the reactions and for the internal force in indicated member $U_{3\text{-}5}$. It is obvious that the influence lines for vertical reactions R_A and R_B are the same as for a simply supported beam.

The construction of the influence line for thrust H is presented here in tabulated form. It is evident that $H_A = H_B = H$.

$P = 1$ left at joint C	$P = 1$ right at joint C
$H \rightarrow \sum M_C^{\text{right}} = 0 \rightarrow -Hf + R_B \frac{l}{2} = 0,$ $H = \frac{R_B l}{2f} \rightarrow \text{IL}(H) = \frac{l}{2f}\text{IL}(R_B)$	$H \rightarrow \sum M_C^{\text{left}} = 0 \rightarrow Hf - R_A \frac{l}{2} = 0,$ $H = \frac{R_A l}{2f} \rightarrow \text{IL}(H) = \frac{l}{2f}\text{IL}(R_A)$

The left portion of influence line H can be obtained from the influence line for R_B by multiplying all ordinates by factor $l/2f$. The maximum thrust caused by concentrated load P is $Pl/4f$. This happens if the force is located at hinge C.

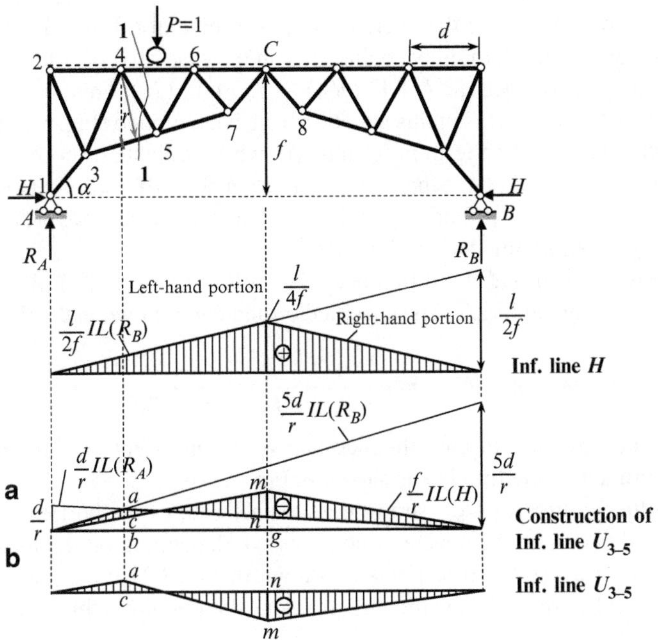

Fig. 3.21 The influence lines of the symmetrical three-hinged truss

Having the influence line for the thrust allows us to determine the internal force in any member due to arbitrary load. Assume for example, that the right half of the truss is loaded by uniformly distributed load q. In this case,

$$H = q\omega = \underbrace{\frac{1}{2}\frac{l}{4f}\frac{l}{2}}_{\omega}q = \frac{ql^2}{16f} \quad \text{and} \quad R_A = \frac{1}{2}\frac{1}{2}\frac{l}{2}q = \frac{ql}{8}.$$

After that, finding the internal force in any member is a matter of elementary calculation. So the influence line for thrust should be regarded as a key influence line.

Force $U_{3\text{-}5}$ (Section 1-1, SPLC Is Panel 4-6; Ritter's Point is 4)

$P = 1$ left at SPLC	$P = 1$ right at SPLC
$U_{3\text{-}5} \rightarrow \sum M_4^{\text{right}} = 0$	$U_{3\text{-}5} \rightarrow \sum M_4^{\text{left}} = 0$
$-U_{3\text{-}5}r + R_B 5d - Hf = 0 \rightarrow U_{3\text{-}5} = \frac{1}{r}(R_B 5d - Hf)$	$U_{3\text{-}5}r - R_A d + Hf = 0 \rightarrow U_{3\text{-}5} = \frac{1}{r}(R_A d - Hf)$
$\text{IL}(U_{3\text{-}5}) = \frac{5d}{r}\text{IL}(R_B) - \frac{f}{r}\text{IL}(H) \quad \text{(i)}$	$\text{IL}(U_{3\text{-}5}) = \frac{d}{r}\text{IL}(R_A) - \frac{f}{r}\text{IL(H)} \quad \text{(ii)}$

The first term in formulas (i) and (ii) corresponds to a common nonthrusted truss (such as a pinned-rolled truss with additional member 7-8). These formulas show that the thrust decreases the internal force in member 3-5; such an influence of the thrust on the distribution of the internal forces is typical for thrusted trusses.

For the construction of influence line $U_{3\text{-}5}$ in the case of $P = 1$ left at SPLC, we need to find the difference $(5d/r)\cdot\text{IL}(R_B) - (f/r)\cdot\text{IL}(H)$. This step is carried out graphically after both terms in this equation have been plotted (Fig. 3.21a). Ordinate ab corresponds to the first term in formula (i), while ordinate bc corresponds to the second term in formula (i). Since $ab > bc$, then the difference between the two functions $(5d/r)\cdot\text{IL}(R_B)$ and $(f/r)\cdot\text{IL}(H)$ leads to the positive ordinates for the final IL $(U_{3\text{-}5})$ within this part of the truss.

Similarly, we can find the difference $(d/r)\cdot\text{IL}(R_B) - (f/r)\cdot\text{IL}(H)$ for the case of $P = 1$ right at SPLC. The specified ordinate nm shown in the final influence line is

$$nm = ng - mg = \frac{d}{2r} - \frac{f}{r}\frac{l}{4f} = \frac{d}{2r} - \frac{3d}{2r} = -\frac{d}{r}.$$

Note that the ordinates of the influence line within the right portion are negative. The final influence line for $U_{3\text{-}5}$ is shown in Fig. 3.21b.

The influence line for $U_{3\text{-}5}$ shows that this element of the bottom chord in this three-hinged truss may be tensile or compressed, depending on the location of the load. This is because the thrust acts as shown in Fig. 3.21, i.e., external force P tends to extend element 3-5, while thrust H tends to compress this element.

3.5.2 Trusses with a Hinged Chain

Some examples of trusses with a hinged chain are presented in Fig. 3.22. In all cases these systems consist of two trusses, AC and CB, connected by hinge C and stiffened by additional structures called a hinged chain. The hinged chain may be located above or below the trusses. Vertical members (hangers or suspensions) connect the hinged chain with the trusses. The connections between the members of the chain and the hangers are hinged. In case (c), all the hinges of the hinged chain are located on one line. In cases (a) and (b), a load is applied to the truss directly, while in case (c), the load is applied to the joint of the hinged chain and then transmitted to the truss.

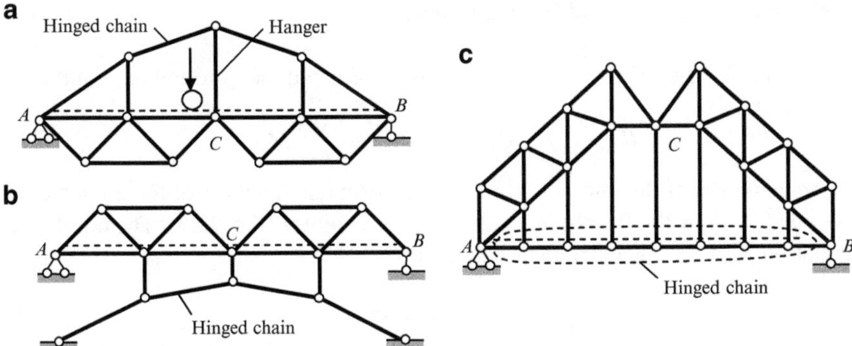

Fig. 3.22 Trusses with hinged chain

The typical truss with a hinged chain located above the truss is shown in Fig. 3.23. Assume that the parameters of the structure are as follows: $d = 3\,\text{m}$, $h = 2\,\text{m}$, $f = 7\,\text{m}$, and $L = 24\,\text{m}$. We need to construct the influence lines for the reactions and the internal forces in hanger, V_n. As usual we start with the kinematical analysis of the structure: it shows that the structure is geometrically unchangeable and statically determinate.

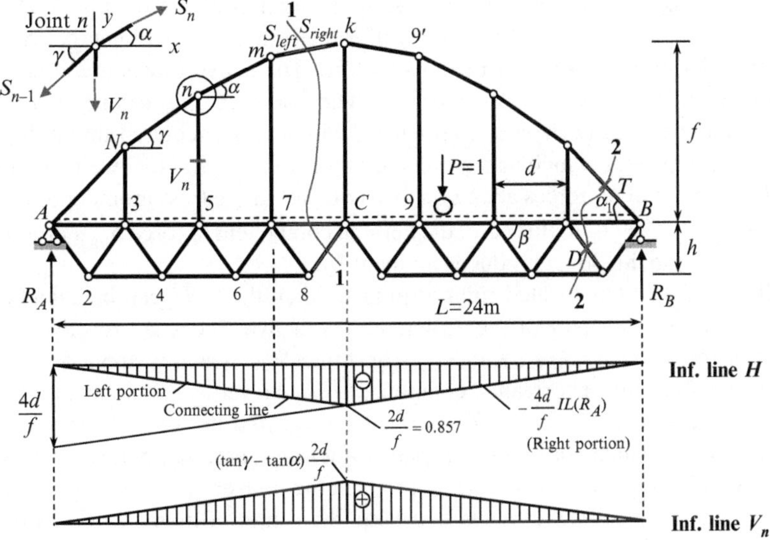

Fig. 3.23 Truss with a hinged chain: design diagram and influence lines

Reaction of Supports and Internal Forces

Reactions R_A and R_B for any load can be calculated using following equilibrium conditions:

$$R_A \to \sum M_B = 0;\ R_B \to \sum M_A = 0.$$

For calculation of the internal forces that arise in the members of the hinged chain, we need to show the free-body diagram for any joint n (Fig. 3.23). The equilibrium equation $\sum X = 0$ leads to the relationship

$$S_n \cos\alpha = S_{n-1}\cos\gamma = H. \tag{3.2}$$

Thus, for any vertical load acting on the given truss, the horizontal component of the forces, which arise in the all members of the hinged chain, are equal. The horizontal component H of the forces S_n, S_{n-1} is called a thrust.

Now we will provide an analysis for the case of a moving load. The influence lines for reactions R_A and R_B are the same as for a simply supported beam. However, the construction of an influence line for thrust H has some special features. Let us consider them.

Thrust *H* (Section 1-1, SPLC Is Panel 7-C; Ritter's Point is *C*)

Internal force S, which arises in the element m-k of the hinged chain, is denoted as S_{left} and S_{right}. The meaning of the subscript notation is clear from Fig. 3.23.

If load $P = 1$ is located to the left of joint 7, then thrust H can be calculated by considering the *right* part of the structure. The active forces are reaction R_B and internal forces $S_{7\text{-}C}$, $S_{8\text{-}C}$, and S_{right}. The last force S_{right} can be resolved into two components: a horizontal component, which is the required thrust H, and a vertical component, which acts along the vertical line C-k. Now we form the sum of the moment of all forces acting on the right part of the truss around point C, i.e., $H \to \sum M_C^{\text{right}} = 0$. In this case, the vertical component of force S_{right} produces no moment, while the thrust produces moment Hf.

If load $P = 1$ is located right at joint C, then thrust H can be calculated by considering the *left* part of the structure. The active forces are reaction R_A and internal forces $S_{7\text{-}C}$, $S_{8\text{-}C}$, and S_{left}. The force S_{left}, which is applied at joint m, can be resolved into a horizontal component H and a vertical component. The latter component acts along vertical line m-7. Now we find the sum of the moment of all the forces, which act on the *left* part of the truss, around point C. In this case, the vertical component of force S_{left} produce the nonzero moment around joint C and thrust H has a new arm (m-7) around the center of moments C. In order to avoid these difficulties we translate the force S_{left} along the line of its action from joint m into joint k. After that we resolve this force into its vertical and horizontal components. This procedure allows us to eliminate the moment due to the vertical component of S, while the moment due to the horizontal component of S is easily calculated as Hf.

Construction of the influence line for H is presented in the table below.

$P = 1$ left at SPLC	$P = 1$ right at SPLC
$H \to \sum M_C^{\text{right}} = 0 \to R_B 4d + Hf = 0,$ $H = -\frac{4d}{f} R_B \to \text{IL}(H) = -\frac{4d}{f} \text{IL}(R_B)$	$H \to \sum M_C^{\text{left}} = 0 \to R_A 4d + Hf = 0,$ $H = -\frac{4d}{f} R_A \to \text{IL}(H) = -\frac{4d}{f} \text{IL}(R_A)$

The left portion of the influence line for H (portion A-7) presents the influence line for R_B multiplied by coefficient $-4d/f$ and the right-hand portion (portion C-B) presents the influence line for R_A multiplied by the same coefficient. The connecting line is between points 7 and C (Fig. 3.23). The negative sign for thrust indicates that all members of the arched chain are in compression.

Force V_n

Equilibrium condition for joint n leads to the following result:

$$\sum Y = 0: \quad -V_n + S_n \sin\alpha - S_{n-1} \sin\gamma = 0 \to V_n = H\left(\tan\alpha - \tan\gamma\right).$$

Therefore,

$$\text{IL}\left(V_n\right) = \left(\tan\alpha - \tan\gamma\right) \cdot \text{IL}(H).$$

Since $\alpha < \gamma$ and H is negative, then all hangers are in tension. The corresponding influence line is shown in Fig. 3.23.

The influence line for thrust H can be considered the key influence line, since thrust H always appears in any cut-section for the entire structure. This influence line allows us to calculate thrust for an arbitrary load. After that, the internal force in any member can be calculated simply by considering all the external loads, the reactions, and the thrust as an additional external force.

Discussion

For any location of a load the hangers are in tension and all members of the chain are compressed. The maximum internal force at *any* hanger occurs if load P is placed at joint C.

To calculate the internal forces in different members caused by an arbitrary fixed load, the following procedure is recommended:

1. Construct the influence line for the thrust
2. Calculate the thrust caused by a fixed load
3. Calculate the required internal force considering thrust as an additional external force

This algorithm combines both approaches: the methods of fixed and of moving loads and so provides a very powerful tool for the analysis of complex structures.

Example 3.2. The structure in Fig. 3.23 is subjected to a uniformly distributed load q within the entire span L. Calculate the internal forces T and D in the indicated elements.

Solution. The thrust of the arch chain equals

$$H = q\Omega_{\mathrm{H}} = -q\frac{1}{2}L\frac{2d}{f} = -\frac{qLd}{f},$$

where Ω_{H} is area of the influence line for H under the load q. After that, the required force T according to (3.2) is

$$T = \frac{H}{\cos\alpha_1} = -\frac{qLd}{f\,\cos\alpha_1}.$$

We can see that in order to decrease the force T we must increase the height f and/or decrease the angle α_1.

To calculate force D, we can use section 2-2 and consider the equilibrium of the right part of the structure:

$$D \to \sum Y = 0:$$

$$D\sin\beta + R_B + T\sin\alpha_1 = 0 \to D = -\frac{1}{\sin\beta}\left(\frac{qL}{2} - \frac{qLd}{f}\tan\alpha_1\right).$$

Thus, this problem is solved using the fixed and moving load approaches: thrust H is determined using corresponding influence lines, while internal forces D and T are computed using H and the classical method of through sections.

3.5.3 Complex Trusses

Complex trusses are generated using special methods to connect rigid discs. These methods are different from those used to create the simple trusses, three-hinged trusses, etc. analyzed in the previous sections of this chapter. An example of a complex truss is a Wichert truss.

Figure 3.24 presents a design diagram of a typical Wichert truss. As before, statical determinacy of the structure can be verified by the formula $W = 2J - S - S_0$, where J, S, and S_0 are the number of hinged joints, members, and constraints of

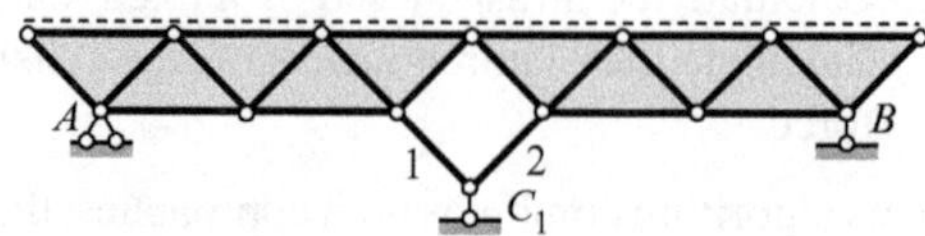

Fig. 3.24 Wichert truss

supports, respectively. For this truss, we have $W = 2 \times 14 - 24 - 4 = 0$. Thus, the truss in Fig. 3.24 is statically determinate.

In this truss two rigid discs are connected using members 1, 2, and the rolled support C. If the constraint C was absent, then this structure would be geometrically changeable. However, the connections of both discs to the ground using constraint C lead to a geometrically unchangeable structure.

A peculiarity of this multispan truss is that even though this structure is statically determinate, its reactions cannot be determined using the three equilibrium equations for the truss as a whole. Therefore, the Wichert truss requires a new approach for its analysis: the replacement bar method, also called the Henneberg method (1886). The main idea behind this method is that the entire structure is transformed into a new structure. For this, the intermediate support is eliminated and a new element is introduced in such a way that the new structure becomes geometrically unchangeable and can be easily analyzed. The equivalent condition of both systems, new and original, allows us to determine the unknown reaction in the eliminated constraint.

As an example, let us consider the symmetrical Wichert truss supported at points A, B, and C and carrying a load P, as shown in Fig. 3.25a. Assume that angle $\alpha = 60°$.

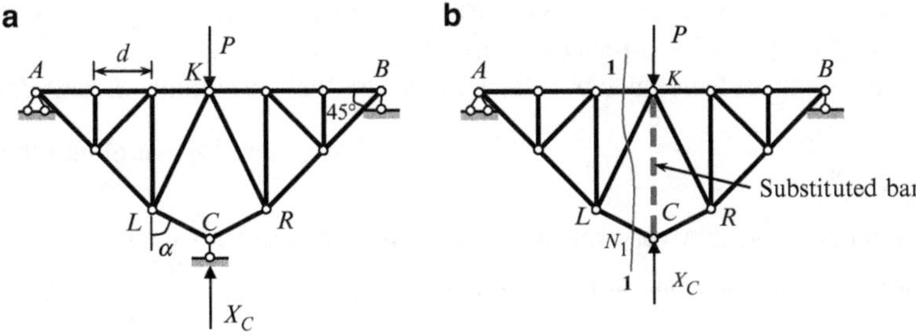

Fig. 3.25 (**a**) Design diagram of a Wichert truss; (**b**) Substituted system

To analyze this system, let us replace support C by an additional vertical member and apply external force X_C, which is equal to the unknown reaction of support C (Fig. 3.25b). The additional element *CK* is called a substituted bar. According to the superposition principle, the internal force in the substituted bar is $F_C + F_P$, i.e., it is the sum of the internal forces due to unknown reaction X_C of support C and given external load P. Both systems (a) and (b) are equivalent if the internal force in the substituted bar is zero, i.e., $F_C + F_P = 0$. In expanded form, this equation may be written as:

$$\overline{F} X_C + F_P = 0, \tag{3.3}$$

where F_P is the internal force in the substituted bar due to external force P; $\overline{F}$ is the internal force in the substituted bar due to unit force $X_C = 1$; and $\overline{F} X_C$ represents the internal force in the substituted bar due to unknown reaction X_C.

Two conditions should be considered. They are P-loading and X_C-loading. Design diagram for both conditions present a truss simply supported at points A and B.

For each condition, the force in the substituted bar can be found in two steps: first, compute the force N_1 in the member LC (using section 1-1) and then find required force in substitute bar, considering joint C. This procedure is presented in Table 3.1.

According to (3.3), the unknown reaction of support C equals

$$X_C = -\frac{F_P}{\overline{F}} = -\frac{-0.672P}{-0.328} = -2.048P. \tag{3.4}$$

After this, the calculation of reactions R_A and R_B presents no further difficulty. As soon as all the reactions are known, the calculation of the internal forces in all the members can be carried out as usual.

Discussion

1. For the given Wichert truss, loaded by vertical force P directed downward, the vertical reaction of the intermediate support is also directed downward. This surprising result can be easily explained. Imagine that support C is removed. Obviously, point K would be displaced downward, rigid disc AKL would rotate clockwise around support A and rigid disc KBR would rotate counterclockwise around support B. Therefore, support C would be displaced upward, so the reaction of support C is directed downward.
2. If the substituted member is a horizontal bar (between joints L and R), then $F_P = \frac{3}{4}P$, $\overline{F} = \frac{\sqrt{3}}{2} - \frac{1}{2} = 0.366$ and for reaction X_C we obtain the same result.
3. A Wichert truss is a very sensitive structure with respect to angle α. Indeed, if angle $\alpha = 45°$, then $F_P = 0.707P$ and $\overline{F} = 0$, so $X_C = -\frac{0.707P}{0}$. This means that the system is simultaneously changeable. If angle $\alpha = 30°$, then reaction X_C is positive and equals $3.55P$.

3.5.4 Summary

Analysis of any truss can be performed using two different analytical methods: the fixed and moving load approaches.

The fixed load approach (methods of joints and through sections) allows us to calculate the internal force in any selected member of a truss.

The moving load approach requires the construction of influence lines. The following approach for the construction of influence lines may be used: the moving load, $P = 1$, is placed at each successive joint of the loading chord and for *each* such loading the required internal force is calculated by the method of joints or method of cuts. This procedure leads to a correct picture of the

Table 3.1 Computation of internal force in substitute bar

State	Design diagram	Free-body diagram	Equilibrium equations
P-loading	Substituted system in Fig. 3.25b subjected to given load $P(\downarrow)$; in this case $R_A = 0.5P(\uparrow)$	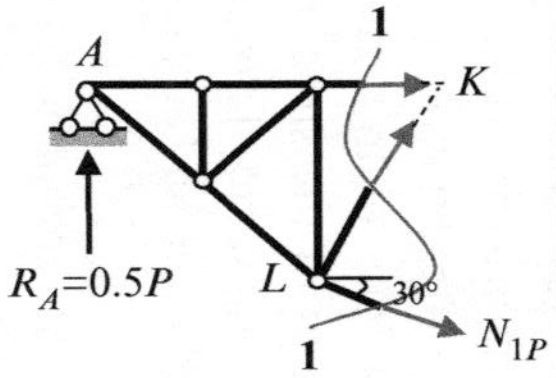 A, K, L, 1, 1, R_A=0.5P, 30°, N_{1P}	$N_{1P} \to \sum M_K^{\text{left}} = 0$ $-R_A 3d + N_{1P} \cos 30° \cdot 2d + N_{1P} \sin 30° \cdot d = 0$ $N_{1P}(2\cos 30° + \sin 30°) = 3R_A$ $N_{1P} = \dfrac{3P/2}{2\sqrt{3}/2 + 1/2} = 0.672P$
		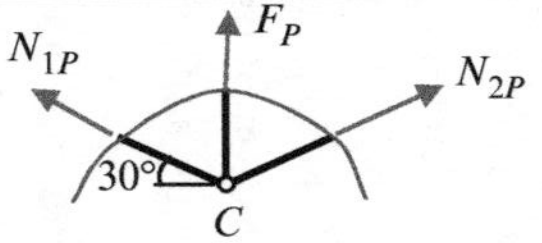 N_{1P}, F_P, N_{2P}, 30°, C	$\sum X = 0 \to N_{1P} = N_{2P} = N$ $\sum Y = 0 \to F_P + 2N \sin 30° = 0$ $F_P = -2N \sin 30° = -0.672P$
X_C-loading	Substituted system in Fig. 3.25b subjected to unit force $X_C = 1(\uparrow)$; in this case $R_A = 0.5(\downarrow)$	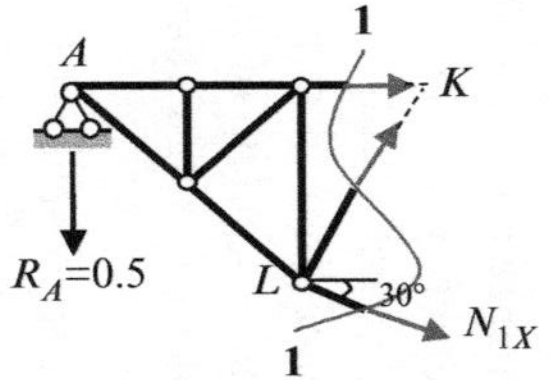 A, K, L, 1, 1, R_A=0.5, 30°, N_{1X}	$N_{1X} \to \sum M_K^{\text{left}} = 0$ $R_A 3d + N_{1X} \cos 30° \cdot 2d + N_{1X} \sin 30° \cdot d = 0$ $N_{1X}(2\cos 30° + \sin 30°) = -3R_A$ $N_{1X} = -\dfrac{3 \times 1/2}{2\sqrt{3}/2 + 1/2} = -0.672$
		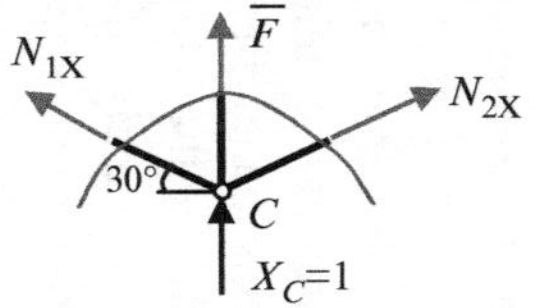 N_{1X}, $\overline{F}$, N_{2X}, 30°, C, X_C=1	$\sum X = 0 \to N_{1X} = N_{2X} = N_X$ $\sum Y = 0 \to \overline{F} + 2N_X \sin 30° + 1 = 0$ $\overline{F} = -2N_X \sin 30° - 1 = -0.328$

influence line, but this procedure is repetitive and bothersome; mainly, it is essentially pointless.

This harmful inefficient approach is based on the calculation of the required force for the different locations of the load *in order to show* the influence lines. In this case the whole necessity of the influence line is lost, since by constructing influence lines this way we *already* determined the internal forces for all positions of the load $P = 1$.

On the contrary, we need the influence line *in order to find* the value of the required internal force for *any* location of the load. By no means should this graph be plotted using repeated computation. In principal, such a flawed approach defeats the purpose of the influence line as a powerful analytical tool.

This book contains an approach to the construction of influence lines, based on accepted methods of truss analysis: using the joint and through section methods we obtain an expression for the required force and after that we transform this expression into an *equation* for the influence line. The algorithm described above allows the influence line to be presented as a *function* (in contrast to a *set of numerical ordinates* for the influence line). Stated as functions, influence lines can realize their full potential as extremely versatile tools for providing different types of analysis.

A very important aspect of this approach is that it allows us to find specific ordinates of the influence lines in terms of the parameters of the structure, such as heights, panel dimensions, angle of inclination of diagonals, etc. This allows us to determine the influence of each parameter on a required force and so provides optimization analysis.

Both fundamental approaches, fixed and moving loads, complement each other. This combination is very effective for truss analysis with some peculiarities (especially for statically indeterminate trusses). Also, the combination of both approaches is an effective way to analyze the same truss in different load cases (snow, dead, live, etc.). For example, for a thrusted truss with supports on different levels, we can find the thrust by considering the system of equilibrium equations. Then, knowing the thrust, we can find all the required internal forces. This is not difficult, but if we need to do it many times for each loading, then it would be much wiser to construct the influence line for the thrust only once. Then we could find its value due to each loading. It is obvious that in this case the influence line for the thrust should be treated as the key influence line. In the case of a truss with a hinged chain, the internal force in the member of the chain under analysis (or better, the thrust as the horizontal component of such an internal force) should be considered to be the key influence line. Clearly, for a complex truss, such as the Wichert truss, the key influence line is the influence line for the reaction of the middle support.

Problems

3.1. Provide kinematical analysis for beams in Fig. P3.1. Construct interaction scheme, point the main and suspended beams, and explain a load path.

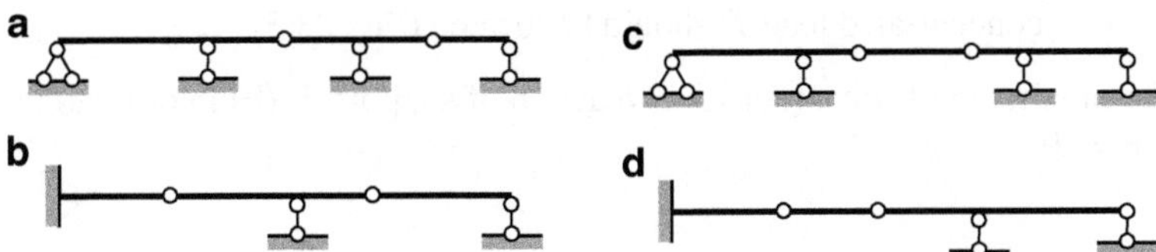

Fig. P3.1

3.2. Vertical settlement of support A of the beam in Fig. P3.2 causes appearance of internal forces in the portion

1. ABC; 2. CDE; 3. Nowhere; 4. Everywhere

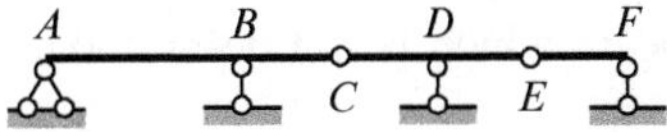

Fig. P3.2

3.3. Beam in Fig. P3.3 is subjected to change of temperature. As a result, internal forces appear in the portion

1. ABC; 2. CDE; 3. Nowhere; 4. Everywhere

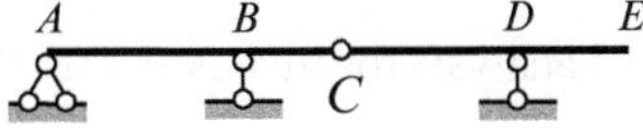

Fig. P3.3

3.4. Find direction of all reactions if uniformly distributed load is located within portion CD (Fig. P3.4)

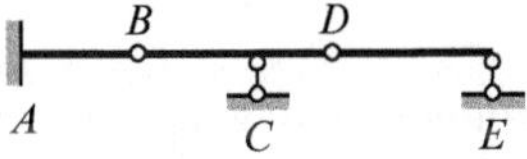

Fig. P3.4

Ans. $R_A < 0\,(\downarrow)$; $R_C > 0\,(\uparrow)$; $R_E = 0$.

3.5. Where concentrated load P should be located (Fig. P3.5)

(a) For maximum positive (negative) reaction of support A; (b) In order to reaction $R_D = 0$

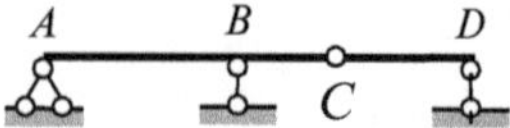

Fig. P3.5

3.6. Analyze the design diagram of two-span Gerber beam is shown in Fig. P3.6.

(a) Find location of distributed load in order to reaction at support C will be equal zero
(b) Find portion of the beam which is not deformable if concentrated load P is located at point B
(c) Where are located extended fibers within the portion AB if concentrated load P is located at point D

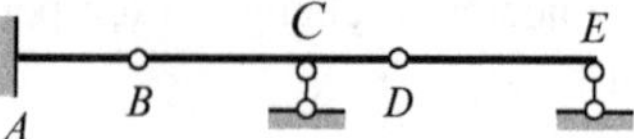

Fig. P3.6

3.7a–b. Provide kinematical analysis for trusses in Fig. P3.7.

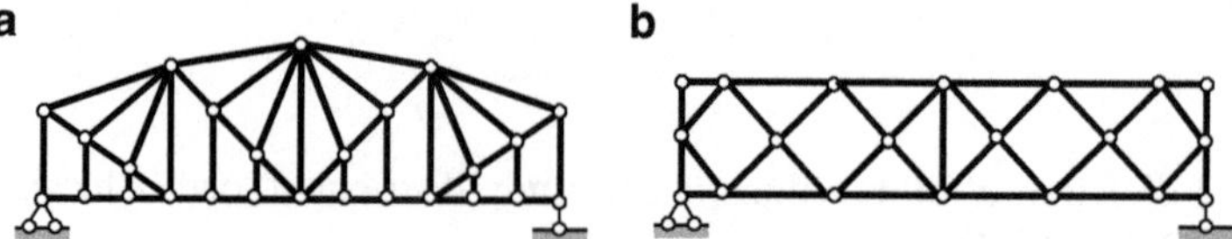

Fig. P3.7

3.8. Design diagram for a simple truss with a lower loaded chord is presented in Fig. P3.8. Consruct the influence lines of the internal forces at the indicated members.

Based on the constructed influence lines check the correct answer:

1. Force V_1 is the maximum tension, if concentrated load P is located at joint 1. E; 2. C; 3. K
2. Force V_2 is always 1. positive; 2. negative; 3. zero

3. Force U_3 is maximum tensile, if distributed load q is located within the portion 1. AC; 2. KB; 3. AB
4. If load P is located at joint K then the force in members 5 and 6 are 1. $D_5 > D_6$; 2. $D_5 < D_6$; 3. $D_5 = D_6$.

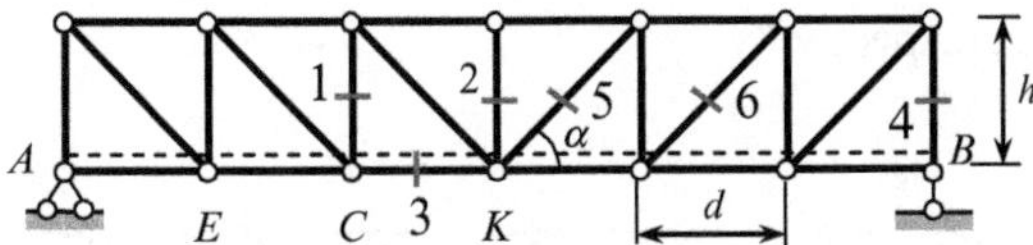

Fig. P3.8

3.9. A single-span K-truss is shown in Fig. P3.9. Construct the influence lines for the internal forces at the indicated members. Using the constructed influence lines, calculate these forces if uniformly distributed load q is distributed within the three panels on the right and concentrated force P is applied at the middle of the span.

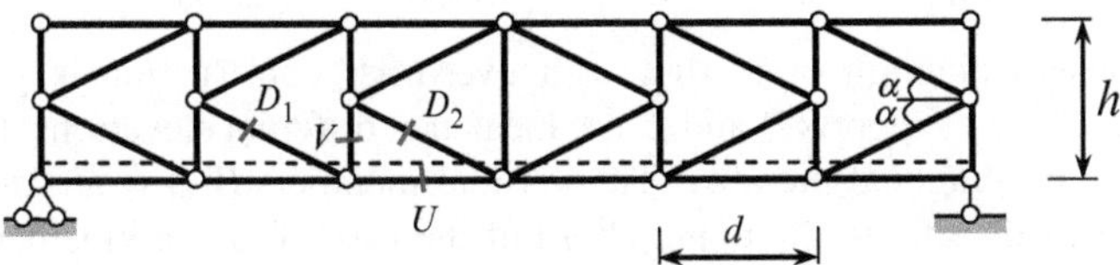

Fig. P3.9

Ans. $U = \frac{3}{2}\frac{qd^2}{h} + P\frac{d}{h}$, $V = -\frac{3}{8}qd - \frac{P}{4}$, $D_1 = D_2 = \frac{3}{8}\frac{qd}{\sin\alpha} + \frac{P}{4\sin\alpha}$.

3.10 a,b. Two trusses with subdivided panels are presented in Fig. P3.10a, b. For case (a) the diagonal members of the main truss are angled downward in the left part of the truss and angled upward in the right part. In case (b) all the diagonal members of the main truss are angled downward. Construct the influence lines for the indicated members and compare the influence lines for the two cases. Using the constructed influence lines, calculate these forces if load P is placed at joint n and uniformly distributed load q is distributed within the panel A-n.

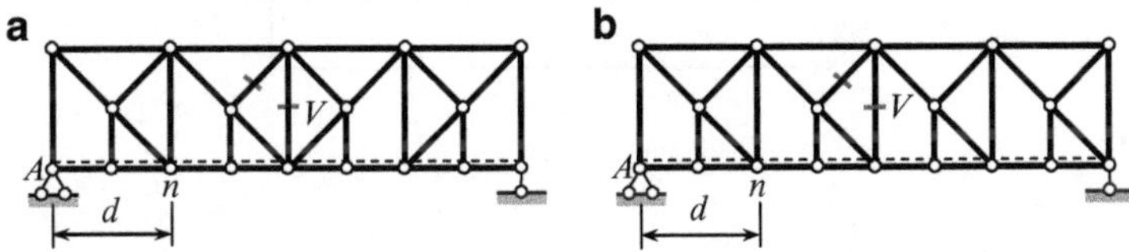

Fig. P3.10

Ans. (b) $V = \frac{qd}{8} + \frac{P}{4}$.

3.11. A design diagram of a truss is presented in Fig. P3.11. Moving loads are applied to the lower chord.

(a) Perform the kinematical analysis; (b) Construct the influence lines for the internal forces in the indicated members; (c) Using the constructed influence lines, compute the indicated internal forces if external forces P_1 and P_2 are placed at joints 7 and 9, respectively.

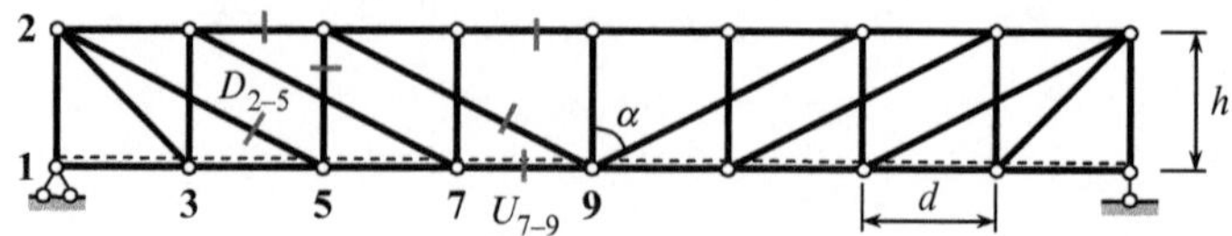

Fig. P3.11

$$\text{Ans. } U_{7\text{-}9} = \frac{9d}{4h}P_1 + \frac{d}{h}P_2;\ D_{2\text{-}5} = -\frac{3}{8\cos\alpha}P_1 + \frac{1}{2\cos\alpha}P_2$$

3.12. A design diagram of a truss with over-truss construction is presented in Fig. P3.12. Pinned supports A and B are located at different elevations. Panel block 1-2-3-4 has no diagonal member. The vertical members (links) are used only to transmit loads directly to the upper chord of the truss. Provide kinematical analysis. Construct the influence line for thrust H. Using the constructed influence line, compute H if forces P_1 and P_2 are placed at joints $1'$ and $2'$, respectively. What happens to the structure if supports A and B are located at one level?

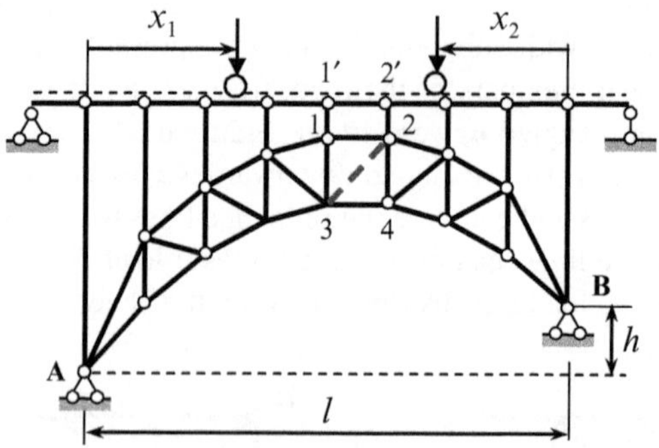

Fig. P3.12

$$\text{Ans. } H = \frac{l}{2h}P_1 - \frac{3l}{8h}P_2$$

Chapter 4
Three-Hinged Arches

The arches are widely used in modern engineering. Arches permit to cover a larger span. The greater is the span than an arch becomes more economical than a truss. From esthetic point of view the arches are more attractive than trusses. Materials of the modern arches are concrete, steel, and wood. The body of the arch may be solid or consist of separate members.

The arches are classified as three-hinged, two-hinged, and arch with fixed supports. A three-hinged arch is geometrically unchangeable statically determinate structure which consists of two curvilinear members, connected together by means of a hinge, with two-hinged supports resting on the abutment. This chapter is devoted to analyze only three-hinged arches. Two-hinged arch and arch with fixed supports will be considered in Part 2.

4.1 Preliminary Remarks

4.1.1 Design Diagram of Three-Hinged Arch

The arch with overarched members is shown in Fig. 4.1a. The arch contains three hinges. Two of them are located at the supports and third is placed at the crown. These hinges are distinguishing features of the three-hinged arch. Design diagram also contains information about the shape of the neutral line of the arch. Usually this shape is given by the expression, $y = f(x)$.

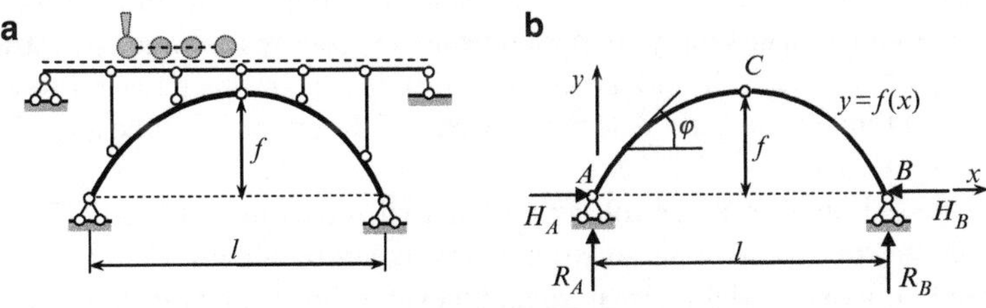

Fig. 4.1 **(a)** Design diagrams for deck-arch bridge; **(b)** Design diagram of three-hinged arch

I.A. Karnovsky and O. Lebed, *Advanced Methods of Structural Analysis*,
DOI 10.1007/978-1-4419-1047-9_4, © Springer Science+Business Media, LLC 2010

The each post which connects the beams of overarched structure with arch itself has the hinges at the ends. It means that in the poles only axial force arises. Idealized design diagram of the three-hinged arch without overarched members is shown in Fig. 4.1b.

Degrees of freedom of three-hinged arch according to Chebushev formula (1.1) is $W = 0$, so this structure is geometrically unchangeable. Indeed, two rigid discs AC and BC are connected with the ground by two hinges A and B and line AB does not pass through the intermediate hinge C.

This structure has four unknown reactions, i.e., two vertical reactions R_A, R_B and two horizontal reactions H_A, H_B. For their determination, three equilibrium equations can be formulated considering the structure in whole. Since bending moment at the hinge C is zero, this provides additional equation for equilibrium of the part of the system. It means that the sum of the moments of all external forces, which are located on the right (or on the left) part of the structure with respect to hinge C is zero

$$\sum_{\text{left}} M_C = 0 \quad \text{or} \quad \sum_{\text{right}} M_C = 0. \tag{4.1}$$

These four equations of equilibrium determine all four reactions at the supports. Therefore, three-hinged arch is a geometrically unchangeable and statically determinate structure.

4.1.2 Peculiarities of the Arches

The fundamental feature of arched structure is that horizontal reactions appear even if the structure is subjected to vertical load only. These horizontal reactions $H_A = H_B = H$ are called a *thrust*; such types of structures are often called as thrusted structures. One type of trusted structures (thrusted frame) was considered in Chap. 3.

It will be shown later that at any cross section of the arch the bending moments, shear, and axial forces arise. However, the bending moments and shear forces are considerably smaller than corresponding internal forces in a simply supported beam covering the same span and subjected to the same load. This is the fundamental property of the arch thanks to thrust. Thrusts in both supports are oriented toward each other and reduce the bending moments that would arise in beams of the same span and load. Therefore, the height of the cross section of the arch can be much less than the height of a beam to resist the same loading. So the three-hinged arch is more economical than simply supported beam, especially for large-span structures.

Both parts of the arch may be connected by a tie. In this case in order for the structure to remain statically determinate, one of the supports of the arch should be rolled (Fig. 4.2a, b).

Prestressed tie allows controlling the internal forces in arch itself. Tie is an element connected by its ends to the arch by mean of hinges, therefore the tie is subjected only to an axial internal force. So even if horizontal reactions of supports equal zero, an extended force (thrust) arises in the tie.

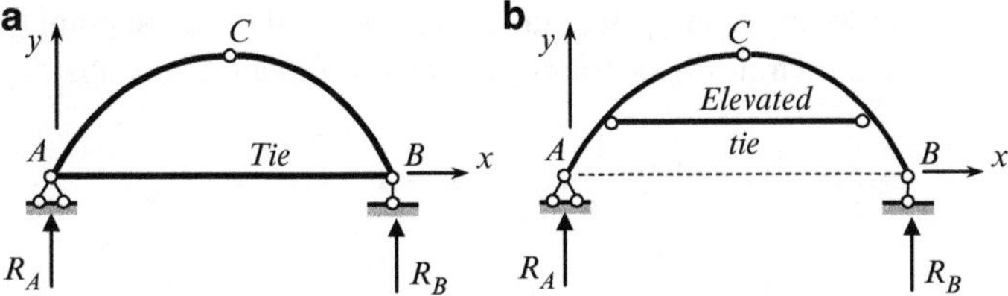

Fig. 4.2 (**a**, **b**) Design diagram of three-hinged arch with tie on the level supports and elevated tie

Thus, the arch is characterized by two fundamental markers such as a curvilinear axis and appearance of the thrust. Therefore the structure in Fig. 4.3 presents the curvilinear *trustless* simply supported element, i.e., this is just members with curvilinear axis, but no arch. It is obvious that, unlike the beam, in this structure the axial compressed forces arise; however, the distribution of bending moments for this structure and for beam of the same span and load will not differ, while the shear forces are less in this structure than that in beam. Thus, the fundamental feature of the arch (decreasing of the bending moments due to the appearance of the thrust) for structure in Fig. 4.3 is not observed.

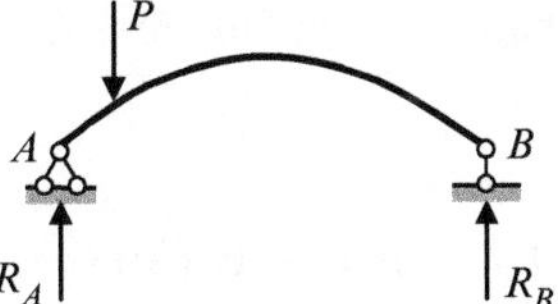

Fig. 4.3 Simply supported thrustless curvilinear member

4.1.3 Geometric Parameters of Circular and Parabolic Arches

Distribution of internal forces in arches depends on a shape of the central line of an arch. Equation of the central line and some necessary formulae for circular and parabolic arches are presented below. For both cases, origin of coordinate axis is located at point *A* (Figs. 4.1b and 4.2).

Circular Arch

Ordinate y of any point of the central line of the arch is defined by formula

$$y = \sqrt{R^2 - \left(\frac{l}{2} - x\right)^2} - R + f; \quad R = \frac{f}{2} + \frac{l^2}{8f}, \tag{4.2}$$

where x is the abscissa of the same point of the central line of the arch; R the radius of curvature of the arch; f and l are the rise and span of the arch.

The angle φ between the tangent to the center line of the arch at point $(x,\ y)$ and horizontal axis is shown in Fig. 4.1b. Trigonometric functions for this angle are as follows:

$$\sin\varphi = (l - 2x)\frac{1}{2R}; \quad \cos\varphi = (y + R - f)\frac{1}{R}. \tag{4.3}$$

Parabolic Arch

Ordinate y of any point of the central line of the arch

$$y = 4fx\,(l - x)\frac{1}{l^2}. \tag{4.4}$$

Trigonometric functions of the angle between the tangent to the center line of the arch at point $(x,\ y)$ and a horizontal axis are as follows

$$\tan\varphi = \frac{dy}{dx} = \frac{4f}{l^2}(l - 2x)\,;\ \cos\varphi = \frac{1}{\sqrt{1 + \tan^2\phi}};\ \sin\varphi = \cos\varphi \cdot \tan\varphi \tag{4.5}$$

For the left-hand half-arch, the functions $\sin\varphi > 0,\ \cos\varphi > 0$, and for the right-hand half-arch the functions $\sin\varphi < 0$ and $\cos\varphi > 0$.

4.2 Internal Forces

Design diagram of a three-hinged symmetrical arch with intermediate hinge C at the highest point of the arch and with supports A and B on one elevation is presented in Fig. 4.4. The span and rise of the arch are labeled as l and f, respectively; equation of central line of the arch is $y = y\,(x)$.

Reactions of Supports

Determination of internal forces, and especially, construction of influence lines for internal forces of the three-hinged arch may be easily and attractively performed using the conception of the "reference (or substitute) beam." The reference beam is a simply supported beam of the same span as the given arch and subjected to the same loads, which act on the arch (Fig. 4.4a).

The following reactions arise in the arch: R_A, R_B, H_A, H_B. The vertical reactions of three-hinged arches carrying the vertical loads have same values as the reactions of the reference beam

$$R_A = R_A^0;\ R_B = R_B^0. \tag{4.6}$$

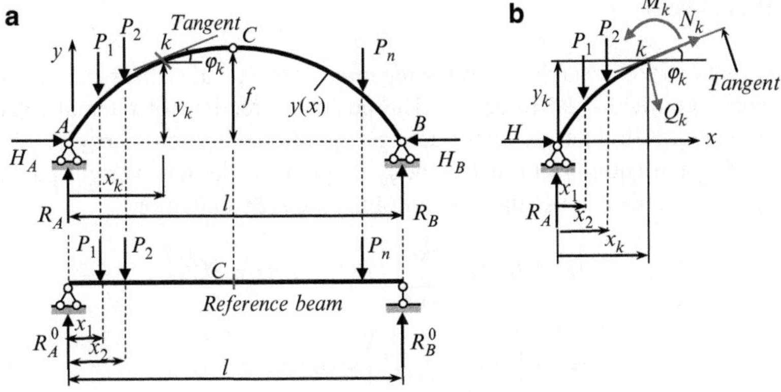

Fig. 4.4 Three-hinged arch. **(a)** Design diagram and reference beam. **(b)** Positive internal forces at any section k

The horizontal reactions (thrust) at both supports of three-hinged arches subjected to the vertical loads are equal in magnitude and opposite in direction

$$H_A = H_B = H. \tag{4.7}$$

Bending moment at the hinge C of the arch is zero. Therefore, by definition of the bending moment

$$M_C = \underbrace{R_A\frac{l}{2} - P_1\left(\frac{l}{2} - x_1\right) - P_2\left(\frac{l}{2} - x_2\right)}_{M_C^0} - H_A \cdot f = 0.$$

Underlined set of terms is the bending moment acting over section C of the reference beam (this section is located under the hinge of the arch). Therefore last equation may be rewritten in the form

$$M_C^0 - H_A f = 0,$$

which allows immediately to calculate the thrust

$$H = \frac{M_C^0}{f}. \tag{4.8}$$

Thus, the thrust of the arch equals to bending moment at section C of the reference beam divided by the rise of the arch.

Internal Forces

In any section k of the arch, the following internal forces arise: the bending moment M_k, shear Q_k, and axial force N_k. The positive directions of internal forces are shown in Fig. 4.4b. Internal forces acting over a cross section k may be obtained considering the equilibrium of free-body diagram of the left or right part of the arch. It is convenient to use the left part of the arch. By definition

$$\begin{aligned} M_k &= R_A x_k - \sum_{\text{left}} P_i\,(x_k - x_i) - H y_k, \\ Q_k &= \left(R_A - \sum_{\text{left}} P\right)\cos\varphi_k - H\sin\varphi_k, \\ N_k &= -\left(R_A - \sum_{\text{left}} P\right)\sin\varphi_k - H\cos\varphi_k, \end{aligned} \tag{4.9}$$

where P_i are forces which are located at the left side of the section k; x_i are corresponding abscises of the points of application; x_k, y_k are coordinates of the point k; and φ_k is angle between the tangent to the center line of the arch at point k and a horizontal.

These equations may be represented in the following convenient form

$$\begin{aligned} M_k &= M_k^0 - H y_k, \\ Q_k &= Q_k^0\cos\varphi_k - H\sin\varphi_k, \\ N_k &= -Q_k^0\sin\varphi_k - H\cos\varphi_k, \end{aligned} \tag{4.10}$$

where expressions M_k^0 and Q_k^0 represent the bending moment and shear force at the section k for the reference beam (beam's bending moment and beam's shear).

Analysis of Formulae [(4.8), (4.10)]

1. Thrust of the arch is inversely proportional to the rise of the arch.
2. In order to calculate the bending moment in any cross section of the three-hinged arch, the bending moment at the same section of the reference beam should be decreased by value Hy_k. Therefore, the bending moment in the arch is less than in the reference beam. This is the reason why the three-hinged arch is more economical than simply supported beam, especially for large-span structures.
   In order to calculate shear force in any cross section of the three-hinged arch, the shear force at the same section of the reference beam should be multiplied by $\cos\varphi_k$ and this value should be decreased by $H\sin\varphi_k$.
3. Unlike beams loaded by vertical loads, there are axial forces, which arise in arches loaded by vertical loads only. These axial forces are always compressed.

Analysis of three-hinged arch subjected to fixed loads is presented below. This analysis implies determination of reactions of supports and construction of internal force diagrams.

Design diagram of the three-hinged circular arch subjected to fixed loads is presented in Fig. 4.5. The forces $P_1 = 10\,\text{kN},\ P_2 = 8\,\text{kN},\ q = 2\,\text{kN/m}$. It is necessary to construct the internal force diagrams M, Q, N.

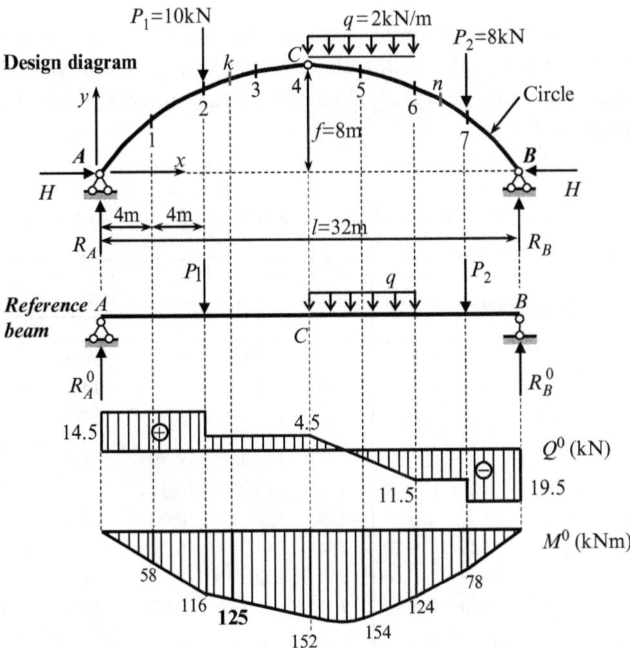

Fig. 4.5 Three-hinged circular arch. Design diagram, reference beam, and corresponding internal force diagrams

Solution.

Reference beam The reactions are determined from the equilibrium equations of all the external forces acting on the reference beam

$$R_A^0 \to \sum M_B = 0:$$

$$-R_A^0 \times 32 + P_1 \times 24 + q \times 8 \times 12 + P_2 \times 4 = 0 \to R_A^0 = 14.5\,\text{kN},$$

$$R_B^0 \to \sum M_A = 0:$$

$$R_B^0 \times 32 - P_1 \times 8 - q \times 8 \times 20 - P_2 \times 28 = 0 \to R_B^0 = 19.5\,\text{kN}.$$

The values of the two reactions just found should be checked using the equilibrium equation $\sum Y = R_A^0 + R_B^0 - P_1 - q \times 8 - P_2 = 14.5 + 19.5 - 10 - 2 \times 8 - 8 = 34 - 34 = 0$.

The bending moment M^0 and shear Q^0 diagrams for reference beam are presented in Fig. 4.5. At point C ($x = 16$ m) the bending moment is $M_C^0 = 152$ kNm.

Three-hinged arch The vertical reactions and thrust of the arch are

$$R_A = R_A^0 = 14.5\,\text{kN},\ R_B = R_B^0 = 19.5\,\text{kN},\ H = \frac{M_C^0}{f} = \frac{152}{8} = 19\,\text{kN}.$$

For construction of internal forces diagrams of the arch, a set of sections has to be considered and for each section the internal forces should be calculated. All computations concerning geometrical parameters and internal forces of the arch are presented in Table 4.1.

Table 4.1 Internal forces in three-hinged circular arch (Fig. 4.5); ($R_A = 14.5$ kN; $R_B = 19.5$ kN; $H = 19$ kN)

Section	x (m)	y (m)	$\sin\varphi$	$\cos\varphi$	M_x^0 (kNm)	H_y (kNm)	M_x (kNm)	Q_x^0 (kN)	Q_x (kN)	N_x (kN)
0	1	2	3	4	5	5′	6	7	8	9
A	0.0	0.0	0.8	0.6	0	0.0	0	14.5	−6.5	−23
1	4	4	0.6	0.8	58	76	−18	14.5	0.2	−23.9
2	8	6.330	0.4	0.9165	116	120.27	−4.27	14.5 / 4.5	5.6892 / −3.4757	−23.213 / −19.213
k	10	7.0788	0.3	0.9539	125	134.497	−9.497	4.5	−1.4074	−19.474
3	12	7.596	0.2	0.9798	134	144.324	−10.324	4.5	0.6091	−19.516
4 (C)	16	8.0	0.0	1.0	152	152	0.0	4.5	4.5	−19.00
5	20	7.596	−0.20	0.9798	154	144.324	9.676	−3.5	0.3707	−19.316
6	24	6.330	−0.40	0.9165	124	120.27	3.73	−11.5	−2.9397	−22.013
n	26	5.3205	−0.50	0.8660	101	101.089	−0.089	−11.5	−0.459	−22.204
7	28	4	−0.6	0.8	78	76	2	−11.5 / −19.5	2.2 / −4.2	−22.1 / −26.9
								−19.5	−4.2	−26.9
B	32	0.0	−0.8	0.6	0	0.0	0	−19.5	3.5	−27

Notes: Values in nominator and denominator (columns 8 and 9) mean value of the force to the left and to the right of corresponding section. Values of discontinuity due to concentrated load equal $P\cos\varphi$ and $P\sin\varphi$ in shear and normal force diagrams, respectively

The column 0 contains the numbers of sections. For specified sections A, 1–7, and B the abscissa x and corresponding ordinate y (in meters) are presented in columns 1 and 2, respectively. Radius of curvature of the arch is

$$R = \frac{f}{2} + \frac{l^2}{8f} = \frac{8}{2} + \frac{32^2}{8 \cdot 8} = 20\,\text{m}.$$

Coordinates y are calculated using the following expression

$$y = y(x) = \sqrt{R^2 - \left(\frac{l}{2} - x\right)^2} - R + f = \sqrt{400 - (16 - x)^2} - 12.$$

Column 3 and 4 contain values of sinφ and cosφ, which are calculated by formulae

$$\sin\varphi = \frac{l-2x}{2R} = \frac{32-2x}{40}, \ \cos\varphi = \frac{y+R-f}{R} = \frac{y+12}{20}.$$

Values of bending moment and shear for reference beam, which are presented in columns 5 and 7, are taken directly from corresponding diagrams in Fig. 4.5. Values for Hy are contained in column 5′. Columns containing separate terms for $Q^0\cos\varphi$, $Q^0\sin\varphi$, $H\cos\varphi$, $H\sin\varphi$ are not presented. Values of bending moment, shear and normal forces for three-hinged arch are tabulated in columns 6, 8, and 9. They have been computed using (4.10). For example, for section A we have

$$Q_A = Q_A^0\cos\varphi_A - H\sin\varphi_A = 14.5\times 0.6 - 19\times 0.8 = -6.5\,\text{kN},$$
$$N_A = -Q_A^0\sin\varphi_A - H\cos\varphi_A = -14.5\times 0.8 - 19\times 0.6 = -23\,\text{kN}.$$

The final internal force diagrams for arch are presented in Fig. 4.6.

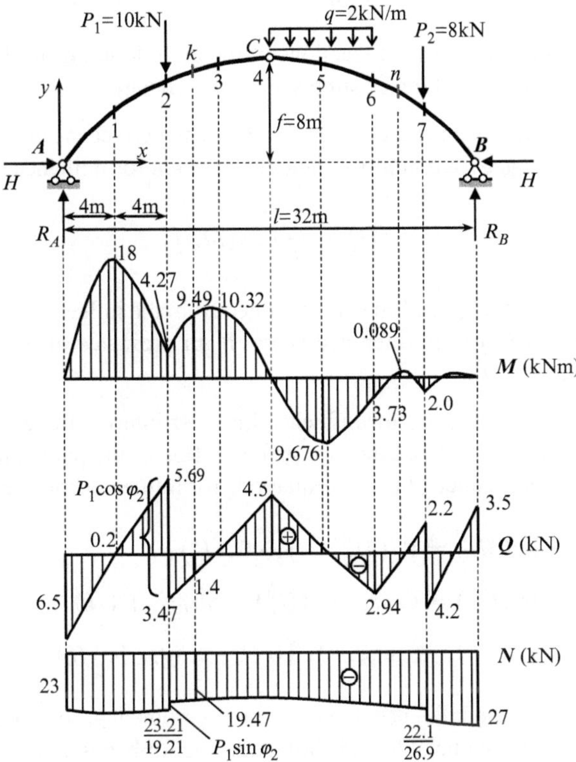

Fig. 4.6 Design diagram of three-hinged circular arch. Internal forces diagrams

Bending moment diagram is shown on the side of the extended fibers, thus the signs of bending moments are omitted. As for beam, the bending moment and shear diagrams satisfy to Schwedler's differential relationships. In particularly, if at any point a shear changes sign, then a slope of the bending moment diagram equals zero, i.e., at this point the bending moment has local extreme (for example, points 2, 7, etc.).

It can be seen that the bending moments which arise in cross sections of the arch are much less than in a reference beam.

4.3 Influence Lines for Reactions and Internal Forces

Equations (4.6), (4.8), and (4.10) can be used for deriving of equations for influence lines.

Vertical reactions The equations for influence lines for vertical reactions of the arch are derived from (4.6). Therefore the equations for influence lines become

$$\mathrm{IL}\,(R_A) = \mathrm{IL}\left(R_A^0\right); \quad \mathrm{IL}\,(R_B) = \mathrm{IL}\left(R_B^0\right). \tag{4.11}$$

Thus, influence lines for vertical reactions of the arch do not differ from influence lines for reactions of the reference simply supported beam.

Thrust The equation of influence lines for thrust is derived from (4.8). Since for given arch a rise f is a *fixed* number, then the equation for influence lines become

$$\mathrm{IL}\,(H) = \frac{1}{f} \cdot \mathrm{IL}\left(M_C^0\right). \tag{4.12}$$

Thus, influence line for trust H may be obtained from the influence line for bending moment at section C of the reference beam, if all ordinates of the latter will be divided by parameter f.

Internal forces The equations for influence lines for internal forces at any section k may be derived from (4.10). Since for given section k, the parameters y_k, $\sin\varphi_k$, and $\cos\varphi_k$ are *fixed* numbers, then the equations for influence lines become

$$\begin{aligned}
\mathrm{IL}\,(M_k) &= \mathrm{IL}\left(M_k^0\right) - y_k \cdot \mathrm{IL}\,(H),\\
\mathrm{IL}\,(Q_k) &= \cos\varphi_k \cdot \mathrm{IL}\left(Q_k^0\right) - \sin\varphi_k \cdot \mathrm{IL}\,(H),\\
\mathrm{IL}\,(N_k) &= -\sin\varphi_k \cdot \mathrm{IL}\left(Q_k^0\right) - \cos\varphi_k \cdot \mathrm{IL}\,(H).
\end{aligned} \tag{4.13}$$

In order to construct the influence line for bending moment at section k, it is necessary to sum two graphs: one of them is influence line for bending moment at section k for reference beam and second is influence line for thrust H with all ordinates of which have been multiplied by a constant factor $(-y_k)$.

Equation of influence lines for shear also has two terms. The first term presents influence line for shear at section k in the reference beam all the ordinates of which have been multiplied by a constant factor $\cos\varphi_k$. The second term presents the influence line of the thrust of the arch, all the ordinates of which have been multiplied by a constant factor $(-\sin\varphi_k)$. Summation of these two graphs leads to the required influence line for shear force at section k. Similar procedure should be applied for construction of influence line for axial force. Note that both terms for axial force are negative.

Analysis of three-hinged arch subjected to moving loads is presented below. This analysis of loads implies the construction of influence lines for reactions and internal forces and their application for analysis in cases of fixed and moving load.

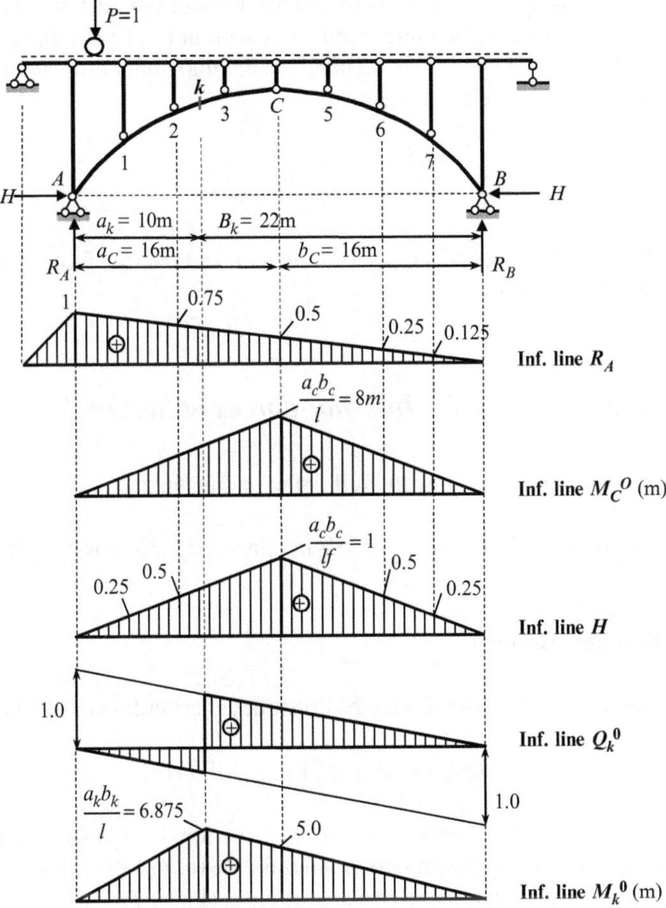

Fig. 4.7 Three-hinged arch. Design diagram and influence lines for reactions of supports and internal forces at section k for substitute beam

Figure 4.7 presents the arched structure consists from the arch itself and overarched construction, which includes the set of simply supported beams and vertical posts with hinged ends. Unit load, which moves along the horizontal beams, is transmitted over the posts on the arch at discrete points. Thus, this design diagram corresponds to indirect load application. Parameters of the arch are same as in Fig. 4.5.

4.3.1 *Influence Lines for Reactions*

According to (4.11), influence lines for vertical reactions R_A and R_B of the arch do not differ from influence lines for reaction of supports of a simply supported beam. Influence line for thrust may be constructed according to (4.12); the maximum ordinate of influence line for bending moment at section C of the reference beam is equal to $(a_C b_C)/l = l/4 = 8$ (m). Therefore the maximum ordinate of influence line for thrust H of the arch becomes

$$\frac{1}{f} \cdot \frac{a_C b_C}{l} = \frac{l}{4f} = \frac{32}{4 \times 8} = 1.$$

Influence lines for reactions of supports of the arch and internal forces for reference beam are shown in Fig. 4.7.

4.3.2 *Influence Lines for Internal Forces at Section k*

The section k is characterized by the following parameters:

$$a_k = 10\,\text{m},\ b_k = 22\,\text{m},\ y_k = 7.0788\,\text{m},\ \sin\varphi_j = 0.30,\ \cos\varphi_j = 0.9539.$$

4.3.2.1 Bending Moment

Influence line for M at section k may be constructed according to (4.13)

$$\text{IL}(M_k) = \text{IL}\left(M_k^0\right) - y_k \cdot \text{IL}(H). \tag{4.13a}$$

Step 1. Influence line for bending moment at section k of reference beam M_k^0 presents the triangle with maximum ordinate $(a_k b_k)/l = (10 \times 22)/32 = 6.875$ m at section k and 5.0 m at section C (Fig. 4.7).

Step 2. Influence line for thrust H presents triangle with maximum ordinate $l/4f = 1$ at the section C. Term $y_k \cdot \text{IL}(H)$ presents the similar graph; the maximum ordinate is $y_k \cdot 1 = 7.0788$ m. So the specified ordinates of graph $y_k \cdot \text{IL}(H)$ at section k and C are 4.42425 m and 7.0788 m, respectively.

Step 3. Procedure (4.13a) is presented in Fig. 4.8, Construction Inf. Line M_k. Since both terms in (4.13a) has *different* signs, then both graphs, $\text{IL}\left(M_k^0\right)$ and

$y_k \cdot \text{IL}(H)$ should be plotted on the *one side* on the basic line. The ordinates of required IL (M_k) will be located *between* these both graphs. Specified ordinates of final graph (4.13a) at section k and C are

$$6.875 - 4.42425 = 2.45075\,\text{m} \quad \text{and} \quad 5.0 - 7.0788 = -2.0788\,\text{m}.$$

Step 4. Influence line between joints 2 and 3 presents a straight line because of indirect load application. Final influence line IL (M_k) is presented in Fig. 4.8; the connected line between joints 2 and 3 is shown by solid line.

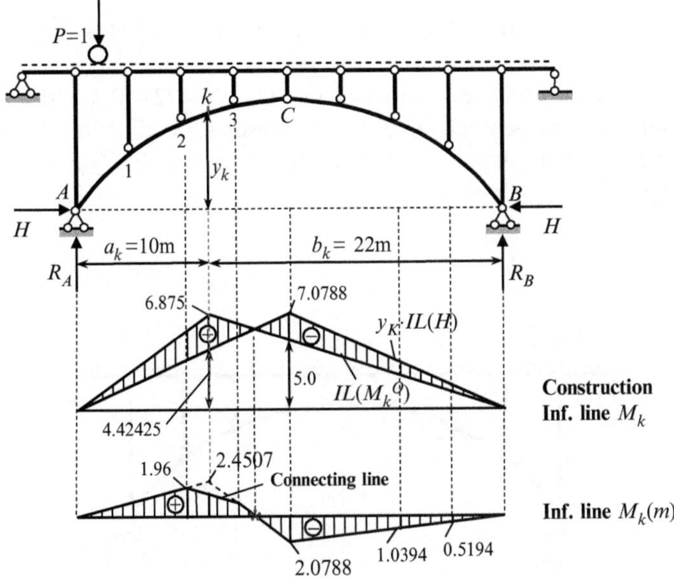

Fig. 4.8 Three-hinged arch. Design diagram and construction of influence line for bending moment at section k of the arch

4.3.2.2 Shear Force

This influence line may be constructed according to equation

$$\text{IL}(Q_k) = \cos\varphi_k \cdot \text{IL}\left(Q_k^O\right) - \sin\varphi_k \cdot \text{IL}(H). \tag{4.13b}$$

Step 1. Influence line for shear at section k for reference beam is shown in Fig. 4.7; the specified ordinates at supports A and B equal to 1.0. The first term $\cos\varphi_k \cdot \text{IL}\left(Q_k^0\right)$ of (4.13b) presents a similar graph with specified ordinates $\cos\varphi_k = 0.954$ at supports A and B, so ordinates at the left and right of section k are -0.298 and 0.656, while at crown C is 0.477.

Step 2. Influence line for thrust is shown in Fig. 4.7; the specified ordinates at crown C equals to 1.0. The second term $\sin\varphi_k \cdot \text{IL}(H)$ of (4.13b) presents a similar graph with specified ordinates $0.3 \times 1.0 = 0.3$ at crown C. Specified ordinate at section k is 0.1875.

Step 3. Procedure (4.13b) is presented in Fig. 4.9. As in case for bending moment, both terms in (4.13b) has *different* signs, therefore both graphs, $\cos\varphi_k \cdot \text{IL}(Q_k^0)$ and $\sin\varphi_k \cdot \text{IL}(H)$ should be plotted on the *one side* on the basic line. Ordinates *between* both graphs present the required ordinates for influence line for shear. Specified ordinates of final graph (4.13b) left and right at section k are

$$0.298 + 0.1875 = 0.4855 \quad \text{and} \quad 0.656 - 0.1875 = 0.4685.$$

At crown C ordinate of influence line Q_k is $0.477 - 0.3 = 0.177$.

Step 4. Influence line between joints 2 and 3 presents a straight line; this connected line is shown by solid line. Final influence line $\text{IL}(Q_k)$ is presented in Fig. 4.9.

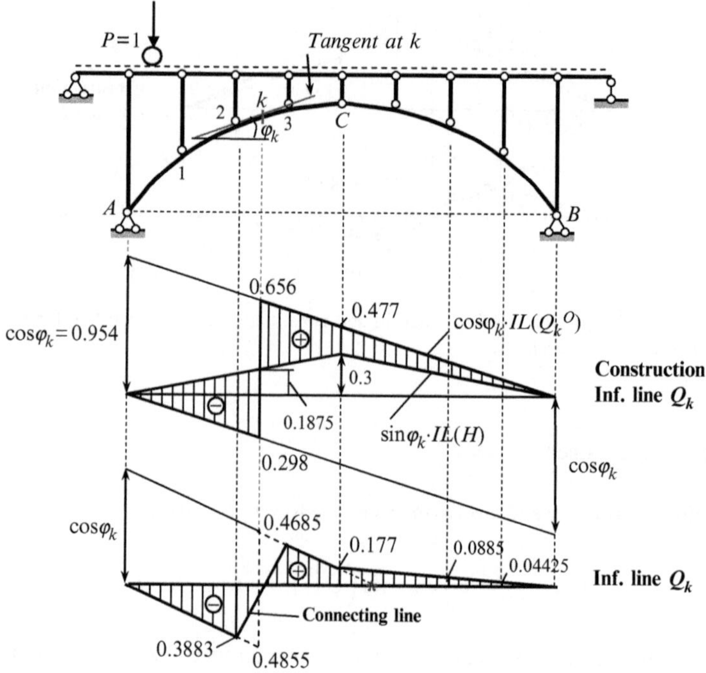

Fig. 4.9 Three-hinged arch. Design diagram and construction of influence line for shear at section k of the arch

4.3.2.3 Axial Force

This influence line may be constructed according to equation

$$\mathrm{IL}\,(N_k) = -\sin\varphi_k \cdot \mathrm{IL}\left(Q_k^0\right) - \cos\varphi_k \cdot \mathrm{IL}\,(H)\,. \qquad (4.13c)$$

Step 1. Influence line for shear at section k for reference beam is shown in Fig. 4.7. The first term $\sin\varphi_k \cdot \mathrm{IL}\left(Q_k^0\right)$ of (4.13c) presents a similar graph with specified ordinates $\sin\varphi_k = 0.30$ at supports A and B, so at the left and right of section $k$ordinates are 0.09375 and -0.20625, while at crown C is -0.15.

Step 2. Influence line for thrust is shown in Fig. 4.7; the specified ordinates at crown C equals to 1.0. The second term $\cos\varphi_k \cdot \mathrm{IL}\,(H)$ of (4.13c) presents a similar graph with specified ordinates $0.9539 \times 1.0 = 0.9539$ at crown C. Specified ordinate at section k is 0.59618.

Step 3. Procedure (4.13c) is presented in Fig. 4.10. Both terms in (4.13c) has *same* signs, therefore both graphs, $\sin\varphi_k \cdot \mathrm{IL}\left(Q_k^0\right)$ and $\cos\varphi_k \cdot \mathrm{IL}\,(H)$, should be plotted on the *different sides* on the basic line. Ordinates for required $\mathrm{IL}\,(N_k)$ are located *between* these both graphs.

Specified ordinates of final graph (4.13c) left and right at section k are

$$-(0.59618 - 0.09375) = -0.50243 \quad \text{and}$$
$$-(0.59618 + 0.20625) = -0.80243.$$

At crown C ordinate of influence line N_k is $-(0.9539 + 0.15) = -1.1039$.

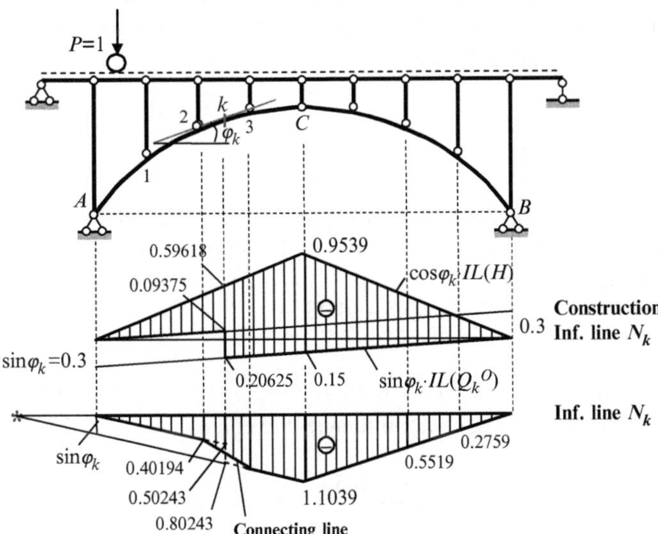

Fig. 4.10 Three-hinged arch. Design diagram and construction of influence line for axial force at section k of the arch

Step 4. Influence line between joints 2 and 3 presents a straight line; this connected line is shown by solid line. Final influence line IL (N_k) is shown in Fig. 4.10.

4.3.2.4 Properties of the Influence Lines for Internal Forces

1. Influence line for bending moment has significantly less ordinates than for reference beam. This influence line contains the positive and negative ordinates. It means that at section k extended fibers can be located below or above the neutral line depending on where the load is placed.
2. Influence line for shear, as in case of reference beam, has two portions with positive and negative ordinates; all ordinates are significantly less than in the reference beam. Influence line for axial force has only negative ordinates. So in case of arbitrary load the axial forces in arch are always compressed.

4.3.3 Application of Influence Lines

Assume that arch is subjected to fixed loads as shown in Fig. 4.5. The reactions of supports and internal forces caused by fixed load may be calculated by formula

$$Z = \sum P_i y_i + \sum q_j \Omega_j,$$

where Z is any force, for which the influence line is constructed; y the ordinate of influence line under the concentrated load P; and Ω is the area of influence line graph under the distributed load q.

Reactions of Supports

Ordinates of influence line for R_A at the points of application the loads P_1 and P_2 are 0.75 and 0.125, respectively. The area of the influence line under the uniformly distributed load is

$$\Omega = \frac{0.5 + 0.25}{2} \times 8 = 3.0.$$

Therefore, the reaction $R_A = P_1 \times 0.75 + q \times 3 + P_2 \times 0.125 = 14.5\,\text{kN}$

The thrust H of the arch, using influence line equals

$$H = P_1 \times 0.5 + q\frac{1 + 0.5}{2} \times 8 + P_2 \times 0.25 = 19\,\text{kN}.$$

These values of reactions coincide with those computed previously in Sect. 4.2.1.

Internal Forces in Section *k*

The internal forces can be found in a similar way, using the relevant influence lines. They are the following:

$$M_k = P_1 \times 1.96 - q\frac{2.0788 + 1.0394}{2} \times 8 - P_2 \times 0.5194 = -9.500\,\text{kN m},$$

$$Q_k = -P_1 \times 0.3883 + q\frac{0.177 + 0.0885}{2} \times 8 + P_2 \times 0.04425 = -1.405\,\text{kN},$$

$$N_k = -P_1 \times 0.40194 - q\frac{1.1039 + 0.5519}{2} \times 8 - P_2 \times 0.2759 = -19.473\,\text{kN}.$$

The magnitudes of just found internal forces M_k, Q_k, and N_k coincide with those computed in Sect. 342 and presented in Table 4.1.

Example 4.1. Let us consider design diagram of the arch in Fig. 4.7. It is necessary to find bending moment in section 3 due to the force $P = 10\,\text{kN}$, applied at point 7.

Solution. The feature of this problem is as follows: we will compute the bending moment at the section 3 *without* influence line for M_3 but using influence line for *reaction*. As the first step, we obtain the vertical reaction and thrust, which are necessary for calculation of internal forces.

Step 1. Find H and R_A from previously constructed influence lines presented in Fig. 4.7

$$R_A = P \times y_R = 10 \times 0.125 = 1.25\,\text{kN};$$
$$H = P \times y_H = 10 \times 0.25 = 2.5\,\text{kN},$$

where y_R and y_H are ordinates of influence line for R_A and H, respectively, under concentrated force P.

Step 2. The bending moment in section 3, considering left forces, becomes

$$M_3 = -H \cdot y_3 + R_A \cdot x_3 = -2.5 \times 7.596 + 1.25 \times 12 = -3.99\,\text{kN m},$$

where x_3 and y_3 are presented in Table 4.1.

This example shows that one of advantages of influence line is that the influence lines for reactions and thrust constructed *once* may be used for their computation for different cases of arbitrary loads. Then, by knowing reactions and thrust, the internal forces at any point of the arch may be calculated by definition without using influence line for that particular internal force.

This idea is the basis of complex usage of influence lines together with fixed load approach, which will be effectively applied for tedious analysis of complicated structures, in particularly for statically indeterminate ones.

4.4 Nil Point Method for Construction of Influence Lines

Each influence line shown in Figs. 4.8–4.10 has the specified points labeled as (*). These points are called nil (or neutral) point of corresponding influence line. Such points of influence lines indicate a position of the concentrated load on the arch, so internal forces M, Q, and N in the given section k would be zero. Nil points may be used for simple procedure for construction of influence lines for internal forces and checking the influence lines which were constructed by analytical approach. This procedure for symmetrical three-hinged arch of span l is discussed below.

4.4.1 Bending Moment

Step 1. Find nil point (NP) of influence line M_k. If load P is located on the left half of the arch, then reaction of the support B pass through crown C. Bending moment at section k equals zero, if reaction of support A pass through the point k. Therefore, NP (M_k) is point of intersection of line BC and Ak (theorem about three concurrent forces). The nil point (*) is always located between the crown C and section k (Fig. 4.11).

Step 2. Lay off along the vertical passing through the support A, the abscissa of section k, i.e., x_k.

Step 3. Connect this ordinate with nil point and continue this line till a vertical passing through crown C and then connect this point with support B.

Step 4. Take into account indirect load application; connecting line between joints 2 and 3 is not shown.

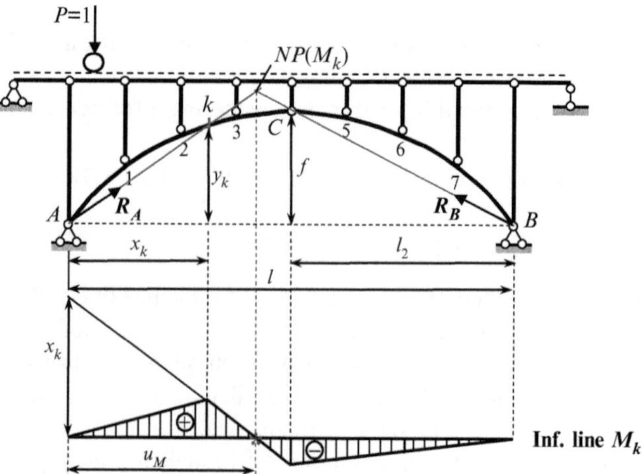

Fig. 4.11 Construction of Influence line M_k using nil point method

Location of NP (M_k) may be computed by formula

$$u_M = \frac{lfx_k}{y_k l_2 + x_k f}.$$

4.4.2 Shear Force

Step 1. Find nil point (NP) of influence line Q_k. If load P is located on the left half of the arch, then reaction of the support B pass through crown C. Shear force at section k equals zero, if reaction of support A will be parallel to tangent at point k. Therefore, NP (Q_k) is point of intersection of line BC and line which is parallel to tangent at point k. For given design diagram and specified section k the nil point (*) is fictitious one (Fig. 4.12a).

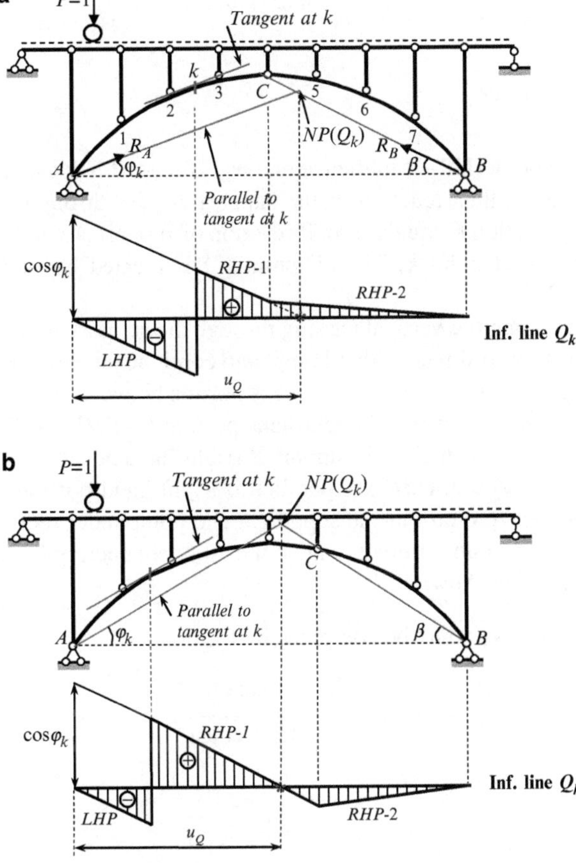

Fig. 4.12 **(a)** Construction of influence line Q_k using nil point method. The case of fictitious nil point. **(b)** Nonsymmetrical three-hinged arch. Construction of influence line Q_k using nil point method. The case of real nil point

Step 2. Lay off along the vertical passing through the support A, the value $\cos\varphi_k$.
Step 3. Connect this ordinate with nil point. A working zone of influence line is portion between section k and vertical passing through crown C – Right-hand portion 1 (*RHP*-1). Then connect the point under crown C with support B – Right-hand portion 2 (*RHP*-2).
Step 4. Left-hand portion (*LHP*) is parallel to right-hand portion 1 and connects two points: zero ordinate at support A and point under section k.
Step 5. Take into account indirect load application; connecting line between joints 2 and 3 is not shown.

Figure. 4.12b presents a nonsymmetrical three-hinged arch with real nil point for influence line Q_k; this point is located within the span of the arch. Therefore, we have one portion with positive shear and two portions with negative shear.

Location of NP (Q_k) (Fig. 4.12a, b) may be computed by formula

$$u_Q = \frac{l \tan\beta}{\tan\beta + \tan\varphi_k}.$$

4.4.3 Axial Force

Step 1. Find nil point (NP) of influence line N_k. If load P is located on the left half of the arch, then reaction of the support B pass through crown C. Axial force at section k equals zero, if reaction of support A will be perpendicular to tangent at point k. The nil point (*) is located beyond the arch span (Fig. 4.13).
Step 2. Lay off along the vertical passing through the support A, the value $\sin\varphi_k$.
Step 3. Connect this ordinate with nil point and continue this line till vertical passes through crown C. A working zone is portion between section k and vertical passing through crown C (right-hand portion 1 – *RHP*-1). Then connect the point under crown C with support B (right-hand portion 2 – *RHP*-2).
Step 4. Left-hand portion (*LHP*) is parallel to Right-hand portion 1 and connects two points: zero ordinate at support A and point under section k.
Step 5. Take into account indirect load application; connecting line between joints 2 and 3 is not shown.

Location of NP (N_k) may be computed by formula

$$u_N = \frac{l \tan\beta}{\tan\beta - \cot\varphi_k}.$$

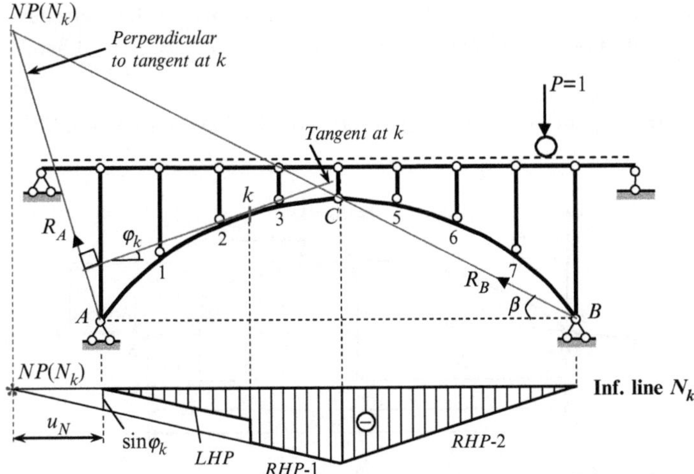

Fig. 4.13 Construction of influence line N_k using nil point method

4.5 Special Types of Arches

This section is devoted to analysis of special types of arches. Among them are arch with support points located on the different levels and parabolic three-hinged arch with complex tie.

4.5.1 Askew Arch

The arch with support points located on the different levels is called askew (or rising) arch. Three-hinged askew arch is geometrically unchangeable and statically determinate structure. Analysis of askew arch subjected to the fixed and moving loads has some features.

Design diagram of three-hinged askew arch is presented in Fig. 4.14. Let the shape of the arch is parabola, span of the arch $l = 42\,\text{m}$ and support B is $\Delta = 3.5\,\text{m}$ higher than support A. The total height of the arch at hinge C is $f + f_0 = 8\,\text{m}$. The arch is loaded by force $P = 10\,\text{kN}$. It is necessary to calculate the reactions and bending moment at section k, construct the influence lines for thrust and bending moment M_k, and apply influence lines for calculation of bending moment and reactions due to fixed load.

Equation of the axis of parabolic arch

$$y = 4\,(f + f_0)\,(L - x)\,\frac{x}{L^2}, \tag{4.14}$$

where span for arch $A - C - B'$ with support points on the same level is $L = l + l_0 = 48$ m. For $x = 42$ m (support B) the ordinate $y = \Delta = 3.5$ m, so

$$\tan\alpha = \frac{\Delta}{l} = \frac{3.5}{42} = 0.0833 \rightarrow \cos\alpha = 0.9965 \rightarrow \sin\alpha = 0.08304.$$

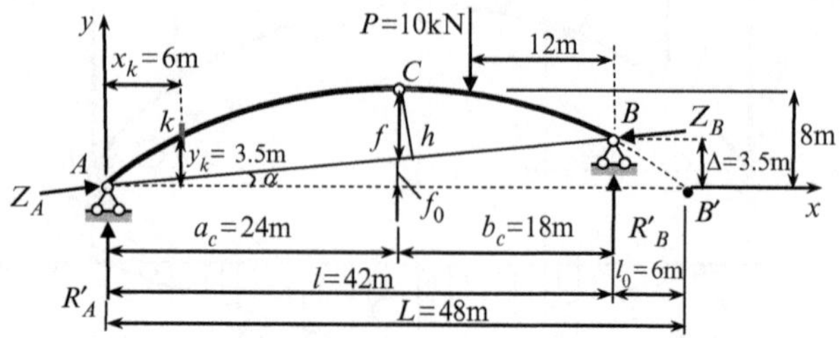

Fig. 4.14 Design diagram of an askew three-hinged arch

Other geometrical parameters are

$$\begin{aligned} f_0 &= a_C \tan\alpha = 24\tan\alpha = 2.0\,\text{m} \rightarrow f = 8 - 2 = 6\,\text{m} \rightarrow \\ h &= f\cos\alpha = 6 \cdot 0.9965 = 5.979\,\text{m}. \end{aligned} \tag{a}$$

For $x = 6$ m (section k), the ordinate $y_k = 3.5$ m.

4.5.1.1 Reactions and Bending Moment at Section *k*

Reactions of supports It is convenient to resolve total reaction at point A into two components. One of them, R'_A, has vertical direction and other, Z_A, is directed along line AB. Similar resolve the reaction at the support B. These components are R'_B and Z_B. The vertical forces R'_A and R'_B present a *part* of the total vertical reactions. These vertical forces may be computed as for reference beam

$$\begin{aligned} R'_A &\rightarrow \sum M_B = 0: \quad -R'_A \cdot 42 + P \cdot 12 = 0 \rightarrow R'_A = 2.857\,\text{kN}, \\ R'_B &\rightarrow \sum M_A = 0: \quad R'_B \cdot 42 - P \cdot 30 = 0 \rightarrow R'_B = 7.143\,\text{kN}. \end{aligned} \tag{b}$$

Since a bending moment at crown C is zero then

$$\begin{aligned} &Z_A \rightarrow \sum M_C^{\text{left}} 0: \; Z_A \cdot h - M_C^0 = 0 \rightarrow Z_A = \frac{M_C^0}{h} = \frac{2.857 \times 24}{5.979} = 11.468\,\text{kN}, \\ &Z_A = Z_B = Z, \end{aligned} \tag{c}$$

where M_C^0 is a bending moment at section C for reference beam.

Thrust H presents the horizontal component of the Z, i.e.,

$$H = Z\cos\alpha = 11.468 \times 0.9965 = 11.428\,\text{kN}. \tag{4.15}$$

The total vertical reactions may be defined as follows

$$R_A = R'_A + Z\sin\alpha = 2.857 + 11.468 \times 0.08304 = 3.809\,\text{kN},$$
$$R_B = R'_B - Z\sin\alpha = 7.143 - 11.468 \times 0.08304 = 6.191\,\text{kN}. \qquad (4.16)$$

Bending moment at section k:

$$M_k = M_k^0 - Hy = 3.809 \times 6 - 11.428 \times 3.5 = -17.144\,\text{kN}. \qquad (4.17)$$

4.5.1.2 Influence Lines for Thrust and Bending Moment M_k.

Thrust Since $H = \left(M_C^0/H\right)\cos\alpha$, then equation of influence line for thrust becomes

$$\text{IL}\,(H) = \frac{\cos\alpha}{h} \cdot \text{IL}\left(M_C^0\right). \qquad (4.18)$$

The maximum ordinate of influence line occurs at crown C and equals

$$\frac{\cos\alpha}{h} \cdot \frac{a_C b_C}{l} = \frac{0.9965}{5.979} \times \frac{24 \times 18}{24 + 18} = 1.71428.$$

Bending moment Since $M_k = M_k^0 - Hy_k$, then equation of influence line for bending moment at section k becomes

$$\text{IL}\,(M_k) = \text{IL}\left(M_k^0\right) - y_k \cdot \text{IL}\,(H). \qquad (4.19)$$

Influence line may be easily constructed using the nil point method. Equation of the line Ak is

$$y = \frac{3.5}{6}x = 0.5833x. \qquad \text{(d)}$$

Equation of the line BC is

$$y - y_C = m\,(x - x_C) \rightarrow y - 8 = -\frac{4.5}{18}\,(x - 24) \rightarrow y = 14 - 0.25x, \qquad \text{(e)}$$

where m is a slope of the line BC.

The nil point NP (M_k) of influence line for M_k is point of intersection of lines Ak and BC. Solution of equations (d) and (e) is $x_0 = 16.8$ m. Influence lines for H and M_k are presented in Fig. 4.15. Maximum positive and negative bending moment at section k occurs if load P is located at section k and hinge C, respectively. If load P is located within portion x_0, then extended fibers at the section k are located below the neutral line of the arch.

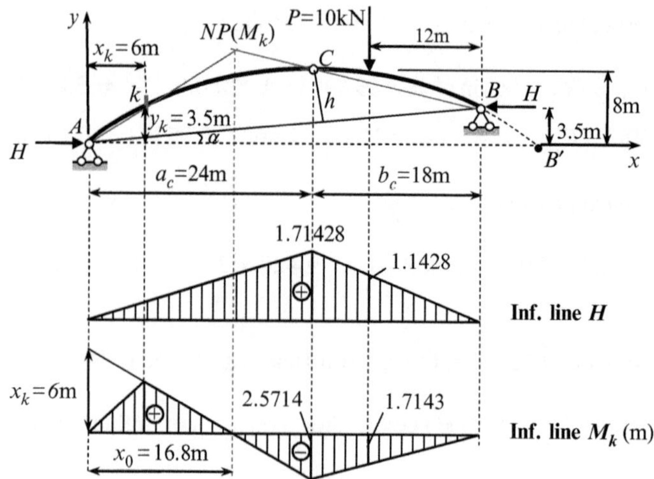

Fig. 4.15 Three-hinged askew arch. Design diagram and influence lines

The thrust and bending moment at the section k may be calculated using the relevant influence lines

$$H = Py = 10 \times 1.1428 = 11.428\,\text{kN},$$

$$M_k = Py = 10 \times (-1.7143) = -17.143\,\text{kN m}.$$

These values coincide exactly with those calculated by formulas (4.15) and (4.17).

As before, the influence lines for thrust constructed *once* may be used for its computation for different cases of arbitrary loads. Then, knowing the vertical reactions and thrust, the internal forces at any point of the arch may be calculated by definition without using influence line for that particular internal force.

4.5.2 Parabolic Arch with Complex Tie

Analysis of such structure subjected to fixed and moving load has some features.

Design diagram of the symmetrical parabolic arch with complex tie is presented in Fig. 4.16. The arch is loaded by vertical uniformly distributed load $q = 2\,\text{kN/m}$. We need to determine the reactions of the supports, thrust, and bending moment at section k ($a_k = 18\,\text{m}$, $y_k = 11.25\,\text{m}$, $\cos\varphi_k = 0.970$, $\sin\varphi_k = 0.2425$) as well as to construct the influence lines for above-mentioned factors.

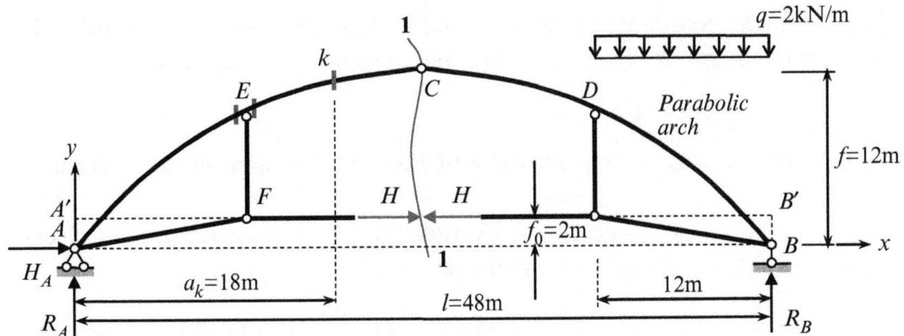

Fig. 4.16 Design diagram of the arch with complex tie

4.5.2.1 Reactions and Bending Moment at Section *k*

The vertical reactions are determined from the equilibrium equations of all the external forces acting on the arch

$$\begin{aligned} R_A \to \sum M_B = 0: &\quad -R_A \times 48 + q \times 12 \times 6 = 0 \to R_A = 3\,\text{kN}, \\ R_B \to \sum M_A = 0: &\quad R_B \times 48 - q \times 12 \times 42 = 0 \to R_B = 21\,\text{kN}. \end{aligned} \tag{a}$$

Horizontal reaction at support A is $H_A = 0$.

The thrust H in the tie (section 1-1) is determined from the following equation

$$H \to \sum M_C^{\text{left}} = 0: \quad -R_A \frac{l}{2} + H\,(f - f_0) = 0 \to H = \frac{M_C^0}{f - f_0} = 7.2\,\text{kN}. \tag{4.20}$$

Equilibrium equations of joint F lead to the axial forces at the members of AF and EF of the tie.

Internal forces at section k for a reference simply supported beam are as follows:

$$\begin{aligned} M_k^0 &= R_A \cdot x_k = 3 \times 18 = 54\,\text{kN m}, \\ Q_K^0 &= R_A = 3\,\text{kN}. \end{aligned} \tag{b}$$

Internal forces at the point k for three-hinged arch are determined as follows:

$$\begin{aligned} M_k &= M_k^0 - H\,(y_k - f_0) = 54 - 7.2\,(11.25 - 2) = -12.6\,\text{kN m}, \\ Q_k &= Q_k^0 \cos\varphi_k - H \sin\varphi_k = 3 \times 0.970 - 7.2 \times 0.2425 = 1.164\,\text{kN}, \\ N_k &= -\left(Q_k^0 \sin\varphi_k + H \cos\varphi_k\right) = -\,(3 \times 0.2425 + 7.2 \times 0.970) = -7.711\,\text{kN}. \end{aligned} \tag{4.21}$$

Note, that the discontinuity of the shear and normal forces at section E left and right at the vertical member EF are $N_{EF}\cos\varphi$ and $N_{EF}\sin\varphi$, respectively.

4.5.2.2 Influence Lines for Thrust and Bending Moment at the Section k

Vertical reactions Influence lines for vertical reactions R_A and R_B for arch and for reference simply supported beam coincide, i.e.,

$$\text{IL}\,(R_A) = \text{IL}\left(R_A^0\right); \quad \text{IL}\,(R_B) = \text{IL}\left(R_B^0\right).$$

Thrust According to expression (4.20), the equation of influence line for thrust becomes

$$\text{IL}\,(H) = \frac{1}{f - f_0} \cdot \text{IL}\left(M_C^0\right). \tag{4.22}$$

The maximum ordinate of influence line for H at crown C

$$\frac{1}{(f - f_0)} \cdot \frac{l}{4} = \frac{48}{4 \times (12 - 2)} = 1.2. \tag{c}$$

Influence line for thrust may be considered as key influence line.

Bending moment According to expression (4.21) for bending moment at any section, the equation of influence line for bending moment at section k becomes

$$\text{IL}\,(M_k) = \text{IL}\left(M_k^0\right) - (y_k - f_0) \cdot \text{IL}\,(H) = \text{IL}\left(M_k^0\right) - 9.25 \cdot \text{IL}\,(H). \tag{4.23}$$

Influence line M_k^0 presents a triangle with maximum ordinate $(a_k b_k)/l = (18 \times 30)/48 = 11.25\,\text{m}$ at section k, so the ordinate at crown C equals to 9 m. Influence line for thrust H presents the triangle with maximum ordinate 1.2 at crown C. Ordinate of the graph $(y_k - f_0) \cdot \text{IL}\,(H)$ at crown C equals $(11.25 - 2) \times 1.2 = 11.1\,\text{m}$, so ordinate at section k equals 8.325 m. Detailed construction of influence line M_k is shown in Fig. 4.17. Since both terms in (4.23) has *different* signs, they should be plotted on the *one side* on the basic line; the final ordinates of influence line are located *between* two graphs $\text{IL}\left(M_k^0\right)$ and $9.25 \cdot \text{IL}\,(H)$.

Maximum bending moment at section k occurs if load P is located above section k and crown C. Bending moment at section k may be positive, negative, and zero. If load P is located within the portion A-NP(M_k), then extended fibers at section k are located below neutral line of the arch.

Figure. 4.17 also presents the construction of influence lines for bending moment using nil points; pay attention that construction of this point must be done on the basis of conventional supports A' and B'.

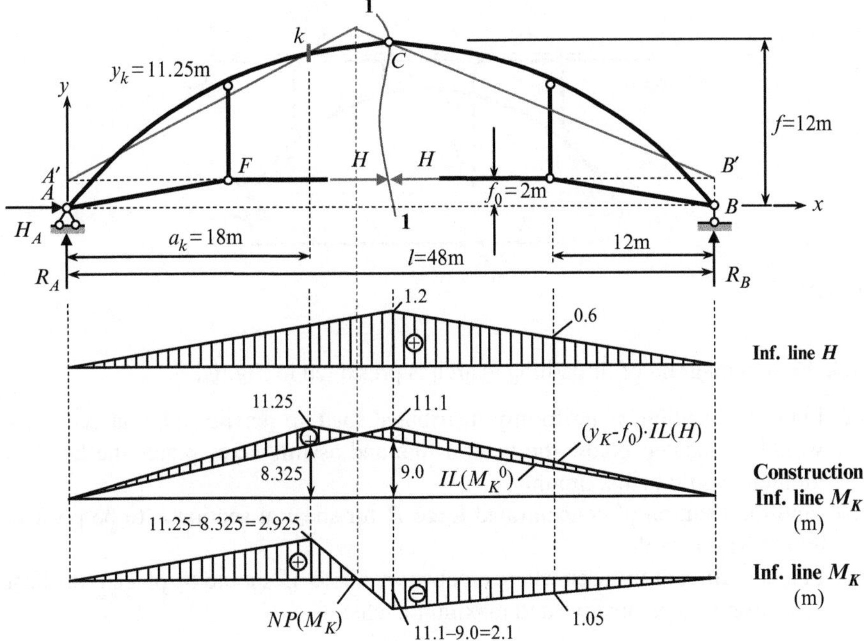

Fig. 4.17 Three-hinged arch with complex tie. Influence lines for H and M_k

Problems

4.1. Design diagram of symmetrical three-hinged arch is presented in Fig. P4.1. Construct the influence lines for internal forces at section n using the nil point method.

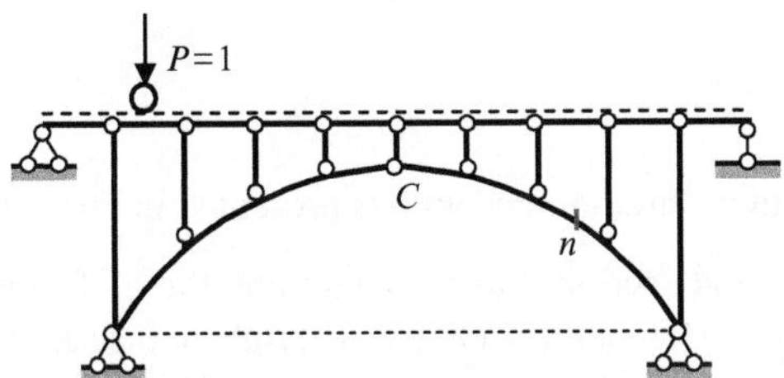

Fig. P4.1

4.2. Design diagram of nonsymmetrical three-hinged arch is presented in Fig. P4.2. Construct the following influence lines (a) for vertical reactions at arch supports A and B; (b) for thrust; and (c) for internal forces at sections k and n. Use the nil point method. Take into account indirect load application.

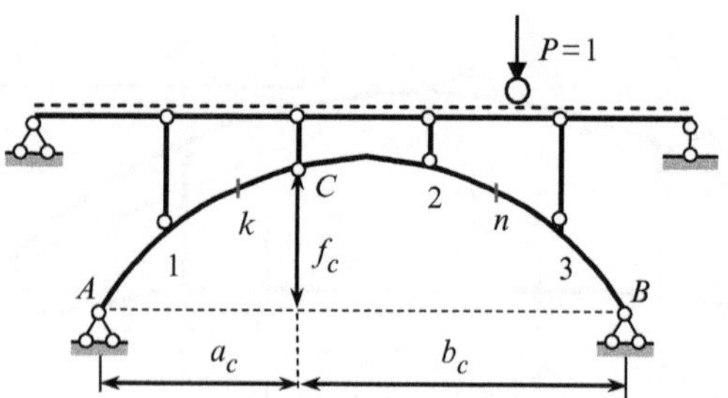

Fig. P4.2

4.3. Design diagram of three-hinged arch is presented in Fig. P4.3.

(a) Find the location of uniformly distributed load so tensile fibers at section k would be located below the neutral line and as this takes place, the bending moment would be maximum
(b) Find the location of concentrated force P for shear at section k to be positive and maximum value
(c) Find a portion of the arch where clockwise couple M should be placed for shear at section k to be positive and maximum value.

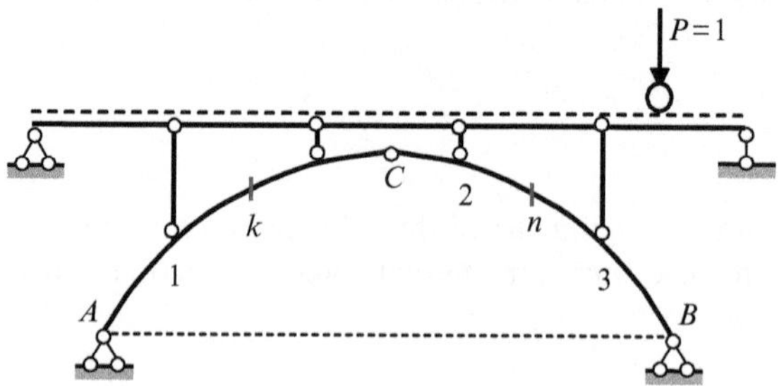

Fig. P4.3

4.4. Design diagram of three-hinged arch is presented in Fig. P4.4.

1. For the axial force at section n to be maximum, the uniformly distrubuted load should be located (a) left at the section n; (b) right at the section n; and (c) within all span AB.
2. For the axial force at section n to be maximum a concentrated force P should be located
   (a) left at the section n; (b) right at the section n; (c) at the hinge C; (d) left at hinge C; (e) right at hinge C.
3. The vertical settlement of support A leads to appearance of internal forces at the following part of the arch: (a) portion AC; (b) portion CB; (c) all arch AB; (d) nowhere.

4. Decreasing of external temperature within the left half-arch AC leads to appearance of internal forces at at the following part of the arch: (a) portion AC; (b) portion CB; (c) all arch AB; (d) nowhere.

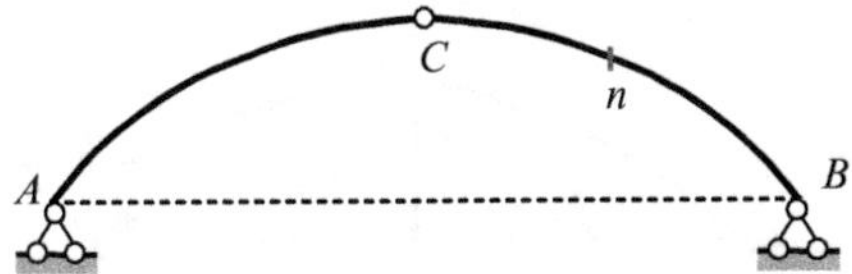

Fig. P4.4

4.5. Three-hinged parabolic nonsymmetrical arch span l is subjected to concentrated force P at the hinge C (Fig. P4.5). Determine the effect of the location of the hinge C on the value of the trrust H. (Hint: $y = 4f_0x(l-x)\frac{1}{l^2}$, $H = \frac{M_C^0}{f_C}$)

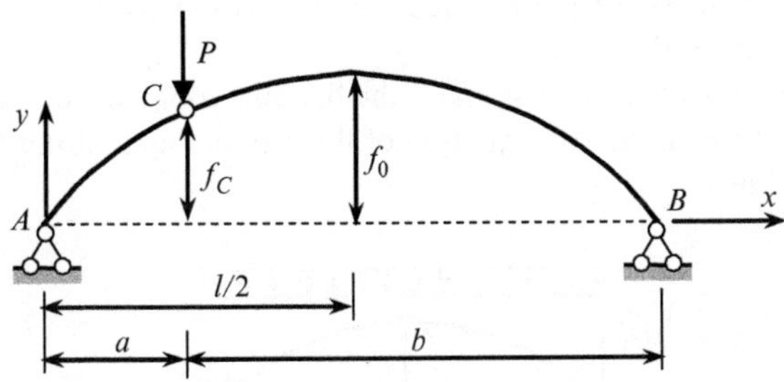

Fig. P4.5

Ans. $H = \dfrac{Pl}{4f_0}$

4.6. Three-hinged arch with tie is subjected to concentrated force P at the hinge C (Fig. P4.6). Determine the effect of the location f_0 of the tie on the axial force H in the tie.

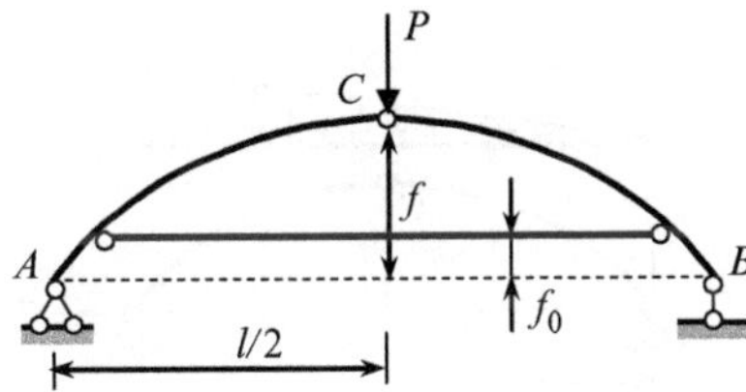

Fig. P4.6

Ans. $H = \dfrac{Pl}{4(f - f_0)}$

4.7. Three-hinged askew arch span l is subjected to concentrated force P at the hinge C (Fig. P4.7). Determine the effect of the parameter f_0 on the thrust of the arch. (Hint: $H = Z\cos\alpha$).

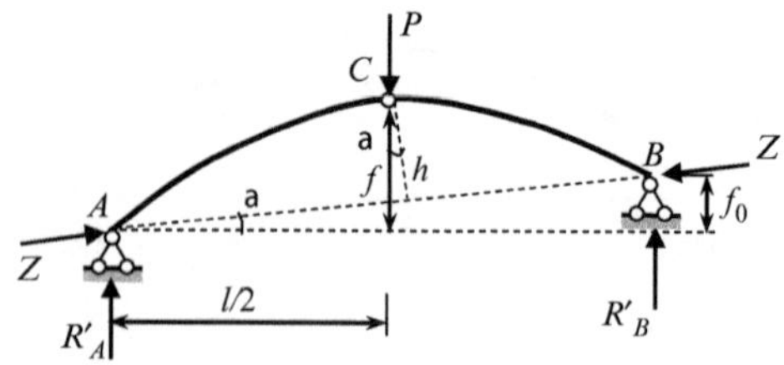

Fig. P4.7

$$\text{Ans. } H = \frac{Pl}{2(2f - f_0)}$$

4.8. Three-hinged symmetrical arch is loaded by uniformly distributed load q (Fig. P4.8). The span and rise of the arch are l and f, respectively. Derive the equation of the rational axis of the arch.

(Note: the arch is called as rational if the bending moments do not arise at all the cross sections of the arch. The equation of the rational axis of the arch depends on the type of loading).

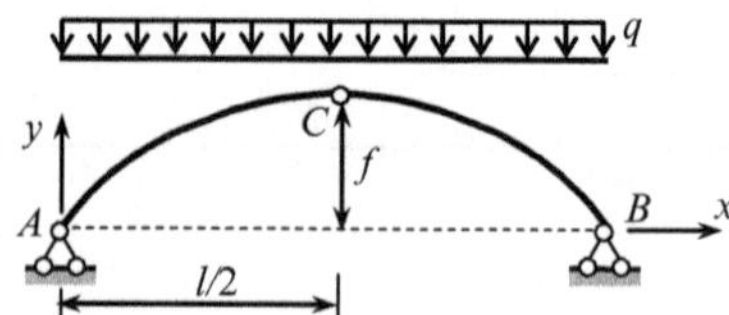

Fig. P4.8

$$\text{Ans. } y = 4f\frac{x}{l}\left(1 - \frac{x}{l}\right)$$

4.9. Three-hinged symmetrical arch is loaded by distributed load q as shown in Fig. P4.9. Derive the equation of the rational axis of the arch.

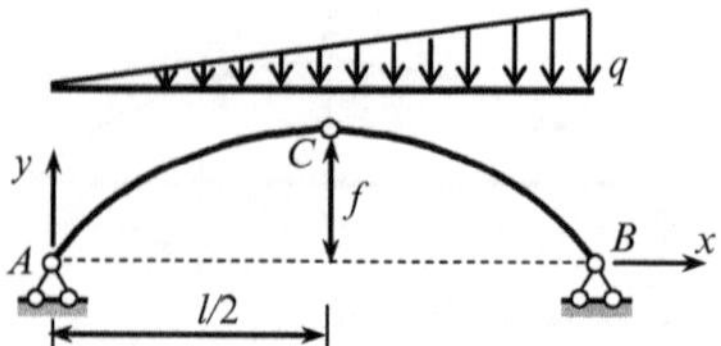

Fig. P4.9

$$\text{Ans. } y = \frac{8}{3}f\frac{x}{l}\left(1 - \frac{x^2}{l^2}\right)$$

4.10. Three-hinged symmetrical arch is loaded by distributed load q as shown in Fig. P4.10.

Find equation of the rational axis of the arch.

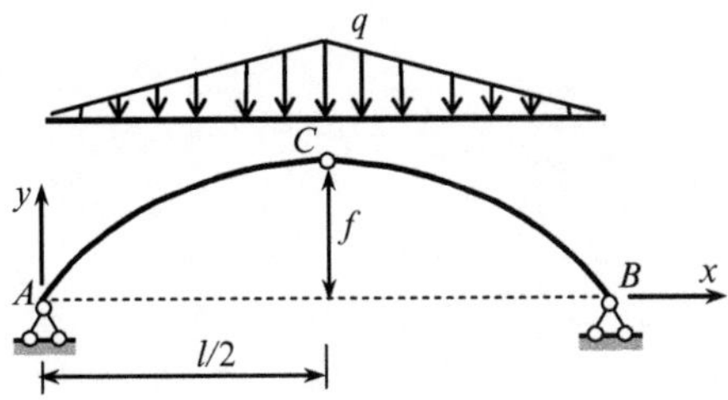

Fig. P4.10

Ans. $y = 3f\frac{x}{l}\left(1 - \frac{4}{3}\frac{x^2}{l^2}\right)$

4.11. Three-hinged symmetrical arch is loaded by the radial distributed load q. Show that the rational axis of the arch presents the circle.

Chapter 5
Cables

This chapter is devoted to analysis of cables under fixed loads of different types. Among them are concentrated loads, uniformly distributed load along the horizontal line and along the cable itself. Important formulas for analysis of the cable subjected to *arbitrary*loads are derived; they allow determining the *changing*of reactions, internal forces and shape due by any additional live loads. Relationships between the thrust, internal forces and total length of a cable are established. The influence of elastic properties of a cable is discussed.

5.1 Preliminary Remarks

The cables as the permanent members of the load-bearing structures are used extensively in modern engineering. Some examples of cabled structures are as follows: suspension bridges, anchoring systems of different objects such as guy-rope of the masts, sea drilling platforms, stadium covering, cableways, floating breakwaters, light-vessel, etc. Cables are also used as the temporary guys during erection of the structures. Suspension bridge is shown in Fig. 5.1.

Cables are made from high-strength steel wires twisted together, and present a flexible system, which can resist only axial tension. The cables allow cover very large spans. This may be explained by two reasons. (1) In axial tension, the stresses are distributed uniformly within all areas of cross section, so the material of a cable is utilized in full measure; (2) Cables are made from steel wires with very high ultimate tensile strength ($\sigma_{\upsilon} \cong 1,860\,\text{MPa}$, while for structural steel ASTM-A36 $\sigma_{\upsilon} \cong 400\,\text{MPa}$). Therefore the own weight of a load-bearing structure becomes relatively small and the effectiveness application of cables increases with the increasing of the spans. Modern suspension bridges permit coverage of spans hundreds of meters in length.

A cable as a load-bearing structure has several features. One of them is the vertical load that gives rise to horizontal reactions, which, as in case of an arch, is called a thrust. To carry out the thrust it is necessary to have a supporting structure. It may be a pillar of a bridge (Fig. 5.1) or a supporting ring for the covering of a stadium. The cost of the supporting structure may be a significant factor in the overall cost of the whole structure.

I.A. Karnovsky and O. Lebed, *Advanced Methods of Structural Analysis*,
DOI 10.1007/978-1-4419-1047-9_5,

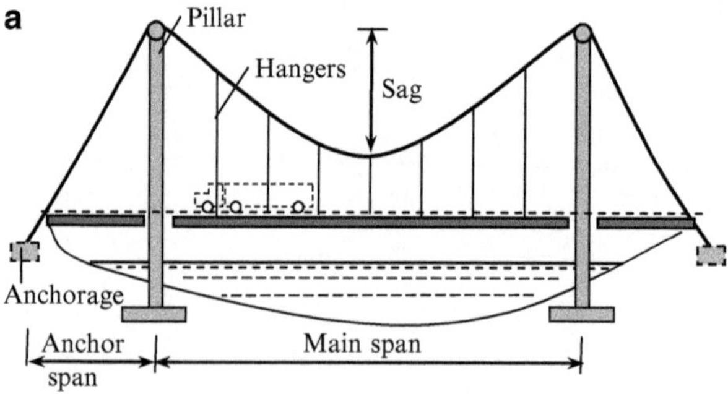

Fig. 5.1 Suspension bridge

Another feature of cables is the high sensitivity of their shape, depending on the total length of the cable, type of the load as well as a load location along the span. The fundamental feature of a cable is its unknown shape in advance. Defining the shape of a cable is one of the important problems.

The following types of external loads will be considered:

1. Cable with self-weight ignored (a cable subjected to concentrated loads, to uniformly distributed load as well as to arbitrary dead load and after that additional any live load).
2. Cable with self-weight.

Under these loads the cable shape takes different forms. Shape of a cable in case of concentrated loads is referred to slopes of straight portions of a cable. Shape of a cable in case of distributed load along horizontal and along cable itself are parabola and catenary, respectively.

Assumptions

1. The cable is inextensible one (elastic properties of the cable will be considered in Sect. 5.6).
2. The cable is perfectly flexible, i.e., the cable does not resist to shear and bending. In this case the internal force in any point of a cable is tensile one, which is directed along the tangent to this point of a cable.

Horizontal component of a tension force is called a thrust. Later it will be shown that thrust remains constant along the length of the cable.

5.1.1 Direct and Inverse Problems

The simplest problem is finding thrust, internal forces, and total length of a cable if the shape of a cable is given. Solution of this problem is well known. However, for practice this type of problem does not have a reasonable sense because shape of the cable cannot be known in advance. Instead, an engineer knows allowable horizontal force, which a tower can resist at the top. Based on this, we will consider the following two fundamental practical problems for cables under fixed loads:

1. Find the shape of a cable and internal forces, if a thrust is given. This type of problem is called the thrust–shape problem.
2. Compute thrust and internal forces, if a total length of a cable is given. This type of problem is called the length–thrust problem.

5.1.2 Fundamental Relationships

Let us consider a cable shown in Fig. 5.2a. The cable is loaded by some vertical concentrated loads P_i (the distance from left support are x_i) and arbitrary load q distributed along horizontal. The self-weight of the cable may be neglected if this weight comprises no more than 10% of an external load on the cable. Since the cable is perfectly flexible, then axial force N_x at any section is directed along tangent at this section. Axial forces N_A and N_B at the supports A and B are resolved into two directions. They are the vertical direction and direction A–B. These forces are

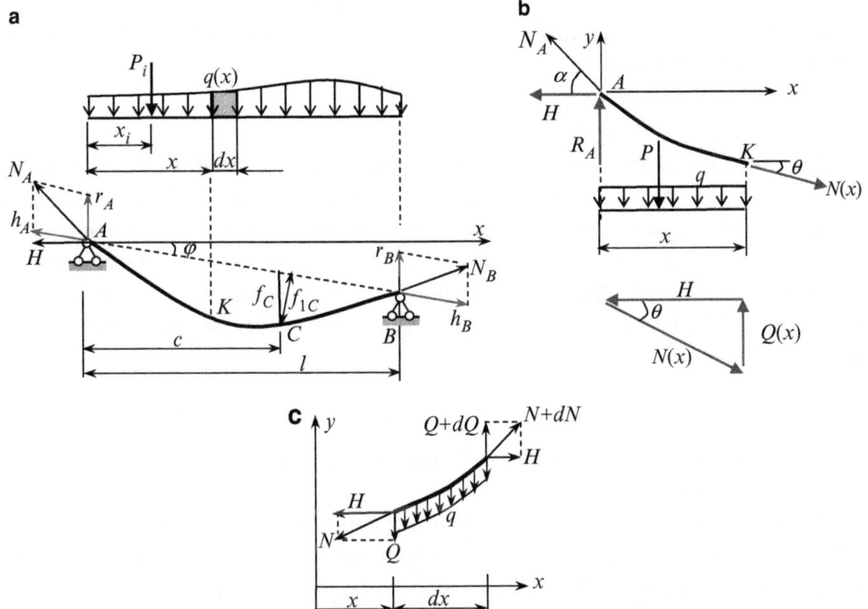

Fig. 5.2 (**a**) Cable subjected to arbitrary load; (**b**) free-body diagram of the part of the cable; (**c**) free-body diagram of the elementary portion of the cable

denoted as r_A and h_A for support A and r_B and h_B for support B. Summing of all forces on the horizontal axis leads to the relationship $h_A = h_B = h$.

The vertical component of the reaction at B

$$r_B \to \sum M_A = 0: \quad r_B l - \sum P_i x_i - \int_l q(x)\mathrm{d}x = 0 \to r_B = \frac{\sum P_i x_i + \int_l q(x)\mathrm{d}x}{l} \tag{5.1}$$

This expression coincides with general formula for reaction of simply supported beam. Therefore if axial forces at supports are resolved as shown in Fig. 5.2a, then a vertical component of axial force at support equals to corresponding reaction of the reference beam.

Now we can calculate ordinate of the cable at the section C with abscissa c. Since the bending moment at any section C is zero, then

$$M_C = r_A \cdot c - h_A \cdot f_{1C} - \sum P_i (c - x_i) - \int_0^c q(x)(c - x)\mathrm{d}x = 0$$

$$f_{1C} = \frac{r_A \cdot c - \sum P_i (c - x_i) - \int_0^c q(x)(c - x)\mathrm{d}x}{h} \tag{5.2}$$

where f_{1C} is a perpendicular to the inclined chord AB.

It is obvious that nominator presents the bending moment at the section C of the reference beam. Therefore

$$f_{1C} = \frac{M_C^0}{h}. \tag{5.3}$$

To determine the vertical coordinate of the point C we need to modify this formula. For this we need to introduce a concept of a thrust. The thrust H is a horizontal component of the axial force at any section. From Fig. 5.2b we can see that equilibrium equation $\sum X = 0$ leads to the formula $H = \text{const}$ for any cable subjected to arbitrary vertical loads. At any support $h = H/\cos\varphi$. Since $f_{1C} = f_C \cos\varphi$, then expression (5.3) can be rewritten as $f_C \cos\varphi = M_C^0 / (H/\cos\phi)$, which leads to the formula:

$$f_C = \frac{M_C^0}{H} \tag{5.4}$$

where f_C is measured vertically from the inclined chord AB. This parameter is called the sag of the cable.

The formula (5.4) defines the shape of the cable subjected to any load. The sag and y-ordinate are equal if and only if supports are located on the same elevation, the origin is placed on the horizontal chord AB and y-axis is directed downward.

Now the total vertical reactions of supports may be determined as follows:

$$\begin{aligned} R_A &= r_A + h \sin\varphi = r_A + H \tan\varphi, \\ R_B &= r_B - h \sin\varphi = r_B - H \tan\varphi. \end{aligned} \tag{5.5}$$

The concept of "shear" at the any section of the reference beam will be helpful. The portion length of x is loaded by uniformly distributed load q, concentrated force P, reaction R_A, thrust H, and axial force $N(x)$ as shown in Fig. 5.2b. The beam shear is $Q(x) = R_A - P - qx$. At the section x (point K) three concurrent forces acts. They are $Q(x)$, $N(x)$, and H. It is obvious that for any section

$$N(x) = \sqrt{H^2 + Q^2(x)}. \tag{5.6}$$

Now let us derive the differential relationships between load q and ordinate of the cable y. The angle θ between the axial line of a cable and x-axis (Fig. 5.2b) obeys to equation

$$\tan\theta = \frac{\mathrm{d}y}{\mathrm{d}x} = \frac{Q}{H}. \tag{5.7}$$

To derive a relationship between Q and intensity of the load q, consider a free body diagram of the elementary portion of the cable (Fig. 5.2c). The axial forces at the left and right ends of a portion are N and $N + \mathrm{d}N$, respectively. Their components are H and Q at the left end, while at the right end H and $Q + \mathrm{d}Q$.

Equilibrium equation $\sum Y = 0$ leads to the following relationship: $\mathrm{d}Q/\mathrm{d}x = q$. Taking into account (5.7), the required relationships between y and q becomes:

$$\frac{\mathrm{d}^2y}{\mathrm{d}x^2} = \frac{q}{H}. \tag{5.8}$$

The first and second integration of (5.8) leads to the expressions for slope and y ordinate, respectively. The constant integration should be determined from the boundary conditions.

5.2 Cable with Neglected Self-Weight

This section contains analysis of the perfectly flexible inextensible cables subjected to concentrated loads, as well as distributed loads along horizontal foot of the cable. For both loadings (concentrated and distributed), the thrust–shape and length–thrust problems are considered.

5.2.1 Cables Subjected to Concentrated Load

In this case of loading each portion of the cable between two adjacent forces presents the straight segment. The simplest design diagram of a cable, subjected to one concentrated load, is presented in Fig. 5.3a. Parameters of the system are $a = 10\,\mathrm{m}$, $l = 25\,\mathrm{m}$, $P = 20\,\mathrm{kN}$.

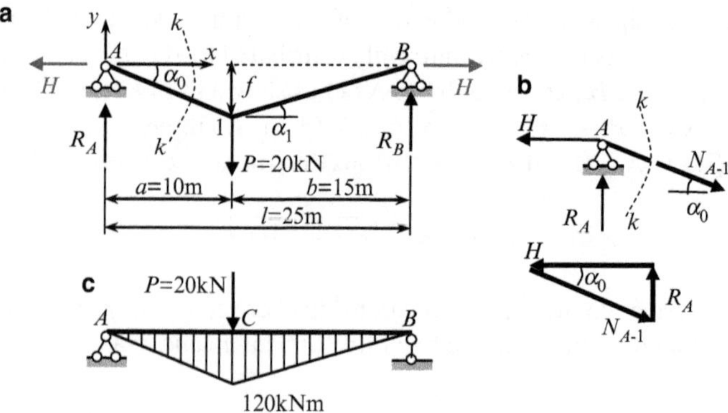

Fig. 5.3 (**a**) Design diagram of the cable; (**b**) free-body diagram; (**c**) reference beam and corresponding bending moment diagram

5.2.1.1 Direct Problem

Determine a shape of the cable, if the thrust of the system is given $H = 24\,\text{kN}$. Vertical reactions of the cable

$$R_A \to \sum M_B = 0: \quad R_A = \frac{P(l-a)}{l} = \frac{20(25-10)}{25} = 12\,\text{kN},$$
$$R_B \to \sum M_A = 0: \quad R_B = \frac{Pa}{l} = \frac{20 \times 10}{25} = 8\,\text{kN}.$$

We can see that the vertical reactions do not depend on the value of the trust H. It happens because supports A and B are located on the same elevation. The forces acting on the segment at support A and corresponding force polygon is shown in Fig. 5.3b. For assumed x–y coordinate system the angles α_0 and α_1 belong to the fourth and first quadrant, respectively, so a shape of the cable is defined as follows

$$\tan\alpha_0 = -\frac{R_A}{H} = -\frac{12}{24} = -\frac{1}{2} \to \cos\alpha_0 = \frac{2}{\sqrt{5}},$$
$$\tan\alpha_1 = \frac{R_B}{H} = \frac{8}{24} = \frac{1}{3} \to \cos\alpha_1 = \frac{3}{\sqrt{10}}.$$

The y-ordinate of the cable at the location of the load P is

$$y = a\tan\alpha_0 = -10 \times 0.5 = -5\,\text{m}.$$

The negative sign corresponds to the adopted x–y coordinate system. The sag at the point C is $f = 5\,\text{m}$.

Tensions in the left and right portions of the cable may be presented in terms of thrust as follows:

$$N_{A-1} = \frac{H}{\cos\alpha_0} = \frac{24 \times \sqrt{5}}{2} = 26.83\,\text{kN}, \; N_{B-1} = \frac{H}{\cos\alpha_1} = \frac{24 \times \sqrt{10}}{3} = 25.30\,\text{kN}.$$

Increasing of the thrust H leads to decreasing of the angle α_0 and α_1; as a result, the sag of the cable is decreasing and tension in both parts of the cable is increasing. In fact, we used, here, the obvious physical statement: in case of concentrated forces the cable presents a set of straight portions.

The tensions in the left and right portions of the cable can be determined using expression (5.6).

Shear forces are $Q_{A-1} = R_A = 12\,\text{kN}, \quad Q_{1-B} = R_A - P = -R_B = -8\,\text{kN}$. Therefore

$$N_{A-1} = \sqrt{Q_{A-1}^2 + H^2} = \sqrt{12^2 + 24^2} = 26.83\,\text{kN};$$

$$N_{1-B} = \sqrt{Q_{1-B}^2 + H^2} = \sqrt{(-8)^2 + 24^2} = 25.30\,\text{kN}.$$

Now let us consider the same problem using the concept of the reference beam (Fig. 5.3c). The bending moment at point C equals $M_C = Pab/l = (20 \times 10 \times 15)/25 = 120\,\text{kNm}$. According to (5.4) the sag f at point C is $f_C = M_C^0/H = 120/24 = 5\,\text{m}$. Obtained result presents the *distance* between chord AB and cable and does not related to adopted x–y coordinate system.

5.2.1.2 Inverse Problem

Determine a thrust of the cable, if total length L of the cable is given.

Length of the cable shown in Fig. 5.3a equals

$$L = \frac{a}{\cos\alpha_0} + \frac{l-a}{\cos\alpha_1} = a\sqrt{1+\tan^2\alpha_0} + (l-a)\sqrt{1+\tan^2\alpha_1}.$$

Length of the cable in terms of active force P and thrust H may be presented as follows

$$L = a\sqrt{1 + \frac{P^2}{H^2}\left(\frac{l-a}{l}\right)^2} + (l-a)\sqrt{1 + \frac{P^2}{H^2}\frac{a^2}{l^2}}.$$

Solving this equation with respect to H leads to following expression for thrust in terms of P, L, and a

$$H = P\frac{2l_0\delta(1-\delta)}{\sqrt{l_0^4 - 2l_0^2(2\delta^2 - 2\delta + 1) + 4\delta^2 - 4\delta + 1}}, \quad l_0 = \frac{L}{l}, \quad \delta = \frac{a}{l}.$$

Let $\delta = 0.4$, $L = 1.2l = 30\,\text{m}$, so $l_0 = 1.2$. In this case the thrust equals

$$H = P\frac{2 \times 1.2 \times 0.4(1-0.4)}{\sqrt{1.2^4 - 2 \times 1.2^2(2 \times 0.4^2 - 2 \times 0.4 + 1) + 4 \times 0.4^2 - 4 \times 0.4 + 1}} = 0.7339P$$

After that, the shape of the cable and internal forces in the cable may be defined easily.

The special case $a = 0.5l$ leads to the following results for thrust and sag

$$H = \frac{P}{2\sqrt{l_0^2 - 1}},$$

$$f = \frac{Pl}{4H} = \frac{l}{2}\sqrt{l_0^2 - 1}.$$

There are some interesting numerical results. Assume that $l_0 = \frac{L}{l} = 1.01$. In this case $f = 0.071l$, so if the total length of the cable L exceeds the span l only by 1%, then sag of the cable comprises 7% of the span.

5.2.2 *Cable Subjected to Uniformly Distributed Load*

Distributed load within the horizontal projection of the cable may be considered for the case when the hangers (Fig. 5.1) are located very often. Design diagram of flexible cable under uniformly distributed load is presented in Fig. 5.4.

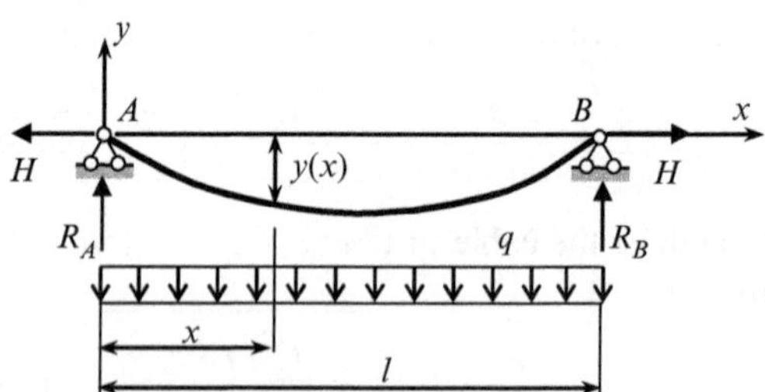

Fig. 5.4 A cable under uniformly distributed load

5.2.2.1 Direct Problem

Thrust H of the cable is given. Determine the shape of the cable and calculate distribution of internal forces. For solving of this problem we will use two approaches.

1. Integration of fundamental equation of a flexible cable (5.8) leads to following expressions for slope and shape of the cable

$$\frac{\mathrm{d}y}{\mathrm{d}x} = \frac{q}{H}x + C_1,$$

$$y = \frac{q}{H}\frac{x^2}{2} + C_1 x + C_2.$$

Constant of integration are obtained from boundary conditions: at $x = 0$ (support A) $y = 0$ and at $x = l$ (support B) $y = 0$. These conditions applied to equation $y(x)$ lead to following constants of integration: $C_2 = 0$ and $C_1 = -ql/2H$. Now the shape of a cable and slope in terms of load q and thrust H is described by equations

$$y(x) = -\frac{ql^2}{2H}\left(\frac{x}{l} - \frac{x^2}{l^2}\right) \tag{5.9a}$$

$$\tan\theta = \frac{\mathrm{d}y}{\mathrm{d}x} = \frac{ql}{2H}\left(2\frac{x}{l} - 1\right). \tag{5.9b}$$

Equation $y(x)$ presents symmetrical parabola. At $x = l/2$ a maximum y-coordinate equals to

$$y_{\max} = -\frac{ql^2}{8H}. \tag{5.10}$$

The tension N at any section x in terms of thrust H is

$$N(x) = \frac{H}{\cos\theta} = H\sqrt{1 + \tan^2\theta} = H\sqrt{1 + \frac{q^2l^2}{4H^2}\left(2\frac{x}{l} - 1\right)^2}. \tag{5.11}$$

This equation may be obtained from (5.6), where shear $Q = R_A - qx$.

At the lowest point ($x = l/2$) a tension $N = H$. The maximum tension occurs at supports

$$N_{\max} = H\sqrt{1 + \frac{q^2l^2}{4H^2}}. \tag{5.12}$$

2. The concept of the reference beam leads to the following procedure. The bending moment for reference beam at any section x is $M_x^0 = (ql/2)x - qx^2/2$. According to (5.4) and taking into account the direction of the y-axis (Fig. 5.4), we immediately get the expression (5.9a) for $y(x)$, and as result, the formulas (5.9b–5.12) for slope, $y_{\max}$, tension $N(x)$, and maximum tension $N_{\max}$.

5.2.2.2 Inverse Problem

Expression for total length L of the cable is

$$L = \int_0^l \sqrt{1 + \left(\frac{\mathrm{d}y}{\mathrm{d}x}\right)^2}\,\mathrm{d}x. \tag{5.13}$$

Since the sag of the cable is $f = ql^2/8H$, then expression (5.9b) for slope at any x in terms of sag f may be presented as

$$\frac{\mathrm{d}y}{\mathrm{d}x} = \frac{ql}{2H}\left(\frac{2x}{l} - 1\right) = \frac{4f}{l}\left(\frac{2x}{l} - 1\right). \tag{5.14}$$

Therefore, the total length L of the cable in terms of sag according to (5.13) becomes

$$L = \int_0^l \sqrt{1 + \left(\frac{4f}{l}\right)^2 \left(\frac{2x}{l} - 1\right)^2} \, \mathrm{d}x. \tag{5.13a}$$

Approximate Solution of Length Determination

The binomial theorem to expand radical in a series $\left(\sqrt{1+\varepsilon} \cong 1 + \frac{1}{2}\varepsilon - \frac{1}{8}\varepsilon^2 + \ldots\right)$ allows presenting (5.13a) as follows:

$$L = \int_0^l \left\{ 1 + \frac{1}{2}\left[\frac{4f}{l}\left(\frac{2x}{l} - 1\right)\right]^2 - \frac{1}{8}\left[\frac{4f}{l}\left(\frac{2x}{l} - 1\right)\right]^4 + \ldots \right\} \mathrm{d}x. \tag{5.15}$$

Integration of this relation leads to the following approximate expression:

$$L = l\left[1 + \frac{8}{3}\left(\frac{f}{l}\right)^2 - \frac{32}{5}\left(\frac{f}{l}\right)^4 + \frac{256}{7}\left(\frac{f}{l}\right)^6 - \ldots\right] \tag{5.16}$$

This expression allows calculating the sag f of the cable in terms of total length L and span l. Since $f = ql^2/8H$, then length of the cable in terms of thrust H may be presented as follows

$$L = l\left[1 + \frac{1}{24}\frac{q^2l^2}{H^2} - \frac{1}{640}\frac{q^4l^4}{H^4} + \ldots\right] \tag{5.16a}$$

This expression allows calculating thrust H in terms of span l, total length of the cable L, and load q.

Gentile Cable

If $f < 0.1l$, then a cable is called the gentile one. Taking into account two terms in equation (5.16a), we get the following equation

$$L = l\left(1 + \frac{1}{24}\frac{q^2l^2}{H^2}\right) \tag{5.16b}$$

Therefore, thrust in terms of total length of the cable L and span l may be presented in the form

$$H = \frac{ql}{2\sqrt{6}} \cdot \frac{1}{\sqrt{l_0 - 1}}, \quad l_0 = \frac{L}{l}. \tag{5.16c}$$

The maximum y-coordinate of the cable in terms of total length L and span l is

$$y_{\max} = -\frac{\sqrt{3}}{2\sqrt{2}} l \sqrt{\frac{L}{l} - 1}. \tag{5.16d}$$

Exact Solution of the Length Determination

Integrating equation (5.13a) leads to the following exact expression for total length in terms of q, l, and H

$$L = \frac{l}{2}\sqrt{1 + \left(\frac{ql}{2H}\right)^2} + \frac{H}{q}\sinh^{-1}\frac{ql}{2H}. \tag{5.16e}$$

Table 5.1 contains results of numerical solution of approximate equation (5.16a) for different total length L of the cable. Three and two terms of this equation have been hold. Results of numerical solution compared with solution of exact equation (5.16e). The problems are solved for $q = 2\,\mathrm{kN/m}$ and $l = 30\,\mathrm{m}$.

Table 5.1 Cable under uniformly distributed load. Thrust H, kN of the cable vs. the total length L

	Total length of a cable L (m)		
	32	34	36
Two terms	47.4342	33.5410	27.3861
Three terms	45.8884	31.1147	23.9464
Exact solution	46.0987	31.7556	25.3087

We can see that even two terms of (5.16a) leads to quite sufficient accuracy. Moreover, since $H_{3\text{terms}} < H_{\text{ex.sol}} < H_{2\text{terms}}$, then two-term approximation is more preferable for design.

Example 5.1. Design diagram of flexible cable with support points A and B on different levels, is presented in Fig. 5.5. The cable is subjected to uniformly distributed load q. Find shape of the cable and determine distribution of internal forces, if thrust H of the cable is given. Parameters of the system are: $l = 30\,\mathrm{m}$, $c = 3\,\mathrm{m}$, $H = 40\,\mathrm{kN}$, $q = 1.8\,\mathrm{kN/m}$. Use two approaches (a) integrating of differential equation (5.8) and (b) the concept of the reference beam.

Solution.

(a) Differential equation of a flexible cable is $\mathrm{d}^2y/\mathrm{d}x^2 = q/H$; its integration leads to following expressions for slope and shape

$$\frac{\mathrm{d}y}{\mathrm{d}x} = \frac{q}{H}x + C_1$$

$$y = \frac{q}{H}\frac{x^2}{2} + C_1 x + C_2$$

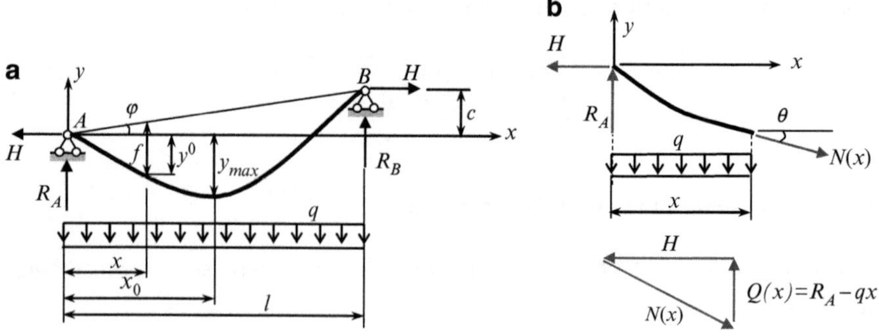

Fig. 5.5 (**a**) Cable under action of uniformly distributed load; (**b**) free-body diagram of the part of the cable

Constants of integration should be calculated from the following boundary conditions: at $x = 0$ (support A) $y = 0$ and at $x = l$ (support B) $y = c$. Constants of integration are $C_2 = 0$ and

$$C_1 = \left(c - \frac{ql^2}{2H}\right)\frac{1}{l}.$$

Now the shape of the cable and its slope for any x may be presented in the form

$$y(x) = \frac{ql^2}{2H}\left(\frac{x^2}{l^2} - \frac{x}{l}\right) + c\frac{x}{l} = \frac{1.8 \times 30^2}{2 \times 40}\left(\frac{x^2}{30^2} - \frac{x}{30}\right) + 3\frac{x}{30} = 0.675\left(\frac{x^2}{30} - x\right) + 0.1x,$$

$$\tan\theta = \frac{\mathrm{d}y}{\mathrm{d}x} = \frac{ql}{2H}\left(2\frac{x}{l} - 1\right) + \frac{c}{l} = \frac{1.8 \times 30}{2 \times 40}\left(2\frac{x}{30} - 1\right) + \frac{3}{30} = 0.675\left(\frac{x}{15} - 1\right) + 0.1.$$

Equation $y(x)$ presents nonsymmetrical parabola. In the lowest point the slope of the cable equals zero

$$\tan\theta = \frac{ql}{2H}\left(2\frac{x}{l} - 1\right) + \frac{c}{l} = 0,$$

This equation leads to corresponding x_0-coordinate

$$x_0 = \frac{H}{q}\left(\frac{ql}{2H} - \frac{c}{l}\right) = \frac{40}{1.8}\left(\frac{1.8 \times 30}{2 \times 40} - \frac{3}{30}\right) = 12.78\,\text{m}.$$

The maximum y occurs at x_0

$$y_{\max} = y(12.78) = 0.675\left(\frac{12.78^2}{30} - 12.78\right) + 0.1 \times 12.78 = -3.674\,\text{m}.$$

Reaction of support A equals

$$R_A \to \sum M_B = 0: \quad -R_A l - Hc + \frac{ql^2}{2} = 0 \to R_A = \frac{ql}{2} - \frac{Hc}{l}$$

Free-body diagram of the left part of the cable and corresponding force triangle are presented in Fig. 5.5b. Since shear $Q(x) = R_A - qx$, then tension at any section according to (5.6) equals

$$N(x) = \sqrt{H^2 + (R_A - qx)^2} = \sqrt{H^2 + \left(\frac{ql}{2} - \frac{Hc}{l} - qx\right)^2}.$$

The tension in the lowest point

$$\begin{aligned} N(x_0) &= \sqrt{H^2 + \left(\frac{ql}{2} - \frac{Hc}{l} - qx_0\right)^2} = \sqrt{H^2 + \left(\frac{1.8 \times 30}{2} - \frac{40 \times 3}{30} - 1.8 \times 12.78\right)^2} \\ &= \sqrt{H^2 + (27 - 27)^2} = H. \end{aligned}$$

Maximum tension occurs at supports

$$N(0) = \sqrt{40^2 + \left(\frac{1.8 \times 30}{2} - \frac{40 \times 3}{30} - 1.8 \times 0\right)^2} = 46.14\,\text{kN},$$

$$N(l) = \sqrt{40^2 + \left(\frac{1.8 \times 30}{2} - \frac{40 \times 3}{30} - 1.8 \times 30\right)^2} = 50.60\,\text{kN}$$

(b) The reaction of the reference beam is $R_A^0 = ql/2$. The bending moment of the reference beam and parameter $f(x)$ measured from the inclined chord AB are

$$\begin{aligned} M^0(x) &= \frac{ql}{2}x - \frac{qx^2}{2}, \\ f(x) &= \frac{M^0(x)}{H} = \frac{1}{H}\left(\frac{ql}{2}x - \frac{qx^2}{2}\right). \end{aligned}$$

Distance y^0 between the horizontal line x and cable becomes

$$y^0(x) = f(x) - x \tan\phi = \frac{1}{H}\left(\frac{ql}{2}x - \frac{qx^2}{2}\right) - \frac{c}{l}x.$$

Condition $\mathrm{d}y^0/\mathrm{d}x = 0$ leads to the parameter $x_0 = 12.78$ m obtained above. The distance between the cable in the lowest point and horizontal line becomes

$$y^0(x_0) = \frac{1}{40}\left(\frac{1.8 \times 30}{2}12.78 - \frac{1.8 \times 12.78^2}{2}\right) - \frac{3}{30} \times 12.78 = 3.674\,\text{m}.$$

5.3 Effect of Arbitrary Load on the Thrust and Sag

So far we considered behavior of the cables that are subjected to concentrated or distributed load only. In case of different loads acting simultaneously formula (5.4) is most appropriate. However in engineering practice another type of loading is possible, mainly: the cable is subjected to any dead load and after that additional live load is applied. How will this change the shape and state of the cable? This problem may be effectively solved by knowing the expression related to the total length of the cable, the thrust, and external loads.

Consider a cable supported at points A and B and subjected to any loads (Fig. 5.6). Assume that a cable is inextensible, and support points of the cable do have the mutual displacements. The covered span is l, while a total unstressed length of a cable is L.

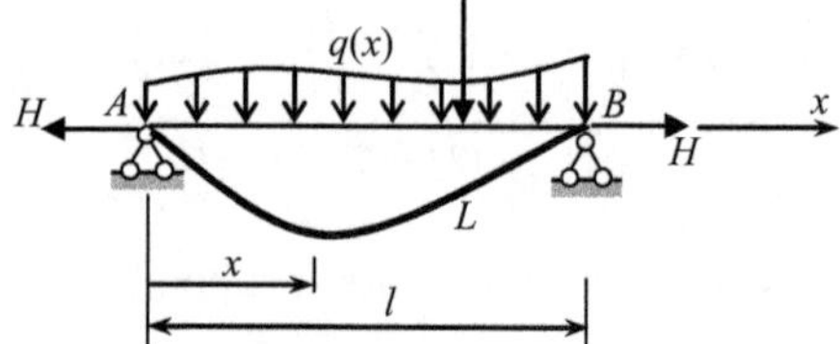

Fig. 5.6 Cable carrying arbitrary dead and live load

The total length of the cable is defined by exact equation (5.13). For gentile cable a slope is small, so

$$\sqrt{1+\left(\frac{\mathrm{d}y}{\mathrm{d}x}\right)^2} \cong 1+\frac{1}{2}\left(\frac{\mathrm{d}y}{\mathrm{d}x}\right)^2 \qquad (5.17)$$

and expression for L becomes

$$L=\int_0^l \sqrt{1+\left(\frac{\mathrm{d}y}{\mathrm{d}x}\right)^2}\,\mathrm{d}x \cong \int_0^l \left[1+\frac{1}{2}\left(\frac{\mathrm{d}y}{\mathrm{d}x}\right)^2\right]\mathrm{d}x. \qquad (5.18)$$

According to (5.4), the shape of the cable is defined by expression $y = M/H$, where M is a bending moment of the reference beam, and H is a thrust. Therefore the slope can be calculated as follows

$$\frac{\mathrm{d}y}{\mathrm{d}x}=\frac{\mathrm{d}}{\mathrm{d}x}\left(\frac{M}{H}\right)=\frac{1}{H}\frac{\mathrm{d}M}{\mathrm{d}x}=\frac{Q(x)}{H},$$

where $Q(x)$ is shear at any section of the reference beam. After that expression (5.18) becomes

$$L=\int_0^l \left[1+\frac{1}{2}\left(\frac{Q(x)}{H}\right)^2\right]\mathrm{d}x = l+\frac{1}{2H^2}\int_0^l Q^2(x)\mathrm{d}x.$$

Usually this formula is written in the following form

$$L = l + \frac{D}{2H^2}, \quad D = \int_0^l Q^2(x)\mathrm{d}x \tag{5.19}$$

Table A.21, contains the load characteristic D for most important cases of loading.

Equation (5.19) allows present the thrust H in terms of load characteristic D as follows

$$H = \sqrt{\frac{D}{2l(l_0 - 1)}}, \quad l_0 = \frac{L}{l} \tag{5.20}$$

It is possible to show that if support points A and B of the cable are located on the different elevation then

$$L = \frac{l}{\cos\varphi} + \frac{D}{2H^2}\cos^3\varphi,$$

where φ is angle between chord AB and x-axis.

Formulas (5.18)–(5.20) entirely solve the problem of determination of the thrust of inextensible gentile cable with supports without their mutual displacements, subjected to *arbitrary* vertical load. These formulas are approximate since for their deriving have been used in approximate relationship (5.17).

Application of expression (5.20) for some classical loading cases is shown below.

1. A cable with total length L and span l carries a concentrated force P throughout distance a starting from the left end of the cable (Fig. 5.3). Let's dimensionless parameters be $l_0 = L/l = 1.2$ and $\xi = a/l = 0.4$. Load characteristic D (Table A.21) is $D = P^2 l\xi(1-\xi) = P^2 l \cdot 0.4(1-0.4) = 0.24P^2 l$.

The thrust $H = \sqrt{\dfrac{D}{2l(l_0 - 1)}} = \sqrt{\dfrac{0.24P^2 l}{2l(1.2 - 1)}} = 0.7746P$. The exact solution is $H = 0.7339P$ (Sect. 5.2, Inverse problem). Error is 5.54%.

2. The cable of total length L and span l carries a uniformly distributed load q (Fig. 5.4). Load characteristic is $D = q^2 l^3/12$ (Table A.21). According to formula (5.16)

$$l_0 = \frac{L}{l} \cong 1 + \frac{8}{3}\frac{f^2}{l^2}.$$

Therefore the thrust becomes

$$H = \sqrt{\frac{D}{2l(l_0 - 1)}} = \sqrt{\frac{q^2 l^3}{12 \times 2l\left(1 + \frac{8}{3}\frac{f^2}{l^2} - 1\right)}} = \frac{ql^2}{8f}.$$

This exact result has been obtained earlier by using formula (5.10).

The formulas above allows considering very important problem determining the *change* of the parameters of the cable if additional live loads are applied.

Example 5.2. The flexible, inextensible, and gentile cable is suspended between two absolutely rigid supports of same elevation. The total length of the cable is L and span equals l. The cable is subjected to uniformly distributed dead load q within the horizontal line; corresponding thrust and sag are H_q and f_q. Calculate a *change* of the thrust and sag of the cable, if the entire cable upon the load q is loaded by additional concentrated live load P at the axis of symmetry of the cable (Fig. 5.7).

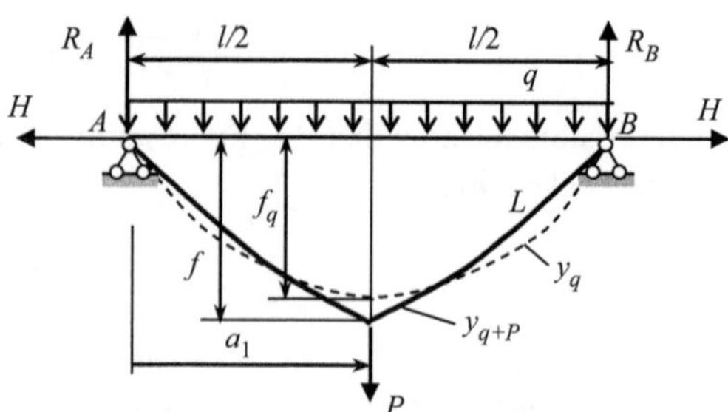

Fig. 5.7 Cable carrying uniformly distributed load q with additional force P

Solution. If a cable is loaded by uniformly distributed load q only then load characteristic is $D_q = q^2l^3/12$.

Corresponding trust and sag are H_q and f_q. In this case (5.19) becomes

$$L = l + \frac{1}{2H_q^2}\frac{q^2l^3}{12} \tag{a}$$

If additional load P is applied at the middle of the cable then load characteristic accordingly (Table A.21, line 5) becomes (the subscript 1 is omitted)

$$D_{q+P} = \frac{q^2l^3}{12}(1 + 3\gamma + 3\gamma^2), \quad \gamma = \frac{P}{ql}.$$

Thus (5.19) becomes

$$L = l + \frac{1}{2H_{q+P}^2}\frac{q^2l^3}{12}(1 + 3\gamma + 3\gamma^2), \tag{b}$$

where H_{q+P} is thrust due to the both loads q and P.

Since the cable is inextensible then left parts of the (a) and (b) are same, therefore

$$l + \frac{1}{2H_q^2}\frac{q^2l^3}{12} = l + \frac{1}{2H_{q+P}^2}\frac{q^2l^3}{12}(1 + 3\gamma + 3\gamma^2),$$

which leads to the required relationship

$$H_{q+P} = H_q\sqrt{1+3\gamma+3\gamma^2}.$$

The sag of the cable

$$f_{q+P} = \frac{M_C^{\text{beam}}}{H_{q+P}} = \frac{\frac{ql^2}{8}+\frac{Pl}{4}}{H_q\sqrt{1+3\gamma+3\gamma^2}} = f_q\frac{1+2\gamma}{\sqrt{1+3\gamma+3\gamma^2}}, \quad f_q = \frac{ql^2}{8H_q}.$$

Let $\gamma = p/ql = 0.2$. In this case the thrust and sag of the cable are

$$H_{q+P} = H_q\sqrt{1+3\times0.2+3\times0.2^2} = 1.31H_q,$$

$$f_{q+P} = f_q\frac{1+2\times0.2}{\sqrt{1+3\times0.2+3\times0.2^2}} = 1.068f_q,$$

i.e., application of additional concentrated force $P = 0.2ql$ leads to increasing of the thrust and sag of the nonextensible cable on 31% and 6.8%, respectively. Timoshenko was the first to solve this problem by other approach (1943).

5.4 Cable with Self-Weight

This section is devoted to analysis of cable carrying a load uniformly distributed along the cable itself. Relationships between parameters of cable, its length, shape, and internal forces are developed. The direct and inverse problems are considered.

5.4.1 Fundamental Relationships

A cable is supported at points A and B and loaded by uniformly distributed load q_0 along the cable *itself.* This load presents not only dead load (self-weight of the cable), but also a live load, such as a glazed ice. Cables of this type are called the *catenary* (Latin word *catena* means a "chain"). It was named by Huygens in 1691. The y-axis passes through the lowest point C. Design diagram is presented in Fig. 5.8. The following notations are used: s is curvilinear coordinate along the cable; $W = q_0 s$ is a total weight of portion s; N is a tension of the cable and H presents a thrust of the cable.

For any point of the cable, relationship between N, H, and W obeys to law of force triangle:

$$N = \sqrt{H^2+W^2}. \tag{5.21}$$

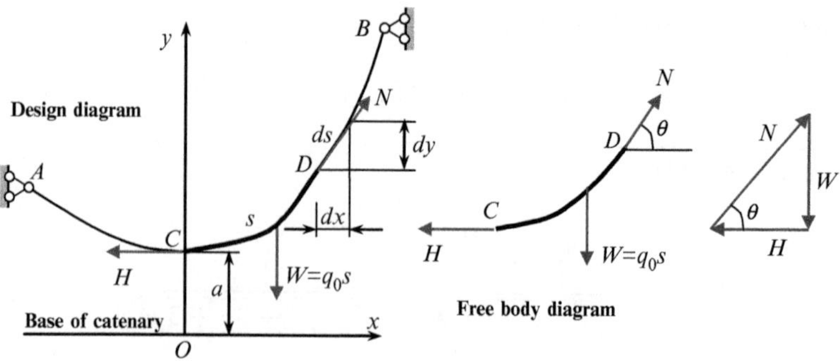

Fig. 5.8 Design diagram of the catenary. Free body diagram for portion *CD* and force triangle

The final results may be conveniently expressed in terms of additional parameter $a = H/q_0$. This parameter is measured from the lowest point C of the cable and defines an origin O and the line O-x called a base of catenary (or chain line). In terms of a and s the normal force, accordingly (5.21) becomes

$$N = q_0\sqrt{a^2 + s^2}. \tag{5.22}$$

The relationship between coordinates x and s in differential form is

$$\mathrm{d}x = \mathrm{d}s\cos\theta = \mathrm{d}s\frac{H}{N} = \frac{aq_0}{q_0\sqrt{a^2 + s^2}}\mathrm{d}s = \frac{a\mathrm{d}s}{\sqrt{a^2 + s^2}}.$$

Integrating this equation from point $C(0, a)$ to $D(x, y)$ allows us to calculate the x-coordinate of any point of the cable in terms of s and $a = H/q_0$

$$x = \int_0^s \frac{a\mathrm{d}s}{\sqrt{a^2 + s^2}} = a\sinh^{-1}\frac{s}{a}. \tag{5.23}$$

The inverse relation, i.e., the solution of (5.23) with respect to s is

$$s = a\sinh\frac{x}{a}. \tag{5.24}$$

This formula allows us to obtain some useful relationships in terms of intensity load q_0 and thrust H.

The length of the cable from lowest point C to $D(x, y)$ is

$$L_{C-D} = \frac{H}{q_0}\sinh\frac{q_0}{H}x. \tag{5.25}$$

Formula (5.22) for tension at any point can be rewritten as follows

$$N(x) = H\sqrt{1 + \sinh^2 \frac{q_0}{H}x} = H \cosh \frac{q_0}{H}x. \tag{5.26}$$

The minimum tension $N = H$ occurs at $x = 0$, to say in the lowest point.

In order to obtain equation of the catenary, we need to express y in terms of x

$$\mathrm{d}y = \mathrm{d}x \tan\theta = \frac{W}{H}\mathrm{d}x = \frac{q_0 s}{a q_0}\mathrm{d}x = \sinh\frac{x}{a}\mathrm{d}x.$$

After integrating this equation from point C to point D, we get

$$\int\limits_a^y \mathrm{d}y = \int\limits_0^x \sinh\frac{x}{a}\mathrm{d}x \to y|_a^y = a\cosh\frac{x}{a}\bigg|_0^x.$$

The equation of the curve, assumed by the cable, and corresponding slope at any point are

$$\begin{aligned} y(x) &= a\cosh\frac{x}{a} = \frac{H}{q_0}\cosh\frac{q_0}{H}x \\ \tan\theta &= \frac{\mathrm{d}y}{\mathrm{d}x} = \sinh\frac{q_0}{H}x. \end{aligned} \tag{5.27}$$

Below is presented analysis of typical cases of cables carrying a uniformly distributed load along the cable itself:

1. Supports located at the same level
2. Supports located at different elevation and saddle point within the span
3. Supports located on the different elevation and cable has not a saddle point within the span

5.4.2 Cable with Supports Located at the Same Level

In this case a cable has the axis of symmetry. It is pertinent here to derive the simple and useful expression for maximum tension for the cable in terms of q_0, f, and H. Maximum tension $N_{\max}$ occurs at the supports ($x = \pm l/2$), then

$$N_{\max} = H\sqrt{1 + \sinh^2\frac{q_0 l}{2H}} \to \sinh^2\frac{q_0 l}{2H} = \frac{N_{\max}^2}{H^2} - 1. \tag{5.28}$$

The sag of the cable f at the axis of symmetry ($x = 0$) in terms of ordinate $y(l/2)$ of the cable at support is

$$f = y\left(\frac{l}{2}\right) - a = \frac{H}{q_0}\cosh\frac{q_0 l}{2H} - \frac{H}{q_0}$$
$$= \frac{H}{q_0}\left(\cosh\frac{q_0 l}{2H} - 1\right) \rightarrow \cosh^2\frac{q_0 l}{2H} = \left(\frac{q_0 f}{H} + 1\right)^2. \quad (5.29)$$

Since $\sinh^2 z = \cosh^2 z - 1$, then expressions (5.28) and (5.29) lead to the following formula

$$N_{\max} = q_0 f + H. \quad (5.30)$$

Example 5.3. A uniform cable weighting $q_0 = 1.25\,\text{kN/m}$ is suspended between two points A and B of equal elevation, which are $l = 20\,\text{m}$ apart as shown in Fig. 5.9.

1. *Thrust–shape problem.* Determine shape of the curve assumed by the cable, distribution of tension, and length of the cable, if a thrust is $H = 5.6\,\text{kN}$
2. *Length–thrust problem.* Determine thrust of the cable and shape of the cable, if a total length $L = 42\,\text{m}$.

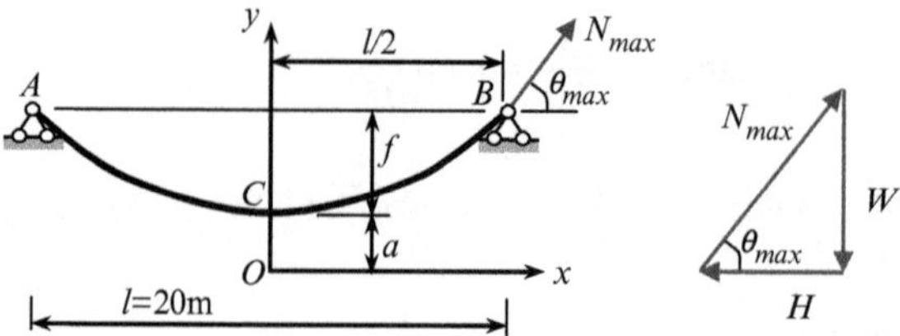

Fig. 5.9 Cable carrying a uniformly distributed load along the cable itself. Design diagram and force triangle for *CB* portion

Solution.

1. *Thrust–shape problem.* Parameter of catenary $a = H/q_0 = 5.6/1.25 = 4.48\,\text{m}$. Equation of the curve assumed by the cable and slope are

$$y(x) = \frac{H}{q_0}\cosh\frac{q_0}{H}x = 4.48\cosh(0.2232x)$$
$$\tan\theta = \sinh\frac{q_0}{H}x = \sinh(0.2232x). \quad \text{(a)}$$

Coordinates of point B are $x = 10\,\text{m}$; $y(10) = 4.48\cosh(0.2232 \times 10) = 21.11\,\text{m}$.

Therefore, sag of the cable at point C becomes $f = y(10) - a = 21.11 - 4.48 = 16.63\,\text{m}$.

Slope at point B is

$$\tan\theta_{\max} = \sinh(0.2232\times 10) = 4.6056 \to$$
$$\sin\theta_{\max} = \sin(\tan^{-1} 4.6056) = 0.9772, \cos\theta_{\max} = \cos(\tan^{-1} 4.6056) = 0.2122.$$

Tension at any point is

$$N = H\sqrt{1+\sinh^2\frac{q_0}{H}x},$$

and maximum tension occurs at point B and equals

$$N_{\max} = 5.6\sqrt{1+\sinh^2(0.2232\times 10)} = 26.39\,\text{kN}.$$

Since shape of the cable is symmetrical, then total length of the cable is

$$L = 2a\sinh\frac{x}{a}\bigg|_{x=\frac{l}{2}} = 2\times 4.48\sinh\frac{10}{4.48} = 41.27\,\text{m}.$$

Control If total length $L = 41.27$m, then total weight of the cable equals $41.27\times 1.25 = 51.58$ kN, and force W, considering portion CB, equals $51.587/2 = 25.79$ kN. From a force triangle we can calculate

$$H = N_{\max}\cos\theta_{\max} = 26.39\times 0.2122 = 5.6\,\text{kN},$$
$$W = N_{\max}\sin\theta_{\max} = 26.39\times 0.9772 = 25.79\,\text{kN},$$
$$\tan\theta_{\max} = \frac{W}{H} = \frac{25.79}{5.6} = 4.605.$$

The maximum tension according to (5.30) equals

$$N_{\max} = q_0 f + H = 1.25(21.11 - 4.48) + 5.6 = 26.39\,\text{kN}.$$

2. *Length–thrust problem.* We assume that a total length L of a cable is given and it is required to find the thrust, shape of the cable, and sag. Since the cable is symmetrical with respect to y-axis, then total length of the cable equals $L = 2a\,\sinh(x_B/a)$, where $L = 42$ m and $x_B = l/2 = 10$ m. Unknown parameter a may be calculated from transcendental equation (5.25), i.e.,

$$42 = 2a\,\sinh\frac{10}{a}.$$

Solution of this equation leads to $a = 4.42$ m. Thrust of the cable is $H = aq_0 = 4.42\times 1.25 = 5.525$ kN.

Equation of the curve and ordinate y for support points are

$$y(x) = \frac{H}{q_0}\cosh\frac{q_0}{H}x = 4.42\,\cosh 0.22642x,$$

$$y_{\max} = 4.42\,\cosh\frac{1.25}{5.525}10 = 21.46\,\text{m}.$$

Sag of the cable at lowest point C is $f = 21.46 - 4.42 = 17.04\,\text{m}$.

5.4.3 Cable with Supports Located on the Different Elevations

In this case the shape of the cable is nonsymmetrical and position of the lowest point C is not known ahead. Availability of only two parameters, such as load per unit length of the cable q_0 and span l, are not enough for cable analysis. For determination of location of point C we need to formulate additional condition for curve $y(x)$ passing through points A and B. One of the additional parameter may be as follows (1) Value of thrust H; (2) Total length L; (3) Coordinates of any additional point of the cable. Two different cases are considered below. They are a saddle point within or outside of the span.

5.4.3.1 Saddle Point Within the Span

The following example illustrates the thrust–length procedure of analysis of the cable and corresponding peculiarities.

Example 5.4. A uniform cable with self-weight $q_0 = 1.25\,\text{kN/m}$ is suspended between two points $l = 20\,\text{m}$ apart horizontally, with one point of support 18 m higher than the other as shown in Fig. 5.10. Thrust of the cable is $H = 5.6\,\text{kN}$. Determine shape of the cable, tension at supports, and total length of the cable. Assume that a saddle point is placed between supports.

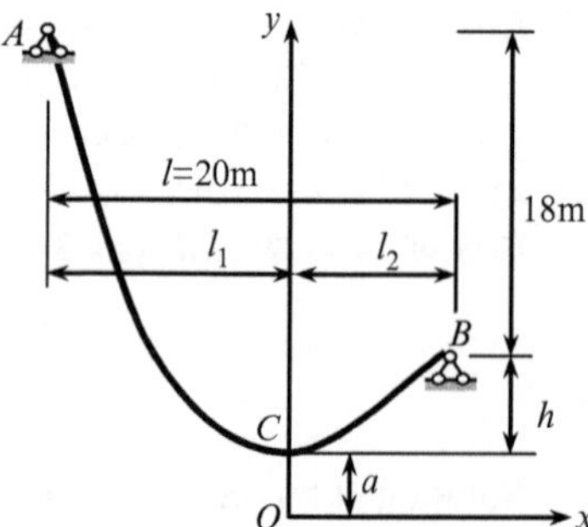

Fig. 5.10 Cable, carrying a uniformly distributed load along the cable itself

Solution. Let a vertical y-axis pass through the lowest point C. Location of this point is defined by parameters l_1 and l_2 which are unknown before head, while $l_1 + l_2 = 20\,\text{m}$. For this direct problem, the parameter of catenary (Fig. 5.10) may be calculated immediately, i.e., $a = H/q_0 = 5.6/1.25 = 4.48\,\text{m}$.

Shape of the cable is defined by equation

$$y(x) = \frac{H}{q_0}\cosh\frac{q_0}{H}x = 4.48\,\cosh(0.2232x). \tag{a}$$

Conditions of the catenary passing through two points A and B are as follows:

At $x = l_2$ (point B) the ordinate $y = a + h = 4.48 + h$.
At $x = -l_1 = -(20 - l_2)$ (point A) the ordinate $y = a + h + 18$.
Therefore, (a) for points B and A becomes
Point B: $4.48\cosh(0.2232l_2) = 4.48 + h$ (b)
Point A: $4.48\cosh\left[-0.2232(20 - l_2)\right] = 4.48 + h + 18$.

These equations contain two unknowns h and l_2. Expressing from both equations' unknown parameter h and after that equating the two expressions, we obtain

$$4.48\cosh(0.2232l_2) = 4.48\cosh(0.2232l_2 - 4.464) - 18.$$

The root of this equation is $l_2 = 8.103\,\text{m}$, which leads to $l_1 = l - l_2 = 11.897\,\text{m}$. Therefore, vertical distance h between points C and B according to formula (b) equals

$$h = 4.48\,\cosh(0.2232 \times 8.103) - 4.48 = 9.555\,\text{m}.$$

Now we can consider the right and left parts of the cable separately.

Curve CB. Equation of the curve and slope are

$$y(x) = 4.48\,\cosh(0.2232x),$$
$$\tan\theta = \frac{dy}{dx} = \sinh(0.2232x).$$

Ordinate y and slope at the point B are

$$y(x = 8.103) = 4.48\cosh(0.2232 \times 8.103) = 14.034\,\text{m},$$
$$\tan\theta(x = 8.103) = \sinh(0.2232 \times 8.103) = 2.969,$$
$$\theta_B = 71.38^\circ; \quad \sin\theta_B = 0.9476.$$

The tension at the point B is

$$N_B = H\sqrt{1 + \sinh^2\frac{q_0}{H}x_B} = 5.6\sqrt{1 + \sinh^2(0.2232 \times 8.103)} = 17.544\,\text{kN}.$$

Length of the portion *CB* is $L_{CB} = a\ \sinh(x_B/\mathrm{a}) = 4.48\ \sinh(8.103/4.48) = 13.303\,\mathrm{m}$

Curve AC. Equation of the curve and slope are

$$y(x) = 4.48\cosh(0.2232x),$$
$$\tan\theta = \frac{\mathrm{d}y}{\mathrm{d}x} = \sinh(0.2232x).$$

Ordinate y and slope at the point A are

$$y(-11.897) = 4.48\cosh(0.2232\times(-11.897)) = 32.034\,\mathrm{m},$$
$$\tan\theta(-11.897) = \sinh(0.2232\times(-11.897)) = -7.0796,$$
$$\theta_A = 81.96^\circ;\quad \sin\theta_A = 0.990.$$

Calculated coordinates y for support points A and B satisfy to given design diagram

$$y_A - y_B = 32.034 - 14.034 = 18\,\mathrm{m}$$

Tension at the point A is

$$N_A = H\sqrt{1+\sinh^2\frac{q_0}{H}x_A} = 5.6\sqrt{1+\sinh^2(0.2232\times 11.897)} = 40.04\,\mathrm{kN}$$

Length of the portion *CA* and total length of the cable are

$$L_{CA} = a\ \sinh\frac{x_A}{a} = 4.48\sinh\frac{11.897}{4.48} = 31.725\,\mathrm{m}$$
$$L = L_{AC} + L_{CB} = 31.725 + 13.303 = 45.028\,\mathrm{m}$$

Control. The tension at the support points and total weight of the cable must satisfy to equation $\sum Y = 0$. In our case we have

$$N_A\ \sin\theta_A + N_B\ \sin\theta_B - Lq_0$$
$$= 40.04\times 0.990 + 17.544\times 0.9476 - 45.028\times 1.25 = 56.264 - 56.285.$$

The relative error equals 0.035%.

5.4.3.2 Saddle Point Outside of the Span

Now let us consider the cable with total length L_0, which is suspended between two points A and B, as shown in Fig. 5.11. Peculiarity of this design diagram, unlike the previous case, is that the curve *AB* of the cable has no point, for which $\tan\theta = 0$, since the lowest point C is located beyond the curve AB. Let the distance between support points in horizontal and vertical directions be D_0 and h_0, respectively; q_0 is weight per unit length. It is required to determine the shape of the cable, thrust H, and tension N_A and N_B at supports.

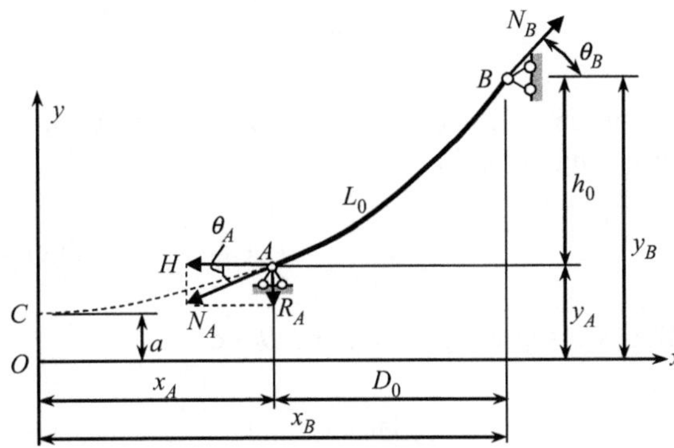

Fig. 5.11 Design diagram of the catenary; saddle point C beyond the span

For this length–thrust problem parameter $a = H/q_0$ cannot be calculated right now because thrust H is unknown yet. However, this parameter may be calculated analytically having the total length of the cable L_0 and dimensions D_0 and h_0. For this some steps should be performed previously.

1. The length of the curve from point C to any point (x, y) according to (5.25) is $L_{C-x} = a \sinh(x/a)$, so the length of curves CB and CA are $L_{CB} = a \sinh(x_B/a)$, $L_{CA} = a \sinh(x_A/a)$. Therefore, the total length AB of a cable is

$$L_0 = L_{CB} - L_{CA} = a \sinh\frac{x_B}{a} - a \sinh\frac{x_A}{a}. \tag{5.31}$$

2. Equation of the curve assumed by cable according to (5.27) is $y(x) = a \cosh\frac{x}{a}$. Therefore, ordinates of points B and A are $y_B = a \cosh\frac{x_B}{a}$, $y_A = a \cosh\frac{x_A}{a}$ and vertical distance between two supports is

$$h_0 = y_B - y_A = a \cosh\frac{x_B}{a} - a \cosh\frac{x_A}{a}. \tag{5.32}$$

Equations (5.31) and (5.32) present relationships between parameters $a = H/q_0$, L_0, h_0, and $D_0 = x_B - x_A$ and contains two unknowns parameters. They are a and x_B (or x_A).

For given geometry parameters L_0, D_0, and h_0 the analytical solution of (5.31) and (5.32) leads to following results.

Parameter a is determined from transcendental equation

$$\cosh\frac{D_0}{a} = 1 + \frac{L_0^2 - h_0^2}{2a^2} \tag{5.33}$$

or in equivalent form

$$\sinh \frac{D_0}{2a} = \frac{1}{2a}\sqrt{L_0^2 - h_0^2} \tag{5.33a}$$

Coordinates x_A and x_A of the points A and B are

$$x_A = a \tanh^{-1}\left(\frac{h_0}{L_0}\right) - \frac{D_0}{2}, \quad x_B = x_A + D_0. \tag{5.34}$$

The thrust of the cable equals $H = q_0 a$. Also, equation of the curve and slope according to (5.27) are

$$y(x) = a \cosh \frac{x}{a}; \quad \tan \theta(x) = \sinh \frac{x}{a}$$

Slopes at support points A and B are

$$\tan \theta_A = \sinh \frac{x_A}{a}; \quad \tan \theta_B = \sinh \frac{x_B}{a} \tag{5.35}$$

Tension at points A and B according (5.1), (5.2), and (5.34) are

$$N_A = H\sqrt{1 + \sinh^2 \frac{x_A}{a}}; \quad N_B = H\sqrt{1 + \sinh^2 \frac{x_B}{a}}. \tag{5.36}$$

Example 5.5. Let us consider the cable, which is presented in Fig. 5.11. The cable has following parameters: $L_0 = 117.7$ m; $D_0 = 100$ m; $h_0 = 57.7$ m; $q_0 = 0.014$ kN/m. Determine the shape of the cable, thrust, and tension.

Solution. Parameter a of the catenary can be calculated from (5.33a)

$$\sinh \frac{100}{2a} = \frac{1}{2a}\sqrt{117.7^2 - 57.7^2} \rightarrow a = 127.41 \text{ m}.$$

Coordinate x of the support A is $x_A = 127.41 \tanh^{-1}(57.7/117.7) - (100/2) = 18.34$ m.

Slopes at supports A and B

$$\tan \theta_A = \sinh \frac{18.34}{127.34} = 0.1446; \quad \sin \theta_A = 0.1431; \quad \cos \theta_A = 0.9897$$

$$\tan \theta_B = \sinh \frac{18.34 + 100}{127.34} = 1.0694; \quad \sin \theta_B = 0.7304; \quad \cos \theta_B = 0.6830.$$

Thrust of the cable is $H = q_0 a = 0.014 \times 127.41 = 1.7837$ kN.

Expressions (5.36) allows calculating the tension at supports A and B

$$N_A = 1.7834\sqrt{1+\sinh^2\frac{18.34}{127.41}} = 1.802\,\text{kN};$$
$$N_B = 1.7834\sqrt{1+\sinh^2\frac{18.34+100}{127.41}} = 2.609\,\text{kN}.$$

Control. Ordinates of support points are

$$y_A = a\cosh\frac{x_A}{a} = 127.41\cosh\frac{18.34}{127.41} = 128.73\,\text{m};$$
$$y_B = 127.41\cosh\frac{18.34+100}{127.41} = 186.43\,\text{m}.$$

Vertical distance between points A and B equals $h_0 = y_B - y_A = 186.43\,\text{m} - 128.73\,\text{m} = 57.7\,\text{m}$.

Equilibrium equation $\sum Y = 0$ for cable in whole leads to the following result

$$\begin{aligned}-N_A\sin\theta_A + N_B\sin\theta_B - L_0 q &= -1.802\times 0.1431 + 2.609\\ &\quad\times 0.7304 - 117.7\times 0.014\\ &= -1.9057 + 1.9056 \cong 0.\end{aligned}$$

Pay attention that vertical reaction of the support A is directed *downward* as shown in Fig. 5.11.

Trust can be calculated by formulas

$$H = N_A\cos\theta_A = 1.802\times 0.9897 = 1.783,$$
$$H = N_B\cos\theta_B = 2.609\times 0.6830 = 1.782.$$

5.5 Comparison of Parabolic and Catenary Cables

Let us compare the main results for parabolic and catenary cables. Assume, that both cables are inextensible, supported on same elevation, and support points does not allow to mutual displacements of the ends of the cable. Both cables have the same span l, subjected to uniformly distributed load q. If the load q is distributed within the horizontal foot then a curve of the cable is parabola (parabolic shape), if the load is distributed within the cable itself, then curve of the cable is catenary. Table 5.2 contains some fundamental parameters for both cables. They are dimensionless sag–span ratio f/l, slope $\tan\theta$ at the supports and dimensionless maximum cable tension–thrust ratio $N_{\max}/H$. Correspondence formulae are presented in brackets. The formulas for parabolic cable may be derived if hyperbolic functions to present as series $\sinh z = z + (z^3/3!) + \ldots$; $\cosh z = 1 + (z^2/2!) + \ldots$.

Table 5.2 Comparison of parabolic and catenary cables

	Parabolic cable	Catenary
f/l	$\frac{ql}{8H}$ (5.10)	$\frac{H}{ql}\left(\cosh\frac{ql}{2H}-1\right)$ (5.29)
$\tan\theta_{\max}$	$\frac{ql}{2H}$ (5.9b)	$\sinh\frac{ql}{2H}$ (5.27)
$\frac{N_{\max}}{H}$	$\sqrt{1+\left(\frac{ql}{2H}\right)^2}$ (5.12)	$\sqrt{1+\sinh^2\frac{ql}{2H}}$ (5.28)

Some numerical results are presented in Fig. 5.12 and 5.13. Comparison is made for two types of cables having the same ql/H ratio.

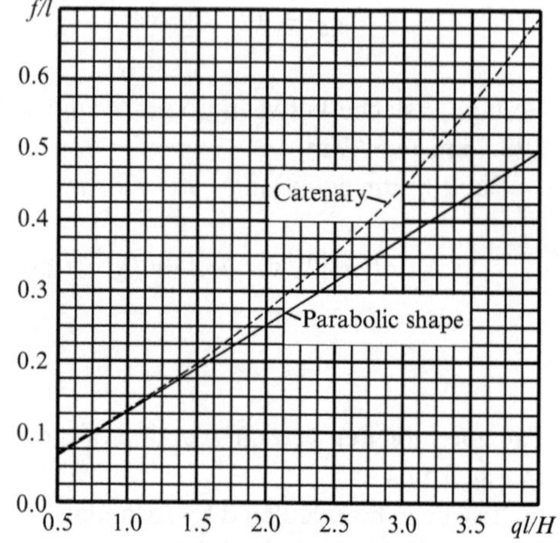

Fig. 5.12 Dimensionless sag–span ratio vs. total load–thrust ratio for parabolic and catenary cables

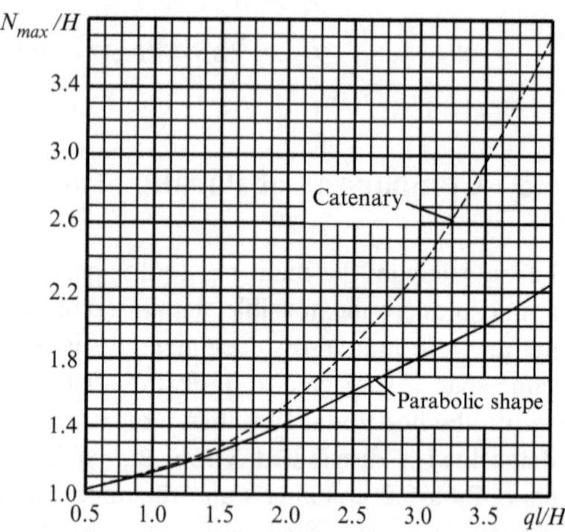

Fig. 5.13 Dimensionless maximum cable tension–thrust ratio vs. total load–thrust ratio for parabolic and catenary cables

For thrust–shape problem the ratio ql/H is known. For relatively small $ql/H(<1.5)$, dimensionless sag f/l and maximum tension $N_{\max}/H$ for parabolic and catenary cables practically coincide. If load q is fixed, then increasing of ql/H

caused by decreasing of the thrust leads to increasing of the sag. If thrust H is fixed, then increasing of ql/H caused by increasing of the load q leads to increasing of the slope of the cable at support point, and as a result, increase in maximum internal force $N_{\max}$ in cable.

5.6 Effect of Axial Stiffness

So far we have considered behavior of a cable without accounting for elastic properties of a cable itself. Elastic properties of a cable have considerable importance in the distribution of tension of a cable and value of sag. Procedure of analysis of elastic cables is presented below. Two cases are considered: cable with concentrated and uniformly distributed loads.

5.6.1 *Elastic Cable with Concentrated Load*

Design diagram of elastic cable is presented in Fig. 5.14. The force P is applied at $x = l/2$. In the deformable state of the cable the sag and inclination are f and α, respectively. Nondeformable length of the cable is L_0. It is necessary to calculate the thrust and sag of the cable taking into account the stiffness of the cable *EA*.

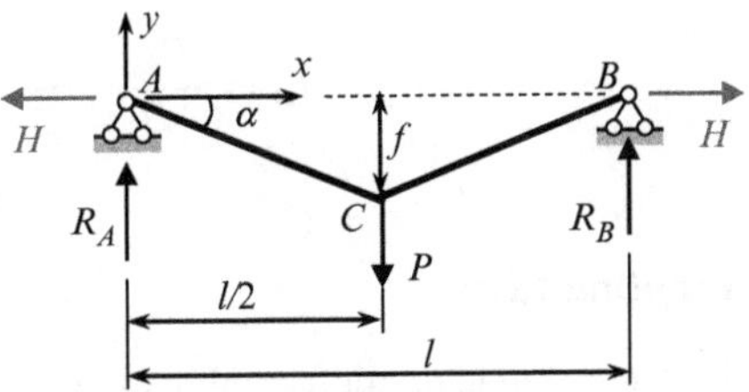

Fig. 5.14 Elastic cable subjected to concentrated load

It is evident that the angle of inclination, y-coordinate, and sag of the cable at point C are

$$\tan\alpha = -\frac{R_A}{H} = -\frac{P}{2H}$$

$$y = \frac{l}{2}\tan\alpha = -\frac{Pl}{4H}, \quad f = \frac{Pl}{4H}.$$

Tension in the left and right portions of a cable may be presented in the form

$$N = \frac{H}{\cos\alpha} = H\sqrt{1+\tan^2\alpha} = H\sqrt{1+\frac{P^2}{4H^2}}.$$

The total length of the cable

$$L = 2\frac{f}{\sin\alpha} = l\sqrt{1+\tan^2\alpha} = l\sqrt{1+\frac{P^2}{4H^2}}. \tag{5.37}$$

Now let us introduce the elastic properties of the cable. Let the stiffness of the cable be *EA*, where *E* is modulus of elasticity; *A* is cross-sectional area of the cable. The strain of the cable

$$\varepsilon = \frac{N}{EA} = \frac{H}{EA}\sqrt{1+\frac{P^2}{4H^2}}. \tag{5.38}$$

If the initial length of the cable is L_0, then the length of the cable upon the load is

$$L = L_0(1+\varepsilon). \tag{5.39}$$

Equations (5.37)–(5.39) lead to one resolving equation

$$P = 2H\sqrt{\frac{\beta^2}{\left(1-\frac{H}{EA}\beta\right)^2} - 1}, \quad \beta = \frac{L_0}{l}. \tag{5.40}$$

This equation allows to calculate unknown thrust H for given P, *EA*, initial length L_0, and span l.

The sag of elastic cable is

$$f_{\text{el}} = \frac{Pl}{4H} = \frac{l}{2}\sqrt{\frac{\beta^2}{\left(1-\frac{H}{EA}\beta\right)^2} - 1}. \tag{5.41}$$

Limiting cases

1. The cable is nondeformable ($EA = \infty$, $L_0 = L$). In this case, the sag

$$f_{\text{nd}} = \frac{l}{2}\sqrt{\frac{L^2}{l^2} - 1}.$$

   This result also may be obtained from geometrical consideration of the design diagram. Let us evaluate increasing of the sag for elastic cable. If $\beta = 1.5$ and $H = 0.01EA$, then $f_{\text{el}} = 1.35 f_{\text{nd}}$, i.e., a sag increased to 35%.
2. The initial length of a cable equals to the span ($L_0 = l$); it means that the cable may be treated as a string. In this case $\beta = 1$ and (5.40) becomes

$$P = 2H\sqrt{\frac{1}{\left(1-\frac{H}{EA}\right)^2} - 1} \tag{5.42}$$

This equation may be rewritten in equivalent form

$$P = \frac{2\sqrt{2}}{1 - \dfrac{H}{EA}} \sqrt{\frac{H^3}{EA}} \sqrt{1 - \frac{H}{2EA}} \tag{5.42a}$$

Since $H/EA \ll 1$, then (5.42a) may be presented as

$$P = \sqrt{\frac{8H^3}{EA}}$$

or

$$H = \frac{1}{2} \sqrt[3]{P^2 EA} \tag{5.43}$$

Corresponding sag is

$$f = \frac{Pl}{4H} = \frac{l}{2} \sqrt[3]{\frac{P}{EA}} \tag{5.44}$$

Equations (5.43) and (5.44) show that the relationships P–H and P–f are nonlinear.

5.6.2 Elastic Cable with Uniformly Distributed Load

Now let us consider a gentile cable of the span l. In this case a distributed load may be considered as sum of external load and the weight of a cable itself. Since the cable is gentile, we assume that the tension in the cable is constant and equal to the thrust

$$N = H\sqrt{1 + \left(\frac{\mathrm{d}y}{\mathrm{d}x}\right)^2} \approx H. \tag{5.45}$$

According to (5.16b) we have

$$\frac{L}{l} = 1 + \frac{1}{24}\frac{q^2 l^2}{H^2}.$$

Solution of this equation leads to the following expression for a thrust (in terms of span l, total length L of the cable, and load q)

$$H = \frac{ql}{2\sqrt{6}} \frac{1}{\sqrt{\dfrac{L}{l} - 1}}. \tag{5.45a}$$

The length L of the cable under load is $L = L_0(1 + \varepsilon)$, where L_0 is initial length of the cable and $\varepsilon = H/EA$, so

$$L = L_0(1 + \varepsilon) = L_0\left(1 + \frac{H}{EA}\right).$$

Thus (5.45a) for thrust H may be presented as

$$H = \frac{ql}{2\sqrt{6}} \frac{1}{\sqrt{\frac{L_0(1+\varepsilon)}{l} - 1}} = \frac{ql}{2\sqrt{6}} \frac{1}{\sqrt{\beta\left(1+\frac{H}{EA}\right) - 1}}, \quad \beta = \frac{L_0}{l}. \quad (5.46)$$

So for computation of the thrust H the equation (5.46) may be rewritten in the following form

$$\frac{\beta}{EA} H^3 + (\beta - 1) H^2 = \frac{q^2 l^2}{24}.$$

This is noncomplete cubic equation with respect to H. Solution of this equation for string ($\beta = 1$) is

$$H = \sqrt[3]{\frac{q^2 l^2 EA}{24}}. \quad (5.46a)$$

The reader is invited to derive the following expression for sag of an gentile cable:

$$f = \frac{\sqrt{3}}{2\sqrt{2}} l \cdot \sqrt{\beta\left(1 + \frac{H}{EA}\right) - 1} \quad (5.47)$$

Hint. Use formula $f = ql^2/8H$ (formula 5.10) and expression (5.46).

We can see that consideration of the elastic properties of a cable leads to the nonlinear relationships H–q and f–H.

Limiting cases

1. In case of nondeformable cable ($EA = \infty$), formulas (5.46) and (5.47) coincide with formulas (5.16c) and (5.16d) for the case of gentile cable carrying uniformly distributed load.
2. If $L_0 = l$ then formulas (5.46a) and (5.47) lead to the following nonlinear relationships f–q :

$$f = \frac{l}{4} \sqrt[3]{\frac{3ql}{EA}} \quad (5.48)$$

Problems

5.1. Design diagram of a cable is presented in Fig. P5.1. Supports A and B are located on different elevations. Parameters of the system are: $a_1 = 10\,\text{m}$, $a_2 = 22\,\text{m}$, $c = 3\,\text{m}$, $l = 30\,\text{m}$, $P_1 = 18\,\text{kN}$, $P_2 = 15\,\text{kN}$.

Determine the thrust H of the cable, if total length of the cable $L = 34\,\text{m}$.

Ans. $H = 23.934\,\text{kN}$

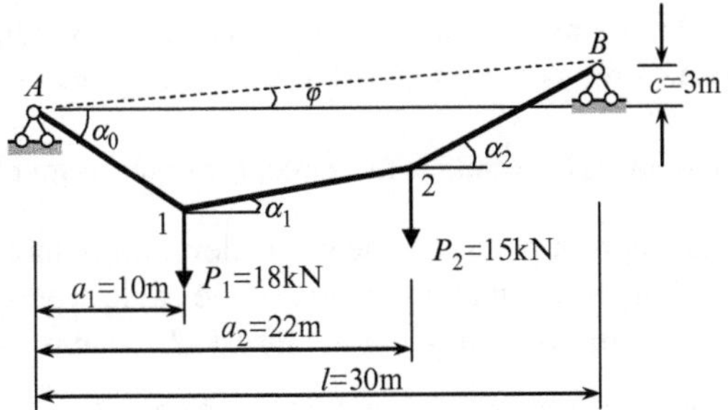

Fig. P5.1

5.2. The cable is subjected to two concentrated forces $P = 30\,\text{kN}$ at joints E and C, and unknown force N at joint D, as shown in Fig. P5.2. The thrust of the cable structure is $H = 60\,\text{kN}$. Determine the force N and corresponding shape of the cable, if the portion CD of the cable is horizontal.

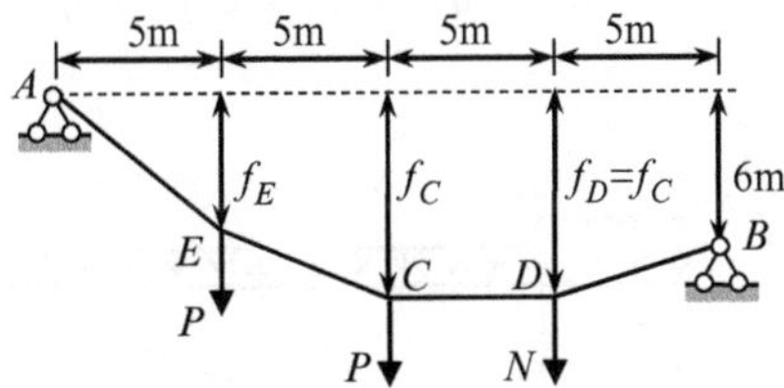

Fig. P5.2

Ans. $N = 18\,\text{kN}$; $f_E = 5.0\,\text{m}$.

5.3. The cable with support points on the same levels has span $l = 36\,\text{m}$ is subjected to uniformly distributed load $q = 2\,\text{kN/m}$ along horizontal projection of the cable. The sag of the cable is $f = 6\,\text{m}$. Determine the thrust of the cable H, maximum axial tension $N_{\max}$ and slope at the support.

Ans. $H = 54\,\text{kN}$; $N_{\max} = 64.89\,\text{kN}$

5.4. The cable with support points on the same levels has span $l = 36\,\text{m}$ and is subjected to uniformly distributed load $q = 2\,\text{kN/m}$ along horizontal projection of the cable. The thrust of the cable is $H = 108\,\text{kN}$. Determine the sag of the cable, maximum axial tension $N_{\max}$, and slope at the support.

Ans. $f = 3\,\text{m}$; $N_{\max} = 113.84\,\text{kN}$

5.5. The cable with support points on the same levels is subjected to uniformly distributed load q along horizontal projection of the cable. The span of the cable is l. If load q increases by two times, but the sag f remains the same, then

(a) Thrust H is (1) remains the same; (2) twice as much; (3) half as much
(b) Maximum axial force is (1) remains the same; (2) twice as much; (3) half as much
(c) Slope at the support is (1) remains the same; (2) twice as much; (3) half as much

5.6. The cable with support points on the same elevation is subjected to uniformly distributed load q along horizontal projection of the cable. The span of the cable is l m. If load q increases by two times, but the thrust H remains the same, then

(a) Sag of the cable is (1) remains the same; (2) greater; (3) twice as much; (4) half as much.
(b) Maximum axial force is (1) remains the same; (2) greater; (3) twice as much; (4) half as much.
(c) Slope at the support is (1) remains the same; (2) greater; (3) twice as much; (4) half as much.

5.7. The flexible inextensible cable with support points on the same levels is subjected to uniformly distributed load q_1 and q_2 within the horizontal line, as shown in Fig. P5.7; the span of the cable is l m and maximum sag is f. Calculate the thrust of the cable. Consider limiting cases $q_1 = q_2$ and $q_2 = q_1$.

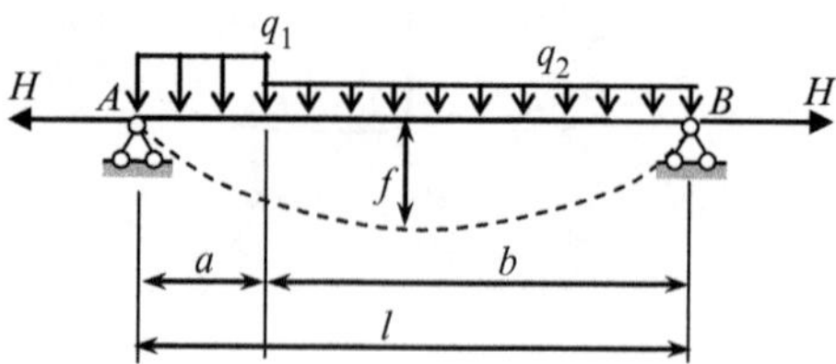

Fig. P5.7

$$\text{Ans. } H = \frac{\left(q_1 a^2 + 2q_2 bl - q_2 b^2\right)^2}{8q_2 f l^2}$$

5.8. The flexible inextensible cable with support points on the same elevation is subjected to uniformly distributed load q within the horizontal line; the span of the cable is l m. Derive expression for maximum tension $N_{\max}$ in terms of sag-to-span ratio $\alpha = f/l$.

$$\text{Ans. } N_{\max} = \frac{ql}{8}\frac{1}{\alpha}\sqrt{1 + 16\alpha^2}$$

5.9. A uniform cable of weight q_0 per unit length is suspended between two points at the same elevation and a distance l apart. Let the maximum axial force $N_{\max}$ and total weight of the cable W be related as $N_{\max} = kW$, where k is any positive number.

(a) Derive equation, which connects parameters q_0, l, H, and k
(b) Calculate the sag–span ratio for which the maximum tension in the cable is equal to 0.75 of the total weight of the entire cable and corresponding value of maximum tension $N_{\max}$.

Ans. (a) $(4k^2 - 1)\sinh^2 \dfrac{q_0 l}{2H} = 1$; (b) 0.2121, $N_{\max} = 0.8337 q_0 l$

5.10. A uniform cable of weight q_0 per unit length, is suspended between two points at the same elevation and a distance l apart. Determine the sag–span ratio, for which the maximum tension is as small as possible.

Ans. $\frac{f}{l} = 0.3377$

5.11. A uniform cable of weight q_0 per unit length is suspended between two points at the same elevation and a distance l apart. The sag of the cable, total length, and thrust are denoted as f, L, and H, respectively. Calculate the maximum tensile force. Present the result in three following forms (1) in terms of H, q_0, l; (2) in terms of H, q_0, f; (3) in terms of H, q_0, L

Ans. (1) $N_{\max} = H \cosh \dfrac{q_0 l}{2H}$; (2) $N_{\max} = q_0 f + H$; (3) $N_{\max} = H\sqrt{1 + \left(\dfrac{q_0 L}{2H}\right)^2}$

5.12. Design diagram of flexible cable with support points A and B on different levels, is presented in Fig. P5.12. The cable is subjected to linearly distributed load q. At the middle point C the sag is $f = 4.5$ m. Find the shape of the cable, thrust, and calculate distribution of internal forces. Parameters of the system are: $l = 60$ m, $c = 12$ m, $q = 2.0$ kN/m. Use the concept of the reference beam.

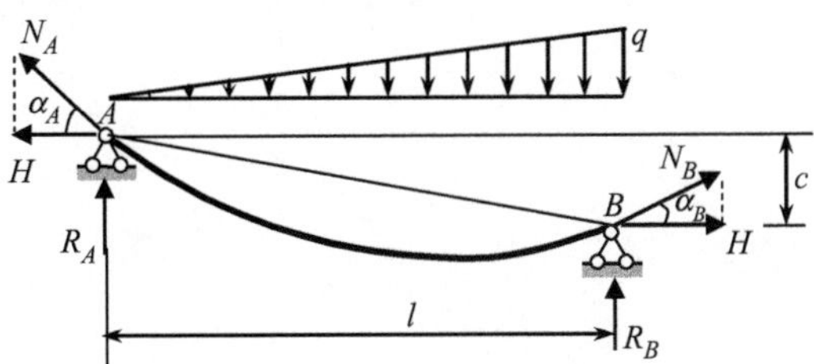

Fig. P5.12

Ans. $H = 100$ kN, $N_A = 107.76$ kN, $N_B = 101.83$ kN, $R_A = 40$ kN, $R_B = 20$ kN

Chapter 6
Deflections of Elastic Structures

This chapter describes some effective methods for computing deflections of deformable structures. The following structures are considered: beams, frames, arches, and trusses subjected to different actions, such as variety of external loads, change of temperature, settlements of supports, and errors of fabrication. Computation of different types of deflections is shown. They are linear, angular, as well as the mutual linear and angular deflections. Advantages and disadvantages of each method and field of their effective application are discussed. Much attention is given to a graph multiplication method which is a most effective method for bending structures. Fundamental properties of deformable structures are described by reciprocal theorems; they will be widely used for analysis of statically indeterminate structures.

6.1 Introduction

Any load which acts on the structure leads to its deformation. It means that a structure changes its shape, the points of the structure displace, and relative position of separate points of a structure changes. There are other reasons of the deformation of structures. Among them is a settlement of supports, change of temperature, etc. Large displacements could lead to disruption of a structure functioning properly and even its collapse. Therefore an existing Building Codes establish limit deflections for different engineering structures. Ability to compute deflections is necessary for estimation of rigidity of a structure, for comparison of theoretical and actual deflections of a structure, as well theoretical and allowable deflections. Beside that, computation of deflections is an important part of analysis of any statically indeterminate structure. Deflections computation is also an integral part of a dynamical analysis of the structures. Thus, the computation of deflections of deformable structures caused by different reasons is a very important problem of Structural Analysis.

Outstanding scientists devoted theirs investigations to the problem of calculation of displacements. Among them are Bernoulli, Euler, Clapeyron, Castigliano, Maxwell, Mohr, etc. They proposed a number of in-depth and ingenious ideas for the

I.A. Karnovsky and O. Lebed, *Advanced Methods of Structural Analysis*,
DOI 10.1007/978-1-4419-1047-9_6,

solution of this problem. At present, methods for computation of the displacements are developed with sufficient completeness and commonness for engineering purposes and are brought to elegant simplicity and perfection.

The deformed shape of a bend structure is defined by transversal displacements $y(x)$ of every points of a structural member. The slope of the deflection curve is given by $\theta(x) = \mathrm{d}y/\mathrm{d}x = y'(x)$. Deflected shapes of some structures are presented in Fig. 6.1. In all cases, elastic curves (EC) reflect the deformable shape of the neutral line of a member; the EC are shown by dotted lines in exaggerated scale.

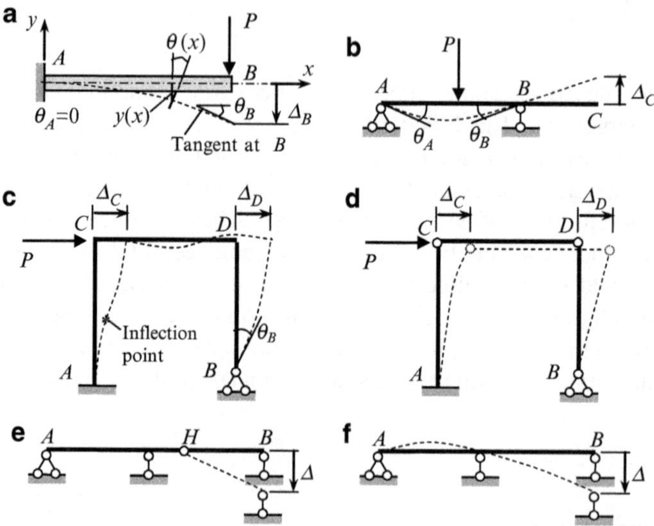

Fig. 6.1 (**a–d**) Deflected shapes of some structures. (**e, f**) Deflected shape of beams caused by the settlement of support B

A cantilever beam with load P at the free end is presented in Fig. 6.1a. All points of the neutral line have some vertical displacements $y(x)$. Equation $y = y(x)$ is the EC equation of a beam. Each section of a beam has not only a transversal (vertical in this case) displacement, but an angular displacement $\theta(x)$ as well. Maximum vertical displacement Δ_B occurs at B; maximum slope θ_B also happens at the same point. At the fixed support A, both linear and angular displacements Δ_A and θ_A are zero.

The simply supported beam with overhang is subjected to vertical load Pas shown in Fig. 6.1b.The vertical displacements at supports A and B are zero. The angles of rotation θ_A and θ_B are maximum, but have different directions. Since overhang BC does not have external loads, the elastic curve along the overhang presents the straight line, i.e., the slope of the elastic curve θ within this portion is constant. The angles of rotation of sections, which are located infinitely close to the left and right of support B are equal.

Figure 6.1c shows the frame due to action of horizontal force P. At fixed support A the linear and angular displacements are zero, while at pinned support B the angle of rotation $\theta_B \neq 0$. The joints C and D have the horizontal displacements Δ_C and Δ_D; under special assumptions these displacements are equal. Joints C and D have angular displacements θ_C and θ_D (they are not labeled on the sketch). The linear and angular displacements of joints C and D lead to deformation of the vertical members as shown on the sketch. Since support A is fixed, then the left member AC has an inflection point.

Figure 6.1d shows the frame with hinged ends of the cross-bar CD; the frame is subjected to horizontal force P. In this case the cross-bar and column BD has a *displacement* but does not have *deflection* and members move as absolutely rigid one – the motion of the member CD is a translation, while the member BD rotates around point B. Thus, it is a possible displacement of the member without the relative displacements of its separate points. So a displacement is not always accompanied by deflections, however, deflections are impossible without displacement of its points.

Figure 6.1e, f shows the shapes of the beams caused by settlement of support. In case 6.1e, a new form of statically determinate beam is characterized by displacement of portion H-B as absolutely rigid body, i.e., without *deflection* of the beam. In case 6.1f, a new form of the beam occurs with deflection of the beam.

There are two principle analytical approaches to computation of displacements. The first of them is based on the integration of the differential equation of the elastic curve of a beam. Modification of this method leads to the initial parameters method. The second approach is based on the fundamental energetic principles. The following precise analytical methods represent the second group: Castigliano theorem, dummy load method (Maxwell–Mohr integral), graph multiplication method (Vereshchagin rule), and elastic load method.

All methods from both groups are exact and based on the following assumptions:

1. Structures are physically linear (material of a structures obey Hook's law);
2. Structures are geometrically linear (displacements of a structures are much less than their overall dimensions).

6.2 Initial Parameters Method

Initial parameters method presents a modification of double integration method in case when a beam has several portions and as result, expressions for bending moments are different for each portion. Initial parameter method allows us to obtain an equation of the elastic curve of a beam with any type of supports (rigid or elastic) and, most important, for any number of portions of a beam.

Fundamental difference between the initial parameter and the double integration method, as it will be shown below, lies in the following facts:

1. Initial parameters method does not require setting up the expressions for bending moments for different portions of a beam, formulating corresponding differential equations and their integration. Instead, the method uses a once-derived

expression for displacement. This expression allows us to calculate slope, bending moments, and shear along the beam and is called the Universal equation of elastic curve of a beam.
2. Universal equation of the elastic curve of a beam contains only two unknown parameters for *any* number of portions.

A general case of a beam under different types of loads and the corresponding notational convention is presented in Fig. 6.2a. The origin is placed at the extreme left end point of a beam, the x-axis is directed along the beam, and y-axis is directed downward. Support A is shown as fixed, however, it can be any type of support or even free end. Load q is distributed along the portion DE. Coordinates of points of application of concentrated force P, couple M, and initial point of distributed load $q$are denoted as a with corresponding subscripts P, M, and q. This beam has five portions (AB, BC, CD, DE, and EL), which leads to the ten constants of integrating using the double integration method.

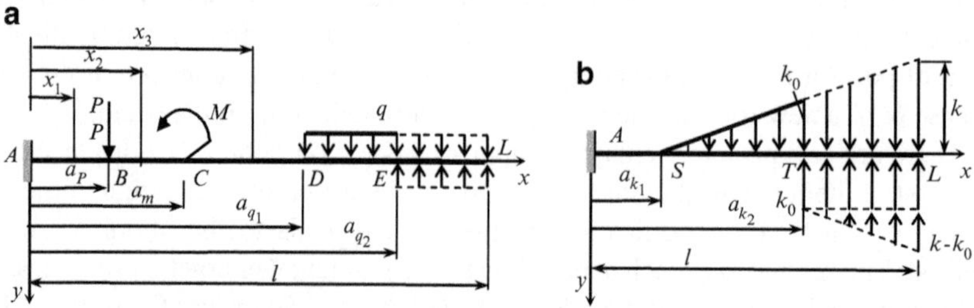

Fig. 6.2 Initial parameters method notation

The initial parameter method requires the following rules to be entertained:

1. Abscises x for all portions should be reckoned from the origin; in this case the bending moment expression for each next portion contains all components related to the previous portion.
2. Uniformly distributed load may start from any point but it must continue to the very right point of the beam. If distributed load is interrupted (point E, Fig. 6.2a), then this load should be continued till the very right point and action of the added load must be compensated by the same but oppositely directed load, as shown in Fig. 6.2a. The same rule remains for load which is distributed by triangle law. If load is located within the portion S-T (Fig. 6.2b), it should be continued till the very right point L of the beam and action of the added load must be compensated by the same but oppositely directed loads (uniformly distributed load with intensity k_0 and load distributed by triangle law with maximum intensity k–k_0 at point L). Both of these compensated loads start at point T and do not interrupt until the extremely right point L.
3. All components of a bending moment within each portion should be presented in unified form using the factor (x–a) in specified power, as shown in Table 6.1. For example, the bending moments for the second and third portions (Fig. 6.2a) caused by the active loads only are

$$M(x_2) = -P(x_2 - a_P),$$
$$M(x_3) = -P(x_3 - a_P) - M(x_3 - a_M)^0.$$

4. Integration of differential equation should be performed *without opening the parenthesis*.

All of these conditions are called Cauchy–Clebsch conditions.

Initial parameters method is based on the equation $EI\,y'' = -M(x)$. Integrating it twice leads to the following expressions for slope and linear displacement

$$EI\,\theta = -\int M(x)\mathrm{d}x + C_1,$$
$$EI\,y = -\int \mathrm{d}x \int M(x)\mathrm{d}x + C_1 x + D_1. \tag{6.1}$$

The transversal displacement and slope at $x = 0$ are $y = y_0$,$\theta = \theta_0$. These displacements are called the initial parameters. Equations (6.1) allow getting the constants in terms of initial parameters

$$D_1 = EIy_0 \text{ and } C_1 = EI\theta_0.$$

Finally (6.1) may be rewritten as

$$EI\theta = EI\theta_0 - \int M(x)\mathrm{d}x,$$
$$EIy = EIy_0 + EI\theta_0 x - \int \mathrm{d}x \int M(x)\mathrm{d}x. \tag{6.2}$$

These equations are called the initial parameter equations. For practical purposes, the integrals from (6.2) should be calculated for special types of loads using the above rules 1–4. These integrals are presented in Table 6.1.

Table 6.1 Bending moments in unified form for different type of loading

	M, a_M, x, y	P, a_P, x, y	q, a_q, x, y	$k=\tan\beta$, β, a_k, x, y
$M(x)$	$\pm M(x-a_M)^0$	$\pm P(x-a_P)^1$	$\pm\dfrac{q(x-a_q)^2}{2}$	$\pm\dfrac{k(x-a_k)^3}{2\cdot 3}$
$\int M(x)\mathrm{d}x$	$\pm M(x-a_M)$	$\pm\dfrac{P(x-a_P)^2}{2}$	$\pm\dfrac{q(x-a_q)^3}{2\cdot 3}$	$\pm\dfrac{k(x-a_k)^4}{2\cdot 3\cdot 4}$
$\int \mathrm{d}x \int M(x)\mathrm{d}x$	$\pm\dfrac{M(x-a_M)^2}{2}$	$\pm\dfrac{P(x-a_P)^3}{2\cdot 3}$	$\pm\dfrac{q(x-a_q)^4}{2\cdot 3\cdot 4}$	$\pm\dfrac{k(x-a_k)^5}{2\cdot 3\cdot 4\cdot 5}$

Combining (6.2) and data in Table 6.1 allows us to write the general expressions for the linear displacements $y(x)$ and slope $\theta(x)$ for a uniform beam:

$$EI\,y(x) = EI\,y_0 + EI\,\theta_0 x - \sum \pm \frac{F(x-a_F)^n}{n!}, \tag{6.3}$$

$$EI\,\theta(x) = EI\,\theta_0 - \sum \pm \frac{F(x-a_F)^{n-1}}{(n-1)!}, \tag{6.4}$$

where *EI* is the flexural rigidity of the beam; F is any load (concentrated, couple, or a distributed one); y_0 and θ_0 are transversal displacement and slope at $x=0$; a_F is the distance from the origin of the beam to the point of application of a concentrated force, couple, or to the starting point of the distributed load and n is the parameter, which depends on the type of the load.

Types of load F and corresponding parameter n are presented in Table 6.2.

Table 6.2 Initial parameters method. Parameter n for the specific loads

Type of load F	F	N
Couple	M	2
Concentrated load	P	3
Uniformly distributed load	q	4
The load distributed by triangle law	k	5

Equation (6.3) is called the Universal equation of elastic curve of a beam. This equation gives an easiest way of deriving the equation of elastic curve of uniform beam and calculating displacement at any specified point. This method is applicable for a beam with arbitrary boundary conditions, subjected to any types of loads.

Notes:

1. The negative sign before the symbol Σ corresponds to the y-axis directed downward.
2. Summation is related only to loads, which are located to the left of the section x. It means that we have to take into account only those terms, for which the difference $(x-a)$ is positive.
3. Reactions of supports and moment of a clamped support must be taken into account as well, like any active force.
4. Consideration of all loads including reactions must start at the very left end and move to the right.
5. Sign of the load factor $\pm(F(x-a_F)^n)/n!$ coincides with the sign of bending moment due to the load, which is located at the left side of the section x.
6. Initial parameters y_0 and θ_0 may be given or be unknown, depending on boundary conditions.
7. Unknown parameters (displacements or forces) are to be determined from the boundary conditions and conditions at specified points, such as the intermediate support and/or intermediate hinge.

For positive bending moments at x due to couple M, force P, and uniformly distributed load q, the expanded equations for displacement and slope are

$$EI\,y(x) = EI\,y_0 + EI\,\theta_0 x - \frac{M(x-a_M)^2}{2!} - \frac{P(x-a_P)^3}{3!} - \frac{q(x-a_q)^4}{4!}, \tag{6.5}$$

$$EI\,\theta(x) = EI\,\theta_0 - M(x-a_M) - \frac{P(x-a_P)^2}{2} - \frac{q(x-a_q)^3}{6}. \tag{6.6}$$

Expressions for bending moment and shear force may be obtained by formula

$$M(x) = -EI\,y''(x), \quad Q(x) = -EI\,y'''(x). \tag{6.7}$$

Advantages of the initial parameters method are as follows:

1. Initial parameters method allows to obtain the *expression* for elastic curve of the beam. The method is very effective in case of large number of portions of a beam.
2. Initial parameters method do not require to form the expressions for bending moment at different portions of a beam and integration of differential equation of elastic curve of a beam; a procedure of integration was once used for deriving the Universal equation of a beam and then only simple algebraic procedures are applied according to expression (6.3).
3. The number of unknown initial parameters is always equals two and does not depend on the number of portions of a beam.
4. Initial parameters method may be effectively applied for beams with elastic supports and beams subjected to displacement of supports. Also, this method may be applied for statically indeterminate beams.

Example 6.1. A simply supported beam is subjected to a uniformly distributed load q over the span (Fig. 6.3). The flexural stiffness EI is constant. Derive the expressions for elastic curve and slope of the beam. Determine the slope at the left and right supports, and the maximum deflection.

Solution. The origin is placed at the left support, the x-axis is directed along the beam and the y-axis is directed downward. The forces that should be taken into

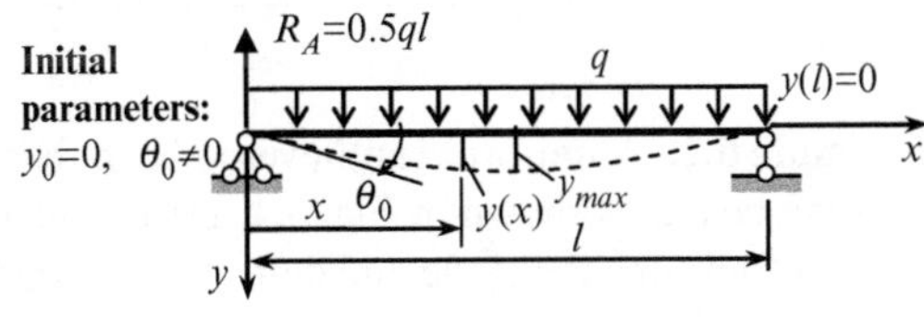

Fig. 6.3 Design diagram of simply supported beam and its deflected shape

account according to the universal equation (6.3) are reaction R_A of the support A and uniformly distributed load q. The expression of elastic curve is

$$EI\,y(x) = EI\,y_0 + EI\,\theta_0 x - \left[\frac{P(x-a_P)^3}{3!} - \frac{q(x-a_q)^4}{4!}\right], \tag{a}$$

where $P = R_A = 0.5ql$, $a_P = 0$, $a_q = 0$; the initial parameters are $y_0 = 0$ and $\theta_0 \neq 0$.

The negative sign before the brackets means that the y-axis is directed downward. The signs before the first and second terms in the brackets indicate that bending moments at x due to a reaction $P = R_A$ and distributed load q are positive and negative, respectively. Equation (a) contains one unknown parameter θ_0, which can be calculated taking into account the boundary condition at the right support. Since the vertical displacement at B is zero, i.e., $y(l) = 0$, then

$$EI\,y(l) = EI\,\theta_0 l - \frac{ql}{2}\frac{l^3}{6} + \frac{ql^4}{24} = 0 \tag{b}$$

This equation leads to $EI\,\theta_0 = ql^3/24$. Therefore, the expression (a) of the elastic curve becomes

$$EI\,y(x) = \frac{ql^3}{24}x - \frac{ql}{2}\frac{x^3}{6} + \frac{qx^4}{24} = \frac{ql^4}{24}\left(\frac{x}{l} - 2\frac{x^3}{l^3} + \frac{x^4}{l^4}\right). \tag{c}$$

An expression for slope of the beam may be derived from (c) by differentiation

$$EI\,\theta(x) = EI\frac{\mathrm{d}y}{\mathrm{d}x} = \frac{ql^3}{24}\left(1 - 6\frac{x^2}{l^2} + 4\frac{x^3}{l^3}\right).$$

The slopes at the left and right supports are $EI\,\theta(0) = ql^3/24$ and $EI\,\theta(l) = -ql^3/24$.

Since the elastic curve is symmetrical with respect to the middle point of the beam, then the maximum displacement occurs at the point $x = 0.5l$. Thus, from (c) we can obtain the following result

$$EI\,y_{\max} = \frac{5}{384}ql^4.$$

The positive sign indicates that the displacement occurs in positive direction of the y-axis.

Example 6.2. A uniform cantilevered beam has a uniform load q over the interval a of the beam, as shown in Fig. 6.4. Derive the equation of the elastic curve of the beam. Determine the slope and deflection of the beam at the free end.

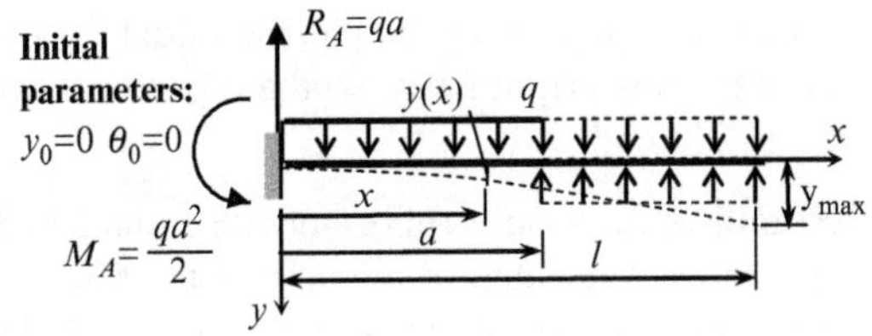

Fig. 6.4 Design diagram of cantilevered beam and its deflected shape

Solution. Since a distributed load is interrupted at $x = a$ then an additional distributed load should be applied from the point $x = a$ until point $x = l$ and this additional load must be compensated by the load of same intensity in the opposite direction. Expression for the elastic curve is

$$EI\,y(x) = EI\,y_0 + EI\,\theta_0 x - \left[-\frac{M_A(x-0)^2}{2!} + \frac{R_A(x-0)^3}{3!} - \frac{q(x-0)^4}{4!} + \frac{q(x-a)^4}{4!}\right],$$

where the initial parameters are $y_0 = 0$ and $\theta_0 = 0$; the third and fourth terms in parentheses take into account the uniformly distributed load over the *all length* of the beam and the compensated distributed load. The reactions are $M_A = qa^2/2$ and $R_A = qa$. So the equations of elastic curve and slope become

$$EI\,y(x) = \frac{qa^2}{2}\frac{x^2}{2} - qa\frac{x^3}{6} + \frac{qx^4}{24} - \frac{q(x-a)^4}{24},$$

$$EI\,\theta(x) = \frac{qa^2x}{2} - \frac{qax^2}{2} + \frac{qx^3}{6} - \frac{q(x-a)^3}{6}.$$

The beam with compensated load has two portions so the last term in both equations must be taken into account only for second portion ($a \le x \le l$), i.e., for positive values of $x - a$.

The transversal deflections of the beam at the point $x = a$ and at the free end are:

$$EI\,y(a) = \frac{qa^2}{2}\frac{a^2}{2} - qa\frac{a^3}{6} + \frac{qa^4}{24} = \frac{qa^4}{8},$$

$$EI\,y(l) = \frac{qa^2l^2}{4} - \frac{qal^3}{6} + \frac{ql^4}{24} - \frac{q(l-a)^4}{24}.$$

The right part of the beam is unloaded and has no constraints, therefore the second portion ($a \le x \le l$) is not deformable and the slopes for any point at this portion are equal

$$EI\,\theta(a) = EI\,\theta(l) = \frac{qa^3}{6}.$$

As can be seen in this example, the initial parameters y_0 and θ_0 for the given structure are known right away, so there is no necessity to use the boundary conditions at the right end.

Example 6.3. A uniform beam with clamped left support and elastic support at the right end is subjected to a concentrated force P at point C, as shown in Fig. 6.5; stiffness parameter of the elastic support is k. Derive the equation of elastic curve. Compute the reaction of a clamped support and consider the special cases $k = 0$ and $k = \infty$.

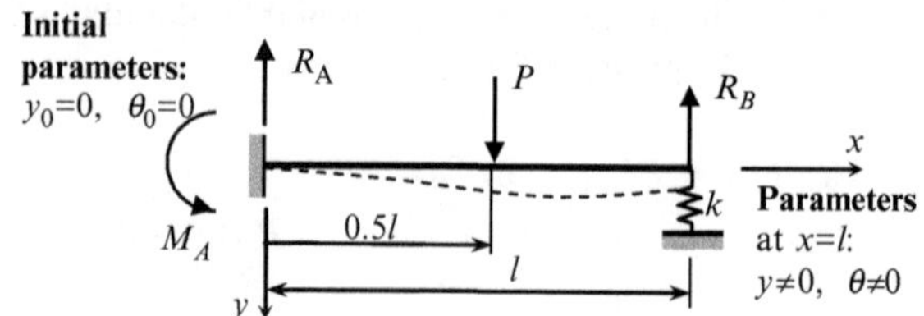

Fig. 6.5 Design diagram of one-span beam with elastic support and its deflected shape

Solution. The general expression for elastic curve according to (6.3) and (6.5) is

$$EI\,y(x) = EI\,y_0 + EI\,\theta_0 x + \frac{M_A(x-0)^2}{2!} - \frac{R_A(x-0)^3}{3!} + \frac{P(x-0.5l)^3}{3!}. \quad \text{(a)}$$

The last term should be taken into account only for positive $(x - 0.5l)$.

Initial parameters are $y_0 = 0$ and $\theta_0 = 0$. These conditions lead to the equation

$$EI\,y(x) = \frac{M_A x^2}{2} - \frac{R_A x^3}{6} + \frac{P(x-0.5l)^3}{6}, \quad \text{(b)}$$

which contains the unknown reaction R_A and reactive moment M_A at the clamped support A. For their determination we have two additional conditions:

1. The bending moment at support B is zero, therefore

$$M(l) = -M_A + R_A l - P\left(l - \frac{l}{2}\right) = 0$$

and the reactive moment M_A in terms of R_A becomes

$$M_A = R_A l - \frac{Pl}{2}. \quad \text{(c)}$$

This expression allows us to rewrite (b) for elastic curve of a beam as follows:

$$EI\,y(x) = \left(R_A l - \frac{Pl}{2}\right)\frac{x^2}{2} - \frac{R_A x^3}{6} + \frac{P}{6}\left(x - \frac{l}{2}\right)^3. \quad \text{(d)}$$

Thus, displacement at point B equals

$$EI\,y(l)=\left(R_A-\frac{P}{2}\right)\frac{l^3}{2}-\frac{R_A l^3}{6}+\frac{P}{6}\left(\frac{l}{2}\right)^3. \tag{e}$$

2. Deflection at point B and stiffness k are related as $y_B=(R_B/k)=(P-R_A)/k$, therefore

$$EI\,y(l)=EI\frac{P-R_A}{k}. \tag{f}$$

Solving system of equations (e, f) leads to an expression for the reaction at support A

$$R_A=P\frac{1+\dfrac{11}{48}\dfrac{kl^3}{EI}}{1+\dfrac{kl^3}{3EI}}.$$

Substitution of the last expression into (d) leads to the equation of elastic curve for the given system.

Special cases:

1. If $k=0$ (cantilever beam), then $R_A=P$
2. If $k=\infty$ (clamped-pinned beam), then $R_A=11P/16$ and $R_B=5P/16$.

Example 6.4. The continuous beam in Fig. 6.6 is subjected to a uniform load $q$in the second span. Flexural stiffness *EI* is constant. Derive the equation of the elastic curve of the beam. Calculate the reactions of supports.

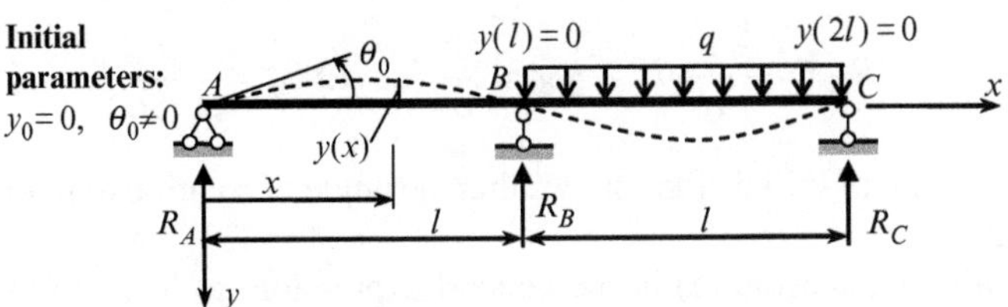

Fig. 6.6 Design diagram of two-span continuous beam and its deflected shape

Solution. The universal equation of elastic curve of a beam is

$$EI\,y(x)=EI\,y_0+EI\,\theta_0 x-\left[\frac{R_A(x-0)^3}{3!}+\frac{R_B(x-l)^3}{3!}-\frac{q(x-l)^4}{4!}\right].$$

Since the initial parameter $y_0=0$ then

$$EI\,y(x)=EI\,\theta_0 x-\frac{R_A x^3}{6}-\frac{R_B(x-l)^3}{6}+\frac{q(x-l)^4}{24}. \tag{a}$$

This equation contains the unknown initial parameter θ_0 and two unknown reactions R_A and R_B. For their calculation we have three additional conditions. They are $EI\,y(l) = 0$, $EI\,y(2l) = 0$, $M(2l) = 0$.

1. Displacement at $x = l$ (support B):

$$EI\,y(l) = EI\,\theta_0 l - \frac{R_A l^3}{6} = 0. \tag{b}$$

2. Displacement at $x = 2l$ (support C):

$$EI\,y(2l) = EI\,\theta_0 2l - \frac{R_A(2l)^3}{6} - \frac{R_B(2l-l)^3}{6} + \frac{q(2l-l)^4}{24} = 0 \quad \text{or}$$

$$EI\,\theta_0 2l - \frac{8R_A l^3}{6} - \frac{R_B l^3}{6} + \frac{ql^4}{24} = 0. \tag{c}$$

3. Bending moment equation may be presented as

$$M(x) = -EI\,y''(x) = R_A x + R_B(x-l) - \frac{q(x-l)^2}{2}.$$

The bending moment at the support C is

$$M(2l) = R_A \cdot 2l + R_B(2l-l) - \frac{q(2l-l)^2}{2} = 0 \quad \text{or} \quad R_A 2l + R_B l - \frac{ql^2}{2} = 0. \tag{d}$$

Solution of (b), (c), and (d) leads to the following results

$$R_A = -\frac{1}{16}ql, \quad R_B = \frac{5}{8}ql, \quad EI\,\theta_0 = -\frac{ql^3}{96}. \tag{e}$$

The negative sign of initial slope shows that the angle of rotation at point A is in the counterclockwise.

Substitution expression (e) in the general expression (a) leads to the following equation for elastic curve

$$EI\,y(x) = -\frac{ql^3}{96}x + \frac{1}{16}ql\frac{x^3}{6} - \frac{5}{8}ql\frac{(x-l)^3}{6} + \frac{q(x-l)^4}{24}.$$

The following terms should be taken into account: for the first span – the first and second terms only and for the second span – all terms of the last equation.

Example 6.5. The beam AB is clamped at the left end and pinned at right end. The beam is subjected to the angular displacement θ_0 at the left end as shown in Fig. 6.7. Derive the equation of the elastic curve and compute the reactions of supports.

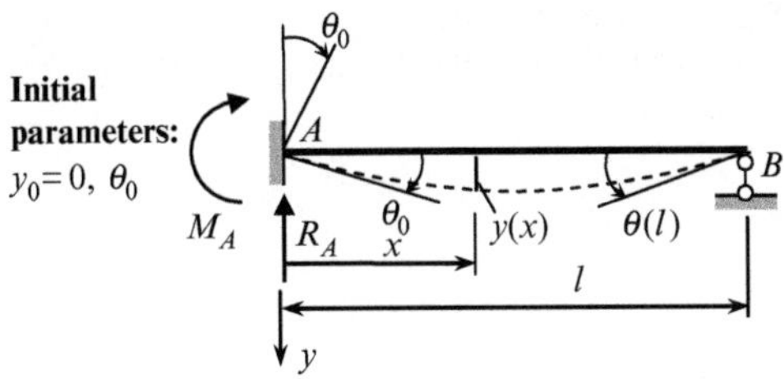

Fig. 6.7 Fixed-rolled beam subjected to the angular settlement of support

Solution. According to (6.5), the expression for elastic curve is

$$EI\,y(x) = EI\,y_0 + EI\,\theta_0 x - \left[\frac{R_A(x-0)^3}{3!} + \frac{M_A(x-0)^2}{2!}\right].$$

Since the initial parameter $y_0 = 0$, then

$$EI\,y(x) = EI\,\theta_0 \cdot x - \frac{R_A x^3}{6} - \frac{M_A x^2}{2}. \tag{a}$$

This equation contains two unknowns M_A and R_A. For their calculation we have two additional equations.

1. The transverse displacement at the right support equals zero

$$EI\,y(l) = EI\,\theta_0 \cdot l - \frac{M_A l^2}{2} - \frac{R_A l^3}{6} = 0. \tag{b}$$

2. Bending moment at support B equals zero. The expression for bending moment may be obtained by twice differentiating (a)

$$M(x) = -EI\,y''(x) = M_A + R_A x. \tag{c}$$

At $x = l$ we have

$$M(l) = M_A + R_A l = 0. \tag{d}$$

Solving (b) and (d) with respect to M_A and R_A leads to the following expressions for reactions

$$M_A = \frac{3EI}{l}\theta_0, \quad R_A = -\frac{3EI}{l^2}\theta_0. \tag{e}$$

The formulas (e) are presented in Table A.3; they are necessary for analysis of statically indeterminate frames by the displacement method (Chap. 8).

Example 6.6. The beam AB is clamped at the left and right ends. Derive equation of the elastic curve for the beam if the vertical relative displacement of supports is Δ_B (Fig. 6.8). Calculate corresponding reactions.

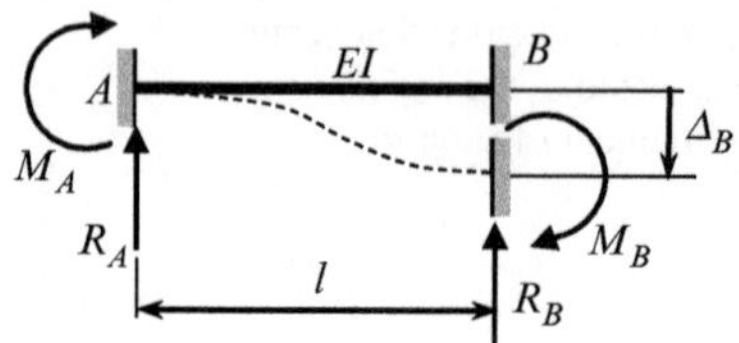

Fig. 6.8 Fixed-fixed beam subjected to linear settlement of support

Solution. According to (6.5), the expression for elastic curve is

$$EI\,y(x) = EI\,y_0 + EI\,\theta_0 x - \left[\frac{R_A(x-0)^3}{3!} + \frac{M_A(x-0)^2}{2!}\right].$$

Since initial parameters are $y_0 = 0$ and $\theta_0 = 0$ then

$$EI\,y(x) = -\frac{R_A x^3}{6} - \frac{M_A x^2}{2}. \tag{a}$$

For right fixed support the displacement is Δ_B, therefore

$$EI\,y(l) = -\frac{R_A l^3}{6} - \frac{M_A l^2}{2} = EI \cdot \Delta_B \tag{b}$$

Expression for slope is $EI\,y'(x) = -(R_A x^2/2) - M_A x$. For right fixed support the slope is zero, so

$$EI\,y'(l) = -\frac{R_A l^2}{2} - M_A l = 0. \tag{c}$$

Solution of (b) and (c) leads to the following results:

$$R_A = \frac{12EI}{l^3}\Delta_B(\uparrow), \quad M_A = -\frac{6EI}{l^2}\Delta_B. \tag{d}$$

Substitution of (d) into (a) allows calculating the transversal displacement of any point of the beam.

Other required reactions may be determined considering the equilibrium equations:

$$R_B = -\frac{12EI}{l^3}\Delta_B(\downarrow) \quad \text{and} \quad M_B = -\frac{6EI}{l^2}\Delta_B. \tag{e}$$

The negative signs show that actual directions for reactions and moments do not coincide with adopted direction in Fig. 6.8.

Formulas (d) and (e) will be used for analysis of statically indeterminate frames by the displacement method (Chap. 8).

6.3 Maxwell–Mohr Method

The Maxwell–Mohr procedure presents a universal method for computation of displacement at any point of any deformable structure. Also, the Maxwell–Mohr procedure allows calculating mutual displacements. Different sources, which may cause displacements of a structure, are considered. They are different types of loads and change of temperature.

6.3.1 Deflections Due to Fixed Loads

For bending systems, the Castigliano's theorem for computation of linear and angular displacements at point k may be presented as follows

$$y_k = \int \frac{M(x)}{EI} \frac{\partial M(x)}{\partial P_k} \mathrm{d}x, \quad \theta_k = \int \frac{M(x)}{EI} \frac{\partial M(x)}{\partial P_k} \mathrm{d}x, \tag{6.8}$$

where $M(x)$ is bending moment at section x, P_k and M_k are force and couple at section k.

Both formulas (6.8) may be simplified. For this purpose let us consider, for example, the simply supported beam subjected to force P and couple M (Fig. 6.9).

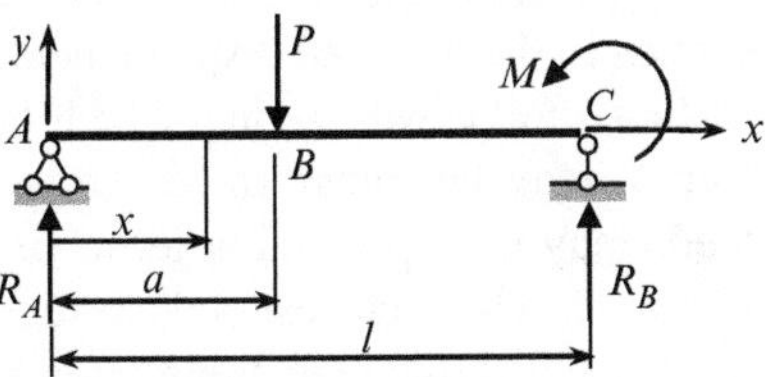

Fig. 6.9 Simply supported beam loaded by P and M

Reaction

$$R_A = P\frac{l-a}{l} + M\frac{1}{l}$$

and the bending moment for the left and right portions of the beam are

$$M(x) = R_A x = P\frac{(l-a)}{l}x + M\frac{x}{l} \quad (x \leq a),$$

$$M(x) = R_A x - P(x-a) = P\frac{a}{l}(l-x) + M\frac{x}{l} \quad (x \geq a).$$

Both expressions present the *linear* functions of the loads P and M. In general case, suppose a structure is subjected to the set of concentrated loads $P_1, P_2, \ldots$, couples $M_1, M_2, \ldots$, and distributed loads $q_1, q_2, \ldots$. This condition of structure is called as P-condition (also known as the actual or loaded condition). In case of P–condition, a bending moment at the any section x is a *linear function* of these loads

$$M(x) = a_1 P_1 + a_2 P_2 + \cdots + b_1 M_1 + b_2 M_2 + \cdots + c_1 q_1 + c_2 q_2 + \cdots, \quad (6.9)$$

where coefficients a_i, b_i, and c_i depend on geometrical parameters of the structure, position of loads, and location of the section x.

If it is required to find displacement at the point of application of P_1, then, as an intermediate step of Castigliano's theorem we need to calculate the partial derivative of bending moment $M(x)$ with respect to force P_1. This derivative is $\partial M(x)/\partial P_1 = a_1$. According to expression for $M(x)$, this parameter a_1 may be considered as the bending moment at section x caused by *unit dimensionless force* ($P_1 = 1$). State of the structure due to action of unit dimensionless load (unit force or unit couple) is called *unit state*. Thus, calculation of partial derivatives in (6.8) may be changed by calculation of a bending moment caused by unit dimensionless load

$$y_k = \int \frac{M(x)}{EI} \frac{\partial M(x)}{\partial P} \mathrm{d}x = \int \frac{M(x)\bar{M}_k}{EI} \mathrm{d}x, \quad (6.10)$$

where $\bar{M}_k$ is bending moment in the unit state. Keep in mind that $\bar{M}_k$ is always a *linear function* and represents the bending moment due to a unit load, which corresponds to the required displacement.

In a similar way, terms, which take into account influence of normal and shear forces, may be transformed. Thus, displacements caused by any combination of loads may be expressed in terms of internal stresses developed by given loads and *unit* load, which corresponds to required displacement. That is the reason why this approach is termed the dummy load method. A general expression for displacement may be written as

$$\Delta_{kp} = \sum \int_0^s \frac{M_p \bar{M}_k}{EI} \mathrm{d}s + \sum \int_0^s \frac{N_p \bar{N}_k}{EA} \mathrm{d}s + \sum \int_0^s \mu \frac{Q_p \bar{Q}_k}{GA} \mathrm{d}s. \quad (6.11)$$

Summation is related to all elements of a structure. Fundamental expression (6.11) is known as Maxwell–Mohr integral. The following notations are used: Δ_{kp} is displacement of a structure in the kth direction in P-condition, i.e., displacement in the direction of unit load (first index k) due to the given load (second index p); M_p, N_p, and Q_p are the internal stresses (bending moment, axial and shear forces) in

P-condition; and $\bar{M}_k$, $\bar{N}_k$, $\bar{Q}_k$ are the internal stresses due to the unit load, which acts in the kth direction and corresponds to the required displacement. GA is transversal rigidity, μ is non-dimensional parameter depends on the shape of the cross-section. For rectangular cross section this parameter equals 1.2, for circular section it equals 10/9. The unit load (force, couple, etc.) is also termed as the dummy load.

For different types of structures, relative contribution of first, second, and third terms of expression (6.11) in the total displacement Δ_{kp} is different. For practical calculation, depending on type and shape of a structure, the following terms from (6.11) should be taken into account:

(a) For trusses – only second term
(b) For beams and frames with ratio of height of cross section to span 0.2 or less – only first term
(c) For beams with ratio of height of cross section to span more than 0.2 – the first and third terms
(d) For gently sloping arches – the first and second terms
(e) For arches with ratio of radius of curvature to height of cross section 5 or more – all terms

In case of trusses, the displacement should be calculated by formula

$$\Delta_{kp} = \sum \int_0^l \frac{N_p \bar{N}_k}{EA} ds. \tag{6.12}$$

Since all elements are straight ones and axial stiffness EA is constant along all length of each element, then this formula may be presented as

$$\Delta_{kp} = \sum \frac{N_p \bar{N}_k}{EA} l. \tag{6.13}$$

Procedure for computation of deflections using Maxwell–Mohr integral is as follows:

1. Express internal forces in P-condition for an arbitrary cross section in terms of its position x
2. Construct the unit condition. For this we should apply unit load (dummy load), which corresponds to the required displacement:

   (a) For linear displacement, a corresponding dummy load represents the unit force, which is applied at the point where displacement is to be determined and acts in the same direction
   (b) For angular displacement, a corresponding dummy load is the unit couple, which is applied at the point where angle of rotation is to be determined

(c) For mutual linear displacement of two sections, a corresponding dummy load represents two unit forces, which are applied at the points where displacement is to be determined and act in the opposite directions
(d) For mutual angular displacement of two sections, a corresponding dummy load represents two unit couples, which are applied at given sections and act in the opposite directions.

3. Express the internal forces in unit condition for an arbitrary cross section in terms of its position x
4. Calculate Maxwell–Mohr integral

Positive sign of displacement means that the real displacement coincides with the direction of the unit load, or work performed by unit load along the actual direction is positive.

Example 6.7. A cantilever uniform beam is subjected to a uniformly distributed load q (Fig. 6.10a). Compute (a) the angle of rotation and (b) vertical displacement at point A. Take into account only bending moments.

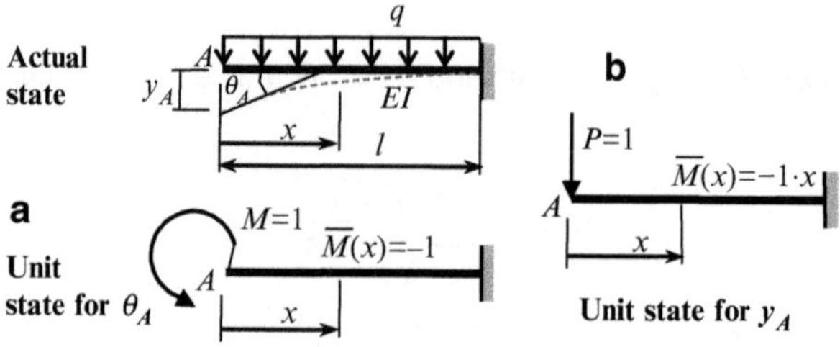

Fig. 6.10 Design diagram of the beam; (**a**) Unit state for θ_A; (**b**) Unit state for y_A

Solution. (a) The angle of rotation may be defined by formula

$$\theta_A = \frac{1}{EI}\int_0^l M_p(x)\bar{M}\,dx. \tag{a}$$

Now we need to consider two states, mainly, the actual and unit ones, and for both of them set up the expressions for bending moments. For actual state, the bending moment is $M_p(x) = -qx^2/2$. Since it is required to determine the slope at point A, then the unit state presents the same structure with unit couple $M = 1$ at point A (Fig. 6.10a); this dummy load may be shown in arbitrary direction. For unit state, the bending moment is $\bar{M} = -1$ for any section x. The formula (a) for required angle of rotation becomes

$$\theta_A = \frac{1}{EI}\int_0^l \left(-\frac{qx^2}{2}\right)\cdot(-1)dx = \frac{ql^3}{6EI}. \tag{b}$$

(b) The vertical displacement at A may be calculated by formula

$$y_A = \frac{1}{EI}\int_0^l M_p(x)\bar{M}\,\mathrm{d}x, \tag{c}$$

where expression for bending moment $M_p(x)$ in the actual state remains without change. In order to construct the unit state, it is necessary to apply unit concentrated force $P = 1$ at the point where it is required to determine displacement (Fig. 6.10b).

For unit state, the bending moment is $\bar{M} = -1 \cdot x$. The formula (c) for vertical displacement becomes

$$y_A = \frac{1}{EI}\int_0^l (-\frac{qx^2}{2}) \cdot (-1 \cdot x)\mathrm{d}x = \frac{ql^4}{8EI}. \tag{d}$$

The positive sign means that adopted unit load make positive work on the real displacement, or other words, actual displacement coincides with assumed one.

Example 6.8. Determine the vertical displacement of joint 6 of symmetrical truss shown in Fig. 6.11. Axial rigidity for diagonal and vertical elements is *EA* and for lower and top chords is 2*EA*.

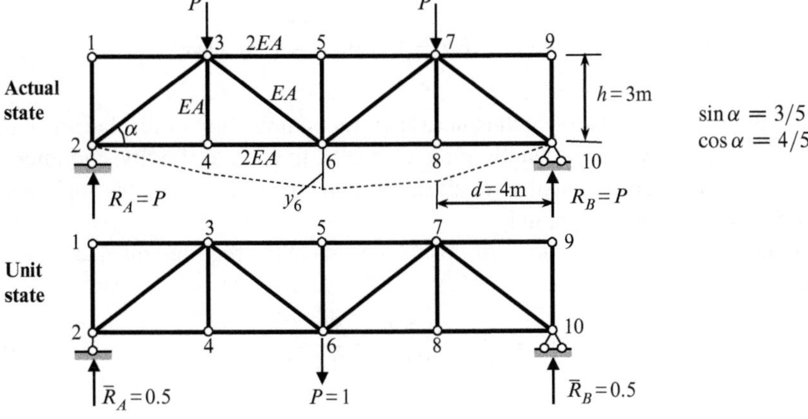

Fig. 6.11 Design diagram of the truss (actual state) and unit state

Solution. All elements of the given structure are subjected to axial loads only, so for required displacement the following formula should be applied:

$$y_6 = \sum \frac{1}{EA} N_p \bar{N}_k l, \tag{a}$$

where N_p and $\overline{N}_k$ are internal forces in actual and unit state, respectively.

Actual state. Reaction of supports $R_A = R_B = P$. As usual, the axial forces in the members are taken positive in tension. Equilibrium equations lead to the following internal forces:

$$\begin{aligned}
O_{35} &\to \sum M_6^{\text{left}} = 0: \quad -R_A \cdot 2d + Pd - O_{35}h = 0 \to O_{35} = -\frac{4}{3}P \\
U_{24} &\to \sum M_3^{\text{left}} = 0: \quad -R_A d + U_{24}h = 0 \to U_{24} = \frac{4}{3}P \\
D_{23} &\to \sum Y^{\text{left}} = 0: \quad D_{23}\sin\alpha + R_A = 0 \to D_{23} = -\frac{5}{3}P \\
D_{36} &\to \sum Y^{\text{left}} = 0: \quad R_A - P - D_{36}\sin\alpha = 0 \to D_{36} = 0
\end{aligned} \tag{b}$$

Unit state. Unit vertical load $P = 1$ is applied at the point where vertical displacement is to be determined. Reaction of supports $\bar{R}_A = \bar{R}_B = 0.5$. Internal forces in unit state are:

$$\begin{aligned}
\bar{O}_{35} &\to \sum M_6^{\text{left}} = 0: \quad -\bar{R}_A \cdot 2d - \bar{O}_{35}h = 0 \to \bar{O}_{35} = -\frac{4}{3} \\
\bar{U}_{24} &\to \sum M_3^{\text{left}} = 0: \quad -\bar{R}_A d + \bar{U}_{24}h = 0 \to \bar{U}_{24} = \frac{2}{3} \\
\bar{D}_{23} &\to \sum Y^{\text{left}} = 0: \quad \bar{D}_{23}\text{sin}\alpha + \bar{R}_A = 0 \to \bar{D}_{23} = -\frac{5}{6} \\
\bar{D}_{36} &\to \sum Y^{\text{left}} = 0: \quad \bar{R}_A - \bar{D}_{36}\text{sin}\alpha = 0 \to \bar{D}_{36} = \frac{5}{6}
\end{aligned} \tag{c}$$

Table 6.3 contains the data which is necessary for computation of displacement according to (a). They are the length of the elements, their axial rigidity, and internal forces in all members in actual and unit states. The last column contains application of (a) for each member separately.

Summation of the last column of this table leads to the required displacement of joint 6

$$y_6 = \sum \frac{N_p \bar{N} l}{EA} = \frac{506P}{18EA}, \tag{d}$$

Note. For calculation of displacement of *all joints* of the bottom (upper) chord of the truss, the different unit states should be considered. For each unit state, the procedure presented in Table 6.3 should be repeated. Thus, the Maxwell–Mohr integral requires calculation of the internal forces in *all* members of the truss for *each* unit state. This procedure is cumbersome and will be extremely simplified using the elastic load method.

Table 6.3 Calculation of vertical displacement of the joint 6 of the truss

Bar	Length of the bar (m)	Axial rigidity	Axial forces		$\frac{N_p \bar{N} l}{EA}$
			Actual state (N_p)	Unit state $\bar{N}$	
O_{1-3}	4	$2EA$	0	0	0
O_{3-5}	4	$2EA$	$-4P/3$	$-4/3$	$64P/18EA$
O_{5-7}	4	$2EA$	$-4P/3$	$-4/3$	$64P/18EA$
O_{7-9}	4	$2EA$	0	0	0
U_{2-4}	4	$2EA$	$4P/3$	2/3	$32P/18EA$
U_{4-6}	4	$2EA$	$4P/3$	2/3	$32P/18EA$
U_{6-8}	4	$2EA$	$4P/3$	2/3	$32P/18EA$
U_{8-10}	4	$2EA$	$4P/3$	2/3	$32P/18EA$
V_{1-2}	3	EA	0	0	0
V_{3-4}	3	EA	0	0	0
V_{5-6}	3	EA	0	0	0
V_{7-8}	3	EA	0	0	0
V_{9-10}	3	EA	0	0	0
D_{2-3}	5	EA	$-5P/3$	$-5/6$	$125P/18E$
D_{3-6}	5	EA	0	5/6	0
D_{6-7}	5	EA	0	5/6	0
D_{7-10}	5	EA	$-5P/3$	$-5/6$	$125P/18E$

6.3.2 *Deflections Due to Change of Temperature*

It is often in engineering practice that the members of a structure undergo thermal effects. In case of statically determinate structures, the change of temperature leads to displacements of points of a structure *without* an appearance of temperature internal forces, while in case of statically indeterminate structures the change of temperature causes an appearance of temperature internal forces. Often these internal forces may approach significant values. Analysis of any statically indeterminate structure subjected to change of temperature is based on calculation of displacement of statically determinate structure. So, calculation of displacements due to change of temperature is a very important problem for analysis of both statically determinate and indeterminate structures.

Two first terms of Maxwell–Mohr's integral (6.11) may be rewritten as follows

$$\Delta_{kp} = \sum \int_0^l \bar{M}_k \Delta_{\theta p} + \sum \int_0^l \bar{N}_k \Delta_{xp}, \tag{6.14}$$

where $\Delta_{\theta p} = (M_p/EI)\mathrm{d}x$ is the mutual angular displacement of both sections faced apart at a distance $\mathrm{d}x$ due to the given load and $\Delta_{xp} = (N_p/EA)\mathrm{d}x$ is the mutual axial displacement of both sections faced apart at a distance $\mathrm{d}x$ due to the given load.

These terms may be easily computed for the case of temperature change. Let us consider elementary part of a structure with length $\mathrm{d}x$. The height of the cross section of the member is h_0. The upper and bottom fibers of the member are subjected to temperature increase t_1 and t_2, respectively, above some reference temperature. Corresponding distribution of temperature (temperature profile) is presented in Fig. 6.12. If the change of temperature for bottom and uppers fibers is equal ($t_1 = t_2$), then this case presents the uniform change of temperature; if $t_1 \neq t_2$ then this case is referred as nonuniform change of temperature.

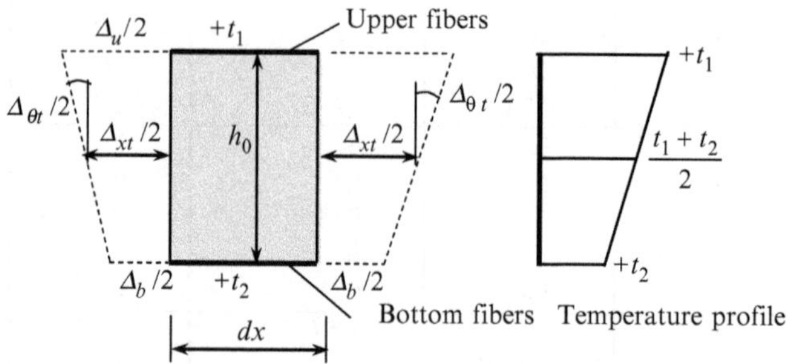

Fig. 6.12 Distribution of temperature and displacements within the height of cross section

The expansion of the upper and bottom fibers equals to $\Delta_u = \alpha t_1 \mathrm{d}x$ and $\Delta_b = \alpha t_2 \mathrm{d}x$, respectively; these expressions contain coefficient of thermal expansion α of member material. In the case of symmetrical cross section, the expansion of the fiber at the mid-height equals to

$$\Delta_{xt} = \alpha \frac{t_1 + t_2}{2} \mathrm{d}x. \tag{a}$$

The mutual angle of rotation of two plane sections, which are located apart from each other on distance $\mathrm{d}x$

$$\Delta_{\theta t} = \alpha \frac{|t_1 - t_2|}{h_0} \mathrm{d}x. \tag{b}$$

Now we can substitute (a) and (b) into (6.14). Finally the displacement in kth direction due to change of temperature may be presented in the following form:

$$\Delta_{kt} = \sum \int_0^s \alpha \frac{t_1 + t_2}{2} \bar{N}_k \mathrm{d}s + \sum \int_0^s \alpha \frac{|t_1 - t_2|}{h_0} \bar{M}_k \mathrm{d}s, \tag{6.15}$$

where $\bar{M}, \bar{N}$ are bending moment and axial force due to the unit generalized force in kth direction; this force should be corresponding to required temperature displacements.

A difference $t_1 - t_2$ is a temperature gradient; a half-sum $(t_1 + t_2)/2$ is a temperature at the centroid of the symmetric cross section (the axis of symmetry coincides with neutral axis). If the cross section is nonsymmetrical about its neutral axis, then the term $(t_1+t_2)/2$ must be replaced by $t_2+((t_1-t_2)/2)y$, where y is the distance of the lower fiber to the neutral axis.

The term $(t_1+t_2)/2$ means that a bar is subjected to uniform thermal effect; in this case all fibers are expanded by the same values. The term $|t_1-t_2|/h_0$ means that a bar is subjected to nonuniform thermal effect; in this case a bar is subjected to bending in such way that the fibers on the neutral line have no thermal elongation. So, the first and second terms in (6.15) present displacements in kth direction due to uniform and nonuniform change of temperature, respectively. Integrals $\int \bar{M}_k \mathrm{d}s$ and $\int \bar{N}_k \mathrm{d}s$ present the areas of bending moment and axial force diagram in unit condition, which corresponds to required displacement.

The presentation of Maxwell–Mohr integral in formula (6.15) allows us to calculate any displacement (linear, angular, mutual linear, mutual angular) caused by uniform or nonuniform change of temperature. This formula does not take into account the influence of shear. The procedure of summation in formula (6.15) must be carried over all members of the system. The signs at all terms in this formula will be obtained as follows: if the displacements of the element induced by both the change of temperature and by the unit load occur at the same direction, then the corresponding term of the equation will be positive.

Procedure for analysis is as follows:

1. Construct the unit state. For this we should apply unit generalized force X, which corresponds to the required displacement
2. Construct the bending moment and axial force diagrams in the unit state
3. For each member of a structure to compute the term $\int \bar{N}_k \mathrm{d}x$, which is the area of axial force diagram in the unit state
4. For each member of a structure to compute the term $\int \bar{M}_k \mathrm{d}x$, which is the area of bending moment diagram in the unit state
5. Apply formula (6.15).

Example 6.9. Determine the vertical displacement of point C at the free end of the knee frame shown in Fig. 6.13, when the indoor temperature rises by 20°C and outdoor temperature remains constant. The height of the element *AB* and *BC* are b and d, respectively.

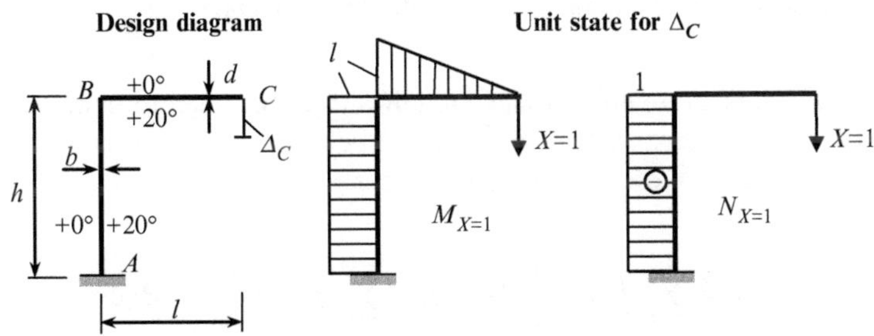

Fig. 6.13 Design diagram of the frame and unit state for Δ_C

Solution. Since the required displacement is Δ_C, then a structure in the unit state is loaded by unit force $X = 1$ at point C. The bending moment and axial force diagrams are shown in Fig. 6.13. General expression for displacements due to change of temperature is

$$\Delta_{kt} = \sum \int_0^s \alpha \frac{t_1 + t_2}{2} \bar{N}_k \mathrm{d}x + \sum \int_0^s \alpha \frac{|t_1 - t_2|}{h_0} \bar{M}_k \mathrm{d}x. \qquad \text{(a)}$$

For member AB of the structure integral $\int \bar{N}_k \mathrm{d}x = h \cdot 1$, while for member BC integral $\int \bar{N}_k \mathrm{d}x = 0$.

For members BC and AB of the frame, the integral $\int \bar{M}_k \mathrm{d}x$ equals to $(1/2)l \cdot l$ and $l \cdot h$, respectively. Therefore, for the given structure the required displacement becomes

$$\Delta_C = \underbrace{-\alpha \frac{0 + 20}{2} \cdot h \cdot 1}_{\text{first term 6.15}} \underbrace{- \alpha \frac{|0 - 20|}{d} \cdot \frac{1}{2} l \cdot l - \alpha \frac{|0 - 20|}{b} \cdot l \cdot h}_{\text{second term 6.15}}. \qquad \text{(b)}$$

The first term of (b), which takes into account axial forces, is negative, since the strains of the element AB induced by the variation in temperature is positive (elongation) and by the unit load is negative (compressed). The second and third terms of (b), which take into account bending moment, are negative, since the tensile fibers of the elements AB and BC induced by the variation in temperature are located inside of the frame, and by the unit load are located outside of the frame.

The final result for the required displacement is

$$\Delta_C = -10\alpha \left(h + \frac{l^2}{d} + 2\frac{lh}{b} \right).$$

Negative sign shows that the actual displacement Δ_C due to the variation in temperature is opposite with induced unit load X.

Example 6.10. Determine the horizontal displacement of point B of the uniform semicircular bar in Fig. 6.14a, if the indoor and outdoor temperature rises by t_1°C and t_2°C, respectively. The height of cross section bar is h_0.

Solution. A temperature effect related to curvilinear bar, therefore the general expression for temperature displacement should be presented in terms of curvilinear coordinate s instead of x as for straight member

$$\Delta_{Bt} = \sum \int_0^s \alpha \frac{t_1 + t_2}{2} \bar{N}_k \mathrm{d}s + \sum \int_0^s \alpha \frac{|t_1 - t_2|}{h_0} \bar{M}_k \mathrm{d}s. \qquad \text{(a)}$$

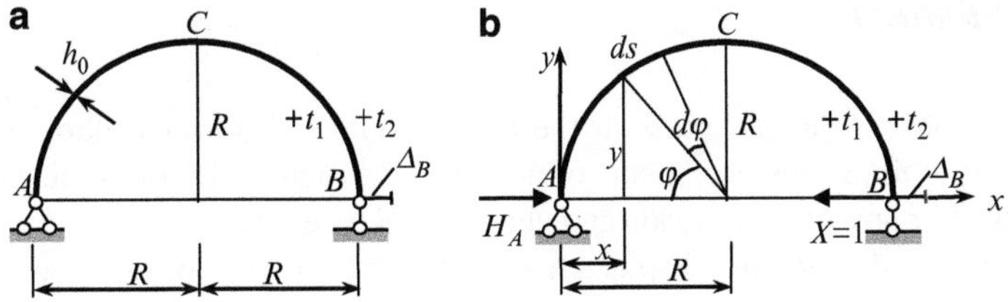

Fig. 6.14 Curvilinear bar. Design diagram and unit state

Unit load $X = 1$ (Fig. 6.14b) corresponds to required horizontal displacement at B. In the unit state reaction $H_A = 1$ and internal forces due to unit load X are as follows

$$\bar{N}_k = -1 \cdot \sin\varphi; \quad \bar{M}_k = -1 \cdot y = -1 \cdot R\sin\varphi. \tag{b}$$

Thus equation (a) for displacement at B becomes

$$\Delta_B = \alpha\frac{t_1 + t_2}{2}\int_0^{\pi R}(-1 \cdot \sin\varphi)\mathrm{d}s + \alpha\frac{|t_1 - t_2|}{h_0}\int_0^{\pi R}(-R \cdot \sin\varphi)\mathrm{d}s. \tag{c}$$

Integration is performed along a curvilinear road of length πR. In the polar coordinates

$$\mathrm{d}\varphi = \frac{\mathrm{d}s}{R}, \quad \sin\varphi = \frac{y}{R},$$

the limits of integration become $0-\pi$

$$\Delta_B = \alpha\frac{t_1 + t_2}{2}\int_0^{\pi}(-1 \cdot \sin\varphi)R\mathrm{d}\varphi + \alpha\frac{|t_1 - t_2|}{h_0}\int_0^{\pi}(-R \cdot \sin\varphi)R\mathrm{d}\varphi. \tag{d}$$

Thus, for required displacement we get the following expression

$$\Delta_B = \alpha\frac{t_1 + t_2}{2}(-2R) + \alpha\frac{|t_1 - t_2|}{h_0}(-2R^2) = -\alpha R(t_1 + t_2) - 2\alpha R^2\frac{|t_1 - t_2|}{h_0}. \tag{e}$$

Negative sign in (e) means that unit force X produces negative work on the real displacement, i.e., the displacement of the point B due to temperature changes is directed from left to right.

For uniform change of temperature (i.e., when gradients for indoor and outdoor temperatures are the same), i.e., $t_1 = t_2$, a difference $t_1 – t_2 = 0$ and only first term of (6.15) or (e) should be taken into account. In this case, the horizontal displacement equals to $\Delta_B = -2\alpha Rt$.

6.3.3 Summary

1. Maxwell–Mohr integral presents the fundamental and power method for calculation of arbitrary displacements of any elastic structure. Displacements may be the result of any types of loads and change of temperature.
2. In order to calculate any displacement, it is necessary to consider two states of a structure, i.e., given and unit states. Unit state presents the same structure, but loaded by unit generalized force corresponding to the required displacement.
3. According to the type of structure, which terms of (6.11) should be taken into account can be decided. For both states, given and unit, it is necessary to set up expressions for corresponding internal forces and calculate the required displacement by the Maxwell–Mohr integral.

6.4 Displacement Due to Settlement of Supports and Errors of Fabrication

A settlement of supports often occurs in engineering practice. If the settlement of support happens in direction of reaction of this support, then in case of statically determinate structures such influence leads to a new position of structure without deformation of its separate members; it means that internal stresses are not induced. Computation of displacement of any point of statically determinate structures due to settlement of supports is considered below.

Let us consider a portal frame; support B settles on Δ as shown in Fig. 6.15a. The new position of the frame members after settlement of support B is shown by dotted line. It is necessary to calculate the linear displacement Δ_k of the point k. Unit state presents the same frame subjected to unit force X, which acts in the direction of the required displacement Δ_k (Fig. 6.15b). This unit force X produces the reaction R at the support B. Assume that direction of this reaction coincides with settlement Δ of support.

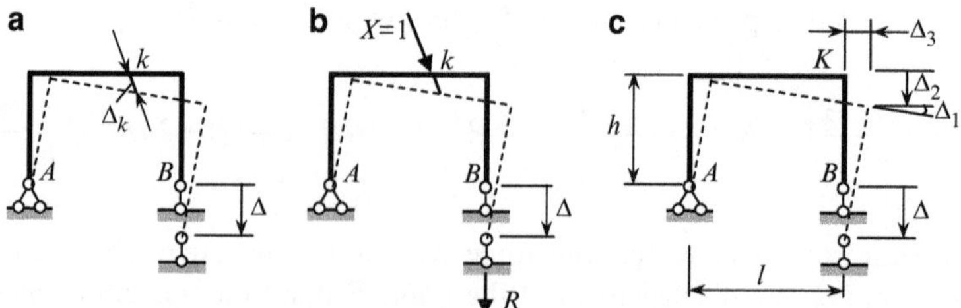

Fig. 6.15 (**a**) Settlement of support B; (**b**) unit state; (**c**) displacements at point K

Effective method for solution of this type of problem is the principle of virtual displacements

$$\sum \delta W_{\text{act}} = 0. \tag{6.16}$$

According to this principle, the elementary work done by all the active forces on any virtual displacement, which is compatible with constraints, is zero. This principle may be applicable for structure with finite displacements. The force R should be considered as active one and (6.16) becomes

$$X \cdot \Delta_k + R \cdot \Delta = 0.$$

Since $X = 1$, then

$$\Delta_k = -R \cdot \Delta. \tag{6.17}$$

The formula (6.17) may be generalized for case of displacements caused by settlements of *several* supports

$$\Delta_{kd} = -\sum R \cdot \Delta, \tag{6.18}$$

where Δ_{kd} is the displacement in kth direction due to settlement of supports, Δ is the given settlement of support, and R is the reaction in the support which is settled; this reaction caused by unit load which corresponds to the required displacement. Summation covers all supports.

Procedure for computation of displacement caused by the settlement of support is as follows:

1. At the point K where displacement should be determined we need to apply unit generalized force $X = 1$ corresponding to the required displacement
2. To show reactions R at the settled support, caused by unit generalized force $X = 1$ and compute these reactions
3. Calculate the work done by these reactions on the displacements of the support; multiply this result by negative unity

Discussion. Equation (6.18) reflects a *kinematical* character of problem; it means that displacements of any point of statically determinate (SD) structure are determined by the geometrical parameters of a structure without taking into account the deformations of its elements. Any settlement of support of SD structure leads to displacement of its separate parts as rigid discs. Displacements of any point of SD structure caused by settlement of supports do not depend on the stiffness of the structure.

Let us consider a procedure (6.18) for portal frame in more detail. The support B is moved on Δ as shown in Fig. 6.15a; it is necessary to calculate the angular and linear displacements of the joint K (Fig. 6.15c). Procedure (6.18) is presented in Table 6.4. The first line of this table presents unit generalized force $X_i = 1$ which corresponds to required displacements. Each unit force may be shown in arbitrary direction. The second line of the table shows the reaction, which arise at the settled support; each of these reactions is caused by unit generalized force $X_i = 1$.

Example 6.11. Two structures are hinged together at C as shown in Fig. 6.16a. Compute the horizontal displacement of point K due to the following settlements of the support A: $\Delta_1 = 0.02\,\text{m}$, $\Delta_2 = 0.03\,\text{m}$, $\theta = 0.01\,\text{rad}$ (scales for displacements Δ_1, Δ_2, and θ and for dimensions of the structure are different).

Table 6.4 Determination of displacements due to settlements of support

	Displacement of the joint K (Fig. 6.15c)		
	Angular displacement Δ_1	Vertical displacement Δ_2	Horizontal displacement Δ_3
The frame under unit load which corresponds to required displacement	$X_1=1$, A, B, R_1	$X_2=1$, A, B, R_2	$X_3=1$, A, B, R_3
Reaction at support B in direction Δ*	$R_1 = -\frac{1}{l}$	$R_2 = -1$	$R_3 = -\frac{1 \cdot h}{l}$
Expression (6.18)	$\Delta_1 = -R_1\,\Delta = -\left(-\frac{1}{l}\right)\Delta = \frac{\Delta}{l}$	$\Delta_2 = -R_2\,\Delta = -(-1)\Delta = \Delta$	$\Delta_3 = -R_3\,\Delta = -\left(-\frac{1 \cdot h}{l}\right)\Delta = \frac{h\Delta}{l}$

*In all cases the negative sign for reaction means that real direction of the reaction R *in unit state* and settlement Δ are oppositely directed. The positive results for required Δ_i means that unit load X on the displacement Δ_i perform the positive work

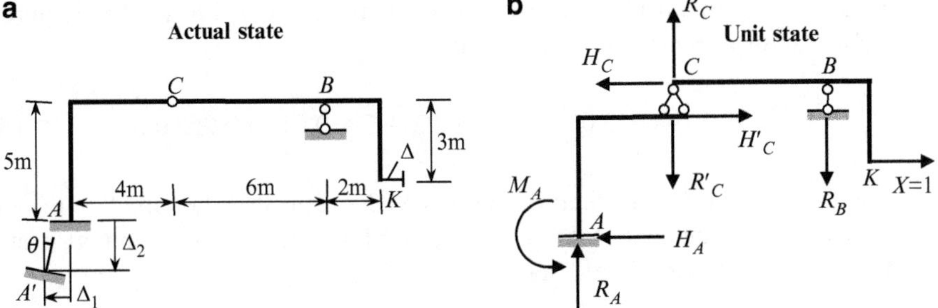

Fig. 6.16 (a) Hinged structure and settlements of support A; (b) Interaction diagram and unit state

Solution. 1. Apply the unit dimensionless horizontal force $X = 1$ corresponding to the required horizontal displacement at point K.

2. Now we have to calculate reactions at the settled support A. Since support A has the horizontal and vertical component of the settlements as well as the angle of rotation, i.e., Δ_1, Δ_2, and θ, then we need to calculate reactions in these directions, i.e., H_A, R_A, and M_A.

   For secondary structure CBK the following reactions arises:

$$R_B \to \sum M_C = 0: \quad X \cdot 3 - R_B \cdot 6 = 0 \to R_B = 0.5,$$

$$R_C = R_B = 0.5; \quad H_C = P = 1.$$

   Reactions H_C and R_C from the secondary structure are transmitted on the primary structure AC as active forces H'_C and R'_C. Reactions, which arise at support A, are

$$H_A = 1; \quad R_A = R'_C = 0.5; \quad M_A = R'_C \cdot 4 + H'_C \cdot 5 = 7.$$

   All reactions are dimensionless, while the M_A is measured in the unit of length; in our case $M_A = 7\,\text{m}$. All these reactions are shown in real direction which corresponds to direction of $X = 1$.

3. The principle of virtual displacements is

$$\sum \delta W_{\text{act}} = 0. \tag{a}$$

   The work done by all active forces on the displacements, which are compatible with constraints, is zero, i.e.,

$$X\Delta + H_A\Delta_1 - R_A \cdot \Delta_2 - M_A\theta = 0. \tag{b}$$

   Since $X = 1$, then

$$\Delta = -H_A\Delta_1 + R_A \cdot \Delta_2 + M_A\theta. \tag{c}$$

In fact the formula (c) is expression (6.18) in the expanded form. The required horizontal displacement at point K equals

$$\Delta = -1 \times 0.02 + 0.5 \times 0.03 + 7 \times 0.01 = 0.065\,\text{m}. \quad \text{(d)}$$

The positive sign for required displacement Δ means that the direction of real horizontal displacement of the point K caused by settlements of support A and direction of unit load X are same.

Example 6.12. The telescope mirror is placed on the inclined CD element of the truss (Fig. 6.17). Determine the angle of rotation of the support bar CD due to the vertical settlements of supports A and B: $\Delta_1 = 0.02\,\text{m}$, $\Delta_2 = 0.03\,\text{m}$.

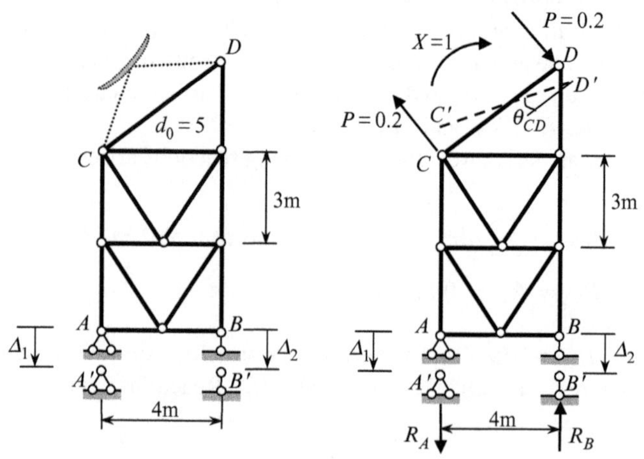

Fig. 6.17 Design diagram of a truss and unit states

Solution. 1. For required angle of rotation θ_{CD} we have to apply unit couple $X = 1$ to the bar CD. This couple is presented as two forces $P = 1/d_0 = 0.2$, where d_0 is the length of the bar CD. Reaction of the supports caused by two forces $P = 0.2$ are:

$$R_B \to \sum M_A = 0: \quad R_B \cdot 4 - 0.2d = 0 \to R_B = 0.25,$$
$$R_A \to \sum Y = 0: \quad R_A = R_B = 0.25.$$

2. Application of principle of virtual displacements leads to following expression

$$X \cdot \theta_{CD} + R_A \cdot \Delta_1 - R_B \cdot \Delta_2 = 0. \quad \text{(a)}$$

However, $X = 1$, and therefore the required displacement, according to (6.18) equals

$$\theta_{CD} = -\sum Rd = -(R_A \cdot \Delta_1 - R_B \cdot \Delta_2). \tag{b}$$

For given $\Delta_i (i = 1, 2)$, the angle of rotation of bar CD becomes

$$\theta_{CD} = -0.25 \times 0.02 + 0.25 \times 0.03 = 0.0025 \text{ rad}$$

Positive sign of the required displacement means that the adopted clockwise couple X within the real angular displacement produce the positive work. It is obvious, that the horizontal displacement at support A does not affect on the angle of rotation of any bar, since horizontal reaction at point A due to applied unit couple is zero.

Deflections of the structural members may occur as a result of the geometric misfit. This topic is sometimes referred to be the name geometric incompatibility.

The following procedure may be applied for this type of problems:

1. At the point K where displacement should be determined we need to apply unit generalized force $X = 1$ corresponding to the required displacement
2. Compute all reactions caused by unit generalized force $X = 1$
3. Calculate the work done by these reactions on the displacements

Example 6.13. The tie AB of the arch ACB in Fig. 6.18 is $\Delta = 0.02$ m longer then required length l. Find the vertical displacement at point C, if $l = 48$ m, $f = 6$ m.

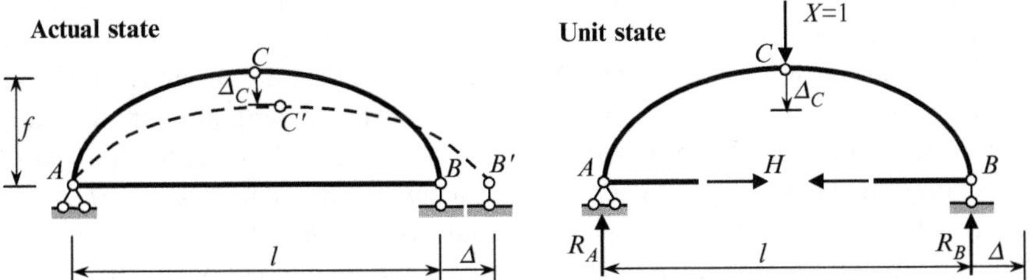

Fig. 6.18 Design diagram of the arch (error fabrication) and unit state

Solution. The actual position of the tie is AB' instead of project AB position. For computation of the vertical displacement Δ_C we have to apply unit vertical force at C. Reactions of the three-hinged arch and thrust in tie caused by force $P = 1$ equals $R_A = R_B = 0.5$, $H = M_C^0 / f = l/4f = 2$.

Application of principle of virtual displacements leads to the following expression

$$X \cdot \Delta_C - H \cdot \Delta = 0.$$

Since $X = 1$, then the required displacement becomes

$$\Delta_C = +H \cdot \Delta = +0.04 \text{ m} \quad \text{(downward)}.$$

It is obvious that the effect of geometric incompatibility may be useful for regulation of the stresses in the structure. Let us consider a three-hinged arch which is loaded by any fixed load. The bending moments are

$$M(x) = M^0 - Hy,$$

where M^0 is a bending moment in the reference beam. If a tie is fabricated longer then required, then the thrust becomes $H = H_1 + H_2$, where H_1 and H_2 are thrust due to fixed load and errors of fabrication, respectively.

Discussion. For computation of displacement due to the settlement of supports and errors of fabrication, we use the principle of virtual work. The common concept for this principle and for Maxwell–Mohr integral is the concept of generalized coordinate and corresponding generalized unit force.

6.5 Graph Multiplication Method

Graph multiplication method presents most effective way for computation of any displacement (linear, angular, mutual, etc.) of bending structures, particularly for framed structures. The advantage of this method is that the integration procedure according to Maxwell–Mohr integral is replaced by elementary algebraic procedure on two bending moment diagrams in the actual and unit states. This method was developed by Russian engineer Vereshchagin in 1925 and is often referred as the Vereshchagin rule.

Let us consider some portion *AB* which is a part of a bending structure; the bending stiffness, *EI*, within of this portion is constant. The bending moment diagrams for this portion in actual and unit state are M_p and $\bar{M}$. Both diagrams for portion *AB* are presented in Fig. 6.19. In general case, a bending moment diagram M_p in the actual state is bounded by curve, but for special cases it may be bounded by straight line (if a structure is subjected to concentrated forces and/or couples). However, it

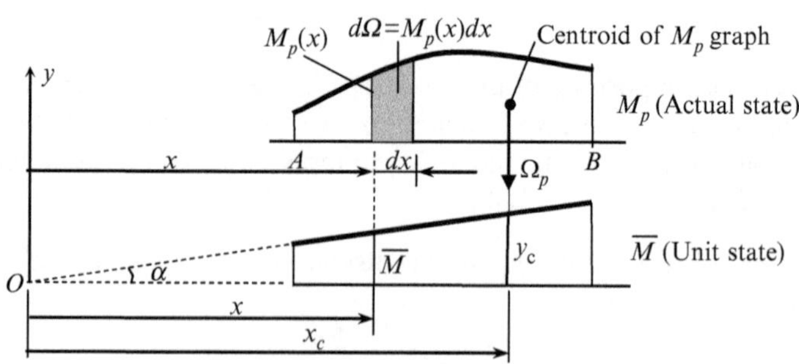

Fig. 6.19 Graph multiplication method. Bending moment diagrams M_p and $\overline{M}$ in actual and unit states

is obvious that in the *unit state* the bending moment diagram $\bar{M}$ is always bounded by the *straight line*. Just this property of unit bending moment diagram allows us to present the Maxwell–Mohr integral for bending systems in the simple form.

Ordinate of the bending moment in actual state at section x is $M_p(x)$. Elementary area of a bending moment diagram in actual condition is $\mathrm{d}\Omega = M_p(x)\mathrm{d}x$. Since $\bar{M} = x\tan\alpha$, then integral in Maxwell–Mohr formula may be presented as (coefficient $1/EI$ by convention is omitted)

$$\int M_p\bar{M}\,\mathrm{d}x = \int (x\tan\alpha)M_p\mathrm{d}x = \tan\alpha\int x\mathrm{d}\Omega. \tag{6.19}$$

Integral $\int x\mathrm{d}\Omega$ represents the static moment of the area of the bending moment diagram in actual state with respect to axis Oy. It is well known that a static moment may be expressed in terms of total area Ω and coordinate of its centroid x_c by formula $\int x\mathrm{d}\Omega = \Omega_p x_c$. It is obvious that $x_c\tan\alpha = y_c$. Therefore, the Maxwell–Mohr integral may be presented as follows

$$\frac{1}{EI}\int M_p\bar{M}\,\mathrm{d}x = \frac{\Omega_p y_c}{EI}. \tag{6.20}$$

The procedure of integration $\int M_p\bar{M}\,\mathrm{d}x = \Omega_p y_c$ is called the "multiplication" of two graphs.

The result of multiplication of two graphs, at least one of which is bounded by a straight line (bending moment diagram in unit state), equals to area Ω of the bending moment diagram M_p in actual state multiplied by the ordinate y_c from the unit bending moment diagram $\bar{M}$, which is located under the centroid of the M_p diagram.

It should be remembered, that *the ordinate y_c must be taken from the diagram bounded by a straight line.* The graph multiplication procedure (6.20) may be presented by conventional symbol $(\times)$ as

$$\Delta_{kp} = \frac{1}{EI}\int M_p\bar{M}_k\mathrm{d}x = \frac{M_p\times\bar{M}_k}{EI}. \tag{6.21}$$

It is obvious that the same procedure may be applicable to calculation of similar integrals, which appear in Maxwell–Mohr integral, i.e., $\int N_p\bar{N}\,\mathrm{d}x$ and $\int Q_p\bar{Q}\mathrm{d}x$.

If the structure in the actual state is subjected to concentrated forces and/or couples, then both the bending moment diagrams in actual and unit states are bounded by the straight lines (Fig. 6.20a). In this case, the multiplication procedure of two diagrams is commutative. It means that the area Ω could be calculated on any of the two diagrams and corresponding ordinate y_c will be measured from the second one, i.e., $\Omega_1 y_1 = \Omega_2 y_2$. This expression may be expressed in terms of specific ordinates, as presented in Fig. 6.20b.

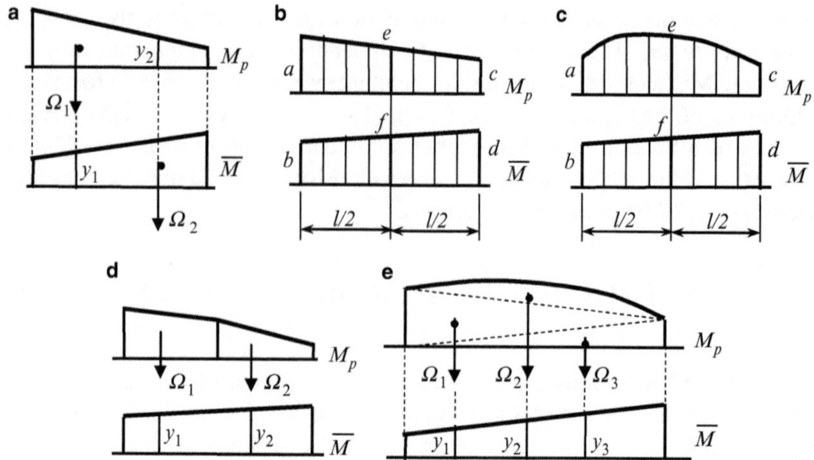

Fig. 6.20 Multiplication of two bending moment diagrams

In this case, the displacement as a result of the multiplication of two graphs may be calculated using two following rules:

1. Trapezoid rule allows calculating the required displacement in terms of *extreme* ordinates

$$\Delta = \frac{l}{6EI}(2ab + 2cd + ad + bc), \tag{6.22}$$

where the crosswise end ordinates has unity coefficients. This formula is precise.
2. Simpson's rule allows calculating the required displacement in terms of *extreme* and *middle* ordinates

$$\Delta = \frac{l}{6EI}(ab + 4ef + cd). \tag{6.23}$$

Equation (6.23) may also be used for calculation of displacements, if the bending moment diagram in the actual condition is bounded by a *curve*line. If the bending moment diagram M_p is bounded by quadratic parabola (Fig. 6.20c), then the result of multiplication of two bending moment diagrams by formula (6.23) is exact; this case occurs if a structure is carrying uniformly distributed load. If the bending moment diagram M_p is bounded by cubic parabola, then the procedure (6.23) leads to the approximate result.

If a graph M_p is bounded by a broken line, then both graphs have to be divided by several portions as shown in Fig. 6.20d. In this case, the result of multiplication of both graphs is

$$\int M_p \bar{M} \, dx = \Omega_1 y_1 + \Omega_2 y_2. \tag{6.24}$$

Sometimes it is convenient to subdivide the curved bending moment diagram by a number of "good" shapes, for example in Fig. 6.20e. In this case

$$\int M_p \bar{M} \, dx = \Omega_1 y_1 + \Omega_2 y_2 + \Omega_3 y_3. \tag{6.25}$$

Signs rule. According to (6.21), the displacement will be positive, when the area of the diagram M_p and the ordinate y_c of the diagram $\bar{M}$ have the same sign. If ordinates in (6.22) or (6.23) of bending moment diagram for actual and unit states are placed on the *different sides* of the basic line, then result of their multiplication is negative. The positive result indicates that displacement occurs in the direction of applied unit load.

Procedure for computation of deflections by graph multiplication method is as follows:

1. Draw the bending moment diagram M_p for the actual state of the structure.
2. Create a unit state of a structure. For this apply a unit load at the point where the deflection is to be evaluated. For computation of linear displacement we need to apply unit force $P = 1$, for angular displacement to apply unit couple $M = 1$, etc.
3. Draw the bending moment diagram $\bar{M}$ for the unit state of the structure. Since the unit load (force, couple) is dimensionless, then the ordinates of unit bending moment diagram $\bar{M}$ in case of force $F = 1$ and moment $M = 1$ are units of length (m) and dimensionless, respectively.
4. Apply the graph multiplication procedure using the most appropriate form: Vereshchagin rule (6.20), trapezoid rule (6.22), or Simpson's formula (6.23).

Graph multiplication method requires the rapid computation of graph areas of different shapes and determination of the position of their centroid. Table A.1 contains the most typical graphs of bending moment diagrams, their areas, and positions of the centroid. Useful formulas for multiplication of two bending moment diagrams are presented in Table A.2.

Example 6.14. A cantilever beam *AB*, length l, carrying a uniformly distributed load q (Fig. 6.21). Bending stiffness *EI* is constant. Compute (a) the angle of rotation θ_A; (b) the vertical displacement Δ_A at the free end.

Solution. Analysis of the structure starts from construction of bending moment diagram M_p due to given external load. This diagram is bounded by quadratic parabola and maximum ordinate equals $ql^2/2$.

(a) *Angle of rotation at point A*. The unit state presents the same beam subjected to unit couple $M = 1$ at the point where it is required to find angular displacement; direction of this couple is arbitrary (Fig. 6.21a). It is convenient that both unit and actual state and their bending moment diagrams locate one under another.

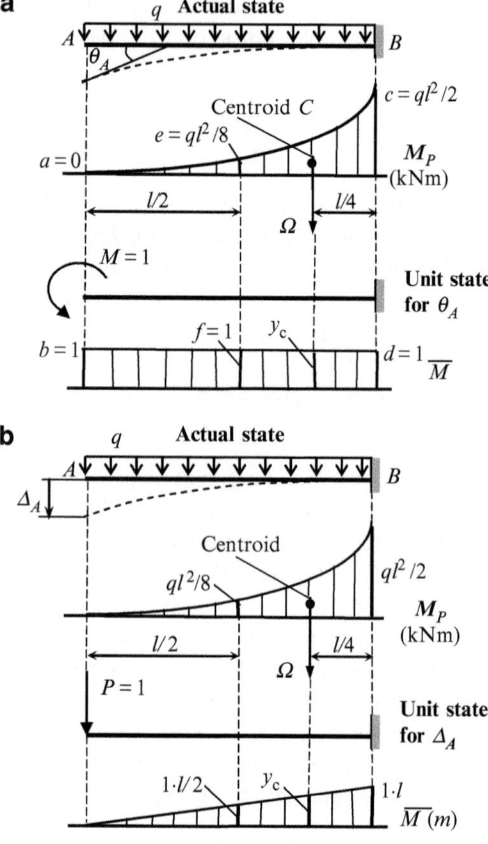

Fig. 6.21 (**a**) Actual state, unit state for θ_A and corresponding bending moment diagrams. (**b**) Actual state, unit state for Δ_A and corresponding bending moment diagrams

The next step is "multiplication" of two bending moment diagrams. The area of square parabola according to Table A.1 is

$$\Omega = \frac{1}{3} \cdot \frac{ql^2}{2} \cdot l.$$

Centroid of this diagram is located on the distance $l/4$ from fixed support. Corresponding ordinate y_c from diagram $\bar{M}$ of unit state is 1. Multiplication procedure is presented in Table 6.5.

This table also contains computation of required displacement using the Simpson rule (6.23). Ordinates a and b are taken from the bending moment diagrams for actual and unit states, respectively, at the left end of a beam (point A); ordinates e and f are taken at the middle of the beam AB, and ordinates c and d at the right end (point B).

Table 6.5 Graph multiplication procedures

Displacement	General formula (6.20) $\Delta = \frac{1}{EI}\Omega y_c$	Simpson rule (6.23) $\Delta = \frac{l}{6EI}(ab + 4ef + cd)$
(a) Angular $\theta_A = \frac{M_p \times \bar{M}}{EI}$	$\theta_A = \frac{1}{EI} \cdot \underbrace{\frac{1}{3}\frac{ql^2}{2}l}_{\Omega} \cdot \underbrace{1}_{y_c} = \frac{ql^3}{6EI}$	$\theta_A = \frac{l}{6EI}\left(\underbrace{0 \cdot 1}_{ab} + \underbrace{4\frac{ql^2}{8} \cdot 1}_{4ef} + \underbrace{\frac{ql^2}{2} \cdot 1}_{cd}\right) = \frac{ql^3}{6EI}$
(b) Linear $\Delta_A = \frac{M_p \times \bar{M}}{EI}$	$\Delta_A = \frac{1}{EI} \cdot \underbrace{\frac{1}{3}\frac{ql^2}{2}l}_{\Omega} \cdot \underbrace{\frac{3}{4} \cdot 1 \cdot l}_{y_c} = \frac{ql^4}{8EI}$	$\Delta_A = \frac{l}{6EI}\left(\underbrace{0 \cdot 0}_{ab} + \underbrace{4\frac{ql^2}{8} \cdot 1 \cdot \frac{l}{2}}_{4ef} + \underbrace{\frac{ql^2}{2} \cdot 1 \cdot l}_{cd}\right) = \frac{ql^4}{8EI}$

(b) *Vertical displacement at point A*. The bending moment diagram M_p for actual state is shown in Fig. 6.21b; this diagram for problems (a) and (b) is same. The unit state presents the same structure with concentrated force $P = 1$, which acts at point A; direction of the unit force is chosen in arbitrary way. The unit state with corresponding bending moment diagram $\bar{M}$ is presented in Fig. 6.21b.

Computation of displacements using Vereshchagin rule in general form and by Simpson rule are presented in Table 6.5.

Discussion:

1. Elastic curve of the beam is shown by dotted line. The tensile fibers for actual and unit states are located above the neutral axis of the beam. Bending moment diagrams are plotted on side of tensile fibers. In the general formula and Simpson rule we use positive sign, because bending moment diagrams for actual and unit states are located on the same side of the basic line. Positive signs in the resulted displacement mean that displacement occurs in the direction of the applied unit load. The units of the ordinates M_p and $\bar{M}$ are (kN m) and (m), respectively.
2. The results, which are obtained by formula (6.20), are precise. Formula (6.23) is approximate one, but for the given problem it leads to the exact result, because the beam is loaded by *uniformly* distributed load, the bending moment diagram presents quadratic parabola, and total order of curves presenting two bending moment diagrams in the actual and unit states is equal to three. If the total orders are more than three, then formula (6.23) leads to the approximate result.
3. The reader is invited to solve the problems above by double integration method, initial parameters method, conjugate beam method, Castigliano theorem, Maxwell–Mohr integral, compare their effectiveness with graph multiplication method, and make personal conclusion about its proficient.

Example 6.15. Design diagram of symmetrical nonuniform simply supported beam of length l is shown in Fig. 6.22. Bending stiffness equals EI for segments AD and EB; while kEI for segment DE; parameter k is any positive number. The beam is carrying force P. Determine the vertical displacement of point C.

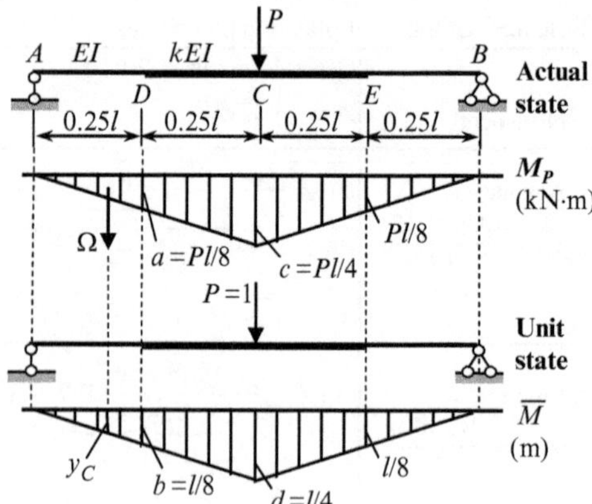

Fig. 6.22 Design diagram of the beam and bending moment diagrams for actual and unit states

Solution. Bending moment diagrams in actual and unit states are presented in Fig. 6.24. For computation $\Delta_C = (M_p \times \bar{M})/EI$, we have to subdivide bending moment diagrams on the parts, within which the bending stiffness is constant. These parts for the left half of the beam are $AD = l/4$ and $DC = l/4$. The Vereshchagin rule for multiplication of diagrams M_p and $\bar{M}$ within portion AD leads to the following result

$$\Delta_{C1} = \frac{1}{EI} \cdot \underbrace{\frac{1}{2} \cdot \frac{Pl}{8} \cdot \frac{l}{4}}_{\Omega} \cdot \underbrace{\frac{2}{3} \cdot \frac{l}{8}}_{y_c} = \frac{Pl^3}{768EI}. \tag{a}$$

For portion CD the trapezoid rule is applied. According to (6.22) we get

$$\Delta_{C2} = \frac{l/4}{6 \cdot k\,EI} \left[\underbrace{2 \cdot \frac{Pl}{8} \cdot \frac{l}{8}}_{2ab} + \underbrace{2 \frac{Pl}{4} \cdot \frac{l}{4}}_{2cd} + \underbrace{\frac{Pl}{8} \cdot \frac{l}{4}}_{ad} + \underbrace{\frac{Pl}{4} \cdot \frac{l}{8}}_{cb} \right] = \frac{7}{768} \frac{Pl^3}{k\,EI}. \tag{b}$$

Finally, the vertical displacement at C becomes

$$\Delta_C = 2\left(\frac{Pl^3}{768EI} + \frac{7}{768} \frac{Pl^3}{k\,EI} \right) = \frac{Pl^3}{48EI}\eta,$$

where $\eta = (1/8) + (7/8k)$; factor 2 takes into account symmetrical part CEB of the beam. If $k = 1$, then $\eta = 1$.

Example 6.16. A portal frame is subjected to horizontal force P as shown in Fig. 6.23. The bending stiffness for each member is shown on design diagram. Calculate (a) the horizontal displacement at the rolled support B and (b) the angle of rotation at the same point B.

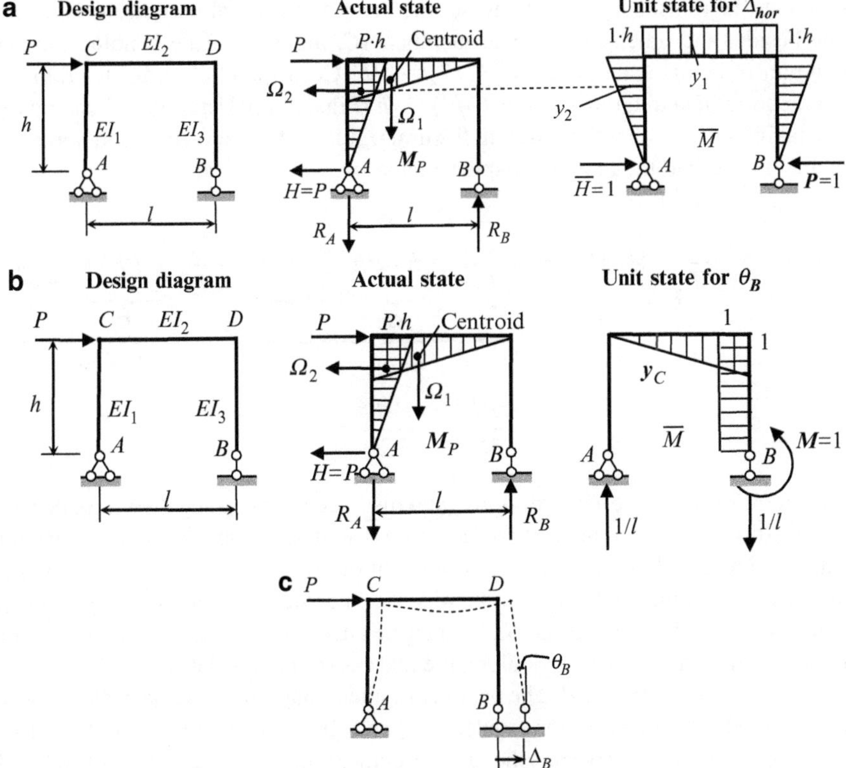

Fig. 6.23 (**a**) Portal frame. Actual and unit states and corresponding bending moment diagrams. (**b**) Portal frame. Actual and unit states and corresponding bending moment diagrams. (**c**) Design diagram and elastic curve for portal frame

Solution. As usual, the analysis starts from construction of bending moment diagram in actual state.

Reactions of supports are $H = P$, $R_A = R_B = Ph/l$. The real directions of reactions are shown in Fig. 6.23a. The tensile fibers on elements *CD* and *AC* are located below and right from the neutral lines of the elements, respectively. Bending moment ordinates at point *C* for vertical and horizontal members are *Ph*.

(a) *Horizontal displacement at B*. For required displacement Δ_{hor}, the unit state presents the same frame with horizontal force $P = 1$, which is applied at point *B*. Direction of the unit force is chosen in arbitrary way. Only horizontal reaction $H = 1$ is induced. The tensile fibers are located outdoor of the frame. The bending moments at rigid joints *C* and *D* and within cross bar equal to $1 \cdot h$.

Multiplication of the bending moment diagrams should be performed for members *AC*, *CD*, and *DB* separately. For horizontal member *CD*, the area of the bending moment diagram in actual state is $\Omega_1 = (1/2)Ph \cdot l$ and corresponding ordinate from unit state is $y_1 = 1 \cdot h$. We assume that horizontal portion *CD* of both bending moment diagrams is located one under the other and vertical portion AC is located one besides the other. For vertical member AC, the area of the bending moment diagram in actual state is $\Omega_2 = (1/2)h \cdot Ph$. Corresponding ordinate from unit bending moment diagram is $y_2 = (2/3) \cdot 1 \cdot h$ (Fig. 6.23a). Using Vereshchagin rule and taking into account the different flexural rigidities for the vertical and horizontal members, we find the required displacement:

$$\Delta_B = \sum \frac{1}{EI}\int_0^s M_p \bar{M}\,ds = \underbrace{-\frac{1}{EI_1} \cdot \underbrace{\frac{1}{2}h \cdot Ph}_{\Omega} \cdot \underbrace{\frac{2}{3} \cdot 1 \cdot h}_{y_c}}_{AC\text{ element}} \underbrace{- \frac{1}{EI_2} \cdot \underbrace{\frac{1}{2}Ph \cdot l}_{\Omega} \cdot \underbrace{1 \cdot h}_{y_c}}_{CD\text{ element}}$$

$$= -\frac{Ph^3}{3EI_1} - \frac{Plh^2}{2EI_2}. \quad \text{(a)}$$

Each term of the expression for horizontal displacement has negative sign, because the bending moment diagrams for actual and unit states are located on different sides of the basic line of the frame. The result of multiplication of diagrams within the vertical member BD equals to zero, since in actual state the bending moments within the member BD are zeros. A final negative sign means that assumed unit force produces a negative work along the real horizontal displacement Δ_B.

(b) *Angle of rotation at B*. The bending moment diagram for actual state is shown in Fig. 6.23b; this diagram for problems (a) and (b) is same. For required displacement θ_B, the unit state presents the same structure with concentrated couple $M = 1$, which acts at point B; direction of the unit couple is chosen in arbitrary way. In the unit state, only vertical reactions $1/l$ arise. The extended fibers are located indoor of the frame. The centroid and area Ω of bending moment diagram in actual state and corresponding ordinate y_c from bending moment diagram for unit state are shown in Fig. 6.23b.

The result of multiplication of diagrams within two vertical members AC and BD equals to zero. Indeed, for these portions the procedure $M_p \times \bar{M} = 0$ because $\bar{M} = 0$ for member AC, and $M_p = 0$ for member BD. For member CD, the Vereshchagin rule leads to the following result

$$\theta_B = \sum \frac{1}{EI}\int_0^s M_p \bar{M}\,ds = \underbrace{\frac{1}{EI_2} \cdot \underbrace{\frac{1}{2}Ph \cdot l}_{\Omega} \cdot \underbrace{\frac{1}{3} \cdot 1}_{y_c}}_{CD\text{ element}} = \frac{Plh}{6EI_2}. \quad \text{(b)}$$

The positive sign is adopted because the bending moment diagrams for actual and unit states are located on one side of the basic line CD of the frame. A final positive sign means that assumed unit couple produce a positive work along the real angular displacement θ_B. Or by other words, the actual direction of angular displacement coincides with the chosen direction for unit couple M, i.e., the section at support B rotates counterclockwise.

Discussion. In actual state, the bending moment along the right vertical bar does not arise; as a result, multiplication of bending moment diagrams for this element for both problems (a) and (b) equals to zero. Therefore, the final result for problems (a) and (b) does not contain a term with stiffness EI_3. This happens because in actual state the element *BD* does not subjected to bending, i.e., this member has displacement as absolutely rigid body. Elastic curve of the frame is shown in Fig. 6.23c.

6.6 Elastic Loads Method

Elastic load method allows *simultaneous* computation of displacements for *set* of points of a structure. This method is based on conjugate beam method. The method especially effective for computation of displacements for set of joints of the truss chord; for trusses this method leads to the precise results.

Elastic loads W are fictitious loads which are applied to the conjugate structure. Bending moments of the conjugate structure and displacements of the real structure at the point of application of elastic loads coincide. A final expression for elastic load at joint n of the truss may be calculated by formula

$$W_n = \sum \frac{\overline{N}_n \cdot N_p \cdot l}{EA} \tag{6.26}$$

This formula uses the following notation: N_p is internal forces due to given load and $\bar{N}_n$ is internal forces in all members of the truss in the unit state.

The right part of the formula (6.26) is similar to formula (6.12), however left part of 6.26 is elastic load, while in Maxwell-Mohr formula - left part is displacement.

Computation of displacements procedure is as follows:

1. Calculate the axial forces N_p in all elements of the truss caused by given load.
2. Calculate the elastic load at a joint n. For this:
   a. Show a fictitious truss. If a real truss is simply supported then the fictitious truss is also simply supported.
   b. Apply two unit couples $M = 1$ to members, which are adjacent to the joint n. Present each couple using forces $F_{n-1} = 1/d_{n-1}$ for span d_{n-1} and $F_n = 1/d_n$ for span d_n, as shown in Fig. 6.24.
   c. Calculate the axial forces $\bar{N}_n$ in all elements of the truss caused by forces in Fig. 6.24.
   d. Calculate the elastic load W_n at the joint n by formula (6.26).
3. Calculate the elastic loads W for remaining joints of the truss chord, as explained in pos. 2.
4. Show the fictitious simply supported beam subjected to all elastic loads W. If the elastic load is positive, then it should be directed downward, i.e., in the same direction as the adjacent forces of neighboring couples.

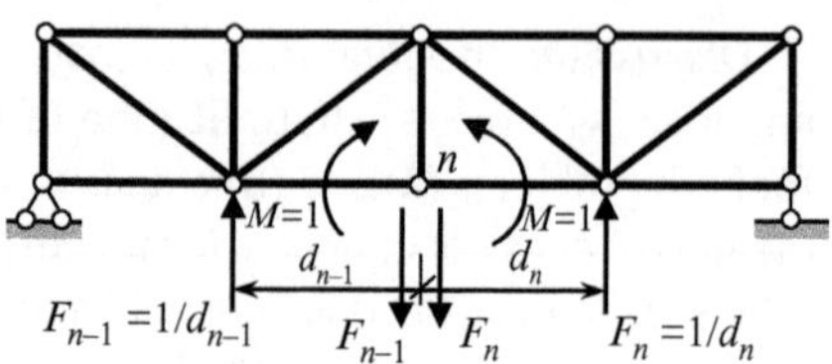

Fig. 6.24 *Unit* state for calculation of elastic load at joint *n*

5. Construct the bending moment diagram for fictitious beam on the tensile fibers. This diagram will present the precise displacements of all joints of the entire real truss.

Example 6.17. Design diagram of the truss is presented in Fig. 6.25a. The truss is carrying two equal forces P. The axial stiffness of the upper and lower chords is $2EA$ and diagonal and vertical members have axial stiffness EA. Compute displacements of the joints of the lower chord.

Solution. Calculation of internal forces in all members of the truss due to applied loads P may be performed by analytical method. Corresponding diagram N_p of internal forces is presented in Fig. 6.25b.

Since it is required to find displacements of the joints of the lower chord, therefore the elastic loads should be applied to the joints of the same chord.

Elastic load W_1 This load is related to joint 1 and represents the mutual angle of rotation of members 0-1 and 1-2. In order to find the elastic load W_1, we need to apply two unit couples (consisting of forces $1/d = 1/4$) to members 0-1 and 1-2 and compute the internal forces in all members of the truss induced by these four forces. Corresponding distribution of internal forces is presented in diagram $\bar{N}_1$ (Fig. 6.25c).

Magnitude of elastic load W_1 is obtained by "multiplication" of two axial force diagrams $\bar{N}_1$ and N_p

$$W_1 = \sum \frac{\bar{N}_1 \cdot N_p \cdot l}{EA}. \quad \text{(a)}$$

The summation should be extended over all members of the truss. For members $0-a$ and $a-2$, internal forces in actual condition are equal $5P/3$ and have opposite signs, while in unit conditions are equal $-5/12$ for both members. Therefore summation within these members equals to zero. Summation within members 0-1and 1-2 of the truss leads to the following result

$$W_1 = \sum \frac{\bar{N}_1 \cdot N_p \cdot l}{EA} = \frac{1}{2EA}\frac{4P}{3} \times \frac{1}{3} \times 4 \times 2 = \frac{64}{36}\frac{P}{EA}. \quad \text{(b)}$$

This procedure explains why the elastic load method is very effective. The four forces present the self-equilibrated set of forces. Therefore the reactions of supports are zero, the members with nonzero internal forces are located within only two panels of the truss, and as a result, multiplication of both axial force diagrams related to only for members which belong to two adjusted panels.

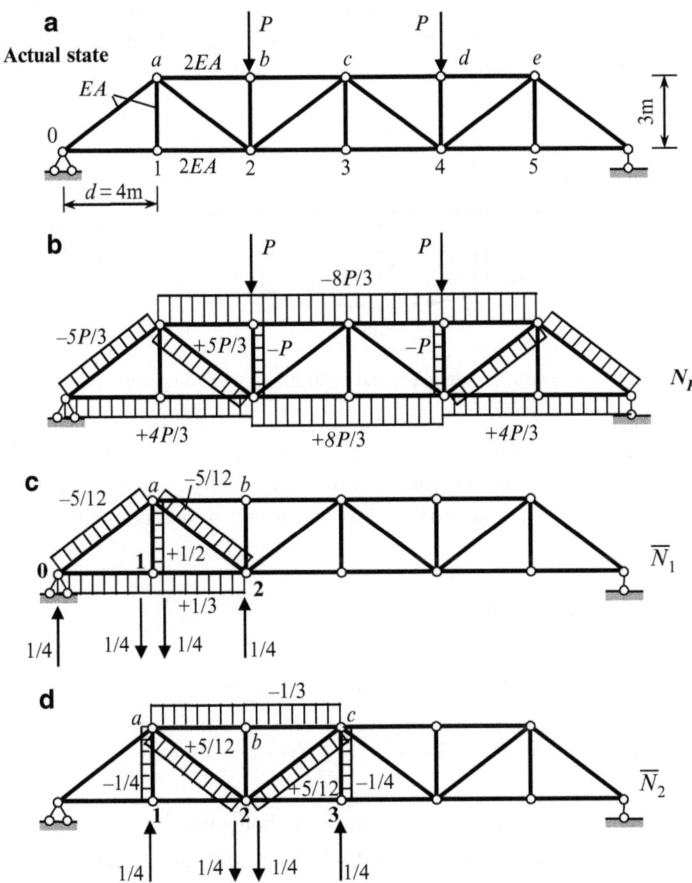

Fig. 6.25 (**a, b**) Design diagram of the truss, and internal forces due to given loads. (**c**) Calculation of elastic load W_1. (**d**) Calculation of elastic load W_2

Elastic load W_2 Two unit couples are applied to members 1-2 and 2-3. Corresponding distribution of internal forces is presented in diagram $\bar{N}_2$ (Fig. 6.25d).

Elastic load W_2 is obtained by "multiplication" of two axial force diagrams $\bar{N}_2$ and N_p. Multiplication of diagrams within only three members ($a-b$, $b-c$, and $a-2$) has nonzero result.

$$W_2 = \sum \frac{\bar{N}_2 \cdot N_p \cdot l}{EA} = \frac{1}{2EA}\frac{8P}{3}\times\frac{1}{3}\times4\times2+\frac{1}{EA}\frac{5P}{3}\times\frac{5}{12}\times5 = \frac{253}{36}\frac{P}{EA}. \qquad \text{(c)}$$

Elastic load W_3 Two unit couples are applied to members 2-3 and 3-4. Performing similar procedure we get

$$W_3 = \frac{128}{36}\frac{P}{EA}.$$

A fictitious beam is a simply supported one, which is loaded at joints 1,...,5 by elastic loads W_i, $i = 1, \ldots, 5$. The elastic loads W_i are positive, so they must be directed downward (Fig. 6.26).

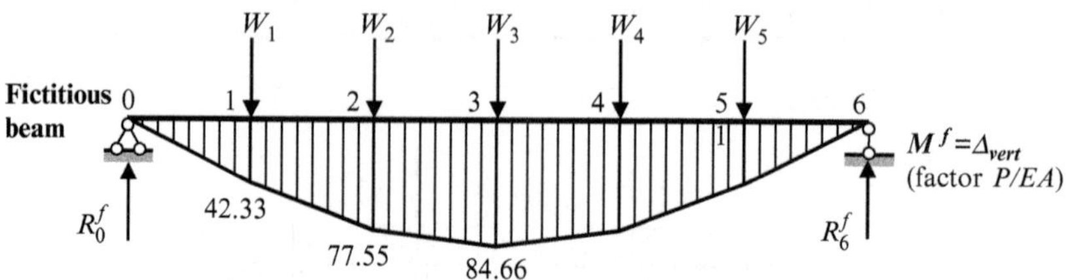

Fig. 6.26 Fictitious beam with elastic loads and bending moment diagram

Since the truss under consideration is symmetrical, then $W_1 = W_5$ and $W_2 = W_4$. Reactions of supports of the fictitious beam are $R_0^{\mathrm{f}} = R_6^{\mathrm{f}} = (381/36) \times (P/EA)$.

Bending moments at the specific points of the fictitious beam caused by elastic loads are

$$\begin{aligned} M_1^{\mathrm{f}} &= R_0^{\mathrm{f}} \cdot 4 = 42.33\frac{P}{EA} \\ M_2^{\mathrm{f}} &= R_0^{\mathrm{f}} \cdot 8 - W_1 \cdot 4 = 77.55\frac{P}{EA} \\ M_3^{\mathrm{f}} &= R_0^{\mathrm{f}} \cdot 12 - W_1 \cdot 8 - W_3 \cdot 4 = 84.66\frac{P}{EA} \end{aligned} \tag{d}$$

Ordinates of the bending moment diagram of the fictitious beam present the vertical displacement of the joints of the lower chord of the truss, $M^{\mathrm{f}} = \Delta_{\mathrm{vert}}$.

$$\Delta_1 = 42.33\frac{P}{EA}, \quad \Delta_2 = 77.55\frac{P}{EA}, \quad \Delta_3 = 84.66\frac{P}{EA}. \tag{e}$$

Discussion. For calculation of displacement at joint 1 by the Maxwell–Mohr integral, it is necessary to apply unit force at this joint and calculate all internal forces (for given truss 21 forces) and then calculate 21 standard term $(\bar{N} \cdot N_p \cdot l)/EA$. For calculation of displacements of joints 1, 2, and 3 the total number of unknown forces in the unit states is $3 \times 21 = 63$ and the number of standard terms according to Maxwell–Mohr integral is 63.

Application of elastic loads to each joint of a truss leads to appearance of internal forces only in the members of two adjacent panels. As a result, for given truss the total number of unknown forces in the all unit states is 16 and the number of terms for elastic loads, according to (6.26), equals 4.

Summary

The elastic load method is based on conjugate beam method. According to this method, displacement of any point of a real structure is proportional to bending moment at the same point of fictitious beam (factor 1/*EI*). Based on this, the elastic load method uses external resemblance of displacements of elastic structure and bending moment diagrams.

Elastic load at joint n presents the mutual angle of rotation of two members of the truss chord, which is adjacent to joint n.

Elastic load method presents very effective way for computation of displacements for *set* of the joints of a truss chord. For trusses this method leads to the precise results. The advantage of the method is that the procedure (6.26) is not related to all members of a truss, but only to small subset of the members.

Elastic load method is also applicable for beams and arches, but for such structures this method is approximate and cumbersome. However, it has advantages over exact methods (for example, Initial Parameters method); this fact can be seen in case of beam with variable cross section.

6.7 Reciprocal Theorems

Reciprocal theorems reflect fundamental properties of any linear statical determinate or indeterminate elastic systems. These theorems will be extensively used for analysis of statically indeterminate structures. Primary investigations were performed by Betti (1872), Maxwell (1864), Lord Rayleigh (1873–1875), Castigliano (1872), and Helmholtz (1886).

6.7.1 Theorem of Reciprocal Works (Betti Theorem)

Let us consider elastic structure subjected to loads P_1 and P_2 separately; let us call it as first and second states (Fig. 6.27). Set of displacements Δ_{mn} for each state are shown below. The first index m indicates the direction of the displacement and the second index n denotes the load, which causes this displacement. Thus

Δ_{11} and Δ_{12} are displacements in the direction of load P_1 due to load P_1 and P_2, respectively

Δ_{21} and Δ_{22} are displacements in the direction of load P_2 due to load P_1 and P_2, respectively.

Let us calculate the strain energy of the system by considering consequent applications of loads P_1 and P_2, i.e., state 1 is *additionally* subjected to load P_2. Total work done by both of these loads consists of three parts:

1. Work done by the force P_1 on the displacement Δ_{11}. Since load P_1 is applied *statically* (from zero to P_1 according to triangle law), then $W_1 = (1/2)P_1\Delta_{11}$.

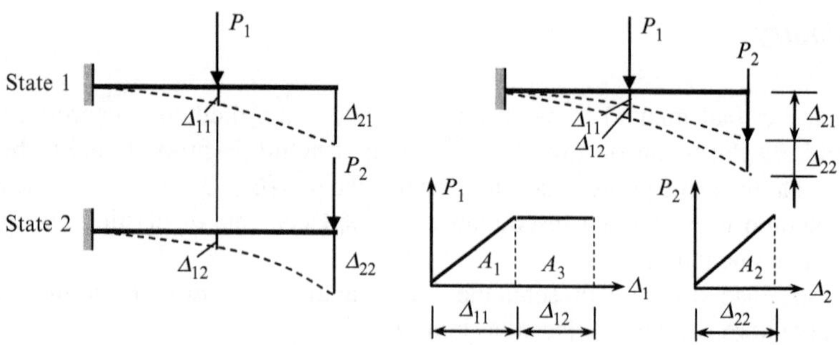

Fig. 6.27 Two state of the elastic structure. Computation of work done by the load P_1 and additional load P_2

2. Work done by the force P_2 on the displacement Δ_{22}. Since load P_2 is applied statically, then $W_2 = (1/2)P_2\Delta_{22}$.
3. Work done by the force P_1 on the displacement Δ_{12}; this displacement is caused by load P_2. The load P_1 approached its maximum value P_1 early (before application of P_2). Corresponding P_1–Δ_1 diagram is shown in Fig. 6.27, so $W_3 = P_1\Delta_{12}$.

Since potential energy U equals to the total work, then

$$U = \frac{1}{2}P_1\Delta_{11} + \frac{1}{2}P_2\Delta_{22} + P_1\Delta_{12}. \tag{6.27}$$

On the other hand, considering of application of load P_2 first and then P_1, i.e., if state 2 is additionally subjected to load P_1, then potential energy U equals

$$U = \frac{1}{2}P_2\Delta_{22} + \frac{1}{2}P_1\Delta_{11} + P_2\Delta_{21}. \tag{6.28}$$

Since strain energy does not depend on the order of loading, then the following fundamental relationship is obtained

$$P_1\Delta_{12} = P_2\Delta_{21} \quad \text{or} \quad W_{12} = W_{21}. \tag{6.29}$$

The theorem of reciprocal works (6.29) said that **in any elastic system the work performed by load of state 1 along displacement caused by load of state 2 equals to work performed by load of state 2 along displacement caused by load of state 1.**

6.7.2 Theorem of Reciprocal Unit Displacements (Maxwell Theorem)

Let us consider two states of elastic structure subjected to *unit loads* $P_1 = 1$ and $P_2 = 1$. Displacement caused by unit loads is called the *unit displacements* and denoted by letter δ_{mn}. The first index m indicates the direction of the displacement and the second index n denotes the unit load, which causes this displacement.

Thus, δ_{11} and δ_{12} are displacements in the direction of load P_1 due to load $P_1 = 1$ and $P_2 = 1$, respectively;

δ_{21} and δ_{22} are displacements in the direction of load P_2 due to load $P_1 = 1$ and $P_2 = 1$, respectively.

In case of unit loads, the theorem of reciprocal works $P_2\Delta_{21} = P_1\Delta_{12}$ leads to the following fundamental relationship $\delta_{12} = \delta_{21}$. In general,

$$\delta_{nm} = \delta_{mn}. \tag{6.30}$$

This equation shows that **in any elastic system, displacement along *n*th load caused by unit *m*th load equals to displacement along *m*th load caused by unit *n*th load**. The term "displacement" refers to linear or angular displacements, and the term "load" means force or moment.

This theorem is demonstrated by the following example. Simply supported beam is subjected to unit load P in the first condition and unit moment M in the second condition (Fig. 6.28). Displacements δ_{11} and δ_{12} are linear displacements along force P in the first and second states; displacements δ_{21} and δ_{22} are angular displacements along moment M in the first and second states.

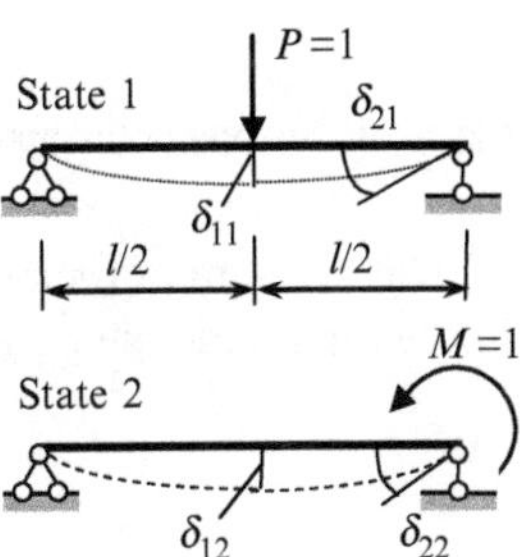

Fig. 6.28 Theorem of reciprocal unit displacements

In the first state, displacement due to load $P = 1$ along the load of the state 2 is

$$\delta_{21} = \theta = \frac{1 \cdot l^2}{16EI}.$$

In the second state, displacement due to load $M = 1$ along the load of the state 1 is

$$\delta_{12} = y = \frac{1 \cdot l^2}{16EI}.$$

Theorem of reciprocal displacements will be widely used for analysis of statically indeterminate structures by the Force method. Theorem of reciprocal unit displacements was proved by Maxwell (1864) before the more general Betti theorem; however, Maxwell's proof was unoticed by scientists and engineers. Mohr proved this theorem in 1864 independently from Betti and Maxwell.

6.7.3 *Theorem of Reciprocal Unit Reactions (Rayleigh First Theorem)*

Let us consider two states of elastic structure subjected to *unit displacements* of supports. They are $Z_1 = 1$ and $Z_2 = 1$ (Fig. 6.29a). Reactions caused by unit displacements are called the *unit reactions* and denoted by letter r_{mn}. The first index m indicates constrain where unit reaction arises and the second index n denotes constrain, which is subjected to unit displacement.

Thus r_{11} and r_{12} are reactions in the constrain 1 due to displacement $Z_1 = 1$ and $Z_2 = 1$, respectively and r_{21} and r_{22} are reactions in the constrain 2 due to displacement $Z_1 = 1$ and $Z_2 = 1$, respectively.

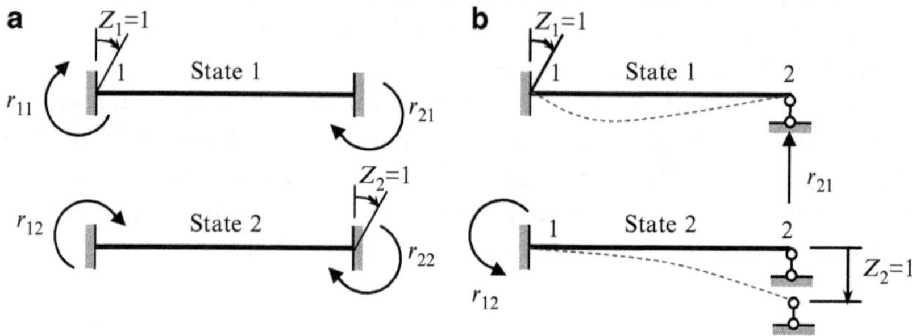

Fig. 6.29 Theorem of reciprocal unit reactions

The theorem of reciprocal works $r_{11} \cdot 0 + r_{21} \cdot 1 = r_{12} \cdot 1 + r_{22} \cdot 0$ leads to the following relationship $r_{21} = r_{12}$. In general,

$$r_{nm} = r_{mn}. \tag{6.31}$$

The theorem of reciprocal reactions said that **in any elastic system reaction r_{nm}, which arises in nth constrain due to unit displacement of constrain m, equals reaction r_{mn}, which arises in mth constrain due to unit displacement of constrain n.**

This is demonstrated by the following example (Fig. 6.29b). Unit displacements of the clamped-pinned beam are $Z_1 = 1$ is a unit angle of rotation of the clamped support and $Z_2 = 1$ is a vertical linear displacement of the pinned support. Unit reactions are as follows: r_{21} is vertical reaction in constrain 2 caused by unit angular displacement of support 1 and r_{12} is moment in constrain 1 caused by unit vertical linear displacement of support 2.

Using Table A.3 (pos. 1 and 2), unit reactions may be written as $r_{21}=r_{12}=3EI/l^2$. Theorem of reciprocal reactions will be widely used for analysis of statically indeterminate structures by the displacement method.

6.7.4 Theorem of Reciprocal Unit Displacements and Reactions (Rayleigh Second Theorem)

Let us consider two states of elastic structure subjected to *unit displacement* $Z_1 = 1$ and *unit load* $P_2 = 1$ (Fig. 6.30a). Reaction r_{12} arises in constrain 1 due to unit load P_2. Displacement δ_{21} occurs in direction of load P_2 due to unit displacement Z_1.

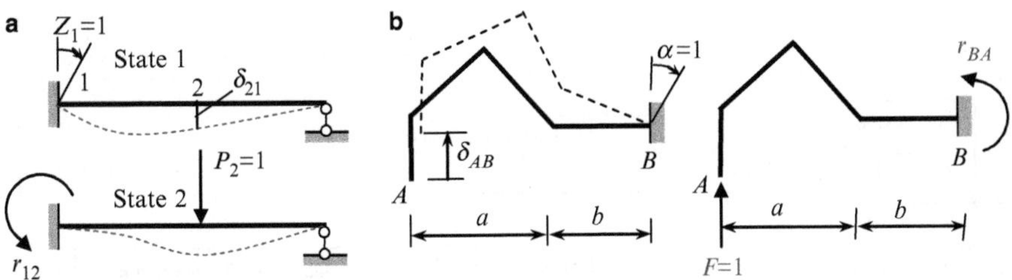

Fig. 6.30 Theorem of reciprocal of unit displacements and reactions

The theorem of reciprocal work in extended form should be presented as follows $-r_{12} \cdot 1 = 1 \cdot \delta_{21}$ so we get that $-r_{12} = \delta_{21}$. In general,

$$r_{jk} = -\delta_{kj}. \tag{6.32}$$

The theorem of reciprocal unit displacements and reactions said that **reaction in *j*th constrain due to unit load of *k*th direction and displacement in *k*th direction due to unit displacement of *j*th constrain are equal in magnitude but opposite in sign**.

This theorem is illustrated in Fig. 6.30b. In order to find a vertical displacement at the point A due to unit rotation of the support B, apply unit force $F = 1$ in the direction δ_{AB}. Moment at fixed support due to force $F = 1$ is $r_{BA} = -F(a + b)$. Since $F = 1$, therefore the vertical displacement is $\delta_{AB} = a + b$.

Theorem of reciprocal reactions and displacements will be used for analysis of statically indeterminate structures by mixed method. In general form, this theorem was considered by Rayleigh (1873–1875). The form (6.32) was presented by Prof. A.A. Gvozdev (1927).

6.7.5 Summary

There are two principle approaches to computation of displacements. The first of them is based on the integration of differential equation of an elastic curve of a beam. The second approach is based on the fundamental energetic principles. Relationships between different methods of calculation displacement and their evolution are presented in Fig. 6.31.

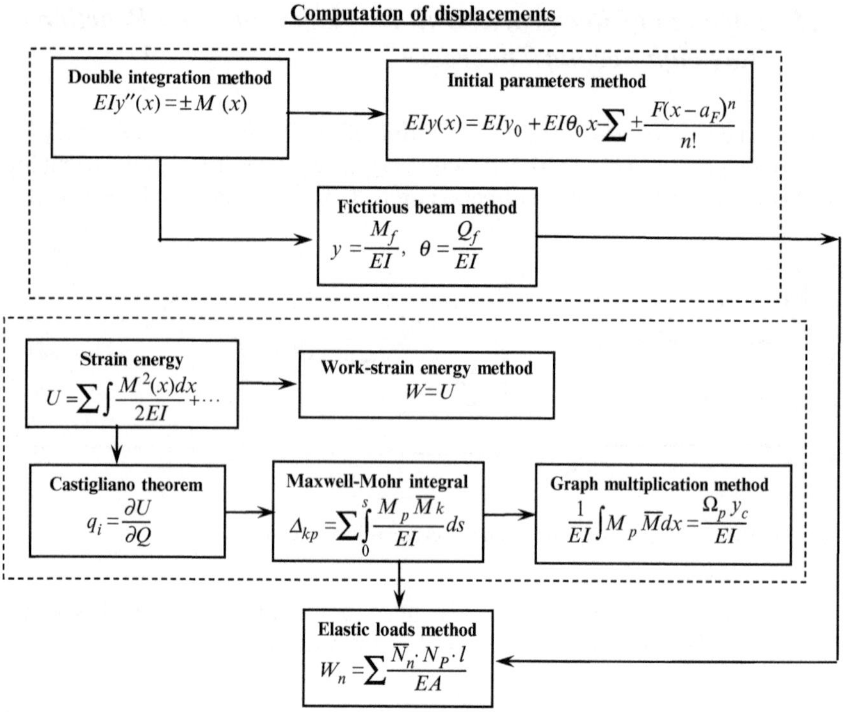

Computation of displacements

Double integration method $EIy''(x) = \pm M(x)$

Initial parameters method $EIy(x) = EIy_0 + EI\theta_0 x - \sum \pm \frac{F(x-a_F)^n}{n!}$

Fictitious beam method $y = \frac{M_f}{EI}$, $\theta = \frac{Q_f}{EI}$

Strain energy $U = \sum \int \frac{M^2(x)dx}{2EI} + \cdots$

Work-strain energy method $W = U$

Castigliano theorem $q_i = \frac{\partial U}{\partial Q}$

Maxwell-Mohr integral $\Delta_{kp} = \sum \int_0^s \frac{M_p \overline{M}_k}{EI} ds$

Graph multiplication method $\frac{1}{EI} \int M_p \overline{M} dx = \frac{\Omega_p y_c}{EI}$

Elastic loads method $W_n = \sum \frac{\overline{N}_n \cdot N_P \cdot l}{EA}$

Fig. 6.31 Two fundamental approaches to calculation of displacements of elastic structures

Group 1 This group contains the double integration method, initial parameters method, and conjugate beam method (fictitious beam method). All methods of this group are based on the differential equation of the elastic curve of the beam.

- Double integration of differential equation allows finding the equation of an elastic curve of a beam. Integration procedure should be performed for each specified problem. This procedure leads to appearance of the constants, which should be determined from the boundary conditions. The number of the constants is twice more than the number of portions of a beam. For beams with two or more portions, this method becomes very cumbersome.
- Initial parameters method is a modification of double integration method. This method is effective for deriving equation of elastic curve for uniform beams with large number of portions and any types of loads. The integrating procedure is performed once at the deriving of universal equation and therefore, practical application of this method reduces only to the algebraic procedures.
- Conjugate beam method also presents the modification of the double integration method and allows computing the linear and angular displacements at specified section of the beam. This method required constructing the fictitious beam and computing the fictitious bending moment and shear at specified section. In case of complex loading, this method leads to the cumbersome computation.

Group 2 This group presents methods, which are based on the concept of the strain energy. The following precise analytical methods are presented in the second group: strain energy method (Clapeyron and Castigliano theorems), Maxwell–Mohr integral (dummy load method), and Vereshchagin rule (graph multiplication method). All methods of this group use a concept of generalized force and corresponding generalized coordinate.

- Work–strain energy method allows calculating displacement at specified points. Even if a numerical procedure of this method is very simple, the area of application of this method is limited.
- Castigliano theorem allows calculating any displacement at specified direction as a partial derivative of the strain energy with respect to generalized force. This theorem has a fundamental character.
- Maxwell–Mohr integral presents the principal formula for computation of displacement at any specified direction. This formula is presented in terms of internal forces caused by a given load (or change of temperature) and unit generalized force, which corresponds to required generalized coordinate. In general, application of this method reduced to integration procedure.
- The graph multiplication method is the modification of the Maxwell–Mohr integral and presents extremely convenient procedure for computation of displacements at specified points for bending structures. This approach allows us to avoid the integration procedure and requires plotting only two bending moment diagrams due to given loads and unit load. After that simple algebraic procedures over them should be performed. In fact, this is the most effective method for calculation of displacement of any nonuniform beams and frames. For trusses, the graph multiplication method coincides with Maxwell–Mohr integral.

The graph multiplication method is so much simple and effective that it is hard to expect that new methods for calculation of displacements of elastic structures may be developed.

- Elastic load method presents the combination of conjugate beam method and Maxwell–Mohr integral. This method allows calculating displacements at the *set of points* simultaneously. The method is especially effective for computation of displacements of the joints of a truss.

Problems

Problems 6.1 through 6.8 should be solved by initial parameter method. The flexural rigidity, EI, is constant for each beam; the span of the beam is l.

6.1. The cantilevered beam is subjected to uniformly varying load as shown in Fig. P6.1. The maximum ordinate of load is q. Derive an equation for the elastic curve of the beam. Determine the vertical displacement and slope at the free end.

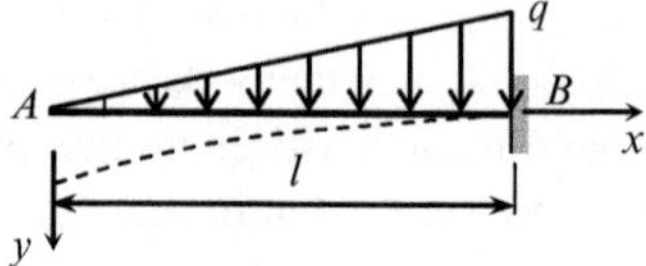

Fig. P6.1

$$\text{Ans. } y(x) = \frac{ql^4}{30EI}\left(1 - \frac{5}{4}\frac{x}{l} + \frac{x^5}{4l^5}\right), \quad y_A = \frac{ql^4}{30EI}, \quad \theta_A = -\frac{ql^3}{24EI}.$$

6.2. The uniform beam is subjected to distributed load q within the portion $a = \xi l$, as shown in Fig. P6.2. Derive an equation for elastic curve of the beam.

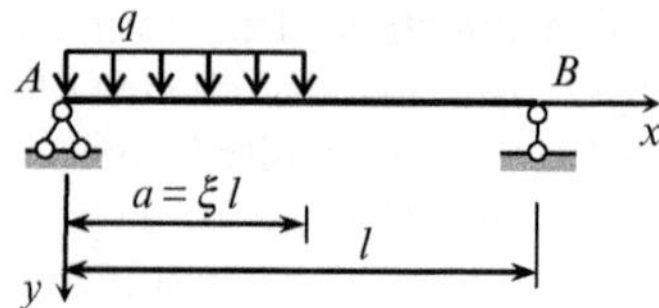

Fig. P6.2

$$\text{Ans. } \theta_A = \frac{qa^2 l}{6EI}(1 - 0.5\xi)^2.$$

6.3. The uniform simply supported beam is subjected to uniformly varying load (Fig. P6.3). The maximum ordinate of load is q. Derive an equation of the elastic curve. Determine the maximum vertical displacement, its location, and slope at the supports A and B.

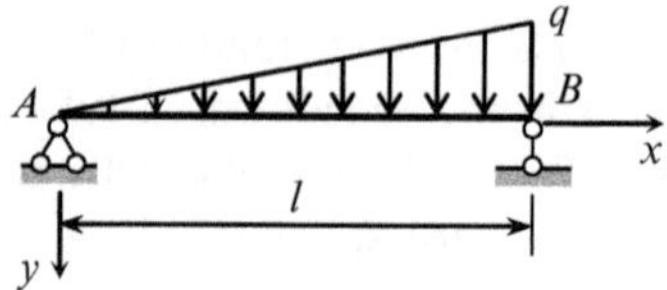

Fig. P6.3

$$\text{Ans. } y_x = \frac{ql^4}{360EI}\left(7\frac{x}{l} - 10\frac{x^3}{l^3} + 3\frac{x^5}{l^5}\right), \quad y_{\max} = \frac{5}{768}\frac{ql^4}{EI},$$

$$\theta_A = \frac{7}{360}\frac{ql^3}{EI}, \quad \theta_B = -\frac{8}{360}\frac{ql^3}{EI}.$$

6.4. Fixed-sliding beam A–B is subjected to action of concentrated force P (Fig. P6.4). Derive an equation of the elastic curve for the beam. Determine the displacement of the beam at the right end.

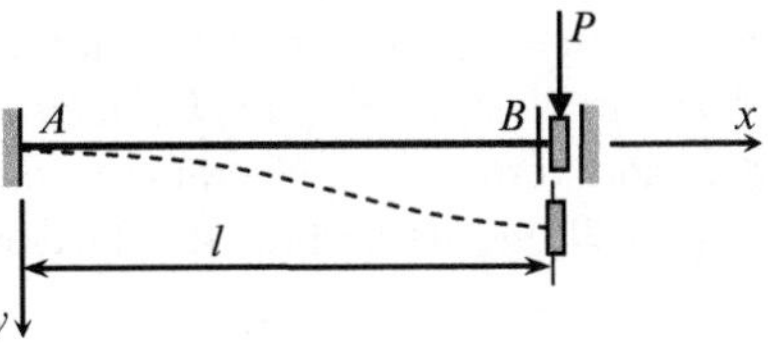

Fig. P6.4

$$\text{Ans. } EI\,y(x) = \frac{Pl^3}{12}\left(3\frac{x^2}{l^2} - 2\frac{x^3}{l^3}\right); \quad EI\,y(l) = \frac{Pl^3}{12}.$$

6.5. A beam pinned with torsion spring at the left end and pinned at the right end is subjected to uniformly distributed load q. Torsion stiffness parameter equals k_{rot}. Calculate the reaction R_A and slope θ_0. Derive an equation of the elastic curve. The reactive moment and slope at A are related by expression $M_A = k_{\text{rot}}\theta(0)$. Consider limiting cases $k_{\text{rot}} = 0$, and $k_{\text{rot}} = \infty$. Find the range of R_A for any stiffness k_{rot}.

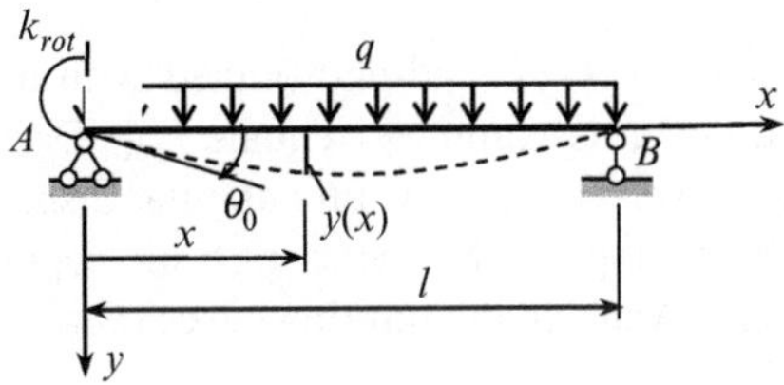

Fig. P6.5

$$\text{Ans. } R_A = \frac{ql}{2}\frac{1 + \dfrac{5}{12}\dfrac{k_{\text{rot}}l}{EI}}{1 + \dfrac{k_{\text{rot}}l}{3EI}}, \quad \theta_0 = R_A\frac{l}{k_{\text{rot}}} - \frac{ql^2}{2k_{\text{rot}}},$$

$0.5ql \le R_A \le 0.625ql$ for any stiffness k_{rot}.

6.6. The fixed-pinned beam is subjected to uniformly distributed load q (Fig. P6.6). Derive an equation of elastic curve of the beam

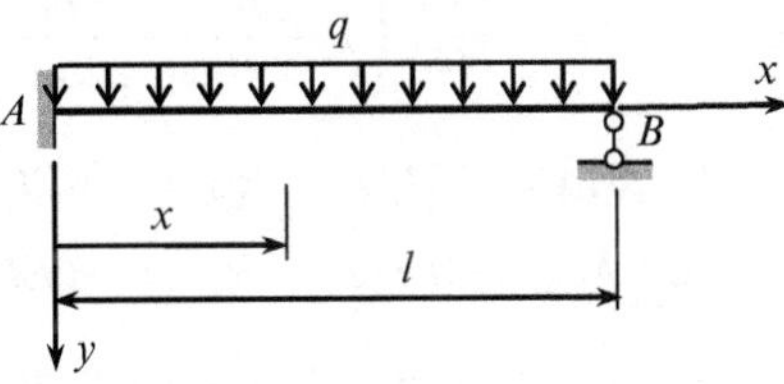

Fig. P6.6

$$\text{Ans. } EI\,y(x) = \frac{ql^2}{8}\frac{x^2}{2} - \frac{5}{8}ql\frac{x^3}{6} + \frac{qx^4}{24}.$$

6.7. The uniform beam is subjected to vertical displacement Δ of support A as shown in Fig. P6.7. Stiffness of elastic support is k. Determine the reactive moment at the clamped support A and vertical displacement at the elastic support B. Reaction R_B of elastic support B and its vertical displacement Δ_B satisfy equation $R_B = k\Delta_B$.

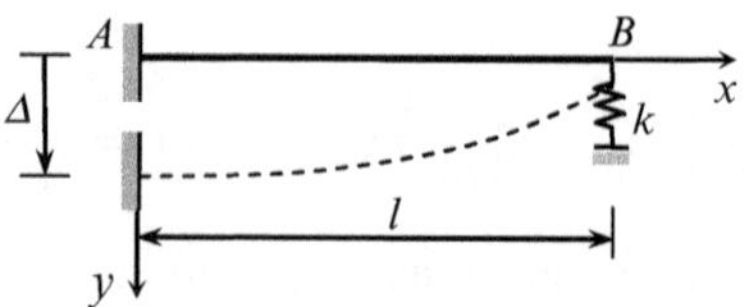

Fig. P6.7

$$\text{Ans. } M = \frac{3EI}{l^2} \cdot \frac{\Delta}{1 + \frac{3EI}{kl^3}}, \quad y(l) = \frac{3EI}{kl^3} \cdot \frac{\Delta}{1 + \frac{3EI}{kl^3}}.$$

6.8. A beam is clamped at the left end and pinned with torsional spring support at right end. Torsional stiffness parameter equals k_{rot}. Derive an equation of the elastic curve for the beam subjected to the unit angular displacement at the left end. Calculate the reactions of supports. Moment at elastic support B and stiffness are related as $M_B = \theta(l)k_{\text{rot}}$. Analyze the limiting cases ($k_{\text{rot}} = 0,\ k_{\text{rot}} = \infty$).

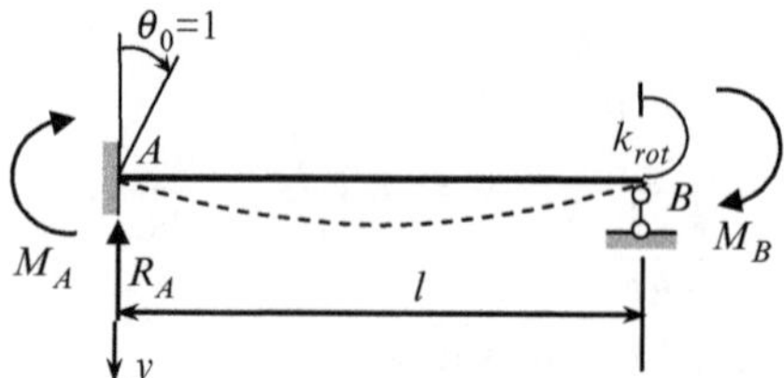

Fig. P6.8

$$\text{Ans. } M_A = \frac{3i + k_{\text{rot}}}{1 + \frac{1}{4}\frac{k_{\text{rot}}}{i}}, \quad R_A = \frac{6EI}{l^2}\left(1 - \frac{1}{2} \cdot \frac{3i + k_{\text{rot}}}{i + \frac{k_{\text{rot}}}{4}}\right), \quad i = \frac{EI}{l}.$$

Limiting cases: $k_{\text{rot}} = 0: \ M_A = 3EI/l; \quad R_A = -3EI/l^2;$

$k_{\text{rot}} = \infty: \ M_A = 4EI/l; \quad R_A = -6EI/l^2.$

Problems 6.9 through 6.16 should be solved by Maxwell–Mohr integral.

6.9. The simply supported beam AB carrying uniformly distributed load q and concentrated force P applied at midspan point C (Fig. P6.9). Determine the vertical displacement at $x = l/2$ and slope at support A.

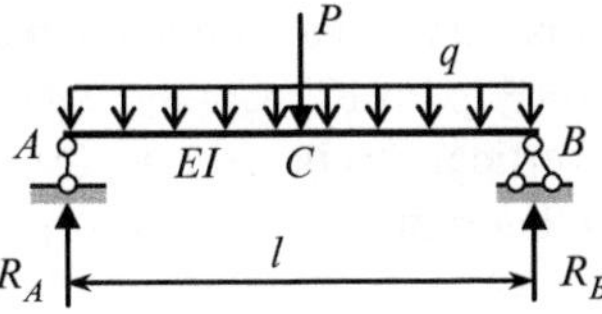

Fig. P6.9

Ans. $y_c = \frac{Pl^3}{48EI} + \frac{5}{384}\frac{ql^4}{EI}; \quad \theta_A = \frac{ql^3}{24EI} + \frac{Pl^2}{16EI}$ (clockwise direction).

6.10. The truss in Fig. P6.10 supports the concentrated force P, which is applied at joint 3. Axial rigidity for diagonals and vertical elements equal EA, for lower and top chords equal $2EA$. Calculate the relative displacement of joints 4 and 7 along the line joining them and the angle of rotation of the element 5–7.

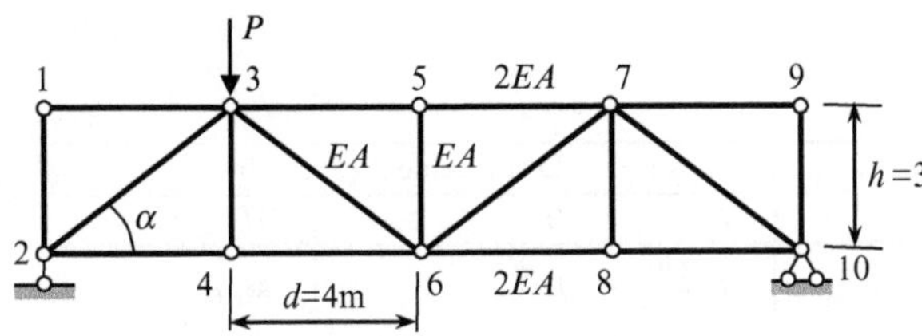

Fig. P6.10

Ans. $\Delta_{4-7} = -3.054\frac{P}{EA}$ (apart); $\quad \theta_{57} = 1.2014\frac{P}{EA}$ (counter clockwise).

6.11. Semicircular simply supported bar of radius R with uniform cross section (EI is a flexural stiffness) is presented in Fig. P6.11. Determine the horizontal displacement of support B.

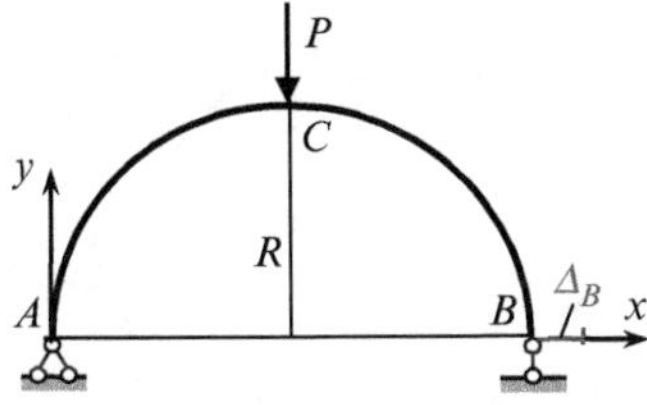

Fig. P6.11

Ans. $\Delta_B = -\frac{PR^3}{2EI}$.

6.12. A uniform circular bar is clamped at point B and carrying the couple M_0, the forces P and H at free end A (Fig. P6.12). Compute the following displacements at the free end A ($\varphi = 0$): (a) vertical; (b) horizontal; (c) angle of rotation. Check the reciprocal unit displacement theorem.

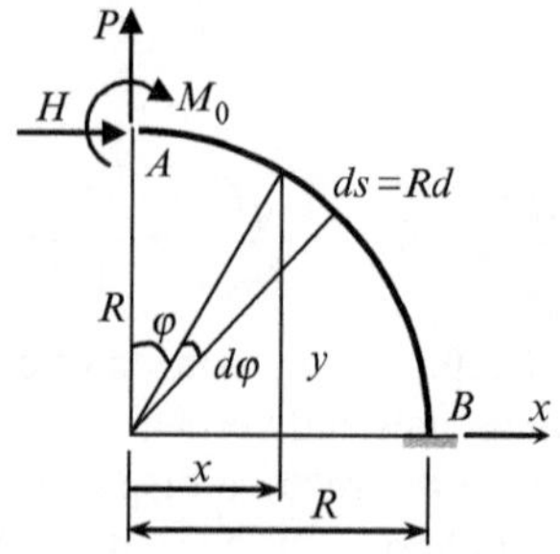

Fig. P6.12

Ans.

	Displacement caused by		
	P	H	M_0
$EI\,\Delta_{vert}$	$P\dfrac{\pi R^3}{4}$	$H\dfrac{R^3}{2}$	$M_0 R^2$
$EI\,\Delta_{hor}$	$P\dfrac{R^3}{2}$	$HR^3\left(\dfrac{3\pi}{4}-2\right)$	$M_0 R^2\left(\dfrac{\pi}{2}-1\right)$
$EI\,\theta$	PR^2	$HR^2\left(\dfrac{\pi}{2}-1\right)$	$M_0\dfrac{\pi R}{2}$

6.13. Three-hinged uniform semicircular arch of radius R carrying the concentrated force P at point C (Fig. P6.13). The flexural stiffness of the arch is EI. Calculate the vertical displacement of the hinge C.

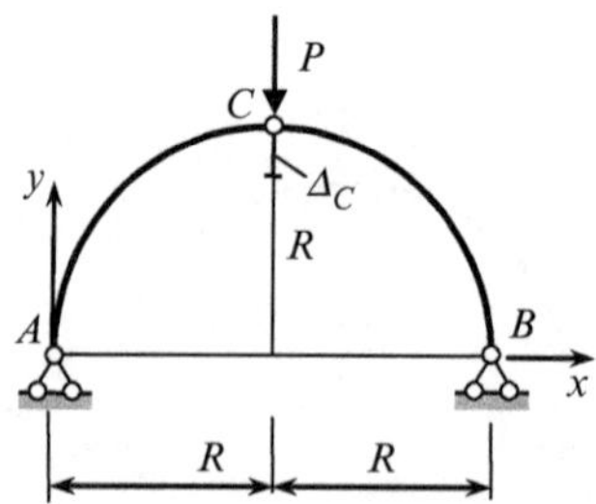

Fig. P6.13

Ans. $\Delta_C = \dfrac{PR^3}{2EI}(\pi - 3)$.

6.14. A circular bar with central angle 80° is clamped at point A and free at point B. The bar is subjected to horizontal force P at the free end B (Fig. P6.14). The area A of cross section of the bar and moment of inertia I are constant. Calculate the horizontal displacement Δ_B at point B. All the three terms of Maxwell–Mohr's integral should be used. Estimate each term for the following data: cross section is rectangular ($h = 2b$), $h/R = 0.1$, the shear modulus $G = E/2(1 + \nu)$, the Poisson's coefficient $\nu = 0.25$ and coefficient $\mu = 1.2$.

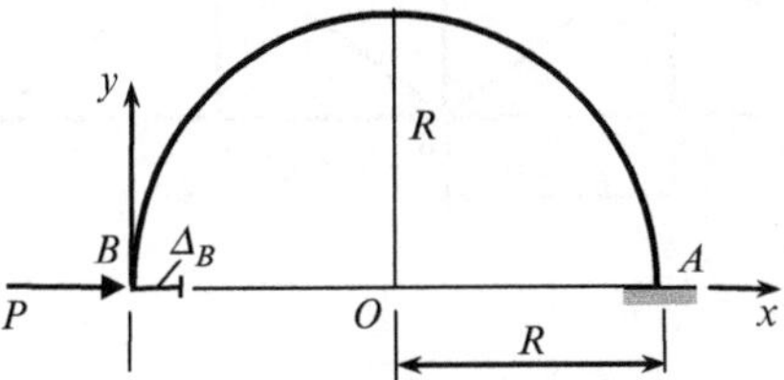

Fig. P6.14

$$\text{Ans. } \Delta_B = \frac{PR^3}{EI}\frac{\pi}{2} + \frac{PR}{EA}\frac{\pi}{2} + \frac{\mu PR}{GA}\frac{\pi}{2}.$$

6.15. Find the horizontal and angular displacement at support C of the frame shown in Fig. P6.15, when the indoor temperature rises by 10°C and outdoor temperature rises by 30°C and 20°C for elements AB and BC, respectively. The height and temperature coefficients of the elements BC and AB are b_1, α_1 and b_2, α_2, respectively.

Fig. P6.15

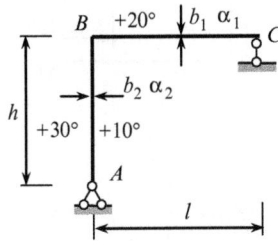

$$\text{Ans. } \Delta_{Ct} = 5\alpha_1 l\left(\frac{h}{b_1} + 3\right) + 10\alpha_2\frac{h^2}{b_2}\left(1 + 2\frac{b_2}{l}\right) \ (\rightarrow),$$

$$\theta_C = 5\alpha_1\frac{l}{b_1} + 20\alpha_2\frac{h}{l} \quad \text{(clockwise)}.$$

6.16. Design diagram of the truss is shown in Fig. P6.16. Temperature of the top chord of the truss decreases by 30°C, and of the bottom chord increases by +45°C; the temperature of diagonals and vertical elements remain constant. The coefficient of thermal expansion of material is α. Compute the

(a) Vertical displacement of joint 4
(b) Mutual (relative) displacement of joints 2 and 3
(c) Angle of rotation of the bar 1–3
(d) Mutual angle of rotation of the bars 1–3 and 3–5

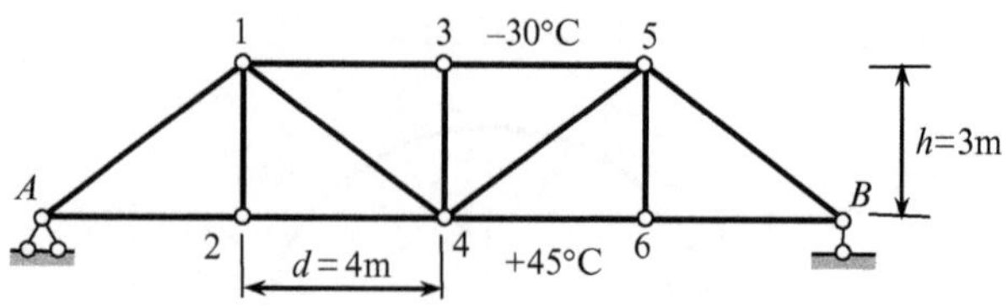

Fig. P6.16

Ans. (a) $\Delta_{4t} = 800\alpha$ (m)(↓); (b) $\Delta_{23} = 321\alpha$(m);
(c)$\theta_{13} = 80\alpha$ (radian) (clockwise); (d)$\theta_{13-35} = 60\alpha$ (rad).

Problems 6.17 through 6.28 are to be solved by graph multiplication method.

6.17. Cantilevered nonuniform beam is subjected to concentrated load P (Fig. P6.17). Calculate the vertical displacement of free end.

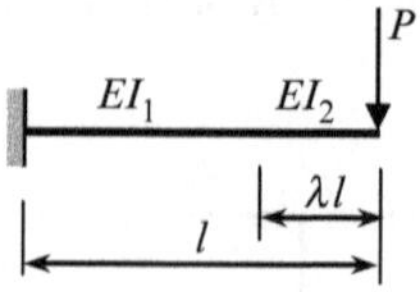

Fig. P6.17

Ans. $y = (1 + \eta\lambda^3)\dfrac{Pl^3}{3EI_1}, \quad \eta = \dfrac{1}{n} - 1, \quad n = \dfrac{I_2}{I_1}$

6.18. A nonuniform cantilevered beam AB carrying the distributed load q (Fig. P6.18). Bending stiffness equals EI on the segment $CB = 0.5l$ and kEI on the segment $AC = 0.5l$; parameter k is any positive number. Calculate the vertical displacement Δ_A at the free end. Analyze the limiting case.

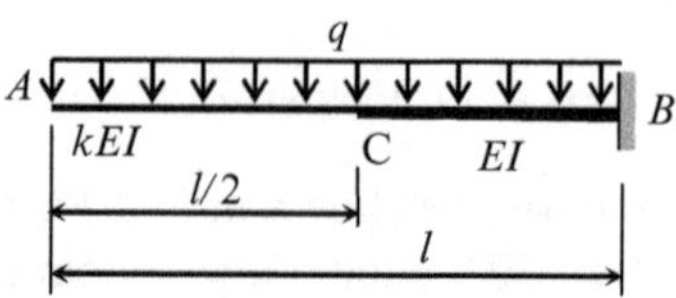

Fig. P6.18

$$\text{Ans. } \Delta_A = \frac{1}{k}\frac{ql^4}{128EI} + \frac{15}{128}\frac{ql^4}{EI} = \frac{ql^4}{8EI}\eta, \quad \eta = \frac{1}{16k} + \frac{15}{16}.$$

6.19. Uniform simply supported beam is subjected to concentrated force P as shown in Fig. P6.19. Determine the vertical displacement at point C and slope at supports A and B.

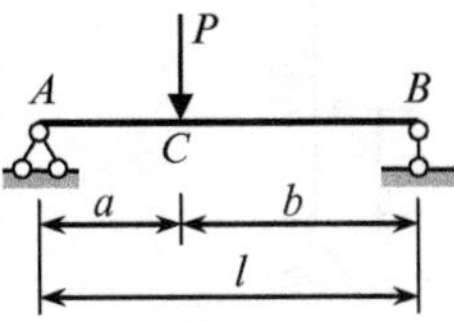

Fig. P6.19

$$\text{Ans. } y_{\text{c}} = \frac{Pa^2b^2}{3EI\,l}, \quad \theta_A = \frac{Pab(l+b)}{6EI\,l}, \quad \theta_B = \frac{Pab(l+a)}{6EI\,l}.$$

6.20. Two span statically determinate beam is subjected to uniformly distributed load $q = 3\text{kN/m}$. Parameters of the beam are $l_1 = 6\,\text{m}$, $l_2 = 5\,\text{m}$, $a = 2\,\text{m}$ (Fig. P6.20). Determine the vertical displacement at point C and slope at supports A and D.

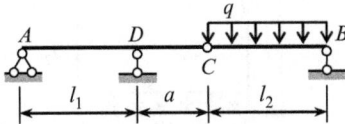

Fig. P6.20

$$\text{Ans. } y_{\text{c}} = \frac{80}{EI}(\text{down}), \quad \theta_A = \frac{15}{EI}(\text{counter clockwise}), \quad \theta_D = \frac{30}{EI}(\text{clockwise}).$$

6.21. Design diagram of the frame is presented in Fig. P6.21. The bending stiffness for all members is EI. Determine the horizontal displacement at points A and B.

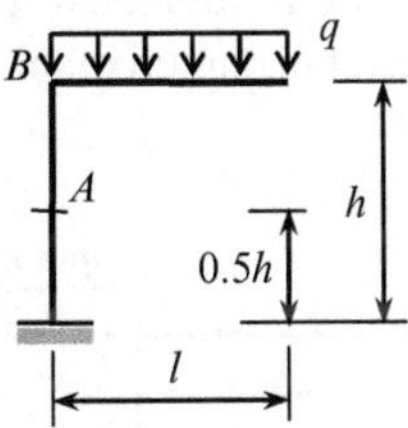

Fig. P6.21

$$\text{Ans. } \Delta_A^{\text{hor}} = \frac{ql^2h^2}{16EI}(\text{right}), \quad \Delta_B^{\text{hor}} = \frac{ql^2h^2}{4EI}(\text{right}).$$

6.22. The frame is subjected to horizontal force P acting at point K (Fig. P6.22). The numbers 1 and 2 in the circles show the relative stiffness, which mean the bending stiffness $1EI$ and $2EI$, respectively. Determine the horizontal displacement of support C.

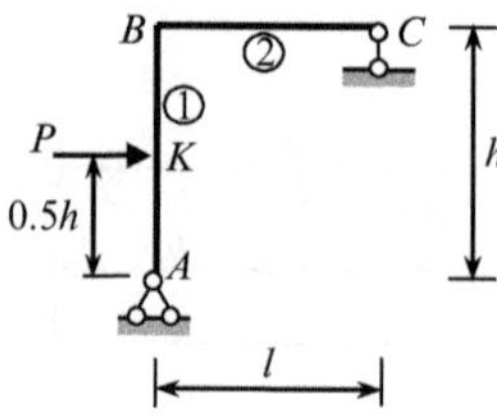

Fig. P6.22

$$\text{Ans. } \Delta_C^{\text{hor}} = \frac{Ph^2 l}{12EI} + \frac{11}{48}\frac{Ph^3}{EI}.$$

6.23. Compute the mutual linear displacement and mutual angular displacement at points A and B for structure shown in Fig. P6.23. Bending stiffness is EI_1 for horizontal element and EI_2 for vertical ones.

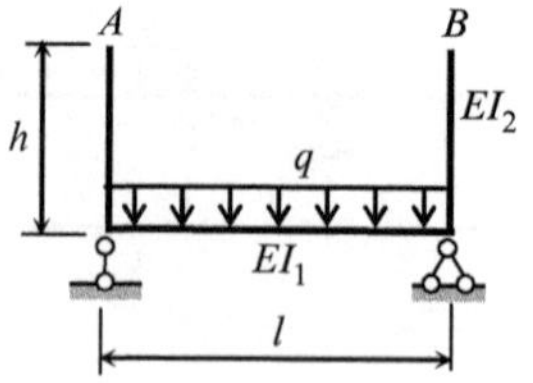

Fig. P6.23

$$\text{Ans. } \Delta_{AB} = \frac{ql^3 h}{12EI_1}; \quad \theta_{AB} = \frac{ql^3}{12EI_1}.$$

6.24. Design diagram of the frame is shown in Fig. P6.24. The relative stiffnesses are shown in the circles. The frame is subjected to uniformly distributed load $q = 4\text{kN/m}$. Compute the vertical and horizontal displacements at point C, and angular displacement at points C and D.

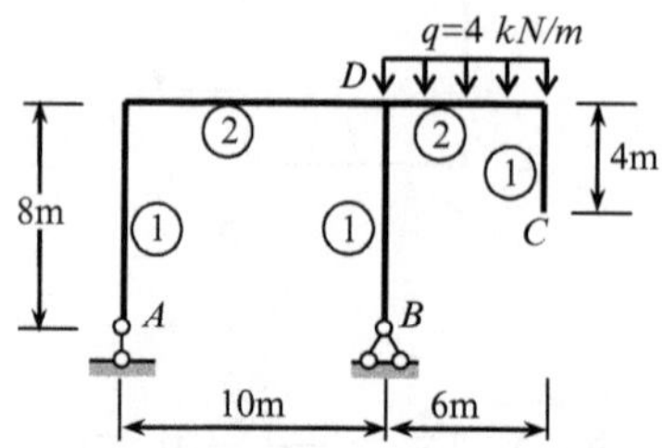

Fig. P6.24

$$\text{Ans. } \Delta_{\text{vert}} = \frac{1024}{EI}(\downarrow); \quad \Delta_{\text{hor}} = -\frac{192}{EI}(\rightarrow); \quad \theta_C = \frac{192}{EI} \text{ (clockwise)};$$

$$\theta_D = -\frac{120}{EI} \text{ (counterclockwise)}.$$

6.25. The portal simply supported frame is subjected to horizontal uniformly distributed load $q = 4\,\text{kN/m}$, as shown in Fig. P6.25. Calculate the horizontal displacement of cross-bar CD.

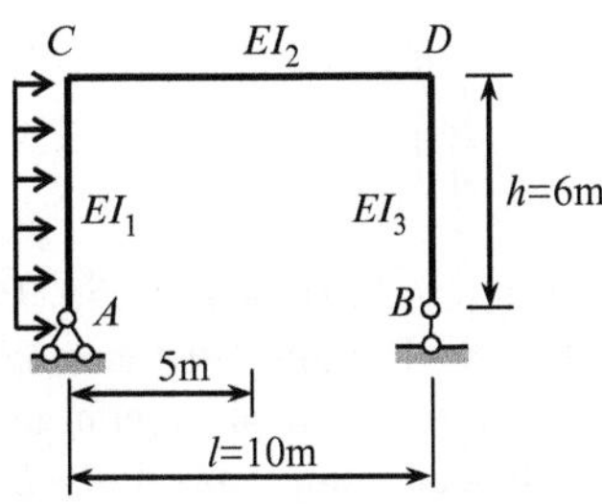

Fig. P6.25

$$\text{Ans. } \Delta_{\text{hor}}^{CD} = \frac{1080}{EI_1} + \frac{1440}{EI_2}.$$

6.26. A three-hinged frame carrying uniformly distributed load $q = 4\,\text{kN/m}$ is shown in Fig. P6.26. The hinge C is located at the middle span AB. The flexural stiffness of each member is shown in design diagram. Compute the mutual angular rotation of points E and D and horizontal displacement of the cross-bar ED.

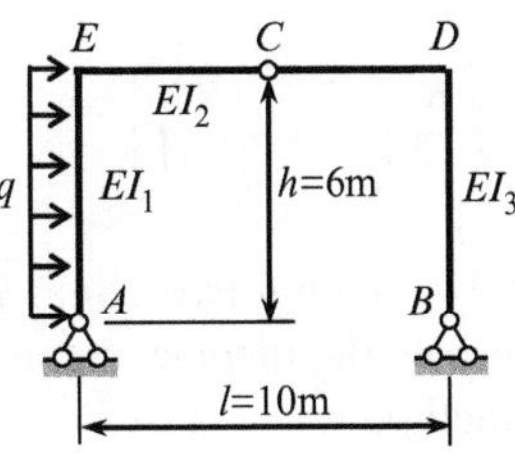

Fig. P6.26

$$\text{Ans. } \theta_{ED} = -\frac{108}{EI_1} + \frac{72}{EI_3}; \quad \Delta_{ED}^{\text{hor}} = \frac{324}{EI_1} + \frac{36}{EI_2} + \frac{21.6}{EI_3}.$$

6.27. Design diagram of the three-hinged frame with elastic tie is presented in Fig. P6.27. The bending stiffness of all members is EI, the axial stiffness of the tie is EA. Determine the vertical displacement at point C. All dimensions are in meters.

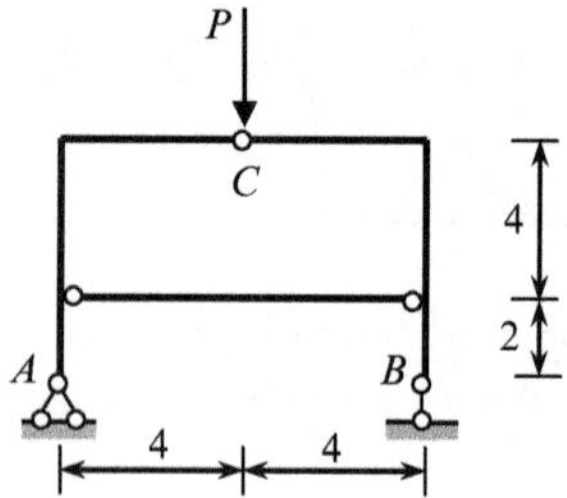

Fig. P6.27

Ans. $\Delta_C = \frac{64}{3} \cdot \frac{P}{EI} \left(1 + \frac{3}{32} \frac{I}{A}\right)$.

6.28. Design diagram of a structure is shown in Fig. P6.28. The top part of the structure presents semicircular bar. Bending stiffness EI = constant for all parts of the structure. Calculate the horizontal displacement at point B due to horizontal uniform distribute load q which acts within a portion AE. Hint: for curvilinear part apply the Maxwell–Mohr integral.

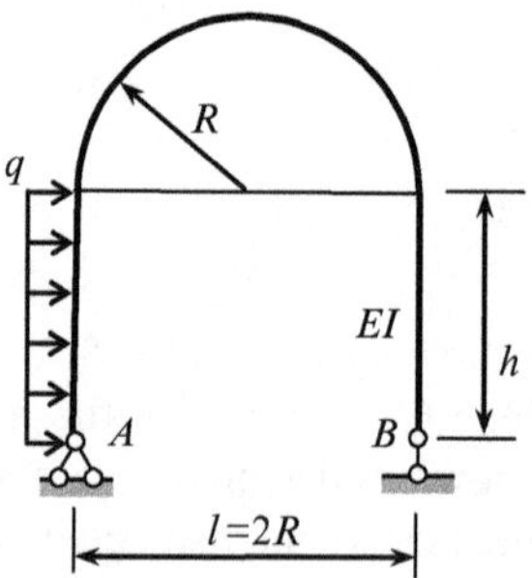

Fig. P6.28

Ans. $\Delta_B^{\text{hor}} = \frac{5}{24} \frac{qh^4}{EI} \left[1 + \frac{6}{5} \frac{R}{h} \left(\pi + 2\frac{R}{h}\right)\right]$.

6.29. The truss in Fig. P6.29 is carrying two equal forces P. The axial stiffness of all the members is EA. Compute the displacements of the all joints of the lower chord. Apply the elastic load method.

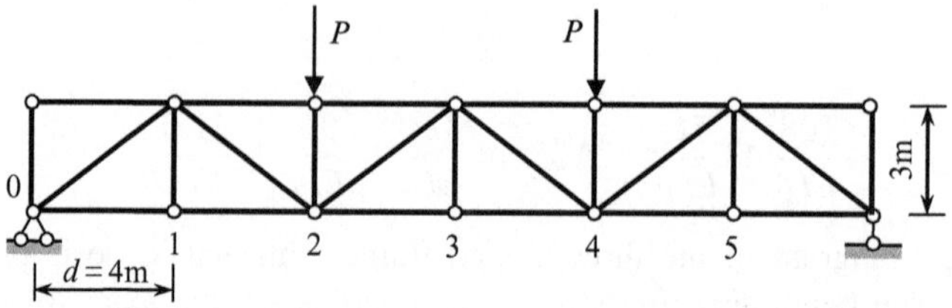

Fig. P6.29

$$\text{Ans. } y_1 = \frac{637}{9}\frac{P}{EA}; \quad y_2 = \frac{1146}{9}\frac{P}{EA}; \quad y_3 = \frac{1274}{9}\frac{P}{EA}.$$

Problems 6.30 through 6.33 are to be solved using the reciprocal theorems

6.30. Calculate the angle of rotation of point A caused by given load P and displacement along the P caused by moment M at point A. The flexural stiffness EI is constant for all members.

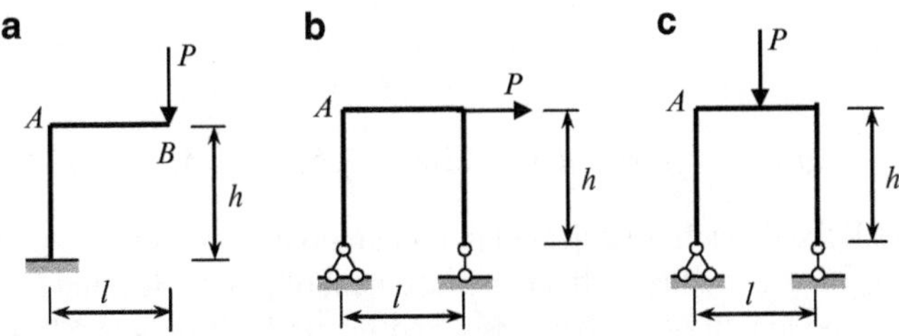

Fig. P6.30

$$\text{Ans. (a) } \theta_A = P\frac{lh}{EI}; \quad y_B = M\frac{lh}{EI}.$$

6.31. Show elastic curves for both states and displacements δ_{21} for first state and δ_{12} for second state. Verify of reciprocal unit displacements theorem (Fig. P6.31).

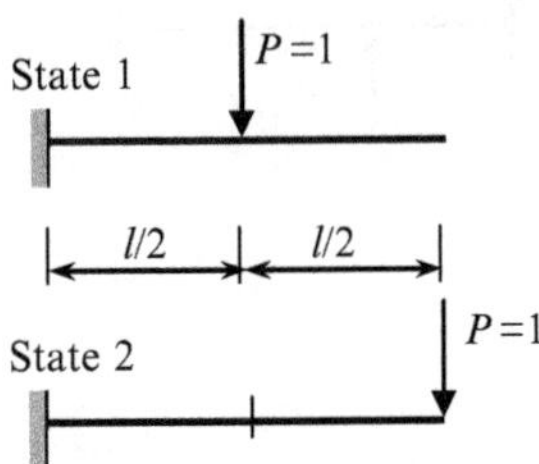

Fig. P6.31

$$\text{Ans. } \delta_{12} = \delta_{21} = \frac{5}{48}\frac{l^3}{EI}$$

6.32. A fixed–fixed beam is subjected to unit vertical displacement of the right support (point j). Calculate the vertical displacement of point k located at $l/4$ from the left support. The length of the beam is l and flexural stiffness is EI. (Hint: use the displacement and reactions reciprocal theorem)

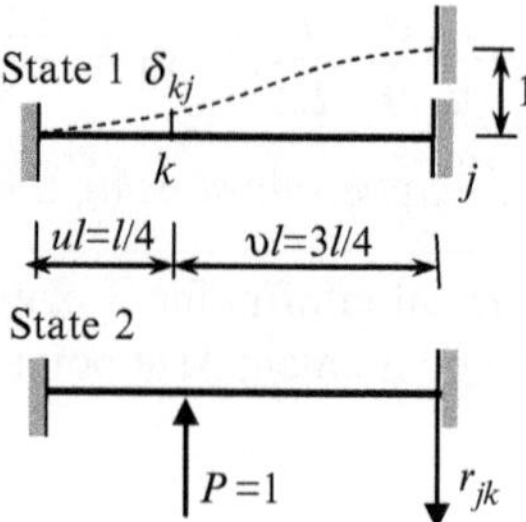

Fig. P6.32

Ans. $\delta_{kj} = -r_{jk}; \quad r_{jk} = -u^2(1 + 2\upsilon) = -5/32, \quad \delta_{kj} = 5/32.$

6.33. A fixed–fixed beam is subjected to unit angular displacement of the right support (point j). Calculate the vertical displacement of point k located at $l/4$ from the left support. The length of the beam is l and flexural stiffness is EI. (Hint: use the displacement and reactions reciprocal theorem).

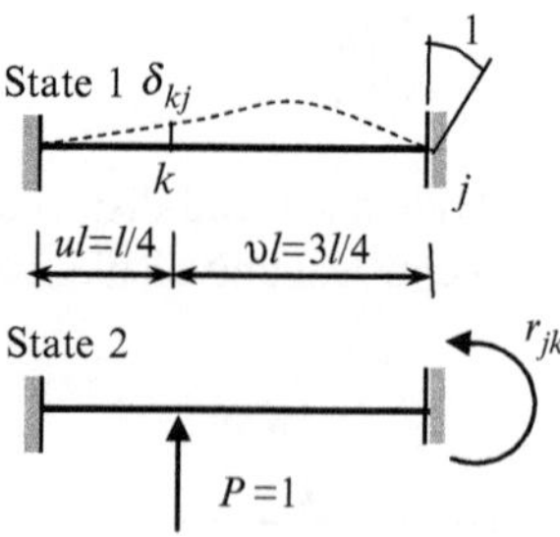

Fig. P6.33

Ans. $\dfrac{3}{64}l.$

Part II
Statically Indeterminate Structures

Chapter 7
The Force Method

The force method presents a powerful method for analyzing linear elastic statically indeterminate structures; this method also has a wide application in problems of stability and dynamics of structures. The method is very attractive because it has clear physical meaning, which is based on a convenient and well-ordered procedure of calculation of displacements of deformable structures, and presently, this method has been brought to elegant simplicity and perfection. In this chapter, the reader find the numerous examples of application of this method for analysis of different structures subjected to external loads, temperature change, settlements of supports, and errors of fabrication.

7.1 Fundamental Idea of the Force Method

Definitions. Statically indeterminate structures are called the structures for which *all* reactions and internal forces cannot be determined solely using equilibrium equations. Redundant constraints (or excess) are constraints, which are not necessary for geometrical unchangeability of a given structure.

7.1.1 Degree of Redundancy, Primary Unknowns and Primary System

Degree of redundancy, or statical indeterminacy, equals to the number of redundant constraints whose elimination leads to the new geometrically unchangeable and statically determinate structure. Thus, degree of statical redundancy is the difference between the number of constraints and number of independent equilibrium equations that can be written for a given structure.

Primary unknowns represent *reactions* (forces and/or moments), which arise in redundant constraints. That is the reason why the method is called the Force Method; this method also is called the flexibility method or the method of consistent deformations. Unknown *internal* forces also may be treated as primary unknowns. Primary

I.A. Karnovsky and O. Lebed, *Advanced Methods of Structural Analysis*,
DOI 10.1007/978-1-4419-1047-9_7, © Springer Science+Business Media, LLC 2010

system (principal or released structure) is such structure, which is obtained from the given one by eliminating redundant constraints and replacing them by primary unknowns.

Let us consider some statically indeterminate structures, the versions of primary systems, and the corresponding primary unknowns. A two-span beam is presented in Fig. 7.1a. The total number of constraints, and as result, the number of unknown reactions, is four. For determination of reactions of this planar set of forces, only three equilibrium equations may be written. Therefore, the degree of redundancy is $n = 4 - 3 = 1$, where four is a total number of reactions, while three is a number of equilibrium equations for given structure. In other words, this structure has one redundant constraint or statical indeterminacy of the first degree.

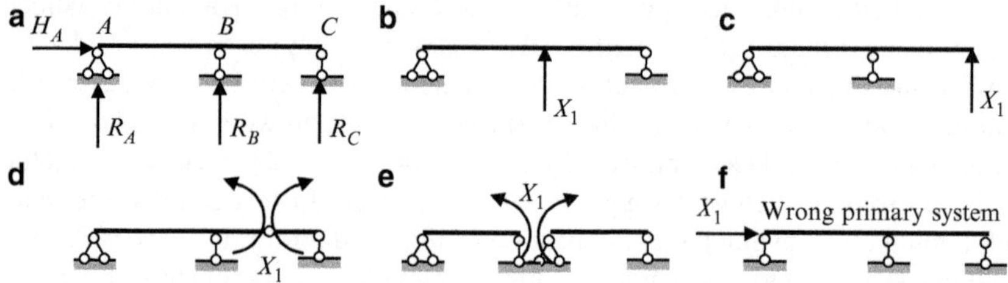

Fig. 7.1 (**a**) Design diagram of a beam; (**b–e**) The different versions of the primary system; (**f**) Wrong primary system

Four versions of the primary system and corresponding primary unknowns are shown in Fig. 7.1b–e. The primary unknown X_1 in cases (b) and (c) are reaction of support B and C, respectively. The primary unknown in cases (d) and (e) are bending moments. In case (d), the primary unknown is the bending moment at any point in the span, while in case (e), the primary unknown is the bending moment at the support B. Each of the primary systems is geometrically unchangeable and statically determinate; the structure in Fig. 7.1d is a Gerber–Semikolenov beam; in case (e), the primary system is a set of simply supported statically determinate beams. The constraint which prevents the horizontal displacement at support A cannot be considered as redundant one. Its elimination leads to the beam on the three parallel constrains, i.e., to the geometrically changeable system. So the structure in Fig. 7.1f cannot be considered as primary system. The first condition for the primary system – geometrically unchangeable – is the necessary condition. The second condition – statical determinacy – is not a necessary demand; however, in this book we will consider only statically determinate primary systems.

The statically indeterminate frame is presented in Fig. 7.2a. The degree of redundancy is $n = 4 - 3 = 1$. The structure in Fig. 7.2b presents a possible version of the primary system. Indeed, the constraint which prevents horizontal displacement at the right support is not a necessary one in order to provide geometrical unchangeability of a structure (i.e., it is a redundant constraint) and it may be eliminated, so the primary unknown presents the horizontal reaction of support. Other version of the primary system is shown in Fig. 7.2c; in this case the primary unknown presents the bending moment at the rigid joint.

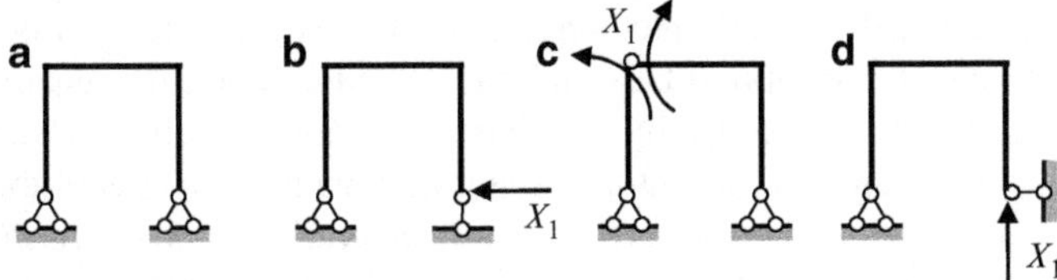

Fig. 7.2 (**a**) Design diagram of a portal frame; (**b–c**) The different versions of the primary system; (**d**) Wrong primary system

The system shown in Fig. 7.2d is geometrically changeable because three remaining support bars would be incapable of preventing rotation of the frame with respect to the left pinned support. Indeed, the constraint which prevents vertical displacement at the right support is a *necessary* constraint in order to provide a geometrical unchangeability of a structure (i.e., it is not a redundant one). It means that the system shown in Fig. 7.2d cannot be accepted as a version of primary system.

Another statically indeterminate frame is presented in Fig. 7.3a. The degree of redundancy is $n = 6 - 3 = 3$.

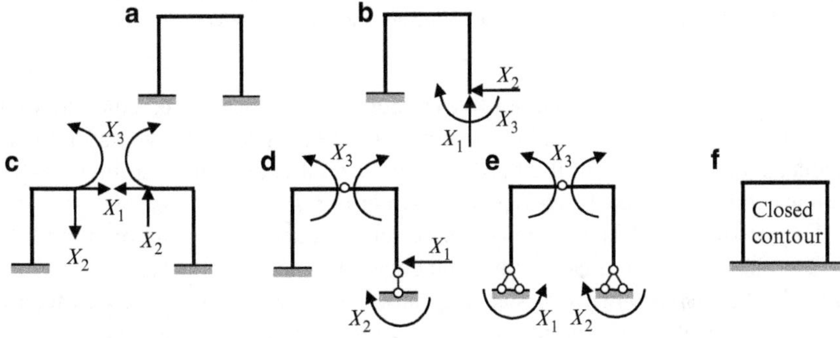

Fig. 7.3 (**a**) Statically indeterminate frame; (**b–e**) The different versions of the primary system; (**f**) A concept of closed contour

One version of the primary system and corresponding primary unknowns is shown in Fig. 7.3b. The primary unknowns are reactions of support. The structure shown in Fig. 7.3c presents the primary system where primary unknowns are internal forces (axial force X_1, shear X_2, and moment X_3), which appear in pairs. Figure 7.3d presents another version of the primary system. In this case, we eliminate two constraints which prevent two displacements (horizontal and angular) at support and one constraint which prevents *mutual* angular displacement, i.e., the primary unknowns are a combination of reactions X_1 and X_2 and internal moment X_3. Is it obvious that three-hinged frame (Fig. 7.3e) can be adopted as the primary system.

The structure shown in Fig. 7.3a can also be considered as a system with closed contour (Fig. 7.3f). One closed contour has three degrees of redundancy, and primary system can be similar as in Fig. 7.3c.

More complicated statically indeterminate frame is presented in Fig. 7.4a. This structure contains three support bars and one closed contour. All reactions of supports may be determined using only equilibrium equations, while the internal forces in the members of the closed contour cannot be obtained using equilibrium equations. Thus, this structure is *externally* statically determinate and *internally* statically indeterminate. The degree of redundancy is $n = 3$. The primary unknowns are internal forces as shown in Fig. 7.4b. They are axial force X_1, shear X_2, and bending moment X_3.

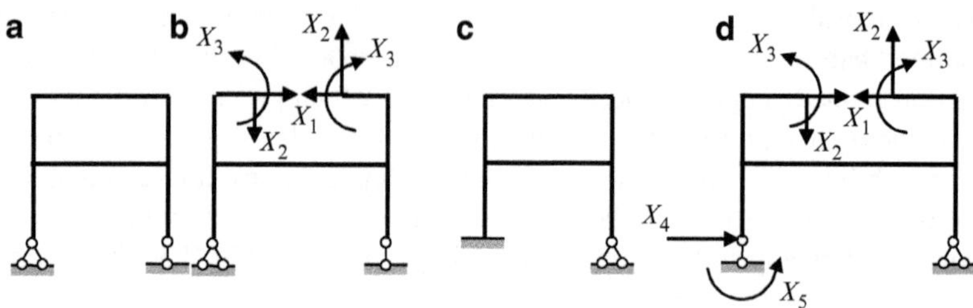

Fig. 7.4 (**a, b**) Internally statically indeterminate structure and primary system and (**c, d**) externally and internally statically indeterminate structure and primary system

The frame in Fig. 7.4c contains five support bars and one closed contour. So this structure is externally statically indeterminate in the second degree and internally statically indeterminate in the third degree. A total statical indeterminacy is five. Primary unknowns may be chosen as shown in Fig. 7.4d.

Degree of statical indeterminacy of structures does not depend on a load. It is evident that inclusion of each redundant constraint increases the rigidity of a structure. So the displacements of statically indeterminate structures are less than the displacements of corresponding structures without redundant constraints.

The different forms of the force method are considered in this chapter. However, first of all we will consider the superposition principle, which is the fundamental basis for the analysis of any statically indeterminate structure. The idea of analysis of a statically indeterminate structure using the superposition principle is presented below.

7.1.2 Compatibility Equation in Simplest Case

Two-span beam subjected to arbitrary load P is shown in Fig. 7.5. This structure is statically indeterminate to the first degree. Two versions of primary system are shown in Table 7.1.

Assume that the middle support B is the redundant one (Version 1 of the primary system). Thus the reaction of this constraint, i.e., $R_B = X$, is a primary unknown. For the given structure, the displacement at point B is zero. The primary unknown

Fig. 7.5 Statically indeterminate beam. Design diagram and reactions of supports

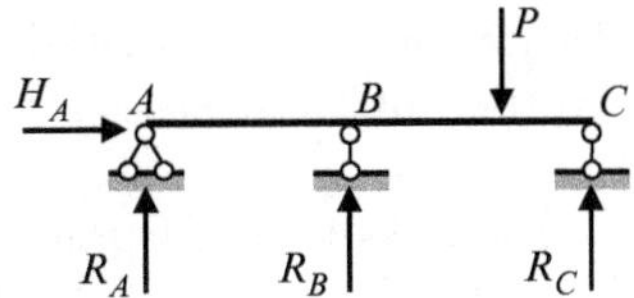

Table 7.1 Analysis of statically indeterminate beam by superposition method

	Version 1	Version 2
Primary system	H_A, A, B, P, C, R_A, $X{=}R_B$, R_C	H_A, A, B, P, C, R_A, R_B, $X{=}R_C$
Primary unknown	$X = R_B$	$X = R_C$
Displacement caused by given loads	A, B, P, C, $y_{B(P)}$	A, B, P, C, $y_{C(P)}$
Displacement caused by primary unknown X	$y_{B(X)}$, A, B, C, $X{=}R_B$	A, B, C, $y_{C(X)}$, $X{=}R_C$
Compatibility equation	$y_B = 0$ $y_B = y_{B(P)} + y_{B(X)} = 0$	$y_C = 0$ $y_C = y_{C(P)} + y_{C(X)} = 0$

X should be determined from the following condition: behavior of the actual beam and primary system must be identical. Since point B has no displacement in the actual state, then the compatibility condition is $y_B = 0$. The displacement of point B in the primary system is caused by given load P, as well as by the primary unknown $R_B = X$. So the compatibility condition may be written in the following form

$$y_B = y_{B(P)} + y_{B(X)} = 0, \tag{7.1}$$

where $y_{B(P)}$ and $y_{B(X)}$ are displacement of point B in the primary system due to given load P, and primary unknown $R_B = X$, respectively. The compatibility equation means that both structures – the given and primary ones – are equivalent. These displacements may be calculated by any method, which are described in Chap. 6. The solution of compatibility equation allows calculating the primary unknown X. The obtained value $X{=}R_B$ should be considered as active external load, which acts (together with given load P) on the statically determinate beam. Analysis of this beam (calculation of all reactions, construction of the internal force diagram, and elastic curve) creates no difficulties at all.

The version 2 of the primary unknown and corresponding compatibility equation is shown also in Table 7.1. The primary systems 1 and 2 are not unique. Using a rational primary system can significantly simplify the analysis of a structure.

The following procedure may be recommended for analysis of statically indeterminate structures by the superposition principle:

1. Determine the degree of statical indeterminacy
2. Choose the redundant unknowns; their number equals to degree of statical indeterminacy
3. Construct the statically determinate structure (primary structure) by eliminating all redundant constraints
4. Replace the eliminated constraints by primary unknowns. These unknowns present reactions of eliminated constraints
5. Form the compatibility equations; their number is equal to degree of statical indeterminacy. Each compatibility equation should be presented in terms of given loads and primary unknowns
6. Solve the system of equations with respect to primary unknowns
7. Since reactions of the redundant constraints are determined, then the computation of all remaining reactions and analysis of the structure may be performed as for the statically determinate structure

Example 7.1. Determine the reactions of the beam shown in Fig. 7.6 and construct the bending moment diagram. Bending stiffness *EI* is constant.

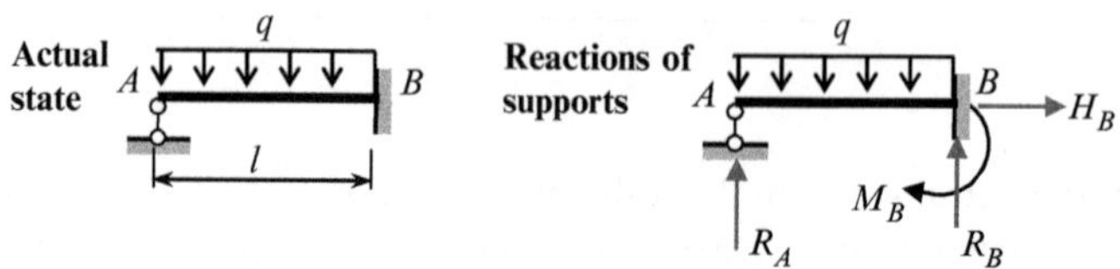

Fig. 7.6 Design diagram of the beam and reactions of supports

Solution. The structure has four unknown reactions (the vertical reactions R_A, R_B, horizontal reaction H_B, and support moment M_B), so the structure is the first degree of statical indeterminacy. A detailed solution for two versions of primary systems is presented in Table 7.2.

Table 7.2 Two versions of primary system and corresponding solution using principle of superposition

	Version 1	Version 2
Primary unknown	Vertical reaction $X = R_A$	Support moment $X = M_B$
Primary system	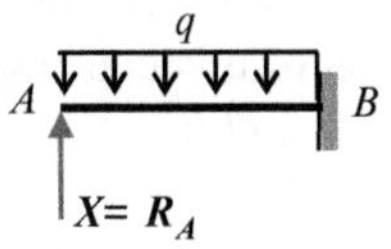	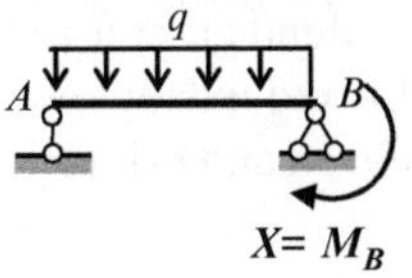
Compatibility condition	$y_A = 0$	$\theta_B = 0$
Displacement due to given load q in primary system	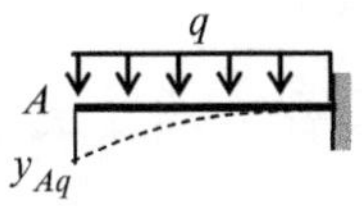	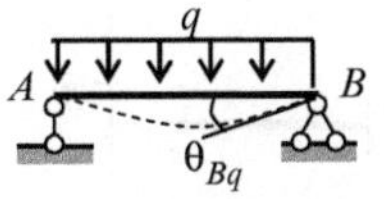
Displacement due to primary unknown X in primary system	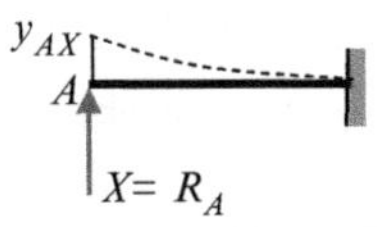	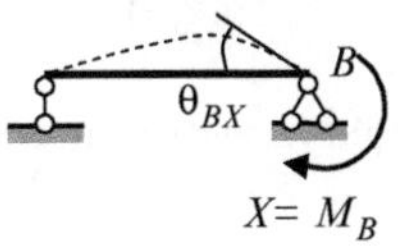
Solution of compatibility equation	$y_A = y_{Aq} - y_{AX} = 0$ $y_{Aq} = \frac{ql^4}{8EI}, \quad y_{AX} = \frac{Xl^3}{3EI}$ $X = R_A = \frac{3}{8}ql$	$\theta_B = \theta_{Bq} - \theta_{BX} = 0$ $\theta_{Bq} = \frac{ql^3}{24EI}, \quad \theta_{BX} = \frac{Ml}{3EI}$ $X = M_B = \frac{ql^2}{8}$
Statically determinate structure and bending moment diagram M for entire structure	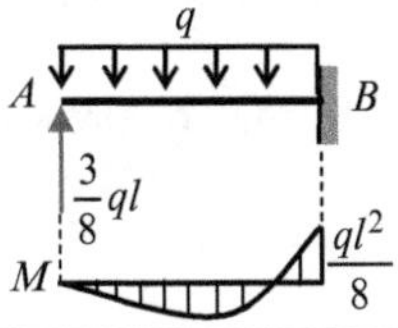	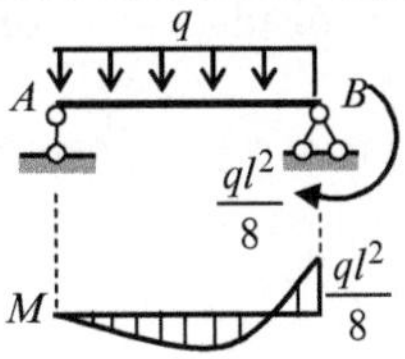

7.2 Canonical Equations of Force Method

Canonical equations of force method offer a unified procedure for analysis of statically indeterminate structures of different types. The word "canonical" indicates that these equations are presented in standard, or in an orderly fashion form. Very important is that canonical equations of the force method may be presented in a matrix form. Thus, this set of equations is a first bridge between classical analytical methods and numerical ones.

7.2.1 The Concept of Unit Displacements

Analysis of any statically indeterminate structure by the force method begins with determination of degree of statical indeterminacy. Primary system is obtained by elimination of redundant constraints and replacing them by reactions of these constrains. Primary unknowns X_i represent reactions (forces or moments) in eliminated redundant constrains.

Let us consider a simple redundant structure, such as clamped-pinned beam. The number of redundant constraints is $n = 4 - 3 = 1$. Assume that the right rolled support is the redundant one. Thus the reaction of this constraint, X_1, is a primary unknown. Given and primary systems are shown in Fig. 7.7.

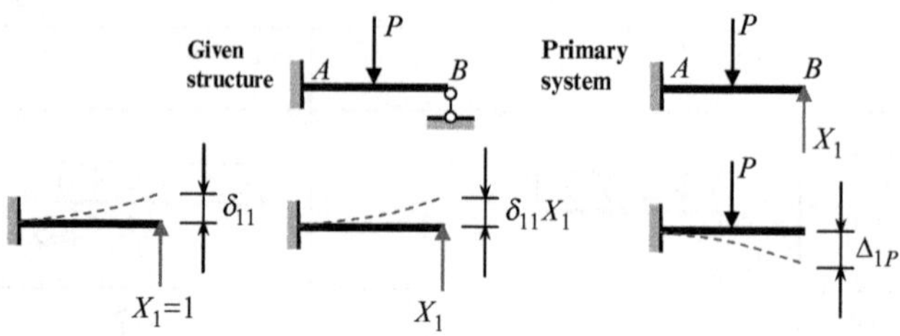

Fig. 7.7 Simple redundant structure. The idea of the force method and the concept of unit displacement

The compatibility condition may be written in the following form

$$y_B = y_{B(P)} + y_{B(X_1)} = 0, \tag{7.2}$$

where $y_{B(P)}$ is displacement of point B in primary system due to given load P, and $y_{B(X_1)}$ is displacement of point B in primary system due to primary unknown $R_B = X_1$.

Displacement $y_{B(X_1)}$ caused by unknown X_1 may be presented as

$$y_{B(X_1)} = \delta_{11} X_1, \tag{7.2a}$$

where δ_{11} presents the displacement in direction 1 (first index) caused by the force $X_1 = 1$ (second index). Coefficient δ_{11} is called the *unit displacement* since it is caused by *unit* primary unknown $X_1 = 1$. The term $\delta_{11} X_1$ presents the displacement in the direction of the eliminated constraint 1 caused by the *actual* primary unknown X_1. If displacement in direction 1 caused by given load $y_{B(P)}$ is denoted as Δ_{1P}, then (7.2) may be rewritten in the following form

$$\delta_{11} X_1 + \Delta_{1P} = 0. \tag{7.3}$$

Left part of equation presents a total displacement in the direction of eliminated constraint 1 (first index) caused by primary unknown X_1 and a given load. If this total displacement is zero, then behavior of both structures (entire structure subjected to given load and primary structure subjected to given load as well as primary unknown X_1) is identical.

The compatibility equation in form (7.4) is called the *canonical equation of the force method* for any structure with one redundant constraint; the free term Δ_{1P} is called the loaded term (loaded displacement, free term). The solution of the canonical equation allows us to calculate the primary unknown X_1, i.e., $X_1 = -\Delta_{1P}/\delta_{11}$.

General case of canonical equations. The canonical equations of the force method for a statically indeterminate structure with n redundant constraints are written as follows

$$\begin{aligned} \delta_{11}X_1 + \delta_{12}X_2 + \cdots + \delta_{1n}X_n + \Delta_{1P} &= 0 \\ \delta_{21}X_1 + \delta_{22}X_2 + \cdots + \delta_{2n}X_n + \Delta_{2P} &= 0 \\ \cdots\cdots\cdots\cdots\cdots\cdots & \\ \delta_{n1}X_1 + \delta_{n2}X_2 + \cdots + \delta_{nn}X_n + \Delta_{nP} &= 0 \end{aligned} \tag{7.4}$$

The form of presentation of the canonical equations as shown in (7.4) is always the same; it does not depend on the type of a structure, its peculiarities, and type of external exposures (forces, support settlements, temperature change, fabrication error). The number n of these equations equals to the degree of statical indeterminacy of a given structure.

All coefficients δ_{ik} of canonical equations represent a *displacement* of the primary structure due to *unit* primary unknowns; these coefficients are called the *unit displacements.*

Coefficient δ_{ik} is the displacement along the direction of unknown X_i due to action of unit unknown X_k; term $\delta_{ik}X_k$ presents displacement along the direction of unknown X_i due to action of real unknown X_k. Coefficients δ_{ik}, which are located on the principal diagonal ($i = k$) are called the principal (main) displacements. All other displacements δ_{ik} ($i \neq k$) are called the secondary unit displacements.

Free term Δ_{iP} presents displacement along the direction of unknown X_i due to action of actual load in primary system. Displacements Δ_{iP} caused by applied loads are called the loaded terms or free terms.

Physical meaning of the canonical equations. The left part of the ith equation presents the total displacement along the direction of unknown X_i due to action of all real unknowns X_k as well as applied load. Total displacement of the primary structure in directions of eliminated restrictions caused by primary unknowns and applied load equals zero. In this case, the difference between the given and primary structures is vanished.

7.2.2 Calculation of Coefficients and Free Terms of Canonical Equations

Computation of coefficients and free terms of canonical equations presents significant and very important part of analysis of any statically indeterminate structure. For their calculation, any methods can be applied. The graph multiplication method is best suited for beams and framed structures. For this, it is necessary in primary system to construct bending moment diagrams $\bar{M}_1, \bar{M}_2, \ldots, \bar{M}_n$ due to unit primary unknowns $X_i, i = 1, \ldots, n$ and diagram M_P^0 due to given load. Unit displacements and loaded terms are calculated by formulas

$$\delta_{ik} = \sum \int \frac{\bar{M}_i \cdot \bar{M}_k}{EI} \mathrm{d}s, \quad \Delta_{iP} = \sum \int \frac{\bar{M}_i \cdot M_P^0}{EI} \mathrm{d}s. \tag{7.5}$$

Accordingly to expression 6.21, these formulas may be presented in conventional forms. Properties of unit coefficients are as follows:

1. Main displacements are strictly positive ($\delta_{ii} > 0$).
2. Secondary displacements δ_{ik}, $i \neq k$ may be positive, negative, or zero.
3. Secondary displacements satisfy the reciprocal displacement theorem

$$\delta_{ik} = \delta_{ki}. \tag{7.5a}$$

It means that unit displacements symmetrically placed with respect to principal diagonal of canonical equations are equal.

The unit of displacements δ_{ik} presents the ratio of unit for displacement according to index i and units for force according to index k.

Construction of internal force diagrams. Solution of (7.4) is the primary unknowns X_i, $i = 1, \ldots, n$. After that the primary system may be loaded by determined primary unknowns and given load. Internal forces may be computed as for usual statically determinate structure. However, the following way allows once again an effective use of the bending moment diagrams in primary system. The final bending moment diagram M_P may be constructed by formula

$$M_P = \bar{M}_1 \cdot X_1 + \bar{M}_2 \cdot X_2 + \cdots + \bar{M}_n \cdot X_n + M_P^0. \tag{7.6}$$

Thus in order to compute the ordinates of the resulting bending moment diagram, it is necessary to multiply each unit bending moment diagrams $\bar{M}_k$ by corresponding primary unknown X_k and summing up with bending moment diagram due to applied load in the primary system M_P^0. This formula expresses the superposition principle. Advantage of formula (7.6) is that it may be effectively presented in tabulated form.

Shear forces may be calculated on the basis of bending moment diagram using Schwedler theorem and axial forces may be calculated on the basis of shear force diagram by considering equilibrium of joints of the structure. Finally, having internal force diagrams, all reactions are easy to determine.

Procedure for analysis The following procedure provides analysis of statically indeterminate beams and frames using the canonical equations of the force method:

1. Provide the kinematical analysis and define the degree of statically indeterminacy n of a structure.
2. Choose the primary system and replace the eliminated redundant constraints by corresponding primary unknowns X_i, $i = 1, \ldots, n$.
3. Formulate the canonical equations of the force method.
4. Apply the successive unit forces $X_1 = 1$, $X_2 = 1, \ldots, X_n = 1$ to primary system and for each unit primary unknown construct corresponding bending moment diagrams $\bar{M}_1, \bar{M}_2, \ldots, \bar{M}_n$.

5. Calculate the unit coefficients δ_{ik}.
6. Construct the bending moment diagram M_P^0 due to applied load in primary system and calculate the load terms Δ_{iP} of (7.4).
7. Solve the system of equations with respect to primary unknowns $X_1, X_2, \ldots, X_n$.
8. Construct the bending moment diagrams by (7.6), next compute the shear and construct corresponding shear force diagram, and lastly compute axial forces and construct the corresponding axial force diagram.
9. Having internal force diagrams, calculate the reactions of supports. Other way is consider primary system subjected to determined primary unknowns and given load and provide computation of all internal forces by definition; this way is less effective.
10. Provide the static control for all structure (or any its part).
11. Provide the kinematical control (7.2) for displacements of an entire structure in direction of primary unknowns.

Intermediate checking of computation These verifications are recommended to be performed *before* solving canonical equations for determining primary unknowns X_i, i.e., on the steps 5 and 6 of the algorithm above. For control of unit displacements and free terms, it is necessary to construct *summary* unit bending moment diagram $\bar{M}_\Sigma = \bar{M}_1 + \bar{M}_2 + \cdots + \bar{M}_n$. The following types of controls are suggested as follows:

(a) Row verification of unit displacements: Multiply summary unit bending moment diagram $\bar{M}_\Sigma$ on a primary bending moment diagram $\bar{M}_i$:

$$\frac{\bar{M}_\Sigma \times \bar{M}_i}{EI} = \sum \int \left(\bar{M}_1 + \bar{M}_2 + \cdots + \bar{M}_n\right) \cdot \bar{M}_i \frac{ds}{EI} = \delta_{i1} + \delta_{i2} + \cdots + \delta_{in}.$$

The result of this multiplication equals to the sum of unit displacements of the i-equation.

(b) Total verification of unit displacements: Multiply the summary unit bending moment diagram $\bar{M}_\Sigma$ on itself.

$$\frac{\bar{M}_\Sigma \times \bar{M}_\Sigma}{EI} = \underbrace{\delta_{11} + \cdots + \delta_{1n}}_{\text{1st equation}} + \underbrace{\delta_{21} + \cdots + \delta_{2n}}_{\text{2nd equation}} + \cdots + \underbrace{\delta_{n1} + \cdots + \delta_{nn}}_{n\text{th equation}} = \sum_{i,k=1}^{n} \delta_{ik}.$$

The result of this multiplication equals to the sum of all unit displacements of canonical equations. It is quite sufficient to perform only total control, however, if errors occur, then the row control should be performed for tracking the wrong coefficient.

(c) Verification of the load displacements: Multiply summary unit bending moment diagram $\bar{M}_\Sigma$ on a bending moment diagram M_P^0.

$$\frac{\bar{M}_\Sigma \times M_P^0}{EI} = \sum \int (\bar{M}_1 + \bar{M}_2 + \cdots + \bar{M}_n) \cdot M_P^0 \frac{ds}{EI} = \Delta_{1P} + \Delta_{2P} + \cdots + \Delta_{nP}.$$

The result of this multiplication equals to the sum of all free terms of canonical equations.

7.3 Analysis of Statically Indeterminate Structures

This section contains application of the force method in canonical form to detailed analysis of different types of structures. Among them are continuous beams, frames, trusses, and arches.

7.3.1 *Continuous Beams*

Let us consider two-span continuous beam on rigid supports (Fig. 7.8a). Figure 7.8b shows one version of the primary system, which presents the set of statically determinate beams. The primary unknown X_1 is bending moment at the intermediate support.

Canonical equation of the force method is $\delta_{11}X_1 + \Delta_{1P} = 0$, where δ_{11} is a displacement in direction of first primary unknown due to unit primary unknown $X_1 = 1$; Δ_{1P} is displacement in the same direction due to applied load. This canonical equation shows that for the adopted primary system the *mutual* angle of rotation at the support 1 caused by primary unknown X_1 and the given load P is zero.

For calculation of displacements δ_{11} and Δ_{1P}, it is necessary to construct the bending moment diagrams in the primary system caused by unit primary unknown $X_1 = 1$ and acting loads; they are shown in Fig. 7.8c, d, respectively.

For calculation of unit displacement, we need to multiply the bending moment diagram $\bar{M}_1$ by itself

$$\delta_{11} = \frac{\bar{M}_1 \times \bar{M}_1}{EI} = 2\frac{1}{EI}\frac{1}{2} \times 1 \times l \times \frac{2}{3} \times 1 = \frac{2l}{3EI}\left(\frac{\text{rad}}{\text{kN m}}\right).$$

For calculation of free term, we need to multiply the bending moment diagram M_P^0 by $\bar{M}_1$

$$\Delta_{1P} = \frac{\bar{M}_1 \times M_P^0}{EI} = \frac{l}{6EI}[1 \times (2 - 0.4) + 0 \times (1 + 0.4)] \times 0.24Pl = 0.064\frac{Pl}{EI} \text{ (Table A.2, line 5)}$$

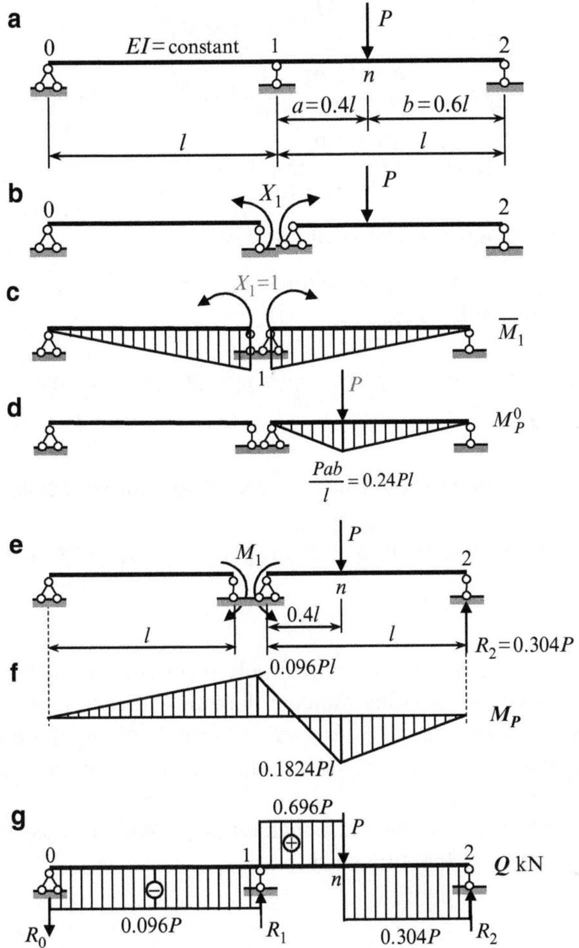

Fig. 7.8 (**a, b**) Design diagram of a beam and primary system. (**c, d**) Bending moment diagrams in the primary system due to unit primary unknown ($\bar{M}_1$) and given load (M_P^0). (**e**) Primary system loaded by primary unknown and given load; (**f**) Final bending moment diagram. (**g**) Shear force diagram and reactions of supports

Primary unknown (bending moment at support 1) is $X_1 = -\Delta_{1P}/\delta_{11} = -0.096Pl$.

Construction of bending moment diagram The bending moment diagram can be constructed considering each simply supported beam separately under action of applied load and obtained bending moments at supports, as shown in Fig. 7.8e; constraints at the support 1 are shown apart for convenient. Final bending moment diagram is shown in Fig. 7.8f.

Computation of shear Having the bending moment diagram we can calculate a shear using differential relationships $Q = \mathrm{d}M/\mathrm{d}x$. In our case we get

$$Q_{0-1} = -\frac{0.096Pl}{l} = -0.096P,$$
$$Q_{1-n} = \frac{0.1824Pl - (-0.096Pl)}{0.4l} = 0.696P,$$
$$Q_{n-2} = \frac{0.1824Pl}{0.6l} = 0.304P.$$

Final shear force diagram is shown in Fig. 7.8g.

Reactions of supports Having shear force diagram we can calculate the reaction of all supports. They are following:

$$R_0 = -0.096P, \quad R_1 = Q_1^{\text{right}} - Q_1^{\text{left}} = 0.696P - (-0.096P) = 0.792P,$$
$$R_2 = 0.304P.$$

Static verification Equilibrium condition for all structure in whole is

$$\sum Y = -0.096P + 0.792P + 0.304P - P = -1.096P + 1.096P = 0.$$

Discussion:

1. Adopted primary system as a set of simply supported beams leads to the simple (triangular) shape of bending moment diagrams in each unit states. The bending moment diagram due to given load is located only within each loaded span. Therefore, computation of coefficients and free terms of canonical equations is elementary procedure.
2. Canonical equations of the force method allow easy to take into account the different bending stiffness for each span.

7.3.2 Analysis of Statically Indeterminate Frames

Design diagram of the simplest frame is presented in Fig. 7.9a. The flexural stiffness for all members is EI. It is necessary to construct the bending moment diagram and calculate a horizontal displacement of the cross bar.

Primary system and primary unknown. The structure has four unknown reactions, so the degree of redundancy is $n = 4 - 3 = 1$. Let us choose the primary unknown X_1 be a vertical reaction at point 1. The primary system is obtained by eliminating support 1 and replacing it by X_1 (Fig. 7.9b).

Canonical equation of the force method is $\delta_{11}X_1 + \Delta_{1P} = 0$. This equation shows that for the adopted primary system the vertical displacements of the left rolled support caused by primary unknown X and the given load q is zero.

Bending moment diagrams in the primary system caused by unit primary unknown $X_1 = 1$ and given load are shown in Fig. 7.9c, d. These graphs also show the displacements along eliminated constraint.

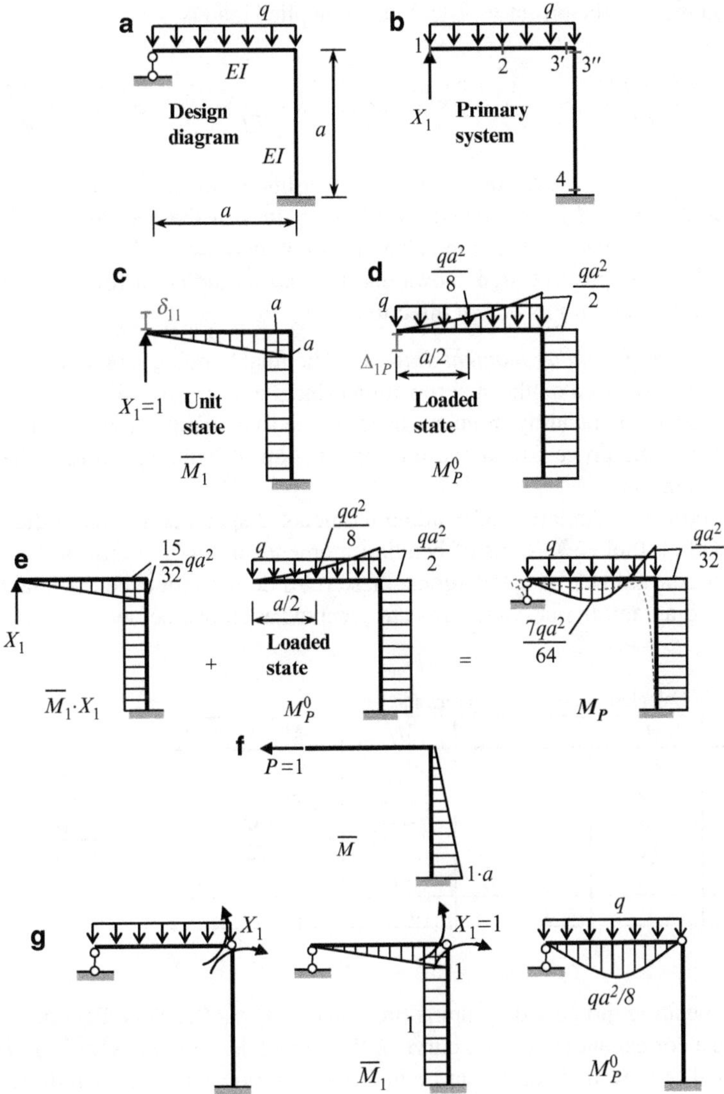

Fig. 7.9 (**a, b**) Design diagram of a frame and primary system. (**c, d**) Unit and loaded states and corresponding bending moment diagrams. (**e**) Construction of bending moment diagram using the superposition principle. (**f**) Unit state for calculation of horizontal displacement. (**g**) New version of primary system and corresponding bending moment diagrams

The unit displacement δ_{11} is obtained by "multiplying" the $\bar{M}_1$ graph by itself, i.e.,

$$\delta_{11} = \frac{\bar{M}_1 \times \bar{M}_1}{EI} = \frac{1}{EI} \times \frac{1}{2} a \times a \times \frac{2}{3} a + \frac{1}{EI} \times a \times a \times a = \frac{4}{3} \frac{a^3}{EI} \left(\frac{\text{m}}{\text{kN}} \right).$$

Displacement in the primary system due to applied load is

$$\Delta_{1P} = \frac{\bar{M}_1 \times M_P^0}{EI} = -\frac{1}{EI} \times \frac{1}{3} \times \frac{qa^2}{2} \times a \times \frac{3}{4}a - \frac{1}{EI} \times \frac{qa^2}{2} \times a \times a = -\frac{5}{8}\frac{qa^4}{EI} \text{ (m)}.$$

The negative sign at each term means that ordinates of the two bending moment diagram $\bar{M}_1$ and M_P^0 are located on different sides of the neutral line of the corresponding members of the frame. The primary unknown is $X_1 = -\Delta_{1P}/\delta_{11} = \frac{15}{32}qa$ (kN). The positive sign shows that the *chosen* direction for the primary unknown coincides with its *actual* direction.

Construction of bending moment diagram The final bending moment diagram will be constructed based on the superposition principle $M = \bar{M}_1 \cdot X_1 + M_P^0$. The first term presents the bending moment diagram due to actual primary unknown $X_1 = (15/32)qa$; the procedure above for construction of bending moment diagram is presented in Fig. 7.9e.

The ordered calculation of bending moments at specified points of the frame is presented in Table 7.3. Signs of bending moments are chosen arbitrarily and used only for convenience of calculations; these rules do not influence the final bending moment diagram. In our case, signs are accepted as shown below.

Table 7.3 Calculation of bending moments

Points	$\bar{M}_1$	$\bar{M}_1 \cdot X_1$	M_P^0	$\bar{M}_1 \cdot X_1 + M_P^0$
1	0	0	0	0
2	$-a/2$	$-15/64$	$+1/8$	$-7/64$
3′	$-a$	$-15/32$	$+1/2$	$+1/32$
3″	$+a$	$+15/32$	$-1/2$	$-1/32$
4	$+a$	$+15/32$	$-1/2$	$-1/32$
Factor		qa^2	qa^2	qa^2

signs of
bending moments
⊕ ⊖ ⊕ ⊖

The bending moment diagram is presented in Fig. 7.9e. This diagram allows us to trace a corresponding elastic curve of the frame; this curve is shown on Fig. 7.9e by dotted line. Rolled support at point 1 does not prevent horizontal displacement of the cross-bar. Therefore, this structure presents the frame with sidesway (cross-bar translation). In this case, all ordinates of bending moment diagram for vertical member 3–4 are located on one side and therefore, elastic curve for this member has no point of inflection.

Kinematical verification Displacement in the direction of constraint 1 in the original system has to be zero. This displacement may be calculated by multiplication of two bending moment diagrams, i.e., one is final bending moment diagram for given structure M_P (Fig. 7.9e), and the second is bending moment diagram $\bar{M}_1$ in unit state (Fig. 7.9c).

$$\Delta_1 = \frac{M_P \times \bar{M}_1}{EI} = \underbrace{\frac{a}{6EI}\left(0\times 0 + 4\times\frac{1}{2}a\times\frac{7}{64}qa^2 - a\times\frac{1}{32}qa^2\right)}_{\text{horizontal portion, Simpson rule}}$$

$$\underbrace{-\frac{1}{EI}\times a\times a\times\frac{1}{32}qa^2}_{\text{vertical element}} = \frac{qa^4}{32EI} - \frac{qa^4}{32EI} = 0.$$

Horizontal displacement of the cross bar Unit state is shown in Fig. 7.9f. Required displacement is

$$\Delta_{\text{hor}} = \frac{M_P \times \bar{M}}{EI} = \frac{1}{EI}\frac{1}{2}a\times 1\times a\times\frac{qa^2}{32} = \frac{qa^4}{64EI}.$$

The positive result means that crossbar shifted from right to left. Corresponding elastic curve for frame in whole is shown in Fig. 7.9e by dashed line. Note that inflection point for vertical member is absent.

Another version of the primary system and corresponding bending moment diagrams $\bar{M}_1$ and M_P^0 are shown in Fig. 7.9g.

In this case

$$\delta_{11} = \frac{4}{3}\frac{a}{EI}, \quad \Delta_{1P} = \frac{qa^3}{24EI}.$$

A primary unknown $X_1 = -qa^2/32$; this is bending moment at the rigid joint, as presented in Fig. 7.9e.

Property of Statically Indeterminate Frames of the First Degree of Redundancy

Let us consider important property of any statically indeterminate frames of the first degree of redundancy. Design diagram of the frame is presented in Fig. 7.10.

Two primary systems and corresponding bending moment diagrams for the unit conditions are presented in Fig. 7.10 and are denoted as versions 1–2. For version 1, the primary system is obtained by eliminating of the support constraint; and for

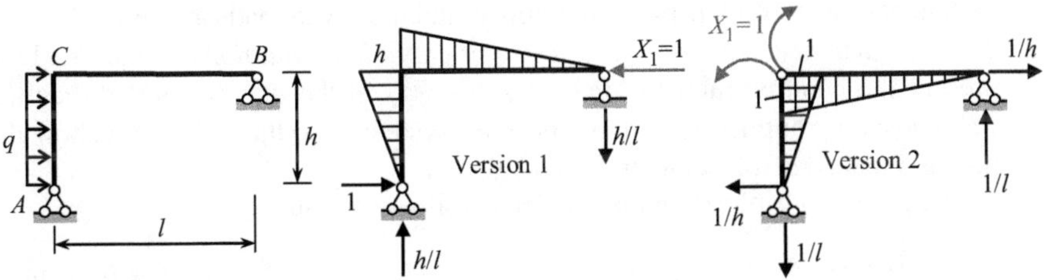

Fig. 7.10 Property of a primary system for structure of the first degree of redundancy

version 2 by introducing a hinge at the joint C. Therefore, the primary unknown in version 1 is reactions of support; primary unknown in the version 2 is bending moment at joint C, so the dimensions of the δ_{11} is m/kN for version 1 and rad/kN m for version 2.

The unit bending moment diagrams for *any* primary system are *similar*. This is a general property for a statically indeterminate structure with first degree of redundancy.

After bending moment diagram is constructed, the kinematical verification must be performed. This procedure involves multiplication of the final and unit bending moment diagram in any primary system. It is obvious that multiplication of the final and unit bending moment diagram for each version will be equal to zero. However, meaning of these multiplications will be different. For example $M \times \bar{M}_{\text{vers1}} = 0$ means that horizontal displacement at point B is zero; $M \times \bar{M}_{\text{vers2}} = 0$ means that mutual angle of rotation of two sections at joint C is zero.

Detailed analysis of statically indeterminate frame using canonical equations of the force method is presented in Example 7.2; this design diagram will be analyzed more detail at a later time (settlements of supports, change of temperature, other methods of analysis).

Example 7.2. A frame is clamped at point A and rolled at points B and C as presented in Fig. 7.11a. The frame is loaded by force $P = 8$ kN and uniformly distributed load, $q =$ kN/m. The relative flexural stiffness of each element is shown in a circle. Construct the bending moment, shear, and normal force diagrams. Determine the reactions of supports.

Solution. 1. *Primary system and primary unknowns.* The structure has five unknown reactions. The degree of indeterminacy is $n = 5 - 3 = 2$. One version of the primary system with primary unknowns X_1 and X_2 (vertical and horizontal reactions at points B and C) is presented in Fig. 7.11b.

Canonical equations of the force method are

$$\begin{aligned} \delta_{11}X_1 + \delta_{12}X_2 + \Delta_{1P} &= 0, \\ \delta_{21}X_1 + \delta_{22}X_2 + \Delta_{2P} &= 0. \end{aligned} \tag{a}$$

These equations show that for the adopted primary system the vertical displacement at support B and horizontal displacement at support C in the primary system caused by both primary unknowns and the given loads are zero.

2. *The unit displacements and free terms of canonical equations.* Figure 7.11c–e present the bending moment diagrams $\bar{M}_1$, $\bar{M}_2$ in the unit states and diagram M_P^0 caused by applied load in the primary system; also these diagram show the unit and loaded displacements δ_{ik} and Δ_{iP}.

The graph multiplication method leads following results:

$$\delta_{11} = \frac{\bar{M}_1 \times \bar{M}_1}{EI} = \frac{1}{1EI} \times 10 \times 5 \times 10 + \frac{1}{2EI} \times \frac{1}{2} \times 10 \times 10 \times \frac{2}{3} \times 10 = \frac{666.67}{EI} \left(\frac{\text{m}}{\text{kN}}\right),$$

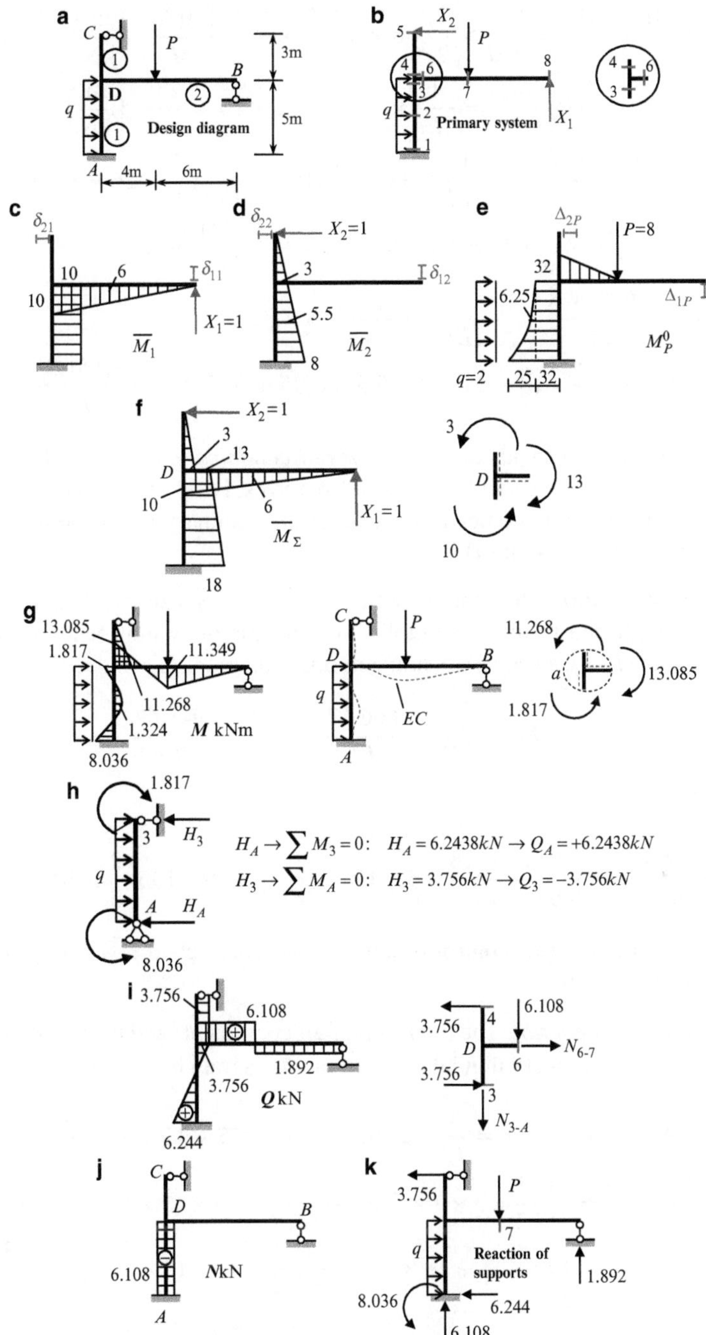

Fig. 7.11 (**a, b**) Design diagram of a frame and primary system. (**c–e**) Bending moment diagrams in primary system caused by unit primary unknowns and the given load (**f**) Summary unit bending moment diagram. (**g**) Final bending moment diagram, elastic curve (EC) and static control. (**h**) Calculation of shear for member 1-3. (**i**) Shear force and free body diagram for joint D. (**j, k**) Axial force diagram and reactions of supports

$$\delta_{22} = \frac{\bar{M}_2 \times \bar{M}_2}{EI} = \frac{1}{1EI} \times \frac{1}{2} \times 8 \times 8 \times \frac{2}{3} \times 8 = \frac{170.67}{EI} \left(\frac{\text{m}}{\text{kN}}\right), \tag{b}$$

$$\delta_{12} = \delta_{21} = \frac{\bar{M}_1 \times \bar{M}_2}{EI} = \frac{1}{1EI} \times \frac{3+8}{2} \times 5 \times 10 = \frac{275}{EI} \left(\frac{\text{m}}{\text{kN}}\right).$$

$$\Delta_{1P} = \frac{\bar{M}_1 \times M_P^0}{EI} = -\frac{1}{1EI} \times 32 \times 5 \times 10 - \frac{1}{1EI} \times \frac{1}{3} \times 25 \times 5 \times 10$$
$$- \frac{1}{2EI} \times \frac{4}{6} \times (2 \times 10 + 6) \times 32 = -\frac{2,294}{EI} \text{ (m)},$$

$$\Delta_{2P} = \frac{\bar{M}_2 \times M_P^0}{EI} = -\frac{1}{1EI} \times \frac{3+8}{2} \times 5 \times 32 -$$
$$\frac{1}{1EI} \times \frac{5}{6} \times (8 \times 25 + 4 \times 5.5 \times 6.25 + 3 \times 0) = -\frac{1,161.25}{EI} \text{ (m)}. \tag{c}$$

3. *Verification of coefficients and free terms of canonical equations.* The unit and loaded displacement should be checked before solving of canonical equations (a). For this purpose, we need to construct the summary unit bending moment diagram $\bar{M}_\Sigma = \bar{M}_1 + \bar{M}_2$ (Fig. 7.11f).

   - *First row control.* The sum of coefficients in the first canonical equation must be equal to the result of multiplication the summary unit bending moment diagram $\bar{M}_\Sigma$ by a primary bending moment diagram $\bar{M}_1$. Indeed,

$$\delta_{11} + \delta_{12} = \frac{666.67}{EI} + \frac{275}{EI} = \frac{941.67}{EI},$$

   while

$$\frac{\bar{M}_\Sigma \times \bar{M}_1}{EI} = \frac{1}{1EI} \frac{13+18}{2} \times 5 \times 10 + \frac{1}{2EI} \frac{1}{2} \times 10 \times 10 \times \frac{2}{3} \times 10 = \frac{941.67}{EI}.$$

   Therefore, the first row control is satisfactory. The second row control may be performed similarly.

   - *Simultaneous control.* The total sum of all coefficients and the result of multiplication of summary unit bending moment diagram by "itself" are

$$\delta_{11} + \delta_{12} + \delta_{21} + \delta_{22} = \frac{1}{EI}(666.67 + 275 + 275 + 170.67) = \frac{1,387.34}{EI},$$

$$\frac{\bar{M}_\Sigma \times \bar{M}_\Sigma}{EI} = \frac{1}{1EI} \frac{1}{2} \times 3 \times 3 \times \frac{2}{3} \times 3 + \frac{1}{1EI} \frac{5}{6} \times (2 \times 13 \times 13 + 2 \times 18 \times 18$$
$$+ 13 \times 18 + 18 \times 13) + \frac{1}{2EI} \frac{1}{2} \times 10 \times 10 \times \frac{2}{3} \times 10 = \frac{1,387.34}{EI}.$$

   - *Free terms control.* The sum of the loaded displacements is

$$\Delta_{1P} + \Delta_{2P} = -\frac{1}{EI}(2,294 + 1,161.25) = -\frac{3,455.25}{EI},$$

while the multiplication of the summary unit bending moment diagram $\bar{M}_{\Sigma}$ by a bending moment diagram M_P^0 due to applied load in the primary structure

$$\frac{\bar{M}_{\Sigma} \times M_P^0}{EI} = -\frac{1}{1EI} \times \frac{5}{6}\left(18 \times 57 + 13 \times 32 + 4 \times 38.25 \times \frac{13+18}{2}\right)$$
$$-\frac{1}{2EI} \times \frac{4}{6}(2 \times 10 + 6) \times 32 = -\frac{3{,}455.25}{EI}.$$

Therefore, the coefficients and free terms of equations (a) are computed correctly.

4. *Primary unknowns.* Canonical equations for primary unknowns X_1 and X_2 becomes

$$\begin{aligned} &666.67X_1 + 275X_2 - 2{,}294 = 0, \\ &275X_1 + 170.67X_2 - 1{,}161.25 = 0. \end{aligned} \qquad \text{(d)}$$

All coefficients and free terms contain factor 1/*EI*, which can be cancelled. It means that primary unknowns of the force method depend only on *relative* stiffnesses of the elements.

The solution of these two equations leads to

$$X_1 = 1.8915\,\text{kN}, \; X_2 = 3.7562\,\text{kN}.$$

5. *Internal force diagrams.* The bending moment diagram of the structure will be readily obtained using the expression

$$M = \bar{M}_1 \cdot X_1 + \bar{M}_2 \cdot X_2 + M_P^0. \qquad \text{(e)}$$

Location of the specified points 1–8 is shown in Fig. 7.11b. Corresponding calculation is presented in Table 7.4. The sign of bending moments is chosen arbitrarily for summation purposes only.

Table 7.4 Calculation of bending moments at the specified points

Points	$\bar{M}_1$	$\bar{M}_1 \cdot X_1$	$\bar{M}_2$	$\bar{M}_2 \cdot X_2$	M_P^0	M
1	−10	−18.915	−8.0	−30.049	+57.0	+8.036
2	−10	−18.915	−5.5	−20.659	+38.25	−1.324
3	−10	−18.915	−3.0	−11.268	+32.0	+1.817
4	0.0	0.0	−3.0	−11.268	+32.0	−11.268
5	0.0	0.0	0.0	0.0	0.0	0.0
6	−10	−18.915	0.0	0.0	+32.0	+13.085
7	−6.0	−11.349	0.0	0.0	0.0	−11.349
8	0.0	0.0	0.0	0.0	0.0	0.0

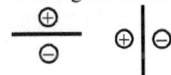

signs of bending moments
⊕ ⊖ ⊕ ⊖

The resulting bending moment diagram and corresponding elastic curve are presented in Fig. 7.11g.

- *Statical verification of bending moment diagram.* The free-body diagram of joint D (closed section a) is shown in Fig. 7.11g. Equilibrium condition of this joint is

$$\sum M = 11.268 + 1.817 - 13.085 = 0.$$

- *Kinematical verification of bending moment diagram.* Displacement in the direction of any primary unknown in the given system must be equal to zero. This condition is verified by multiplying bending moment diagram M_P in the actual state by bending moment diagram $\bar{M}_i$ in *any* primary system.

Displacement in the direction of the first primary unknown is

$$\begin{aligned}\Delta_1 = \frac{\bar{M}_1 \times M_P}{EI} &= \frac{1}{1EI} \times \frac{5}{6}(-8.036\times10-1.817\times10+4\times1.324\times10)\\ &+\frac{1}{2EI} \times \frac{4}{6}(-2 \times 13.085 \times 10+2 \times 11.349 \times 6-13.085 \times 6+10\times11.349)\\ &+\frac{1}{2EI} \times \frac{1}{2} \times 11.349 \times 6 \times \frac{2}{3} \times 6 = \frac{1}{EI}(195.453 - 195.511) \approx 0.\end{aligned}$$

The relative error is 0.029%. Similarly it is easy to check that displacement in the direction of the second primary unknown is zero, i.e., $\Delta_2 = \bar{M}_2 \times M_{P/EI} = 0$.

Shear forces may be calculated using differential relationships $Q = \mathrm{d}M/\mathrm{d}x$. This formula leads to the following results:

$$Q_{6-7} = \frac{13.085 - (-11.349)}{4} = 6.1085\,\text{kN}$$

$$Q_{7-8} = -\frac{11.349}{6} = -1.8915\,\text{kN}; \quad Q_{4-5} = -\frac{11.268}{3} = -3.756\,\text{kN}.$$

The portion A-3 subjected to load q and couples at points A and 3 is shown in Fig. 7.11h.

$$H_A \to \sum M_3 = 0: \quad H_A = 6.2438\,\text{kN} \to Q_A = +6.2438\,\text{kN},$$

$$H_3 \to \sum M_A = 0: \quad H_3 = 3.756\,\text{kN} \to Q_3 = -3.756\,\text{kN}.$$

The final shear force diagram is shown in Fig. 7.11i.

Axial forces may be derived from the equilibrium of rigid joint D; a corresponding free-body diagram is presented in Fig. 7.11i. The shear at point 3 is negative, so this force, according to the sign law, should rotate the body counterclockwise. It is assumed that unknown forces N_{6-7} and N_{3-4} are tensile.

$$N_{6-7} \to \sum X = 0: \quad N_{6-7} = 0,$$

$$N_{3-A} \to \sum Y = 0: \quad N_{3-A} = -6.108\,\text{kN}.$$

Final diagram for axial force N is presented in Fig. 7.11j. Since the member DB has no constraints for its horizontal displacement, a normal force in this member is zero. The similar conclusion related to the member DC.

6. *Reactions of supports.* Internal force diagrams M, Q, and N allow us to show all reactions of the support. The negative shear 3.756 means that horizontal reaction at support C equals 3.756 kN and directed from right to left. The positive shear 6.444 means that horizontal reaction at clamped support A equals 6.444 kN and this direction of the reaction R_A produces the positive shear, etc. All reactions of supports are shown in Fig. 7.11k.

Now we can perform controls of the obtained reactions of supports. Equilibrium equations for the frame in whole are

$$\sum X = q \cdot 5 - 6.244 - 3.756 = 10 - 10 = 0$$

$$\sum Y = -P + 6.108 + 1.892 = -8 + 8 = 0$$

$$\begin{aligned}\sum M_7 &= 2 \times 5 \times 2.5 + 8.036 - 6.108 \times 4 - 6.244 \times 5 + 3.756 \times 3 + 1.892 \times 6 \\ &= 55.656 - 55.652 \cong 0\end{aligned}$$

Therefore, equilibrium of the structure in a whole is held. It is left as an exercise for the reader to check the equilibrium of some parts of the system (for example, use a cut section through left/or right at point 7) and considering the equilibrium of the either part of the structure.

Discussion:

1. For any statically indeterminate structure subjected to action of arbitrary external load, the distribution of internal forces (bending moment, shear and normal forces), as well as reactions of supports depend only on relative stiffnesses of the elements, and does not depend on their absolute value of flexural stiffness, EI.
2. Two free-body diagrams for joint D takes into account different internal forces: the Fig. 7.11g contains only the bending moments, while Fig. 7.11i contains the shear and axial forces. It happens because the section is passed *infinitely close* to joint D.
3. The unit bending moment diagrams are used at different steps of analysis, so the effectiveness of these diagrams is very high.

7.3.3 Analysis of Statically Indeterminate Trusses

Statically indeterminate trusses are geometrically unchangeable structures for which all reactions and all internal forces cannot be determined using only the equilibrium of equations. Figure 7.12 presents three types of statically indeterminate trusses. They are (a) externally, (b) internally, and (c) mixed statically indeterminate trusses.

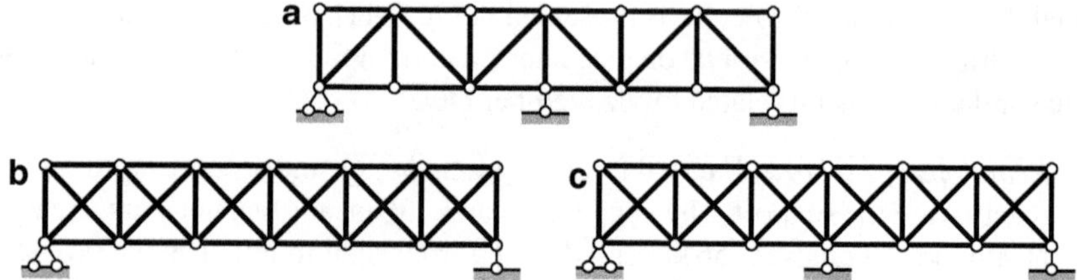

Fig. 7.12 Trusses. Types of statically indeterminacy

The truss (7.12a) contains one redundant *support* member and does not contain the redundant members in the web. This structure presents the first degree of *externally* statically indeterminate truss. In cases (b) and (c), the diagonal members do not have a point of intersection, i.e., these members are not connected (nor hinged, nor fixed) with each other. Case (7.12b) presents the *internally* statically indeterminate truss to six degrees of redundancy. Case (7.12c) presents the *internally* statically indeterminate truss to six degrees and *externally* to the first degree.

The analysis of statically indeterminate trusses may be effectively performed by the force method in canonical form. The primary system is obtained by elimination of redundant constraints. As in the general case (7.4) of canonical equations, X_n are primary unknowns; δ_{ik} are unit displacements of the primary system in the direction of i th primary unknown due to unit primary unknown $X_k = 1$; Δ_{iP} are displacements of the primary system in the direction of ith primary unknown $X_k = 1; \Delta_{ip}$ due to acting load.

For computation of coefficient and free terms of canonical equations, we will use the second term of the Maxwell–Mohr integral (6.11).

$$\delta_{ik} = \sum_n \frac{\bar{N}_i \cdot \bar{N}_k \cdot l}{EA},$$

$$\Delta_{iP} = \sum_n \frac{\bar{N}_i \cdot N_P^0 \cdot l}{EA}, \tag{7.7}$$

where l is length of nth member of the truss; $\bar{N}_i$, $\bar{N}_k$ are axial forces in nth member due to unit primary unknowns $X_i = 1$, $X_k = 1$; and N_P^0 is axial force in nth member of the primary system due to acting load.

Summation procedure in (7.7) should be performed on all members of the truss (subscript n is omitted).

Solution of (7.4) is the primary unknowns X_i $(i = 1, \ldots, n)$. Internal forces in the members of the truss may be constructed by the formula

$$N = \bar{N}_1 \cdot X_1 + \bar{N}_2 \cdot X_2 + \cdots + N_P^0. \tag{7.8}$$

Kinematical verification of computed internal forces may be done using the following formula

$$\sum \bar{N} N \frac{l}{EA} = 0. \tag{7.9}$$

If a primary unknown is the *reaction* of support, then this equation means that displacement in the direction of the primary unknown due to primary unknown and given loads is zero. If the primary unknown is the internal force in any redundant member of a web, then (7.9) means that a *mutual* displacement in the direction of the primary unknown due to this unknown and given loads is zero.

Let us consider the symmetrical truss that carries two equal forces $P = 120\,\text{kN}$ (Fig. 7.13a). The axial stiffness for all members is *EA*.This structure is first degree of statically indeterminacy. The reaction of the intermediate support is being as the primary unknown. The primary system is shown in Fig. 7.13b.

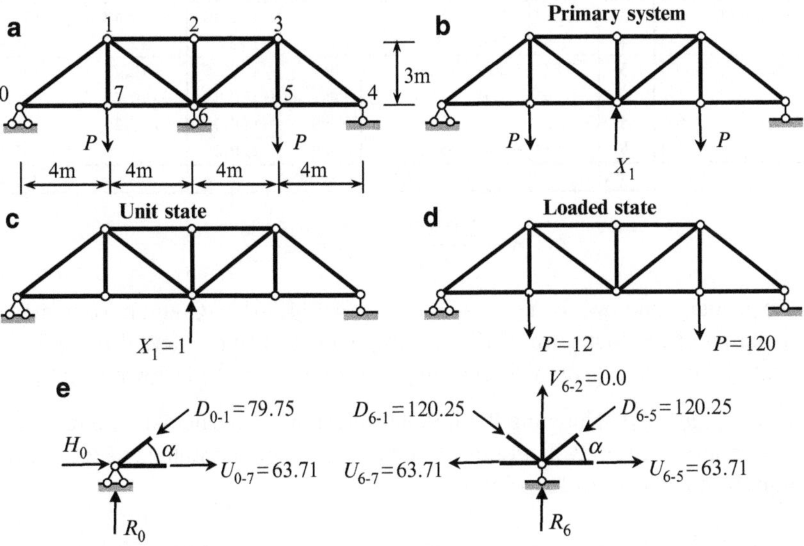

Fig. 7.13 (**a**, **b**) Redundant truss and primary system. (**c**, **d**) Unit and loaded states. (**e**) The free-body diagrams of joint 0 and 6

Canonical equation of the force method and primary unknown are

$$\delta_{11} X_1 + \Delta_{1P} = 0 \rightarrow X_1 = -\frac{\Delta_{1P}}{\delta_{11}}.$$

For calculation of δ_{11}, Δ_{1P}, it is necessary to show the unit and loading states (Fig. 7.13c, d).

The analysis of this statically indeterminate truss is presented in the tabulated form (Table 7.5). Column 1 contains the flexibility for each member; the factor 1/*EA* is omitted. Internal forces for all members in unit and loaded states are presented in columns 2 and 3, respectively. For computation of unit displacement δ_{11}, the entries of the column 5 should be summated. Similar procedure should be performed for computation of loaded displacement Δ_{1P} (column 4).

Table 7.5 Calculation of internal forces in the members of a statically indeterminate truss

Members		$\frac{l}{EA}$	$\bar{N}_1$	N_P^0	$\bar{N}_1 \cdot N_P^0 \frac{l}{EA}$	$\bar{N}_1^2 \cdot \frac{l}{EA}$	$\bar{N}_1 \cdot X_1$	$N = N_P^0 + \bar{N}_1 X_1$	$\bar{N}_1 \cdot N \frac{l}{EA}$
0		1	2	3	4	5	6	7	8
Vertical members	1–7	3	0	120	0	0	0	120	0
	2–6	3	0	0	0	0	0	0	0
	3–5	3	0	120	0	0	0	120	0
Lower chord	0–7	4	−0.667	160	−426.88	1.774	−96.288	63.71	−169.98
	7–6	4	−0.667	160	−426.88	1.774	−96.288	63.71	−169.98
	6–5	4	−0.667	160	−426.88	1.774	−96.288	63.71	−169.98
	5–4	4	−0.667	160	−426.88	1.774	−96.288	63.71	−169.98
Upper chord	1–2	4	1.333	−160	−853.12	7.107	192.43	32.43	172.92
	2–3	4	1.333	−160	−853.12	7.107	192.43	32.43	172.92
Diagonals	0–1	5	0.833	−200	−833	3.469	120.25	−79.75	−332.16
	1–6	5	−0.833	0	0	3.469	−120.25	−120.25	500.84
	6–3	5	−0.833	0	0	3.469	−120.25	−120.25	500.84
	3–4	5	0.833	−200	−833	3.469	120.25	−79.75	−332.14
Factor		$1/EA$			$1/EA$	$1/EA$			$1/EA$

$$\Delta_{1P} = -\frac{5,079.76}{EA}; \quad \delta_{11} = \frac{35.186}{EA}$$

The primary unknown is $X_1 = -\Delta_{1P}/\delta_{11} = 144.36\,\text{kN}$. Column 6 contains the computation of the first term of (7.8). Computation of final internal force in each element of the truss is provided according to formula (7.8) and shown in column 7.

Reaction of supports Knowing the internal forces we can calculate the reaction R of any support. Free-body diagrams of joints 0 and 6 are presented in Fig. 7.13e.

Equilibrium equations for joint 0 is

$$\sum Y = 0: \quad R_0 - 79.75 \sin\alpha = 0 \rightarrow R_0 = 47.85\,\text{kN}.$$

It is obvious that $R_4 = R_0 = 47.85\,\text{kN}$. Equilibrium equation for joint 6

$$\sum Y = -2 \times 120.25 \sin\alpha + R_6 = 0$$

leads to the following reaction at the intermediate support of the truss: $R_6 = 2 \times 120.25 \times 0.6 = 144.3\,\text{kN}$. Pay attention, this result has been obtained early as the primary unknown X_1.

Static verification For truss in whole the equilibrium equations

$$\sum Y = 2 \times 47.85 + 144.3 - 2 \times 120 = 240 - 240 = 0,$$

$$\sum X = 0: \quad H_0 = 0.$$

The last result may be checked. For joint 0, the equilibrium equation $\sum X = 0$ (Fig. 7.13e) leads to the following result: $H_0 + 63.71 - 79.75 \times 0.8 = 0 \rightarrow H_0 = 63.8 - 63.71$. The relative error is 0.14%.

Kinematical verification Displacement in the direction of the primary unknown due to primary unknown and given loads is zero. According to (7.9), for computation of this displacement we need to multiply the entries of column 7, 2, and 1 (column 8). The sum of entries of this column equals

$$\sum \bar{N}_1 \cdot N \frac{l}{EI} = -1{,}344.24 + 1{,}347.52 = 3.28.$$

The relative error is

$$\frac{3.28 \times 100\%}{1{,}344} = 0.24\%.$$

Discussion:

1. If statically indeterminate truss is subjected to any loads, then distribution of internal forces depends on *relative* axial stiffness *EA* of the members and does not depend on their absolute value *EA*.
2. Fundamental property of statically indeterminate trusses: any change of rigidity *EA* of specified element of the truss cannot change the *sign* of internal force in this element.

7.3.4 Analysis of Statically Indeterminate Arches

Different types of statically indeterminate arches are presented in Fig. 7.14. They are two-hinged arches with or without tie (Fig. 7.14a, b), one-hinged arch (Fig. 7.14c), and arch with fixed ends (Fig. 7.14d).

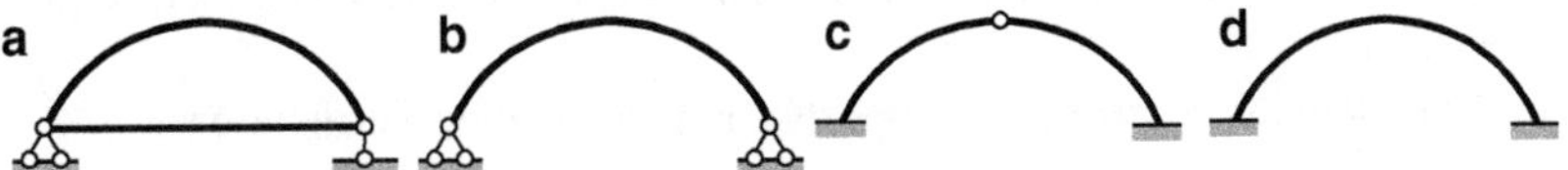

Fig. 7.14 Types of the redundant arches

The most effective method for analytical analysis of statically indeterminate arches is the force method in canonical form (7.4); for two-hinged arches we have one primary unknown, two unknowns for one-hinged arch, and three unknowns for hingeless arch.

As for three-hinged arch, the distribution of internal forces depends on a shape of the neutral line (parabola, circular, etc.). This should be taken into account when calculating unit coefficients and free terms of canonical equations. In general case, these coefficients and free terms depend on bending moments, shear, and axial forces. In calculating displacements, we will take into account only bending moments for the arch itself and an axial force for the tie, while the shear and axial forces for arch may be neglected. Since the axis of the arch is curvilinear, then the graph multiplication method leads to approximate results.

Unlike three-hinged arches, in redundant arches, as for any statically indeterminate structure, the internal forces arise in case of displacements of supports, changes of temperature, and errors of fabrication. For masonry or concrete arches, a concrete shrinkage should be taken into account, since this property of material leads to the appearance of additional stresses.

Procedure for analysis of statically indeterminate arches is as follows:

1. Choose the primary system of the force method.
2. Accept the simplified model of the arch, i.e., the arch is divided into several portions and each curvilinear portion is changed by straight member. Calculate the geometrical parameters of the arch at specified points.
3. Calculate the unit and loaded displacements, neglecting the shear and axial forces in arch:

$$\delta_{ik} = \int\limits_{(s)} \frac{\bar{M}_i \cdot \bar{M}_k}{EI} ds; \quad \Delta_{iP} = \int\limits_{(s)} \frac{\bar{M}_i \cdot M_P^0}{EI} ds.$$

Computation of these displacements may be performed using the graph multiplication method.

4. Find the primary unknown using canonical equation (7.4) of the force method.
5. Construct the internal force diagrams; the following formulas may be applied:

$$\begin{aligned} M &= \bar{M}_1 X_1 + \bar{M}_2 X_2 + \cdots + M_P^0 \\ Q &= \bar{Q}_1 X_1 + \bar{Q}_2 X_2 + \cdots + Q_P^0 \\ N &= \bar{N}_1 X_1 + \bar{N}_2 X_2 + \cdots + N_P^0, \end{aligned}$$

   where X_i are primary unknowns; $\bar{M}_i$, $\bar{Q}_i$, $\bar{N}_i$ are bending moment, shear, and axial force caused by unit ith primary unknown $X_i = 1$; and M_P^0, Q_P^0, N_P^0 are bending moment, shear, and axial force caused by given load in primary system.
6. Calculate the reactions of supports and provide their verifications.

Let us show this procedure for analysis of the parabolic two-hinged uniform arch shown in Fig. 7.15a. The flexural stiffness of the cross section of the arch is *EI*. The equation of the neutral line of the arch is $y = (4f/l^2)x(l - x)$. The arch is subjected to uniformly distributed load q within the all span. It is necessary to find the distribution of internal forces.

The arch under investigation is statically indeterminate of the first degree. The primary system is shown in Fig. 7.15b; the primary unknown X_1 is the horizontal reaction of the right support. Canonical equation of the force method is $\delta_{11} X_1 + \Delta_{1P} = 0$. The primary unknown $X_1 = \Delta_{1P}/\delta_{11}$.

Specified points of the arch The *span* of the arch is divided into eight equal parts; the specified points are labeled 0-8. Parameters of the arch for these sections are

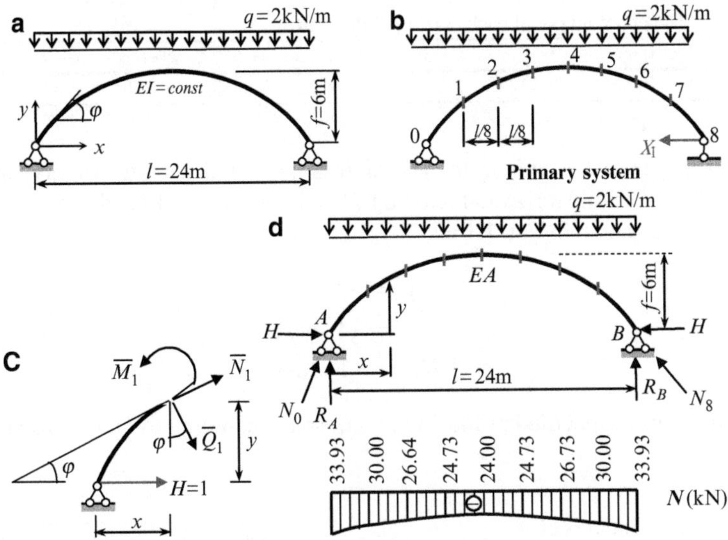

Fig. 7.15 (**a, b**) Two-hinged parabolic arch. Design diagram and primary system, (**c**) Positive directions of internal forces, (**d**) Final axial force diagram and reaction of supports

presented in Table 7.6; the following formulas for calculation of trigonometric functions of the angle φ between the tangent to the arch and x-axis have been used:

$$\tan\varphi = y' = \frac{4f(l-2x)}{l^2}, \quad \cos\varphi = \frac{1}{\sqrt{1+\tan^2\varphi}}, \quad \sin\varphi = \cos\varphi\,\tan\varphi.$$

Table 7.6 Geometrical parameters of parabolic arch

Points	Coordinates (m)				
	x	y	$\tan\varphi$	$\cos\varphi$	$\sin\varphi$
0	0	0.0	1.00	0.7070	0.7070
1	3	2.625	0.75	0.800	0.6000
2	6	4.500	0.50	0.8944	0.4472
3	9	5.625	0.25	0.9701	0.2425
4	12	6.000	0.0	1.0	0.0
5	15	5.625	−0.25	0.9701	−0.2425
6	18	4.500	−0.5	0.8944	−0.4472
7	21	2.625	−0.75	0.800	−0.6000
8	24	0.0	−1.00	0.7070	−0.7070

The length of the chord between points n and n-1 equals

$$l_{n,n-1} = \sqrt{(x_n - x_{n-1})^2 + (y_n - y_{n-1})^2}. \tag{7.10}$$

The chord lengths of each portion of the arch are presented in Table 7.7.

Table 7.7 The chord length of each portion of the arch

Portion	0–1	1–2	2–3	3–4	4–5	5–6	6–7	7–8
Length (m)	3.9863	3.5377	3.2040	3.0233	3.0233	3.2040	3.5377	3.9863

Internal forces in the unit state The arch is subjected to unit primary unknown $X_1 = 1$ (Fig. 7.15b). Horizontal reaction $H = 1$ and the positive directions of internal forces are shown in Fig. 7.15c.

$$\begin{aligned} \bar{M}_1 &= -1 \cdot y \\ \bar{Q}_1 &= -1 \cdot \sin \varphi \\ \bar{N}_1 &= -1 \cdot \cos \varphi. \end{aligned} \tag{7.11}$$

Internal forces at specified points in the unit state according to (7.11) are presented in Table 7.8.

Table 7.8 Internal forces of the arch in the unit state

Points	$\bar{M}_1$	$\bar{Q}_1$	$\bar{N}_1$
0	0.0	−0.7070	−0.7070
1	−2.625	−0.6000	−0.8000
2	−4.50	−0.4472	−0.8944
3	−5.625	−0.2425	−0.9701
4	−6.00	0.0	−1.0000
5	−5.625	0.2425	−0.9701
6	−4.50	0.4472	−0.8944
7	−2.625	0.6000	−0.8000
8	0.0	0.7070	−0.7070

The unit displacement caused by primary unknown $X = 1$ equals

$$\delta_{11} = \int_{(s)} \frac{\bar{M}_1 \cdot \bar{M}_1}{EI} \mathrm{d}s. \tag{a}$$

Thus the *only* column $\bar{M}_1$ (Table 7.8) will be used for calculation of unit displacement; the columns $\bar{Q}_1$ and $\bar{N}_1$ will be used for computation of final shear and axial forces as indicated at step 5 of procedure.

Internal forces in the loaded state Displacement in the primary system caused by applied load equals

$$\Delta_{1P} = \int_{(s)} \frac{\bar{M}_1 \cdot M_P^0}{EI} \mathrm{d}s, \tag{b}$$

where M_P^0 is bending moments in the arch in the primary system due to given load q. Thus, as in case of unit displacement, for computation of loaded displacement we will take into account only bending moment.

The reactions of supports of the primary system in the loaded state are $R_A^0 = R_B^0 = 24\,\mathrm{kN}$; this state is not shown. Expressions for internal forces are follows $(0 \le x \le 24)$

$$
\begin{aligned}
M_P^0 &= R_A^0 x - q\frac{x^2}{2} = 24x - x^2 \\
Q_P^0 &= \left(R_A^0 - qx\right)\cos\varphi = (24 - 2x)\cos\varphi \\
N_P^0 &= -\left(R_A^0 - qx\right)\sin\varphi = -(24 - 2x)\sin\varphi.
\end{aligned}
\tag{c}
$$

Internal forces at specified points in the loaded state of the primary system are presented in Table 7.9.

Table 7.9 Internal forces of the arch in the loaded state

Points	M_P^0	Q_P^0	N_P^0
0	0.0	16.968	−16.968
1	63	14.400	−10.800
2	108	10.7328	−5.178
3	135	5.8206	−1.455
4	144	0.0	0.0
5	135	−5.8206	−1.455
6	108	−10.7328	−5.178
7	63	−14.400	−10.800
8	0.0	−16.968	−16.968

Computation of unit and loaded displacements For calculation of displacements, the Simpson's formula is applied. Unit and loaded displacements are

$$
\begin{aligned}
\delta_{11} &= \frac{\bar{M}_1 \times \bar{M}_1}{EI} = \sum_1^n \frac{l_i}{6EI}\left(a_1^2 + 4c_1^2 + b_1^2\right), \\
\Delta_{1P} &= \frac{\bar{M}_1 \times M_P^0}{EI} = \sum_1^n \frac{l_i}{6EI}\left(a_1 a_P + 4c_1 c_P + b_1 b_P\right),
\end{aligned}
\tag{d}
$$

where l_i is the length of the ith straight portion of the arch (Table 7.7); n number of the straight portions of the arch; a_1, a_P ordinates of the bending moment diagrams $\bar{M}_1$ and M_P^0 at the extreme left end of the portion; b_1, b_P ordinates of the same bending moment diagrams at the extreme right end of the portion; and c_1, c_P are the ordinates of the same bending moment diagrams at the middle point of the portion.

Calculation of the unit and loaded displacements is presented in Table 7.10. Section "Unit state," columns a_1, c_1, andb_1 contain data from column $\bar{M}_1$ (Table 7.8). Section "Loaded state," columns a_P, c_P, and b_P contain data from column M_P^0 (Table 7.9). As an example for portion 1–2($l_{1-2}/6 = 3.5377/6 = 0.5896$), the entries of columns 6 and 10 are obtained by following way

$$
\begin{aligned}
&\frac{0.5896}{EI}\left[(-2.625)^2 + 4\,(-3.5625)^2 + (-4.50)^2\right] = 45.9335/EI, \\
&\frac{0.5896}{EI}\left[(-2.625) \times 63 + 4\,(-3.5625) \times 85.5 + (-4.50) \times 108\right] \\
&= -1{,}102.40/EI.
\end{aligned}
$$

Table 7.10 Calculation of coefficient and free term of canonical equation

Portion	$\frac{l_i}{6EI}$	Unit state a_1	c_1	b_1	$\frac{\bar{M}_1 \times \bar{M}_1}{EI}$	Loaded state a_P	c_P	b_P	$\frac{\bar{M}_1 \times M_P^0}{EI}$
1	2	3	4	5	6	7	8	9	10
0-1	0.6644	0.0	−1.3125	−2.625	9.1563	0.0	31.5	63	−219.75
1-2	0.5896	−2.625	−3.5625	−4.500	45.9335	63	85.5	108	−1, 102.40
2-3	0.5340	−4.500	−5.0625	−5.625	82.4529	108	120.5	135	−1, 978.87
3-4	0.5039	−5.625	−5.8125	−6.000	102.1815	135	139.5	144	−2, 452.35
4-5	0.5039	−6.000	−5.8125	−5.625	102.1815	144	139.5	135	−2, 452.35
5-6	0.5340	−5.625	−5.0625	−4.500	82.4529	135	121.5	108	−1, 978.87
6-7	0.5896	−4.500	−3.5625	−2.625	45.9335	108	85.5	63	−1, 102.40
7-8	0.6644	−2.625	−1.3125	0.0	9.1563	63	31.5	0.0	−219.40
Factor	$1/EI$				$1/EI$				$1/EI$

$$\delta_{11} = \frac{479.4484}{EI} \text{ (m/kN)} \quad \Delta_{1P} = -\frac{11{,}506.74}{EI} \text{ (m)}$$

Canonical equation and primary unknown Canonical equation and primary unknown (thrust) are

$$\frac{479.4484}{EI} X_1 - \frac{11{,}506.74}{EI} = 0 \rightarrow X_1 = 24.00\,\overrightarrow{\text{kN}}.$$

Construction of internal force diagrams Internal forces, which arise in the entire structure, may be calculated by formulas

$$\begin{aligned} M &= \bar{M}_1 X_1 + M_P^0 \\ Q &= \bar{Q}_1 X_1 + Q_P^0 \\ N &= \bar{N}_1 X_1 + N_P^0. \end{aligned} \tag{e}$$

Calculation of internal forces in the arch due to given fixed load is presented in Table 7.11; internal forces $\bar{M}_1$,$\overline{Q}_1$, and $\bar{N}_1$ due to unit primary unknown $X_1 = 1$ are presented earlier in Table 7.8.

Table 7.11 Calculation of internal forces at specified points of the arch

Points	$\overline{M}_1 X_1$	$\overline{Q}_1 X_1$	$\overline{N}_1 X_1$	M_P^0	Q_P^0	N_P^0	M	Q	N
	1	2	3	4	5	6	1 + 4	2 + 5	3 + 6
0	0.0	−16.968	−16.968	0.0	16.968	−16.968	0.0	0.0	−33.936
1	−63	−14.400	−19.2	63	14.400	−10.80	0.0	0.0	−30.0
2	−108	−10.733	−21.466	108	10.733	−5.178	0.0	0.0	−26.644
3	−135	−5.820	−23.282	135	5.820	−1.455	0.0	0.0	−24.737
4	−144	0.0	−24.00	144	0.0	0.0	0.0	0.0	−24.0
5	−135	5.820	−23.282	135	−5.820	−1.455	0.0	0.0	−24.737
6	−108	10.733	−21.466	108	−10.733	−5.178	0.0	0.0	−26.674
7	−63	14.40	−19.20	63	−14.40	−10.800	0.0	0.0	−30.0
8	0.0	16.968	−16.968	0.0	−16.968	−16.968	0.0	0.0	−33.936

Corresponding axial force diagram is presented in Fig. 7.15d.

Knowing the internal forces at points 0 and 8 we can calculate the reactions at supports A and B. Axial force $N_0 = 33.936\,\text{kN}$ is shown at support A (Fig. 7.15d). Reactions of this support are $R_A = N_0 \sin\varphi_0 = 33.936 \times 0.707 = 24\,\text{kN}$ and $H = N_0 \cos\varphi_0 = 24\,\text{kN}$. For primary unknown X_1 we have obtained the same result.

Verification of results.

(a) For arch in whole $\sum Y = R_A + R_B - q \cdot 24 = 0$.
(b) The bending moment at any point of the arch

$$M(x) = R_A x - Hy - \frac{qx^2}{2} = 24x - 24\frac{4f}{l^2}x(l-x) - \frac{qx^2}{2}$$

for given parameters f, l, and q indeed equals to zero for *any* x.

Discussion:

1. If two-hinged uniform parabolic arch is subjected to uniformly distributed load within all span, then this arch is rational since the bending moments and shear forces are equal to zero in all sections of the arch. In this case, only axial compressed forces arise in all section of the arch.
2. Procedure for analysis of nonuniform arch remains same. However, in this case the Table 7.6 must contain additional column with parameter EI for *each point* 0–8, Table 7.7 must contain parameter EI_i for *middle point* of each portion, and column 2 of the Table 7.10 should be replaced by column $l_i/6EI_i$.

7.4 Computation of Deflections of Redundant Structures

As known, for calculation of deflection of any bending structure it is necessary to construct the bending moment diagram in the actual state, then construct unit state and corresponding bending moment diagram, and finally, both diagrams must be multiplied

$$\Delta_k = \int_s \frac{M_P \bar{M}}{EI} \mathrm{d}s = \frac{M_P \times \bar{M}}{EI}. \tag{7.12}$$

Here M_P is a bending moment diagram of a given statically indeterminate structure due to applied load, while a bending moment diagram $\bar{M}$ is pertaining to unit state.

The construction of bending moment diagram M_P in the entire state is discussed above using the superposition principle. Other presentation of superposition principle will be considered in the following sections of the book. Now the following principal question arises: *how to construct the unit state*? It is obvious that unit load must correspond to the required deflection. But which structure must carry this unit load? It is obvious that unit load may be applied to the given statically indeterminate structure. For construction of bending moment diagram in unit state for statically indeterminate structure the additional analysis is required. Therefore computation of deflections for statically indeterminate structure becomes cumbersome. However, solution of this problem can be significantly simplified, taking into account a following fundamental concept.

Bending moment diagram M_P of any statically indeterminate structure can be considered as a result of application of two types of loads to a *statically determinate structure*. They are the given external loads and primary unknowns. It means that a given statically indeterminate structure may be replaced by *any statically determinate structure* subjected to a given load and primary unknowns, which are treated now as *external* forces. It does not matter which primary system has been used for final construction of bending moment diagram, since on the basis of *any* primary system the *final* bending moment diagram will be the same. Therefore, the unit load (force, moment, etc.), which corresponds to required displacement (linear, angular, etc.) should be applied in *any statically determinate (!) structure*, obtained from a given structure by elimination of *any* redundant constraints.

This fundamental idea is applicable for arbitrary statically indeterminate structures. Moreover, this concept may be effectively applied for verification of the resulting bending moment diagram. Since displacement in the direction of the *primary unknown* is zero, then

$$\Delta_k = \int_s \frac{M_P \bar{M}}{EI} \mathrm{d}s = \frac{M_P \times \bar{M}}{EI} = 0, \tag{7.12a}$$

where $\bar{M}$ is the bending moment diagram due to unit primary unknown. This is called a *kinematical control* of the resulting bending moment diagram. Equations (7.12) and (7.12a) are applicable for determination of deflections and kinematical verification for *any* flexural system.

Kinematical verification for structure in Table 7.2 is shown below. For given structure the vertical displacement of support A is zero. We can check this fact using above theory. Unit state is constructed as follows: the support A is eliminated and unit load $P = 1$ is applied at point A. Two bending moment diagrams, M_P and $\bar{M}$, are shown in Fig. 7.16a. Their multiplication leads to the following results:

$$\Delta_A^{\text{ver}} = \frac{\bar{M}_P \times M_P}{EI_i} = \underbrace{\frac{l}{6EI}\left(0 \times 0 + 4 \times \frac{ql^2}{16} \times \frac{l}{2} - \frac{ql^2}{8} \times l\right)}_{\text{Simpson rule}} = 0.$$

Indeed, the vertical displacement of the support A is zero. Now let us calculate the slope at support A. For this we need to show bending moment diagram M_P in entire structure (Fig. 7.16a) and apply unit moment at support A in *any* statically determinate structure. Two versions of unit states are shown in Fig. 7.16b. Computation of the slope leads to the following results:

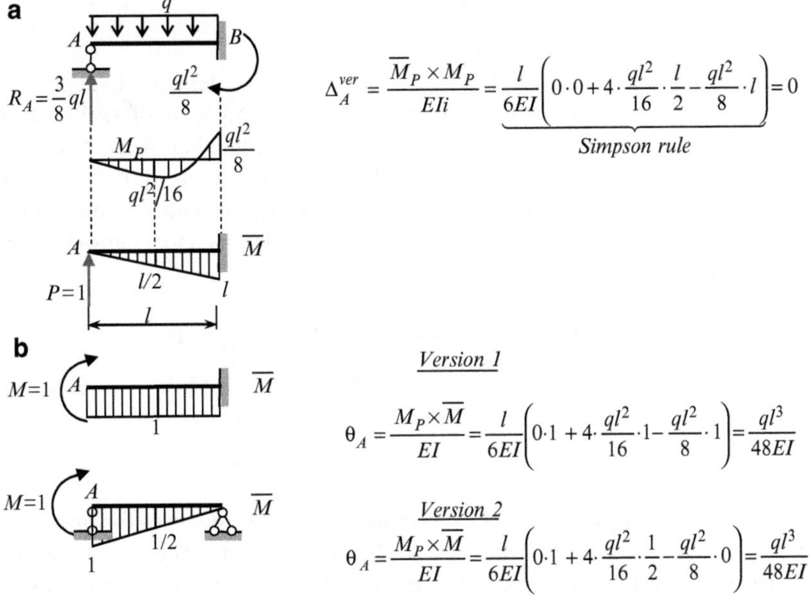

Fig. 7.16 (**a**) Computation of vertical displacement at support A. (**b**) Computation of slope at support A. Two versions of unit state

Version 1:

$$\theta_A = \frac{M_P \times \bar{M}}{EI} = \frac{l}{6EI}\left(0 \times 1 + 4 \times \frac{ql^2}{16} \times 1 - \frac{ql^2}{8} \times 1\right) = \frac{ql^3}{48EI}.$$

Version 2:

$$\theta_A = \frac{M_P \times \bar{M}}{EI} = \frac{l}{6EI}\left(0 \times 1 + 4 \times \frac{ql^2}{16} \times \frac{1}{2} - \frac{ql^2}{8} \times 0\right) = \frac{ql^3}{48EI}.$$

It easy to check that superposition principle leads to the same result. Indeed, in case of simply supported beam subjected to uniformly distributed load q and support moment $M_B = ql^2/8$, the slope at support A equals

$$\theta_A = \theta_A^q + \theta_A^{M_B} = \frac{ql^3}{24EI} - \frac{M_B l}{6EI} = \frac{ql^3}{24EI} - \frac{ql^2}{8}\frac{l}{6EI} = \frac{ql^3}{48EI}.$$

The reader is invited to check that slope at support B is zero. For multiplication of both bending moment diagrams, it is recommended to apply the Simpson's rule.

7.5 Settlements of Supports

If any statically indeterminate structure is subjected to settlement of supports, then internal forces arise in the members of the structure. Analysis of such structures may be effectively performed by the force method in canonical form. The primary system and primary unknowns should be adopted as in the case of the fixed loads.

Let us consider any statically indeterminate structure with n redundant constraints. Some of the supports have linear and/or angular displacements d_i. Canonical equations are

$$\begin{aligned}
&\delta_{11}X_1 + \delta_{12}X_2 + \cdots + \delta_{1n}X_n + \Delta_{1s} = 0\\
&\delta_{21}X_1 + \delta_{22}X_2 + \cdots + \delta_{2n}X_n + \Delta_{2s} = 0\\
&\cdot \quad \cdot \quad \cdot \quad \cdot \quad \cdot \quad \cdot \quad \cdot \quad \cdot \quad \cdot \quad \cdot \quad \cdot \quad \cdot\\
&\delta_{n1}X_1 + \delta_{n2}X_2 + \cdots + \delta_{nn}X_n + \Delta_{ns} = 0,
\end{aligned} \tag{7.13}$$

where free terms Δ_{ks} $(k = 1, 2, \ldots, n)$ represent displacements of the primary system in the direction of primary unknowns X_k due to *settlements* of the supports. For calculation of these terms, we need to use the theorem of reciprocal unit displacements and reactions (Rayleigh second theorem).

Let the support i has the unit displacement δ_i. The displacement at any point k may be calculated using the above-mentioned theorem, i.e., $\delta_{ki} = -r_{ik}$. So, the displacement in direction k due to unit displacement at direction i may be calculated as reaction at support i caused by unit load at direction k.

If the support i has nonunity displacement d_i, then the displacement Δ_k at any point k may be calculated using formula $\Delta_{ki} = -\bar{R}_{ik}d_i$, where $\bar{R}_{ik}$ presents the reaction at support i due to unit load at direction k. In fact, it means that both parts of formula $\delta_{ki} = -r_{ik}$ are multiplied by d_i. In case of several displacements d_i of the supports, the free terms of canonical equation of the force method are calculated using the following expressions (index s – "settlements" – at Δ_{ks} is omitted):

$$\Delta_k = -\sum_i \bar{R}_{ik} \cdot d_i. \tag{7.14}$$

In this formula $\bar{R}_{ik}$ is reaction of the constraint in the direction of a given displacement d_i due to unit primary unknown $X_k = 1$. In other words, $\bar{R}_{i1}$ and $\bar{R}_{i2}$ are found as reactions in the primary system due to primary unknowns $X_1 = 1$ and $X_2 = 1$, respectively; these reactions are determined in supports, which are subjected to displacement. After calculation of the primary unknowns X_i construction of internal forces is performed as usual.

The final bending moment diagram can be constructed by formula

$$M_s = \bar{M}_1 \cdot X_1 + \bar{M}_2 \cdot X_2 + \cdots + \bar{M}_n \cdot X_n. \tag{7.15}$$

Note that unlike the analysis of structures subjected to loads, the term M_s^0 in (7.15) is absent.

Kinematical verification of the final bending moment diagram may be performed by the following expression

$$\sum \int \frac{M \cdot \bar{M}_{\Sigma}}{EI} ds + \sum \Delta_{is} = 0, \tag{7.16}$$

where $\bar{M}_{\Sigma}$ is a summary unit bending moment diagram.

Procedure for analysis of redundant structures subjected to the settlement of supports is as follows:

1. Provide the kinematical analysis, determine the degree of redundancy, choose the primary system of the force method, and formulate the canonical equations (7.13)
2. Construct the unit bending moment diagrams and calculate the unit displacements
3. Calculate the free terms of canonical equations
4. Solve the canonical equation with respect to primary unknowns X_i
5. Construct the internal force diagrams
6. Calculate the reactions of supports and provide their verifications

Let us consider a two-span uniform beam with equal spans. The middle support 1 is shifted by Δ as shown in Fig. 7.17a. It is necessary to calculate the bending moment at support 1.

The degree of static indeterminacy of the structure equals one. The primary system is the set of two simply supported beams (Fig. 7.17b). Bending moment diagram due to unit primary unknown is shown in Fig. 7.17c.

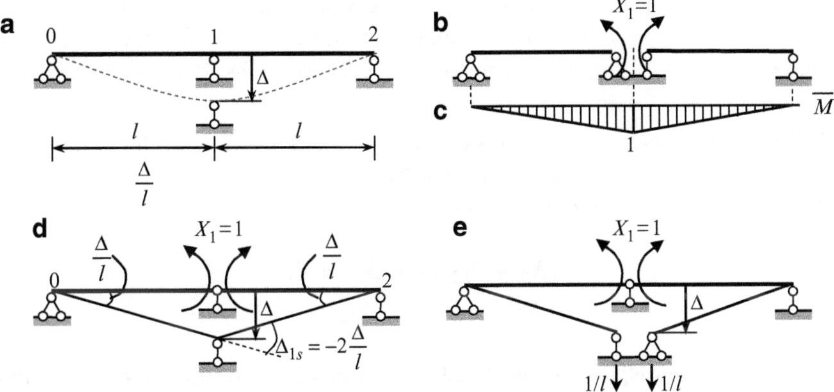

Fig. 7.17 **(a–c)** Two-span beam. Design diagram, primary system and bending moment diagram due to $X_1 = 1$. **(d, e)** Two approaches for computation of free term of canonical equation

The canonical equation and primary unknown are $\delta_{11}X_1 + \Delta_{1s} = 0$, $X_1 = -\Delta_{1s}/\delta_{11}$.

The unit displacement

$$\delta_{11} = \frac{\bar{M}_1 \times \bar{M}_1}{EI} = 2 \times \frac{1}{2}l \times 1 \times \frac{2}{3} \times \frac{1}{EI} = \frac{2l}{3EI}.$$

The free term of canonical equation Δ_{1s} is a displacement in direction of the primary unknown X_1 (i.e., the *mutual* angle of rotation at support 1) caused by the given settlements of support). In this simple case, the Δ_{1s} can be calculated by procedure shown in Fig. 7.17d. If support 1 is shifted, then slope at supports 0 and 2 are Δ/l, therefore the mutual angle of rotation $\Delta_{1s} = -\left(\frac{\Delta}{l} + \frac{\Delta}{l}\right) = -2\frac{\Delta}{l}$. Negative sign means that each moment X produce the negative work on the angular displacement Δ/l due to the settlement of support.

The free term can be calculated using a general algorithm (7.14) as follows: let us apply $X_1 = 1$ and compute the *reaction in direction of given displacement*. As a result, two reactions each $1/l$ arises at the support 1, as shown in Fig. 7.14e. Expression (7.14) leads to the same result. Indeed, $\Delta_{1s} = -2\,(1/l) \times \Delta$.

Canonical equation becomes $(2l/3EI)\,X_1 - 2\,(\Delta/l) = 0$. The primary unknown, i.e., the bending moment at support 1 equals $X_1 = M_1 = 3\left(EI/l^2\right)\Delta$. A positive sign shows that extended fibers are located below the longitudinal axis of the beam. Note that bending moments at supports of uniform continuous beams with equal spans caused by vertical displacements of one of its supports are presented in Table A.18.

Conclusion: In case of settlements supports, the distribution of internal forces depends not only on relative stiffness, but on absolute stiffness *EI* as well.

Example 7.3. The design diagram of the redundant frame is the same as in Example 7.2. No external load is applied to the frame, but the frame is subjected to settlement of fixed support *A* as presented in Fig. 7.18a. Construct the internal force diagrams and calculate reactions of supports. Assume that the vertical, horizontal, and angular settlements are $a = 2$ cm, $b = 1$ cm, and $\varphi = 0.01$ rad $= 34'30''$, respectively.

Solution. Let the primary system is chosen as in Example 7.2, so the primary unknowns are reactions X_1 and X_2 (Fig. 7.18b).

Canonical equations of the force method are

$$\begin{aligned} \delta_{11}X_1 + \delta_{12}X_2 + \Delta_{1s} &= 0, \\ \delta_{21}X_1 + \delta_{22}X_2 + \Delta_{2s} &= 0, \end{aligned} \tag{a}$$

where δ_{ik} are unit displacements, which have been obtained in Example 7.2. They are equal to

$$\delta_{11} = \frac{666.67}{EI}, \quad \delta_{12} = \delta_{21} = \frac{275}{EI}, \quad \delta_{22} = \frac{170.67}{EI}. \tag{b}$$

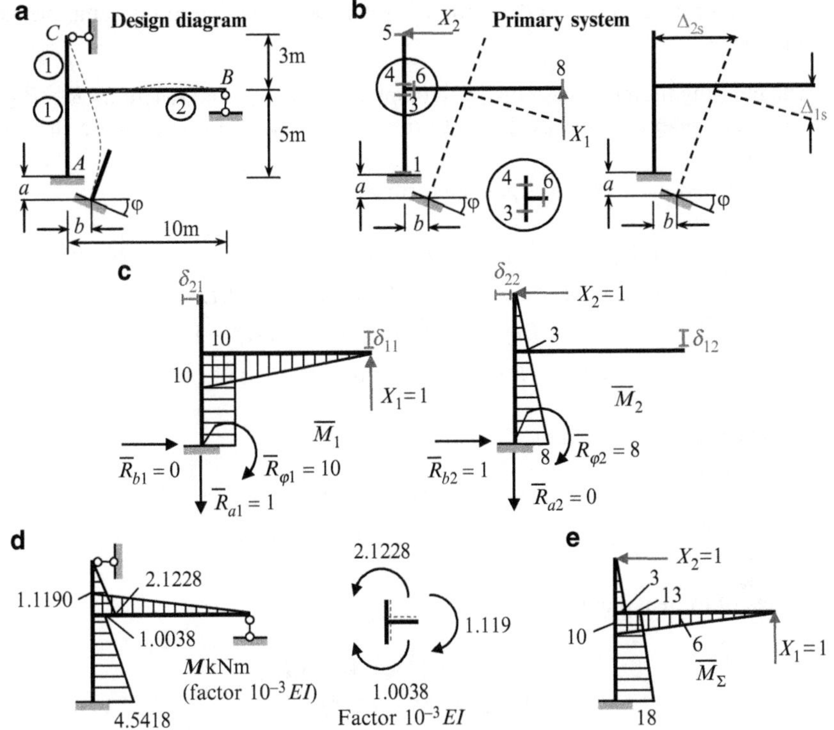

Fig. 7.18 (**a, b**) Design diagram, primary system and corresponding displacements. (**c**) Bending moment diagrams in unit conditions and reactions at the support A. (**d**) Resulting bending moment diagram; (**e**) Summary unit bending moment diagram

The free terms Δ_{1s} and Δ_{2s} present the displacements in the direction of primary unknowns X_1 and X_2 in a primary system due to settlement of support (Fig. 7.18b). According to (7.14), we have:

$$\begin{aligned}\Delta_{1s} &= -\sum \bar{R}_{i1} \cdot d_i = -(\bar{R}_{a1} \cdot a + \bar{R}_{b1} \cdot b + \bar{R}_{\varphi1} \cdot \varphi),\\ \Delta_{2s} &= -\sum \bar{R}_{i2} \cdot d_i = -(\bar{R}_{a2} \cdot a + \bar{R}_{b2} \cdot b + \bar{R}_{\varphi2} \cdot \varphi),\end{aligned} \quad (7.3)$$

where a, b, and φ are the vertical, horizontal, and angular displacements of support A. The unit reactions $\bar{R}$ related to support, which is subjected to the given settlement. Each reaction has two indexes. The first index (a, b, or φ) corresponds to given direction of the settlement of the support. The second index (1 and 2) means the primary unknown (X_1 or X_2) which leads to appearance of unit reactions. Thus,

$\bar{R}_{a1}, \bar{R}_{a2}$ are reactions in direction of a-displacement due to primary unknowns $X_1 = 1$ and $X_2 = 1$
$\bar{R}_{b1}, \bar{R}_{b2}$ are reactions in direction of b-displacement due to primary unknowns $X_1 = 1$ and $X_2 = 1$

$\bar{R}_{\varphi 1}$, $\bar{R}_{\varphi 2}$ are reactions in direction of φ-displacement due to the same primary unknowns

The bending moment diagrams, unit displacements δ_{ik}, and reactions $\bar{R}_{i1}$ $\bar{R}_{i2}$ in the primary system caused by the unit primary unknowns X_1 and X_2 are shown in Fig. 7.18c.

The expressions (c) in an expanded form are

$$\begin{aligned}\Delta_{1s} &= -(\bar{R}_{a1} \cdot a + \bar{R}_{b1} \cdot b + \bar{R}_{\varphi 1} \cdot \varphi) = -(1 \cdot a + 0 \cdot b + 10 \cdot \varphi) \\ &= -(1 \times 0.02 + 0 \times 0.01 + 10 \times 0.01) = -0.12\,\text{m}, \\ \Delta_{2s} &= -(\bar{R}_{a2} \cdot a + \bar{R}_{b2} \cdot b + \bar{R}_{\varphi 2} \cdot \varphi) = -(0 \cdot a + 1 \cdot b + 8 \cdot \varphi) \\ &= -(0 \times 0.02 + 1 \times 0.01 + 8 \times 0.01) = -0.09\,\text{m}.\end{aligned} \tag{7.4}$$

Canonical equations become

$$\begin{aligned}666.67X_1 + 275X_2 - 0.12EI &= 0, \\ 275X_1 + 170.67X_2 - 0.09EI &= 0.\end{aligned} \tag{7.5}$$

The roots of these equations are

$$\begin{aligned}X_1 &= -1.119 \times 10^{-4}EI, \\ X_2 &= 7.076 \times 10^{-4}EI.\end{aligned} \tag{7.6}$$

The bending moment diagram is constructed using the superposition principle according to (7.15)

$$M = \bar{M}_1 \cdot X_1 + \bar{M}_2 \cdot X_2. \tag{7.7}$$

The corresponding calculations are presented in Table 7.12.

Table 7.12 Calculation of bending moments

Points	$\bar{M}_1$	$\bar{M}_1 \cdot X_1$	$\bar{M}_2$	$\bar{M}_2 \cdot X_2$	M
1	−10	1.119	−8.0	−5.6608	−4.5418
3	−10	1.119	−3.0	−2.1228	−1.0038
4	0.0	0.0	−3.0	−2.1228	−2.1228
5	0.0	0.0	0.0	0.0	0.0
6	−10	1.119	0.0	0.0	1.1190
8	0.0	0.0	0.0	0.0	0.0
Factor		$10^{-3}EI$		$10^{-3}EI$	$10^{-3}EI$

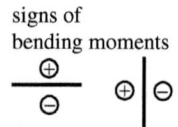

signs of bending moments

⊕ ⊖ ⊕ ⊖

The resulting bending moment diagram is presented in Fig. 7.18d.

Static verification A free body diagram for rigid joint of the frame is shown in Fig. 7.18d. The extended fibers are shown by dotted lines. Equilibrium of the rigid joint

$$\sum M = (2.1228 - 1.0038 - 1.119) \times 10^{-3}EI = 0.$$

Kinematical verification The summary unit bending moment diagram $\bar{M}_{\Sigma} = \bar{M}_1 + \bar{M}_2$ is shown in Fig. 7.18e. The formula (7.16) leads to the following result

$$\frac{M \times \bar{M}_{\Sigma}}{EI} + \sum \Delta_{is}$$
$$= \left[\underbrace{\frac{1}{1EI} \times \frac{1}{2} \times 3 \times 3 \times \frac{2}{3} \times 2.1228}_{\text{portion 4-5}} - \underbrace{\frac{1}{2EI} \times \frac{1}{2} \times 10 \times 10 \times \frac{2}{3} \times 1.119}_{\text{portion 6-8}}\right] \times 10^{-3} EI$$
$$+ \underbrace{\frac{1}{1EI} \times \frac{5}{6}(2 \times 13 \times 1.0038 + 2 \times 18 \times 4.5418 + 13 \times 4.5418 + 18 \times 1.0038) \times 10^{-3} EI}_{\text{portion 1-3}}$$
$$+ \underbrace{(-0.12)}_{\Delta_{1s}} + \underbrace{(-0.09)}_{\Delta_{2s}} = 0.2286 - 0.2286 = 0.$$

After verification for bending moment diagram, we can construct the shear and axial force diagrams and find reactions of supports. These procedures have been described in Example 7.2.

Discussion: In the case of settlements of supports, the primary unknowns as well as the reactions and internal forces (bending moment, shear, and normal force) depends on both the relative and absolute stiffnesses *EI*. This is the common property of any statically indeterminate structure subjected to settlement of supports.

7.6 Temperature Changes

If an arbitrary statically indeterminate structure is subjected to change of temperature, then the internal forces arises in the members of the structure. Analysis of such structure may be effectively performed on the basis of a canonical equation of the force method. A primary system and the primary unknowns of the force method are chosen as usual. The canonical equations for structure with n redundant constraints are

$$\begin{aligned}
\delta_{11}X_1 + \delta_{12}X_2 + \cdots + \delta_{1n}X_n + \Delta_{1t} &= 0 \\
\delta_{21}X_1 + \delta_{22}X_2 + \cdots + \delta_{2n}X_n + \Delta_{2t} &= 0 \\
\cdots\cdots\cdots\cdots\cdots\cdots \\
\delta_{n1}X_1 + \delta_{n2}X_2 + \cdots + \delta_{nn}X_n + \Delta_{nt} &= 0,
\end{aligned} \tag{7.17}$$

where X_n are primary unknowns and δ_{ik} and Δ_{it} are displacements of the primary system in the direction of ith primary unknown caused by the unit primary unknown $X_k = 1$ and by the change of temperature, respectively.

The unit displacements should be calculated as usual. The free terms Δ_{it} $(i = 1, 2, \ldots, n)$ should be calculated using the following expression

$$\Delta_{it} = \sum \int \alpha t_{av} \bar{N}_i \mathrm{d}s + \sum \int \alpha \frac{\Delta t}{h} \bar{M}_i \mathrm{d}s, \tag{7.18}$$

where α is the coefficient of thermal expansion; h is a height of a cross section of a member; $\bar{N}_i$, $\bar{M}_i$ are normal force and bending moment in a primary system due to action of unit primary unknown X_i.

In case of constant α, t_{av}, Δt, and h within the each member

$$\Delta_{it} = \sum \alpha t_{\text{av}} \int \bar{N}_i \mathrm{d}s + \sum \alpha \frac{\Delta t}{h} \int \bar{M}_i \mathrm{d}s. \tag{7.18a}$$

Thus for computation of Δ_{1t}, the unit primary unknown $X_1 = 1$ should be applied to the primary system and then procedure (7.18a) should be performed; the procedure of summation is related to all members.

The temperature at axial line (average temperature) and temperature gradient are

$$t_{\text{av}} = \frac{t_1 + t_2}{2}, \quad \Delta t = |t_1 - t_2|, \tag{7.19}$$

where t_1 and t_2 are changes of temperature on the top and bottom fibers of the member; the average temperature t_{av} and temperature gradient Δt are related to uniform and nonuniform temperature changes, respectively.

The integrals $\int \bar{N}_i \mathrm{d}s$, $\int \bar{M}_i \mathrm{d}s$ present area of corresponding diagram in the primary system.

Solution of (7.17) is the primary unknowns X_i. Bending moment diagram is constructed using the formula

$$M_t = \bar{M}_1 \cdot X_1 + \bar{M}_2 \cdot X_2 + \cdots + \bar{M}_n \cdot X_n. \tag{7.20}$$

Kinematical control of the final bending moment diagram may be performed using the following expression

$$\sum \int \frac{M_t \cdot \bar{M}_\Sigma}{EI} \mathrm{d}s + \sum \Delta_{it} = 0, \tag{7.21}$$

where $\bar{M}_\Sigma$ is the summary unit bending moment diagram and M_t is the resultant bending moment diagram caused by change of temperature.

Procedure for analysis of redundant structures subjected to the change of temperature is as follows:

1. Provide the kinematical analysis, determine the degree of redundancy, choose the primary system of the force method, and formulate the canonical equations (7.17)

2. Construct the unit bending moment and axial force diagrams
3. Calculate the unit displacements. In case of bending structures take into account only bending moments
4. Calculate the free terms of canonical equations. For this:

   (a) Calculate the average temperature and temperature gradient for each member of a structure
   (b) Apply formula (7.18a)

5. Solve the canonical equation (7.17) with respect to primary unknowns X_i.
6. Construct the internal force diagrams.
7. Calculate the reactions of supports and provide their verifications.

Let us consider fixed-rolled uniform beam (the height of the cross section of the beam is h), subjected to following change of temperature: the temperature of the above beam is increased by t_1°, while below of the beam increased by the t_2°, $t_1 > t_2$ (Fig. 7.19a).

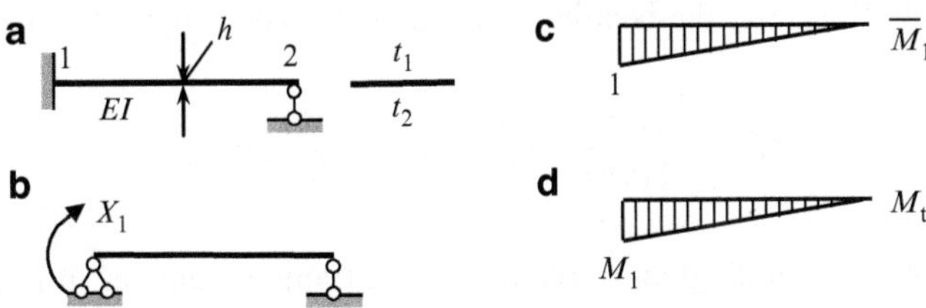

Fig. 7.19 (**a**) Design diagram, (**b**) Primary system; (**c**) Unit bending moment diagram; (**d**) Final bending moment diagram

The primary system is the simply supported beam and the primary unknown is the moment at the fixed support (Fig. 7.19b). Bending moment diagram for unit state in primary system is shown in Fig. 7.19c.

The canonical equation is $\delta_{11}X_1 + \Delta_{1t} = 0$. Bending moment diagram due to unit primary unknown is shown in Fig. 7.19c. The unit displacement is

$$\delta_{11} = \frac{\bar{M}_1 \times \bar{M}_1}{EI} = \frac{1}{2}l \times 1 \times \frac{2}{3} \times \frac{1}{EI} = \frac{l}{3EI}.$$

The axial force $\bar{N}_1$ in primary system due to the primary unknown is zero, so the first term of (7.18a) is zero. The second term of (7.18a) contains the expression $\int \bar{M}_1 \mathrm{d}s$, which is the area of the bending moment diagram in the unit state. In our case the free term may be presented as

$$\Delta_{1t} = \alpha\frac{\Delta t}{h}\int \bar{M}_1 \mathrm{d}s = -a\frac{t_1 - t_2}{h} \times \underbrace{\frac{1}{2} \times 1 \times l}_{\text{area } \bar{M}_1} = -\frac{al}{2h}(t_1 - t_2).$$

The negative sign means that bending moment diagram is plotted on the more cold fibers.

Canonical equation becomes

$$\frac{l}{3EI}X_1 - \frac{al}{2h}(t_1 - t_2) = 0.$$

So, the bending moment at the clamped support is $X_1 = M_1 = (3aEI/2h)(t_1 - t_2)$.

Final bending moment diagram M_t is constructed on the basis of (7.20) and shown in Fig. 7.19d.

Kinematical control leads to the following result

$$\int \frac{M_t \bar{M}_1}{EI} ds + \Delta_{1t} = \frac{1}{EI} \cdot \underbrace{\frac{1}{2} \times 1 \times l}_{\Omega} \cdot \underbrace{\frac{2}{3} \times \frac{3aEI}{2h}(t_1 - t_2)}_{y_c} - \frac{al}{2h}(t_1 - t_2) = 0.$$

Discussion: Let the height of rectangular cross section of the beam is increased by n times. In this case, the bending moment at clamped support increases by n^2 because

$$M_1 = \frac{3aE}{2(nh)} \frac{b(nh)^3}{12} \Delta t = \frac{3aEI}{2h} \Delta t \cdot n^2.$$

Example 7.4. Design diagram of the redundant frame is same as in Example 7.2. Stiffness for vertical and horizontal members are $1EI$ and $2EI$, respectively. Heights of the cross section h for vertical and horizontal members are 0.60 m. The frame is subjected to temperature changes as presented in Fig. 7.20a. Construct the internal force diagrams and calculate reactions of supports.

Solution. Let us accept a primary system as shown in Fig. 7.20b.

Canonical equations of the force method are

$$\begin{aligned} \delta_{11} X_1 + \delta_{12} X_2 + \Delta_{1t} &= 0, \\ \delta_{21} X_1 + \delta_{22} X_2 + \Delta_{2t} &= 0, \end{aligned} \tag{a}$$

where δ_{ik} are unit displacements and Δ_{1t} and Δ_{2t} are displacements in a primary system in the direction of primary unknowns X_1 and X_2, respectively, due to change of temperature.

The unit states and corresponding bending moment and axial force diagrams caused by unit primary unknowns are presented in Fig. 7.20c. All normal forces due to $\bar{X}_2 = 1$ are zero, i.e., $\bar{N}_2 = 0$.

The unit displacements δ_{ik} for this primary system have been obtained in Example 7.2; they are

$$\delta_{11} = \frac{666.67}{EI}, \quad \delta_{12} = \delta_{21} = \frac{275}{EI}, \quad \delta_{22} = \frac{170.67}{EI}. \tag{b}$$

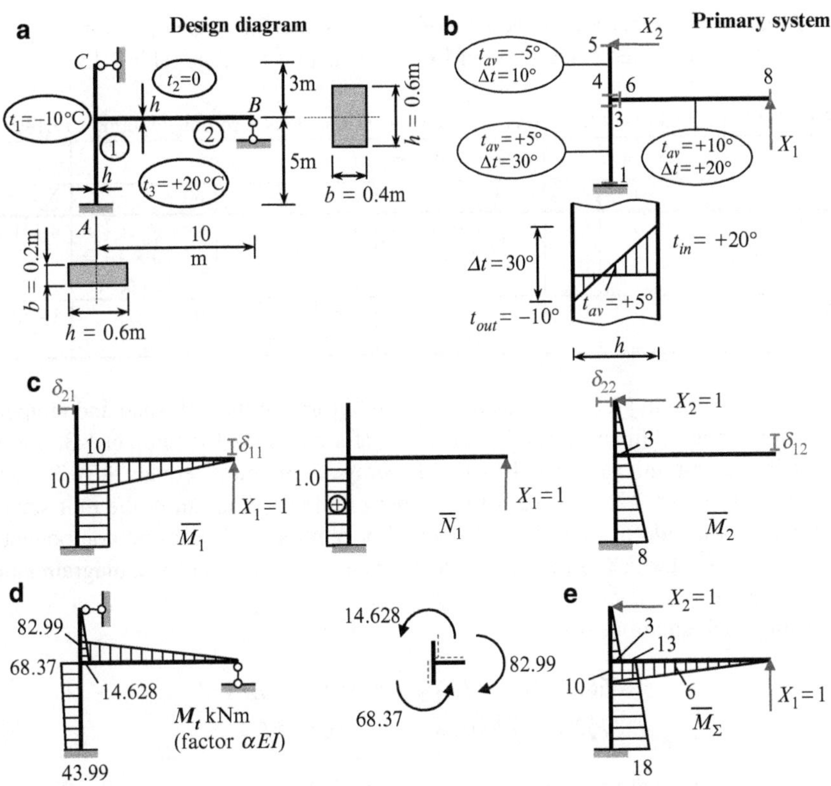

Fig. 7.20 (**a**)Design diagram and temperature distribution; (**b**) Primary system. (**c**) Bending moment and axial force diagrams in unit states. (**d**) Final bending moment diagram and static verification. (**e**) Summary unit bending moment diagram

For calculation of free terms we will use expressions (7.18a). These expressions contain the average temperature t_{av} and temperature gradient Δt for each member; they are

For portion 1-3

$$t_{av} = \frac{20 + (-10)}{2} = 5°; \quad \Delta t = |-10 - (+20)| = 30°,$$

for portion 4-5

$$t_{av} = \frac{0 + (-10)}{2} = -5° \quad \Delta t = |-10 - 0| = 10°, \tag{c}$$

for portion 6-8

$$t_{av} = \frac{0+20}{2} = 10° \quad \Delta t = |0-(+20)| = +20°.$$

These parameters for each element of the frame are shown on the primary system in Fig. 7.20b. Detailed calculation of free terms is presented in Table 7.13.

Table 7.13 Calculation of free terms Δ_{it} of canonical equations according to formula (7.18a)

	$\alpha t_{av} \int \bar{N}_i ds$ for portion			$\alpha \frac{\Delta t}{h} \int \bar{M}_i ds$ for portion			
	1–3	4–5	6–8	1–3	4–5	6–8	$\sum$
Δ_{1t}	$\alpha \times 5 \times 1 \times 5 = 25\alpha$	0	0	$\alpha \frac{30}{0.6} \times 10 \times 5 = 2{,}500\alpha$	0	$\alpha \times \frac{20}{0.6} \times \frac{1}{2} \times 10 \times 10 = 1{,}666.7\alpha$	4,191.7 α
Δ_{2t}	0	0	0	$\alpha \times \frac{30}{0.6} \times \frac{3+8}{2} \times 5 = 1{,}375\alpha$	$\alpha \times \frac{10}{0.6} \times \frac{1}{2} \times 3 \times 3 = 75\alpha$	0	1,450α

The first term in (7.18a) is positive if normal force in the unit state and temperature of the neutral fiber t_{av} has the same sign (for example, for element 1-3, normal force in the first unit state equals +1 and average temperature equals +5). The second term in (7.18a) is positive if the bending moment diagram in the unit state is located on the side of more heated fibers. For member 1-3, the first component is $\Delta_{1t} = \alpha \times 5 \times 1 \times 5$, where 1×5 presents the area of the axial force diagram along this member.

Canonical equations become

$$\begin{aligned} 666.67X_1 + 275X_2 + 4,191.7\alpha \cdot EI &= 0, \\ 275X_1 + 170.67X_2 + 1,450\alpha \cdot EI &= 0. \end{aligned} \qquad \text{(d)}$$

Roots of these equations are $X_1 = -8.2988\alpha EI$, $X_2 = 4.8759\alpha EI$. Bending moment diagram is constructed using formula (7.20). The corresponding calculation is presented in Table 7.14. Columns $\bar{M}_1 \cdot X_1$, $\bar{M}_2 \cdot X_2$, and M_t have to be multiplied by factor αEI.

Table 7.14 Calculation of bending moments

Points	$\bar{M}_1$	$\bar{M}_1 \cdot X_1$	$\bar{M}_2$	$\bar{M}_2 \cdot X_2$	M_t
1	−10	+82.998	−8.0	−39.008	+43.99
3	−10	+82.998	−3.0	−14.628	+68.37
4	0.0	0.0	−3.0	−14.628	−14.628
5	0.0	0.0	0.0	0.0	0.0
6	−10	+82.998	0.0	0.0	+82.99
8	0.0	0.0	0.0	0.0	0.0

signs of bending moments
⊕ ⊖ ⊕ ⊖

The resulting bending moment diagram is presented in Fig. 7.20d.

Static verification.

$$\sum M = (68.37 + 14.628 - 82.99) \cdot \alpha EI = (82.998 - 82.99) \cdot \alpha EI \approx 0.$$

Kinematical verification The summary unit bending moment diagram $\bar{M}_{\Sigma}$ is shown in Fig. 7.20e.

The expression (7.21) leads to the following result:

$$\sum\int\frac{M_t\cdot\bar{M}_{\Sigma}}{EI}ds+\sum\Delta_{it}=-\frac{5}{1EI\times 6}\left(18\times 43.99+4\times\frac{13+18}{2}\times\frac{43.99+68.37}{2}\right.$$

$$\left.+13\times 68.37\right)\cdot\alpha\,EI$$

$$\underbrace{-\frac{1}{1EI}\times\frac{1}{2}\times 3\times 3\times\frac{2}{3}\times(-14.628)\,\alpha\,EI}_{\text{portion 4-5}}-\underbrace{\frac{1}{2EI}\times\frac{1}{2}\times 10\times 10\times\frac{2}{3}\times 82.99\alpha\,EI}_{\text{portion 6-8}}$$

$$+\underbrace{4{,}191.7\alpha}_{\Delta_{1t}}+\underbrace{1{,}450\alpha}_{\Delta_{2t}}=-5{,}686.32\alpha+5{,}685.58\alpha\cong 0.$$

Summary: Distribution of internal forces (bending moment, shear force and normal force) as well as reactions depends on both relative and absolute stiffnesses, as well as on coefficient of thermal expansion. This is the property of any statically indeterminate structure subjected to change of temperature.

Statically indeterminate trusses If any members of a truss are subjected to change of temperature, then the unit displacements and free terms of canonical equations of the force method (7.17) should be calculated by the formulas

$$\delta_{ik}=\sum_j\frac{\bar{N}_i\cdot\bar{N}_k\cdot l}{EA},$$

$$\Delta_{it}=\alpha\sum_j\bar{N}_i\cdot\Delta t\cdot l, \tag{7.22}$$

where l is a length of jth member of the truss; $\bar{N}_i, \bar{N}_k$ are internal forces in jth member due to unit primary unknowns $X_i=1$, $X_k=1$; Δt is a thermal gradient; and α is a coefficient of thermal expansion.

In expressions (7.22), summation is done on all members of the truss (subscripts j are not shown).

Axial force in members of the truss using the superposition principle is determined by the formula

$$N=\bar{N}_1\cdot X_1+\bar{N}_2\cdot X_2+\cdots+\bar{N}_n\cdot X_n, \tag{7.23}$$

where n is degree of statical indeterminacy.

Example 7.5. Design diagram of the truss is presented in Fig. 7.21. Axial stiffness for diagonal and vertical members equals EA, for upper and lower chords equal $2EA$. Determine the reaction of the middle support and internal forces in all members of the truss caused by temperature changes on Δt degrees. Consider two cases: the temperature gradient is applied to (a) all members of the truss and (b) the lower chord only.

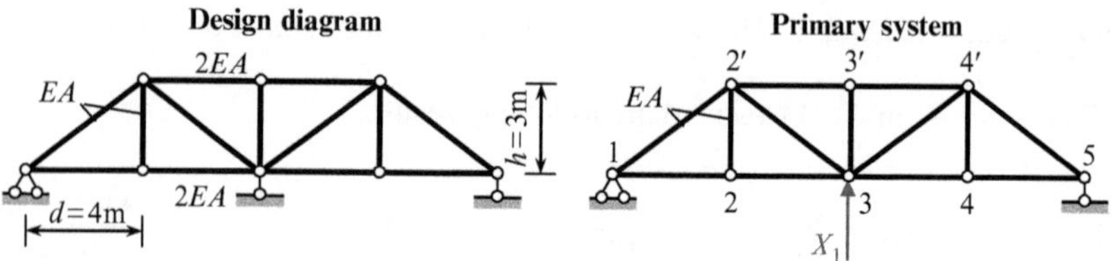

Fig. 7.21 Externally statically indeterminate truss. Design diagram and primary system

Solution. The primary system is obtained by eliminating the middle support. The primary unknown X_1 is the reaction of the middle support (Fig. 7.21).

Canonical equation of the force method is $\delta_{11}X_1 + \Delta_{1t} = 0$. The primary unknown is $X_1 = -\Delta_{1t}/\delta_{11}$. Final internal forces may be calculated by formula $N_t = \bar{N}_1 \cdot X_1$

Results of the analysis are presented in Table 7.15. The column 1 contains lengths of all members of the truss. Internal forces in the primary system caused by unit primary unknown $X_1 = 1$ are presented in column 2. The column 3 serves for calculation of the free term of the canonical equation for the case when *all members of the truss* are subjected to temperature gradient Δt. In this case, $\Delta_{1t} = 18.994 - 19.002 \cong 0$ (relative error equals 0.04%), and for primary unknown we get $X_1 = 0$. It means that if the change of temperature is related to *all* members of the truss, then internal forces in the members of the truss do not arise.

Table 7.15 Statically indeterminate truss subjected to temperature change – calculation of internal forces in all members of the truss

Members		$L\,(m)$	$\bar{N}_1$	$\alpha\bar{N}_1 \cdot \Delta t \cdot l$ Δt related to *all* members	$\alpha\bar{N}_1 \cdot \Delta t \cdot l$ Δt related to *lower chord only*	$\bar{N}_1^2 \cdot \dfrac{l}{EA}$	$N_t = \bar{N}_1 \cdot X_1$	$\bar{N}_1 \cdot N_t \dfrac{l}{EA}$
		1	2	3	4	5	6	7
Upper chord	2′-3′	4	1.333	5.332	0	3.554	0.580	1.546
	3′-4′	4	1.333	5.332	0	3.554	0.580	1.546
Lower chord	1-2	4	−0.667	−2.668	−2.668	0.890	−0.290	0.387
	2-3	4	−0.667	−2.668	−2.668	0.890	−0.290	0.387
	3-4	4	−0.667	−2.668	−2.668	0.890	−0.290	0.387
	4-5	4	−0.667	−2.668	−2.668	0.890	−0.290	0.387
Vertical members	2-2′	3	0	0	0	0	0	0
	3-3′	3	0	0	0	0	0	0
	4-4′	3	0	0	0	0	0	0
Diagonals	1-2′	5	0.833	4.165	0	3.470	0.362	1.508
	3-2′	5	0.833	4.165	0	3.470	0.362	1.508
	3-4′	5	0.833	4.165	0	3.470	0.362	1.508
	5-4′	5	0.833	4.165	0	3.470	0.362	1.508
				$\Delta_{1t} \approx 0$	$\Delta_{1t} = -10.672$	$\delta_{11} = 24,548$		10.672
	Factor		–	$\alpha \cdot \Delta t$	$\alpha \cdot \Delta t$	$1/EA$	$\alpha \cdot \Delta t \cdot EA$	$\alpha \cdot \Delta t$

The column 4 serves for calculation of free term of the canonical equation if *only members of the lower chord* of the truss are subjected to temperature change Δt.

In this case,

$$\Delta_{1t} = -10.672\alpha \times \Delta t.$$

Column 5 is used for calculation of unit displacement of the canonical equation. In our case,

$$\delta_{11} = \frac{24.548}{EA}.$$

These results yield the primary unknown

$$X_1 = -\frac{\Delta_{1t}}{\delta_{11}} = \frac{10.672\alpha \times \Delta t \times EA}{24.548} = 0.435\alpha \times \Delta t \times EA.$$

Column 6 contains internal forces in all members of the truss caused by temperature changes of the lower chord only. These forces, according to (7.22), are equal to $N = \bar{N}_1 \cdot X_1$. Column 7 serves for control of analysis: the sum of all terms of this column equals to Δ_{1t} with the opposite sign.

Discussion: 1. The structure under consideration has one absolutely necessary constraint, the reaction of which may be determined from the equilibrium equation. This constraint is the support bar at the joint 1, which prevents horizontal displacement. If the truss is *externally* statically indeterminate, then the temperature gradient, which is related to *all members* of the truss, induces a displacement in the direction of that absolutely necessary constraint, and internal forces in all members of the truss induced by temperature gradient are equal to zero.

2. If any member of the truss has been made by Δ units longer than required, then this error of fabrication may be treated as a thermal expansion, i.e $\Delta = \alpha tl$, where l is a length of a member; for all other members $\alpha tl = 0$. Canonical equation becomes $\delta_{11}\mathrm{X}_1 + \Delta_{1\mathrm{t}} = 0$, $\Delta_{1\mathrm{t}} = \alpha Ntl$, where N is the stress induced in the same member by a unit force X_1.

Problems

Problems 7.1 through 7.6 are to be solved by superposition principle. The flexural rigidity, EI, is constant for each beam.

7.1. Continuous two-span beam supports the uniformly distributed load q (Fig. P7.1). Find the reaction of supports and construct the bending moment diagram.

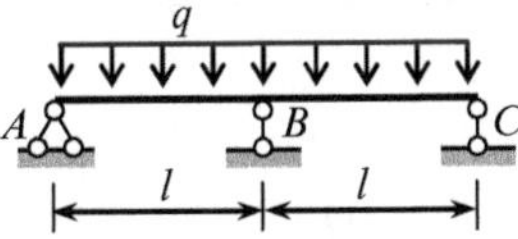

Fig. P7.1

Ans. $R_B = \frac{5}{4}ql$; $M_B = -0.125ql^2$; $M_{\max}(0.375l) = 0.0703ql^2$.

7.2. A clamped-pinned beam is subjected to couple M_0 at the roller support B (Fig. P7.2). Find the reaction of supports and construct the bending moment diagram.

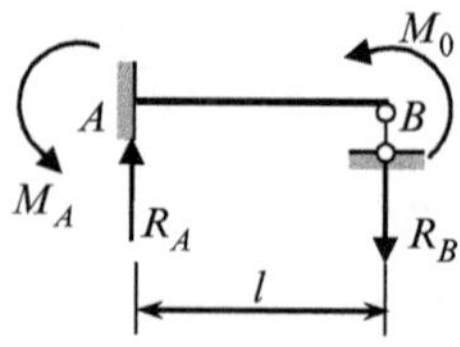

Fig. P7.2

$$\text{Ans. } M_A = -0.5M_0; \quad R_A = R_B = \frac{3}{2}\frac{M_0}{l}$$

7.3. Fixed-pinned beam is subjected to concentrated force P at the midspan (Fig. P7.3). Find the reaction of supports and construct the bending moment diagram.

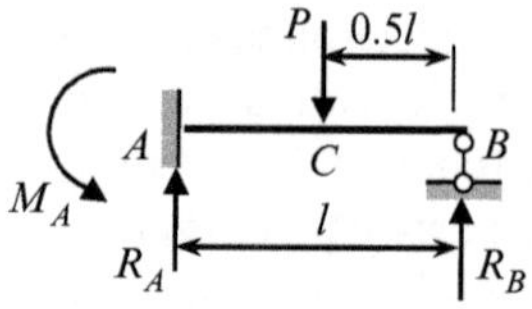

Fig. P7.3

$$\text{Ans. } M_A = -\frac{3}{16}Pl; \quad M_C = \frac{5}{32}Pl; \quad R_B = \frac{5}{16}P.$$

7.4. The beam AB with clamped support A and elastic support B is subjected to uniformly distributed load q; the stiffness coefficient of elastic support is k (Fig. P7.4). The deflection λ of support B, reaction R_B of this support and stiffness coefficient k are related as $\lambda = R_B/k$. Determine the reaction of supports. Consider two special cases: $k = \propto$, $k = 0$.

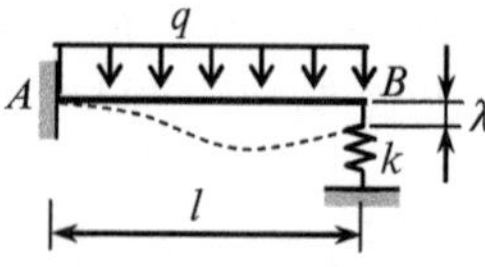

Fig. P7.4

$$\text{Ans. } R_B = \frac{3}{8}ql \cdot \frac{1}{1+\alpha}, \quad \alpha = \frac{3EI}{kl^3}$$

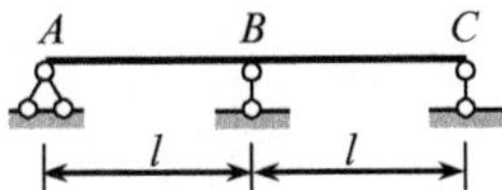

Fig. P7.5

7.5. Calculate the reaction and bending moment at the support B (Fig. P7.5), if the vertical settlement of this support Δ.

Ans. $R_B = \dfrac{6EI}{l^3}\Delta; \quad M_B = R_A l = \dfrac{3EI}{l^2}\Delta$

7.6. Pinned–pinned–pinned uniform beam is subjected to distributed load q (Fig. P7.6). Construct the bending moment diagram. Consider three versions of primary system. Primary unknown is: (1) Reaction R_1; (2) Reaction R_2; (3) Bendig moment at support 1. Calculate the deflection at the middle of the first span.

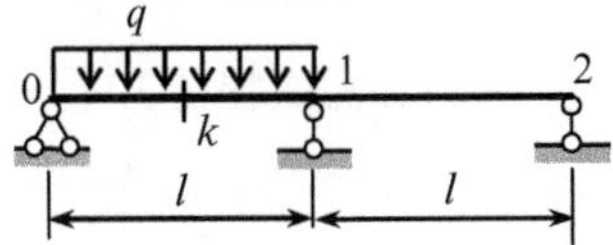

Fig. P7.6

Ans. $M_1 = -\dfrac{ql^2}{16}, \quad y_k = \dfrac{7}{768}\dfrac{ql^4}{EI}$

For next problems the Canonical equations of the force method should be used

7.7–7.8. The uniform beam is subjected to concentrated force P (Figs. P7.7 and P7.8). Calculate the reaction of supports and construct the internal force diagrams. Show the elastic curve of the beam.

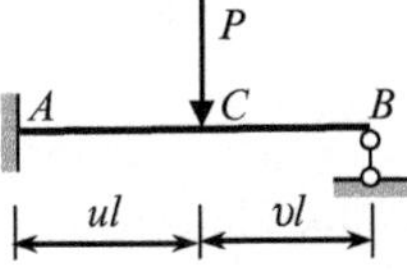

Fig. P7.7

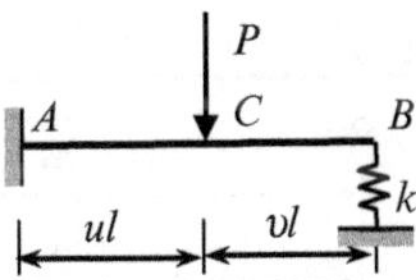

Fig. P7.8

Ans.

$$(7.7)\ R_A = \frac{P\upsilon}{2}\left(3-\upsilon^2\right), \quad M_C = \frac{Plu^2\upsilon}{2}(3-u), \quad M_A = -\frac{Pl}{2}\upsilon\left(1-\upsilon^2\right).$$

$$(7.8)\ R_B = \frac{Pu^2}{2}(3-u)\frac{1}{1+\alpha}, \quad \alpha = \frac{3EI}{kl^3}.$$

7.9–7.10. The uniform beam is subjected to uniformly distributed load q Figs. P7.9 and P7.10). Calculate the reaction of supports and construct the internal force diagrams. Show the elastic curve. For problem (7.10) use the following relationship $R = \lambda k$, where k is a stiffness coefficient of elastic support and R and λ are eaction and deflection of support B.

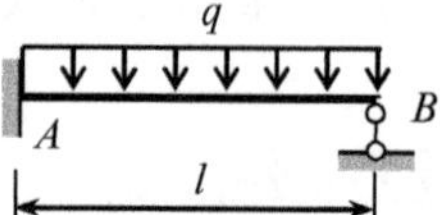

Fig. P7.9

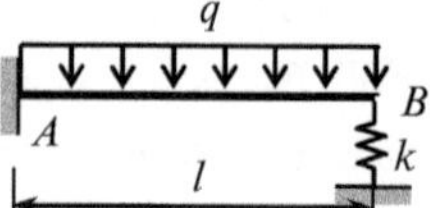

Fig. P7.10

Ans. (7.9) $R_A = \frac{5}{8}ql$; $M_A = -\frac{ql^2}{8}$; (7.10) $R_B = \frac{3}{8}ql\frac{1}{1+\alpha}$, $\alpha = \frac{3EI}{kl^3}$.

7.11. Continuous beam with clamped left support and cantilever at the right is presented in Fig. P7.11. Compute the bending moments at the supports 1 and 2.

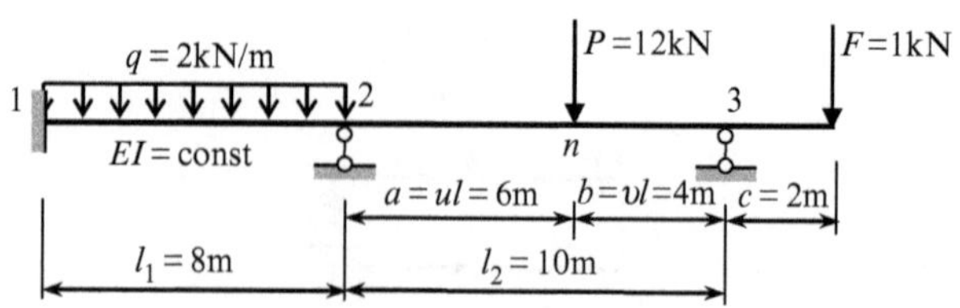

Fig. P7.11

Ans. $M_1 = -8.013\,\text{kN m}$, $M_2 = -15.975\,\text{kN m}$

7.12. The frame is subjected to uniformly distributed load q, as shown in Fig. P7.12. Construct the internal force diagrams. Calculate the reactions of supports. Show the elastic curve of the frame. Determine the horizontal displacement of the joint C.

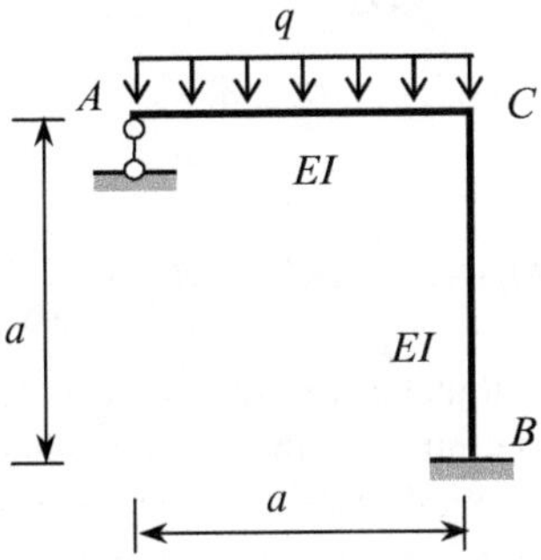

Fig. P7.12

Ans. $M_B = \dfrac{qa^2}{32}, \quad R_A = \dfrac{15}{32}qa, \quad \Delta_C^{\text{hor}} = \dfrac{qa^4}{64EI}\,(\leftarrow)$.

7.13. The frame is subjected to uniformly distributed load q, as shown in Fig. P7.13; parameter n is any positive number. Construct the bending moment diagram. Explain relationship between moments at extreme left and right points of the cross-bar. Show the elastic curve of the frame. Explain the influence of parameter n.

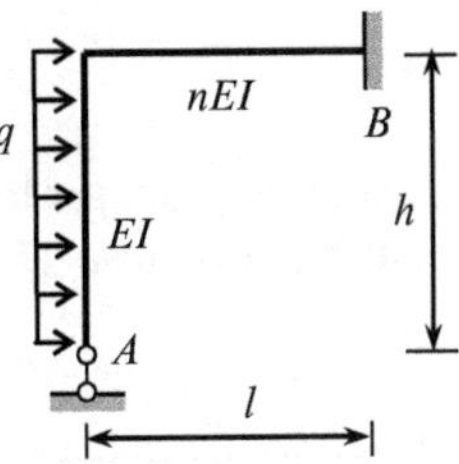

Fig. P7.13

Ans. $R_A = \dfrac{3}{4}\dfrac{qh^2}{l}$.

7.14. The portal frame is loaded by uniformly distributed load q, as shown in Fig. P7.14. Calculate the axial force X_1 in the member AB. Construct the bending moment diagram. Determine the horizontal displacement of the cross-bar.

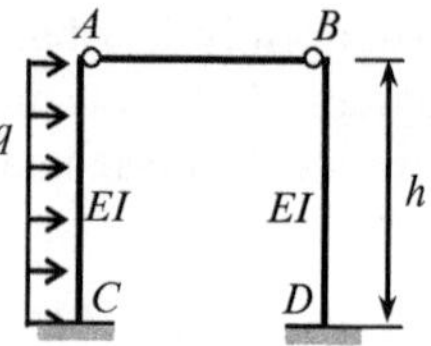

Fig. P7.14

Ans. $X_1 = -\dfrac{3}{16}qh, \quad M_C = \dfrac{5}{16}qh^2, \quad M_D = \dfrac{3}{16}qh^2, \quad \Delta = \dfrac{qh^4}{16EI}.$

7.15. The portal frame is subjected to horizontal force P. The stiffness of vertical members is *EI*, and horizontal member is *nEI*, where n is any positive number (Fig. P7.15). Calculate horizontal reaction H at support B. Construct the internal force diagrams. Analyze the influence parameter n on distribution of internal forces.

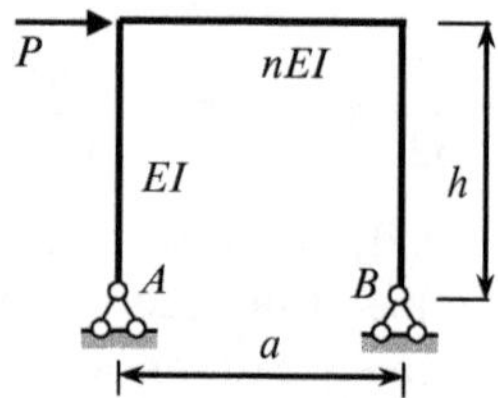

Fig. P7.15

Ans. $H = p/2.$

7.16. Construct the bending moment diagram for beam ($EI = const$) with clamped ends, subjected to uniformly distributed load q (Fig. P7.16). Calculate maximum deflection.

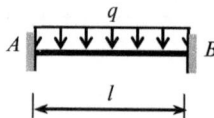

Fig. P7.16

Ans. $M_A = -\dfrac{ql^2}{12}, \quad y\left(\dfrac{l}{2}\right) = \dfrac{ql^4}{384EI}.$

7.17. The beam with clamped supports is subjected to force P as shown in Fig. P7.17. Construct the bending moment diagram. Calculate the deflection at point C. The bending stiffness is *EI*.

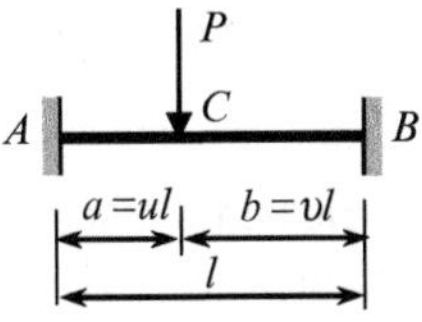

Fig. P7.17

Ans. $R_A = \frac{Pb^2}{l^3}(3a+b)$, $M_A = -Pluv^2$, $M_C = 2Plu^2v^2$,

$f_C = P\frac{a^3b^3}{3EI\,l^3}$.

7.18. The frame in Fig. P7.18 is subjected to concentrated force P and uniformly distributed load q within the vertical element. The relative flexural stiffnesses are shown in circle; $h = 5\,\text{m}$, $l = 10\,\text{m}$, $a = 4\,\text{m}$, $b = 6\,\text{m}$, $P = 8\,\text{kN}$, $q = 4\,\text{kN/m}$. Construct the bending moment diagram using different primary systems. Provide the kinematical verification. Calculate shear and axial forces and determine the reactions of supports.

Trace the elastic curve and show inflection points.

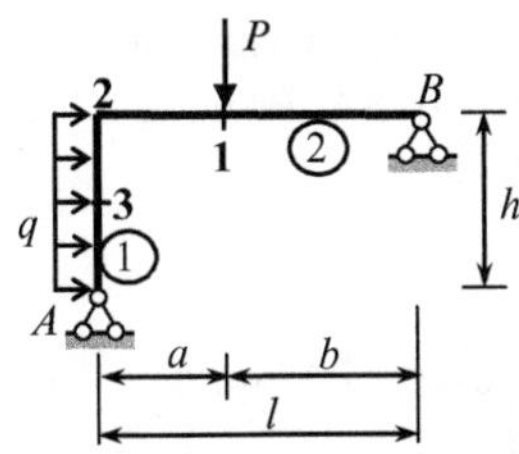

Fig. P7.18

Ans. $M_1 = 10.842\,\text{kN m}$, $M_2 = -13.93\,\text{kN m}$.

7.19. Design diagram of the frame is shown in Fig. P7.19. The flexural stiffnesses EI for all members are equal. Construct the internal force diagrams and trace the elastic curve.

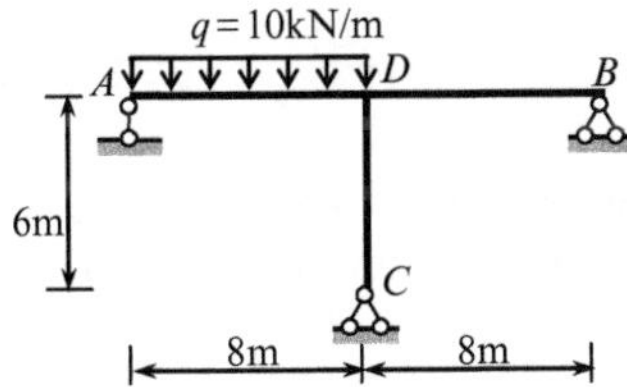

Fig. P7.19

Ans. $Q_A = 32.99\,\text{kN}, \quad Q_B = 3.02\,\text{kN}, \quad Q_C = 5.32\,\text{kN},$
$N_{CD} = -50.03\,\text{kN}.$

7.20. Symmetrical structure is subjected to load P, as shown in Fig. P7.20. The cross-sectional area of the vertical member is A, and for inclined members are kA, where k is any positive number. The length of inclined members is l. Calculate the internal force in the vertical member.

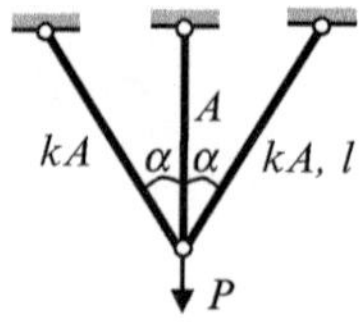

Fig. P7.20

$$\text{Ans. } N_{\text{vert}} = \frac{P}{1 + 2k\cos^3\alpha}.$$

7.21. The structure consists of the truss, which is supported at points 1 and 7, and the above truss construction. The structure carries five concentrated loads P, which are applied at joints 2-6 of the upper chord of the truss. Axial stiffness for upper and lower chords of the truss is 2*EA*, for vertical and diagonal elements of the truss and the above truss construction is *EA*. Provide a kinematical analysis. Determine internal force in the member 4-c.

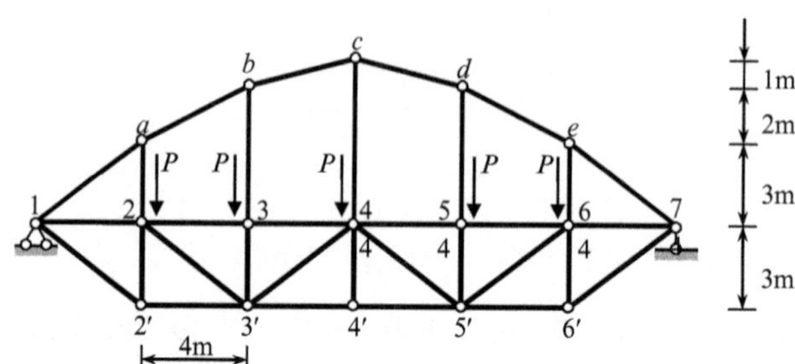

Fig. P7.21

Ans. $X_1 = 0.916P$.

7.22. Uniform semicircular arch with fixed ends is loaded by uniform load q (Fig. P7.22). Calculate reactions of supports. Construct the internal force diagrams

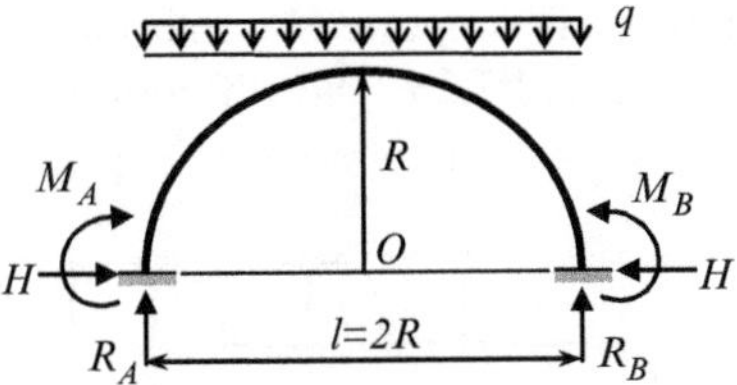

Fig. P7.22

Ans. $R_A = qR, \quad H = 0.56012qR, \quad M_A = 0.10658qR^2$

7.23. Design diagram of a two-hinged parabolic arch with the tie is presented in Fig. P7.23. The equation of the neutral line of the arch is $\left(4f/l^2\right)x\,(l-x)$. The cross-sectional moments of inertia vary by law $I_x = I_C \cos\varphi_x$ (m^4). The flexural stiffness in the highest section of the arch is EI_C. The cross-sectional area of the tie is $A_\mathrm{t}(\mathrm{m}^2)$. Modulus of elasticity of the material of the arch and the tie are $E(\mathrm{kN/m}^2)$ and E_t, respectively. Calculate internal axial force in the tie.

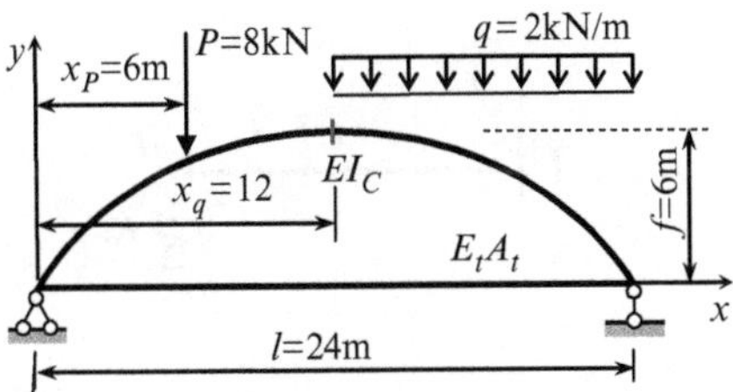

Fig. P7.23

$$\text{Ans. } X_1 = \frac{8558.67}{517.082 + 24\beta}, \quad \beta = \frac{EI_C}{E_\mathrm{t}A_\mathrm{t}}.$$

7.24. A rigid horizontal weightless beam is supported by three identical vertical rods at points 1-3. The beam supports a load P, as shown in Fig. P7.24. The stiffness EA is constant for all vertical rods. Determine the axial forces in members 1-3.

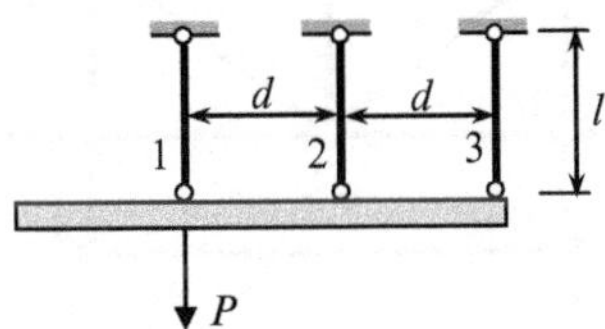

Fig. P7.24

Ans. $N_1 = 0.833P; \quad N_2 = 0.333P; \quad N_3 = -0.167P.$

7.25. A rigid horizontal beam is supported by four identical vertical rods and supports a load P, as shown in Fig. P7.25. The stiffness *EA* is constant for all vertical rods. Determine the axial forces in members 1-4. Neglect the weight of the beam.

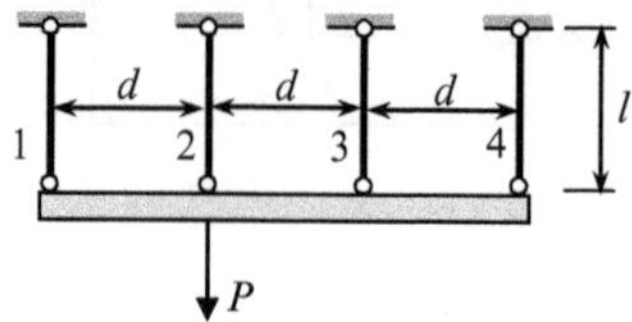

Fig. P7.25

Ans. $N_1 = 0.4P; \quad N_2 = 0.3P; \quad N_3 = 0.2P; \quad N_4 = 0.1P$

7.26. A beam with clamped support *A* and elastic support *B* is subjected to unit angular displacement of clamped support (Fig. P7.26). The flexural stiffness *EI* is constant; stiffness of elastic support is k $(R_B = k\Delta_B)$; and the length of the beam is l. Construct the bending moment diagram. Consider limiting cases ($k = 0$, and $k = \infty$).

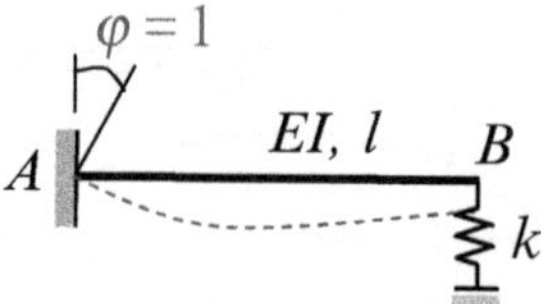

Fig. P7.26

$$\text{Ans. } M_A = \frac{3EI}{l} \cdot \frac{1}{1+\alpha}, \quad \alpha = \frac{3EI}{kl^3}.$$

7.27. The tie *ab* of a truss is subjected to thermal gradient Δt, degrees. Axial stiffness of the upper and lower chords of the truss is 2*EA*; for all other members of the structure axial stiffness is *EA*. Calculate the axial force at the member *ab*.

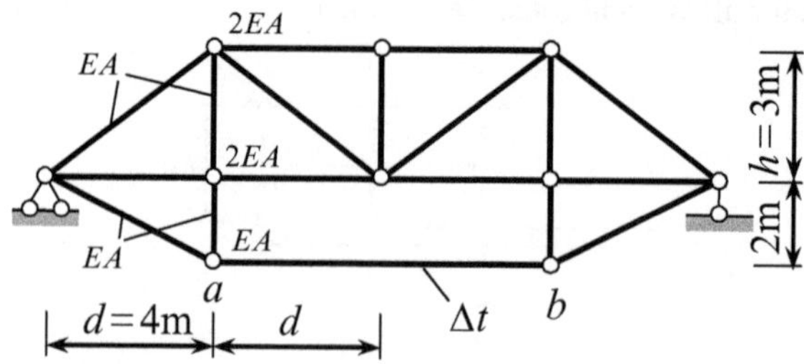

Fig. P7.27

Ans. $X_1 = -0.152\alpha \times \Delta t \times EA$

7.28. The portal frame is subjected to settlements a, b, and φ of support B. Calculate the free terms of canonical equation of the force method for three different versions of a primary system.

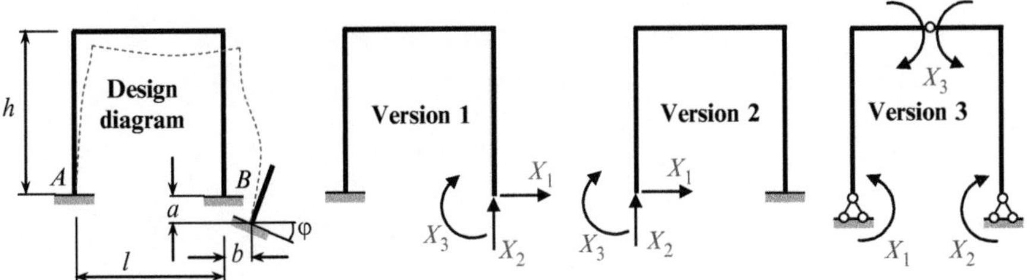

Fig. P7.28

Ans.

Version of primary system	Δ_{1d}	Δ_{2d}	Δ_{3d}
1	$-b$	$+a$	$-\varphi$
2	$+b$	$-a + l\varphi$	$+\varphi$
3	$-\frac{b}{2h} - \frac{a}{l}$	$-\frac{b}{2h} + \frac{a}{l}$	$+\frac{b}{h}$

Chapter 8
The Displacement Method

The displacement method is a powerful method for analysis of statically indeterminate frames subjected to different external exposures.

The displacement method is effective especially for analyzing sophisticated structures with a large number of redundant constraints. Two forms of the displacement method are applied in structural engineering. One uses the expanded form and the other the canonical form equations. In this book only canonical form will be considered.

The displacement method was introduced in 1826 by C. L. Navier (1785–1836). A. Clebsch (1833–1872) presented the displacement method for trusses in expanded form (1862). This method was further developed by H. Manderla (1880), E. Winkler (1892), and O. Mohr (1892). The application of the displacement method to frames was presented by A. Bendixen (1914) and was called the slope-deflection method. Later this method was developed and presented in canonical form by A. Ostenfeld (1926). A significant contribution in the development of the canonical form was added by S. Leites (1936).

The displacement method in canonical form is brought to elegant simplicity and contains a deep fundamental idea. This form offers a unified and rigorous convenient algorithm for analysis of different statically indeterminate structures. Among them are the single- or multispan as well as multileveled frames with deformable or infinitely rigid crossbars, etc. Canonical form can be effectively applied for the analyses of structures subjected to different loads, temperature changes, settlements of supports, and errors of fabrication. Application of the displacement method in canonical form for the construction of influence lines is very effective and leads to a simple and clear procedure.

Moreover, the displacement method in canonical form is a very effective tool for the special parts of structural analysis, such as stability and structural dynamics.

8.1 Fundamental Idea of the Displacement Method

Analysis of any statically indeterminate structure by the displacement method in the canonical form begins with determining the degree of kinematical indeterminacy.

I.A. Karnovsky and O. Lebed, *Advanced Methods of Structural Analysis*,
DOI 10.1007/978-1-4419-1047-9_8, © Springer Science+Business Media, LLC 2010

8.1.1 Kinematical Indeterminacy

In the case of the force method, the unknowns are forces at the redundant constraints. Knowing these forces we can find the distribution of internal forces and after that, displacements at any point of a structure.

The fundamental approach in the displacement method is the opposite: initially we calculate the *displacements* at the ends of the members and then the internal forces in the members. Thus, the primary unknowns in the displacement method are the *displacements*.

Analysis of a structure by the displacement method is based on the following assumptions:

1. The deformations of the members caused by axial and shear forces can be neglected.
2. The difference between the length of the deformable element and its initial length can be neglected.

Analysis of any statically indeterminate structure by the displacement method begins with determining the degree of *kinematical* indeterminacy. Generally, the degree of *kinematical* indeterminacy n of a structure is determined by the formula

$$n = n_{\mathrm{r}} + n_{\mathrm{d}}, \tag{8.1}$$

where n_{r} is the number of unknown angles of rotation of the rigid joints of a structure and n_{d} is the number of independent linear displacements of the joints.

Note that, in general, the degrees of *kinematical* and statical indeterminacy are not equal.

To calculate the number of linear displacements, n_{d}, we need to introduce the concept of the "hinged system or scheme." A hinged system is obtained from the original structure by introducing hinges at all rigid joints and supports while considering all members of the hinged scheme to be absolutely rigid. The degree of mobility of a hinged system is determined by the number of additional members, which would transform a hinged system into a geometrically unchangeable structure. The degree of mobility in turn determines the number of independent linear joint displacements, n_{d}.

Let us consider the structure shown in Fig. 8.1a; the elastic curve is shown by a dotted line. Since axial deformations are neglected and support 1 is *unmovable*, the joints of the entire structure have no horizontal displacements. Since the number of rigid joints equals two, then $n_{\mathrm{r}} = 2$. To show the hinged scheme, we introduce hinges at all rigid joints; this scheme represents a geometrically unchangeable structure. Indeed, the structure 1-2-3 presents a rigid disc and joint 4 is connected with this rigid disc and with the ground at support 5. So $n_{\mathrm{d}} = 0$ and the total degree of kinematical indeterminacy equals two.

Due to the action of the external load, rigid joints 2 and 4 rotate by angles φ_2 and φ_4, respectively. These angular displacements, φ_2 and φ_4, determine completely the deformable shape of the structure and represent the unknowns of the displacement method.

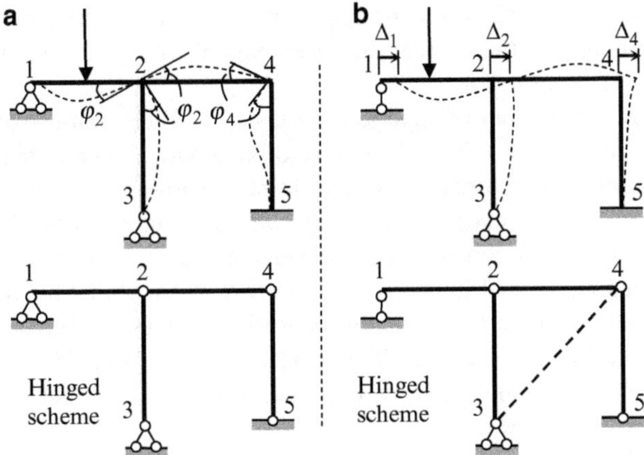

Fig. 8.1 (**a, b**) Design diagrams of frames, deflected shapes and hinged schemes

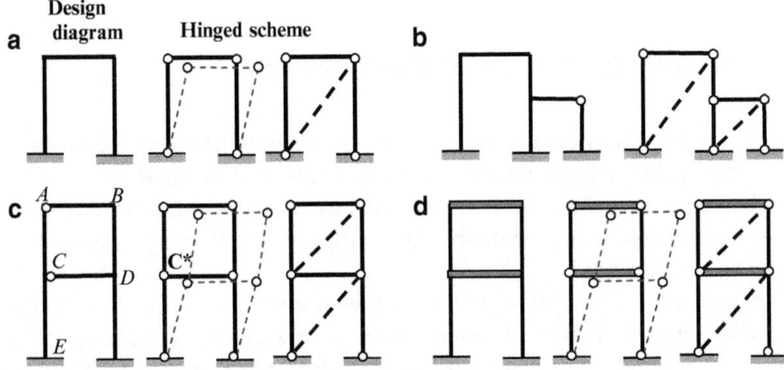

Fig. 8.2 Calculation of independent linear joints displacements

Figure 8.1b presents the same frame, but with a *movable* support 1. The number of unknown angles of rotation of rigid joints is, as before, $n_r = 2$; angular displacements φ_2 and φ_4 are not shown. In the case of rolled support 1 the structure has the linear displacements Δ of joints 1, 2, and 4 as well. According to conditions 1 and 2 of the assumptions, these linear displacements Δ are the same for all joints. So the number of independent linear displacements is $n_d = 1$. Indeed, the hinged structure presents a geometrically changeable system because support 1 is movable. To obtain a geometrically unchangeable structure, only one additional member needs to be introduced into this system (as shown by line 3-4), thus $n_d = 1$. Therefore, the total degree of kinematical indeterminacy equals three. The unknowns of the displacement method are angular displacements φ_2 and φ_4 and lateral displacement Δ.

Some statically indeterminate frames and their hinged schemes are presented in Fig. 8.2. In case (a), only one additional member (shown by the dashed line) transforms the hinged system into a geometrically unchangeable structure, so $n_d = 1$; the number of rigid joints is 2, hence the degree of kinematical indeterminacy of the entire structure equals 3: the angular displacements of rigid joints and the horizontal displacement of the crossbar.

In case (b), the number of rigid joints equals 3 and the number of independent linear displacements, $n_{\mathrm{d}} = 2$, so the degree of kinematical indeterminacy equals 5.

Case (c) presents a multistorey frame. Joint C shows that member CD is connected with vertical member AE by a hinge, while the member ACE does not have a hinge at point C. To construct the hinged scheme, it is necessary to introduce hinges at joints B, D, and joint C for member ACE (note that hinge C^* includes the original hinge at joint C and the introduced hinge). For the given structure, the number of rigid joints is 3 (joints B, D, and C for member ACE) and the number of independent linear joint displacements is 2, so the total number of unknowns by the displacement method is 5.

Case (d) also presents a multistorey frame, in which the crossbars are absolutely rigid. Since the joints cannot rotate, the number of unknown angles of rotation is zero and the number of independent linear joint displacements is 2.

8.1.2 Primary System and Primary Unknowns

The primary system of the displacement method is obtained from the given one by *introducing* additional constraints to prevent rotation of all rigid joints and all independent displacements of various joints. These introduced constraints are shown by the shaded squares and the double lines, respectively. Primary systems for some structures are shown in Fig. 8.3.

Primary unknowns Z_i $(i = 1, 2, \ldots, n)$ represent *displacements* of introduced constraints (angles of rotations and/or linear displacements of various joints of a frame). The number of primary unknowns, n, for each structure equals to the degrees of its kinematical indeterminacy.

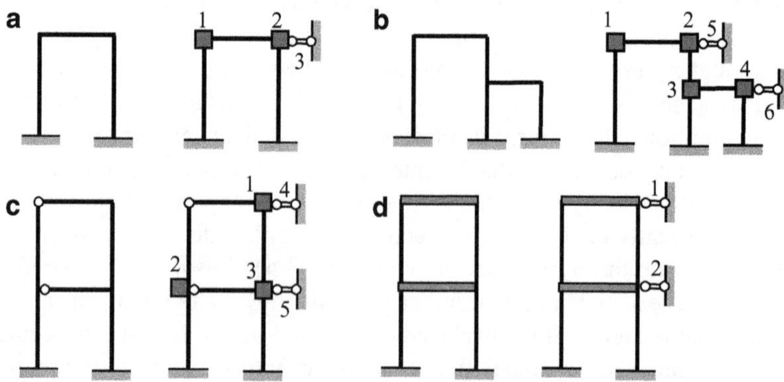

Fig. 8.3 Design diagrams and primary systems of the displacement method

In case 8.3a, the primary unknowns Z_1 and Z_2 are angular displacements of introduced rigid joints 1 and 2; the primary unknown, Z_3, is the linear displacement of introduced constraint 3. In case 8.3b, the primary unknowns are angular displacements $Z_1, \ldots, Z_4$ and linear displacements Z_5, Z_6. In case 8.3c, the primary unknowns are angular displacements Z_1, Z_2, Z_3 and linear displacements Z_4 and Z_5.

It can be seen that a primary system of the displacement method consists of a number of single-span redundant beams. In most cases, they are fixed-fixed and fixed-pinned uniform beams; fixed-guided beams and beams with elastic support as well as nonuniform beams are also possible. The next sections show how to use the primary system and tabulated data (Tables A.3–A.8) for the analysis of statically indeterminate structures in canonical form.

8.1.3 Compatibility Equation. Concept of Unit Reaction

The fundamental idea of the displacement method presentation in canonical form is explained below by considering the simplest frame in Fig. 8.4a.

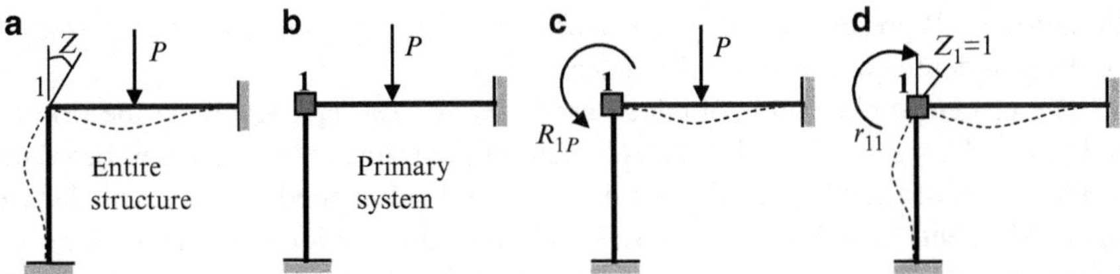

Fig. 8.4 (**a, b**) Design diagram and primary system; (**c, d**) The primary system is subjected to given load P and unit rotation of introduced constrain 1

1. The elastic curve of the frame caused by given load P is shown by the dashed line; rigid joint 1 rotates in a clockwise direction by some angle Z.
2. The primary system is obtained by introducing additional constraint 1 (Fig. 8.4b). This additional constraint prevents angular displacement. Therefore, the reactive moment due to external load P arises in the introduced constraint and deformation occurs in the horizontal element only. The elastic curve and reactive moment R_{1P} at constraint 1 in the primary system are shown in Fig. 8.4c. The difference between the given structure and the primary one is obvious: in the primary system the reactive moment arises at the additional constraint.
3. In order to eliminate the difference between the given and primary structures, constraint 1 must be rotated by some angle Z_1. In this case, the reactive moment arises in constraint 1. If the angle of rotation is a unit angle ($Z_1 = 1$), then the reactive moment is called a *unit reaction* and is denoted as r_{11}. The term "unit" means that this moment is caused by a *unit displacement* of the introduced constraint. The first numeral in r_{11} is the label of the constraint where the reactive

moment arises and the second numeral is the label of the constraint, which is rotated (in this example, they are the same).

Sign rule. If constraint 1 is rotated clockwise and *after that it is released*, then this constraint tends to rotate back counterclockwise. Therefore, the reactive moment will act in an opposite direction, i.e., clockwise. Thus, the positive reaction *coincides* with the positive displacement of the introduced constraint. Figure 8.4d shows the positive directions for displacement Z_1 and unit reaction r_{11}.

4. The total reaction caused by rotation of the introduced constraint and the given load is $r_{11}Z_1 + R_{1P}$, where the first term $r_{11}Z_1$ is the reactive moment in constraint 1 due to the real angle of rotation Z_1. The second term represents the reactive moment in constraint 1 due to the actual load. Lowercase letter r means that this reaction is caused by the *unit* rotation while capital letter R means that this reaction is caused by the *real* external load. If this total reactive moment (due to both the given load and the angular displacement of the included constraint) is equal to zero, then the behavior of the given and primary structures is identical. This statement may be written in the following form:

$$r_{11}Z_1 + R_{1P} = 0. \tag{8.2}$$

Equation (8.2) presents the *displacement method in canonical form* for a structure of the first degree of kinematical indeterminacy.

The primary unknown is obtained as $Z_1 = -R_{1P}/r_{11}$. Knowing the primary unknown Z_1, we can consider each element of the frame to be a standard beam due to the action of found angle Z_1 and the external load applied to this particular element. Moments of supports for these standard beams are found in Tables A.3–A.6. Using the principle of superposition, the final bending moment diagram for the entire frame is constructed by the formula

$$M_P = \bar{M}_1 \cdot Z_1 + M_P^0, \tag{8.3}$$

where $\bar{M}_1$ is the bending moment diagram caused by the unit primary unknown; term $\bar{M}_1 \cdot Z_1$ is the bending moment diagram caused by the *actual* primary unknown Z_1; and the bending moment diagram in the primary system caused by the given load is denoted as M_P^0.

8.2 Canonical Equations of Displacement Method

8.2.1 Compatibility Equations in General Case

Now we consider an arbitrary n-times kinematically indeterminate structure. The primary unknowns Z_i $(i = 1,\ldots,n)$ are displacements (linear and/or angular) of introduced constraints. The canonical equations of the displacement method will be written as follows:

$$\begin{aligned} r_{11}Z_1 + r_{12}Z_2 + \cdots + r_{1n}Z_n + R_{1P} &= 0 \\ r_{21}Z_1 + r_{22}Z_2 + \cdots + r_{2n}Z_n + R_{2P} &= 0 \\ \cdot \quad \cdot \quad \cdot \quad \cdot \quad \cdot \quad \cdot \quad \cdot \quad \cdot \quad \cdot \quad \cdot \quad \cdot \quad \cdot \quad & \cdot \\ r_{n1}Z_1 + r_{n2}Z_2 + \cdots + r_{nn}Z_n + R_{nP} &= 0. \end{aligned} \tag{8.4}$$

The number of canonical equations is equal to number of primary unknowns of the displacement method.

Interpretation of the canonical equations Coefficient r_{ik} represents the unit reaction, i.e., the reaction (force or moment), which arises in the ith introduced constraint (first letter in subscript) caused by *unit* displacement $Z_k = 1$ of kth introduced constraint (second letter in subscript). The term $r_{ik}Z_k$ represents the reaction, which arises in the ith introduced constraint due to the action of *real* unknown displacement Z_k. Free term R_{iP} is the reaction in the ith introduced constraint due to the action of the applied loads. Thus, the left part of the ith equation represents a total reaction, which arises in the ith introduced constraint due to the actions of all real unknowns Z as well as the applied load.

The total reaction in each introduced constraint in the primary system caused by all primary unknowns (the linear and angular displacements of the introduced constraints) and the applied loads is equal to zero. In this case, the difference between the given structure and the primary system vanishes, or in other words, the behavior of the given and the primary systems is the same.

8.2.2 Calculation of Unit Reactions

The frame presented in Fig. 8.5a allows angular displacement of the rigid joint and horizontal displacement of the crossbar. So the structure is twice kinematically indeterminate. The primary system of the displacement method is presented in Fig. 8.5b.

Constraints 1 and 2 are additional introduced constraints that prevent angular and linear displacements. In constraint 1, which prevents angular displacement, only the reactive moment arises; in constraint 2, which prevents only linear displacement, only the reactive force arises. The corresponding canonical equations are

$$\begin{aligned} r_{11}Z_1 + r_{12}Z_2 + R_{1P} &= 0, \\ r_{21}Z_1 + r_{22}Z_2 + R_{2P} &= 0. \end{aligned}$$

To determine coefficients r_{ik} of these equations, we consider two states. State 1 presents the primary system subjected to unit angular displacement $Z_1 = 1$. State 2 presents the primary system subjected to unit horizontal displacement $Z_2 = 1$. For both states, we will show the bending moment diagrams. These diagrams caused by the unit displacements of introduced constraints $Z_1 = 1$ and $Z_2 = 1$ are shown in Fig. 8.5c, d, respectively. The elastic curves are shown by dashed lines; the asterisk

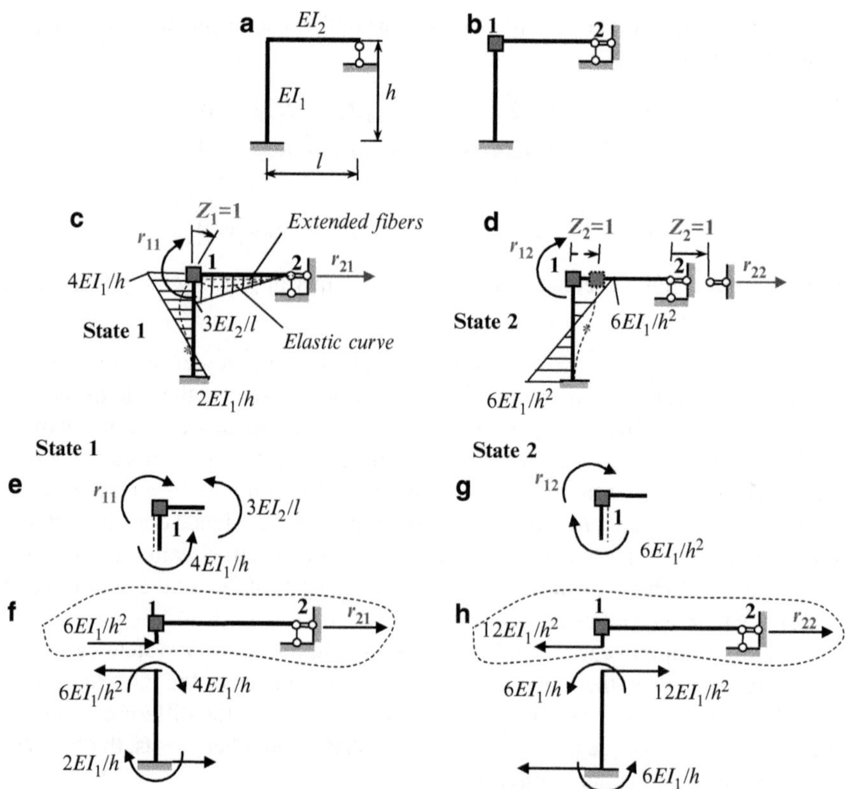

Fig. 8.5 (**a**) Design diagram; (**b**) Primary system of the displacement method; (**c, d**) Unit states and corresponding bending moment diagrams. Calculation of unit reactions: (**e, f**) Free-body diagrams for joint 1 and crossbar, state 1; (**g, h**) Free-body diagrams for joint 1 and crossbar, state 2

(*) denotes the inflection points of the elastic curves. For member 1-2, the extended fibers are located below the neutral line. Bending moment diagrams are plotted on the extended fibers.

First unit state ($Z_1 = 1$) Rigid joint 1 rotates clockwise (positive direction) through angle $Z_1 = 1$. The reactions for both members (fixed-pinned and fixed-fixed) in the case of angular displacement of the fixed supports are presented in Tables A.3 and A.4. In this case, the bending moment at joint 1 of the horizontal member equals $3EI_2/l$ (Table A.3, row 1); for the vertical member, the specified ordinates are $4EI_1/h$ for joint 1 and $2EI_1/h$ at the bottom clamped support (Table A.4). As a result of the angular displacement, unit reactive moment r_{11} arises in constraint 1 and reactive force r_{21} arises in constraint 2; all unit reactions are shown in the positive direction.

Second unit state ($Z_2 = 1$) If constraint 2 has horizontal displacement $Z_2 = 1$ from left to right (positive direction), then introduced joint 1 has the same displacement and, as a result, the vertical member is subjected to bending. Unit reactive moment r_{12} and unit reactive force r_{22} arise in constraints 1 and 2, respectively. The specified ordinates for the vertical member at the bottom and at point 1 are equal to $6EI_1/h^2$.

Unit reactions r_{11}, r_{12} represent the reactive *moments* in constraint 1 for both states and unit reactions r_{21}, r_{22} represent the reactive *forces* in constraint 2 for both states. Unit reactions r_{ii}, located on the main diagonal of the canonical equations, are called the main reactions (r_{11}, r_{22}); other unit reactions are called secondary ones (r_{21}, r_{12}).

To calculate all the unit reactions, we need to consider the free-body diagrams for joint 1 and for crossbar 1-2 for each state. The free-body diagram for joint 1 in state 1 is shown in Fig. 8.5e. The direction of moments $4EI_1/h$ and $3EI_2/l$ corresponds to the location of the extended fibers in the vicinity of joint 1 (Fig. 8.5e); the extended fibers are shown by dashed lines. Positive unit reactive moment r_{11} is shown by the direction of unit displacement Z_1.

The equilibrium condition for joint 1 is $\Sigma M = 0$, therefore we get

$$r_{11} = \frac{4EI_1}{h} + \frac{3EI_2}{l}.$$

To calculate unit reaction r_{21}, we need to consider the free-body diagram for the crossbar. The shear force infinitely close to joint 1 is found by considering the equilibrium of the vertical element through the following steps. Moments at the end points of the vertical element shown in Fig. 8.5f are taken from the bending moment diagram in Fig. 8.5c. The moments $4EI_1/h$ at the top and $2EI_1/h$ at the support are equilibrated by two equal forces, $6EI_1/h^2$. The upper force is transmitted to the crossbar. After that, unit reaction r_{21} is found by considering the equilibrium equation of the crossbar, ($\Sigma F_x = 0$), so

$$r_{21} = -\frac{1}{h}\left(\frac{4EI_1}{h} + \frac{2EI_1}{h}\right) = -\frac{6EI_1}{h^2}.$$

It is obvious that this reaction can also be taken directly from Table A.4, row 1.

Similarly, equilibrium equations $\Sigma M = 0$ for the free-body diagrams for joint 1 in state 2 (Fig. 8.5g) and equilibrium $\Sigma F_x = 0$ for the crossbar in state 2 (Fig. 8.5h) lead to the following unit reactions:

$$r_{12} = -\frac{6EI_1}{h^2}, \; r_{22} = \frac{12EI_1}{h^2}.$$

8.2.3 Properties of Unit Reactions

1. Main reactions are strictly positive ($r_{ii} > 0$). Secondary reactions r_{ik} may be positive, negative, or zero.

2. According to the reciprocal reactions theorem, the secondary reactions satisfy the symmetry condition $r_{ik} = r_{ki}$, i.e., $r_{12} = r_{21}$.
3. The dimensions of unit reaction r_{ik} are determined by the following rule: the dimension of *reaction* (force or moment) at index i is divided by the dimension of the *displacement* (linear or angular) at index k. In our case $[r_{11}] = \frac{\text{kN m}}{\text{rad}}$, $[r_{12}] = \frac{\text{kN m}}{\text{m}} = \text{kN}$, $[r_{21}] = \frac{\text{kN}}{\text{rad}} = \text{kN}$, $[r_{22}] = \frac{\text{kN}}{\text{m}}$.

8.2.4 Procedure for Analysis

For analysis of statically indeterminate continuous beams and frames by the displacement method in canonical form the following procedure is suggested:

1. Define the degree of kinematical indeterminacy and construct the primary system of the displacement method
2. Formulate the canonical equations of the displacement method (8.4)
3. Apply successively unit displacements $Z_1 = 1, Z_2 = 1, \ldots, Z_n = 1$ to the primary structure. Construct the corresponding bending moment diagrams $\bar{M}_1, \bar{M}_2, \ldots, \bar{M}_n$ using Tables A.3–A.8
4. Calculate the main and secondary unit reactions r_{ik}
5. Construct the bending moment diagram M_P^0 due to the applied load in the primary system and calculate the free terms R_{iP} of the canonical equations
6. Solve the system of equations with respect to unknown displacements $Z_1, Z_2, \ldots, Z_n$
7. Construct the bending moment diagrams by formula

$$M_P = \bar{M}_1 \cdot Z_1 + \bar{M}_2 \cdot Z_2 + \cdots + \bar{M}_n \cdot Z_n + M_P^0. \tag{8.5}$$

   The term $\bar{M}_n \cdot Z_n$ represents a bending moment diagram due to actual displacement Z_n. The term M_P^0 represents a bending moment diagram in the primary system due to actual load
8. Compute the shear forces using the Schwedler theorem considering each member due to the given loads and end bending moments and construct the corresponding shear diagram
9. Compute the axial forces from the consideration of the equilibrium of joints of the frame and construct the corresponding axial force diagram
10. Calculate reactions of supports and check them using the equilibrium conditions for an entire structure as a whole or for any separated part

Let us show the application of this algorithm to the analysis of a uniform continuous two-span beam *A*-1-*B* (Fig. 8.6a). According to (8.1), this continuous beam is kinematically indeterminate to the first degree. Indeed, support *A* prevents linear displacement of the joints and there is one rigid joint at support 1, therefore there is only angular displacement at this support. Thus, the primary unknown is the angular displacement Z_1 at support 1. The primary system is obtained from a given structure by introducing constraint 1 at middle support 1 (Fig. 8.6a); this constraint prevents angular displacement at support 1.

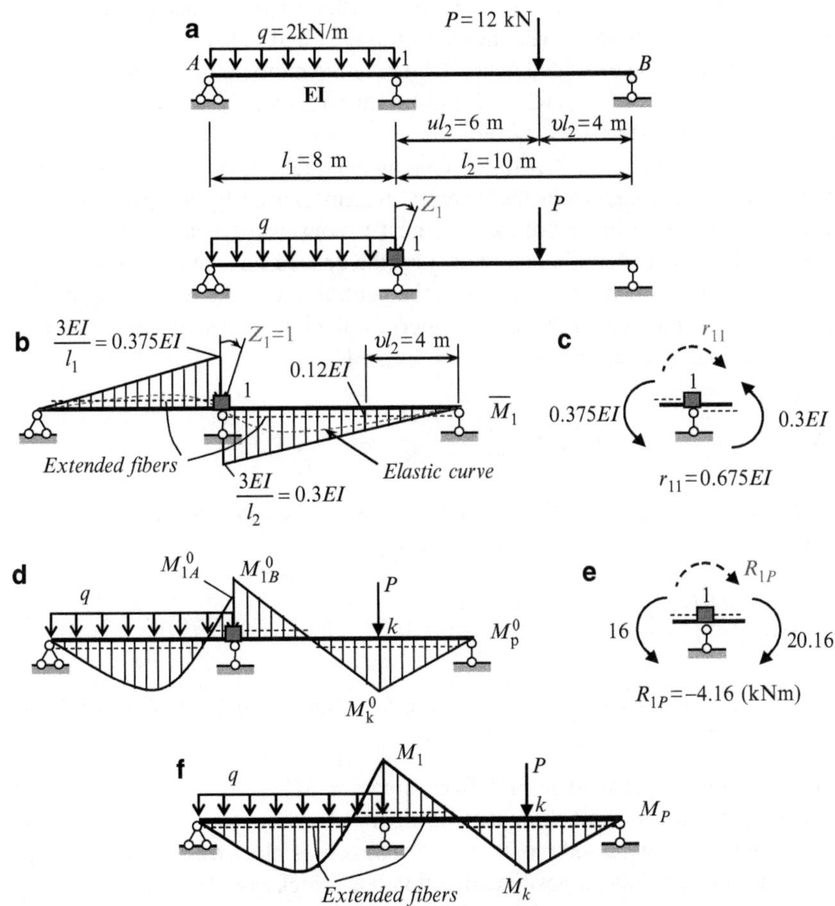

Fig. 8.6 (**a**) Design diagram of the beam and primary system. (**b**) Bending moment diagram caused by unit angular displacement; (**c**) Calculation of r_{11}. (**d**) Bending moment diagram of a primary system caused by a given load; (**e**) Calculation of free term R_{1P}. (**f**) Final bending moment diagram

The canonical equation of the displacement method is

$$r_{11}Z_1 + R_{1P} = 0. \tag{a}$$

To calculate unit reaction r_{11}, we need to rotate the introduced constraint clockwise by angle $Z_1 = 1$. The corresponding elastic curve, the location of the extended fibers, and the bending moment diagram caused by the unit primary unknown are shown in Fig. 8.6b. The bending moment at the fixed support for a fixed-pinned beam is $3EI/l$ (Table A.3). The free-body diagram of joint 1 from diagram $\bar{M}_1$ is shown in Fig. 8.6c. According to the elastic curve, the extended fibers in the vicinity of joint 1 are located above the neutral line to the left of point 1 and below the neutral line to the right of point 1. The moments $0.375EI$ and $0.3EI$ are shown according to the location of the extended fibers. Unit reactive moment r_{11} is shown assuming its positive direction (clockwise). Equilibrium condition $\sum M = 0$ leads to unit reaction $r_{11} = 0.675EI$ (kN m/rad).

To calculate free term R_{1P} of the canonical equation, we need to construct the bending moment diagram in the primary system caused by the given load. This diagram is shown in Fig. 8.6d; each element is considered as a separate beam; the location of the extended fibers is shown by the dashed line. The extended fibers in the vicinity of joint 1 are located above the neutral line to the left and right of joint 1. The bending moment at the fixed support for the left span subjected to uniformly distributed load q, according to Table A.3, equals

$$M_{1A}^0 = \frac{ql_1^2}{8} = \frac{q \cdot 8^2}{8} = 16 \text{ (kN m)}.$$

The bending moment at the specified points for the right span subjected to concentrated force P, according to Table A.3, equals

$$M_{1B}^0 = \frac{Pl_2}{2}\upsilon\left(1-\upsilon^2\right) = \frac{12 \times 10}{2} \times 0.4 \times \left(1-0.4^2\right) = 20.16 \text{ (kN m)},$$

$$M_k^0 = \frac{Pl_2}{2}u^2\upsilon\left(3-u\right) = \frac{12 \times 10}{2} \times 0.6^2 \times 0.4\left(3-0.4\right) = 20.736 \text{ (kN m)}.$$

The free-body diagram of joint 1 from diagram M_P^0 is shown in Fig. 8.6e. According to the location of the extended fibers, the moment of 16 kN m is shown to be counterclockwise and moment of 20.16 kN m is clockwise. Reactive moment R_{1P} is assumed to have a positive direction, i.e., clockwise. Equilibrium condition $\sum M = 0$ leads to $R_{1P} = -4.16$ kN m.

Canonical equation (a) becomes $0.675EI \cdot Z_1 - 4.16 = 0$. The root of this equation, i.e., the primary unknown is

$$Z_1 = \frac{6.163}{EI} \text{ (rad)}. \tag{b}$$

The bending moments at the specified points can be calculated by the following formula

$$M_P = \bar{M}_1 \cdot Z_1 + M_P^0. \tag{c}$$

In our case we have

$$M_{1A} = -0.375EI \times \frac{6.163}{EI} - 16 = -18.31 \text{ kN m},$$
$$M_{1B} = 0.3EI \times \frac{6.163}{EI} - 20.16 = -18.31 \text{ kN m},$$
$$M_k = 0.12EI \times \frac{6.163}{EI} + 20.736 = 21.475 \text{ kN m}.$$

Of course, M_{1A} and M_{1B} are equal. The negative sign indicates that the extended fibers at support 1 are located above the neutral line. The final bending moment diagram M_P is presented in Fig. 8.6f; the location of the extended fibers is shown by the dotted line.

Now we will consider the analysis of some frames by the displacement method in canonical form.

Example 8.1. The crossbar of the frame in Fig. 8.7a is connected with vertical members by means of hinges. The bending stiffness is *EI* for all vertical members and 2*EI* for the crossbar; their relative stiffnesses, 1 and 2, are shown in the circles. Concentrated force P acts horizontally at the level of the crossbar. Construct the internal force diagrams.

Solution. The system has two unknowns of the displacement method: the angular displacement of joint 1 and the linear displacement of the crossbar. The primary system is shown in Fig. 8.7b.

The introduced constraint 1 is related only to the horizontal member, but not the vertical one. The primary unknowns of the displacement method are angular displacement Z_1 of constraint 1 and linear displacement Z_2 of constraint 2. The bending stiffness per unit length for the vertical and horizontal members are $i_{\text{vert}} = \frac{1EI}{5} = 0.2EI$, $i_{\text{hor}} = \frac{2EI}{6} = 0.333EI$. The canonical equations of the displacement method are:

$$\begin{aligned} r_{11}Z_1 + r_{12}Z_2 + R_{1P} &= 0, \\ r_{21}Z_1 + r_{22}Z_2 + R_{2P} &= 0. \end{aligned} \tag{a}$$

Calculation of unit reactions To calculate coefficients r_{11} and r_{21}, we need to construct the bending moment diagram $\bar{M}_1$ in the primary system due to the rotation of induced constraint 1 (Fig. 8.7c).

The bending moment at the clamped support is $3i = (3 \times 2EI)/6 = 1EI$ (Table A.3, row 1). Because of the hinged connections of the vertical elements with the horizontal bar, the bending moments in the vertical elements do not arise. The positive unknown unit reactions are shown by the dotted arrow. The equilibrium of joint 1 from bending moment diagram $\bar{M}_1$ leads to $r_{11} = 2EI$ (kN m/rad).

The equilibrium of the crossbar, considering the bending moment diagram $\bar{M}_1$, leads to $r_{21} = 0$. Indeed, within the vertical members the bending moments do not arise. Therefore, shear forces within these members also do not arise and equation $\sum X = 0$ for the crossbar leads to the above result.

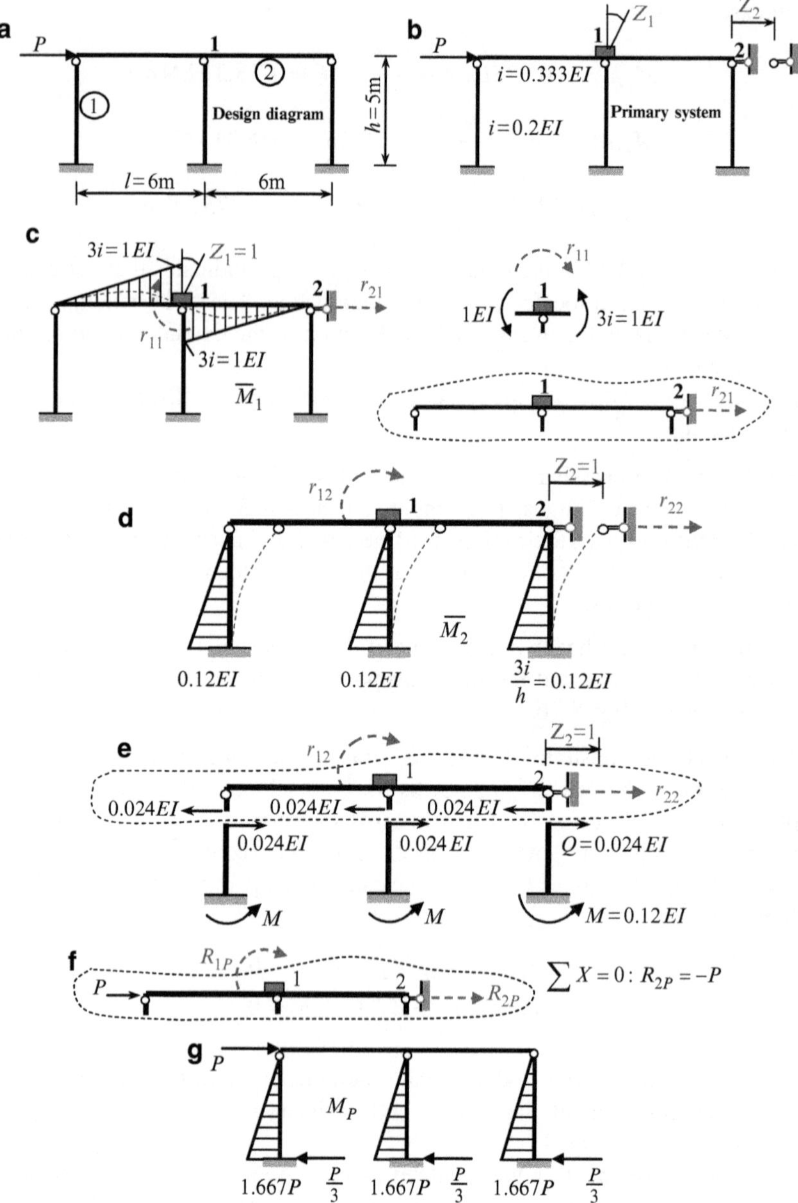

Fig. 8.7 (**a, b**) Design diagram of the frame and primary system. (**c**) Unit bending moment diagram due to $Z_1 = 1$ and calculation of r_{11} and r_{21}. (**d, e**) Unit bending moment diagram due to $Z_2 = 1$ and calculation of r_{12} and r_{22}. (**f**) Free-body diagram of the crossbar in the loaded condition. (**g**) Final bending moment diagram and reactions of supports

To calculate coefficients of r_{12} and r_{22} it is necessary to construct the bending moment diagram $\bar{M}_2$ in the primary system due to the linear displacement of induced constraint 2 (Fig. 8.7d). The bending moment at the bottom of the vertical members is $3i/h = (3 \times 0.2EI)/5 = 0.12EI$ (Table A.3, row 2). The equilibrium equation of joint 1 leads to $r_{12} = 0$.

The free-body diagram for the crossbar is presented in Fig. 8.7e, where shear in each vertical member due to unit displacement $Z_2 = 1$ is

$$Q = \frac{0.12EI}{h} = \frac{0.12EI}{5} = 0.024EI.$$

Therefore, the equilibrium equation $\sum X = 0$ for the crossbar leads to the following result: $r_{22} = 3 \times 0.024EI = 0.072EI$ (kN/m).

Calculation of free terms. The force P is applied at the level of the crossbar, therefore there is no bending of the horizontal elements in the primary system. Since constraint 2 prevents displacement in a horizontal direction then bending of the vertical members of the frame does not occur either. Thus, in the primary system there are no elements subject to bending. But this does not imply that all reactions in induced restrictions 1 and 2 are zero. Indeed, $R_{1P} = 0$, but $R_{2P} = -P$. The last expression is obtained from the equilibrium of the crossbar subjected to the given load P (Fig. 8.7f).

The canonical equations become

$$\begin{aligned} &2EI \cdot Z_1 + 0 \cdot Z_2 = 0, \\ &0 \cdot Z_1 + 0.072EI \cdot Z_2 - P = 0. \end{aligned} \tag{b}$$

The roots of these equations are $Z_1 = 0$, $Z_2 = P/0.072EI$. The result $Z_1 = 0$ means that the crossbar is not deformed, but displaced in the horizontal direction as an absolutely rigid element. This is because the crossbar is connected to the vertical elements by means of hinges.

The bending moment diagram can be constructed using the principle of superposition:

$$M_P = \bar{M}_1 \cdot Z_1 + \bar{M}_2 \cdot Z_2 + M_P^0. \tag{c}$$

Since $Z_1 = 0$ and acting load P does not cause bending of the members, then formula (c) becomes $M_P = \bar{M}_2 \cdot Z_2$. The resulting bending moment diagram is presented in Fig. 8.7g.

The bending moments at the clamped supports are

$$0.12EI \times \frac{P}{0.072EI} = 1.667P \text{ (kN m)}.$$

The shear force at vertical members

$$Q = \frac{1.667P}{h} = \frac{1.667P}{5} = \frac{P}{3} \text{ (kN)}.$$

Discussion:

1. The active load acts at the level of the crossbar and the bending moments in the primary system due to this load do not arise. However, this is not to say that all free terms of canonical equations are zero. Introduced constraints prevent angular and linear displacements of the frame. So free terms present the reactive moment and force in the introduced constraints. The reactive moment $R_{1P} = 0$, while the reactive force of introduced constraint $R_{2P} \neq 0$.
2. Even if the primary system has a nonzero bending moment diagram within the crossbar (diagram $\bar{M}_1$), the resulting bending moment diagram along the crossbar is zero. This happens because $Z_1 = 0$, $\bar{M}_2 = 0$, and $M_P^0 = 0$.
3. If the flexural stiffness of the crossbar is increased n times, then unit reaction r_{11} becomes $r_{11} = 2EI \cdot n$, while all other coefficients and free terms remain the same. So equations (b) lead to the same primary unknowns and the resulting bending moment diagram remains same. This happens because for given design diagram the bending deformations for crossbar are absent.
4. The length l of the span has no effect on the bending moment diagram.
5. If the frame, shown in Fig. 8.7a, is modified (the number of the vertical members being k), then the reaction of each support is P/k.

The next example presents a detailed analysis of a frame by the displacement method in canonical form. This frame was analyzed earlier by the force method. Therefore, the reader can compare the different analytical approaches to the same structure and see their advantages and disadvantages.

Example 8.2. A design diagram of the frame is presented in Fig. 8.8a. The flexural stiffnesses for the vertical member and crossbar are EI and $2EI$, respectively; their relative flexural stiffnesses are shown in circles. The frame is loaded by force $P = 8\,\text{kN}$ and uniformly distributed load $q = 2\,\text{kN/m}$. Construct the bending moment diagram.

Solution. It is easy to check the number of independent linear displacements $n_d = 1$ (the hinged scheme is not shown) and the total number of unknowns of the displacement method is 2. The primary system is obtained by introducing two additional constraints, labeled 1 and 2 (Fig. 8.8b). Constraint 1 prevents only angular displacement of the rigid joint and constraint 2 prevents only linear displacement of the crossbar.

The flexural stiffness per unit length for each element of the structure is as follows:

$$i_{1\text{-}3} = \frac{1EI}{5} = 0.2EI;\; i_{6\text{-}8} = \frac{2EI}{10} = 0.2EI;\; i_{4\text{-}5} = \frac{1EI}{3} = 0.333EI,$$

where the subscript of each parameter i indicates an element of the frame (Fig. 8.8c).

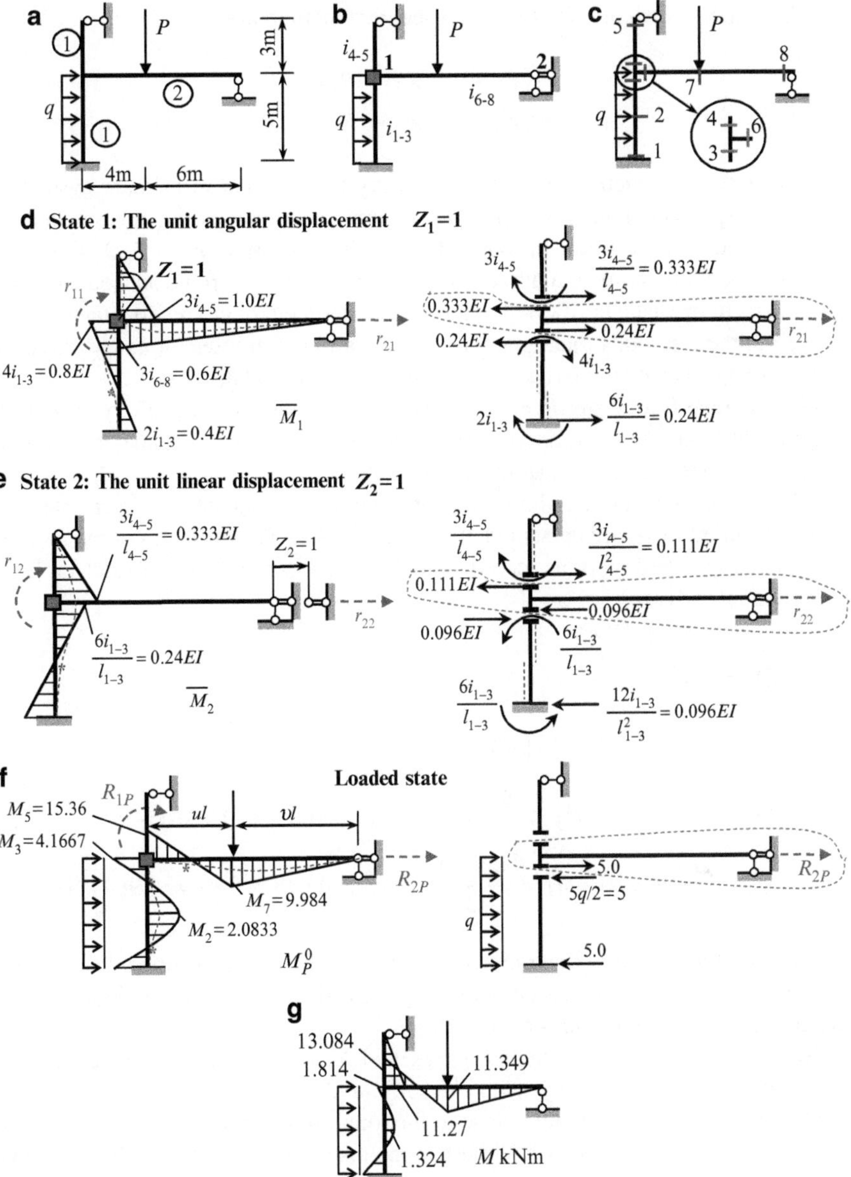

Fig. 8.8 (**a**) Design diagram; (**b**) Primary system; (**c**) Specified sections; (**d**) State 1. Bending moment diagram due to unit angular displacement $Z_1 = 1$ and free-body diagram for the calculation of r_{21}. (**e**) State 2. Bending moment diagram due to unit linear displacement $Z_2 = 1$ and free-body diagram for the calculation of r_{22}. (**f**) Bending moment diagram in the loaded state and free-body diagram for the calculation of load reaction R_{2P}. (**g**) Final bending moment diagram M (kN m)

The canonical equations of the displacement method are

$$r_{11}Z_1 + r_{12}Z_2 + R_{1P} = 0,$$
$$r_{21}Z_1 + r_{22}Z_2 + R_{2P} = 0.$$

To calculate unit reactions r_{ik}, it is necessary to construct in primary system the unit bending moment diagrams $\bar{M}_1$, $\bar{M}_2$. To construct unit bending moment diagram $\bar{M}_1$, it is necessary to *rotate* introduced constraint 1 by angle $Z_1 = 1$ clockwise. To construct bending moment diagram $\bar{M}_2$, it is necessary to *shift* introduced constraint 2 to the right by distance $Z_2 = 1$. The bending moment diagrams for states 1 and 2 as well as the free-body diagrams for the calculation of unit reactions r_{ik} are shown in Fig. 8.8d,e. The positive reactions are shown by dashed arrows.

To calculate free terms R_{iP}, it is necessary to construct in primary system the diagram M_P^0 caused by the applied load (loaded state). This state and corresponding bending moment diagram-is shown in Fig. 8.8f.

The ordinates of the bending moment diagrams for standard uniform elements due to different loads are presented in Tables A.3–A.6. The ordinates of the M_P^0 diagram at the specified sections according to Table A.4 (for element 1-3) and Table A.3 (for element 6-8) are

$$M_1 = M_3 = \frac{ql_{1\text{-}3}^2}{12} = 4.1667\,\text{kN m};$$
$$M_2 = \frac{ql_{1\text{-}3}^2}{24} = 2.0833\,\text{kN m};$$
$$M_6 = \frac{Pl}{2}\upsilon\left(1 - \upsilon^2\right) = \frac{8 \times 10}{2} \times 0.6\left(1 - 0.6^2\right) = 15.36\,\text{kN m};$$
$$M_7 = \frac{Pl}{2}u^2\upsilon\,(3 - u) = \frac{8 \times 10}{2} \times 0.4^2 \times 0.6\,(3 - 0.4) = 9.984\,\text{kN m}.$$

All the bending moment diagrams are plotted on the extended fibers of the frame (Fig. 8.8d–f). Elastic curves are shown by dashed lines. The asterisks (*) on the elastic curves show the points of inflection.

To calculate reactive moments r_{11}, r_{12}, and R_{1P}, it is necessary to consider the free-body diagrams of joint 1 using $\bar{M}_1$, $\bar{M}_2$, and M_P^0 diagrams. The ordinates of the bending moments infinitely close to the joint are taken from the corresponding diagrams. The direction of each of these moments must correspond to the location of the extended fibers in each diagram. The calculations of unit reactions r_{11}, r_{12}, and R_{1P} are presented in Table 8.1. Positive reactive moments are directed clockwise; locations of extended fibers are shown by dashed lines.

To calculate reactive *forces* r_{21}, r_{22}, and R_{2P}, it is necessary to consider the equilibrium of horizontal member 6-8. For this we need to cut off element 6-8 from diagrams $\bar{M}_1$, $\bar{M}_2$, and M_P^0 by sections infinitely close to joint 1 from above and below. The calculations of unit reactions r_{21}, r_{22}, and loaded reaction R_{2P} are presented in Table 8.2.

Table 8.1 Calculation of unit and loaded reactions at introduced constraint 1

Diagram	Free-body diagram of joint 1	Equilibrium equation	Reaction
$\bar{M}_1$	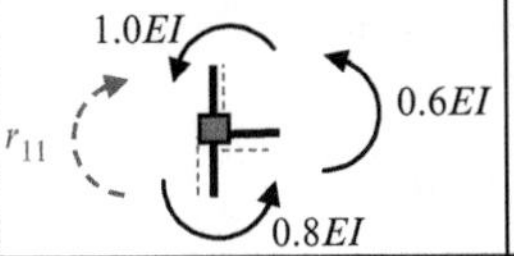 $1.0EI$, $0.6EI$, r_{11}, $0.8EI$	$\sum M = 0$ $-r_{11} + (1.0 + 0.6 + 0.8)\,EI = 0$	$r_{11} =$ $2.4EI$ (kN m/rad)
$\bar{M}_2$	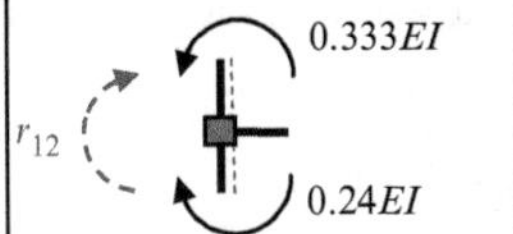 $0.333EI$, r_{12}, $0.24EI$	$\sum M = 0$ $-r_{12} - 0.24EI + 0.333EI = 0$	$r_{12} =$ $0.093EI$ (kN m/m)
M_P^0	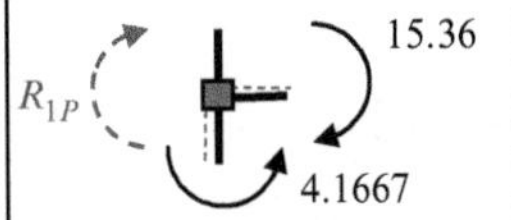 15.36, R_{1P}, 4.1667	$\sum M = 0$ $-R_{1P} - 15.36 + 4.1667 = 0$	$R_{1P} =$ -11.1933 kNm

Table 8.2 Calculation of unit and loaded reactions at introduced constraint 2

Diagram	Portion (crossbar) of the structure	Equilibrium equation	Reaction
$\bar{M}_1$	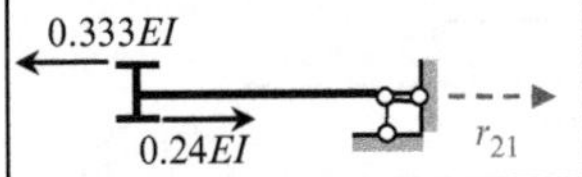 $0.333EI$, $0.24EI$, r_{21}	$\sum X = 0$ $r_{21} + 0.24EI - 0.333EI = 0$	$r_{21} =$ $0.093EI$ (kN/rad)
$\bar{M}_2$	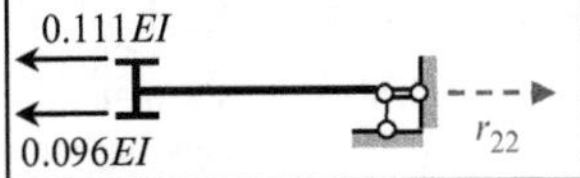 $0.111EI$, $0.096EI$, r_{22}	$\sum X = 0$ $r_{22} - 0.111EI - 0.096EI = 0$	$r_{22} =$ $0.207EI$ (kN/m)
M_P^0	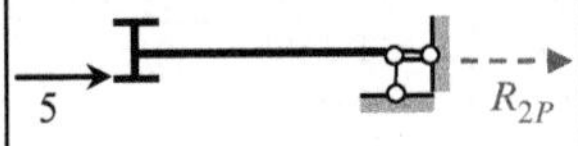 5, R_{2P}	$\sum X = 0$ $R_{2P} + 5 = 0$	$R_{2P} = -5$ kN

Let us consider in detail the procedure for the calculation of reactive forces r_{21}. When a section is passed infinitely close to joint 1 from above (Fig. 8.8d), the moment $3i_{4\text{-}5}$ is applied. Since the extended fibers are located to the right of member 4-5, then moment $3i_{4\text{-}5}$ is directed clockwise considering the top part of the portion 4-5. Moment $3i_{4\text{-}5}$ is equilibrated by two forces, $3i_{4\text{-}5}/l_{4\text{-}5} = 0.333EI$. (The force at support 5 is not shown.) Force $0.333EI$ is transmitted to the horizontal member and has an opposite direction.

Member 1-3 should be considered in a similar way: pass a section infinitely close to joint 1 from below, apply two bending moments $4i_{1\text{-}3}$ (the top of the member) and $2i_{1\text{-}3}$ (the bottom of the member) to correspond to the location of the extended fibers and then equilibrate them by two forces, $6i_{1\text{-}3}/l_{1\text{-}3} = 0.24EI$. As the last step, this force is transmitted to the horizontal member in the opposite direction.

The canonical equations of the displacement method become:

$$2.4Z_1 + 0.093Z_2 - \frac{11.1933}{EI} = 0,$$

$$0.093Z_1 + 0.207Z_2 - \frac{5}{EI} = 0. \tag{a}$$

The roots of these equations present the primary unknowns and they are equal to

$$Z_1 = \frac{3.794}{EI}\ \text{(rad)}\,,\ Z_2 = \frac{22.450}{EI}\ \text{(m)}\,.$$

The final bending moment diagram is constructed by a formula using the principle of superposition:

$$M_P = \bar{M}_1 \cdot Z_1 + \bar{M}_2 \cdot Z_2 + M_P^0.$$

The corresponding calculations for the specified points (Fig. 8.8c) are presented in Table 8.3. The signs of the bending moments in the unit conditions are conventional.

Table 8.3 Calculation of bending moments

Points	$\bar{M}_1$	$\bar{M}_1 \cdot Z_1$	$\bar{M}_2$	$\bar{M}_2 \cdot Z_2$	M_P^0	M (kN m)
1	−0.4	−1.5176	+0.24	+5.388	+4.1667	+8.037
2	+0.2	+0.7588	0.0	0.0	−2.0833	−1.324
3	+0.8	+3.0352	−0.24	−5.388	+4.1667	+1.814
4	−1.0	−3.794	−0.333	−7.476	0.0	−11.27
5	0.0	0.0	0.0	0.0	0.0	0.0
6	−0.6	−2.2764	0.0	0.0	+15.36	+13.084
7	−0.36	−1.3658	0.0	0.0	−9.984	−11.349
8	0.0	0.0	0.0	0.0	0.0	0.0
Factor	EI		EI			

signs of bending moments

⊕ ⊖

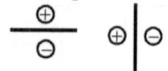

The final bending moment diagram M is presented in Fig. 8.8g. The same diagram was obtained by the force method (Example 7.2).

Construction of shear and axial force diagrams, as well as computation of the reactions is described in detail in Example 7.2.

Summary. The presentation of the equation of the displacement method in canonical form is conveniently organized and also prescribes a well-defined algorithm for the analysis of complex structures. Special parts of structural analysis can be carried out more easily with the canonical form of the displacement method. These include the construction of influence lines (Chap. 10), Stability (Chap. 13), and Vibration (Chap. 14).

8.3 Comparison of the Force and Displacement Methods

The force and displacement methods are the principal analytical methods in structural analysis. Both of these methods are widely used, not only for static analysis, but also for stability and dynamical analysis. Below we will provide a comparison pertinent to the two methods' presentation in their canonical forms.

Both methods require construction of primary systems. Both methods require construction of bending moment diagrams for unit exposures (forces or displacements). In both methods, a difference between the primary system and the original one is eliminated using the set of canonical equations.

Fundamental differences between these methods are presented in Table 8.4. It can be easily seen that these methods are duel, i.e., one column of the table can be obtained from the other by linguistic restatement.

Table 8.4 Fundamental differences between the force and displacement methods

Comparison criteria	Force method	Displacement method
Primary system (PS)	Obtained by **eliminating redundant** constraints **from** a structure	Obtained by **introducing additional** constraints **to** a structure
Primary unknowns (PU)	Uses **forces** (forces and moments), which **simulate** the actions of **eliminated** constraints **Reactions of eliminated constrains are PU**	Uses **displacements** (linear and angular), which **neutralize** the actions of **introduced** constraints **Displacements of induced constraints are PU**
Number of PU	Equals the degree of **statical** indeterminacy	Equals the degree of **kinematical** indeterminacy
Number of PS and way of obtaining PS	**Nonunique**. PS **can be chosen** so that all redundant constraints must be eliminated and replaced by corresponding reactions (forces and/or moments) Which redundant constraints should be eliminated is a matter of choice, but the obtained PS should be statically determinate	**Unique**. PS **must be constructed** so that in every rigid joint an additional constraint is introduced to prevent angular rotation; and for every independent linear displacement an additional constraint is introduced to prevent linear displacement PS presents a set of standard statically indeterminate beams
Canonical equations	$\delta_{11}X_1 + \delta_{12}X_2 + \cdots + \Delta_{1P} = 0$ $\delta_{21}X_1 + \delta_{22}X_2 + \cdots + \Delta_{2P} = 0$ . . . . Number of canonical equations equals the number of PU	$r_{11}Z_1 + r_{12}Z_2 + \cdots + R_{1P} = 0$ $r_{21}Z_1 + r_{22}Z_2 + \cdots + R_{2P} = 0$ . . . . Number of canonical equations equals the number of PU

(continued)

Table 8.4 (continued)

Meaning of equations	Total **displacement** in the direction of **eliminated** constraints caused by the action of all primary unknowns **(forces or moments)** and applied forces is zero	Total **reaction** in the direction of **introduced** constraints caused by the action of all primary unknowns **(linear or angular displacements)** and applied forces is zero
Character of canonical equations	**Kinematical:** the left part of canonical equations represents **displacements**	**Statical:** the left part of canonical equations represents **reactions**
Matrix of coefficients of canonical equations	$A = \begin{bmatrix} \delta_{11} & \delta_{12} & \cdots & \delta_{1n} \\ \delta_{21} & \delta_{22} & \cdots & \delta_{2n} \\ \cdots & \cdots & \cdots & \cdots \\ \delta_{n1} & \delta_{n2} & \cdots & \delta_{nn} \end{bmatrix}$, det $A > 0$, where A is the **flexibility** matrix	$R = \begin{bmatrix} r_{11} & r_{12} & \cdots & r_{1n} \\ r_{21} & r_{22} & \cdots & r_{2n} \\ \cdots & \cdots & \cdots & \cdots \\ r_{n1} & r_{n2} & \cdots & r_{nn} \end{bmatrix}$, det $R > 0$, where R is the **stiffness** matrix
Meaning of unit coefficients	Unit **displacement** $\boldsymbol{\delta_{ik}}$ presents **displacement** in the direction of ith **eliminated** constraints due to primary unknown **(force)** $X_k = 1$	Unit **reaction** $\boldsymbol{r_{ik}}$ presents **reaction** in the ith **introduced** constraints due to primary unknown **(displacement)** $Z_k = 1$
Meaning of free terms	**Displacement** $\boldsymbol{\Delta_{iP}}$ presents **displacement** in the direction of ith **eliminated** constraint due to applied forces	**Reaction** $\boldsymbol{R_{iP}}$ presents **reaction** in the ith **introduced** constraint due to applied forces
Dimensions of unit coefficients	$\boldsymbol{\delta_{ik}}$ – Dimension of **displacement** at i is divided by dimension of action **(force or moment)** at k	$\boldsymbol{r_{ik}}$ – Dimension of **force** at i is divided by dimension of action **(linear or angular displacement)** at k

We are providing only the classical approach, however, an experienced reader may choose to designate the primary system of the force method as statically indeterminate, providing that he/she has all the necessary formulas for calculating the accepted statically indeterminate primary system

8.3.1 Properties of Canonical Equations

1. The main coefficients of the canonical equations of the force and displacement methods are strictly positive: $\delta_{ii} > 0$; $r_{ii} > 0$.
2. The matrix of coefficients of the canonical equations is symmetrical with respect to the main diagonal: $\delta_{ik} = \delta_{ki}$; $r_{ik} = r_{ki}$. These coefficients may be positive, negative, or zero.
3. The coefficients of the canonical equations depend only on the type of structure, but do not depend on external loads, settlements of supports, temperature changes, or errors of fabrication.

4. The determinant of the matrix of coefficients of the canonical equations is strictly positive. This condition describes an internal property of structures based on a fundamental law of elastic systems: the potential energy of a structure subjected to any load is positive. Since $D \neq 0$, the solution to the canonical equations of any redundant structure, subjected to any load, change of temperature, or settlements of supports, is unique.

It is time to ask a question: when is it more convenient to apply the force method and when the displacement method? The general answer is the following: the more rigid the system due to given constraints, then the more efficient the displacement method will be. This can be illustrated by considering different structures. Figure 8.9a presents a frame with fixed supports. The frame, according to the force method, has nine unknowns, while by the displacement method it has only one unknown: the angle of rotation of the only rigid joint. Analysis of this structure by the displacement method is a very simple problem. Should this structure be modified by adding more elements connected at the rigid joint, nevertheless, the number of unknowns by the displacement method is still the same, while the number of unknowns by the force method is increased with the addition of each new element.

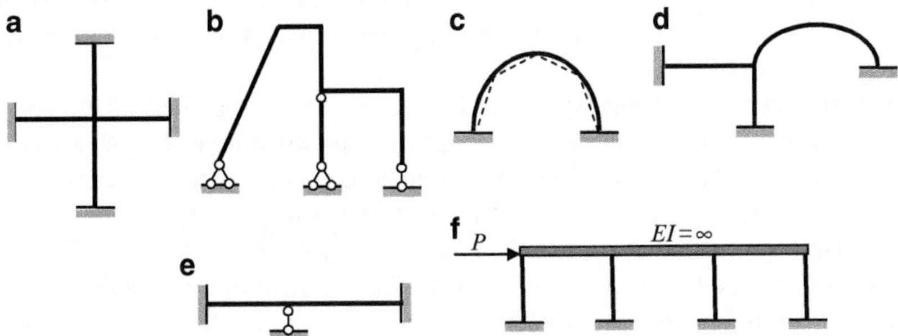

Fig. 8.9 Different types of structures that can be analyzed by either the force method or the displacement method

Another frame is shown in Fig. 8.9b. The number of unknowns by the force method is one, while by the displacement method it is six (four rigid joints and two linear independent displacements). Also, it is important to note that this frame contains inclined members, which lead to additional cumbersome computations of the reactions in the introduced constraints. Therefore the force method is more preferable.

A statically indeterminate arch with fixed supports can also be analyzed by both the force and displacement methods. The number of unknowns by the force method is three. In order to analyze this kind of arch by the displacement method, its curvilinear axis must be replaced by a set of straight members (since the standard elements of the displacement method are straight members). One version of such segmentation of the arch is shown in Fig. 8.9c by dashed lines; in this case, the number of unknowns by the displacement method is six. Obviously, the force method is more preferable here. If the arch is part of any complex structure, as shown in Fig. 8.9d, then such a structure can be easily analyzed by the displacement method in canonical form. Indeed, the number of unknowns by the force method is six, while by the displacement method it is just one (the angular displacement of the rigid joint). Detailed tables for parabolic uniform and nonuniform arches are presented in Tables A.19 and A.20.

Figure 8.9e shows a statically indeterminate beam. The number of unknowns by the force method is four. The number of unknowns by the displacement method is one. It is obvious that the displacement method is more preferable in this case.

Figure 8.9f shows a statically indeterminate frame with an absolutely rigid crossbar. By the displacement method this structure has only one unknown no matter how many vertical elements it has; analysis of frames of this type is considered in Sect. 8.4.

8.4 Sidesway Frames with Absolutely Rigid Crossbars

So far it has been assumed that the each rigid joint of the frame can rotate. Such joint corresponds to one unknown of the displacement method. This section describes analysis of a special type of frame: sidesway frames with absolutely rigid crossbars (flexural stiffness $EI = \infty$).

Let us consider the frame shown in Fig. 8.10a. The connections of the crossbar and the vertical members are rigid. A feature of this structure is that the crossbar is an absolutely rigid body; therefore, even if the joints are rigid, there are no angles of rotation of the rigid joints. Thus, the frame has only one primary unknown, i.e., the linear displacement Z_1 of the crossbar. The primary system is shown in Fig. 8.10b. Introduced constraint 1 prevents horizontal displacement of the crossbar. Flexural stiffness per unit length is $i = EI/h$.

The canonical equation of the displacement method is $r_{11}Z_1 + R_{1P} = 0$.

The bending moment diagram caused by unit linear displacement Z_1 is shown in Fig. 8.10c.

Unit reaction r_{11} is calculated using the equilibrium equation for the crossbar. The bending moment for the clamped-clamped beam due to lateral unit displacement is $M = 6i/h$; therefore the shear force for vertical elements is $Q = 2M/h = 12i/h^2$. This force is transmitted to the crossbar (Fig. 8.10d) and after that the unit reaction can be calculated as follows:

$$r_{11} \to \sum X = -3\frac{12i}{h^2} + r_{11} = 0 \to r_{11} = \frac{36i}{h^2}.$$

Applied load P does not produce bending moments in the primary system. Nevertheless, the free term is not zero and should be calculated using the free-body diagram for the crossbar (Fig. 8.10e).

Equilibrium condition $\sum X = 0$ leads to the following result: $R_{1P} = -P$.

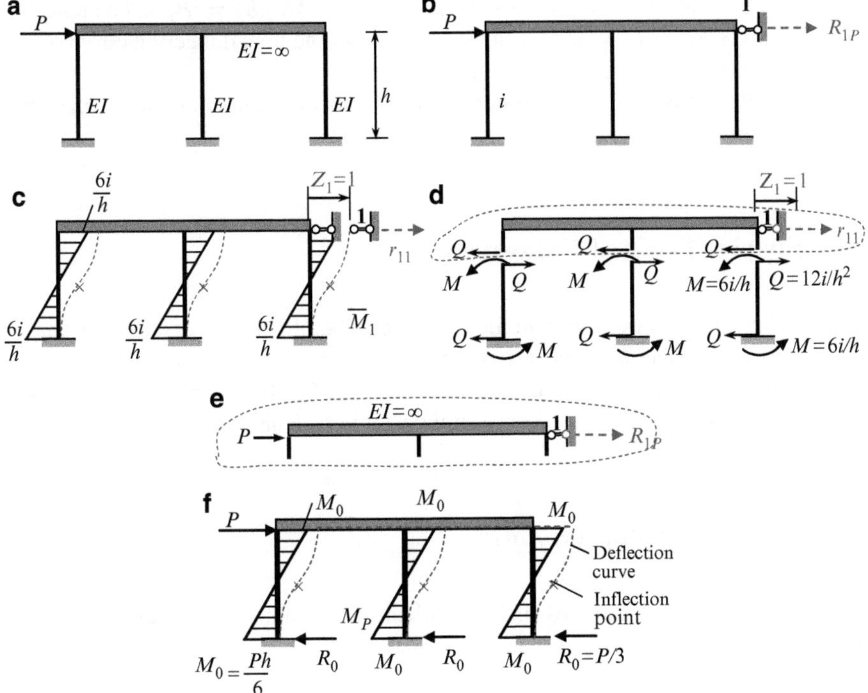

Fig. 8.10 Frame with infinitely rigid crossbar: (**a**) Design diagram; (**b**) Primary system; (**c**) Bending moment diagram in unit state; (**d**) Free-body diagram of the crossbar. (**e**) Free-body diagram of the crossbar in loaded state; (**f**) Final bending moment diagram

The primary unknown becomes

$$Z_1 = -\frac{R_{1P}}{r_{11}} = \frac{Ph^3}{36EI}.$$

The bending moment at each of the specified sections of the frame is calculated by the formula

$$M_P = \bar{M}_1 \cdot Z_1 + M_P^0.$$

Since applied load P does not produce bending moments in the primary system, then $M_P^0 = 0$ and the resulting bending moments due to applied load P are determined as $M_P = \bar{M}_1 \cdot Z_1$. The corresponding bending moment diagram and the reactions of supports are presented in Fig. 8.10f.

The bending moment at the top and bottom of each vertical member is

$$M_0 = \frac{6i}{h}\frac{Ph^3}{36EI} = \frac{Ph}{6}.$$

Shear within all vertical members is $Q = (M_0 + M_0/h) = P/3$. The reaction of all supports $R_0 = Q = P/3$. Figure 8.10f shows the final deflection curve of the frame and inflection point of all vertical members.

Note that a frame with an absolutely rigid crossbar has only one unknown of the displacement method for any number of vertical members.

8.5 Special Types of Exposures

The displacement method in canonical form can be effectively applied for analysis of statically indeterminate frames subjected to special types of exposures such as settlements of supports and errors of fabrication. For both types of problems, a primary system of the displacement method should be constructed as usual.

8.5.1 *Settlements of Supports*

For a structure with n degrees of kinematical indeterminacy, the canonical equations are

$$\begin{aligned}
&r_{11}Z_1 + r_{12}Z_2 + \cdots + r_{1n}Z_n + R_{1s} = 0\\
&r_{21}Z_1 + r_{22}Z_2 + \cdots + r_{2n}Z_n + R_{2s} = 0\\
&\cdot \quad \cdot \quad \cdot \quad \cdot \quad \cdot \quad \cdot \quad \cdot \quad \cdot \quad \cdot \quad \cdot \quad \cdot\\
&r_{n1}Z_1 + r_{n2}Z_2 + \cdots + r_{2n}Z_n + R_{ns} = 0,
\end{aligned} \tag{8.6}$$

where the free terms R_{js} $(j = 1, 2, \ldots, n)$ represent the reaction in the jth introduced constraint in the primary system due to the *settlements of a support*. These terms are calculated using Tables A.3–A.6, taking into account the *actual* displacements of the support. Unit reactions r_{ik} should be calculated as before. The final bending moment diagram is constructed by the formula

$$M_s = \bar{M}_1 \cdot Z_1 + \bar{M}_2 \cdot Z_2 + \cdots + \bar{M}_2 \cdot Z_2 + M_s^0, \tag{8.7}$$

where M_s^0 is the bending moment diagram in the primary system caused by the given *settlements of the support.*

Example 8.3. The redundant frame in Fig. 8.11 is subjected to the following settlement of fixed support A: $a = 2\,\text{cm}$, $b = 1\,\text{cm}$, and $\varphi = 0.01\,\text{rad} = 34'30''$. Construct the bending moment diagram.

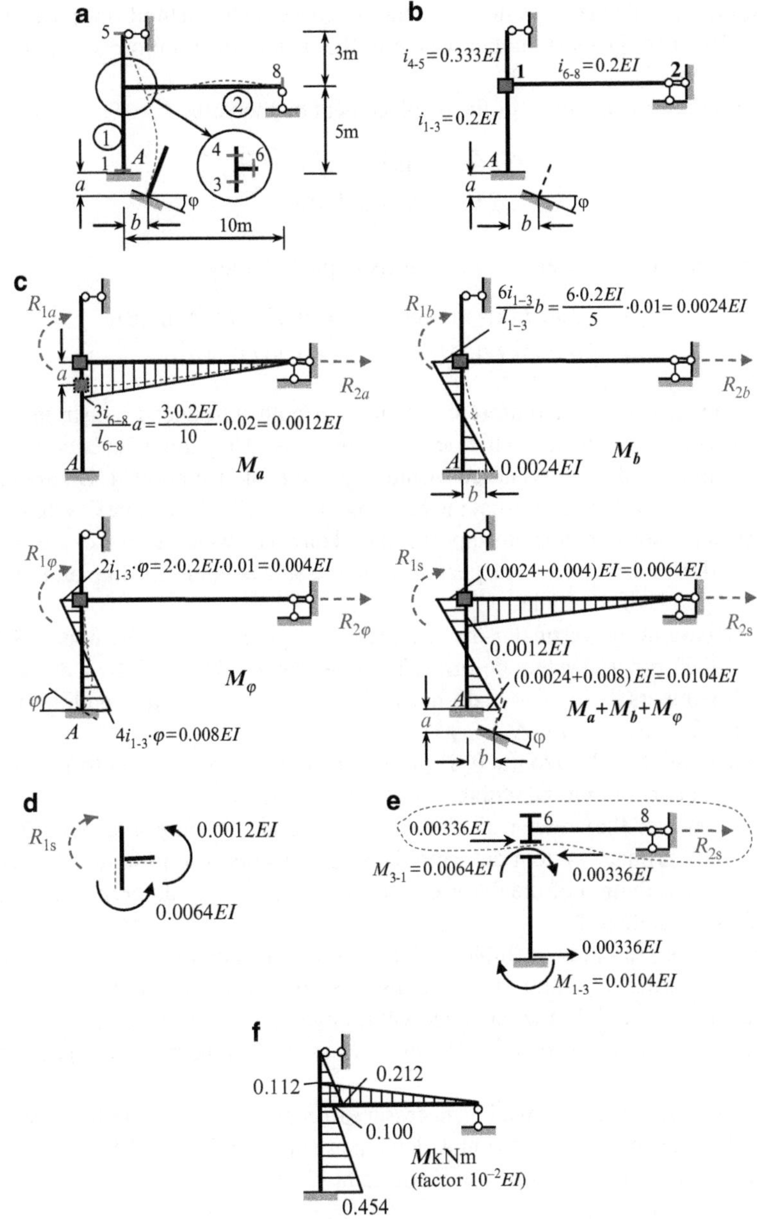

Fig. 8.11 Settlements of support: (**a**) Design diagram; (**b**) Primary system; (**c**) Bending moment diagram in primary system caused by settlements of support A; (**d, e**) Free-body diagrams for rigid joint and crossbar; (**f**) Final bending moment diagram

Solution. The primary system of the displacement method is presented in Fig. 8.11b. The primary unknowns are the angle of rotation Z_1 and linear displacement Z_2.

The canonical equations of the displacement method are

$$r_{11}Z_1 + r_{12}Z_2 + R_{1s} = 0,$$
$$r_{21}Z_1 + r_{22}Z_2 + R_{2s} = 0.$$

The unit reactions r_{ik} were obtained in Example 8.2; they are

$$r_{11} = 2.4EI\,(\text{kN m/rad})\,, r_{12} = 0.093EI\,(\text{kNm/m})\,,$$
$$r_{21} = 0.093EI\,(\text{kN/rad})\,, r_{22} = 0.207\,(\text{kN/m})\,.$$

Free terms R_{1s} and R_{2s} represent reactions in both introduced constraints in the primary system due to the settlement of support A. They are calculated using the bending moment diagrams caused by the displacement of support A. Diagrams M_a, M_b, and M_φ (Fig. 8.11c) present the bending moments in the primary system due to the separate displacements of support A based on given values $a = 2\,\text{cm}$, $b = 1\,\text{cm}$, and $\varphi = 0.01\,\text{rad} = 34'30''$, respectively. Figure 8.5a contains the numeration of the specified points of the frame.

In the case of the vertical a-displacement of support A, only member 6-8 of the frame in the primary system deforms. The elastic curve is shown by a dashed line; vertical members 1-3 and 4-5 do not deform. To construct the bending moment diagram M_a, we use Table A.3.

In the case of the horizontal b-displacement, only member 1-3 deforms; in order to construct bending moment diagram M_b, we use Table A.4.

In the case of the angular displacement of support A, only member 1-3 of the frame in the primary system deforms. This happens because introduced constraint 1 prevents distribution of the deformation. Bending moment diagram M_φ is shown according to Table A.4.

The values presented in Tables A.3 and A.4 correspond to unit displacements, therefore in order to obtain the reactions according to the given displacements, it is necessary to multiply the tabulated values by corresponding values a, b, and φ. Bending moment diagrams M_a, M_b, and M_φ as well as their sum are presented in Fig. 8.11c.

To calculate the free terms, it is necessary to consider the free-body diagrams for the rigid joint and for the horizontal element as shown in Fig. 8.11d, e.

The equilibrium condition of the rigid joint yields

$$R_{1s} = 0.0064EI + 0.0012EI = 0.0076EI.$$

The shear force within portion 1-3 is equals to

$$\frac{M_{3-1} + M_{1-3}}{l_{1-3}} = \frac{0.0064 + 0.0104}{5}EI = 0.00336EI.$$

These forces rotate the member counterclockwise. Force 0.00336EI is transmitted to horizontal member 6-8 in the opposite direction, i.e., from left to right.

The equilibrium condition of the horizontal element is used to calculate R_{2s}

$$\sum X = 0 \rightarrow R_{2s} = -0.00336EI.$$

The canonical equations become

$$2.4Z_1 + 0.093Z_2 + 0.0076 = 0,$$
$$0.093Z_1 + 0.207Z_2 - 0.00336 = 0.$$

The roots of these equations are

$$Z_1 = -0.386 \times 10^{-2}\,\text{rad},$$
$$Z_2 = 1.797 \times 10^{-2}\,\text{m}.$$

The bending moment diagram is constructed by the following expression using the principle of superposition

$$M_s = \bar{M}_1 \cdot Z_1 + \bar{M}_2 \cdot Z_2 + M_s^0,$$

where $M_s^0 = M_a + M_b + M_\phi$ is the bending moment in the primary system caused by the settlements of the support. The corresponding computation is presented in Table 8.5.

Table 8.5 Calculation of bending moments

Points	$\bar{M}_1$	$\bar{M}_1 \cdot Z_1$	$\bar{M}_2$	$\bar{M}_2 \cdot Z_2$	$M_a + M_B + M_\varphi$	M_s
1	−0.4	0.155	+0.24	0.431	−1.04	−0.454
3	+0.8	−0.309	−0.24	−0.431	+0.64	−0.100
4	−1.0	0.386	−0.333	−0.598	0.0	−0.212
5	0.0	0.0	0.0	0.0	0.0	0.0
6	−0.6	0.232	0.0	0.0	−0.12	0.112
8	0.0	0.0	0.0	0.0	0.0	0.0
Factor	EI	$10^{-2}EI$	EI	$10^{-2}EI$	$10^{-2}EI$	$10^{-2}EI$

signs of bending moments

⊕ ⊖ ⊕ ⊖

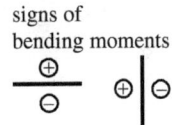

The final bending moment diagram due to the settlements of the support is presented in Fig. 8.11f. This same diagram was obtained previously by the force method (Example 7.3).

Discussion. In case of the structure subjected to external loading, the internal force distribution depends only on the relative stiffness of the elements, while in case of the settlements of support the distribution of internal forces depends on both the relative and the absolute stiffness of the elements.

In the case of settlements of a support, the calculation of free terms is an elementary procedure using the displacement method, while by the force method free terms are more difficult to calculate. Therefore, for settlement of support problems the displacement method is preferable to the force method.

8.5.2 Errors of Fabrication

In cases of fabrication errors, analysis of the frame can be effectively performed by the displacement method in canonical form. Construction of the primary system and calculation of unit reactions should be performed as usual. Canonical equations should be presented in form (8.4), but the free terms R_{iP} should be replaced by R_{ie}. These free terms present the reactions of the introduced constraints caused by the errors of fabrication.

Example 8.4. A design diagram of the frame is presented in Fig. 8.12a. Member *AB* has been fabricated $\Delta = 0.8$ cm too long. The moment of inertia of all cross sections is $I = 9.9895 \times 10^{-5}\,\mathrm{m}^4$, and the modulus of elasticity is $E = 2 \times 10^8\,\mathrm{kN/m}^2$ $\left(EI = 19{,}979\,\mathrm{kN\,m}^2\right)$. Calculate the angle of rotation of joint *B* and construct the bending moment diagram.

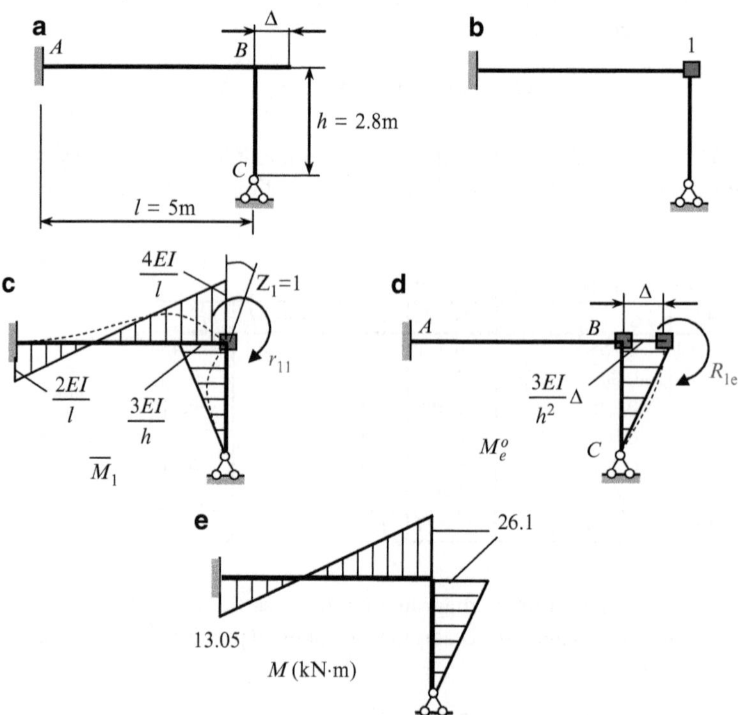

Fig. 8.12 (**a, b**) Errors of fabrication: Design diagram and primary system. (**c, d**) Bending moment diagrams for unit and loading states. (**e**) Resulting bending moment diagram

Solution. The primary system of the displacement method is shown in Fig. 8.12b.

The primary unknown is the angle of rotation of included constraint 1. The canonical equation of the displacement method is $r_{11}Z_1 + R_{1e} = 0$. The primary unknown becomes $Z_1 = -R_{1e}/r_{11}$.

The bending moment diagram in unit state is presented in Fig. 8.12c. The loaded state (Fig. 8.12d) presents Δ-displacement of joint B. The member AB is not subjected to bending; corresponding elastic curve is shown by a dashed line. The moment R_{1e} is a reactive moment at the introduced constraint caused by displacement Δ. Table A.3 presents the reactions caused by the *unit* displacement, therefore the specified ordinate of the bending moment diagram is $\left(3EI/h^2\right)\Delta$.

It is obvious that

$$r_{11} = \frac{4EI}{l} + \frac{3EI}{h}, \; R_{1e} = -\frac{3EI}{h^2}\Delta.$$

So the angle of rotation of joint B becomes

$$Z_1 = \frac{3\Delta}{\left(\frac{4h^2}{l} + 3h\right)} = \frac{3 \cdot 0.008\,\text{m}}{4 \times \frac{2.8^2}{5}\,(\text{m}) + 3 \times 2.8\,(\text{m})} = 0.001635\,\text{rad}.$$

The resulting bending moments at the specified points are

$$M_{BA} = \frac{4EI}{l}Z_1 = \frac{4 \times 19979\,\text{kN}\,\text{m}^2}{5.0\,\text{m}} \times 0.001635\,\text{rad} = 26.1\,\text{kN}\,\text{m};$$

$$M_{AB} = 0.5M_{BA} = 13.05\,\text{kN}\,\text{m};$$

$$M_{BC} = \frac{3EI}{h}Z_1 - \frac{3EI}{h^2}\Delta = \frac{3 \times 19979\,\text{kN}\,\text{m}^2}{2.8\,\text{m}} \times 0.001635\,\text{rad} - \frac{3 \times 19979\,\text{kN}\,\text{m}^2}{2.8^2\,\text{m}^2} \times 0.008\,\text{m} = -26.1\,\text{kN}\,\text{m}.$$

The corresponding bending moment diagram is presented in Fig. 8.12e.

Construction of shear and axial forces and computation of reactions of supports should be performed as usual.

Discussion. We can see that an insignificant error of fabrication leads to significant internal forces in the structure. This fact lays the basis of the inverse problem, which can be formulated as follows: find the initial displacement of specified points of a structure for obtaining the required distribution of internal forces. For example, let us consider the two-span continuous beam subjected to a uniformly distributed load shown in Fig. 8.13.

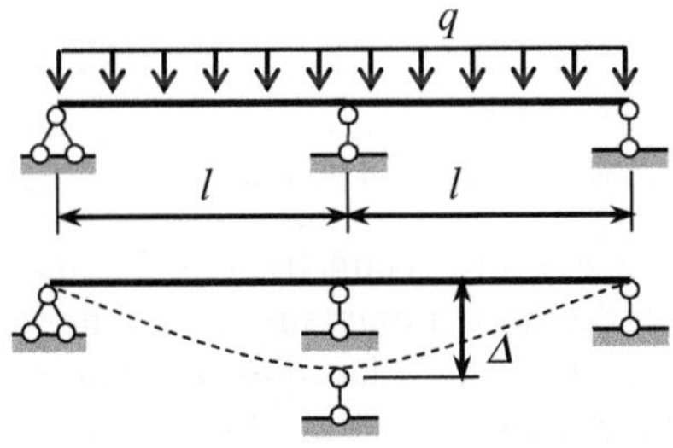

Fig. 8.13 Controlling of the stresses in the beam by initial displacement of middle support

In this case, the extended fibers at the center support are located *above* the neutral line and the corresponding moment at the support is $ql^2/8$. If the external load is absent and the center support is placed below the beam level on Δ, then the extended fibers at the support will be located *below* the neutral line and the corresponding moment at the support will be $\left(3EI/l^2\right)\Delta$. If beam is subjected to load q and displacement Δ simultaneously, then we can determine parameter Δ for the required distribution of internal forces. For example, what Δ should be in order that the bending moment at the center support be zero, or what Δ should be in order that the maximum bending moment at the span and bending moment at the center support be equal, etc.

8.6 Analysis of Symmetrical Structures

Symmetrical structures are used often in structural engineering. *Symmetrical structures* mean their geometrical symmetry, symmetry of supports, and stiffness symmetry of the members. For analysis of symmetrical statically indeterminate frames subjected to *any loads*, the Combined Method (the combination of the force and displacement methods) can be effectively applied.

8.6.1 Symmetrical and Antisymmetrical Loading

We start consideration of the combined method from concept of resolving a total load on symmetrical and antisymmetrical components. Symmetrical frame subjected to horizontal load P is shown in Fig. 8.14a. This load may be presented as a sum of symmetrical and antisymmetrical components (Fig. 8.14b, c). In general, *any load* may be presented as the sum of symmetrical and antisymmetrical components.

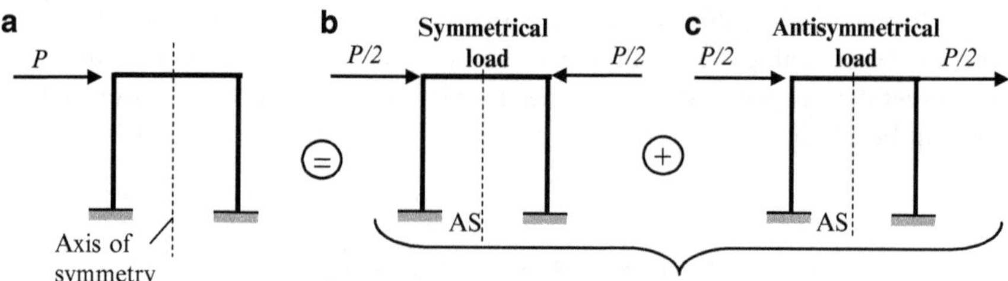

Fig. 8.14 Presentation of the load P as a sum of the symmetrical and antisymmetrical components

Symmetrical frame subjected to symmetrical and antisymmetrical components allows simplifying the entire design diagram. This simplification is based on the change of the entire design diagram by its equivalent *half-frame* (the physical basis of such changing will be discussed below). This procedure leads to the following fact: the degree of statical (or kinematical) indeterminacy for a half-frame is less than for an entire frame. Equivalent half-frames should be analyzed by classical methods. It will be shown below that for analysis of the half-frame, different methods (Force and Displacement methods) should be applied for symmetrical and antisymmetrical components. Therefore, the method under consideration is called the combined method.

8.6.2 Concept of Half-Structure

In case of symmetrical structure subjected to symmetrical and antisymmetrical loads, the elastic curve at the point on the axis of symmetry (AS) has specific properties. Mainly these properties allow an entire frame to replace by an equivalent half-frame, separately for symmetrical and antisymmetrical loading. In case of nonsymmetrical change of temperature or settlements of supports, these exposures could also be replaced by symmetrical and antisymmetrical components.

At any section of a member the following *internal forces* arise: *symmetrical* unknowns, such as bending moment M and axial force N and *antisymmetrical* unknown shear force Q (Fig. 8.15).

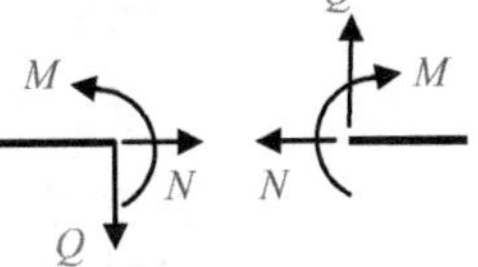

Fig. 8.15 Symmetrical and antisymmetrical internal forces

At the point of the axis of symmetry (AS) the following *displacements* arise: the vertical Δ_v, horizontal Δ_h, and angular φ. Considering of these displacements at the AS (point A) allow us to construct an equivalent half-frame. Depending on loading (symmetrical or antisymmetrical), the different displacements will be at the point A and, as result, the different support conditions for equivalent half-frame.

Let us consider the arbitrary symmetrical frames (Table 8.6). Elastic curves in the neighborhood of the point A on the axis of symmetry are shown by dotted line. Assume that number of spans is odd. In the case of symmetrical load, the horizontal and angular displacements at the point A are zero, while the vertical displacement occurs. Therefore, an equivalent half-frame must contain a support at point A, which would model corresponding displacements, i.e., which would allow the vertical displacement and does not allow the horizontal and angular displacements. Only the slide support corresponds to these types of displacements.

Table 8.6 Symmetric frame and corresponding half-frame for symmetric and antisymmetric loading

Number of spans	Symmetrical loading		Antisymmetrical loading	
1,3,5,...	A Half-frame A $\Delta_v \neq 0$ $\Delta_h = 0$ $\varphi = 0$		A Half-frame A $\Delta_v = 0$ $\Delta_h \neq 0$ $\varphi \neq 0$	
	Number of unknowns		Number of unknowns	
	Entire frame	Half-frame	Entire frame	Half-frame
	FM: 9	FM: 5	FM: 9	**FM: 4**
	DM: 9	**DM: 4**	DM: 9	DM: 5
2,4,6,...	A Half-frame A $\Delta_v = 0$ $\Delta_h = 0$ $\varphi = 0$		A EI Half-frame A $0.5EI$ $\Delta_v = 0$ $\Delta_h \neq 0$ $\varphi \neq 0$	
	Number of unknowns		Number of unknowns	
	Entire frame	Half-frame	Entire frame	Half-frame
	FM: 12	FM: 6	FM: 12	FM: 6
	DM: 10	**DM: 4**	DM: 10	DM: 6
Internal force diagrams	Bending moment diagram – symmetrical Normal force diagram – symmetrical Shear force diagram – antisymmetrical		Bending moment diagram – antisymmetrical Normal force diagram – antisymmetrical Shear force diagram – symmetrical	

In the case of antisymmetrical load, the horizontal and angular displacements at the point A exist, while a vertical displacement is zero. It means that an equivalent half-frame at point A must contain a support, which would allow such displacements, i.e., which would allow the horizontal and angular displacements and does not allow a vertical one. Only roller support corresponds to these types of displacements. Table 8.6 also contains the mathematical conditions for replacing the given frame by its equivalent half-frame. In all cases, the support at the point A for equivalent half-frame simulates the displacement at the point A for entire frame.

Now assume that the number of spans is even. It means that additional vertical member is placed on the axis of symmetry. In the case of symmetrical load, the horizontal and angular displacements at the point A, as in case of odd spans, are zero, while the vertical displacement of A is zero because the vertical member is at axis of symmetry. Therefore, an equivalent half-frame must contain a support at point A, which would model corresponding displacements. Only clamped support is related with displacements described above. In case of antisymmetrical loading, the half frame contains the member at the axis of symmetry with bending stiffness $0.5EI$.

Fundamental properties of internal force diagrams for symmetrical structures are as follows:

1. In case of symmetrical loading, the internal force diagrams for symmetrical unknowns (M, N) are symmetrical and for antisymmetrical unknown (Q) they are antisymmetrical.
2. In case of antisymmetrical loading, the internal force diagrams for symmetrical unknowns (M, N) are antisymmetrical and for antisymmetrical unknown (Q) they are symmetrical.

Table 8.6 also contains the number of unknowns for entire frame and for half-frame by Force (FM) and Displacement (DM) methods. The rational method is shown by bold. We can see that in case of symmetrical loading, the more effective is displacement method, while in the case of antisymmetrical loading, the more effective is the force method. From this table, we can see advantages of the combined method. Bending moment diagram is constructed for each case and then summated (on the basis of superposition principle) to obtain final bending moment diagram for original frame.

The following procedure may be recommended for analysis of symmetrical structures:

1. Resolve the entire load into symmetrical and antisymmetrical components
2. Construct the equivalent half-frame for both types of loading
3. Provide the analysis of each half-frame using the most appropriate method
4. Find the final distribution of internal forces using superposition principle

Problems

In Problems 8.1 through 8.6 provide complete analysis by the displacement method, including the following:

1. Determine the bending moment at the support and construct the internal force diagrams
2. Calculate the reactions of the supports and provide static control
3. Provide kinematical control (check the slope and vertical displacement at the intermediate support)
4. Compute the vertical displacement at the specific points and show the elastic curve

8.1–8.2. A two-span uniform beam with pinned and rolled supports is subjected to a fixed load (Figs. P8.1 and P8.2). Provide complete analysis for each design diagram.

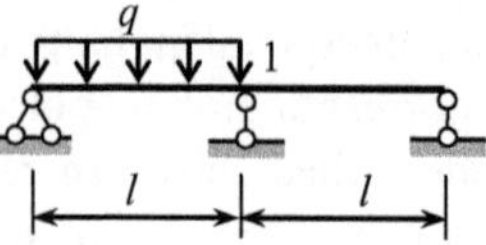

Fig. P8.1

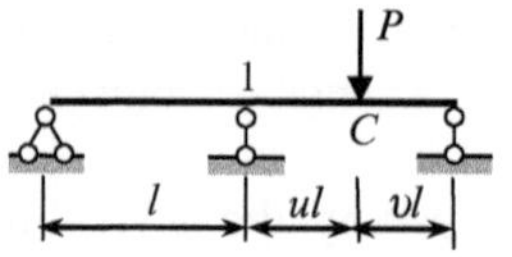

Fig. P8.2

Ans. (8.1) $M_1 = -\dfrac{ql^2}{16}$; (8.2) $M_1 = -\dfrac{Pl}{4}\upsilon\left(1-\upsilon^2\right)$

8.3–8.4. A two-span uniform beam with fixed left and rolled right support is subjected to fixed load (Figs. P8.3 and P8.4). Provide complete analysis.

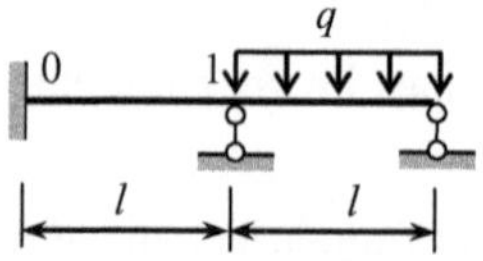

Fig. P8.3

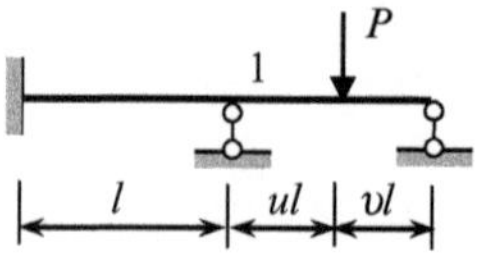

Fig. P8.4

Ans. (8.3) $M_0 = \dfrac{ql^2}{28}$; $M_1 = -\dfrac{ql^2}{14}$; (8.4) $M_1 = -\dfrac{2}{7}Pl\upsilon\left(1-\upsilon^2\right)$.

8.5. A two-span beam uniform beam with two fixed ends is subjected to fixed load (Fig. P8.5). Provide complete analysis.

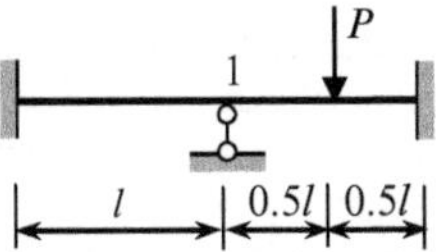

Fig. P8.5

$$\text{Ans. } M_1 = -\frac{Pl}{16}.$$

8.6. A two-span nonuniform beam with fixed ends is loaded as shown in Fig. P8.6. Parameter n is any positive number. Construct the bending moment and shear force diagrams. Show the elastic curve. Prove that $M_0 = M_1/2$.

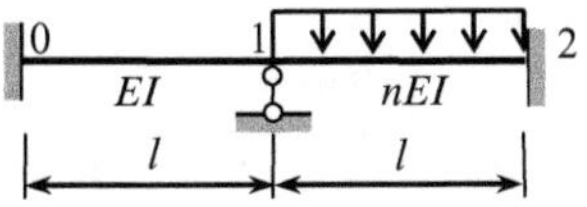

Fig. P8.6

$$\text{Ans. } M_1 = \frac{ql^2}{12\,(1+n)}, \quad M_2 = \frac{ql^2}{24} \cdot \frac{2+3n}{1+n}.$$

8.7. A design diagram of a two-span uniform beam is shown in Fig. P8.7. Calculate the slope at the support 1 and compute the bending moments at the supports.

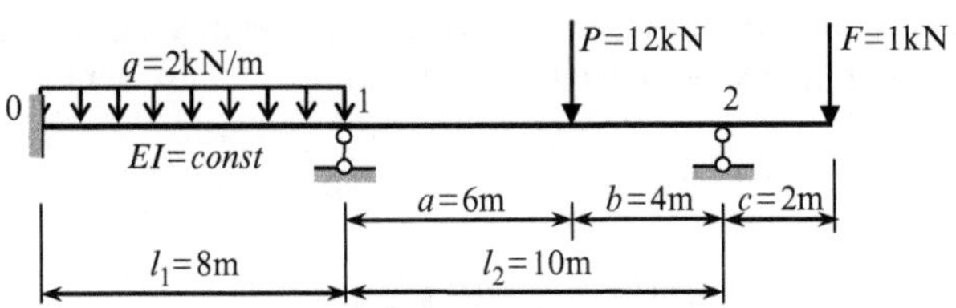

Fig. P8.7

$$\text{Ans. } Z_1 = \frac{10.616}{EI}\text{ rad}, \quad M_0 = 8.013\,\text{kN m}, \quad M_1 = 15.975\,\text{kN m}.$$

8.8. A two-span uniform beam with fixed and pinned end supports is subjected to vertical settlements of support, as shown in Fig. P8.8. Provide complete analysis for both design diagrams.

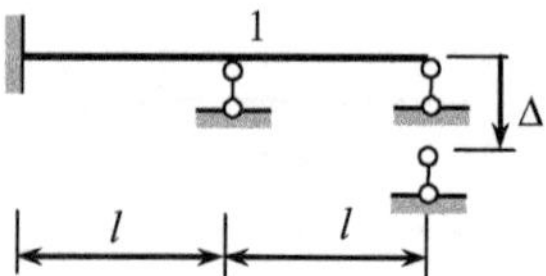

Fig. P8.8

Ans. $M_1 = -\dfrac{12}{7}\dfrac{EI}{l^2}\Delta.$

8.9. A two-span uniform beam with fixed ends is subjected to uniformly distributed load q and vertical settlements of the intermediate support, as shown in Fig. P8.9. Define the value of the settlement in order that: (a) the bending moment at the intermediate support equals zero; (b) the bending moment at the fixed supports equals zero; (c) the bending moment at the midpoint of the span equals zero.

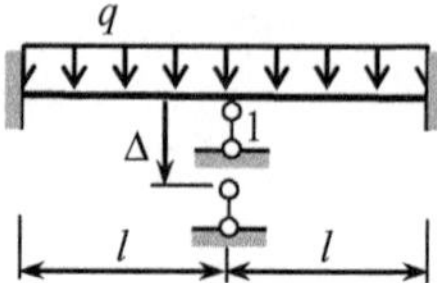

Fig. P8.9

Ans. (a) $\Delta = \dfrac{ql^4}{72EI}\,(\downarrow).$

8.10. A design diagram of a frame is presented in Fig. P8.10. Supports 2, 3, and 4 are fixed, sliding, and rolled, respectively. Bending stiffnesses are EI and $2EI$ for vertical and horizontal members, respectively; their relative stiffnesses are shown in circles. Compute the angle of rotation of the joint 1. Construct the internal force diagrams.

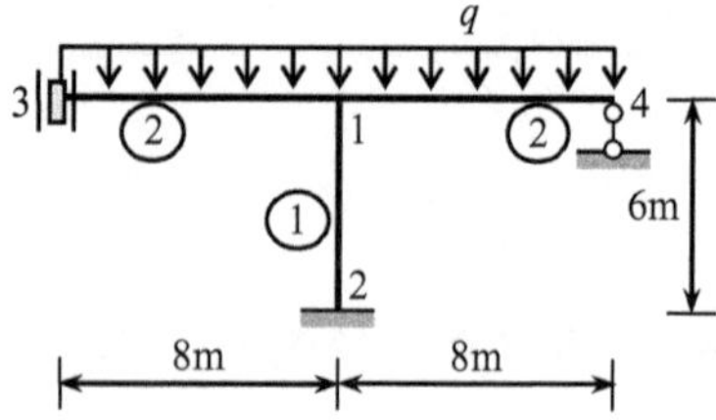

Fig. P8.10

Ans. $Z_1 = -\dfrac{8q}{EI}$ (rad), $M_{13} = \dfrac{58q}{3}$, $M_{31} = -\dfrac{38q}{3}$, $M_{14} = 14q$

8.11. A frame is subjected to uniformly distributed load q, as shown in Fig. P8.11. The stiffness of the right elastic support is k(kN/m). The relative stiffnesses for all members are shown in the circles. Calculate the angle of rotation of the joint 1. Construct the bending moment diagram. Check the limiting cases $k = \infty$ and $k = 0$. Trace the elastic curves for both limiting cases. Is it possible to find a stiffness of elastic support such that rigid joint 1 would not rotate under the given load?

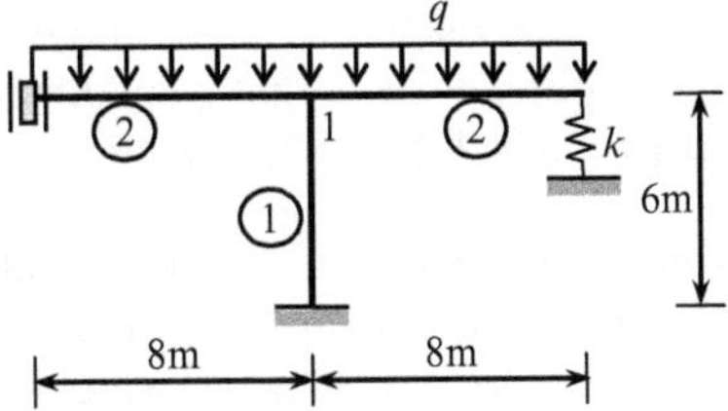

Fig. P8.11

Ans. $Z_1 = -\dfrac{12q}{EI} \cdot \dfrac{24\alpha - \frac{32}{3}}{11 + 9\alpha}, \quad \alpha = \dfrac{1}{1 + \frac{6EI}{512k}}.$

8.12a–d. Design diagrams for four Γ-frames with different boundary conditions at support 3 are shown in Fig. P8.12. All frames are subjected to uniform distributed load q. For all frames the span is l and height $h = l$. The flexural rigidity is EI for all members. Construct the internal force diagrams. Trace the elastic curves and show the inflection points. Explain the differences in structural behavior caused by the different types of supports 3.

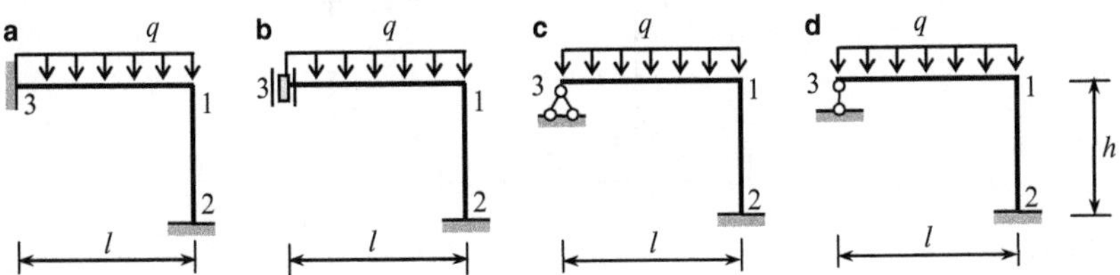

Fig. P8.12

Ans. (a) $Z_1 = -\dfrac{ql^3}{96EI}$; (b) $Z_1 = -\dfrac{ql^3}{15EI}$; (c) $Z_1 = -\dfrac{ql^3}{56EI}$

8.13. A sidesway frame is subjected to horizontal load P along the crossbar (Fig. P8.13). The connections of the crossbar with the left and right vertical members are hinged while the connection with the intermediate vertical member is rigid. The bending stiffness of all vertical members is EI and for the crossbar it is $2EI$. Will the axial forces in the crossbar be equal to zero or not? Verify your answer

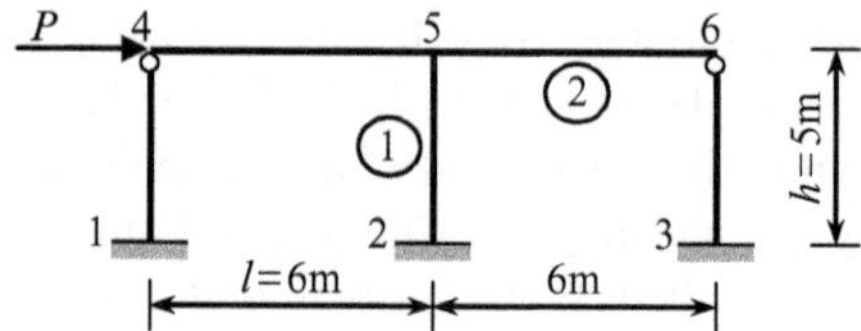

Fig. P8.13

by constructing the internal force diagrams. Trace the elastic curve of the frame. Calculate the reactions of supports. Provide static and kinematical verification.

Ans. $Q_1 = Q_3 = +0.1945P,\ Q_2 = +0.611P.$

8.14. The structure shown in Fig. P8.14 contains a nonuniform parabolic arch. The moment of inertia at the arbitrary section of the arch is $I_x = I_C / \cos\alpha$, where I_C is the moment of inertia of the cross section at the crown C and α is the angle between the horizontal line and the tangent at any section of the arch. Uniformly distributed load $q = 4kN/m$ is placed within the left half-span of the arch. Construct the bending moment diagram. Solve this problem in two versions: (1) The axial forces in the cross section of the arch are ignored and (2) the axial forces in the cross section of the arch are taken into account. The moment of inertia and the area of the cross section at the crown C are $I_C = 894 \times 10^6\,\mathrm{mm}^4$, $A_C = 33,400\,\mathrm{mm}^2$. Hint: The reader can find all the required data for parabolic arches as standard members in Table A.19 (uniform arches) and Table A.20 (non-uniform arches). Compare the results obtained by each version.

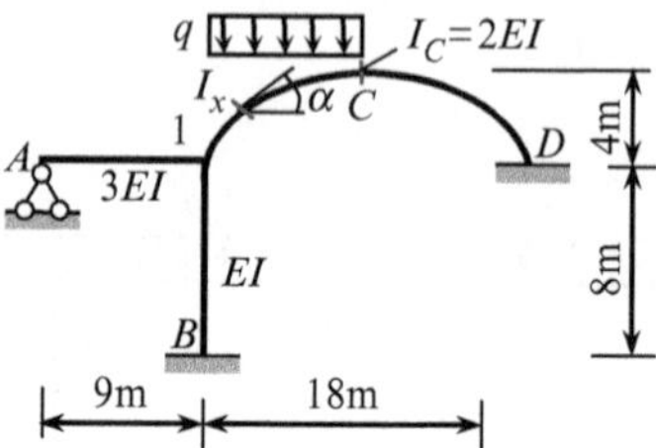

Fig. P8.14

$$\text{Ans. (1) } Z_1 = \frac{8.1}{EI}\ \text{(rad)}, \quad M_{1A} = 8.1\,\mathrm{kN\,m}, \quad M_{1B} = 4.05\,\mathrm{kN\,m},$$

$$M_{1C} = 12.15\,\mathrm{kN\,m}; \quad (2)\ Z_1 = \frac{8.4996}{EI}\ \text{(rad)}.$$

8.15. A frame with an infinitely rigid crossbar is subjected to horizontal load P. The connections between the vertical members and the crossbar are hinged (Fig. P8.15). Calculate the support moments and the reactions. Construct the internal force diagram. Show the elastic curve.

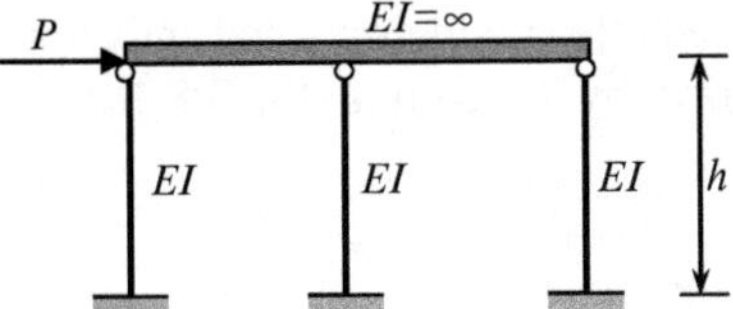

Fig. P8.15

Ans. $M_0 = Ph/3$, $R_0 = P/3$.

8.16. A design diagram of a frame with columns having a step-variable cross section is presented in Fig. P8.16. The frame is subjected to uniformly distributed load $q = 5\,\text{kN/m}$ within the first span. Construct the bending moment diagram using the displacement method in canonical form (use Tables A.7 and A.8).

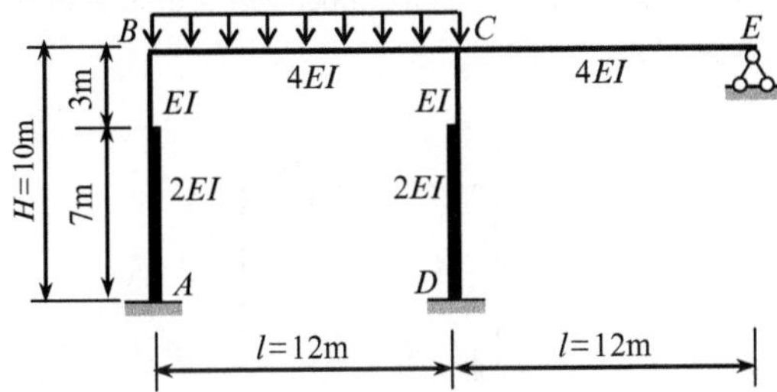

Fig. P8.16

Ans. $M_A = 12.41\,\text{kN m}$, $M_B = 21.02$, $M_D = 8.83$, $M_{CB} = 47.24$, $M_{CD} = 14.95$, $M_{CE} = 32.29$.

8.17. A design diagram of a two-storey frame is presented in Fig. P8.17. The crossbars are absolutely rigid. Connections between the vertical members and the crossbars are rigid. Construct the internal force diagrams. Determine the reactions of support.

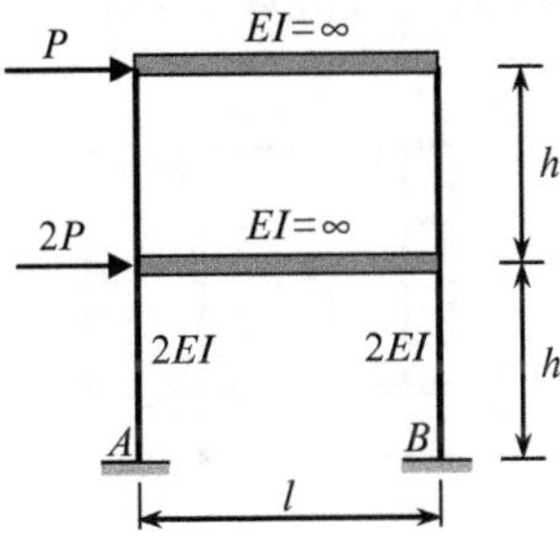

Fig. P8.17

Ans. $Q_A = 1.5P$, $N_{\text{left}} = -N_{\text{right}} = \dfrac{2.5Ph}{l}$, $M_A = 0.75Ph$.

8.18. Resolve a given load into symmetrical and antisymmetrical components. Axis of symmetry (AS) is shown by dotted line (Fig. P8.18).

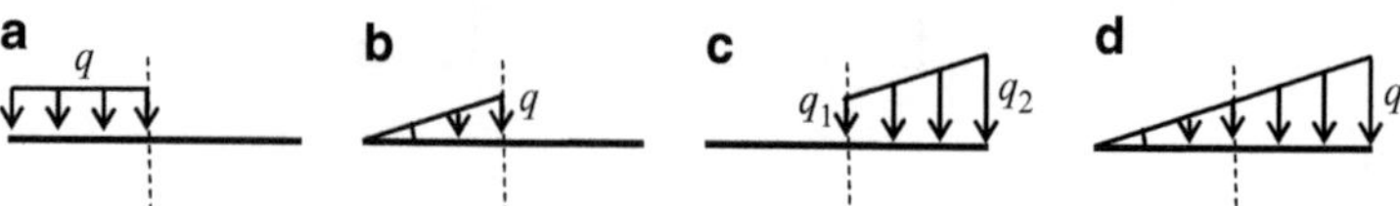

Fig. P8.18

8.19. Symmetrical portal frame is loaded by the force P, as shown in Fig. P8.19. Resolve the given load into symmetrical and antisymmetrical components and calculate the bending moments at specified points for corresponding loading. Calculate the horizontal reaction H. Determine the horizontal displacement of the crossbar and show the elastic curve.

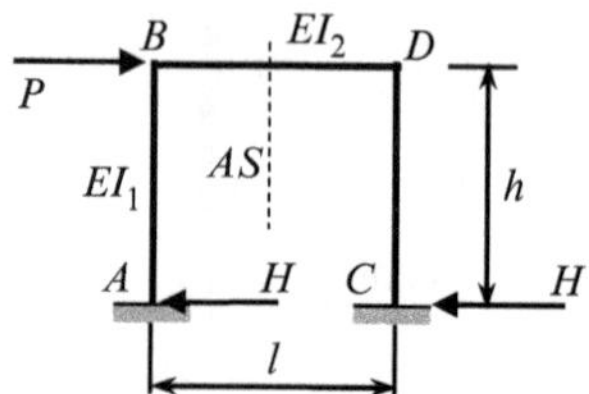

Fig. P8.19

$$\text{Ans. } M_B = \frac{3k}{2(1+6k)}Ph;\ M_A = \frac{1+3k}{2(1+6k)}Ph;\quad H = \frac{P}{2};\quad k = \frac{I_2}{I_1}\cdot\frac{h}{l}.$$

8.20. The portal frame in Fig. P8.20 is subjected to horizontal uniformly distributed load q. Construct the bending moment diagram by the Combined method. Determine the horizontal displacement of the crossbar and show the elastic curve. EI = constant for all members; $l = 4\,\text{m}$, $h = 8\,\text{m}$.

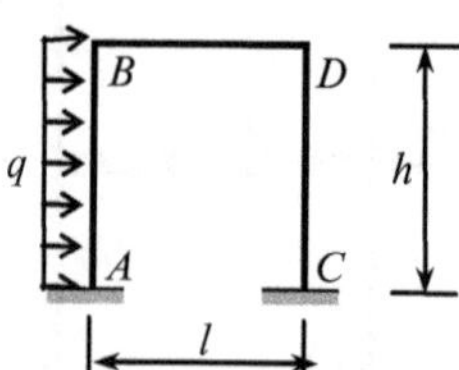

Fig. P8.20

$$\text{Ans. } M_A^{\text{sym}} = 3.335q;\quad M_B^{\text{sym}} = 1.335q;\ M_C^{\text{total}} = 7.746q;\quad M_D^{\text{total}} = 6.257q.$$

Chapter 9
Mixed Method

This chapter is devoted to the analysis of special types of structures. They are the frames with different flexibility ranges of their separate parts. These structures have specific properties which allow simplify their analysis.

9.1 Fundamental Idea of the Mixed Method

In engineering practice the statically indeterminate frames with specified features may be found: one part of a structure contains a small number of reactions and a large number of the rigid joints, while another part contains a large number of reactions and a small number of the rigid joints. Two examples of this type of structures are presented in Fig. 9.1a, b. Part 1 of a structure (a) contains only one support constraint (one vertical reaction at point A) and three rigid joints, while the part 2 contains nine unknown reactions and two rigid joints. We can say that parts 1 and 2 are "soft" and "rigid", respectively. The frame (b) may be considered also as a structure with different ranges of flexibility of their separate parts 1 and 2. The part 1 is soft while the part 2 is rigid one.

For analysis of these types of structures the mixed method should be applied. In this method, some unknowns represent unknowns of the force method and some represent unknowns of the displacement method. The mixed method was introduced and developed by Prof. A. Gvozdev in 1927.

9.1.1 Mixed Indeterminacy and Primary Unknowns

It is convenient to apply the force method to the "soft" part of the structure and the displacement method to the "rigid" part of the structure. The primary system of the mixed method is obtained from a given structure by *eliminating* the redundant constraints in the "soft" part of a structure and *introducing* additional constraints at rigid joints in the "rigid" part of a structure (Fig. 9.1c, d). Thus, the primary unknowns of the mixed method are *forces and displacements simultaneously*. For scheme (a) the primary unknowns are the *force* X_1 and *displacements* Z_2 and Z_3. For scheme

I.A. Karnovsky and O. Lebed, *Advanced Methods of Structural Analysis*,
DOI 10.1007/978-1-4419-1047-9_9,

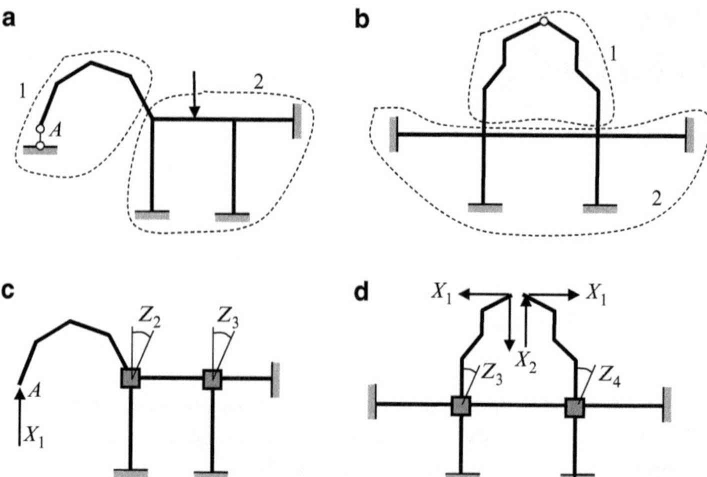

Fig. 9.1 **(a, b)** Design diagrams of frames with different ranges of flexibility of their separate parts; **(c, d)** primary systems of the mixed method

(b) the primary unknowns of the force method are labeled as X_1, X_2, X_3, while the primary unknowns of the displacement method are Z_3, Z_4. A numbering of the unknowns is through sequence.

For frames with different ranges of flexibility of their separate parts the number of unknowns of the mixed method is less than a number of unknowns of the force or displacement methods. This is an advantage of the mixed method. The mixed method may also be presented in canonical form.

Let us consider a frame shown in Fig. 9.2a; the flexural rigidity of vertical and horizontal elements are *EI* and 2*EI*, respectively. For given structure the number of unknowns by the force method equals 4, and the number of unknowns by the displacement method also equals 4 (angles of rotation of rigid joints 2, 4, 5 and linear displacement of the cross bar 4-5).

Both parts of the frame have different ranges of flexibility, mainly, the right part is "rigid" and left part is "soft." Indeed, the right part 1-2-3 has six constraints of supports at the points 1 and 3, while the left part 2-4-5-*A* of the frame has only one constraint support *A*. Therefore, for the left and right parts of the frame it is convenient to apply the force and displacement methods, respectively.

9.1.2 Primary System

The primary system of the mixed method is obtained from the given structure by *eliminating* the left vertical constraint *A* and *introducing* the additional constraint at rigid joint 2, simultaneously (Fig. 9.2b). Primary unknowns are X_1 and Z_2, where X_1 is unknown of the force method and Z_2 is unknown of the displacement method.

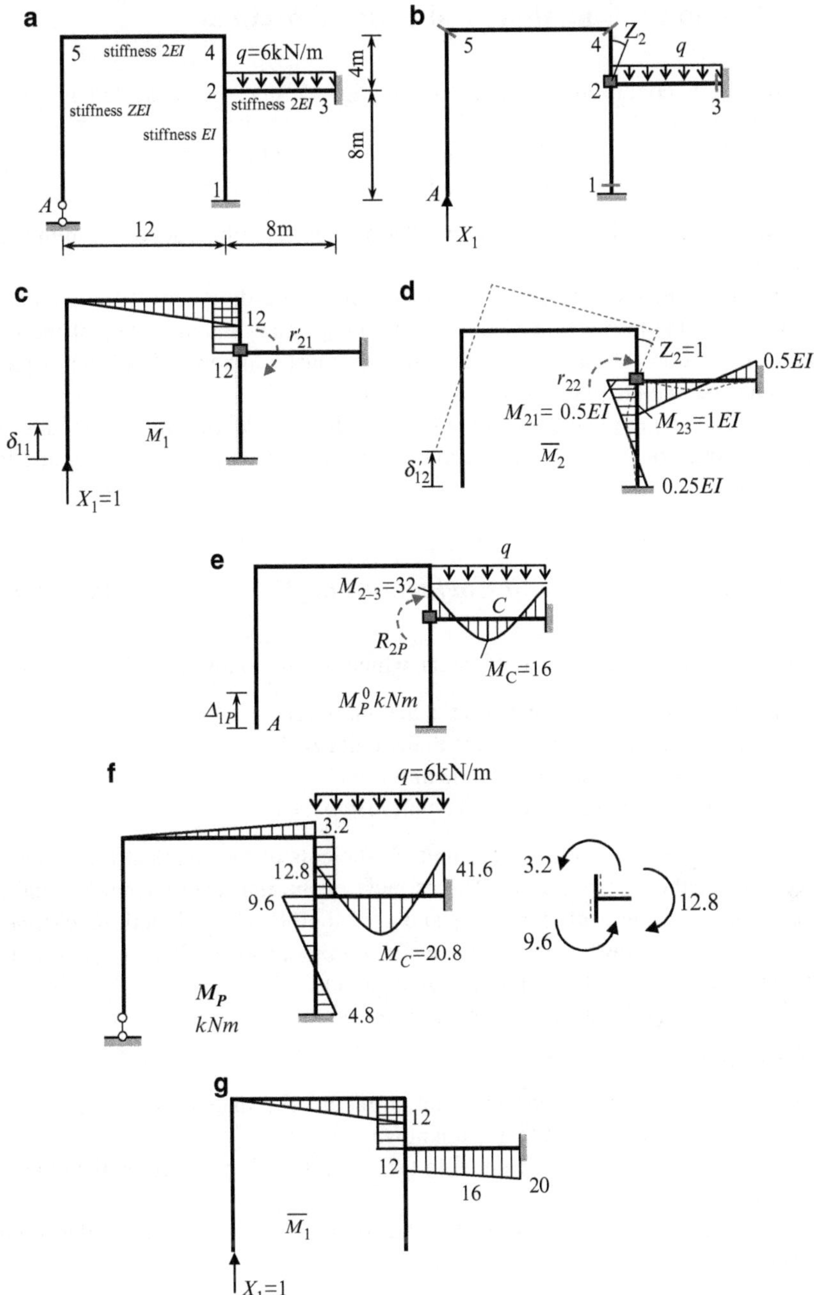

Fig. 9.2 **(a)** Design diagram; **(b)** primary system of the mixed method; **(c)** bending moment diagrams caused by force $X_1 = 1$; **(d)** bending moment diagrams caused by displacement $Z_2 = 1$. **(e)** Bending moment diagram in the primary system due to given load. **(f)** Resulting bending moment diagram and free body diagram for joint 2. **(g)** Unit bending moment diagram in primary system of the force method

9.2 Canonical Equations of the Mixed Method

Compatibility conditions for structure in Fig. 9.2 may be presented in canonical form

$$\begin{aligned} \delta_{11} X_1 + \delta'_{12} Z_2 + \Delta_{1P} &= 0, \\ r'_{21} X_1 + r_{22} Z_2 + R_{2P} &= 0. \end{aligned} \tag{9.1}$$

Thus the mixed method for this particular design diagram reduces the number of unknowns to two.

The first equation means that *displacement* in the direction of the *eliminated* constraint 1, due to reaction of this constraint (primary unknown X_1 of the force method), rotation of the introduced joint 2 (primary unknown Z_2 of the displacement method), and applied load must be zero.

The second equation means that *reaction* in the *introduced* joint 2 due to reaction X_1 of eliminated constraint 1, rotation Z_2 of introduced joint 2 and applied load must be zero.

9.2.1 *The Matter of Unit Coefficients and Canonical Equations*

These equations contain the coefficients, which belong to four different groups, i.e.:

- δ_{11} represents a *displacement* due to the unit *force*
- δ'_{12} represents a *displacement* due to the unit *displacement*
- r'_{21} represents a *reaction* due to the unit *force*
- r_{22} represents a *reaction* due to the unit *displacement*

So, the coefficients δ_{11} and r_{22} are unit coefficients of the classical force and displacement method, respectively. The essence of the unit coefficients δ'_{12} and r'_{21} is different from the coefficients δ_{11} and r_{22}: coefficient δ'_{12} is unit *displacement* caused by the unknown of the *displacement* method, and coefficient r'_{21} is unit *reaction* caused by the unknown of the *force* method.

Each term of (9.1) has the following meaning:

First equation:

- $\delta'_{11} X_1$ = *displacement* in the direction of the eliminated constraint (point A, vertical direction) caused by the unknown *force* X_1
- $\delta'_{12} Z_2$ = *displacement* of point *A* in the same direction caused by the unknown *angle of rotation* Z_2
- Δ_{1P} = *displacement* of point *A* in the same direction caused by the *applied load.*

Second equation:

- $r'_{21} X_1$ = *reaction* of the introduced constraint 2 due to the unknown *force* X_1
- $r_{22} Z_2$ = *reaction* of the same constraint 2 due to the unknown *angle of rotation* Z_2
- R_{2P} = *reaction* of the same constraint 2 due to the *applied loads*

Thus, the character of each equation is *mixed* because it is the displacement (or reaction) caused by primary unknowns of the force method X *and* the displacement method Z simultaneously. Moreover, a meaning of each equation in (9.1) is mixed because they represent *displacement* along of eliminated constraint A *and reaction* of introduced constraint 2.

9.2.2 Calculation of Coefficients and Free Terms

The general properties of elastic structures and reciprocal theorems provide the following properties and relationships between coefficients of canonical equation of the mixed method

$$\delta_{nn} > 0, \quad r_{kk} > 0, \quad \delta_{nm} = \delta_{mn}, \quad r_{nm} = r_{mn}, \quad r_{nm} = -\delta_{mn}. \tag{9.2}$$

Bending moment diagrams due to the primary unknowns $X_1 = 1$ and $Z_2 = 1$ are presented in Fig. 9.2c, d. Construction of bending moment diagram caused by primary unknown $X_1 = 1$ is obvious. For construction of bending moment diagram caused by primary unknown $Z_2 = 1$ we need to rotate introduced constraint 2 through angle $Z_2 = 1$. In this case, members of the right part only will sustain deformation, while left part will move as absolutely rigid body. Corresponding deformable position is shown by dotted line. Unit displacements (δ_{11}, δ_{12}) and unit reactions (r_{22}, r'_{21}) are shown in the corresponding diagrams.

Ordinates of bending moment diagram for clamped–clamped beam due to the unit rotation of joint 2 are

$$M_{23} = \frac{4EI_{2-3}}{l_{2-3}} = 4\frac{2EI}{8} = 1.0EI, \quad M_{21} = \frac{4EI_{2-1}}{l_{2-1}} = 4\frac{1EI}{8} = 0.5EI.$$

Having the unit bending moment diagrams due to $X_1 = 1$ and $Z_2 = 1$ we can calculate the coefficients and free terms of canonical equations of the mixed method.

Multiplication of bending moment diagram $\overline{M}_1$ by itself leads to coefficient of the force method

$$\delta_{11} = \sum \int \frac{\overline{M}_1 \cdot \overline{M}_1}{EI} ds = \frac{1}{2EI} \cdot \frac{12 \times 12}{2} \cdot \frac{2}{3} \cdot 12 + \frac{1}{1EI} \cdot 12 \times 4 \times 12 = \frac{864}{EI} \text{ (m/kN)}$$

Equilibrium condition of induced constraint 2 from bending moment diagram $\overline{M}_1$ leads to

$$r'_{21} = -12\,\text{m}$$

The units of coefficients of canonical equations of the mixed method can be defined as for the force and displacement methods. In general, for r'_{ik}, the unit of *reaction* which corresponds to the index i, should be divided by unit of factor, which corresponds to the index k. In our case we get

$$\left[r'_{21}\right] = \left[\frac{\text{kN m}}{\text{kN}}\right] = [\text{m}].$$

Equilibrium condition of induced constraint 2 from bending moment diagram $\overline{M}_2$ leads to coefficient of the displacement method

$$r_{22} = 1EI + 0.5EI = 1.5EI(\text{kN m/rad}).$$

Theorem of reciprocal displacements and reactions leads to $\delta'_{12} = -r'_{21} = 12\,\text{m/rad}$.

The unit of δ'_{ik} is defined by following rule: the unit of *displacement*, which corresponds to the index i, should be divided by unit of factor, which corresponds to the index k. In our case we get $\left[\delta'_{21}\right] = [\text{m/rad}]$.

For calculation of free terms of canonical equations we need to construct the bending moment diagram in the primary system due to distributed load q; this diagram is presented in Fig. 9.2e. Specific ordinates of M_P^0 diagram are follows: $M_{2-3} = M_{3-2} = ql^2\,/\,12 = (6\times 8^2)/12 = 32\,\text{kN m}$, $M_C = ql^2\,/\,24 = 16\,\text{kN m}$.

From this diagram we get $R_{2P} = -32$ (kN m) and $\Delta_{1P} = 0$.

Canonical equations of the mixed method becomes:

$$\frac{864}{EI}X_1 + 12Z_2 = 0,$$
$$-12X_1 + 1.5EI\,Z_2 - 32 = 0.$$

Roots of these equations are $X_1 = -0.266\,\text{kN}, \quad Z_2 = 19.2/EI(\text{rad})$

9.2.3 Computation of Internal Forces

Resulting bending moments acting at different cross sections of the frame are calculated by formula

$$M_P = \overline{M}_1 X_1 + \overline{M}_2 Z_2 + M_P^0. \tag{9.3}$$

Corresponding calculation for specified points is presented in Table 9.1. Resulting bending moment diagram is presented in Fig. 9.2f.

Verification. The static and kinematical verifications may be considered. They are:

1. Free body diagram for joint 2 is presented in Fig. 9.2f. Direction of the moments is shown according to location of the extended fibers, which are shown by dotted lines. Equilibrium equation $\sum M_2 = 3.2 + 9.6 - 12.8 = 0$ is satisfied.
2. Displacement in the direction of eliminated constraint A in the original system must be zero. This displacement may be computed by multiplication of two bending moment diagrams: one of them is the resultant bending moment diagram M_P for the entire given structure and second is bending moment diagram caused by the $X_1 = 1$ in *any primary system of the force method*. One version of the

primary system of the force method and corresponding unit bending moment diagram $\overline{M}_1$ are presented in Fig. 9.2g.

Table 9.1 Calculation of bending moments

Points	$\overline{M}_1$	$\overline{M}_1 Z_1$	$\overline{M}_2$	$\overline{M}_2 Z_2$	M_P^0	M_P (kNm)
1	0.0	0.0	−0.25	−4.8	0.0	−4.8
2–1	0.0	0.0	+0.5	+9.6	0.0	+9.6
2–3	0.0	0.0	−1.0	−19.2	+32	+12.8
3–2	0.0	0.0	+0.5	+9.6	+32	+41.6
2–4	+12.0	−3.2	0.0	0.0	0.0	−3.2
4–2	+12.0	−3.2	0.0	0.0	0.0	−3.2
4–5	−12.0	+3.2	0.0	0.0	0.0	+3.2
5	0.0	0.0	0.0	0.0	0.0	0.0
C	0.0	0.0	−0.25	−4.8	−16	−20.8
Factor			EI			

The vertical displacement of point A for entire structure equals

$$\Delta_A = \sum \int \frac{M_P \overline{M}_1}{EI} ds = -\frac{1}{2EI} \frac{1}{2} \times 12 \times 12 \times \frac{2}{3} \times 3.2 - \frac{1}{1EI} \times 3.2 \times 4 \times 12$$
$$+ \frac{8}{6 \times 2EI} (-12.8 \times 12 + 4 \times 20.8 \times 16 - 41.6 \times 20)$$
$$= -887.46 + 887.47 \cong 0$$

Construction of shear and axial force diagrams, computation of all reactions and their verifications should be performed as usual.

Problems

9.1. Design diagram of the frame is shown in Fig. P9.1.

1. Determine the number of unknowns by the force method, displacement, and mixed methods.
2. Show the primary system of the mixed method and set up of corresponding canonical equations;
3. Explain the meaning of primary unknowns and canonical equations of the mixed method.
4. Define the unit of coefficients and free term of canonical equations.
5. Describe the way of calculation of all coefficients of canonical equations and free terms.
6. Explain the advantage of the mixed method with comparison of the force and displacement methods.

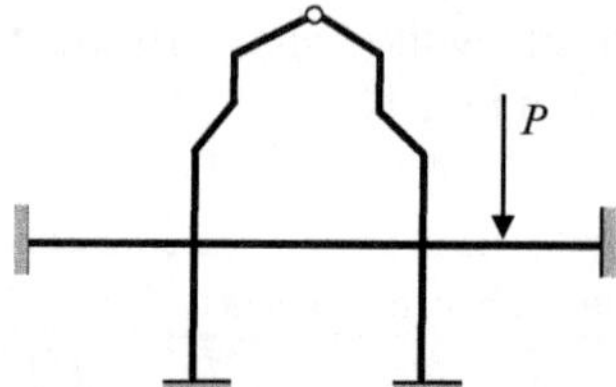

Fig. P9.1

9.2. Design diagrams of the nonsymmetrical frames are presented in Fig. P9.2a–e. Choose the best method (force, displacement, or mixed) of analysis for each frame. Explain your decision. Choose primary unknowns, show primary system, write the canonical equations, and trace way for calculation of coefficients and free terms.

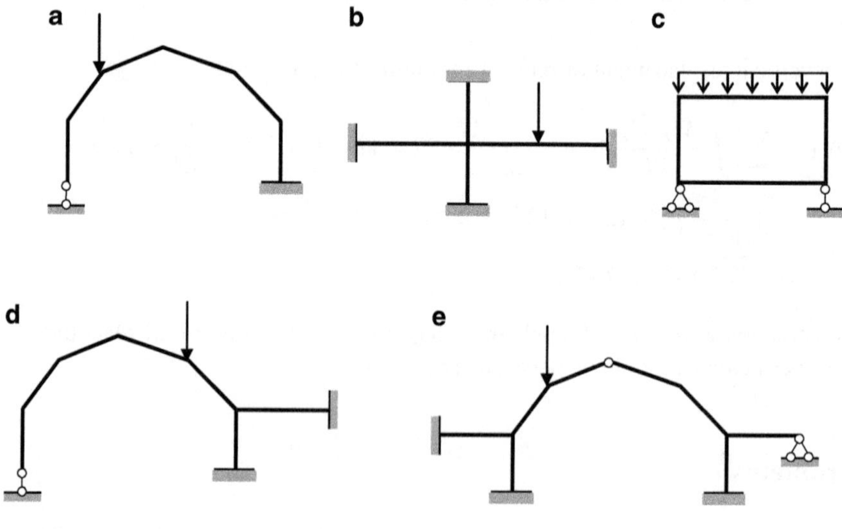

Fig. P9.2

Ans. (a) FM-1; (b) DM-1; (c) FM-3; (d) MM-2; (e) MM-4

9.3. The frame is shown in Fig. P9.3. The bending stiffness for each member is EI. Construct the bending moment diagram and perform its statical and kinematical verifications.

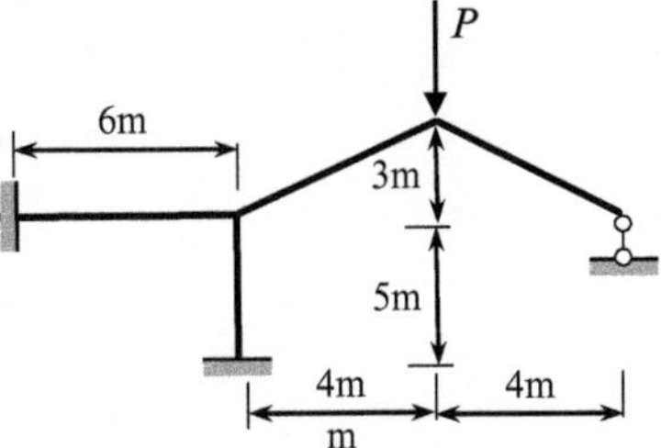

Fig. P9.3

9.4. Construct the bending moment diagram for structure in Fig. P9.4. Perform the static and kinematical control of resulting bending moment diagram.

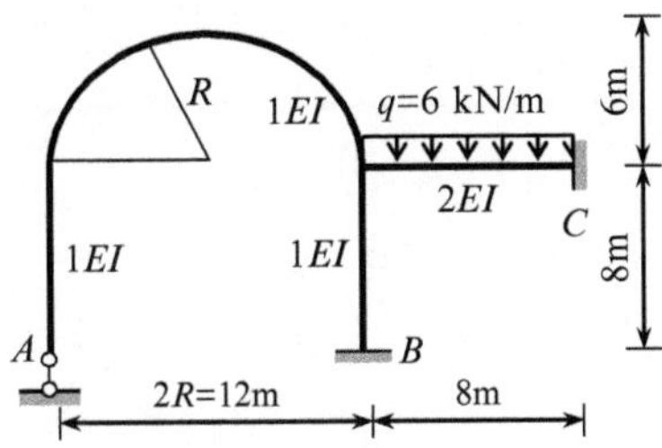

Fig. P9.4

Ans. $M_B = 4.84\,\text{kNm},\ M_C = 41.75\,\text{kNm}$

Chapter 10
Influence Lines Method

This chapter is devoted to construction of influence lines for different statically indeterminate structures. Among them are continuous beams, frames, nonuniform arches, and trusses. Analytical methods based on the force and the displacement methods are applied. Also, kinematical method of construction of influence lines is discussed. This method allows tracing the *models* of influence lines.

10.1 Construction of Influence Lines by the Force Method

Let us consider a first-degree statically indeterminate structure. In case of a *fixed* load, the canonical equation of the force method is

$$\delta_{11}X_1 + \Delta_{1P} = 0. \tag{10.1}$$

In this equation, the free term Δ_{1P} represents displacement caused by given fixed load. Now we need to transform (10.1) for the case of *moving* load. Since moving load is *unit* one, then (for the sake of consistency notations) let us replace free term Δ_{1P} by the *unit* free term δ_{1P}; this free term presents a displacement in primary system in the direction of X_1 caused by load $P = 1$. The primary unknown

$$X_1 = -\frac{\delta_{1P}}{\delta_{11}}.$$

Unit displacement δ_{11} presents displacement caused by primary unknown $X_1 = 1$. Therefore, δ_{11} is some *number*, which depends on the type of a structure, its parameters and chosen primary system and does not depend on the position of the acting load. However, δ_{1P} depends on unit load location. Since load $P = 1$ is traveling, δ_{1P} becomes a *function* of position of this load, and as result, the primary unknown becomes a *function* as well:

$$\text{IL}\left(X_1\right) = -\frac{1}{\delta_{11}}\text{IL}\left(\delta_{1P}\right). \tag{10.2}$$

I.A. Karnovsky and O. Lebed, *Advanced Methods of Structural Analysis*,
DOI 10.1007/978-1-4419-1047-9_10, © Springer Science+Business Media, LLC 2010

Influence line for bending moment In case of *fixed* load, a bending moment at any section k equals

$$M_k = \overline{M}_k \cdot X_1 + M_k^0 \tag{10.3}$$

where X_1 is the primary unknown of the force method; $\overline{M}_k$ is the bending moment at section k in a primary system due to unit primary unknown $X = 1$; and M_k^0 is the bending moment at section k in a primary system due to given load.

Now we need to transform (10.3) for the case of *moving* load. The bending moment $\overline{M}_k$ at any section k presents the *number*, because this moment is caused by the unit primary unknown X_1; the second component X_1 according to (10.2) presents a function. The last component, the bending moment M_k^0, is caused by the given load, which is considered as moving load now; therefore, the bending moment M_k^0 also becomes a *function* of the position of this load. As a result, the bending moment at any section k becomes a *function*:

$$\text{IL}\,(M_k) = \overline{M}_k \cdot \text{IL}\,(X_1) + \text{IL}\left(M_k^0\right). \tag{10.4}$$

Influence line for shear force In case of fixed load, the shear at any section k is a number

$$Q_k = \overline{Q}_k \cdot X_1 + Q_k^0. \tag{10.5}$$

Similarly in case of traveling load, the shear at any section k becomes a function, therefore

$$\text{IL}\,(Q_k) = \overline{Q}_k \text{IL}\,(X_1) + \text{IL}\left(Q_k^0\right). \tag{10.6}$$

For the n-times statically indeterminate structure, the canonical equations of the force method in case of a fixed load $P = 1$ are

$$\begin{array}{l} \delta_{11}X_1 + \delta_{12}X_2 + \cdots + \delta_{1n}X_n + \delta_{1P} = 0 \\ \cdot \quad \cdot \quad \cdot \quad \cdot \quad \cdot \quad \cdot \quad \cdot \quad \cdot \quad \cdot \quad \cdot \quad \cdot \quad \cdot \quad \cdot \\ \delta_{n1}X_1 + \delta_{n2}X_2 + \cdots + \delta_{nn}X_n + \delta_{nP} = 0 \end{array} \tag{10.7}$$

Unit displacements δ_{ik} which are caused by unit primary unknowns present *numbers*.

Displacements δ_{iP} are caused by unit moving load. Fundamental feature of system (10.7) is that the free terms δ_{iP} are some *functions* of position of unit load $P = 1$. Therefore, a solution of the system (10.7) leads to the primary unknowns X_i as the *functions* of the load position, in fact, to $IL\,(X_i), i = 1, \ldots, n$. In this case the bending moment at the any specified section k becomes function, so influence line should be constructed by formula

$$\text{IL}\,(M_k) = \overline{M}_{k1} \cdot \text{IL}\,(X_1) + \overline{M}_{k2} \cdot \text{IL}\,(X_2) + \cdots + \text{IL}\left(M_k^0\right). \tag{10.8}$$

Influence lines expressions for shear and axial force at any section may be constructed similarly.

In case of statically indeterminate truss of the first degree of redundancy, the expression for influence line of internal force, which is induced at any member k of the truss, may be constructed by formula

$$\mathrm{IL}\,(S_k) = \overline{S_k} \cdot \mathrm{IL}\,(X_1) + \mathrm{IL}\left(S_k^0\right), \tag{10.9}$$

where $\overline{S_k}$ is the internal force in kth member of a truss caused by the primary unknown $X_1 = 1$ in the primary system and $\mathrm{IL}\left(S_k^0\right)$ is the influence line for kth member of a truss in the primary system.

Construction of influence lines for internal forces in a statically indeterminate structure starts from construction of influence lines for primary unknowns $\mathrm{IL}\,(X_1)$, $\mathrm{IL}\,(X_2)\,,\ldots$.

Construction of influence lines for statically indeterminate structures by the force method is recommended to perform using the following procedure:

1. To adopt the primary unknowns and primary system. For continuous beams it is recommended to choose a primary system as a set of simply supported beams.
2. To write the canonical equations of the force method.
3. Compute the coefficients δ_{ik} of canonical equations. These unit displacements present some *numbers.*
4. Compute the free terms δ_{iP} of canonical equations. These load terms presents the displacement in the direction of ith primary unknown due to the load which depends on the location of unit load and therefore presents some *functions.*
5. Solve the canonical equations and construct the influence lines for primary unknowns.
6. Construct the influence lines for reactions and internal forces at specified section.

Below we will show detailed procedure for construction of influence lines by the force method for continuous beam, hingeless nonuniform parabolic arch, and truss. As will be shown below, the influence lines for primary unknowns, internal forces, and reactions for statically indeterminate structures are bounded by the curved lines, unlike statically determinate structures.

10.1.1 Continuous Beams

Let us demonstrate the above procedure for uniform two-span continuous beam ABC with equal spans l. It is necessary to construct the influence line for internal forces at a specified section k (Fig. 10.1a).

Primary System

The structure is the statical indeterminate of first degree. The primary system of the force method presents two simply supported beams (Fig. 10.1b). The primary unknown X_1 is the bending moment at the middle support.

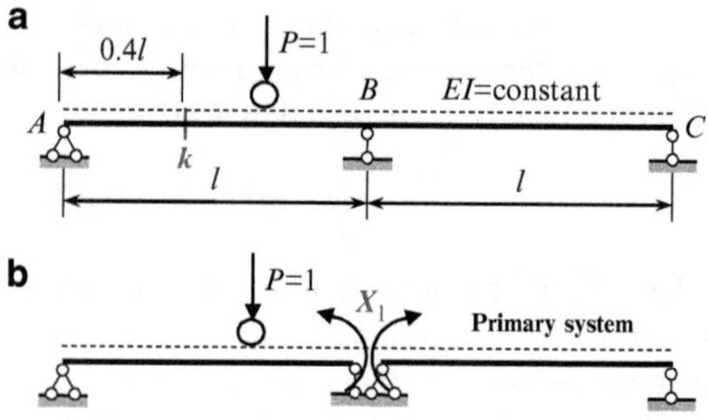

Fig. 10.1 Design diagram of the beam and primary system

Influence Line for Primary Unknown X_1

Equation of influence line for primary unknown is presented by formula IL $(X_1) = -(1/\delta_{11})$ IL (δ_{1P}), where δ_{11} is a mutual angle of rotation in direction of X_1 due to primary unknown $X_1 = 1$ and δ_{1P} is a slope at the middle support in primary system due to traveling load $P = 1$.

1. For calculation of unit displacement δ_{11} we need to construct the bending moment diagram $\overline{M}_1$ in primary system due to primary unknown $X_1 = 1$ (Fig. 10.2a).

   Graph multiplication method leads to the following result:

$$\delta_{11} = \frac{\overline{M}_1 \times \overline{M}_1}{EI} = 2 \cdot \frac{1}{2} l \cdot 1 \cdot \frac{2}{3} \cdot 1 \cdot \frac{1}{EI} = \frac{2l}{3EI}.$$

2. For calculation of displacement δ_{1P} we need to place the moving load at the left and right spans. Let the load $P = 1$ be located within the *left* span of the primary system. The angle of rotation δ_{1P} at the right support B of the simply supported beam is presented in terms of dimensionless parameter u, which defines the position of the load (Fig. 10.2b). If the load $P = 1$ travels along the *right* span, then the angle of rotation δ_{1P} at the left support of the simply supported beam (i.e., the same support B) is presented in terms of dimensionless parameter υ (Fig. 10.2b). Expressions for δ_{1P} in terms of parameters u and υ are shown in Fig. 10.2b; they are taken from Table A.9, pinned–pinned beam. Each span is subdivided into five equal portions and the angle of rotation δ_{1P} is calculated for location of the P $= 1$ at the each point ($u = 0.0, 0.2, 0.4, 0.6, 0.8, 1.0$) (Table 10.1). Parameter u is reckoned from the left support of each span, parameters u and υ satisfy the following condition: $u + \upsilon = 1$.

   Corresponding influence line for δ_{1P} is shown in Fig. 10.2c.
3. Influence line of primary unknown X_1 is obtained by dividing the ordinates of influence line for δ_{1P} by $-\delta_{11} = -2l/3EI$. Corresponding influence line for X_1 is presented in Fig. 10.2d; all ordinates must be multiplied by parameter l.

We can see that for any position of the load the bending moment at support B will be negative, i.e., the extended fibers in vicinity of support B are located above the neutral line.

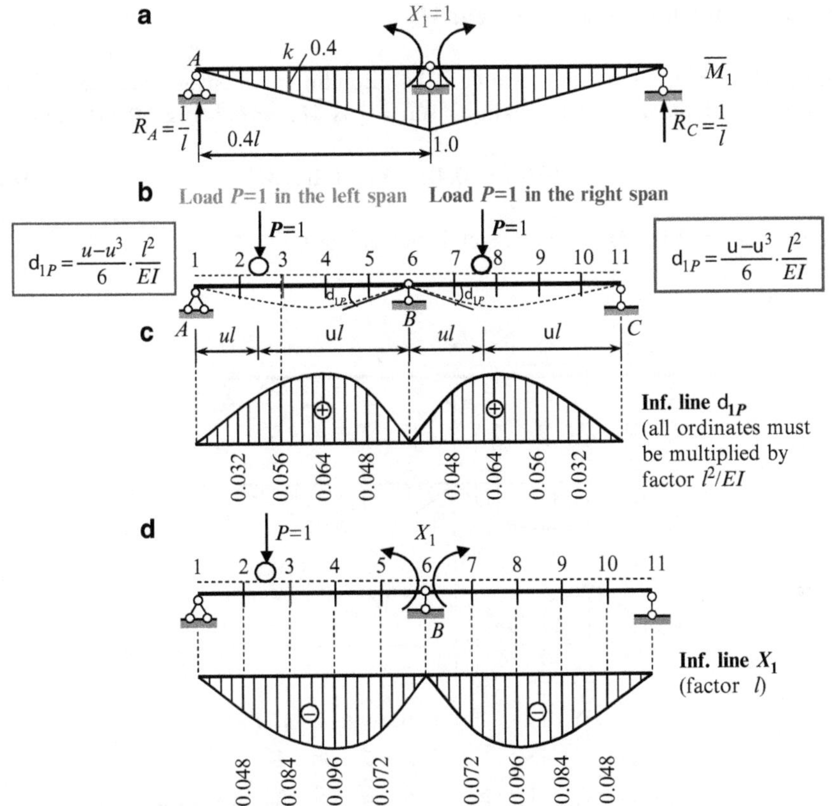

Fig. 10.2 (**a**) Bending moment diagram in primary system due to $X_1 = 1$. (**b**) Location of the load $P = 1$ in the left and right span; (**c**) Influence line for δ_{1P}. (**d**) Primary system and influence line for primary unknown X_1

Table 10.1 Calculation of δ_{1P} (coefficient l^2/EI is omitted)

$P = 1$ in the left span			$P = 1$ in the right span		
Point	Parameter	$\delta_{1P} = \frac{u-u^3}{6}\frac{l^2}{EI}$	Point	Parameter	$\delta_{1P} = \frac{\upsilon-\upsilon^3}{6}\frac{l^2}{EI}$
1 (*A*)	$u = 0.0$	0.0	6 (*B*)	$\upsilon = 1.0$	0.0
2	$u = 0.2$	0.032	7	$\upsilon = 0.8$	0.048
3 (*k*)	$u = 0.4$	0.056	8	$\upsilon = 0.6$	0.064
4	$u = 0.6$	0.064	9	$\upsilon = 0.8$	0.056
5	$u = 0.8$	0.048	10	$\upsilon = 0.2$	0.032
6 (*B*)	$u = 1.0$	0.0	11 (*C*)	$\upsilon = 0.0$	0.0

Influence line for bending Moment M_k

Should be constructed using formula (10.4). Bending moment at section k in the primary system due to primary unknown $X_1 = 1$ is $\overline{M}_k = 0.4$ (Fig. 10.2a), therefore

$$\text{IL}\,(M_k) = 0.4\text{IL}\,(X_1) + \text{IL}\left(M_k^0\right). \tag{10.10}$$

Construction of IL (M_k) step by step is presented in Fig. 10.3.

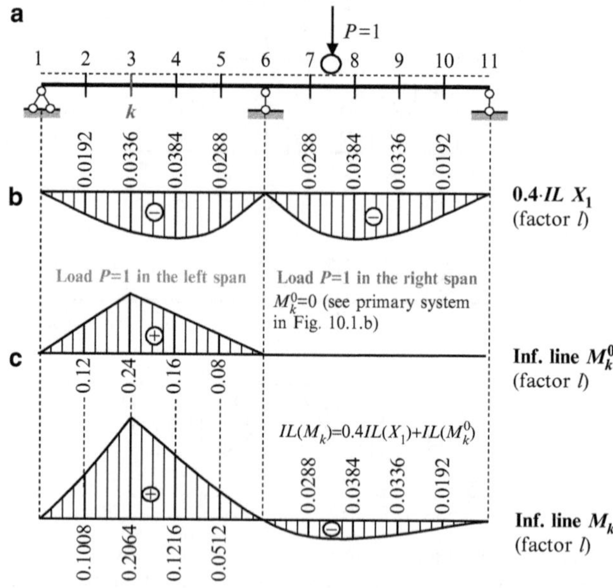

Fig. 10.3 Construction of influence line for bending moment at section k

The first term 0.4IL (X_1) of (10.10) is presented in Fig. 10.3a. The second term IL $\left(M_k^0\right)$ is the influence line of bending moment at section k in the primary system. If load $P = 1$ is located in the left span, then the influence line presents the triangle with maximum ordinate at the section k; this ordinate equals to $ab/l = (0.4l \times 0.6l)\,/\,l = 0.24l$. If load $P = 1$ is located in the right span, then the bending moment at section k does not arise, and the influence line has zeros ordinates. It happens because the primary system presents two separate beams. Influence line for M_k^0 is shown in Fig. 10.3b. Summation of two graphs 0.4IL (X_1) and IL $\left(M_k^0\right)$ leads to the final influence line for bending moment at section k; this influence line is presented in Fig. 10.3c. The maximum bending moment at the section k occurs when the load P is located at the same section. The positive sign means that if a load is located within the left span, then extended fibers at the section k will be located below the neutral line. It is easy to show the following important property of influence line for M_k. The sum of the absolute values of the slopes at the left and right of section k is equal to unity.

Influence line for shear force Q_k

Should be constructed using formula (10.6). Reaction at support A in primary system caused by primary unknown $X_1 = 1$ equals $\overline{R}_A = 1/l\,(\uparrow)$ (Fig. 10.2a), so the shear in primary system at section k due to primary unknown $X_1 = 1$ is $\overline{Q}_k = \overline{R}_A = 1/l$. Therefore

$$\mathrm{IL}\,(Q_k) = \frac{1}{l}\mathrm{IL}\,(X_1) + \mathrm{IL}\,\left(Q_k^0\right). \tag{10.11}$$

Construction of IL (Q_k) step by step is presented in Fig. 10.4.

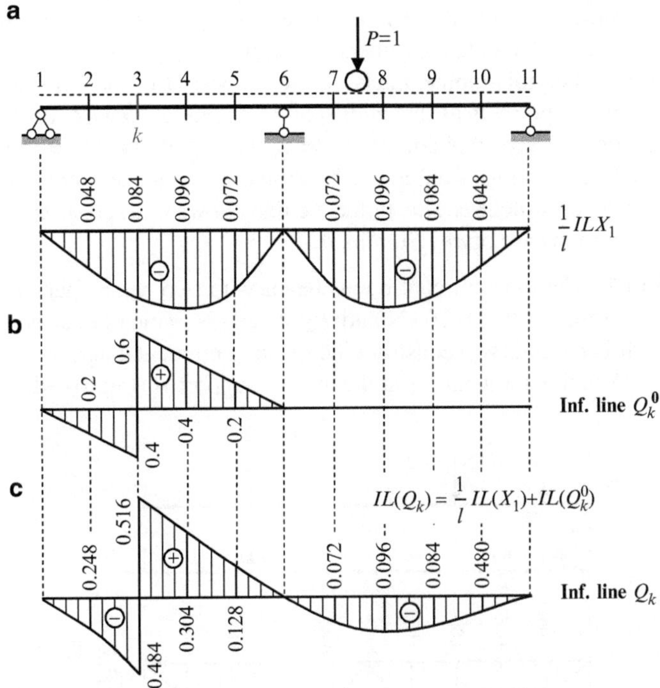

Fig. 10.4 Construction of influence lines for shear force at section k

The first term of (10.11) is the influence line for X_1 scaled by $1/l$, so ordinates of this graph (Fig. 10.4a) are dimensionless. The second term IL $\left(Q_k^0\right)$ is the influence line of shear force at section k in the primary system. If load $P = 1$ is located in the left span, then influence line of shear force for simply supported beam is shown in Fig. 10.4b. If load $P = 1$ is located in the right span, then shear at section k does not arise, and influence line has zeros ordinates. Summation of two graphs $(1/l)\,\mathrm{IL}\,(X_1)$ and IL $\left(Q_k^0\right)$ leads to the required influence line for shear force at section k; this influence line is presented in Fig. 10.4c.

Discussion. Obtained influence lines allow calculate bending moment and shear for *any* section of the beam. Let $P = 10\,\text{kN}$ is located at section 9. In this case, the moment

$$M_k = -0.0336 \times l \times 10 = -0.336l \text{ (kNm)}.$$

It means that reaction of the left support is $R_A = M_k/0.4l = -(0.336l/0.4l) = -0.84\,\text{kN}$. Negative sign shows that reaction is directed downward. Now construction of bending moment and shear diagrams has no problem. For example, the shear is $Q_1 = \ldots = Q_6^{\text{left}} = -0.84\,\text{kN}$. The bending moments at specified points are

$$M_2 = -0.84 \times 0.2l = -0.168\,\text{kNm}, \ldots, \quad M_5 = -0.84 \times 0.8l = -0.672\,\text{kNm}$$

It is obvious that same results may be obtained using influence line for X_1. If load $P = 10\,\text{kN}$ is located at section 9 then moment at support B is $X_1 = -0.84l$. Negative sign indicates that extended fibers in vicinity of support B are located above the neutral line, i.e., this moment acts on the support B of the *left* span *clockwise* and on the *same* support B of the *right* span in counterclockwise direction. In this case the reaction of support A equals $R_A = X_1/l = -(0.84l/l) = -0.84\,\text{kN}$.

Influence line for primary unknown X_1 should be treated as a *fundamental characteristic* for given structure. This influence line allows us to construct the bending moment diagram in case of *any fixed* load.

Example 10.1. The uniform continuous beam with two equal spans $l = 10\,\text{m}$ is loaded by two forces $P_1 = 10\,\text{kN}$ and $P_2 = 20\,\text{kN}$, which act at points 2 and 4 (Fig. 10.5). It is necessary to construct the bending moment diagram. Use the influence line for bending moment X_1 at the middle support B (Fig. 10.2d).

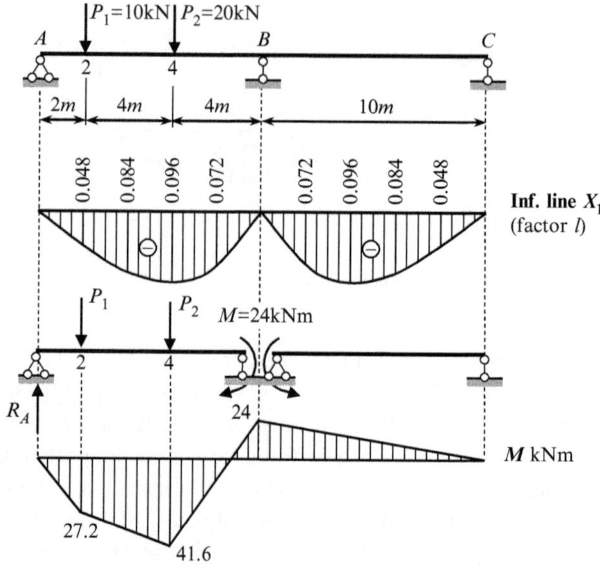

Fig. 10.5 Construction of bending moment diagram using influence line for primary unknown X_1

Solution. The bending moment at the middle support B using corresponding influence line is

$$X_1 = M_B = l \sum Py = 10\,(-10 \times 0.048 - 20 \times 0.096) = -24\,\text{kNm}.$$

After that the initial statically indeterminate continuous beam may be considered as a set of two *statically determinate* simply supported beams subjected to given load and moment at the support B. In other words, *the statical indeterminacy* has been *disclosed.*

Reaction at support A is

$$R_A = \frac{10 \times 8 + 20 \times 4 - 24}{10} = 13.6\,\text{kN}.$$

Bending moment at specified points are

$$\begin{aligned} M_2 &= 13.6 \times 2 = 27.2\,\text{kN m}, \\ M_4 &= 13.6 \times 6 - 10 \times 4 = 41.6\,\text{kN m}, \\ M_B &= 13.6 \times 10 - 10 \times 8 - 20 \times 4 = -24\,\text{kNm}. \end{aligned}$$

Discussion. Influence line for X_1 should be considered as a reference data for analysis of beams subjected to different set of fixed loads. If we need to analyze a structure once due to given set of loads, we can use any classical method or use influence line as a referred data. If we need to analyze the same structure many times, and each time the structure is subjected to different set of loads, then it is much more convenient to construct influence line once and then use it as reference data for all other sets of loadings. Thus, combination of two approaches, i.e., moving and fixed load, is extremely effective for analysis of structures.

10.1.2 Hingeless Nonuniform Arches

Let us apply the general procedure for analysis of symmetrical parabolic nonuniform arch with clamped ends shown in Fig. 10.6a. The equation of the neutral line is

$$y = \frac{4f}{l^2}\,(l - x)\,x.$$

Assume that the cross-sectional moments of inertia varies by law

$$I_X = \frac{I_c}{\cos \varphi_x},$$

where I_C corresponds to the highest point of the arch (crown C); this law corresponds to increasing the moment of inertia from crown to supports. It is necessary to construct the influence lines for reactions of the support A and bending moment at the crown C.

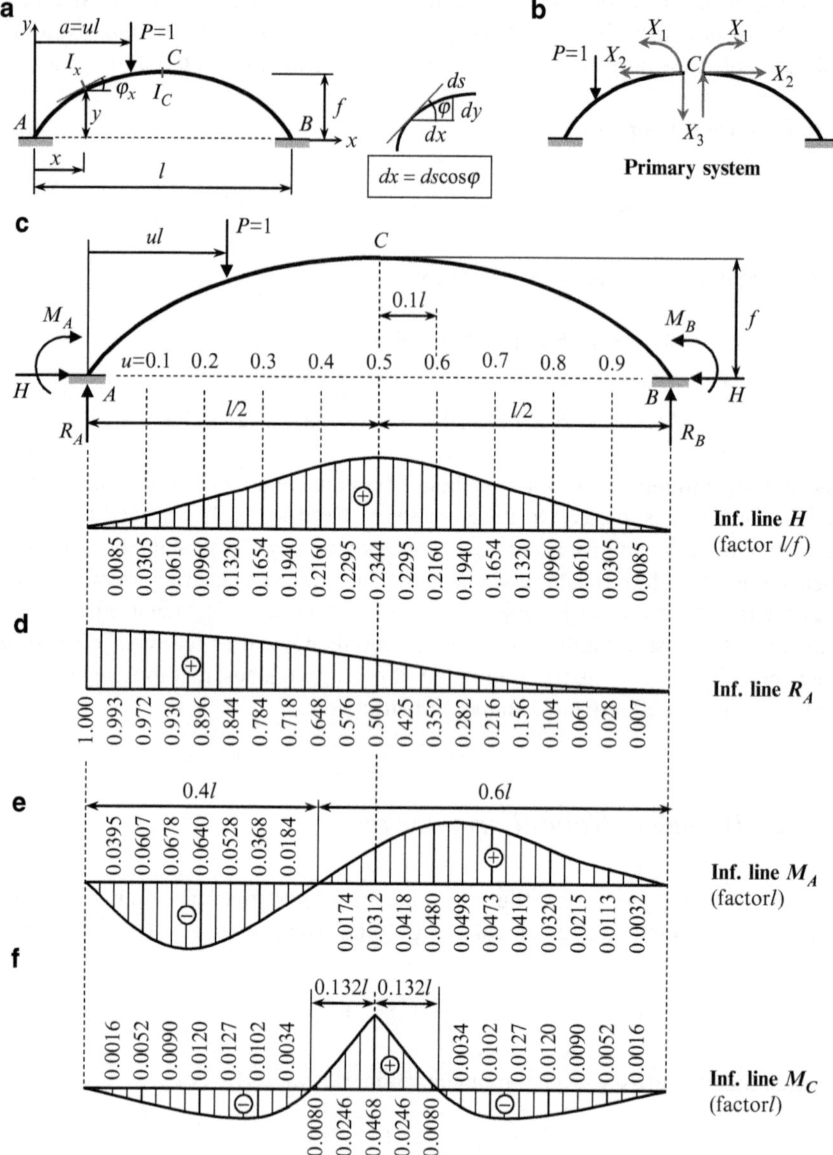

Fig. 10.6 (**a, b**) Parabolic arch with clamped ends. (**a**) Design diagram; (**b**) Primary system. (**c–f**) Parabolic non-uniform arch. Design diagram and influence lines

This arch is statically indeterminate to the third degree. Let us accept the primary system presented in Fig. 10.6b, so the primary unknowns are the bending moment X_1, the normal force X_2, and shear X_3 at the crown C of the arch.

Canonical equations of the force method are

$$\begin{aligned} \delta_{11}X_1 + \delta_{12}X_2 + \delta_{13}X_3 + \Delta_{1P} &= 0, \\ \delta_{21}X_1 + \delta_{22}X_2 + \delta_{23}X_3 + \Delta_{2P} &= 0, \\ \delta_{31}X_1 + \delta_{32}X_2 + \delta_{33}X_3 + \Delta_{3P} &= 0. \end{aligned} \tag{10.12}$$

The unit coefficients can be calculated by formula

$$\delta_{ik} = \int_0^l \frac{\overline{M}_i \cdot \overline{M}_k}{EI_x} \mathrm{d}s.$$

Since X_1 and X_2 are symmetrical unknowns, the unit bending moment diagrams $\overline{M}_1$ and $\overline{M}_2$ are symmetrical, while $\overline{M}_3$ diagram is antisymmetrical. Is obvious that all displacements computed by multiplying symmetrical diagram by antisymmetrical ones equal to zero. Therefore, $\delta_{13} = \delta_{31} = 0$, $\delta_{23} = \delta_{32} = 0$, and the canonical equations (10.12) fall into two independent systems

$$\begin{cases} \delta_{11}X_1 + \delta_{12}X_2 + \Delta_{1P} = 0 \\ \delta_{21}X_1 + \delta_{22}X_2 + \Delta_{2P} = 0 \end{cases} \quad \text{and} \quad \delta_{33}X_3 + \Delta_{3P} = 0. \tag{10.13}$$

Note that it is possible to find a special type of primary system in order that *all* secondary coefficients will be equal to zero. Corresponding primary system is called a rational one. We will not discuss this question.

Coefficients and free terms of canonical equations will be calculated taking into account only bending moments, which arise in the arch. The expression for bending moments in the left part of the primary system for unit and loaded states (the force $P = 1$ is located within the left part of the arch) are presented in Table 10.2.

Table 10.2 Bending moments due to unit primary unknowns and given unit load

	$\overline{M}_1$	$\overline{M}_2$	$\overline{M}_3$	$M^0_{P=1}$
Bending moment expression	1	$1\,(f-y)$	$-1\left(\frac{l}{2}-x\right)$	$M^0_P = -1\,(a-x)$

10.1.2.1 Unit Coefficients

$$\delta_{11} = \int_0^l \frac{\bar{M}_1 \cdot \bar{M}_1}{EI_x} ds = \int_0^l \frac{1 \times 1 \times \cos\varphi_x}{EI_C} \frac{dx}{\cos\varphi_x} = \int_0^l \frac{dx}{EI_C} = \frac{l}{EI_C}$$

$$\delta_{12} = \int_0^l \frac{\bar{M}_1 \cdot \bar{M}_2}{EI_x} ds = \int_0^l 1\,(f - y) \frac{dx}{EI_C}$$

$$= \int_0^l \left[f - \frac{4f}{l^2}(l - x)x \right] \frac{dx}{EI_C} = \frac{fl}{3EI_C} \quad \delta_{12} = \delta_{21} = \frac{fl}{3EI_C}$$

$$\delta_{22} = \int_0^l \frac{\bar{M}_2 \cdot \bar{M}_2}{EI_x} ds = \int_0^l (f - y)^2 \frac{dx}{EI_C}$$

$$= \int_0^l \left[f - \frac{4f}{l^2}(l - x)x \right]^2 \frac{dx}{EI_C} = \frac{f^2 l}{5EI_C}$$

$$\delta_{33} = 2 \int_0^{\frac{l}{2}} \frac{\bar{M}_3 \cdot \bar{M}_3}{EI_x} ds = 2 \int_0^{\frac{l}{2}} \left(\frac{l}{2} - x \right)^2 \frac{dx}{EI_C} = \frac{l^3}{12EI_C}$$

10.1.2.2 Free Terms

Since $P = 1$, the free terms are denoted through δ_{iP}

$$\delta_{1P} = \int_0^a \frac{\bar{M}_1 \cdot \bar{M}_P^0}{EI_x} ds = -\int_0^a 1\,(a - x) \frac{dx}{EI_C} = -\frac{a^2}{2EI_C}$$

$$\delta_{2P} = \int_0^a \frac{\overline{M}_2 \cdot M_P^0}{EI_x} ds = -\int_0^a 1\,(f - y) \times 1 \times (a - x) \frac{dx}{EI_C}$$

$$= -\int_0^a 1 \left[f - \frac{4f}{l^2}(l - x)\,x \right] (a - x) \frac{dx}{EI_C} = \frac{a^2 f}{EI_C} \left(-\frac{1}{2} + \frac{2}{3}u - \frac{1}{3}u^2 \right)$$

$$\delta_{3P} = \int_0^a \frac{\overline{M}_3 \cdot M_P^0}{EI_x} ds = \int_0^a \left(\frac{l}{2} - x \right) (a - x) \frac{dx}{EI_C} = \frac{l^3}{EI_C} u^2 \left(\frac{1}{4} - \frac{u}{6} \right)$$

Canonical equations (10.13) become

$$lX_1 + \frac{fl}{3}X_2 = \frac{a^2}{2}$$

$$\frac{fl}{3}X_1 + \frac{1}{5}f^2lX_2 = -fa^2\left(-\frac{1}{2} + \frac{2}{3}u - \frac{1}{3}u^2\right)$$

and $\frac{l^3}{12}X_3 + l^3u^2\left(\frac{1}{4} - \frac{u}{6}\right) = 0.$

The solution of these equations leads to the following expressions for the primary unknowns in terms of dimensionless parameter $u = a/l$, which defines the location of the unit force P:

$$\begin{aligned} X_1 &= u^2\left(-\frac{3}{4} + \frac{5}{2}u - \frac{5}{4}u^2\right)l \\ X_2 &= \frac{15}{4}u^2(1-u)^2\frac{l}{f} \\ X_3 &= 12u^2\left(-\frac{1}{4} + \frac{u}{6}\right) \end{aligned} \tag{10.14}$$

These formulas should be applied for $0 \le u \le 0.5$. Since X_1 and X_2 are symmetrical unknowns, the expressions for these unknowns for the right part of the arch $(0.5 \le u \le 1)$ may be obtained from expressions (10.14) by substitution $u \to 1-u$. Since X_3 is antisymmetrical unknown, the sign of expressions for X_3 should be changed and parameter u should be substituted by $1-u$. Influence lines for the primary unknowns X_1, X_2, and X_3 may be constructed very easily.

After computation of the primary unknowns we can calculate the reaction and internal forces at any section of the arch.

10.1.2.3 Reactions of Support *A*

The following reactions should be calculated: thrust, vertical reaction, and moment.

Thrust:

$$H = X_2 = \frac{15}{4}u^2(1-u)^2\frac{l}{f} \quad \text{for } 0 \le u \le 1.0$$

This formula presents the thrust of the arch as the function of the dimensionless parameter u, i.e., this expression is the influence line for H (Fig. 10.6c). Maximum thrust is $H_{max} = 0.2344\,(Pl/f)$ and it occurs, when the force P is located at the crown C. This formula shows that decreasing of the rise f leads to increasing of the thrust H.

Vertical Reaction:

$$R_A = X_3 + 1 = 12u^2\left(-\frac{1}{4} + \frac{u}{6}\right) + 1 \quad \text{for } 0 \le u \le 0.5$$

Since X_3 is antisymmetrical unknown, for the right part of the arch it is necessary to change sign on the opposite and make the change $u \to 1-u$. Therefore, if unit load P is located on the right part of the arch then reaction R_A is

$$R_A = X_3 = -12\,(1-u)^2\left(-\frac{1}{4}+\frac{1-u}{6}\right) \quad \text{for } 0.5 \le u \le 1.0.$$

Corresponding influence line is presented in Fig. 10.6d.

Moment at Support A:

$$M_A = -\,1ul + X_1 + X_2 f - X_3\frac{l}{2} = u\left(-1+\frac{9}{2}u-6u^2+\frac{5}{2}u^3\right)l \quad \text{for } 0 \le u \le 0.5$$

$$M_A = X_1 + X_2 f - X_3\frac{l}{2} = (1-u)^2\left(\frac{5}{2}u^2-u\right)l \quad \text{for } 0.5 \le u \le 1.0.$$

Corresponding influence line is presented in Fig. 10.6e.

10.1.2.4 Bending Moment at Crown *C*

$$M_C = X_1 = u^2\left(-\frac{3}{4}+\frac{5}{2}u-\frac{5}{4}u^2\right)l \quad \text{for } 0 \le u \le 0.5.$$

Since X_1 is symmetrical unknown, for the right part of the arch it is necessary to make the change $u \to 1-u$. Therefore, if unit load P is located on the right part of the arch, then bending moment at crown is

$$M_C = X_1 = (1-u)^2\left(-\frac{3}{4}+\frac{5}{2}(1-u)-\frac{5}{4}(1-u)^2\right)l \quad \text{for } 0.5 \le u \le 1.0.$$

Influence line for M_C is presented in Fig. 10.6f.

Conclusions If load P is placed in the portion of $0.132l$ in both sides from crown C, then the extended fibers at C are located below the neutral line of the arch. The direction of the support moment M_A depends on the location of the load: if load P is placed within o.4l from the left support, then extended fibers in vicinity of the A are located outside of the arch.

Discussion

1. For given parabolic nonuniform arch we obtained the precise results. It happens because a cross moment of inertia is increasing from crown to supports according to formula $I_x = I_c/\cos\varphi_x$. Since $\mathrm{d}x = \mathrm{d}s\cos\varphi_x$, $\mathrm{d}s/EI_x = \mathrm{d}x/EI_c$ and all integrals are presented in exact form.
2. In arch with clamped supports subjected to distributed load along half-span the maximum bending moments arise at supports. For this case, the following law

for moment of inertia of cross section may be assumed: $I_x \cos\varphi_x = I_c$. This expression corresponds to increasing of bending stiffness of the arch from crown to supports.

3. In arch with pinned supports, the zeros bending moments arise at supports. For these cases, the following law for moment of inertia of cross section may be taken as: $I_c \cos\varphi_x = I_x$. This expression corresponds to decreasing of bending stiffness of the arch from crown to supports. Both types of arches are presented in Fig. 10.7.

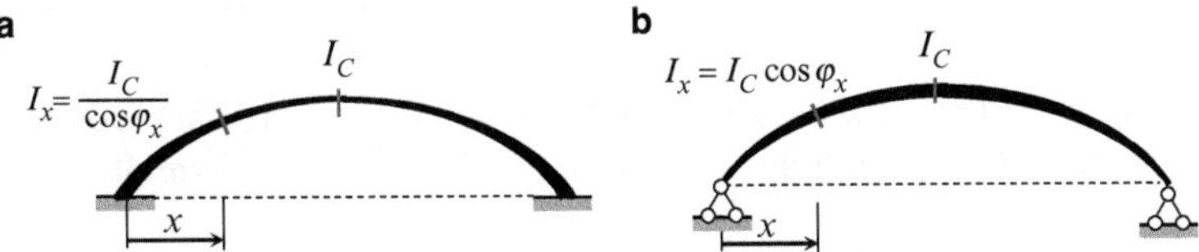

Fig. 10.7 Types of nonuniform arches

Thus, it can be observed, that shape (10.7a) is not wise to use for pinned type of supports, while the shape (10.7b) is dangerous to use in case of clamped supports. It is obvious, that the laws for moment of inertia of cross section in real structures are not limited to two considered cases above.

Example 10.2. Design diagram of symmetric nonuniform parabolic arch with clamped ends is presented in Fig. 10.8. The cross-sectional moments of inertia varies by law $I_x = I_C / \cos\varphi_x$ as considered above. The arch is subjected to concentrated load $P = 30\,\text{kN}$ and uniformly distributed load $q = 2\,\text{kN/m}$, as shown on the design diagram. Calculate the reactions of support A and internal forces (shear and bending moment) at the crown C. Use the influence lines obtained above.

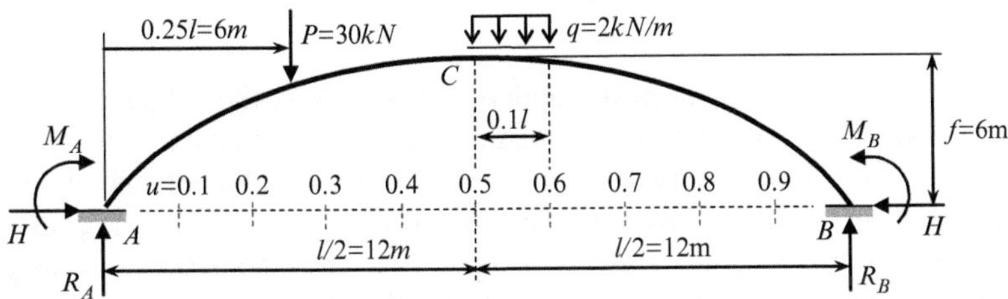

Fig. 10.8 Design diagram of parabolic nonuniform arch

Solution. Influence lines for reactions and bending moment at the crown C are shown in Fig. 10.6c–f. For design diagram in Fig. 10.8, the following factors should be taken into account: $l/f = 4$, and $l = 24\,\text{m}$.

Internal forces may be defined using the corresponding influence lines by formula $S = P \cdot y + q \cdot \Omega$, where y is the ordinate of influence line under concentrated force, Ω is area of influence line within acting distributed load. The area of curvilinear influence line may be calculated approximately by replacing curvilinear segments between two neighboring ordinates by straight lines (Fig. 10.9).

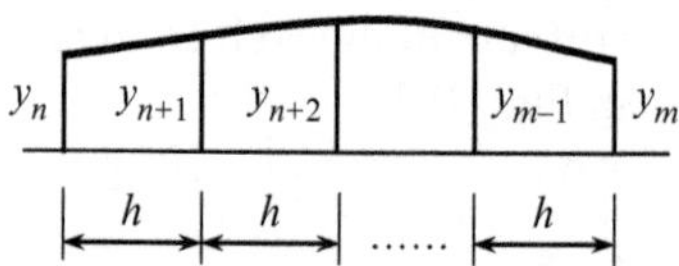

Fig. 10.9 Approximation for calculation of area of curvilinear influence line

If a horizontal distance h, which separates these ordinates, remains constant, then the area bounded by two ordinates y_n and y_m will be given by formula

$$\Omega_n^m = h\left(\frac{y_n}{2} + y_{n+1} + y_{n+2} + \ldots + y_{m-1} + \frac{y_m}{2}\right). \tag{a}$$

Ordinates of influence lines in Fig. 10.6c–f are presented over $0.05l = 1.2\,\text{m}$. Now we can calculate all reactions at support A due to fixed load P and q.

Thrust

$$H = \frac{l}{f}\left[P \times 0.1320 + q\left(\frac{0.2344}{2} + 0.2295 + \frac{0.2160}{2}\right)1.2\right] = 20.20\,\text{kN}$$

Vertical reaction

$$R_A = \left[P \times 0.844 + q\left(\frac{0.5}{2} + 0.425 + \frac{0.352}{2}\right)1.2\right] = 27.36\,\text{kN}$$

Moment at support

$$M_A = l\left[-P \times 0.0528 + q\left(\frac{0.0312}{2} + 0.0418 + \frac{0.048}{2}\right)1.2\right] = -33.32\,\text{kNm}.$$

Obtained values of reactions at support A (as well as the influence lines for primary unknowns X_i) allow for calculating all internal forces at any section of the arch. For this it is necessary to eliminate all constraints at the left end of the arch and replace them by the reactive forces just founded, i.e., to consider the given arch as statically determinate one, which is clamped at B only and is subjected to given load and reactions at support A. For example, bending moment at crown C, by definition, equals

$$\begin{aligned} M_C &= R_A\frac{l}{2} - H \cdot f + M_A - P\left(\frac{l}{2} - 0.25l\right) \\ &= 27.36 \times 12 - 20.20 \times 6 - 33.32 - 30\,(12 - 6) = -6.2\,\text{kNm}. \end{aligned}$$

Here we again use the fixed and moving load approaches in parallel way.

The bending moment at crown C using the influence line

$$M_C = l\left[-P \times 0.0127 + q\left(\frac{0.0468}{2} + 0.0246 + \frac{0.008}{2}\right)1.2\right] = -6.15\,\text{kNm}.$$

Relative error is

$$\frac{6.2 - 6.15}{6.175}100\% = 0.8\%.$$

This error is due to approximate calculation of the area of influence lines.

Shear force at crown C is obtained by projecting all forces, located to the left of this section, on the vertical:

$$Q_C = R_A - P = 27.36 - 30 = -2.64\,\text{kN}.$$

Discussion. Influence lines for reactions of supports have the fundamental meaning, since they allow easy calculations of reactions of statically indeterminate arch, subjected to any fixed load. After that, calculating internal forces at any section of the arch is performed as for statically determinate structure.

10.1.3 Statically Indeterminate Trusses

In case of trusses the general procedure for construction of influence lines has the fundamental features. Let us consider the statically indeterminate truss shown in Fig. 10.10a. The axial stiffness of all members equals *EA*. It is required to construct the influence line for reaction at the middle support. The structure under consideration is externally statically indeterminate truss of the first degree of redundancy. Let the primary unknown X_1 be a reaction of the middle support; the primary system is shown in Fig. 10.10b.

Force method leads to the following expression for the primary unknown

$$X_1 = -\left(\delta_{1P}/\delta_{11}\right),$$

where δ_{1P} is the displacement in the direction of 1-th primary unknown due to traveling load $P = 1$.

Principal concept Expression for X_1 may be modified. According to the reciprocal displacements theorem, $\delta_{1P} = \delta_{P1}$ and expression for X_1 becomes

$$X_1 = -\frac{\delta_{1P}}{\delta_{11}} = -\frac{\delta_{P1}}{\delta_{11}} \tag{10.15}$$

This equation contains the in-depth fundamental concept: *instead of calculation of displacement δ_{1P} in the direction of primary unknown X_1 due to the moving load $P = 1$, we will calculate displacements δ_{P1} at points of application of load $P = 1$ due to unit primary unknown $X_1 = 1$* (Fig. 10.10c). This fundamental idea also will be used in the last section of this chapter for construction of *models* of influence lines.

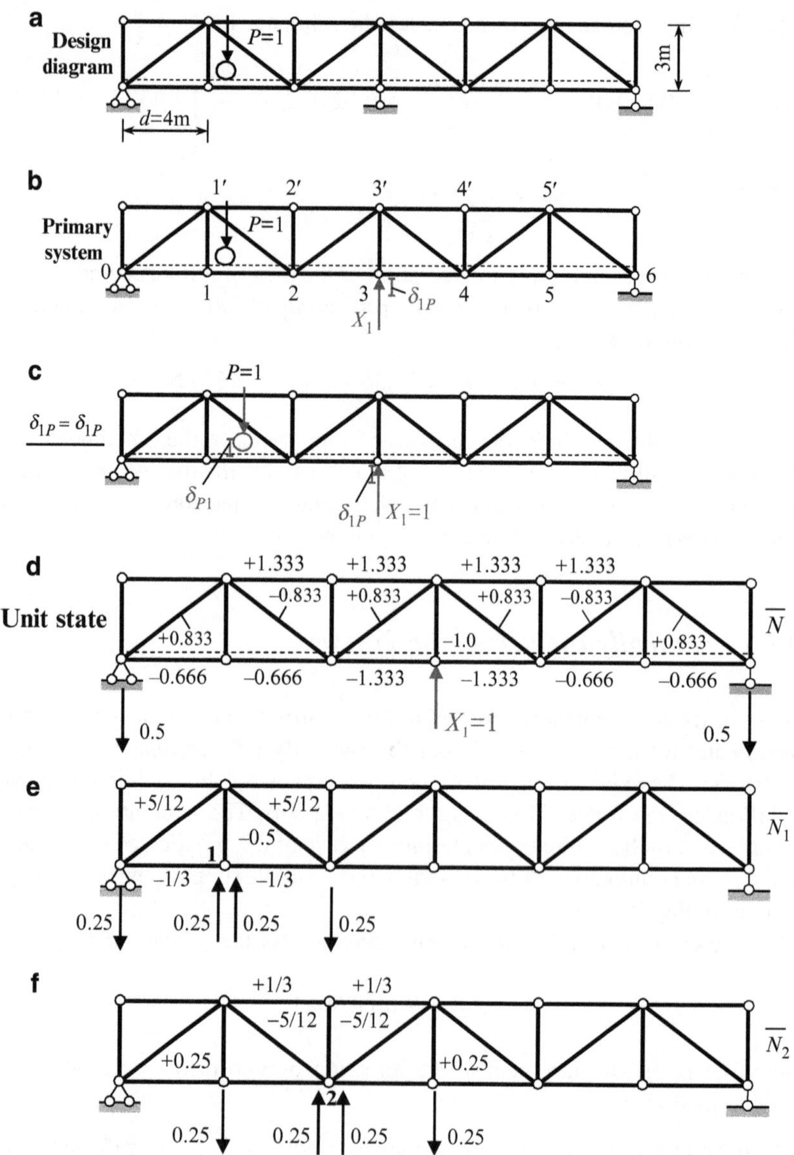

Fig. 10.10 (**a, b**) Design diagram of a statically indeterminate truss and primary system. (**c**) Reciprocal displacements theorem. (**d**) Internal forces $\overline{N}$ caused by unit primary unknown $X_1 = 1$. (**e**) Design diagram for calculation of the first elastic load. (**f**) Design diagram for calculation of the second elastic load

Unit state and corresponding internal forces in all members of the truss are shown in Fig. 10.10d. These internal forces will be used for calculation of δ_{P1} and δ_{11}.

Calculation of δ_{P1} Since the lower chord of the truss is loaded by traveling load, we have to find the vertical displacements of the joints of the lower chord. The most effective procedure for calculation of δ_{P1} is the elastic load method.

Elastic load W_1 According to this method we need to apply two unit couples of opposite directions to the members, which are located to the left and to the right from joint 1. Let the left couple rotate counterclockwise and right clockwise; the couples are not shown. The each couple should be presented as two forces at joints 0 and 1, 1, and 2. These forces have a vertical direction and their values are $1/d = 1/4 = 0.25$. The group of four loads at joints 0, 1, 2 is shown in Fig. 10.10e. These four forces present a self-equilibrate set of forces, therefore the reactions of supports are zeros. Corresponding internal forces in each element of the truss in the primary system caused by these four applied forces 0.25 each are shown in diagram $\overline{N}_1$.

The first elastic load is determined by formula

$$W_1 = \sum \frac{\overline{N}_1 \cdot \overline{N} \cdot l}{EA}, \tag{10.16}$$

where $\overline{N}$ is the internal force in each element of the truss in the primary system caused by the unit primary unknown $X_1 = 1$; these forces are shown in Fig. 10.10d. Summation is performed by all elements of the truss. The first elastic load according to (10.16) becomes

$$W_1 = \underbrace{2\left(-\frac{1}{3}\right)(-0.666)\,4\frac{1}{EA}}_{\text{members } 0-1,\ 1-2} + \underbrace{\frac{5}{12}\times 0.833\times 5\times\frac{1}{EA}}_{\text{member } 0-1'} + \underbrace{\frac{5}{12}(-0.833)\times 5\times\frac{1}{EA}}_{\text{member } 2-1'} = \frac{1.776}{EA}$$

Elastic load W_2 The loads $1/d = 1/4 = 0.25$ are applied at joints 1 and 2, 2 and 3 as shown in Fig. 10.10f; corresponding internal forces are shown on the diagram $\overline{N}_2$.

The second elastic load becomes

$$W_2 = \sum \frac{\overline{N}_2 \cdot \overline{N} \cdot l}{EA} = \underbrace{2\times\frac{1}{3}\times 1.333\times 4\frac{1}{EA}}_{\text{members } 1'-2',\ 2'-3'} + \underbrace{\left(-\frac{5}{12}\right)(-0.833)\times 5\times\frac{1}{EA}}_{\text{members } 1'-2} + \underbrace{\left(-\frac{5}{12}\right)0.833\times 5\times\frac{1}{EA}}_{\text{members } 2-3'} = \frac{3.555}{EA}.$$

Elastic load W_3 Similar procedure leads to the following result for third elastic load $W_3 = 8.525/EA$.

Since the structure is symmetric, the elastic loads $W_4 = W_2$ and $W_5 = W_1$.

Fictitious beam presents the simply supported beam, which is loaded by elastic loads (Fig. 10.10g). The elastic load W_1 is positive, so this load must be directed as forces at support 1, i.e., upward (Fig. 10.10e). Similarly, all the elastic loads should be directed upward (Fig. 10.10g).

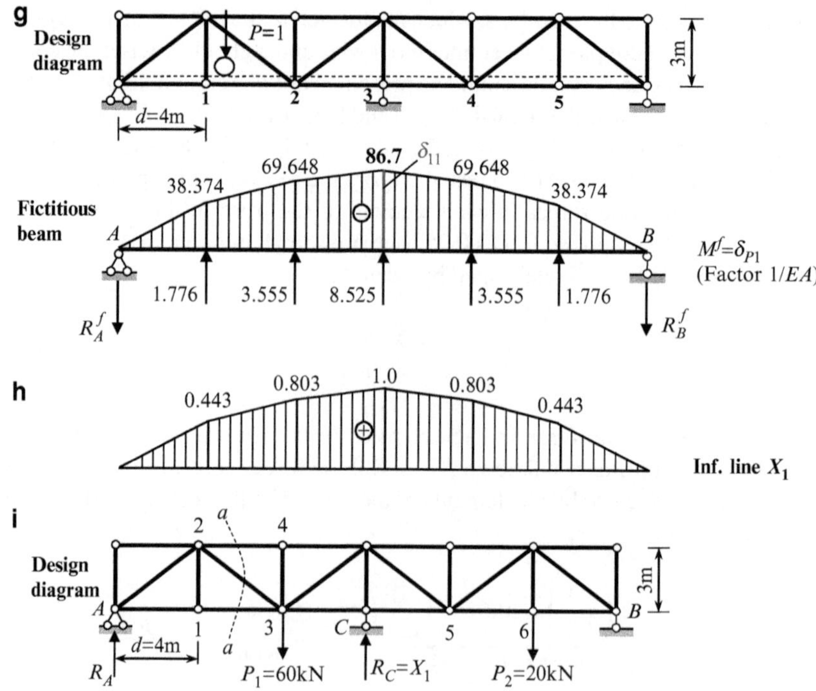

Fig. 10.10 (**g**) Fictitious beam subjected to elastic loads and corresponding bending moment diagram; (**h**) Influence line for primary unknown X_1. (**i**) Truss subjected to fixed load $P_1 = 60$ kN and $P_2 = 20$ kN

Reactions at the left and right supports of fictitious beam are $9.5935/EA$. The corresponding bending moment diagram is presented in Fig. 10.10g; factor $1\,/\,EA$ for all ordinates is not shown. The ordinates of this diagram present the displacements of corresponding joints of the truss due to unit primary unknown $X_1 = 1$. It is obvious that the unit displacement becomes $\delta_{11} = 86.7/EA$.

10.1.3.1 Influence Line for Primary Unknown X_1

This key influence line is obtained by dividing all ordinates of diagram δ_{P1} (Fig. 10.10g) for fictitious beam by $(-\delta_{11})$. Corresponding influence line X_1 is presented in Fig. 10.10h.

In case of arbitrary external load the internal force in any member can be calculated using the influence line for primary unknown X_1. For example, the truss is loaded by concentrated force $P_1 = 60\,\text{kN}$ at joint 3 and $P_2 = 20\,\text{kN}$ at joint 6 as shown in Fig. 10.10i. For this loading the primary unknown X_1 which is the reaction at support C becomes $R_C = X_1 = \sum P_i y_i = P_1 0.803 + P_2 0.443 = 57.04\,\text{kN}$.

Now we can calculate the reaction at support A:

$$R_A \to \sum M_B = 0: \; -R_A 6d + P_1 4d - R_C 3d + P_2 d = 0 \to R_A = 34.81\,\text{kN}.$$

After that we can calculate any internal force. For example, for force U_{1-3} (section $a-a$) we get:

$$U_{1-3} \to \sum M_2^{\text{left}} = 0: \; -R_A d + U_{1-3} h = 0 \to U_{1-3} = \frac{34.81 \times 4}{3} = 46.41\,\text{kN}.$$

Summary

1. The reciprocal displacements theorem allows to consider δ_{P1} as *model* of vertical displacements of joints of the truss (within a constant multiplier which equals $-1/\delta_{11}$).
2. For the calculation of δ_{P1} the elastic loads method have been applied. This method is more effective in comparison of Maxwell–Mohr formula by the following reason. When a set of four loads is applied, then the internal forces arise only in the members, which form the two adjacent panels of the truss. Therefore, procedure of summation is related only to elements, which *belongs to these two panels* of the truss, while using the Maxwell–Mohr integral, a summation is related to *all elements* of the truss.
3. The elastic loads method also allows calculating unit displacement δ_{11}. It is obvious that δ_{11} could be calculated using Maxwell–Mohr formula

$$\delta_{11} = \sum \frac{\overline{N} \cdot \overline{N} \cdot l}{EA}.$$

However, since in the unit state practically in all members of the truss arise internal forces (Fig. 10.10d), then this procedure becomes cumbersome. In fact, the elastic load method allows calculating δ_{P1} for all joints and δ_{11} in one step, as shown in Fig. 10.10g.
4. Influence line of internal force, which is induced at any member k of the truss, may be constructed by formula (10.9).
5. Influence line for primary unknown X_1 should be treated as key influence line. Indeed, using this influence line we can calculate X_1 in case of arbitrary fixed load. After that, the truss should be considered as statically determinate one, which is subjected to the given loads and the force X_1.

10.2 Construction of Influence Lines by the Displacement Method

Let us consider a first-degree kinematically indeterminate structure. In case of a *fixed* load, the canonical equation of the displacement method is

$$r_{11}Z_1 + R_{1P} = 0.$$

Now we need to transform this equation for the case of *moving* load. In this equation the free term R_{1P} represents reaction caused by given actual load. Since moving load is *unit* one, then (for the sake of consistency notations) let us replace a free term R_{1P} by the *unit* free term r_{1P}; this free term presents a reaction in primary system in the introduced constraint 1 caused by load $P = 1$. The primary unknown

$$Z_1 = -\frac{r_{1P}}{r_{11}}.$$

The unit reaction r_{11} presents reaction in the introduced constraint caused by unit displacement of this constraint. Therefore, r_{11} is some specific *number*, which depends on the type of a structure and its parameters and does not depend on the position of the acting load. However, r_{1P} depends on unit load location. Since load $P = 1$ is traveling, r_{1P} becomes a *function* of position of this load, and as result, the primary unknown becomes a *function* as well:

$$\text{IL}\,(Z_1) = -\frac{1}{r_{11}}\text{IL}\,(r_{1P})\,. \tag{10.17}$$

The function IL (r_{1P}) may be constructed using Tables A.3–A.6; nondimensionless parameters u and υ $(u + \upsilon = 1)$ denote the position of load P.

In case of *fixed* load, a bending moment at any section k may be calculated by formula

$$M_k = \overline{M}_k Z_1 + M_k^0.$$

where Z_1 is the primary unknown of the displacement method; $\overline{M}_k$ is the bending moment at section k in a primary system due to unit primary unknown $Z = 1$; and M_k^0 is the bending moment at section k in a primary system due to given load.

Now we need to transform this equation for the case of *moving* load. The bending moment $\overline{M}_k$ at any section k presents the *number*, because this moment is caused by the unit primary unknown Z_1; the second component Z_1 presents a function. The last component moment M_k^0 is caused by the moving load, therefore the bending moment M_k^0 also becomes a *function* of the position of unit load. As a result, the bending moment at any section k becomes a *function*, which can be presented as follows

$$\text{IL}\,(M_k) = \overline{M}_k \cdot \text{IL}\,(Z_1) + \text{IL}\left(M_k^0\right). \tag{10.18}$$

In case of fixed load, the shear at any section k is a number

$$Q_k = \overline{Q}_k \cdot Z_1 + Q_k^0.$$

Similarly in case of traveling load, the shear at any section k becomes a function, therefore

$$\mathrm{IL}\,(Q_k) = \overline{Q}_k \cdot \mathrm{IL}\,(Z_1) + \mathrm{IL}\left(Q_k^0\right). \tag{10.19}$$

For the n-times kinematically indeterminate structure, the canonical equations of the Displacement method in case of a fixed load $P = 1$ are

$$\begin{array}{c} r_{11}Z_1 + r_{12}Z_2 + \cdots + r_{1n}Z_n + r_{1P} = 0 \\ \cdot \quad \cdot \quad \cdot \quad \cdot \quad \cdot \quad \cdot \quad \cdot \quad \cdot \quad \cdot \quad \cdot \quad \cdot \quad \cdot \quad \cdot \\ r_{n1}Z_1 + r_{n2}Z_2 + \cdots + r_{nn}Z_n + r_{nP} = 0 \end{array} \tag{10.20}$$

Unit reaction r_{ik} are caused by unit primary unknowns. They can be calculated using the typical procedures of the displacement method; these reactions are presented as the specific *numbers*.

Reaction r_{iP} can be calculated using Tables A.3–A.6. Fundamental feature of system (10.20) is that the free terms r_{iP} are some *functions* of position of unit load $P = 1$. Therefore, a solution of the system (10.20) leads to the primary unknowns Z_i as the *functions* of the load position, in fact, to $\mathrm{IL}\,(Z_i)$, $i = 1, \ldots, n$.

Bending moment at any section k in case of fixed load may be calculated by formula

$$M_k = \overline{M}_1 \cdot Z_1 + \overline{M}_2 \cdot Z_2 + \ldots + M_P^0.$$

In case of moving load, the bending moment at the any specified section k becomes function, so influence line should be constructed by formula

$$\mathrm{IL}\,(M_k) = \overline{M}_{k1} \cdot \mathrm{IL}\,(Z_1) + \overline{M}_{k2} \cdot \mathrm{IL}\,(Z_2) + \ldots + \mathrm{IL}\left(M_k^0\right). \tag{10.21}$$

Influence lines for shear and axial force at any section can be constructed similarly.

Similarly to the force method, the construction of influence lines for internal forces in statically indeterminate structures by the displacement method starts from construction of influence lines for primary unknowns $\mathrm{IL}\,(Z_1)$, $\mathrm{IL}\,(Z_2)$, ….

The following procedure may be recommended for construction of influence lines for statically indeterminate structures by the displacement method:

1. Determine the degree of kinematical indeterminacy, construct the primary system and formulate the canonical equations of the displacement method.
2. Compute the coefficients r_{ik} of canonical equations; they are unit reaction at introduced constraint i caused by unit *displacements* of introduced constraint k. These unit reactions present the some specific *numbers*.
3. Construct the expressions for free terms r_{iP} of canonical equations. These load terms presents reaction at introduced constraint i due to load, depends of location of unit load and therefore present some *functions*.

4. Solve the canonical equations; since the free terms are functions, the primary unknowns will be also present the *functions* of location of unit load.
5. Construct the influence lines for primary unknowns. This step presents the central and most important part of the procedure.
6. Construct the influence lines for reactions and internal forces (bending moment and shear) at specified section.

The detailed procedure of construction of influence line for primary unknown and internal forces for continuous beam and frame will be illustrated below.

10.2.1 Continuous Beams

Let us demonstrate the above procedure for uniform two-span continuous beam ABC with equal spans l. The purpose of analysis is construction of influence lines for bending moment and shear at a specified section k (Fig. 10.11a).

Primary system The structure is kinematically indeterminate of the first degree. The primary system of the displacement method is shown in Fig. 10.11b. The primary unknown Z_1 is the angle of rotation of additional constraint at the middle support.

10.2.1.1 Influence Line for Primary Unknown Z_1

For construction of this influence line we will have to apply the formula (10.17)

1. For calculation of r_{11} it is necessary to plot the bending moment diagram due to primary unknown $Z_1 = 1$ (Fig. 10.11c). It is obvious that $r_{11} = 6EI/l$.
2. For calculation of r_{1P} it necessary to consider the load $P = 1$ located within the left and right span separately. Bending moment diagrams due to traveling load $P = 1$ are shown in Fig. 10.11d.

   Expressions for reactive moment r_{1P} for different beams are presented in Table A.3. Note that length υl is measured from the pinned (roller) support. Ordinates of influence line for r_{1P} in terms of position υ of the load $P = 1$ in the left and right spans are presented in Table 10.3.

Table 10.3 Calculation of r_{1P}

Moving load $P = 1$ at the left span			Moving load $P = 1$ at the right span		
Point	υ	$r_{1P} = \frac{l}{2}\upsilon\left(1-\upsilon^2\right)$	Point	υ	$r_{1P} = -\frac{l}{2}\upsilon\left(1-\upsilon^2\right)$
1	0	0	6	1.0	0
2	0.2	$0.096l$	7	0.8	$-0.144l$
3	0.4	$0.168l$	8	0.6	$-0.192l$
4	0.6	$0.192l$	9	0.4	$-0.168l$
5	0.8	$0.144l$	10	0.2	$-0.096l$
6	1.0	0	11	0	0

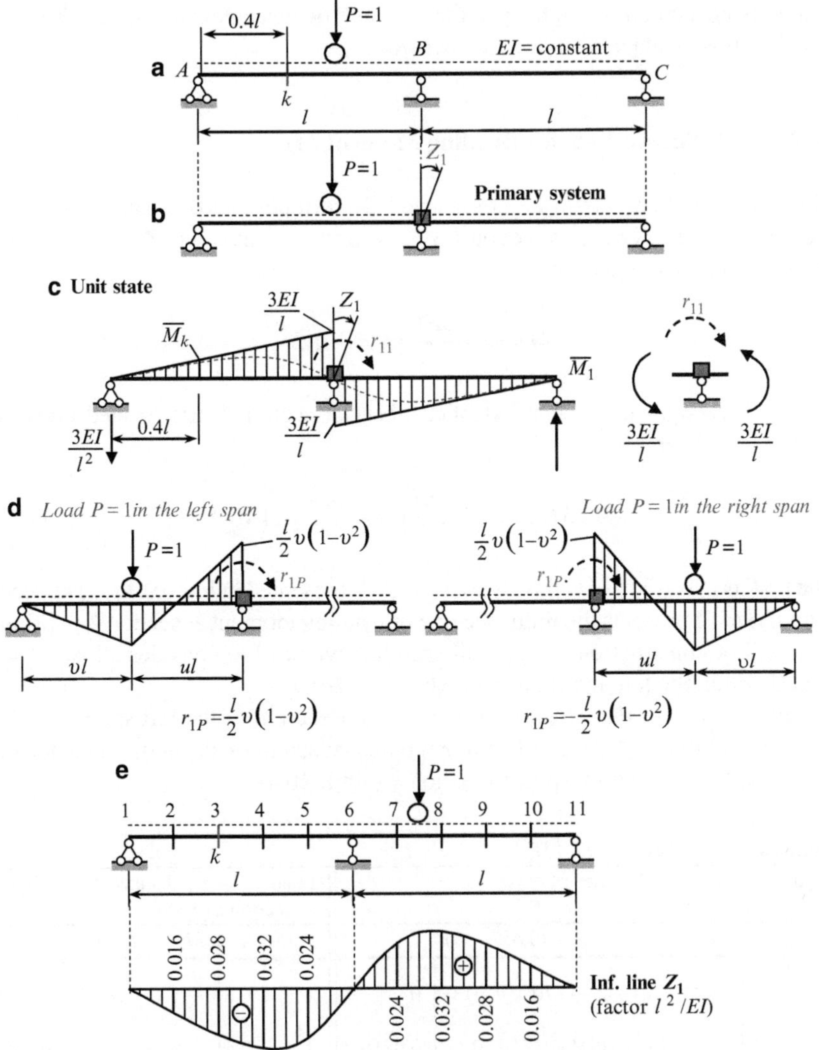

Fig. 10.11 (**a, b**) Design diagram of the beam and primary system. (**c**) Bending moment diagram caused by unit primary unknown and calculation of r_{11}. (**d**) Calculation of r_{1P}. (**e**) Continuous beam. Influence line for primary unknown Z_1

3. Influence line of primary unknown Z_1 is obtained according to formula (10.17): all ordinates of influence line for r_{1P} (Table 10.3) should be divided by

$$-r_{11} = -6\,(EI/l)\,.$$

Corresponding influence line for Z_1 is presented in Fig. 10.11e; all ordinates must be multiplied by parameter l^2/EI.

If load is located on the right span then angular displacement at support B is positive, i.e., this displacement occurs clockwise.

10.2.1.2 Influence Line for Bending Moment M_k

This influence line should be constructed by formula (10.18). The bending moment in primary system at section k due to primary unknown $Z_1 = 1$ is shown in Fig. 10.11c and equals to

$$\overline{M}_k = -\frac{3EI}{l}0.4 = -1.2\frac{EI}{l};$$

the negative sign means that extended fibers at section k are located above the neutral line. Therefore

$$\mathrm{IL}(M_k) = -1.2\frac{EI}{l}\mathrm{IL}(Z_1) + \mathrm{IL}\left(M_k^0\right). \tag{10.22}$$

The first term $-1.2\,(EI/\,l)\,\mathrm{IL}\,(Z_1)$ of (10.22) is presented in Fig. 10.12a. The second term $\mathrm{IL}\,\left(M_k^0\right)$ presents the influence line of bending moment at section k in primary system. For construction of this influence line we need to consider a load $P = 1$ placed within the left and right spans (Fig. 10.12b).

Load $P = 1$ in the left span (Fig. 10.12b). Reaction of the left support is $R = u^2/2\,(3-u)$ (Table A.3). The bending moment at section k depends on the location of the load $P = 1$ with respect to section k (Table 10.4).

Table 10.4 Calculation of $IL\left(M_k^0\right)$

Load $P = 1$ is located to the left of section $k (u \geq 0.6)$				Load $P = 1$ is located to the right of section $k (u \leq 0.6)$			
Point	u	$M_k^0 = R \cdot 0.4l - P(0.4l - vl)$		Point	u	$M_k^0 = R \cdot 0.4l$	
1	1	$\frac{u^2}{2}(3-u)\,0.4l - l\,(0.4-v) =$	0	$k, 3$	0.6	$\frac{u^2}{2}(3-u)\,0.4l$	$0.1728l$
2	0.8	$\frac{u^2}{2}(3-u)\,0.4l - l\,(u-0.6)$	$0.08161l$	4	0.4		$0.0832l$
k,3	0.6		$0.1728l$	5	0.2		$0.0224l$
				6	0		0

Load $P = 1$ in the right span. In this case the bending moment at section k does not arise (because the introduced constraint), and influence line has zeros ordinates. Influence line M_k^0 is shown in Fig. 10.12b.

The final influence line for bending moment at section k is constructed using expression (10.18) and is shown in Fig. 10.12c. The same result had been obtained early by the force method.

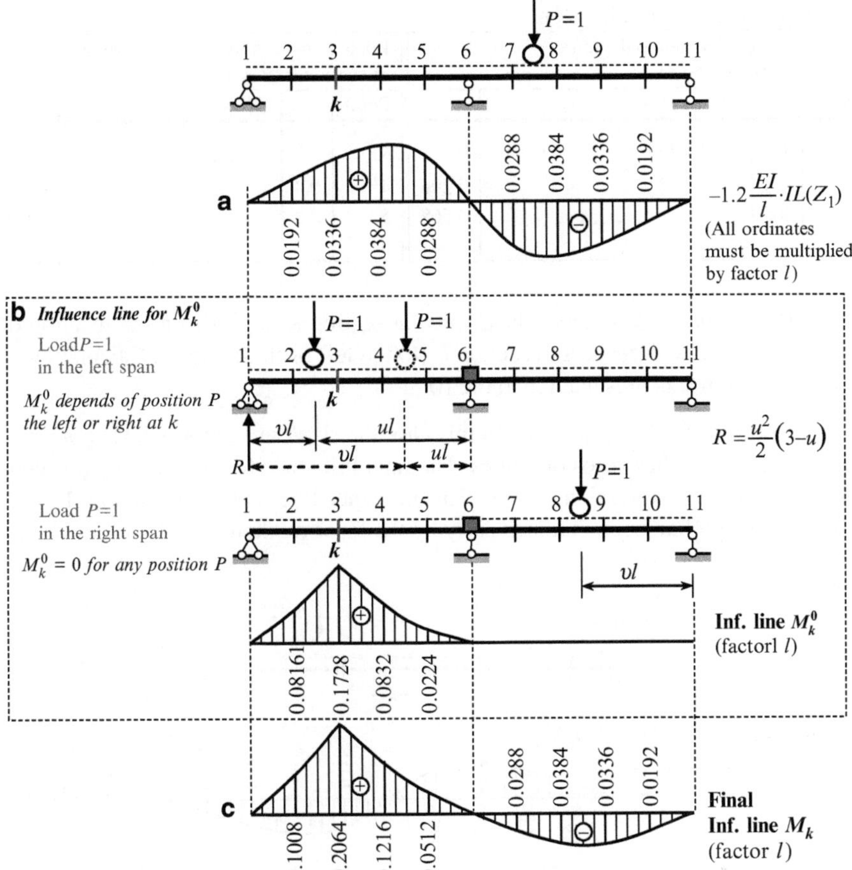

Fig. 10.12 (**a–c**) Construction of influence line for bending moment M_k

10.2.1.3 Influence Line for Shear Force Q_k

This influence line should be constructed by formula (10.19).

According to Fig. 10.11c, the reaction at the support 1 due to $Z_1 = 1$ is directed downward, so the shear in primary system at section k due to primary unknown $Z_1 = 1$ is $\overline{Q_k} = 3EI/l$. Therefore, (10.19) becomes

$$\mathrm{IL}\,(Q_k) = -\frac{3EI}{l^2}\mathrm{IL}\,(Z_1) + \mathrm{IL}\left(Q_k^0\right). \tag{10.23}$$

The first term $-\left(3EI/l^2\right)\cdot \mathrm{IL}\,(Z_1)$ of the (10.23) is presented in Fig. 10.13a. The second term of the (10.23) presents influence line of shear at section k in the primary system. For calculation of this term we need to consider location of the load $P = 1$ in both spans separately.

Table 10.5 Calculation of IL (Q_k^0)

Load $P = 1$ is located to the left of section k, $(u \geq 0.6)$				Load $P = 1$ is located to the right of section k $(u \leq 0.6)$			
Point	u	$Q_k^0 = R - P$		Point	u	$Q_k^0 = R$	
1	1	$\text{IL}(Q_k) = \frac{u^2}{2}(3-u) - 1$	0	k,3	0.6	$\text{IL}(Q_k^0) = \frac{u^2}{2}(3-u)$	0.432
2	0.8		0.296	4	0.4		0.208
k,3	0.6		0.568	5	0.2		0.056
				6	0		0

Load $P = 1$ in the left span Shear force at section k depends on the location of the load $P = 1$ with respect to section k (to the left or to the right). Calculation of shear at section k is presented in Table 10.5.

Load $P = 1$ is in the right span In this case the shear at section k does not arise, and influence line has zeros ordinates. The final influence line for shear Q_k is constructed using the expression (10.23). This influence line is presented in Fig. 10.13c. The same result had been obtained early by the force method.

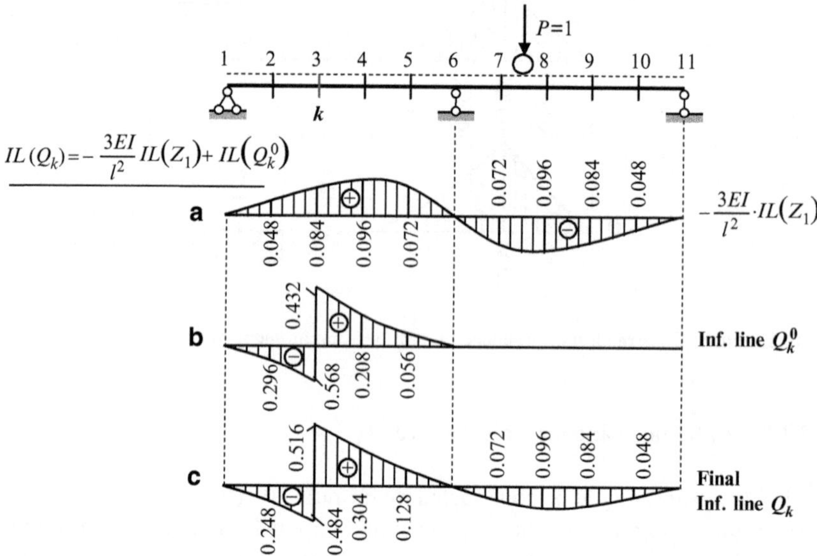

Fig. 10.13 **(a–c)** Construction of influence line for shear Q_k

Discussion

1. Influence lines for M_k^0 and Q_k^0 for the force method are bounded by the *straight* lines, because the primary system is a set of statically *determinate* simply supported beams. The same influence lines in the displacement method are bounded by the *curved* lines, because the primary system is a set of statically *indeterminate* beams.

2. If a load is located in the right span, then M_k^0 and Q_k^0 are zero. In the force method it happens because the primary system presents two *separate* beams; therefore, the load from one beam cannot be transmitted to another. In the displacement method it happens because the introduced support in the primary system does not allow transmitting of internal forces from the right span to the left one.

Example 10.3. The two-span beam with equal spans l is subjected to force P as shown in Fig. 10.14a. The beam is divided into ten equal portions; the number of joints is presented in Fig. 10.11e. Find the reaction at the middle support B. Solve this problem by three different ways (a) Use the influence lines for M_k and Q_k; (b) use the influence lines for primary unknown of the displacement method; (c) use the influence lines for primary unknown of the force method.

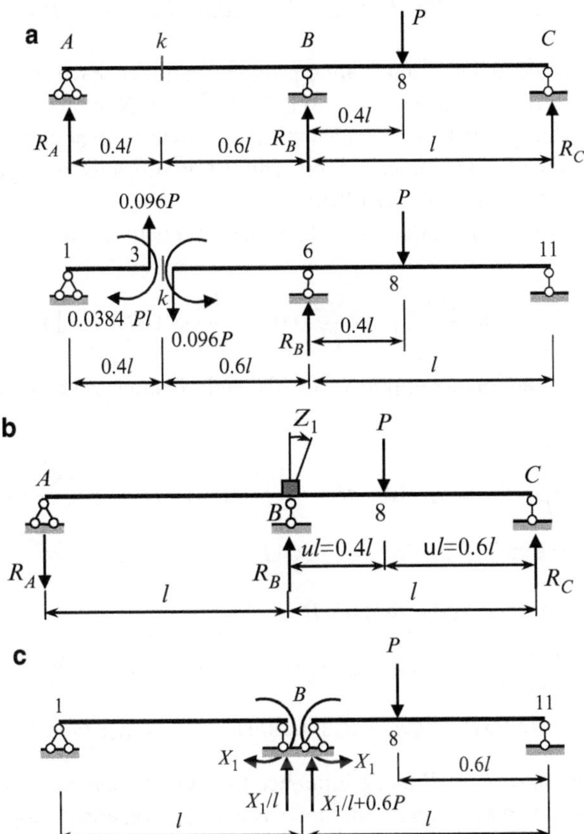

Fig. 10.14 (**a**) Design diagram of the beam and calculation of reaction at support B using the influence line for internal forces at section k. (**b**) Calculation of reaction at support B using the influence line Z_1 of the displacement method. (**c**) Calculation of reaction at support B using the influence line X_1 of the force method

Solution. (a) Since a load P is located at point 8, the internal forces at section k according to Figs. 10.12c and 10.13c are $M_k = -0.0384Pl$ and $Q_k = -0.096P$. Now we can make a cut section of the beam at the point 3 (k) and show corresponding internal forces at this section (Fig. 10.14a). Direction of the bending moment and shear are shown according to their signs.

The free-body diagram for part 1-3 allows us to find the reaction of support A ($0.096P$ downward) and to perform checking of calculation

$$\sum M_A = 0.096P \cdot 0.4l - 0.0384Pl = 0.$$

Free-body diagram for part 3-11 allows calculating all reactions and constructing internal force diagrams, which correspond to given location of the force P. For example

$$R_B \rightarrow \sum M_C = 0$$

$$-R_B l + P \cdot 0.6l + 0.096P\,(0.6l + l) + 0.0384Pl = 0 \rightarrow R_B = 0.792P\ (\uparrow).$$

(b) If force P is located at point 8, then primary unknown Z_1 of the displacement method (Fig. 10.11e) equals $Z_1 = 0.032\left(Pl^2/EI\right)$. Now we can consider two beams. The left pinned–clamped beam AB is subjected to angular displacement Z_1 only and the right clamped–rolled beam BC is subjected to angular displacement Z_1 and force P (Fig. 10.14b).

The vertical reaction at support A for the left beam (according to Table A.3) equals

$$R_A = \frac{3EI}{l^2}Z_1 = \frac{3EI}{l^2}0.032\frac{Pl^2}{EI} = 0.096P\ (\downarrow).$$

The vertical reaction at support C for the right beam (according to Table A.3) equals

$$R_C = \frac{3EI}{l^2}Z_1 + \frac{Pv}{2}(3-u) = \frac{3EI}{l^2}0.032\frac{Pl^2}{EI} + \frac{P0.4^2}{2}(3-0.4)$$
$$= 0.096P + 0.208P = 0.304P\ (\uparrow)$$

Required reaction at the support B

$$R_B \rightarrow \sum Y = 0:\ -R_A + R_B - P + R_C = 0$$
$$\rightarrow -0.096P + R_B - P + 0.304P = 0 \rightarrow R_B = 0.792P\ (\uparrow).$$

(c) Reaction of the support B using influence line X_1 of the force method is calculated by the next way. Primary unknown (bending moment at support B) equals $X_1 = M_B = -P \cdot 0.096l$ (Fig. 10.2d). Now we can consider two simply supported beams (left beam is subjected to moment $M_B = X_1$ only and the right beam is subjected to $M_B = X_1$ and force P) as shown in Fig. 10.14c.

All reactions should be calculated as follows

$$R_A=\frac{X_1}{l}=0.096P(\downarrow),\ R_B=2\frac{X_1}{l}+0.6P=0.792P,\ R_C=-\frac{X_1}{l}+0.4P=0.304P$$

Using the above reactions, we can calculate the bending moments for all sections of the beam

$$M_2=-0.096P\frac{l}{5}=-0.0192Pl,$$
$$M_3=-0.096P\frac{2l}{5}=-0.0384Pl,\ldots$$

Thus, having the influence lines we can easily construct the internal force diagrams for any fixed loads.

10.2.2 Redundant Frames

Construction of influence lines for statically indeterminate frames may be effectively performed using the displacement method. As in case of the fixed load, the displacement method is more effective than the force method for framed structures with high degree of the static indeterminacy. Construction of influence lines for internal forces at any section of the frame starts from construction of influence lines for primary unknowns. The following example illustrates the construction of influence line for primary unknown of redundant frame. As have been shown early, this influence line should be treated as key influence line.

Figure 10.15a presents a design diagram of statically indeterminate frame. Bending stiffness *EI* is constant for all portions of the frame. We need to construct the influence line for angle of rotation of the rigid joint.

The primary system is presented in Fig. 10.15b. The primary unknown Z_1 is the angle of rotation of the rigid joint.

Equation of influence line for primary unknown Z_1 is $\text{IL}\,(Z_1)=-\frac{1}{r_{11}}\text{IL}\,(r_{1P})$.

1. *Calculation of unit reaction* r_{11}. Bending moment diagram $\overline{M}_1$ caused by unit rotation of introduced constraint 1 and free body diagram of this constraint is shown in Fig. 10.15c. Equilibrium condition of rigid joint leads to $r_{11}=10i$.
2. *Calculation of free term* r_{1P}. It is necessary to consider two positions of moving load $P=1$: the load is traveling along the left and right spans. The position of load $P=1$ in the each span is indicated by parameters u and v $(u+v=1)$. Bending moment diagrams are shown in Fig. 10.15d. Ordinates of bending moment diagrams for pinned–fixed and fixed–fixed beams are taken from Tables A.3 and A.4, respectively. Note, in our case, the left span of the frame is a mirror of those presented in Table A.3, however a distance from the left pinned support is labeled as *ul*. This should be taken into account, i.e., in formula for M_A

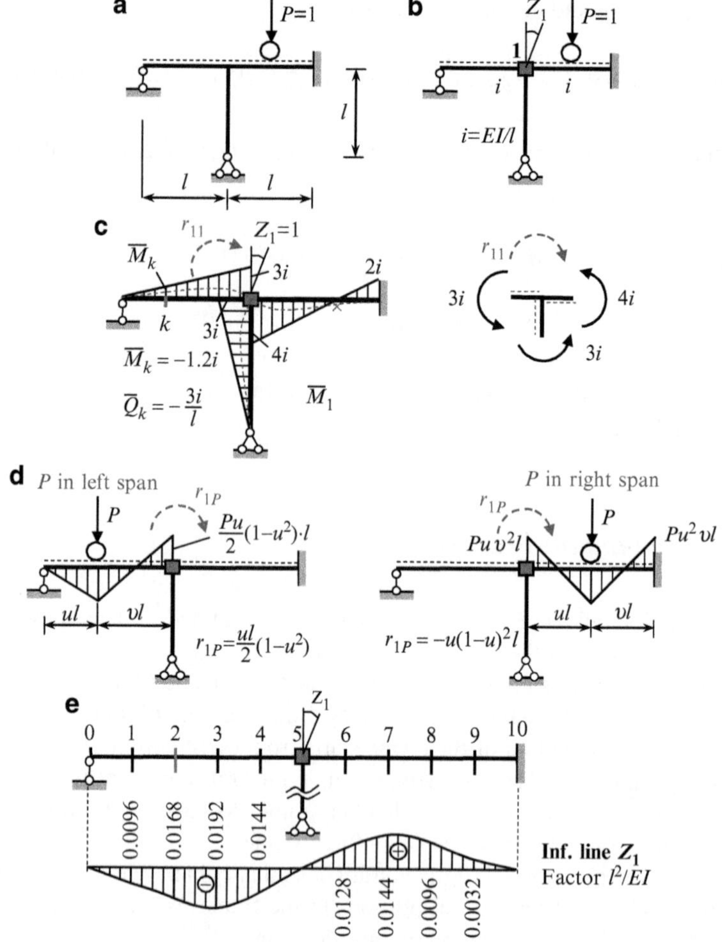

Fig. 10.15 (**a, b**) Design diagram and primary system. (**c**) Unit state and corresponding bending moment diagram. (**d**) Calculation of free term r_{1P}. (**e**) Influence line for primary unknown Z_1

(Table A.3), instead of parameter υ we must write the parameter u. Therefore, if load $P = 1$ is located in the left span, then the free term of canonical equation is $r_{1P} = \frac{ul}{2}\left(1 - u^2\right)$. If load $P = 1$ is located in the right span, then we can directly apply formula from Table A.4, so $r_{1P} = -u\upsilon^2 l = -u(1-u)^2 l$.

3. *Influence line for primary unknown.* Having expressions for r_{1P} in terms of position P and $r_{11} = 10i$, the expressions for primary unknown Z_1 may be presented as follows:

$P = 1$ in the left span

$$\mathrm{IL}\left(Z_1\right) = -\frac{1}{10i}\frac{ul}{2}\left(1 - u^2\right) = -\frac{u}{20}\left(1 - u^2\right)\frac{l^2}{EI}.$$

$P = 1$ in the right span

$$\text{IL}\,(Z_1) = \frac{1}{10i}u\,(1-u)^2\,l = \frac{u}{10}\,(1-u)^2\,\frac{l^2}{EI}.$$

Left and right spans are divided for five equal portions. Calculation of ordinates of influence line of primary unknown at specified points 1–10 is presented in Table 10.6.

Table 10.6 Ordinates of influence line of primary unknown (factor l^2/EI)

Load $P = 1$ in left span			Load $P = 1$ in right span		
Points	u	$\text{IL}(Z_1) = -\frac{u}{20}(1-u^2)\frac{l^2}{EI}$	Points	u	$\text{IL}(Z_1) = \frac{u}{10}(1-u)^2\frac{l^2}{EI}$
0	0.0	0.0	5	0.0	0.0
1	0.2	−0.0096	6	0.2	0.0128
2	0.4	−0.0168	7	0.4	0.0144
3	0.6	−0.0192	8	0.6	0.0096
4	0.8	−0.0144	9	0.8	0.0032
5	1.0	0.0	10	1.0	0.0
Factor		l^2/EI			l^2/EI

The final influence line for primary unknown Z_1 is presented in Fig. 10.15e.

Once constructed influence line for primary unknown presents the fundamental data, because it carries comprehensive and important information about structure. This key influence line can be used for analysis of structure subjected to *arbitrary* load placed along the loaded counter.

10.3 Comparison of the Force and Displacements Methods

For summarizing, let us compare two fundamental analytical methods of structural analysis for construction of influence lines; some results are presented in Table 10.7. As in the case of the dead loads, construction of influence lines for any statically indeterminate structures starts from determining of number and types of unknown and presentation of corresponding primary system of the force and displacement methods.

Notes:

1. In case of continuous beams with pinned end supports both methods lead to the approximately same time consuming.
2. In case of continuous beams with one or two fixed end supports the displacement method is more preferable.
3. In case of frames the preferable method depends on the number of the primary unknowns. For frames similar to those in Sect. 10.2.2, the displacement method is

Table 10.7 Comparison of the force and displacement methods for construction of influence lines

Comparison criteria	Force method	Displacement method
Primary system (PS)	Obtained by *eliminating redundant* constraints *from* a structure	Obtained by *introducing additional* constraints *to* a structure
Primary unknowns (PU)	*Forces* (forces and moments), which *simulate* action of *eliminated* constraints	*Displacements* (linear and angular), which *neutralize* action of *introduced* constraints
Canonical equations in general form	$\delta_{11}X_1 + \delta_{12}X_2 + \ldots + \delta_{1n}X_n + \Delta_{1P} = 0$ $\delta_{21}X_1 + \delta_{22}X_2 + \ldots + \delta_{2n}X_n + \Delta_{2P} = 0$ . . . Number of canonical equations equals to the number of PU	$r_{11}Z_1 + r_{12}Z_2 + \ldots + r_{1n}Z_n + R_{1P} = 0$ $r_{21}Z_1 + r_{22}Z_2 + \ldots + r_{2n}Z_n + R_{2P} = 0$ . . . Number of canonical equations equals to the number of PU
Canonical equations in case of unit moving load $P = 1$	$\delta_{11}X_1 + \delta_{12}X_2 + \ldots + \delta_{1n}X_n + \delta_{1P} = 0$ $\delta_{21}X_1 + \delta_{22}X_2 + \ldots + \delta_{2n}X_n + \delta_{2P} = 0$	$r_{11}Z_1 + r_{12}Z_2 + \ldots + r_{1n}Z_n + r_{1P} = 0$ $r_{21}Z_1 + r_{22}Z_2 + \ldots + r_{2n}Z_n + r_{2P} = 0$ . . . . . .
Features of coefficients and free terms	δ_{ik} are numbers; δ_{iP} are functions of load location	r_{ik} are numbers; r_{iP} are functions of load location
Primary unknowns	X_i are functions of load location	Z_i are functions of load location

Case of $n = 1$	$\delta_{11}X_1 + \delta_{1P} = 0$	$r_{11}Z_1 + r_{1P} = 0$
Primary unknown	$X_1 = -\frac{\delta_{1P}}{\delta_{11}}$ δ_{1P} is function, δ_{11} is number	$Z_1 = -\frac{r_{1P}}{r_{11}}$ r_{1P} is function, r_{11} is number
Influence line for primary unknown	$\text{IL}\,(X_1) = -\frac{1}{\delta_{11}}\text{IL}\,(\delta_{1P})$	$\text{IL}\,(Z_1) = -\frac{1}{r_{11}}\text{IL}\,(r_{1P})$
Influence line for internal forces at section k	$M_k = \overline{M}_k \cdot \text{IL}\,(X_1) + \text{IL}\left(M_k^0\right)$ $Q_k = \overline{Q}_k \cdot \text{IL}\,(X_1) + IL\left(Q_k^0\right)$ $N_k = \overline{N}_k \cdot \text{IL}\,(X_1) + \text{IL}\left(N_k^0\right)$	$M_k = \overline{M}_k \cdot \text{IL}\,(Z_1) + \text{IL}\left(M_k^0\right)$ $Q_k = \overline{Q}_k \cdot \text{IL}\,(Z_1) + \text{IL}\left(Q_k^0\right)$ $N_k = \overline{N}_k \cdot \text{IL}\,(Z_1) + \text{IL}\left(N_k^0\right)$
Unit internal forces $\overline{M}_k\,(\overline{Q}_k, \overline{N}_k)$	Bending moment (shear, axial forces) at section k of primary system of the *Force* method caused by primary unknown $X_1 = 1$	Bending moment (shear, axial forces) at section k of primary system of the *Displacement* method caused by primary unknown $Z_1 = 1$
$IL\left(M_k^0\right)$ $IL\left(Q_k^0\right)$ $IL\left(N_k^0\right)$	Influence line for bending moment (shear, axial forces) at section k in primary system of the *Force* method. These influence lines are *linear*	Influence line for bending moment (shear, axial forces) at section k in primary system of the *Displacement* method. These influence lines are *curvilinear*

more preferable; if the rolled and pinned supports of this frame will be substituted by fixed ones, then advantages of the displacement method are obvious.
4. For construction of influence lines for arches and trusses the force method is more preferable.

10.4 Kinematical Method for Construction of Influence Lines

The shape of influence lines allows finding the most unfavorable position of the load. The shape of influence line is often referred as a model of influence line. The model of influence lines may be constructed by kinematical method using Müller–Breslau principle, which is considered below.

Let us consider n times statically indeterminate continuous beam. It is required to construct an influence line for any reaction (or internal force) X at any section of a beam. Primary system of the force method is obtained by eliminating constraint, which corresponds to required force X and replacing this constrain by force X. Primary system presents $(n-1)$ times statically indeterminate structure. Canonical equation for specified unknown X_1 in case of unit load P is presented in the form $\delta_{11}X_1 + \delta_{1P} = 0$. Influence line for primary unknown X becomes

$$\mathrm{IL}\,(X_1) = -\frac{1}{\delta_{11}}\mathrm{IL}\,(\delta_{1P}),$$

where δ_{11} is the displacement in the direction of primary unknown X_1 caused by unit primary unknown $X_1 = 1$; δ_{1P} presents displacement in the direction of primary unknown caused by moving unit load P.

According to reciprocal displacements theorem, $\delta_{1P} = \delta_{P1}$, where δ_{P1} presents displacement in the direction of moving load P caused by unit primary unknown X_1. Therefore, the influence line for primary unknown may be constructed by formula

$$\mathrm{IL}\,(X_1) = -\frac{1}{\delta_{11}}\mathrm{IL}\,(\delta_{P1}). \tag{10.24}$$

The ordinates of influence line for any function X (reaction, bending moment, etc.) are proportional to ordinates of the elastic curve due to unit force X, which replaces the eliminated constraint where the force X arises (Müller–Breslau principle). This principle with elastic loads method was effectively applied previously for analytical construction of influence lines for statically indeterminate truss. Now we illustrate the Müller–Breslau principle for two types of problems. They are analytical computation of ordinates of influence lines and construction of models of influence lines. Both of these problems are referred as kinematical method. Our consideration of this method will be limited only the continuous beams.

In order to construct the model of influence line for a certain factor X (reaction, bending moment, shear force) by a kinematical method, the following steps must be performed:

1. Indicate the constraint or section, in which factor X arises.
2. Show the new system by eliminating constraint where factor X arises.
3. Apply the $X = 1$ instead of eliminated constraint.
4. Show the elastic curve due to $X = 1$ in new system. This curve is a model for influence line of factor X.

Figure 10.16 presents elimination of constraint k where factor X arises and replacing this constraint by corresponding force X; case (a) should be used for construction of influence line for reaction at any support of continuous beam; the cases (b) and (c) for construction of influence lines for bending moment and shear at any section k, respectively.

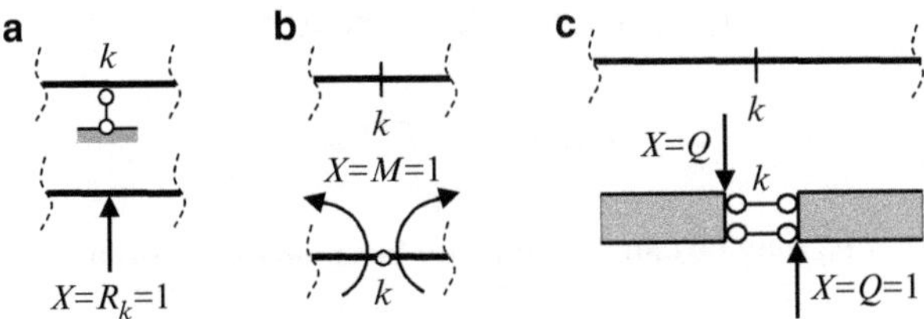

Fig. 10.16 Elimination of constraint and replacing it by corresponding force $X = 1$

In case (a) a support is eliminated and unit reaction is applied. In case (b) we introduce hinge k. Now both parts of the beam can rotate at section k respect to each other, but relative horizontal and relative vertical displacements are absent. In case (c) we introduce special device that allows for the displacements of each part in vertical direction, but the relative horizontal and relative angular displacements are absent.

Figure 10.17a presents two-span uniform continuous beam. The required bending moment at support 1 is considered as a primary unknown X_1. The primary system presents two separate simply supported beams (Fig. 10.17b). Equation of influence line for X_1 is described by (10.24).

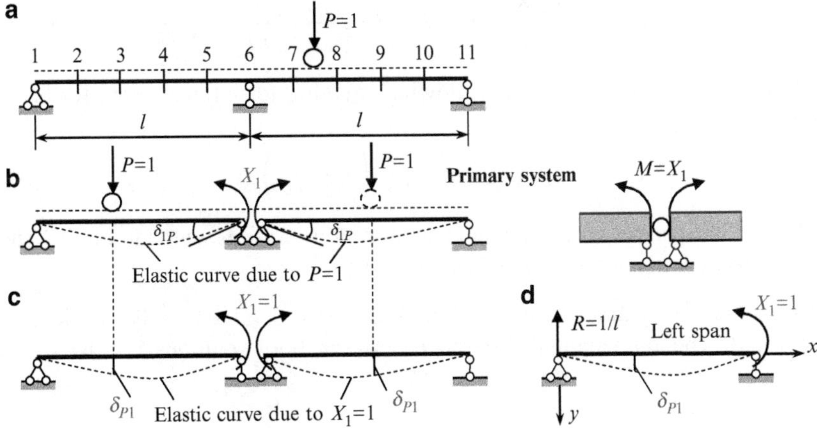

Fig. 10.17 Design diagram, primary system and illustration of reciprocal displacement theorem

The elastic curve caused by $P = 1$ in the primary system is presented in Fig. 10.17b; δ_{1P} is angle of rotation at support 1 caused by moving force $P = 1$. Instead of calculation of δ_{1P} for different position of $P = 1$, we will calculate δ_{P1}; this displacement occurs under the force P caused by unit primary unknown X_1, as shown in Fig. 10.17c.

Elastic curve δ_{P1} caused by fixed unit couple $X_1 = 1$ may be easily constructed by the initial parameter method (Fig. 10.17d). For the left span

$$EI\,y = EI\,\delta_{P1} = EI\,y_0 + EI\,\theta_0 - \frac{R(x-0)^3}{3!},$$

where R is the reaction of the left support. Since $R = 1/l$, and vertical displacement at initial point $y_0 = 0$, the equation of elastic curve becomes

$$EI\,y = EI\,\delta_{P1} = EI\,\theta_0 - \frac{1}{l}\frac{x^3}{6}.$$

Initial parameter θ_0 may be calculated using boundary condition at the right support (point 6):

$$EI\,y\,(l) = EI\,\delta_{P1}\,(l) = EI\,\theta_0 - \frac{1}{l}\frac{l^3}{6} = 0 \rightarrow \theta_0 = \frac{l}{6EI}.$$

Finally, displacement in the direction P caused by unit primary unknown $X_1 = 1$, may be written as follows

$$\delta_{P1} = \frac{l^2}{6EI}\frac{x}{l}\left(1 - \frac{x^2}{l^2}\right) \tag{10.25}$$

Now we need to compute the unit displacement δ_{11}. Bending moment diagram caused by unit primary unknown X_1 is shown in Fig. 10.18b. Using the graph multiplication method we get

$$\delta_{11} = \frac{\overline{M}_1 \times \overline{M}_1}{EI} = 2 \cdot \frac{1}{2} l \cdot 1 \cdot \frac{2}{3} \cdot 1 \frac{1}{EI} = \frac{2}{3}\frac{l}{EI}.$$

Equation (10.24) leads to the following equation for influence line for bending moment at support 1:

$$\text{IL}\,(X_1) = -\frac{l}{4}\frac{x}{l}\left(1 - \frac{x^2}{l^2}\right). \tag{10.26}$$

Influence line for bending moment at support point 6 is presented in Fig. 10.18c. It is obvious that this influence line is symmetrical.

Discussion. The same influence line had been obtained early in Sect. 10.1, Fig. 10.2d. The fundamental difference between both solutions is related to the

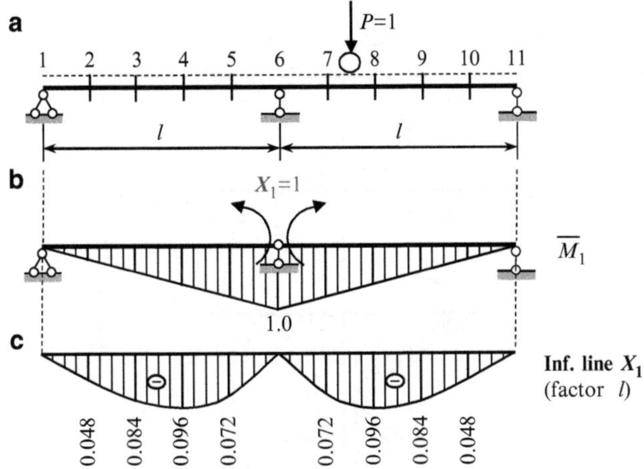

Fig. 10.18 Unit bending moment diagram and influence line for bending moment at support 1

free term of canonical equation of the force method. In Sect. 10.1 we calculated a free term δ_{1P}, which is the angle of rotation at support 1, caused by moving unit load. In Sect. 10.4 we calculated a free term δ_{P1}, which is the vertical displacement at the point of application of the moving load caused by moment $X_1 = 1$ at the middle support 1. Now it is evident that for the construction of influence line for truss, as shown in Sect. 10.3, we have also used the Müller–Breslau principle; however for computation of displacement of the joints, which belong to the loaded contour, the elastic load method was applied.

As shown above for calculation of δ_{P1} the precise analytical method has been applied. As a result, influence line was presented with *computed ordinates* at specified points using formula (10.24). Without consideration of constant factor $(-1/\delta_{11})$, this formula allows constructing the *model* of influence line. Thus, the model of influence line is constructed using only function δ_{P1}, which presents the elastic curve.

In other words, if we apply at point *a* the unit factor *X*, then the elastic curve presents the model of the influence line *X* at point *a*. If we apply at point *a* the unit displacement, which corresponds to the required factor *X*, then the elastic curve presents genuine influence line *X* at point *a*.

Example 10.4. Design diagram for continuous beam is presented in Fig. 10.19. Construct the models of influence lines for reactions and internal forces.

Solution.

Influence line for reaction R_1 In this case we need to eliminate a constraint, where the vertical reaction R_1 arises and apply the positive reaction R_1. Elastic curve due to unit reaction R_1 in the new system is a model of influence line for R_1. If load $P = 1$ is located above support 1, then reaction $R_1 = 1$, therefore ordinate of

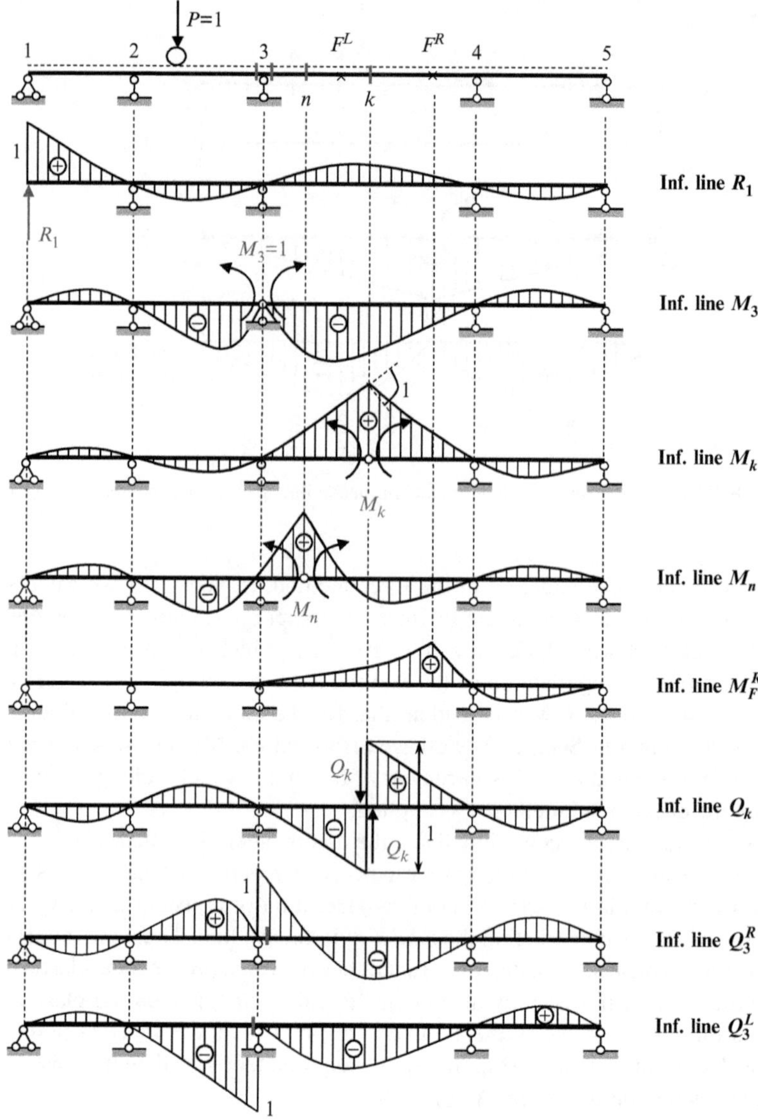

Fig. 10.19 Continuous beam. Design diagram and models of influence lines for reactions and internal forces

influence line at this support equals to 1. If load is located on portions 1-2 and 3-4, then ordinates of influence lines for R_1 are positive, i.e., the vertical reaction R_1 is directed upward.

Influence line for bending moment at the support M_3 In this case we need to include the hinge at the support 3, to apply two positive unit couples M_3, and show the corresponding elastic curve which is a model of required influence line. If a load is located within portions 2-3 and 3-4, then ordinates of influence line are negative, i.e., the extended fibers at the support 3 is located above the longitudinal axis of the beam.

Influence line for bending moment M_k in the span. In this case we need to include the hinge at the section k, to apply two positive couples M_k and show the elastic curve; for this curve a mutual angle of rotation at section k is equal to unity (see Sect. 10.1.1). The positive ordinates of influence line show that extended fibers at the section k are located under of neutral axis of the beam.

For each span we can determine two specific points is called the left and right foci points. They are labeled as F^{left} and F^{right}. Figure 10.19 presents these points for span 3-4 only. Formulas for computation of location F^{L} and F^{R} are presented in Appendix, Section A. Foci Points.

Influence lines for bending moment at the span may be of three different shapes depending on where section is located, i.e., between two foci points, between support and focus, and for section which coincides with focus.

If the section k is located between two foci F^{L} and F^{R}, then ordinates of influence line within corresponding span are positive. If a section n is located between left support and focus F^{L}, then ordinates of influence line within the corresponding span are positive and negative (influence line for M_n). The same conclusion will be done if the section is located between F^{R} and right support. If the section under consideration coincides with F^{R} then bending moment *does not arise* in the section F^{R}, when load P is located in the first and second spans (influence line for $M_{\text{F}}{}^{\text{R}}$). Therefore, for construction of the model of influence line for bending moment at the any section within the span it is necessary first of all, to find location of foci points for given span and then to define which case takes place.

Influence line for shear force Q_k. In this case we need to eliminate the constraint, which corresponds to the shear force at the section k and apply two positive shears Q_k. Elastic curve in the new system due to forces $Q_k = 1$ is a model of influence line for Q_k. Shape of influence line for shear forces at the sections, which are infinitely closed to the support, may be obtained as limiting cases, when the section is located within the span.

It is obvious that the construction of influence lines for statically determinate multispan beams (Ch. 3) using interaction scheme reflects the Müller–Breslau principle.

Summary

The purpose of influence lines and their application for statically determinate structures in case of the fixed and moving load have been discussed in Chaps. 2–4. Of course, this remains also for statically indeterminate structures. However, in case of the statically indeterminate structures, the importance of influence lines and their convenience are sharply increased. Although the construction of influence lines for statically indeterminate structures is not as simple as for statically determinate ones, the consumption of a time is paid off by their advantages. Additional and very important advantage of influence lines is as follows: Influence lines for primary unknown and any factor (reaction, bending moment, the angle of rotation, etc.) for statically indeterminate structure allows calculating not only these unknown and corresponding factor, but also finding a *distribution* of internal forces for any types of fixed loads. It may be done combining the fixed and moving load approaches. It is a principal feature of influence lines for statically indeterminate structures. Ability of an engineer to apply both methods separately and together increases his opportunity of analysis and allows performing in-depth qualities and quantities investigation of structural behavior.

Problems

10.1. Continuous beam is shown in Fig. P10.1. Trace the models of the following influence lines:

(a) Reaction of all supports
(b) Bending moments at all supports and at sections n, and k
(c) Bending moment for sections which coincide with the left and right foci points $(F^{\mathrm{L}}, F^{\mathrm{R}})$
(d) Shear for sections$n$ and k and for sections 1 and 4
(e) Shear for sections which are placed infinitely close left and right to support 2

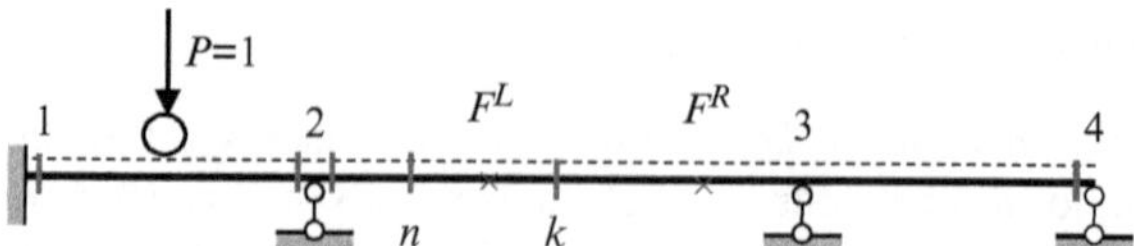

Fig. P10.1

10.2. Design diagram of a structure is shown in Fig. P10.2. Trace the models of the following influence lines:

(a) Reaction of all supports
(b) Bending moments at all supports and at section k
(c) Shear for section k and Shear for sections 1 and 4
(d) Shear for sections which are placed infinitely close to left and right to support 2

For problems (a)–(d) take into account indirect load application.

(e) Will it be changed according to the shape of the influence line models in case of nonuniform continuous beams?
(f) Is it possible to change the sign of influence line as the result of the changing of the stiffness of the structure?

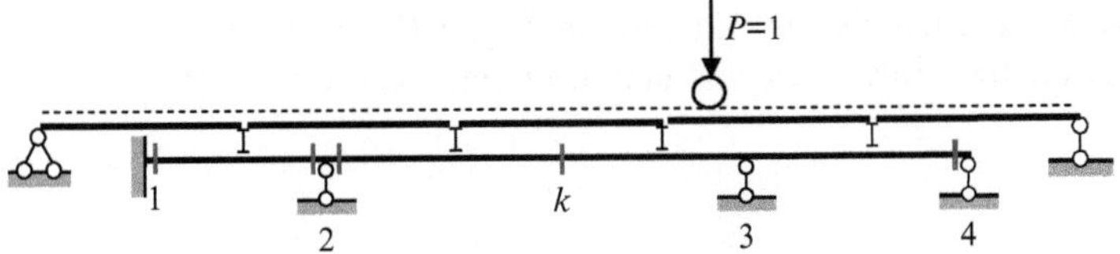

Fig. P10.2

10.3. Uniform clamped–pinned beam is shown in Fig. P10.3.

(a) Construct the influence lines for reactions R_A and R_B, moment at clamped support A, bending moment, and shear at section $k(u = 0.4)$;
(b) Construct the bending moment diagram if force $P = 10\,\mathrm{kN}$ is placed at point $u = 0.6$. Use the above constructed influence lines.

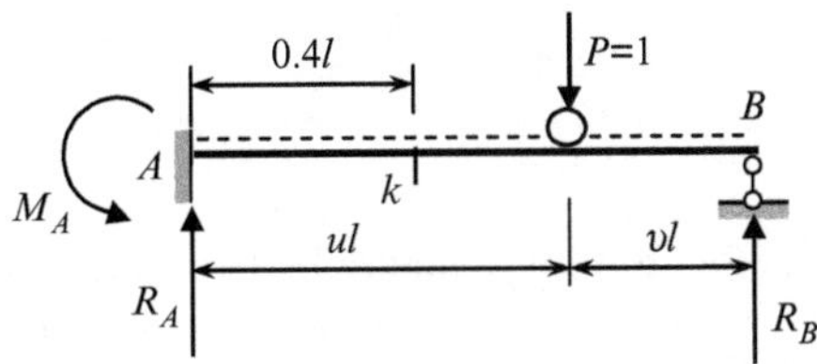

Fig. P10.3

Ans. (b) $R_B = 4.32\,\mathrm{kN}$, $M_A = 1.68l\,\mathrm{kNm}$

10.4. Design diagram of a frame is presented in Fig. P10.4. The relative flexural stiffness are shown in the circle; $a = b = 0.5l$.

1. Construct the influence line for horizontal reaction H at support A and bending moment at section k
2. Construct the bending moment diagram if force $P = 100\,\mathrm{kN}$ is placed at point $x/l = 0.75$. Use the influence line for primary unknown.

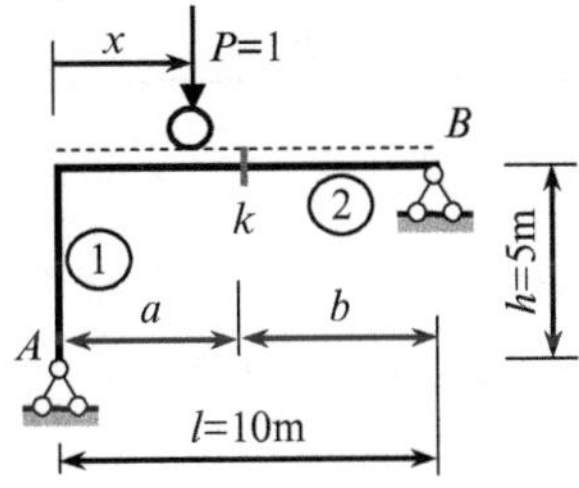

Fig. P10.4

Ans. $H = 11.718\,\mathrm{kN}$; $R_A = 30.859\,\mathrm{kN}$, $M_k = 95.7\,\mathrm{kN\,m}$

10.5. Analyze each design diagram in Fig. P10.5 and choose the most effective method for analytical construction of influence line for bending moment at section k.

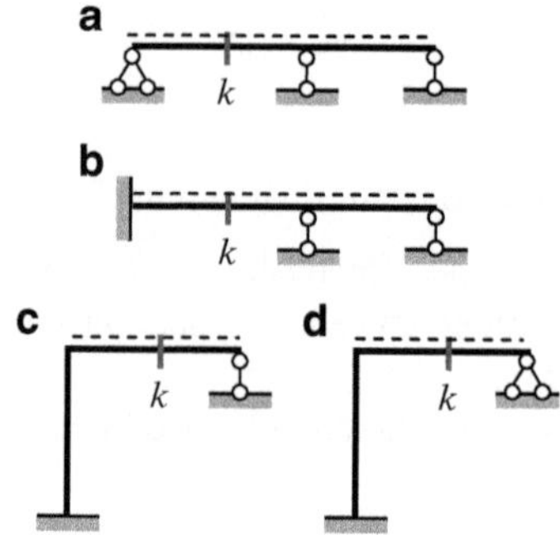

Fig. P10.5

10.6. Design diagram of a uniform semicircular two-hinged arch subjected to traveling load $P = 1$ is presented in Fig. P10.6. Flexural stiffness of the arch is *EI*, axial stiffness of the tie is $E_t A_t$. Derive expression for truss H in terms of the angle φ. Take into account displacement of the arch due to bending deformation of the arch itself and axial deformation of the tie. Analyze limiting case ($E_t A_t = 0$, and $E_t A_t = \infty$).

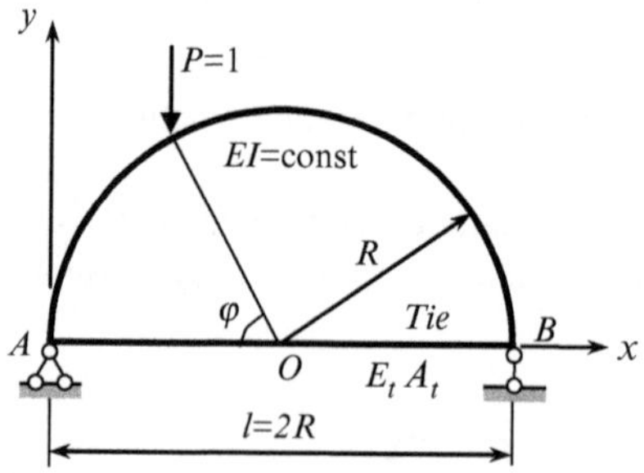

Fig. P10.6

Ans. $H = \dfrac{1}{\pi + 4\dfrac{E}{E_t}\dfrac{I}{A_t R^2}} f(\varphi_P),\; f(\varphi_P) = 2\left(\dfrac{3}{4} - \cos^2\varphi_P + \dfrac{1}{4}\cos 2\varphi_P\right) z$

10.7. Uniform two-span beams with equal spans are presented in Fig. P10.7; the flexural stiffness *EI* is constant. Construct the influence line for angle of rotation at the middle support. Construct the bending moment diagram, if force $P = 100\,\text{kN}$ is located at point 8; use the influence line for primary unknown.

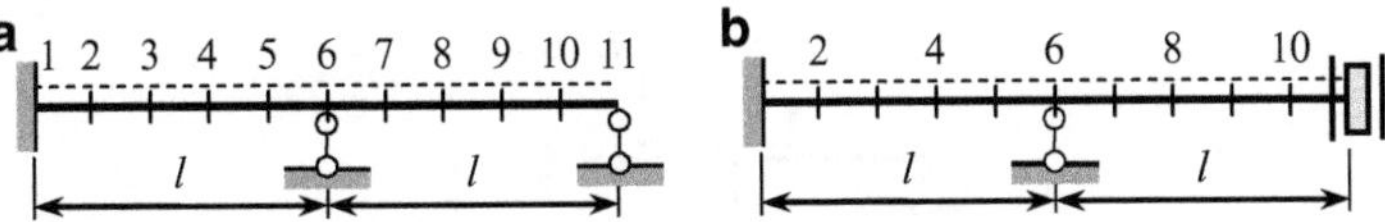

Fig. P10.7

Ans. (a) Ordinate of IL (Z) at point 8 is $0.0274\left(l^2/EI\right)$; $M_1 = 5.48l$, $M_6 = -10.96l$.

10.8. Pinned–pinned–pinned beam is subjected to concentrated load P; $ul = 0.4l$ (Fig. P10.8).

(a) Construct the the influence line for angle of rotation at the middle support; (b) Calculate the bending moment at the support 1; (c) Calculate the vertical displacement at section $k$and show the elastic curve.

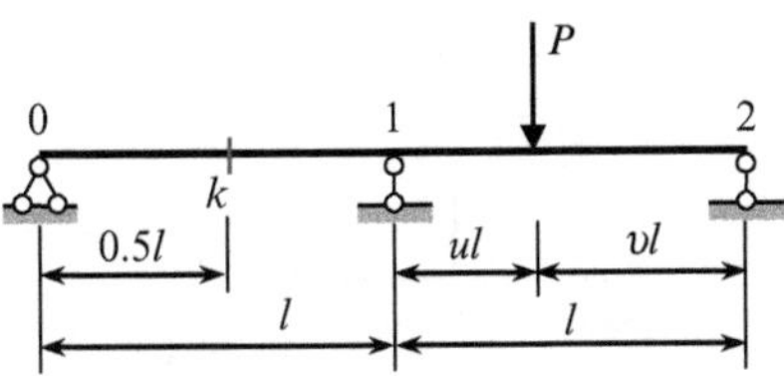

Fig. P10.8

Ans. $M_1 = -0.096Pl$, $f_k = 0.006\left(Pl^3/EI\right)$

10.9. Design diagram of the frame is shown in Fig. P10.9. Bending stiffness for all members is EI.

(a) Construct the influence lines for bending moment and shear at section k. (b) Construct the bending moment diagram if the fixed load $P = 10\,\text{kN}$ is located on distance $0.6l$ from the left support, i.e. in the section 3; (c) construct the bending moment diagram if the fixed load $P = 100\,\text{kN}$ is located on distance $0.6l$ from the rigid joint, i.e. in the section 8. *Hint*: Divide each span by five portions. Joints numeration is shown in Fig. 10.4. For problems (b) and (c) use the influence line for primary unknown.

Ans.
(a) $M_1 = 0.09312l$; $M_k = 0.19296l\ldots$, $Q_1 = -0.2672$; $Q_k^{\text{left}} = -0.5176\ldots$;
(b) $M_3 = 1.5936l$ (kN m); $M_5^{\text{left}} = -1.344l$; $M_5^{\text{right}} = -0.768l$; $M_{10} = 0.384l$;
(c) $M_5^{\text{left}} = -2.88l$; $M_5^{\text{right}} = -5.76l$; $M_8 = 11.904l$; $M_{10} = -16.32l$

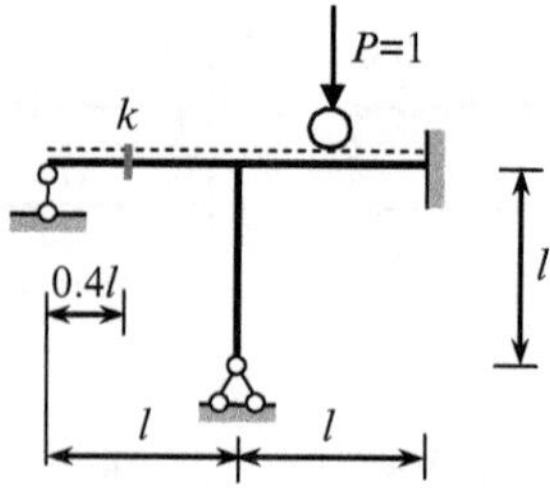

Fig. P10.9

10.10. A frame is subjected to fixed load P (Fig. P10.10). The bending stiffness for each member is EI = const. Construct the influence line for vertical reaction at point A. Use this influence line for constructing the bending moment diagram. Calculate the horizontal displacement at support A and angle of rotation of rigid joint due to fixed load P.

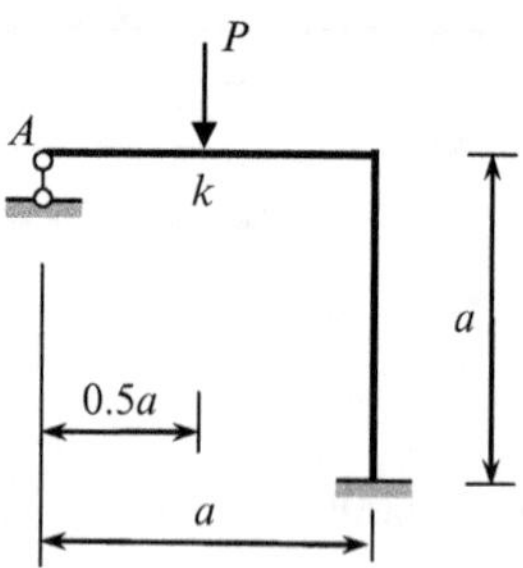

Fig. P10.10

Ans.

$$\text{IL}\,(R_A) = \frac{3}{4}\left[\frac{1}{6}\left(1-\frac{x}{a}\right)^2\left(2+\frac{x}{a}\right)+\left(1-\frac{x}{a}\right)\right]$$

$$M_k = \frac{29}{128}Pa; \quad \Delta_A^{\text{hor}} = \frac{3Pa^3}{128EI}; \quad \theta = \frac{3Pa^2}{64EI}$$

Chapter 11
Matrix Stiffness Method

Matrix stiffness method (MSM) is a modern powerful method of analysis of engineering structures. Its effective and widespread application is associated with availability of modern computers and effective computer programs. The MSM allows performing detail analysis of any sophisticated 2D and 3D engineering structure and takes into account different features of a structure and loading.

The method demands a set of new concepts. They are finite element, global and local coordinate systems, possible displacements of the ends, ancillary diagrams, initial matrices, stiffness matrix of separate element and structure in whole, etc.

This method uses the idea of the displacement method and contains its further development: arbitrary structure should be presented as a set of finite elements and three aspects of any problem – a static, geometrical, and physical – should be presented in matrix form. The MSM does not demand of constructing bending moment diagram caused by unit primary unknowns in the primary system. Instead it is necessary to prepare few initial matrices according to strong algorithms and perform matrix procedures by computer using the standard programs.

This chapter contains the detailed discussion of MSM, all ancillary diagrams and initial matrices are constructed by hand. Mainly such presentation of material allows reader to see the internal logic and the features of the method, to understand physical meaning of each step, and to find the corresponding result in analysis of a structure by displacement method in canonical form.

At the present time, MSM is developed with great detail. The reader can find out in literature the different presentations of MSM. In our book, we will consider this method in simplest form and apply it for analysis of the planar bar structures only. Among them are the statically indeterminate beams, frames, trusses, and combined structures subjected to different external exposures (loads, settlements of supports, change of temperature, influence lines).

11.1 Basic Idea and Concepts

The fundamental concepts of MSM are the following: the finite elements, the possible displacements of the ends and degree of kinematical indeterminacy (degree of freedom), the global and local coordinate systems.

I.A. Karnovsky and O. Lebed, *Advanced Methods of Structural Analysis*,
DOI 10.1007/978-1-4419-1047-9_11, © Springer Science+Business Media, LLC 2010

11.1.1 Finite Elements

Each structure may be subdivided into separate elements of simple geometrical configuration called the finite elements. This step has no theoretical justification. For each finite element, the stress–strain analysis is preliminarily investigated in detail; the result of such analysis is presented in existing handbooks. For presentation of the given structure as a set of the finite elements, the features of the structure as well as required accuracy of analysis should be taken into account. Engineering experience is an important factor for choosing the type and number of the finite elements.

In case of a truss, the separate members of the truss may be adopted as finite elements. Therefore, the discrete model of the truss in terms of finite elements coincides with design diagram of the truss.

In case of the frame with uniform members, separate members of the frame also may be adopted as the finite elements (Fig. 11.1a). If the frame contains the member with variable cross-section, then this member may be divided into several portions with constant stiffness along each element (Fig. 11.1b).

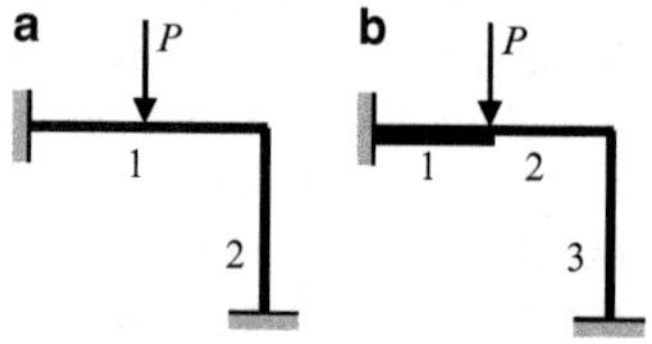

Fig. 11.1 Frames and their presentation by the set of finite elements

The uniform beams in Tables A.3–A.8, subjected to displacements of supports may be considered as the simplest finite elements. The finite elements can be one, two, or three-dimensional. This chapter deals with planar bar structures only, so the finite elements are straight thin bars with three types of constraints at the ends. They are hinged-hinged (truss member), fixed-pinned, and fixed-fixed (frame member).

General idea of MSM. At the end points of each finite element, the some displacements and interaction forces arise. For structure in whole these forces are internal, while for each finite element these forces should be considered as the external loads. For all finite elements, we can write three groups of equations. They are the (1) equilibrium equations, (2) physical equations, and (3) geometrical ones. Equilibrium equations take into account external forces for each finite element. Physical equations relate forces and displacements at end points of each element. Geometrical equations describe continuity conditions between ends of the elements. Solving of these equations allows determining displacements and forces at end points of each element.

11.1.2 Global and Local Coordinate Systems

The local coordinate system is referred to as the specified element, while global system is related to the whole structure. To understand these concepts, let us consider a truss, subjected to force P (Fig. 11.2).

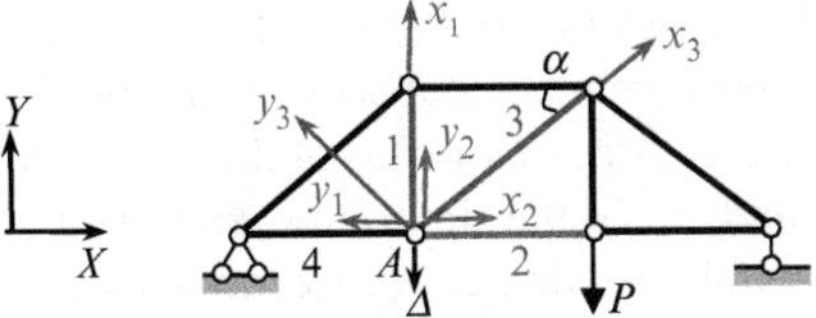

Fig. 11.2 Local and global coordinate systems

The members 1, 2, 3, 4 met at the joint A. Internal forces N_i of the elements 1–4 does not coincide with given external load P. Axial deformation Δ_i of each element does not coincide with vertical displacement Δ of joint A. Therefore, we need to distinguish between the internal and external forces, deflection of separate members and displacements of joints of the structure. For this purpose, we introduce two coordinate systems. *Each element* of the structure, corresponding *internal* force and *deformation* may be referred to the *local* coordinate system x-y. The origin of this system coincides with initial point of this element and one of the axes is directed along the element itself as shown for members 1–3; for member 4, the local coordinates are not shown. The *structure in whole,* the *external* force, and *displacements* of the joints are referenced to *global* coordinates X-Y.

The difference between the local displacements Δ_i of each element and the global vertical displacement Δ of joint A can be seen from the follows relationships:

$$\begin{aligned} x_1 &= -\Delta, & y_1 &= 0; \\ x_2 &= 0, & y_2 &= -\Delta; \\ x_3 &= -\Delta \sin\alpha, & y_3 &= -\Delta \cos\alpha, \ldots \end{aligned}$$

11.1.3 Displacements of Joints and Degrees of Freedom

Once a structure is presented as a set of finite elements, we need to identify the possible displacements of the ends of each member. In fact these displacements present unknowns of the displacement method (angular displacements of the rigid joints and their independent linear displacements). These displacements are called the *possible* angular and linear displacements of the joints. The term *possible* means that in given structure such displacement is possible, but not necessarily utilized. For example, in case of two span continuous beam, a section at the middle support generally rotates; however, if the beam and its loading are symmetrical then this angle of rotation is zero. In summary, we can see that any possible displacement of the joint is a displacement in *global* coordinates.

Degree of kinematical indeterminacy n of a structure is determined by the formula $n = n_r + n_d$, where n_r is a number of unknown angles of rotation of the rigid joints of a structure and n_d is a number of independent linear displacements of joints. Parameter n defines the degrees of freedom of a structure.

The beam in Fig. 11.3a has one unknown of the displacement method, i.e., the degree of kinematical indeterminacy equals one. So the degree of freedom $n = 1$. The frame in Fig. 11.3b has three unknowns of the displacement method (two angular displacements Z_1 and Z_2 of the rigid joints and one linear displacement Z_3 of the cross bar). So the number of degrees of freedom $n = 3$.

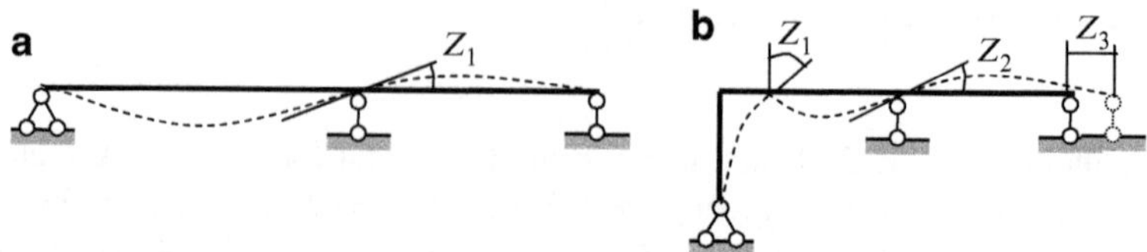

Fig. 11.3 Elastic curves and concept of degrees of freedom

Thus the term "degrees of freedom" implies possibility of angular, and linear displacements of the joints are caused by *deformation* of the structure. It can be seen that for kinematical analysis of a structure (Chap. 1) the term "degrees of freedom" had other meaning; in this case this concept implies independent displacements of the system which contains the *absolutely rigid* discs.

11.2 Ancillary Diagrams

To present a structure in such form which can be accepted by a modern computer, the entire design diagram should be expanded by three ancillary diagrams. They are joint-load (***J-L***), displacement-load (***Z-P***), and internal forces-deformation (***S-e***) diagrams.

11.2.1 Joint-Load (J-L) Diagram

This diagram presents the transformation of the arbitrary load into the *equivalent joint loads*. For construction of joint-load diagram, the followings are necessary:

1. Identify the possible angular and linear displacements of the joints
2. Construct the bending moment diagram M_P^0 in the primary system of the displacement method (first state) due to external exposures (loads, settlements of supports, temperature change)
3. Present the moments and forces as the *joint load* in direction of the possible displacements (second state).

Continuous beam in Fig. 11.4 has one unknown *angular* displacement at support 1. We need to convert the given load to the equivalent *moment* at the support 1. The first state is the bending moment diagram M_P^0 in primary system. The fixed-end

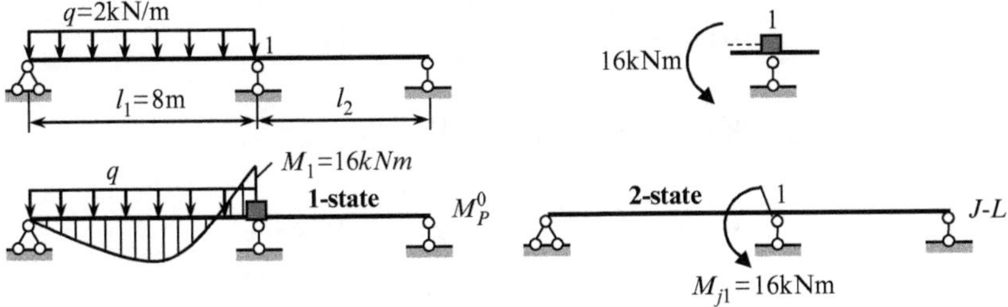

Fig. 11.4 Transformation of external load to the equivalent joint load

moment (FEM) equals to $M_1 = \frac{ql_1^2}{8} = 16$ (kNm). We can show the joint 1 with joint moment 16 kNm counterclockwise according a location of extended fibers. This moment should be transported on the joint-load diagram counterclockwise (second state).

Figure 11.5a presents design diagram of the frame; $P_1 = 10$ kN, $P_2 = 16$ kN, $q = 3$ kN/m. Degree of kinematical indeterminacy equals 4, where $n_r = 3$ and $n_d = 1$. The primary system is shown in Fig. 11.5b; all introduced constraints are labeled as 1–4 and the specified sections as 5–8. Also this figure contains the bending moment diagram in primary system M_P^0 (first state). Bending moments which act on joints 1, 2, 3 are shown in Fig. 11.5c.

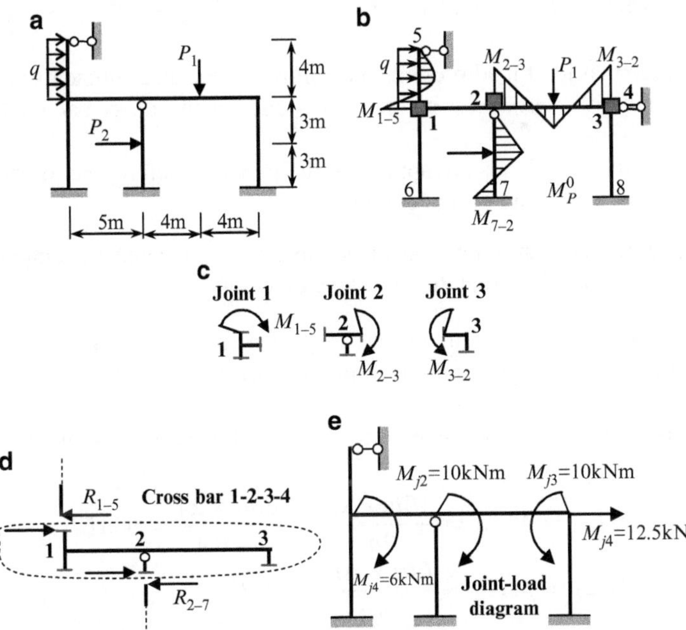

Fig. 11.5 (**a–c**) Design diagram of the frame and computation of the fixed-end moments; (**d**) Computation of equivalent joint load P_{j4}; (**e**) Joint-load diagram (state 2)

$$M_{1-5} = \frac{ql_{1-5}^2}{8} = \frac{3 \times 4^2}{8} = 6\,\text{kNm}$$

$$M_{7-2} = \frac{3}{16} P_2 l_{2-7} = 18\,\text{kNm}$$

$$M_{2-3} = \frac{P_1 l_{2-3}}{8} = \frac{10 \times 8}{8} = 10\,\text{kNm}$$

$$M_{3-2} = M_{2-3} = 10\,\text{kNm}$$

Figure 11.5d shows the horizontal forces which arise in the vertical loaded members 1–5 and 2–7. Both of these forces are transmitted on the cross bar, so final horizontal joint load equals $P_{j4} = R_{1-5} + R_{2-7} = 7.5 + 2.5 = 12.5\,\text{kN}$. This force may be applied at any joint of the frame (1, 2, or 3).

$$R_{1-5} = \frac{5ql_{1-5}}{8} = \frac{5 \times 3 \times 4}{8} = 7.5\,\text{kN};$$

$$R_{2-7} = \frac{5P_2}{16} = \frac{5 \times 16}{16} = 5\,\text{kN}.$$

Thus the entire load may be presented as equivalent moments M_{j1}, M_{j2}, M_{j3} at joints 1–3 and force P_{j4} in horizontal direction; subscript j means that entire loads are transformed to joint load. The final ***J-L*** diagram is shown in Fig. 11.5e

$$M_{j1} = 6\,\text{kNm};\ M_{j2} = 10\,\text{kNm};\ M_{j3} = 10\,\text{kNm};\ P_{j4} = 7.5 + 5 = 12.5\,\text{kN}$$

Note again that joint load presents the *equivalent* bending moments and forces, which are merely *transported* on the joints and on the cross bar on the *same* direction.

In case of truss, all loads are applied at the joints; therefore, the joint-load diagram coincides with entire design diagram.

Example 11.1. The continuous beam is subjected to change of temperature as shown in Fig. 11.6a. Construct the joint-load diagram.

Solution. The primary system of the displacement method and bending moment diagram caused by given temperature exposure is shown in Fig. 11.6b (first state).

The bending moments for pinned-fixed beam and fixed-fixed beam are

$$M_{1-0} = M_{3-4} = \frac{3EI\alpha \cdot \Delta t}{2h} = \frac{3EI\alpha \times 24}{2 \times 0.4} = 90\alpha EI;$$

$$M_{1-2} = M_{3-2} = \frac{EI\alpha \times \Delta t}{h} = 60\alpha EI.$$

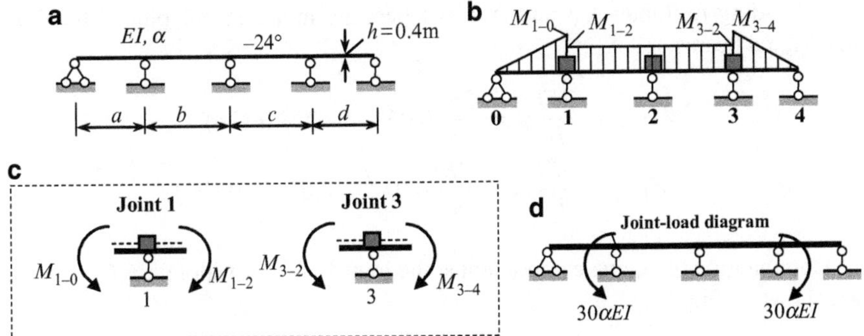

Fig. 11.6 (**a, b**) Design diagram of the beam and computation of the fixed-end moments; (**c, d**) Joint-load diagram in case of temperature change

Figure 11.6c presents the bending moments vicinity the joints 1 and 3. The joint moments equal

$$M_{j1} = M_{1-0} - M_{1-2} = 90\alpha EI - 60\alpha EI = 30\alpha EI \text{ (kNm) counterclockwise;}$$
$$M_{j2} = 0,\ M_{j3} = 90\alpha EI - 60\alpha EI = 30\alpha EI \text{ (kNm) clockwise.}$$

The final joint-load diagram is shown in Fig. 11.6d.

Example 11.2. The continuous beam is subjected to the angular displacement φ of fixed support 0 and vertical displacement Δ of rolled support 2 (Fig. 11.7a). Construct the joint-load diagram.

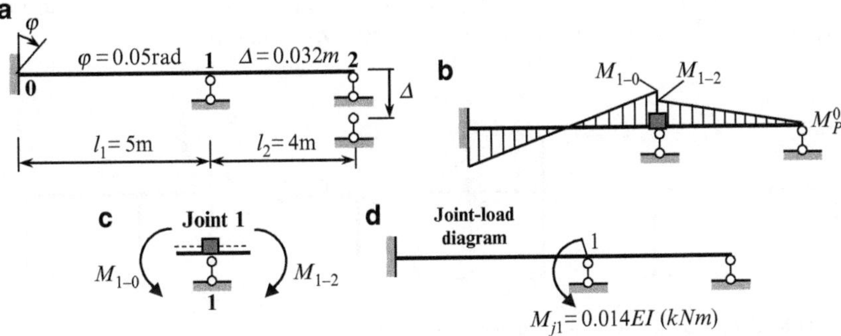

Fig. 11.7 (**a, b**) Design diagram of the beam and computation of the fixed-end moments; (**c, d**) Joint-load diagram in case of settlements of supports

Solution. The primary system of the displacement method and bending moment diagram caused by given settlements of supports is shown in Fig. 11.7b (first state).

According to Tables A.3 and A.4, the bending moments for pinned-fixed and fixed-fixed beams are

$$M_{1-0} = \frac{2EI}{l_1}\varphi = \frac{2EI}{5}0.05 = 0.02EI\,(\text{kNm}),$$
$$M_{1-2} = \frac{3EI}{l_2^2}\Delta = \frac{3EI}{4^2}0.032 = 0.006EI\,(\text{kNm}).$$

Computation of equivalent moment at the joint 1 and corresponding ***J-L*** diagram are shown in Fig. 11.7c, d.

11.2.2 Displacement-Load (Z-P) Diagram

This diagram shows a numeration of angular and independent linear displacements Z at the joints and their positive direction. Also this diagram contains *type* load, which corresponds to displacements Z and their positive directions.

Sign rule. Assume that the positive angular displacement Z occurs clockwise; the positive horizontal linear displacement occurs from left to right, and vertical linear displacement occurs upward. Direction of the positive load coincides with positive direction of the displacement.

Design diagram frame and corresponding primary system of the displacement method are shown in Fig. 11.8a, b. Introduced constraints 1–3 prevent angular displacements Z_1–Z_3; corresponding loads are moments M_1–M_3. Constraint 4 prevents linear displacement Z_4, so corresponding load is P_4.; these loads and all displacements are shown in positive directions. The ***Z-P*** diagram contains only sceleton of the entire scheme of the structure, the type of end displacements, its positive direction and corresponding loads (Fig. 11.8c). Geometrical parameters, stiffnesses, and loading are are not shown on the ***Z-P*** diagram.

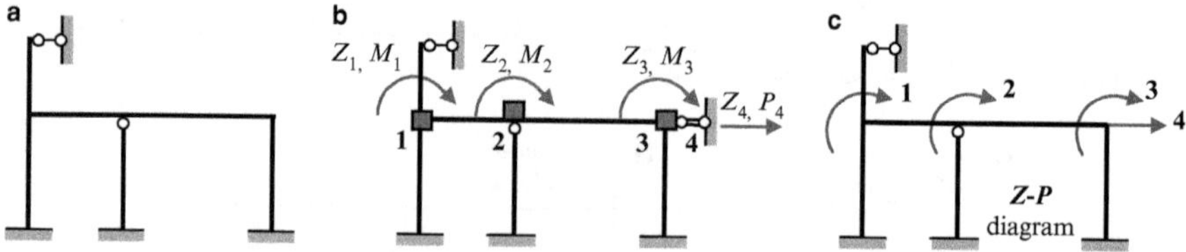

Fig. 11.8 Displacement-load (***Z-P***) diagram for frame

Let us show the construction of ***Z-P*** diagram for truss in Fig. 11.9a. Support A prevents two displacements; therefore, this joint has not possible displacement and corresponding joint loads. Joint C prevents horizontal displacement, so possible displacement of this joint and corresponding load is directed vertically (labeled 1). The possible displacement of joint B and corresponding load are directed horizontally

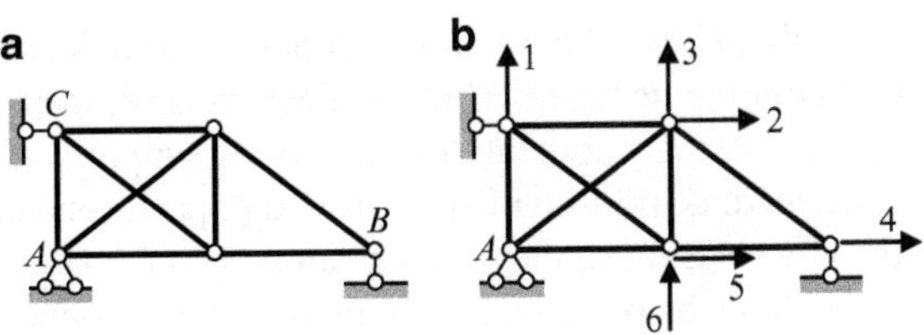

Fig. 11.9 ***Z-P*** diagram for truss

(4). All other joints of frame have two possible displacements (vertical and horizontal) and two corresponding joint loads.

It is obvious that ***Z-P*** diagram (Fig. 11.9b) shows the positive possible displacements and corresponding external loads in global coordinates.

11.2.3 Internal Forces-Deformation (S-e) Diagram

This diagram shows a numeration and positive directions of all unknown internal forces S, which corresponds to deformation e. Figure 11.10a shows three bar elements from any truss. Since all connections are hinged, then for all members only axial deformations e in local coordinates is possible; corresponding internal forces are axial forces. The concept of ***S-e*** diagram for truss is shown in Fig. 11.10b.

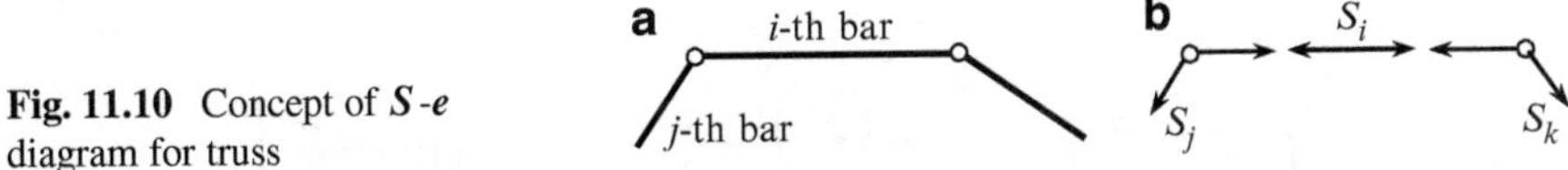

Fig. 11.10 Concept of S-e diagram for truss

Figure 11.11a presents the design diagram of the truss (loading is not shown) and numeration 1–8 of the members. The corresponding ***S-e*** diagram is shown in Fig. 11.11b. Thus, we show the arrows related to the intermediate part of the truss member. This diagram contains the positive axial deformation e_i and force S_i in i-th member; the force S_i is constant along the i-th element.

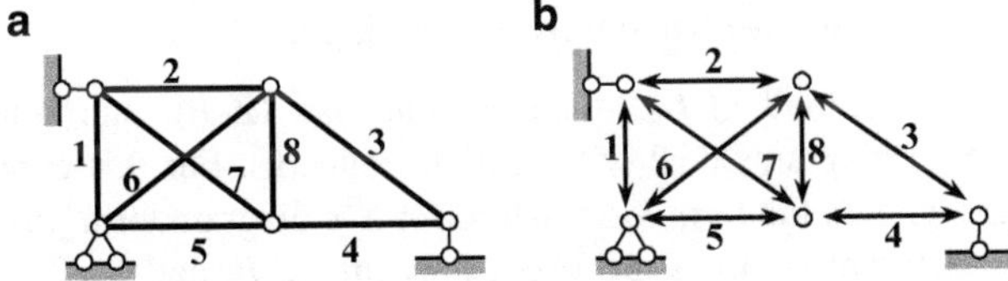

Fig. 11.11 ***S-e*** diagram for truss

For beams we will denote the *sections* with unknown internal forces. Figure 11.12a shows a continuous beam. The ***S-e*** diagram is presented in Fig. 11.12b; this diagram contains internal moments in vicinity of supports. The positive moment rotates the intermediate portion of span around opposite end clockwise. These moments are shown by solid lines; they are transmitted on the adjacent portion with opposite directions (shown by dotted arrows). Figure 11.12c contains a diagram ***S-e*** of positive internal forces.

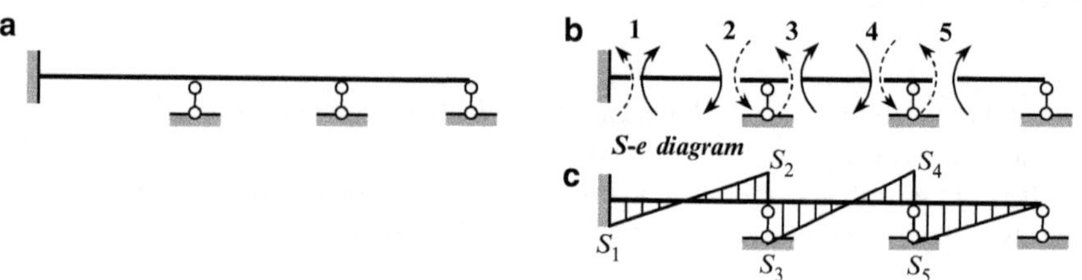

Fig. 11.12 Concept of ***S-e*** diagram for continuous beam

For frames, as for beams, we will denote the *sections* with unknown internal forces. The required internal forces are bending moments at the ends of each member. Figure 11.13a shows the frame; the joints are labeled as i, j, k. The ***S-e*** diagram is presented in Fig. 11.13b; this diagram contains internal moments infinitely close to the fixed support and rigid joint. The positive moment rotate the *intermediate* portion of member around opposite end clockwise. These moments are shown by solid lines; they transmitted on the adjacent portion with opposite directions (shown by dotted arrows). The diagram ***S-e*** of positive internal forces is shown in Fig. 11.13c.

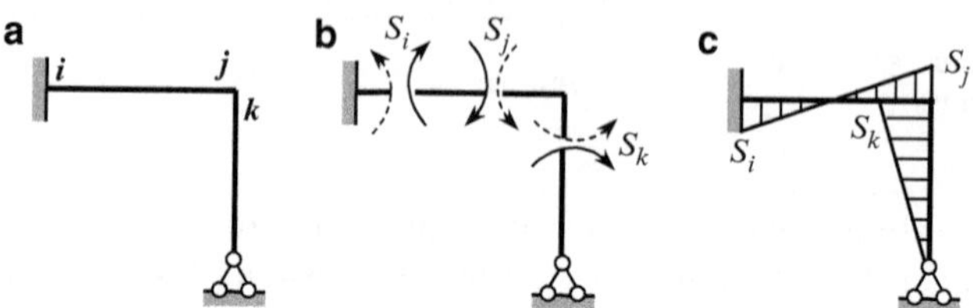

Fig. 11.13 ***S-e*** diagram for frame

We can see that diagram (b) shows the type of internal loads, while the diagram (c) shows positive directions of unknown internal forces. Later both diagram (b) and diagram (c) are referred together simply as ***S-e*** diagram.

Summary The joint-load (***J-L***), displacement-load (***Z-P***), and internal forces-deformation (***S-e***) diagrams deal with similar concepts. For this concept, the ***J-L*** and ***Z-P*** diagrams uses the term *load*, while for ***S-e*** diagram term *forces*. For these concepts, the different symbols are used; they are ***L***, ***P***, and ***S***. The symbol ***L*** is used for presentation of the given loading by equivalent joint loads. The symbol ***P*** is used for presentation of the *type* of the load, which corresponds to possible displacements Z. The symbol ***S*** is used for presentation of the *type* of the unknown internal forces S.

11.3 Initial Matrices

The first group of initial matrices describes the loading of structure and unknown internal forces.

11.3.1 Vector of External Joint Loads

The arbitrary exposure on a structure (fixed or moving load, settlement of supports, change of temperature) should be presented in mathematics form. For this purpose it serves a vector **P** of the external joint loads. This vector may be constructed on the basis of the joint-load (***J-L***) and the displacement-load (***Z-P***) diagrams.

The truss is loaded as shown in Fig. 11.14a; the ***Z-P*** diagram is shown in Fig. 11.14b.

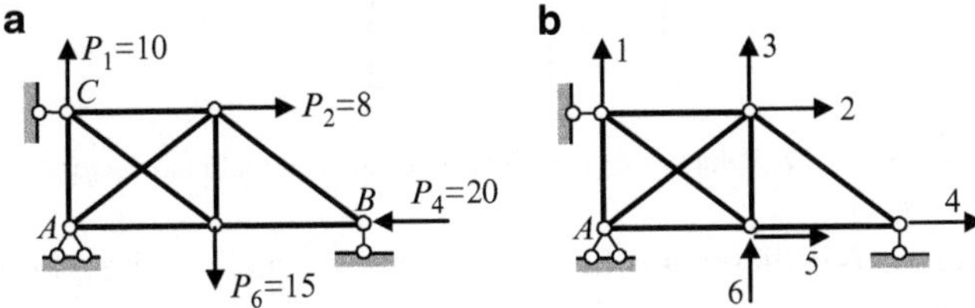

Fig. 11.14 Design diagram and ***Z-P*** diagram for truss

Since the truss has 6 possible displacements, then vector of external load will contains 6 entries, i.e.,

$$\vec{\mathbf{P}} = \begin{bmatrix} P_1 \\ \cdots \\ P_6 \end{bmatrix}$$

For convenience, this vector we will present in transposed form as

$$\vec{\mathbf{P}} = \lfloor P_1 \ P_2 \ \cdots \ P_6 \rfloor^{\mathrm{T}};$$

the symbol ^{T} is reserved for the matrix transpose procedure. Considering Fig. 11.14a, b we can compile the vector $\vec{\mathbf{P}}$ of joint loads:

$$\vec{\mathbf{P}} = \lfloor 10 \quad 8 \quad 0 \quad -20 \quad 0 \quad -15 \rfloor^{\mathrm{T}}$$

The first entry 10 means that in the first direction (Fig. 11.14b) there acts the force $P_1 = 10$. The negative sign at the fourth entry indicates that external force $P_4 = 20$ is directed opposite the positive fourth possible displacement. Zeros for third and fifth entries show that in directions 3 and 5 of ***Z-P*** diagram the external forces are absent.

Now we compile the vector of external forces for frame shown in Fig. 11.15a. For this frame, the joint-load diagram (Fig. 11.5e) is repeated in Fig. 11.15b. The ***Z-P*** diagram is shown in Fig. 11.15c.

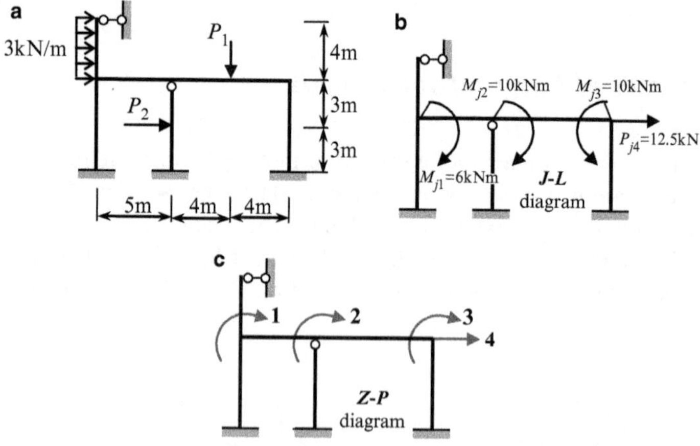

Fig. 11.15 Joint-load and ***Z-P*** diagrams. Formulation of the external load vector

Comparing the ***J-L*** diagram with ***Z-P*** diagram we can form the following vector of external loads

$$\vec{\mathbf{P}} = \lfloor\, 6 \quad 10 \quad -10 \quad 12.5 \,\rfloor^{\mathrm{T}}$$

11.3.2 Vector of Internal Unknown Forces

The required internal forces may be presented in the ordered mathematics form. For this purpose serves the matrix-vector of unknown internal forces **S**. The entries of this vector are the axial forces for end-hinged members and the bending moments at the ends of the bending members. Generally, the vector $\vec{S}$ may be presented as a sum of the vectors of internal forces at the first and second states, i.e. $\vec{S} = \vec{S}_1 + \vec{S}_2$.

Let us consider the frame in Fig. 11.5. This scheme without external load, numeration of sections, and positive directions of the end moments for each element of the frame are shown in Fig. 11.16a–c. Pay attention that the sections where the moments are zeros have been eliminated from consideration. They are point at the upper rolled support (5) and at hinge H that belongs to the second column.

The bending moment diagram in primary system M_P^0 (Fig. 11.5b) and ***S-e*** diagram allow constructing the vector of the moment at the indicated sections in primary system (state 1) due to given load. This vector becomes

$$\vec{\mathbf{S}}_1 = \lfloor\, 0 \;\; 0 \;\; -6 \;\; 0 \;\; 0 \;\; -18 \;\; -10 \;\; 10 \;\; 0 \;\; 0 \,\rfloor^{\mathrm{T}}$$

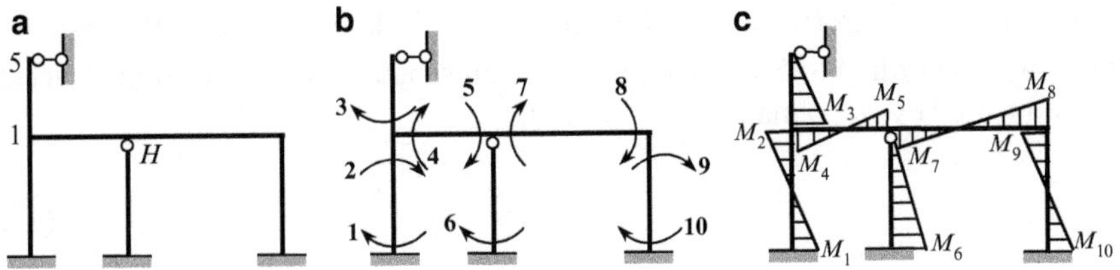

Fig. 11.16 Formulation of the internal force vector

The signs of the moments are established on the basis of the M_P^0 and ***S-e*** diagrams. For example, since a moment $M_{1-5} = 6\,\text{kNm}$ on the bending moment diagram M_P^0 is plotted left (Fig. 11.5b), and the positive moment M_3 at the same section on the ***S-e*** diagram is plotted right (Fig. 11.16c), then third entry of the **S** vector have a negative sign.

The vector of required internal moments at the specified sections 1, …, 10 is

$$\vec{S} = \lfloor M_1 \; M_2 \cdots M_9 \; M_{10} \rfloor^{\text{T}}$$

This vector will present a result of analysis of the frame by computer. If that occurs, for example, the first entry (bending moment M_1) will be positive, then according to ***S-e*** diagram (Fig. 11.16c), the extended fibers at the section 1(bottom of the left column) will be located at the right. Therefore, the internal moment M_1 should acts as shown in Fig. 11.16b, i.e., clockwise. Computation of the vector of internal forces in the second state will be discussed later.

Summary The following combination of two diagrams leads to the initial vectors:

1. Joint-load (***J-L***) diagram + (***Z-P***) diagram → Vector of external joint loads $\vec{\mathbf{P}}$.
2. M_P^0 diagram + (***S-e***) diagram → Vector of internal forces $\vec{\mathbf{S}}_1$ in the first state.

11.4 Resolving Equations

MSM considers three sides of problem: they are static, geometrical, and physical.

On their basis the second group of initial matrices may be constructed. They are static matrix, deformation matrix, and stiffness matrix in local coordinates. These matrices describe the different features of structure: configuration of structure, its geometry and supports, stiffness of each element, and the order of theirs connection.

11.4.1 Static Equations and Static Matrix

Assume that structure is m-times kinematically indeterminate. It means that structure has m external joint loads. These loads, according to ***Z-P*** diagram, may be presented as vector $\vec{\mathbf{P}}$. Let the structure has n unknown internal forces; these unknowns according to ***S-e*** diagram may be presented as vector $\vec{\mathbf{S}}$. Both of vectors are connected by static matrix **A** by formula

$$\vec{\mathbf{P}} = \mathbf{A}\vec{\mathbf{S}}. \tag{11.1}$$

This equation is called the static matrix equation. The number of columns n of the matrix **A** equals to the number of the unknown internal forces; the number of rows m of the static matrix **A** equals to the number of the possible displacements. If $m > n$, then a structure is geometrically changeable; if $m = n$, then a structure is statically determinate; if $m < n$, then a structure is statically indeterminate. In fact, the matrix equation (11.1) describes the structure, its supports, type of joints, and order of the elements connections.

The static matrix $\mathbf{A}_{(m \times n)}$ may be constructed on the basis of a set of equilibrium conditions for specific parts of a structure in ordered form. Equilibrium equations for frames must be constructed for each joint which have the angular displacement and for part of the frame which contains the joints with linear displacements. In case of possible angular displacement, the equation $\sum M = 0$ should be used; if a displacement is a linear one, then uses equation $\sum X = 0$.

In case of trusses the equilibrium of each joint in form

$$\sum X = 0 \text{ and/or } \sum Y = 0$$

should be considered.

Each possible load (each component of the vector **P**) must be presented in terms of all unknowns (all components of the vector $\vec{\mathbf{S}}$). Left part of each equation of equilibrium should contains only a possible load, while the right part unknown internal forces. The type of possible load corresponds to the type of possible displacement according to diagram ***Z-P***. If a possible displacement is the angle of rotation then corresponding load is a moment. If possible displacement is linear one, then a corresponding load is force.

Figure 11.17a presents the continuous beam. Unknown angular displacements **Z** of intermediate supports A and B and corresponding moments $\boldsymbol{P}_1$ and $\boldsymbol{P}_2$ are labeled on the ***Z-P*** diagram by 1 and 2 (Fig. 11.17b). Positive unknown internal moments M_i (i = 1–5) at the ends of each portions are labeled on the ***S-e*** diagram by 1–5 (Fig. 11.17c).

Equilibrium equations for intermediate supports can be rewritten as follows

$$P_1 = 0 \cdot M_1 + 1 \cdot M_2 + 1 \cdot M_3 + 0 \cdot M_4 + 0 \cdot M_5,$$
$$P_2 = 0 \cdot M_1 + 0 \cdot M_2 + 0 \cdot M_3 + 1 \cdot M_4 + 1 \cdot M_5.$$

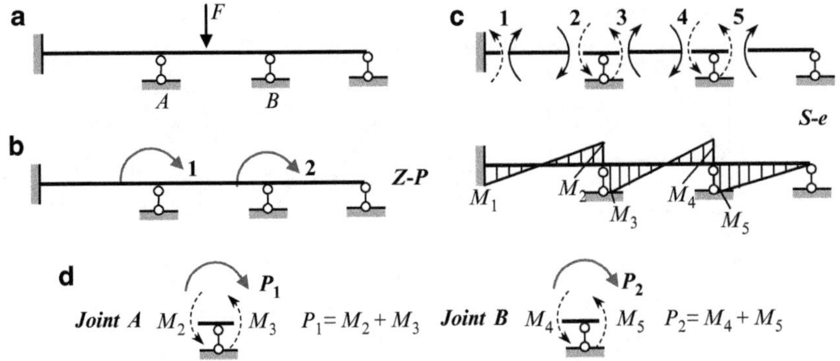

Fig. 11.17 (**a–c**) Continuous beam and corresponding ***Z-P*** and ***S-e*** diagrams; (**d**) Construction of static matrix **A** for beam

In matrix form, these equations can be rewritten as follows

$$\begin{bmatrix} P_1 \\ P_2 \end{bmatrix} = \begin{bmatrix} 0 & 1 & 1 & 0 & 0 \\ 0 & 0 & 0 & 1 & 1 \end{bmatrix} \cdot \begin{bmatrix} M_1 \\ M_2 \\ M_3 \\ M_4 \\ M_5 \end{bmatrix}, \text{ where static matrix } \mathbf{A}_{(2\times 5)} = \begin{bmatrix} 0 & 1 & 1 & 0 & 0 \\ 0 & 0 & 0 & 1 & 1 \end{bmatrix}$$

Each row of matrix **A** presents the corresponding equilibrium equation; the entries of the matrix are coefficients at M_i. Note again that equilibrium equations are formulated for *possible loads P*, which corresponds *to possible displacements, but not for given load F*.

The truss in Fig. 11.18a contains four *possible* linear displacements and corresponding external loads ***P***, labeled as **1–4** (Fig. 11.18b). Support A has no linear displacements; therefore, ***Z-P*** diagram does not contain the vectors of displacement at support A. The possible displacements **1** and **4** describe the rolled supports B and D as well as their orientation.

Equilibrium equations should be formulated for each joint with *possible* load

$$\begin{aligned}
&\text{Joint B} \quad \sum Y = 0: \quad && P_1 = S_1 \\
&\text{Joint C} \quad \sum X = 0: \quad && P_2 = S_2 - S_3 \cos\alpha + S_5 \cos\alpha \\
& && P_2 = S_2 - 0.6S_3 + 0.6S_5 \\
& \qquad\qquad \sum Y = 0: \quad && P_3 = S_3 \sin\alpha + S_5 \sin\alpha \\
& && P_3 = 0.8S_3 + 0.8S_5 \\
&\text{Joint D} \quad \sum X = 0: \quad && P_4 = S_4 + S_3 \cos\alpha \\
& && P_4 = 0.6S_3 + S_4
\end{aligned}$$

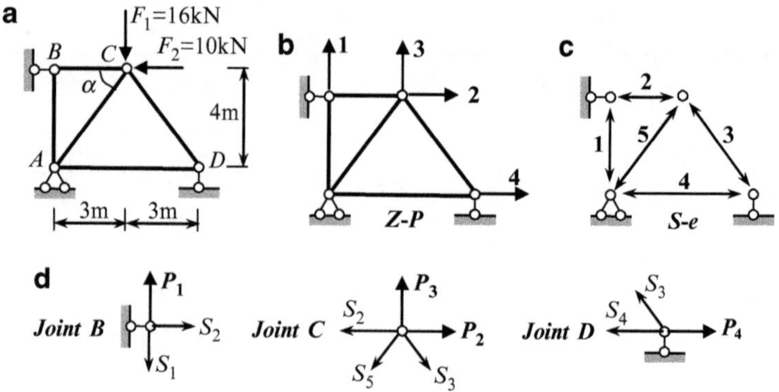

Fig. 11.18 (**a–c**) Design diagram of truss and corresponding ***Z-P*** and ***S-e*** diagrams; (**d**) Construction of static matrix **A** for truss

These equations in the ordered form can be rewritten as follows

$$\begin{aligned} P_1 &= 1 \cdot S_1 + 0 \cdot S_2 + 0 \cdot S_3 + 0 \cdot S_4 + 0 \cdot S_5, \\ P_2 &= 0 \cdot S_1 + 1 \cdot S_2 - 0.6 \cdot S_3 + 0 \cdot S_4 + 0.6 \cdot S_5, \\ P_3 &= 0 \cdot S_1 + 0 \cdot S_2 + 0.8 \cdot S_3 + 0 \cdot S_4 + 0.8 \cdot S_5, \\ P_2 &= 0 \cdot S_1 + 0 \cdot S_2 + 0.6 \cdot S_3 + 1 \cdot S_4 + 0 \cdot S_5. \end{aligned}$$

These equations allows presenting the vector of possible joint forces $\vec{\mathbf{P}}_{(4\times1)} = \lfloor P_1\ P_2\ P_3\ P_4 \rfloor^{\mathrm{T}}$ in terms of the vector of unknown internal forces $\overline{\mathbf{S}}_{(5\times1)} = \lfloor S_1\ S_2\ S_3\ S_4\ S_5 \rfloor^{\mathrm{T}}$ using the static matrix $\mathbf{A}_{(4\times5)}$

$$\begin{bmatrix} P_1 \\ P_2 \\ P_3 \\ P_4 \end{bmatrix} = \underbrace{\begin{bmatrix} 1 & 0 & 0 & 0 & 0 \\ 0 & 1 & -0.6 & 0 & 0.6 \\ 0 & 0 & 0.8 & 0 & 0.8 \\ 0 & 0 & 0.6 & 1 & 0 \end{bmatrix}}_{\mathbf{A}_{(4\times5)}} \cdot \begin{bmatrix} S_1 \\ S_2 \\ S_3 \\ S_4 \\ S_5 \end{bmatrix}$$

Each row of this matrix presents the corresponding equilibrium equation; the entries of the matrix are coefficient at S_i. We can see that static matrix describes the structure. First, this structure is statically indeterminate, since the number of rows $m = 4$ while the number of columns $n = 5$. Moreover, the matrix **A** describes the ways of connections of different members. For example, second row shows that members 2, 3, and 5 are connected together, while the last row shows that members 3 and 4 are connected. At last, this matrix describes the inclination of different members. Therefore, the static matrix **A** is very informative matrix.

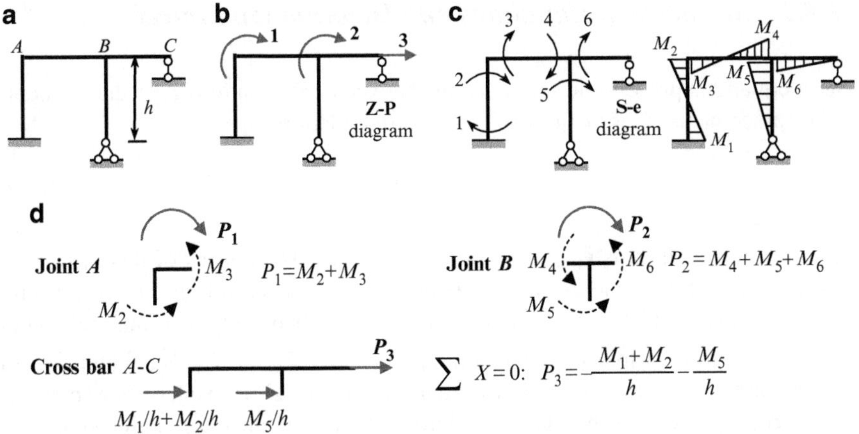

Fig. 11.19 (**a–c**) Design diagram of the frame and ancillary diagrams; (**d**) Construction of static matrix for frame

Example 11.3. The frame is shown in Fig. 11.19. Construct the static matrix.

Solution. This structure has two angular displacements of the rigid joints and one linear displacement. The arrows 1 and 2 show possible angular displacements and corresponding possible external loads (moments, kNm); the arrow 3 shows possible linear displacements and corresponding possible external load (force, kN). The vector of possible joint loads is $\vec{\mathbf{P}} = \lfloor P_1\ P_2\ P_3 \rfloor^{\mathrm{T}}$. The vector of unknown internal moments is $\vec{\mathbf{S}} = \lfloor M_1\ M_2\ \cdots\ M_6 \rfloor^{\mathrm{T}}$.

Equilibrium equations for rigid joints are shown in Fig. 11.19d.

In matrix form the static equations can be presented as follows

$$\begin{bmatrix} P_1 \\ P_2 \\ P_3 \end{bmatrix} = \underbrace{\begin{bmatrix} 0 & 1 & 1 & 0 & 0 & 0 \\ 0 & 0 & 0 & 1 & 1 & 1 \\ -1/h & -1/h & 0 & 0 & -1/h & 0 \end{bmatrix}}_{\mathbf{A}(3\times 6)} \cdot \begin{bmatrix} M_1 \\ M_2 \\ M_3 \\ M_4 \\ M_5 \\ M_6 \end{bmatrix}$$

Note again, that the static matrix defines the structure itself (supports, connection of the members, etc.) and does not depend on the type of external exposures.

11.4.2 Geometrical Equations and Deformation Matrix

These equations present the relationships between deformations e of the elements and displacements Z of the joints. The required relationships is

$$\vec{\mathbf{e}} = \mathbf{B}\vec{\mathbf{Z}}. \tag{11.2}$$

Vector of deformation is $\vec{\mathbf{e}}_{(n\times 1)} = \lfloor e_1\ e_2\ \cdots\ e_n \rfloor^{\mathrm{T}}$. The entries of this vector are deformation of the elements at the sections with unknown internal forces. Therefore, a dimension of this vector equals to the number of unknown internal forces $\mathbf{S}$. Vector of joint displacement is $\vec{\mathbf{Z}}_{(m\times 1)} = \lfloor Z_1\ Z_2\ \cdots\ Z_m \rfloor^{\mathrm{T}}$; dimension of this vector equals to the number of primary unknowns Z of displacement method. Matrix $\mathbf{B}_{(m\times n)}$ presents the matrix of deformation. The entry b_{ij} (i-th row and j-th column) means deformation in direction unknown internal force S_i caused by displacement Z_j. In fact, the matrix equation (11.2) describes the conditions of the deformation *continuity* of the elements.

Let us show the construction of matrix deformation $\mathbf{B}$ for continuous beam (Fig. 11.20a). The primary system, the positive angular displacements Z and positive direction of the unknown bending moments M_i ($i = 1,\ldots,5$) at the ends of each member are shown in Fig. 11.20b.

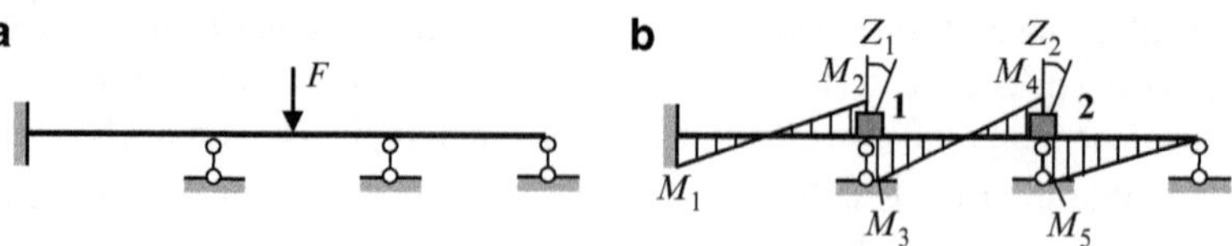

Fig. 11.20 Construction of the deformation matrix

If introduced constraint 1 has angular displacement Z_1 then deformation (angle of rotation) at the section 1 will be zero, i.e., $e_1 = 0 \cdot Z_1$. If introduced constraint 2 has angular displacement Z_2 then deformation at the section 1 will be zero, also. The deformations in direction M_i are

$$\begin{aligned} e_1 &= 0\cdot Z_1 + 0\cdot Z_2 \\ e_2 &= 1\cdot Z_1 + 0\cdot Z_2 \\ e_3 &= 1\cdot Z_1 + 0\cdot Z_2 \\ e_4 &= 0\cdot Z_1 + 1\cdot Z_2 \\ e_5 &= 0\cdot Z_1 + 1\cdot Z_2 \end{aligned} \quad \text{or in matrix form} \quad \begin{bmatrix} e_1 \\ e_2 \\ e_3 \\ e_4 \\ e_5 \end{bmatrix} = \underbrace{\begin{bmatrix} 0 & 0 \\ 1 & 0 \\ 1 & 0 \\ 0 & 1 \\ 0 & 1 \end{bmatrix}}_{\mathbf{B}} \cdot \begin{bmatrix} Z_1 \\ Z_2 \end{bmatrix}$$

The second equation shows that if introduced constraint 1 has angular displacement Z_1 then corresponding deformation (angle of rotation) at the section 2 will be same, i.e., $e_2 = 1 \cdot Z_1$. If introduced constraint 2 has angular displacement Z_2 then corresponding deformation at the section 2 will be zero, because introduced constraint 1.

Note that deformation matrix **B** and static matrix **A** are connected as follows: $\mathbf{B} = \mathbf{A}^{\mathrm{T}}$. This is a general rule, so for calculation of matrix **B** we can apply two approaches.

11.4.3 Physical Equations and Stiffness Matrix in Local Coordinates

These equations present the relationships between unknown internal forces S and deformations e of the elements. The required relationships is

$$\vec{\mathbf{S}} = \tilde{\mathbf{k}}\vec{\mathbf{e}}, \tag{11.3}$$

where $\vec{\mathbf{S}}_{(m\times 1)} = \lfloor S_1 \ S_2 \ \cdots \ S_m \rfloor^{\mathrm{T}}$ is a vector of unknown internal forces; $\vec{\mathbf{e}}_{(m\times 1)} = \lfloor e_1 \ e_2 \ \cdots \ e_m \rfloor^{\mathrm{T}}$ is a vector of deformation; $\tilde{\mathbf{k}}$ is a stiffness matrix of the system.

In general form $\tilde{\mathbf{k}}$ is diagonal matrix

$$\tilde{\mathbf{k}} = \begin{bmatrix} k_1 & 0 & \cdots & 0 \\ 0 & k_2 & \cdots & 0 \\ \cdots & \cdots & \cdots & \cdots \\ 0 & 0 & \cdots & k_m \end{bmatrix} \tag{11.4}$$

The diagonal entry k_i is a stiffness matrix of i-th finite element of a structure. The each diagonal entry k_i is called the *internal stiffness matrix* or stiffness matrix in local coordinates for specified member i; matrix (11.4) in whole is internal stiffness matrix or stiffness matrix in local coordinates for all structure.

For truss element (bar with hinged at the ends), a deformation is

$$e = \frac{Sl}{EA}, \text{ so } S = \frac{EA}{l}e.$$

Thus internal stiffness matrix for truss element contains only one entry and presented as

$$k = \frac{EA}{l}[1]. \tag{11.5}$$

This expression allows to determine the axial force S if axial deformation of element $e = 1$. The symbol [1] means a matrix with the sole entry equals 1.

Let us form the stiffness matrix for truss shown in Fig. 11.21; this figure contains the numeration of the members. Assume that for all members EA is constant.

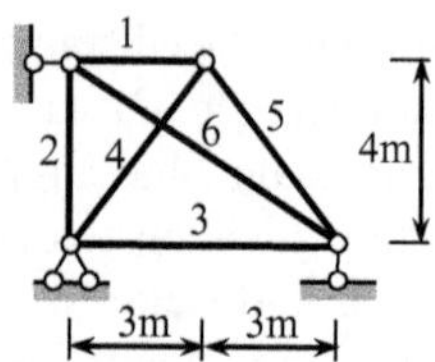

Fig. 11.21 Construction of the internal stiffness matrix for truss

Stiffness matrices for each member are shown below.

$$\mathbf{k}_1 = \frac{EA}{l_1}[1] = \frac{EA}{3}[1]; \quad \mathbf{k}_2 = \frac{EA}{4}[1]; \quad k_3 = \frac{EA}{6}[1];$$
$$\mathbf{k}_4 = \mathbf{k}_5 = \frac{EA}{5}[1]; \quad \mathbf{k}_6 = \frac{EA}{\sqrt{6^2+4^2}}[1] = \frac{EA}{\sqrt{52}}[1]$$

The internal stiffness matrix of the truss becomes

$$\tilde{\mathbf{k}} = EA \begin{bmatrix} 1/3 & 0 & 0 & 0 & 0 & 0 \\ 0 & 1/4 & 0 & 0 & 0 & 0 \\ 0 & 0 & 1/6 & 0 & 0 & 0 \\ 0 & 0 & 0 & 1/5 & 0 & 0 \\ 0 & 0 & 0 & 0 & 1/5 & 0 \\ 0 & 0 & 0 & 0 & 0 & 1/\sqrt{52} \end{bmatrix}$$

For bending elements, we can use the Tables A.3–A.6. If a uniform fixed-pinned beam is subjected to angular displacement e of the fixed support, then a bending moment at this support equals $M = \frac{3EI}{l}e$, so the stiffness matrix of such element in local coordinates is

$$\mathbf{k}_{f-p} = \frac{EI}{l}[3]. \tag{11.6}$$

If a fixed-fixed uniform beam is subjected to unit angular displacements e_1 and e_2 of the fixed ends, then the following bending moments arise at the both supports:

$$M_1 = \frac{EI}{l}(4e_1 + 2e_2),$$
$$M_2 = \frac{EI}{l}(2e_1 + 4e_2).$$

These formulas may be presented in matrix form (11.3), i.e.,

$$\begin{bmatrix} M_1 \\ M_2 \end{bmatrix} = \frac{EI}{l}\begin{bmatrix} 4 & 2 \\ 2 & 4 \end{bmatrix} \cdot \begin{bmatrix} e_1 \\ e_2 \end{bmatrix} \tag{11.7}$$

where the vector of internal forces is $\vec{\mathbf{S}} = \lfloor M_1 \; M_2 \rfloor^{\mathrm{T}}$, vector of angular displacements at the end of element $\vec{e} = \lfloor e_1 \; e_2 \rfloor^{\mathrm{T}}$ and the stiffness matrix in local coordinates for uniform beam with fixed ends becomes

$$\mathbf{k}_{f-f} = \frac{EI}{l}\begin{bmatrix} 4 & 2 \\ 2 & 4 \end{bmatrix} \tag{11.7a}$$

Let us form the stiffness matrix in local coordinates for frame shown in Fig. 11.22. Assume that bending stiffness equals EI for horizontal members 2 and 4, and for vertical members $3EI$.

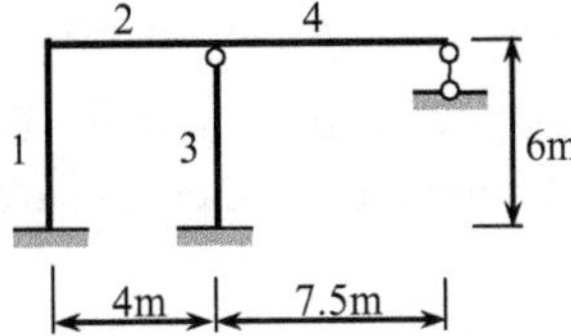

Fig. 11.22 Design diagram of the frame

According to primary system of the displacement method, we have the fixed-fixed members 1 and 2, and fixed-pinned members 3 and 4. Stiffness matrices for each member in local coordinates are

$$\mathbf{k}_1 = \frac{EI_1}{l_1}\begin{bmatrix} 4 & 2 \\ 2 & 4 \end{bmatrix} = \frac{3EI}{6}\begin{bmatrix} 4 & 2 \\ 2 & 4 \end{bmatrix} = EI\begin{bmatrix} 2 & 1 \\ 1 & 2 \end{bmatrix},$$

$$\mathbf{k}_2 = \frac{EI_2}{l_2}\begin{bmatrix} 4 & 2 \\ 2 & 4 \end{bmatrix} = \frac{EI}{4}\begin{bmatrix} 4 & 2 \\ 2 & 4 \end{bmatrix} = EI\begin{bmatrix} 1 & 0.5 \\ 0.5 & 1 \end{bmatrix},$$

$$\mathbf{k}_3 = \frac{EI_3}{l_3}[3] = \frac{3EI}{6}[3] = EI\,[1.5],$$

$$\mathbf{k}_4 = \frac{EI_4}{l_4}[3] = \frac{EI}{7.5}[3] = EI\,[0.4].$$

Each of these stiffness matrices is presented in the form that contains general multiplier EI. The internal stiffness matrix of the frame becomes

$$\tilde{\mathbf{k}} = EI\begin{bmatrix} 2 & 1 & 0 & 0 & 0 & 0 \\ 1 & 2 & 0 & 0 & 0 & 0 \\ 0 & 0 & 1 & 0.5 & 0 & 0 \\ 0 & 0 & 0.5 & 1 & 0 & 0 \\ 0 & 0 & 0 & 0 & 1.5 & 0 \\ 0 & 0 & 0 & 0 & 0 & 0.4 \end{bmatrix}$$

Internal stiffness matrix for combined structure (e.g., the frame with tie) can be constructed by the similar way.

11.5 Set of Formulas and Procedure for Analysis

The behavior of any structure can be described by following three groups of equations:

Equilibrium equations This matrix equation establish relationships between external possible joint loads $\vec{\mathbf{P}}$ and unknown internal forces $\vec{\mathbf{S}}$

$$\vec{\mathbf{P}} = \mathbf{A}\vec{\mathbf{S}}, \tag{11.8}$$

where **A** is a static matrix.

Geometrical equations This matrix equation establish relationships between deformation of elements $\vec{\mathbf{e}}$ and possible global displacements $\vec{\mathbf{Z}}$ of the joints

$$\vec{\mathbf{e}} = \mathbf{B}\vec{\mathbf{Z}} = \mathbf{A}^{\mathrm{T}}\vec{\mathbf{Z}}, \tag{11.9}$$

where **B** is a matrix of deformation.

Physical equations This matrix equation establish relationships between required internal forces $\vec{\mathbf{S}}$ and deformation of elements $\vec{\mathbf{e}}$

$$\vec{\mathbf{S}} = \tilde{\mathbf{k}}\vec{\mathbf{e}}, \tag{11.10}$$

where $\tilde{\mathbf{k}}$ is a stiffness matrix of a structure in local coordinates (internal stiffness matrix).

These three groups of equations describe completely any structure (geometry, distribution of stiffness of separate members, types of connections of the members), types of supports, and external exposure.

11.5.1 Stiffness Matrix in Global Coordinates

Rearrangement of (11.8)–(11.10) allows to obtain the equation for vector of unknown internal forces **S**. For this purpose, let us apply the following procedure.

The vector ***e*** may be eliminated from (11.9) and (11.10). For this (11.9) should be substituted into (11.10). This procedure allows us to express the vector unknown internal forces in term of vector of unknown displacements **Z**

$$\vec{\mathbf{S}} = \tilde{\mathbf{k}}\vec{\mathbf{e}} = \tilde{\mathbf{k}}\mathbf{A}^{\mathrm{T}}\vec{\mathbf{Z}}. \tag{11.11}$$

Now vector ***S*** can be eliminated from (12.8) and (12.11). For this (12.11) should be substituted into (11.8); this procedure allows us to express the vector of possible joint external loads in term of unknown displacements

$$\vec{\mathbf{P}} = \mathbf{A}\vec{\mathbf{S}} = \mathbf{A}\tilde{\mathbf{k}}\mathbf{A}^{\mathrm{T}}\vec{\mathbf{Z}}. \tag{11.12}$$

This equation may be rewritten in the form

$$\vec{\mathbf{P}} = \mathbf{K}\vec{\mathbf{Z}}. \tag{11.13}$$

The matrix **K** presents the stiffness matrix of a structure in global coordinates (external stiffness matrix)

$$\mathbf{K} = \mathbf{A}\tilde{\mathbf{k}}\mathbf{A}^{\mathrm{T}}. \tag{11.14}$$

This matrix is a symmetrical one and it has the strictly positive entries on the main diagonal. Dimension of this matrix is $(n \times n)$, where n is a number of unknown end displacements. Thus, for calculation of stiffness matrix **K** in global coordinates, we need to know a static matrix **A** and stiffness matrix $\tilde{\mathbf{k}}$ (11.4) of a structure in local coordinates.

11.5.2 Unknown Displacements and Internal Forces

Equation (11.13) allows to calculate the vector of ends displacements

$$\vec{\mathbf{Z}} = \mathbf{K}^{-1}\vec{\mathbf{P}}, \tag{11.15}$$

where $\mathbf{K}^{-1}$ is inverse stiffness matrix in global coordinates.

For truss the vector of possible joint external loads **P** is formed on the basis of the entire design diagram and ***Z-P*** diagram. For beams and frames, the vector **P** is formed on the basis of the joint-load and ***Z-P*** diagrams. If a structure is subjected to m different groups of loading, then matrix **P** contains m columns. Each column corresponds to specified group loading.

Knowing vector **Z** we can calculate, according to (11.11), the unknown internal forces of the second state

$$\vec{\mathbf{S}}_2 = \tilde{\mathbf{k}}\mathbf{A}^{\mathrm{T}}\vec{\mathbf{Z}}. \tag{11.16}$$

The formula (11.8) may be used for verification of obtained internal forces, i.e., $\mathbf{A}\vec{\mathbf{S}}_2 = \vec{\mathbf{P}}$.

Final internal forces may be calculated by the formula

$$\vec{\mathbf{S}}_{\text{fin}} = \vec{\mathbf{S}}_1 + \vec{\mathbf{S}}_2, \tag{11.17}$$

where the vector $\vec{\mathbf{S}}_1$ presents the internal forces at specified sections in the first state. It is obvious that for trusses $\vec{\mathbf{S}}_1$ is a zero-vector. For bending elements, the entries of $\vec{\mathbf{S}}_1$ are bending moments at the end sections. This vector forms on the basis of the M_P^0 diagram (1 state) and ***S-e*** diagram. If final ordinate M at any section is positive, then this ordinate should be plotted at the same side of the member as on the ***S-e*** diagram.

11.5.3 Matrix Procedures

The following procedure for analysis of any structure by MSM may be proposed:

1. Define the degree of kinematical indeterminacy and type of displacement for each joint.
2. Calculate the fixed end moments to construct the ***J-L*** diagram.
3. Numerate the possible displacements of the joints, then construct the ***Z-P*** diagram and form the vector $\vec{\mathbf{P}}$ of external joint loads.
4. Numerate the unknowns internal forces S (for truss it is a number of the elements; for frame it is a nonzero bending moments M at the ends of each elements) to construct the ***S-e*** diagram and to form the vector $\vec{\mathbf{S}}_1$ of internal forces. For this use the first state and ***S-e*** diagram; for truss $\vec{\mathbf{S}}_1 = \mathbf{0}$.
5. Consider the equilibrium conditions for each possible displacement of the joint and construct the static matrix **A**; the number of the rows of this matrix equals to degree of kinematical indeterminacy and the number of the columns equals to the number of the unknown internal forces.
6. Construct the stiffness matrix for each member and for all structure in local coordinates.
7. Perform the following matrix procedures:

   Compute the intermediate matrix complex $\tilde{\mathbf{k}}\mathbf{A}^{\mathrm{T}}$ (this complex will be used in the next steps)

   Compute the stiffness matrix in global coordinates $\mathbf{K} = \mathbf{A}\tilde{\mathbf{k}}\mathbf{A}^{\mathrm{T}}$ and its inverse matrix $\mathbf{K}^{-1}$

   Calculate the vector of joint displacements $\vec{\mathbf{Z}} = \mathbf{K}^{-1}\vec{\mathbf{P}}$

   Calculate the vector of unknown internal forces of the second state $\mathbf{S}_2 = \tilde{\mathbf{k}}\mathbf{A}^{\mathrm{T}} \cdot \mathbf{Z}$

   Calculate the vector of final internal forces $\vec{\mathbf{S}}_{\text{fin}} = \vec{\mathbf{S}}_1 + \vec{\mathbf{S}}_2$

All matrix procedures (11.14)–(11.17) may be performed by standard programs using computer.

For trusses the procedure (11.17) leads to the axial forces at the each member. For frames this formula leads to nonzero bending moments at the ends of each element. To plot the final bending moment diagram, the signs of obtained final moments should be consistent with **S-e** diagram.

The shear force can be calculated on the basis of the bending moment diagram considering each member subjected to given loads and the end bending moments; the axial forces can be calculated on the basis of the shear diagram consideration of equilibrium of joints of the frame. Finally, having all internal force diagrams, we can show the reactions of supports and check them using equilibrium conditions for an entire structure as a whole or for any separated part.

Notes:

1. In different textbooks the algorithm above (or slightly modified algorithm) is referred differently. They are the matrix displacement method (do not confuse with the displacement method in matrix form), finite element method, stiffness method. All of them realize the one general idea: presentation of the structure as a set of separate elements with necessary demands about consistent of deformation.
2. The entries of the stiffness matrix $\mathbf{K} = \mathbf{A}\tilde{\mathbf{k}}\mathbf{A}^{\mathrm{T}}$ present the unit reactions of the displacement method in canonical form.
3. Generally speaking, a vector of internal unknown forces **S** may be constructed by different forms. In this textbook, the vector **S** is presented in the *simplest form*. This vector contains only *nonzero* bending moments (for beams and frames) for each separate member at the ends. This choice of the vector **S** leads to the very compact stiffness matrix. Indeed, for fixed-pinned beam, this matrix contains only one entry, for fixed-fixed standard member, this matrix has dimension (2×2). Certainly, we can expand the vector state **S** including, for example, the shear and axial force. However, this leads to the stiffness matrix with expanded dimensions. Even if the final result would contain more complete information, observing over all matrices is difficult. Therefore, we limited our consideration of the MSM only for *simplest* presentation of vector **S**. This leads to the simple and vivid of intermediate and final results, and significant simplification of numerical procedures.

11.6 Analysis of Continuous Beams

This section presents a detailed analysis of statically indeterminate continuous beams subjected to different types of exposures.

Design diagram of the uniform two-span beam subjected to fixed load is shown in Fig. 11.23a. This structure may be presented as a set of two finite elements: they are A-1 and 1-B. To transmit the given load to the joint load, we need to calculate the fixed end moments at support 1 (Fig. 11.23b). They are equal to

$$\begin{aligned} M_{1A}^0 &= \frac{ql_1^2}{8} = \frac{q \times 8^2}{8} = 16\ (\text{kNm}) \text{ and } M_{1B}^0 = \frac{Pl_2}{2}\upsilon\left(1-\upsilon^2\right) \\ &= \frac{12 \times 10}{2} \times 0.4\left(1-0.4^2\right) = 20.16\ (\text{kNm}), \end{aligned}$$

so the equivalent moment $20.16 - 16 = 4.16$ acts clockwise (Fig. 11.23c) and corresponding joint-load diagram (first state) is shown in Fig. 11.23d.

The beam has one unknown angular displacement at support 1 and corresponding one possible external joint load. The displacement-load (***Z-P***) diagram is presented in Fig. 11.23e. Having the joint-load and ***Z-P*** diagrams, we can construct the vector

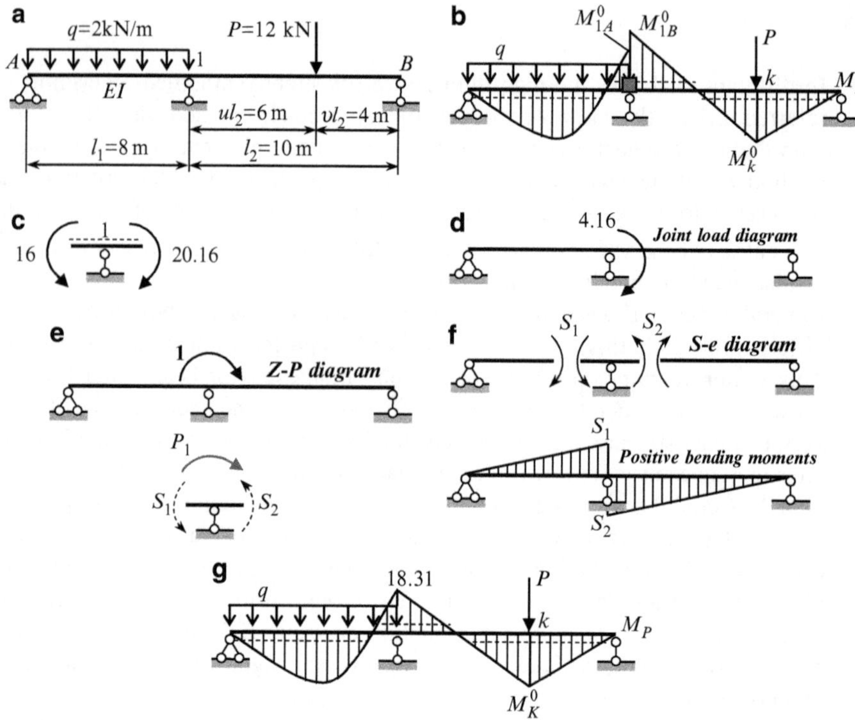

Fig. 11.23 (**a–d**) Fixed load. Continuous beam and corresponding joint-load diagram; (**e, f**) Continuous beam and corresponding ***Z-P*** and ***S-e*** diagrams; (**g**) Design diagram and final bending moment diagram

of external joint moments; in this simplest case, the vector $\vec{\mathbf{P}} = [4.16]$, so this vector have only one entry. Positive sign means that moment at joint load and ***Z-P*** diagrams act at one direction.

Unknown internal forces (moments S_1 and S_2) and their positive directions are shown on ***S-e*** diagram (Fig. 11.23f). To construct the vector **S** of internal forces in the state 1, we need to take into account bending moment diagram M_P^o (Fig. 11.23b). This vector is

$$\vec{S}_1 = \begin{bmatrix} 16 \\ -20.16 \end{bmatrix}.$$

The signs of the entries correspond to ***S-e*** diagram.

A static matrix **A** can be constructed on the basis of the ***Z-P*** and ***S-e*** diagrams. Figure 11.23e shows free body diagram for joint 1 subjected to unknown internal "forces" S_1 and S_2 in vicinity of joint 1 and load P_1, which corresponds to possible displacement of the joint 1. Since this displacement is angle of rotation then this load is a moment. Equilibrium condition leads to the equation $P_1 = S_1 + S_2$. So the static matrix becomes $\mathbf{A} = \lfloor 1\ 1 \rfloor$.

Stiffness matrices for left and right spans are

$$k_1 = \frac{3EI_1}{l_1}[1] = \frac{3EI_1}{8}[1] = \frac{3EI_1}{40}[5],$$

$$k_2 = \frac{3EI_2}{l_2}[1] = \frac{3EI_1}{10}[1] = \frac{3EI_1}{40}[4].$$

Internal stiffness matrix for all beam in local coordinates is

$$\tilde{\mathbf{k}} = \begin{bmatrix} k_1 & 0 \\ 0 & k_2 \end{bmatrix} = \frac{3EI}{40}\begin{bmatrix} 5 & 0 \\ 0 & 4 \end{bmatrix}$$

Stiffness matrix for all structure in global coordinates and its inverse are

$$\mathbf{K} = \mathbf{A}\tilde{\mathbf{k}}\mathbf{A}^{\mathrm{T}} = \lfloor 1\ 1 \rfloor \cdot \frac{3EI}{40}\begin{bmatrix} 5 & 0 \\ 0 & 4 \end{bmatrix} \cdot \begin{bmatrix} 1 \\ 1 \end{bmatrix} = \frac{27}{40EI} \rightarrow \mathbf{K}^{-1} = \frac{40EI}{27}$$

Displacement of the joint 1

$$\vec{\mathbf{Z}} = \mathbf{K}^{-1}\vec{\mathbf{P}} = \frac{40EI}{27} \times 4.16 = \frac{6.163}{EI}$$

Vector of internal forces of the second state

$$\vec{\mathbf{S}}_2 = \tilde{\mathbf{k}}\mathbf{A}^{\mathrm{T}}\vec{\mathbf{Z}} = \frac{3EI}{40}\begin{bmatrix} 5 & 0 \\ 0 & 4 \end{bmatrix} \cdot \begin{bmatrix} 1 \\ 1 \end{bmatrix} \frac{6.163}{EI} = \begin{bmatrix} 2.3111 \\ 1.8489 \end{bmatrix}$$

Final internal forces

$$\vec{\mathbf{S}}_{fin} = \vec{\mathbf{S}}_1 + \vec{\mathbf{S}}_2 = \begin{bmatrix} 16 \\ -20.16 \end{bmatrix} + \begin{bmatrix} 2.3111 \\ 1.8489 \end{bmatrix} = \begin{bmatrix} 18.31 \\ -18.31 \end{bmatrix}$$

The negative sign means that moment S_2, according to ***S-e*** diagram, is directed in opposite direction, so external fibers right at the joint 1 are located above the neutral line. Final bending moment diagram is shown in Fig. 11.23g.

Equilibrium condition $\sum M_1 = 0$ for joint 1 is satisfied. The bending moment at point k may be calculated considering second span as simply supported beam subjected to force P and moment 18.31 kNm counterclockwise.

Notes:

1. Analysis of this beam has been performed early by the displacement method (Chap. 8) so the reader has an opportunity to compare analysis of the same beam by different methods.

2. Computation of shear may be performed on the basis of the M_P diagram. Reactions of supports can be calculated on the basis of the shear diagram. Reaction of intermediate support is $R_1 = Q_1^{\text{right}} - Q_1^{\text{left}}$.

Example 11.4. Design diagram of a nonuniform continuous beam is presented in Fig. 11.24a. The angle of rotation of the clamped support is $\varphi_0 = \varphi = 0.01$ rad and the vertical displacement of the support 2 is $\Delta_2 = \Delta = 0.04$ m. Construct the bending moment diagram.

Solution. Degree of kinematical indeterminacy equals two. Both joints at supports B and C have angular possible displacements. Now we need to present the effect of the settlements of supports in form of moments at the joints B and C. The fixed end moments are

$$M_{AB} = \frac{4EI}{l_{AB}}\varphi = \frac{4EI_0}{6} \times 0.01 = 0.00667EI_0$$

$$M_{BC} = M_{CB} = \frac{6EI}{l_{BC}^2}\Delta = \frac{6 \cdot 2EI_0}{4^2} \times 0.04 = 0.03EI_0$$

$$M_{BA} = \frac{2EI}{l_{AB}}\varphi = \frac{2EI_0}{6} \times 0.01 = 0.00333EI_0$$

$$M_{CD} = \frac{3EI}{l_{CD}^2}\Delta = \frac{3 \cdot 2EI_0}{4^2} \times 0.04 = 0.015EI_0.$$

Corresponding bending moment diagram M_Δ^0 in the primary system of the displacement method and the joint-load diagram are shown in Fig. 11.24b, c.

The positive possible angular displacements of the joints and corresponding possible external forces are presented on the ***Z-P*** diagram (Fig. 11.24d). The ***S-e*** diagram and positive bending moments diagram are shown in Fig. 11.24e. The vector of external joint loads is

$$\vec{\mathbf{P}} = EI_0 \lfloor 0.02667 \quad 0.015 \rfloor^{\mathrm{T}}.$$

Vector of fixed end moments in the first state is

$$\vec{\mathbf{S}}_1 = EI_0 \lfloor 0.00667 \; 0.00333 \; -0.03 \; -0.03 \; 0.015 \rfloor^{\mathrm{T}}$$

Static matrix Considering the ***Z-P*** and ***S-e*** diagrams we get

$$P_1 = S_2 + S_3$$

$$P_2 = S_4 + S_5$$

so the static matrix becomes

$$\mathbf{A}_{(2\times5)} = \begin{bmatrix} 0 & 1 & 1 & 0 & 0 \\ 0 & 0 & 0 & 1 & 1 \end{bmatrix}$$

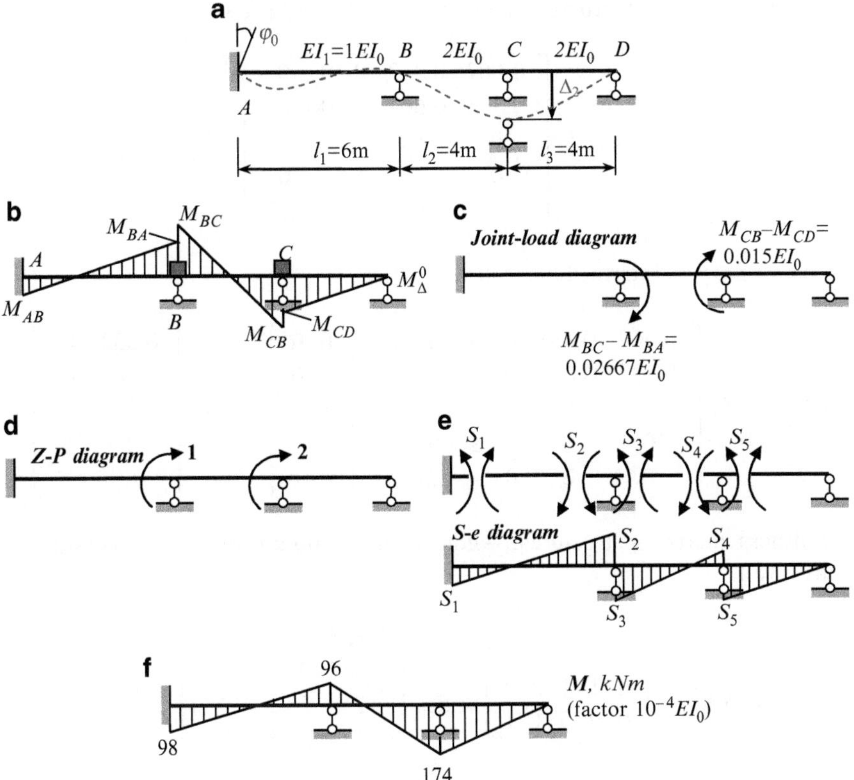

Fig. 11.24 (**a**) Settlements of supports. Design diagram of continuous beam; (**b**, **c**) Conversion of effect due to settlement of support to the joint-load moments; (**d**, **e**) Continuous beam; (**f**) Final bending moment diagram and corresponding ***Z-P*** and ***S-e*** diagrams

Stiffness matrix Stiffness matrices for finite elements *AB*, *BC*, and *CD* in local coordinates are

$$\mathbf{k}_1 = \frac{EI_1}{l_1}\begin{bmatrix} 4 & 2 \\ 2 & 4 \end{bmatrix} = \frac{EI_0}{6}\begin{bmatrix} 4 & 2 \\ 2 & 4 \end{bmatrix} = EI_0\begin{bmatrix} 0.667 & 0.333 \\ 0.333 & 0.667 \end{bmatrix},$$

$$\mathbf{k}_2 = \frac{EI_2}{l_2}\begin{bmatrix} 4 & 2 \\ 2 & 4 \end{bmatrix} = \frac{2EI_0}{4}\begin{bmatrix} 4 & 2 \\ 2 & 4 \end{bmatrix} = EI_0\begin{bmatrix} 2 & 1 \\ 1 & 2 \end{bmatrix},$$

$$\mathbf{k}_3 = \frac{EI_3}{l_3}[3] = \frac{2EI_0}{4}[3] = EI_0[1.5].$$

Stiffness matrix for structure in whole in local coordinates is

$$\tilde{\mathbf{k}}_{(5\times5)} = \begin{bmatrix} 0.667 & 0.333 & 0 & 0 & 0 \\ 0.333 & 0.667 & 0 & 0 & 0 \\ 0 & 0 & 2 & 1 & 0 \\ 0 & 0 & 1 & 2 & 0 \\ 0 & 0 & 0 & 0 & 1.5 \end{bmatrix} EI_0$$

Matrix procedures Intermediate matrix complex

$$\tilde{\mathbf{k}}\mathbf{A}^{\mathrm{T}} = EI_0 \begin{bmatrix} 0.667 & 0.333 & 0 & 0 & 0 \\ 0.333 & 0.667 & 0 & 0 & 0 \\ 0 & 0 & 2 & 1 & 0 \\ 0 & 0 & 1 & 2 & 0 \\ 0 & 0 & 0 & 0 & 1.5 \end{bmatrix} \times \begin{bmatrix} 0 & 0 \\ 1 & 0 \\ 1 & 0 \\ 0 & 1 \\ 0 & 1 \end{bmatrix} = EI_0 \begin{bmatrix} 0.333 & 0 \\ 0.667 & 0 \\ 2 & 1 \\ 1 & 2 \\ 0 & 1.5 \end{bmatrix}$$

Stiffness matrix for structure in whole in global coordinates and inverse stiffness matrix are

$$\mathbf{K} = \mathbf{A} \cdot \tilde{\mathbf{k}}\mathbf{A}^{\mathrm{T}} = \begin{bmatrix} 0 & 1 & 1 & 0 & 0 \\ 0 & 0 & 0 & 1 & 1 \end{bmatrix} \times EI_0 \begin{bmatrix} 0.333 & 0 \\ 0.667 & 0 \\ 2 & 1 \\ 1 & 2 \\ 0 & 1.5 \end{bmatrix} = EI_0 \begin{bmatrix} 2.667 & 1 \\ 1 & 3.5 \end{bmatrix},$$

$$\mathbf{K}^{-1} = \frac{1}{EI_0} \begin{bmatrix} 0.42 & -0.12 \\ -0.12 & 0.32 \end{bmatrix}$$

Vector of joint displacements

$$\vec{\mathbf{Z}} = \mathbf{K}^{-1}\vec{\mathbf{P}} = \frac{1}{EI_0} \begin{bmatrix} 0.42 & -0.12 \\ -0.12 & 0.32 \end{bmatrix} \times EI_0 \begin{bmatrix} 0.02667 \\ 0.015 \end{bmatrix} = \begin{bmatrix} 0.0094 \\ 0.0016 \end{bmatrix}$$

Vector of unknown bending moments of the second state is

$$\vec{\mathbf{S}}_2 = \tilde{\mathbf{k}}\mathbf{A}^{\mathrm{T}} \cdot \mathbf{Z} = EI_0 \begin{bmatrix} 0.333 & 0 \\ 0.667 & 0 \\ 2 & 1 \\ 1 & 2 \\ 0 & 1.5 \end{bmatrix} \times \begin{bmatrix} 0.0094 \\ 0.0016 \end{bmatrix} = EI_0 \begin{bmatrix} 0.00313 \\ 0.00627 \\ 0.0204 \\ 0.0126 \\ 0.0024 \end{bmatrix}$$

The final vector of required bending moments (kNm) is

$$\vec{\mathbf{S}} = \vec{\mathbf{S}}_1 + \tilde{\mathbf{k}}\mathbf{A}^{\mathrm{T}}\mathbf{Z} = EI_0 \begin{bmatrix} 0.00667 \\ 0.00333 \\ -0.03 \\ -0.03 \\ 0.015 \end{bmatrix} + EI_0 \begin{bmatrix} 0.00313 \\ 0.00627 \\ 0.0204 \\ 0.0126 \\ 0.0024 \end{bmatrix} = EI_0 \begin{bmatrix} 0.0098 \\ 0.0096 \\ -0.0096 \\ -0.0174 \\ 0.0174 \end{bmatrix}$$

Corresponding final bending moment diagram is presented in Fig. 11.24f.

Example 11.5. Design diagram of the uniform three-span continuous beam is presented in Fig. 11.25a. Construct the influence lines for bending moments at the supports B and C (sections 6 and 12, respectively).

Solution. Each span of the beam is divided in equal portions and specified sections are numerated (0–18). Next we need to show the displacement-load (***Z-P***) and ***S-e*** diagrams (Fig. 11.25a). Unknown moments S_1, S_2 arise at support B and S_3, S_4 at support C.

Static matrix The ***Z-P*** and ***S-e*** diagrams allow us to constructing the following equilibrium equations:

$$P_1 = S_1 + S_2$$
$$P_2 = S_3 + S_4$$

so the static matrix of the structure is

$$\mathbf{A}_{(2\times4)} = \begin{bmatrix} 1 & 1 & 0 & 0 \\ 0 & 0 & 1 & 1 \end{bmatrix}$$

Stiffness matrix Stiffness matrices for each finite element are

$$\mathbf{k}_1 = \frac{EI}{l}[3], \quad \mathbf{k}_2 = \frac{EI}{l}\begin{bmatrix} 4 & 2 \\ 2 & 4 \end{bmatrix}, \quad \mathbf{k}_3 = \frac{EI}{l}[3].$$

Stiffness matrix of all structure in local coordinates and intermediate complex $\tilde{\mathbf{k}}\mathbf{A}^{\mathrm{T}}$ are

$$\tilde{\mathbf{k}} = \frac{EI}{l}\begin{bmatrix} 3 & 0 & 0 & 0 \\ 0 & 4 & 2 & 0 \\ 0 & 2 & 4 & 0 \\ 0 & 0 & 0 & 3 \end{bmatrix}, \quad \tilde{\mathbf{k}}\mathbf{A}^{\mathrm{T}} = \frac{EI}{l}\begin{bmatrix} 3 & 0 & 0 & 0 \\ 0 & 4 & 2 & 0 \\ 0 & 2 & 4 & 0 \\ 0 & 0 & 0 & 3 \end{bmatrix} \cdot \begin{bmatrix} 1 & 0 \\ 1 & 0 \\ 0 & 1 \\ 0 & 1 \end{bmatrix} = \frac{EI}{l}\begin{bmatrix} 3 & 0 \\ 4 & 2 \\ 2 & 4 \\ 0 & 3 \end{bmatrix}$$

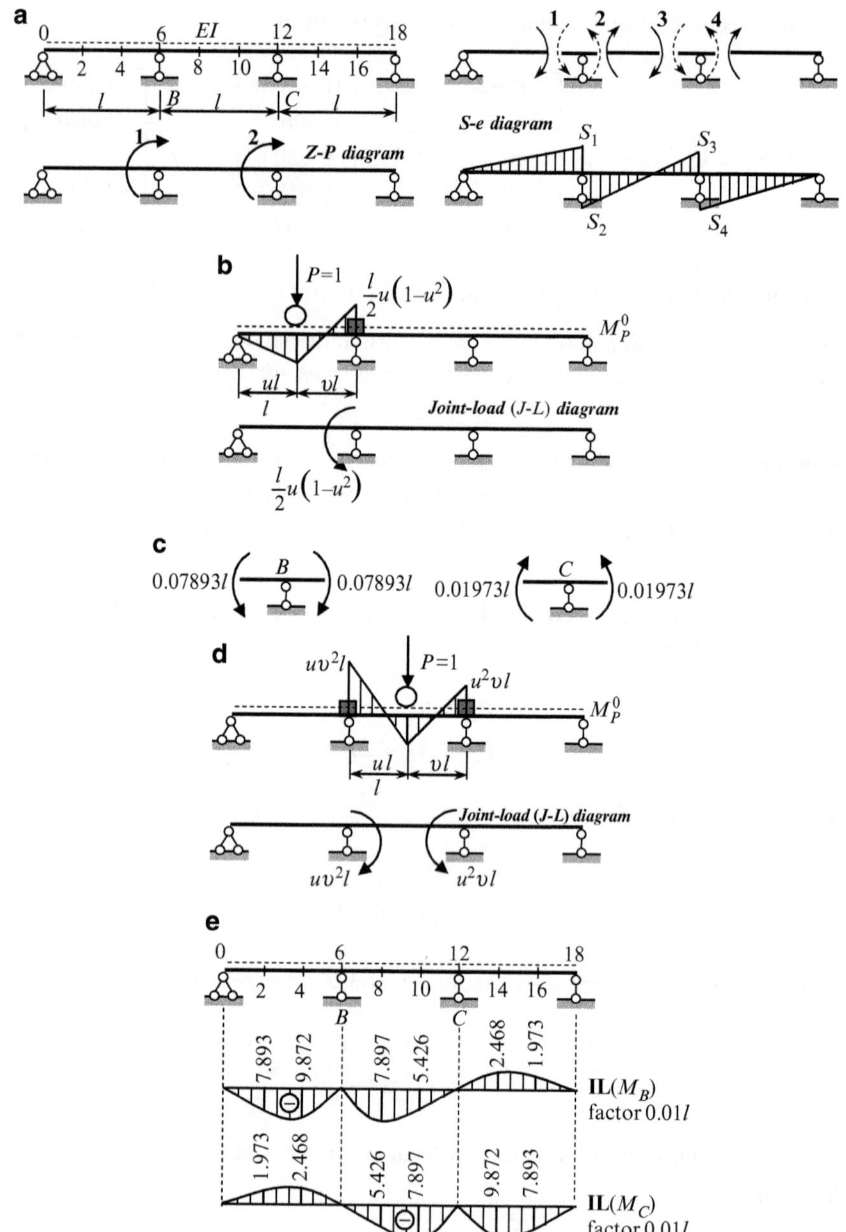

Fig. 11.25 (**a**) Moving load. Design diagram of continuous beam, ***Z-P*** and ***S-e*** diagrams; (**b**) Load $P = 1$ in the first span and corresponding joint-load diagram; (**c**) Ordinates of influence lines at sections 6 and 12; the load $P = 1$ is placed at point 2; (**d**) Load $P = 1$ in the second span; (**e**) Influence lines for bending moments at supports B and C and corresponding joint-load diagram

The stiffness matrix of all structure in global coordinates and inverse stiffness matrix are

$$\mathbf{K} = \mathbf{A}\cdot\tilde{\mathbf{k}}\mathbf{A}^{\mathrm{T}} = \begin{bmatrix} 1 & 1 & 0 & 0 \\ 0 & 0 & 1 & 1 \end{bmatrix}\frac{EI}{l}\begin{bmatrix} 3 & 0 \\ 4 & 2 \\ 2 & 4 \\ 0 & 3 \end{bmatrix} = \frac{EI}{l}\begin{bmatrix} 7 & 2 \\ 2 & 7 \end{bmatrix},$$

$$\mathbf{K}^{-1} = \frac{l}{45EI}\begin{bmatrix} 7 & -2 \\ -2 & 7 \end{bmatrix}$$

To construct the matrices **P** and $\mathbf{S}_1$ we need to consider the unit moving load in all spans separately. Next for each loading we need to construct the bending moment diagram in primary system of displacement method and show the corresponding Joint-load diagram.

Load P* = 1 *in the first span The bending moment diagram in the primary system and equivalent joint-load diagram are shown in Fig. 11.25b.

On the basis of the joint-load and ***Z-P*** diagrams, the vector **P** of the external joint loads for any position u of unit load P in the first span is

$$\vec{\mathbf{P}} = l\begin{bmatrix} -0.5u\left(1-u^2\right) \\ 0 \end{bmatrix}$$

On the basis of the M_P^0 and ***S-e*** diagrams the vector $\mathbf{S}_1$ of unknown internal forces $S_1 - S_4$ in the first state for any position P becomes

$$\vec{\mathbf{S}}_1 = l\begin{bmatrix} 0.5u\left(1-u^2\right) \\ 0 \\ 0 \\ 0 \end{bmatrix}$$

Now we can determine the entries of both of these matrices when moving load P is placed at the sections 2 and 4 and perform corresponding matrix procedures.

$P = 1$ **at the section 2** ($u = 0.333,\ \upsilon = 0.667$)	$P = 1$ **at the section 4** ($u = 0.667, \upsilon = 0.333$)
$\vec{\mathbf{P}} = l\begin{bmatrix} -0.1480 \\ 0 \end{bmatrix}, \vec{\mathbf{S}}_1 = \begin{bmatrix} M_1 \\ M_2 \\ M_3 \\ M_4 \end{bmatrix}_1 = l\begin{bmatrix} 0.1480 \\ 0 \\ 0 \\ 0 \end{bmatrix}$	$\vec{\mathbf{P}} = l\begin{bmatrix} -0.1851 \\ 0 \end{bmatrix}, \vec{\mathbf{S}}_1 = \begin{bmatrix} M_1 \\ M_2 \\ M_3 \\ M_4 \end{bmatrix}_1 = l\begin{bmatrix} 0.1851 \\ 0 \\ 0 \\ 0 \end{bmatrix}$

(continued)

P = 1 at the section 2 ($u = 0.333,\ \upsilon = 0.667$)	**P = 1 at the section 4** ($u = 0.667,\ \upsilon = 0.333$)
Matrix procedures	
$\vec{\mathbf{Z}} = \begin{bmatrix} Z_1 \\ Z_2 \end{bmatrix} = \mathbf{K}^{-1}\vec{\mathbf{P}} = \frac{l}{45EI}\begin{bmatrix} 7 & -2 \\ -2 & 7 \end{bmatrix} \times l \begin{bmatrix} -0.1480 \\ 0 \end{bmatrix} = \frac{l^2}{304.05EI}\begin{bmatrix} -7 \\ 2 \end{bmatrix}$	$\vec{\mathbf{Z}} = \begin{bmatrix} Z_1 \\ Z_2 \end{bmatrix} = \mathbf{K}^{-1}\vec{\mathbf{P}} = \frac{l}{45EI}\begin{bmatrix} 7 & -2 \\ -2 & 7 \end{bmatrix} \times l \begin{bmatrix} -0.1851 \\ 0 \end{bmatrix} = \frac{l^2}{243.11EI}\begin{bmatrix} -7 \\ 2 \end{bmatrix}$
$\vec{\mathbf{S}}_2 = \begin{bmatrix} M_1 \\ M_2 \\ M_3 \\ M_4 \end{bmatrix}_2 = \tilde{\mathbf{k}}\mathbf{A}^{\mathrm{T}} \cdot \vec{\mathbf{Z}} = \frac{EI}{l}\begin{bmatrix} 3 & 0 \\ 4 & 2 \\ 2 & 4 \\ 0 & 3 \end{bmatrix} \times \frac{l^2}{304.05EI}\begin{bmatrix} -7 \\ 2 \end{bmatrix} = l\begin{bmatrix} -0.06907 \\ -0.07893 \\ -0.01973 \\ 0.01973 \end{bmatrix}$	$\vec{\mathbf{S}}_2 = \begin{bmatrix} M_1 \\ M_2 \\ M_3 \\ M_4 \end{bmatrix}_2 = \tilde{\mathbf{k}}\mathbf{A}^{\mathrm{T}} \cdot \vec{\mathbf{Z}} = \frac{EI}{l}\begin{bmatrix} 3 & 0 \\ 4 & 2 \\ 2 & 4 \\ 0 & 3 \end{bmatrix} \times \frac{l^2}{243.11EI}\begin{bmatrix} -7 \\ 2 \end{bmatrix} = l\begin{bmatrix} -0.08638 \\ -0.09872 \\ -0.02468 \\ 0.02468 \end{bmatrix}$
Bending moment at the sections 6 and 12 $\vec{\mathbf{S}} = \vec{\mathbf{S}}_1 + \vec{\mathbf{S}}_2$	
$\vec{\mathbf{S}} = l\begin{bmatrix} 0.1480 \\ 0 \\ 0 \\ 0 \end{bmatrix} + l\begin{bmatrix} -0.06907 \\ -0.07893 \\ -0.01973 \\ 0.01973 \end{bmatrix} = l\begin{bmatrix} 0.07893 \\ -0.07893 \\ -0.01973 \\ 0.01973 \end{bmatrix}$	$\vec{\mathbf{S}} = l\begin{bmatrix} 0.1851 \\ 0 \\ 0 \\ 0 \end{bmatrix} + l\begin{bmatrix} -0.08638 \\ -0.09872 \\ -0.02468 \\ 0.02468 \end{bmatrix} = l\begin{bmatrix} 0.09872 \\ -0.09872 \\ -0.02468 \\ 0.02468 \end{bmatrix}$

The signs of the bending moments should be treated according to ***S-e*** diagram and general rules of the bending moments. If the load $P = 1$ is placed at point 2, then ordinates of influence lines for bending moments at sections $6(B)$ and $2(C)$ are given in Fig. 11.25c.

If the load $P = 1$ is placed at point 4, then ordinates of influence lines for bending moments at the same sections are following: $M_4^B = -0.09872l$; $M_4^C = 0.02468l$.

Load P $= 1$ ***in the second span*** The bending moment diagram in the primary system of displacement method and equivalent joint-load diagram are shown in Fig. 11.25d.

Now for any position u of unit load P in the second span, we can compile the vector $\mathbf{P}$ of the external joint loads (using the ***J-L*** and ***Z-P*** diagrams) and vector $\mathbf{S}_1$ of unknown internal forces in the first state (using the M_P^0 and **S-e** diagrams).

They are

$$\vec{\mathbf{P}} = l\begin{bmatrix} uv^2 \\ -u^2v \end{bmatrix}, \vec{\mathbf{S}}_1 = l\begin{bmatrix} 0 \\ -uv^2 \\ u^2v \\ 0 \end{bmatrix}$$

After that we will determine the entries of these both matrices when moving load P is placed at the sections 8 and 10 and perform corresponding matrix procedures.

Load P $=$ **1 *in the third span*** This case can be considered elementary by taking into account the symmetry of the beam and loading in the first span. The final influence lines for moments at the supports B and C are shown in Fig. 11.25e.

If load $P = 1$ is placed at sections with odd numbers $1, 3, \ldots$ then ordinates of influence lines M_B and M_C may be computed by similar way.

Discussion

Dimension of each submatrix depends on the type of element. For truss member and bending fixed-pinned member, each submatrix is a scalar. For fixed-fixed member, a stiffness matrix is (2×2); for this element the relationships between the bending moments at the ends and angular displacements at the ends according Table A.4 is described by (11.7), so

$$k_{f-f} = \frac{EI}{l}\begin{bmatrix} 4 & 2 \\ 2 & 4 \end{bmatrix}.$$

It is obvious that it is possible to *expand* the number of unknown forces at the ends of the element and consider the displacements at the ends. For example, we can consider four components for unknown forces (not only the bending moments, but the reactions also) and four components for end displacements (not only the angular displacements, but the linear displacements too). According Table A.4, the force–displacement relationships can be presented in the following form

$$\begin{bmatrix} R_1 \\ M_1 \\ R_2 \\ M_2 \end{bmatrix} = \underbrace{\frac{2EI}{l}\begin{bmatrix} 6/l^2 & 3/l & -6/l^2 & 3/l \\ 3/l & 2 & -3/l & 1 \\ -6/l^2 & -3/l & 6/l^2 & -3/l \\ 3/l & 1 & -3/l & 2 \end{bmatrix}}_{\mathbf{k}_{f-f}} \cdot \begin{bmatrix} \Delta_1 \\ \varphi_1 \\ \Delta_2 \\ \varphi_2 \end{bmatrix}$$

In this case we need to use for analysis the (4×4) – stiffness matrix. Similarly for pinned-fixed member the stiffness matrix becomes

$$\mathbf{k}_{p-f} = \frac{3EI}{l}\begin{bmatrix} 1/l^2 & -1/l^2 & 1/l \\ -1/l^2 & 1/l^2 & -1/l \\ 1/l & -1/l & 1 \end{bmatrix}.$$

It is possible to expand more the number of internal forces at the ends considering also the axial forces (and corresponding axial displacements). It is obvious that dimension of all initial and intermediate matrices become very large. So in this chapter the adopted stiffness matrix should be considered as a truncated matrix. Such form leads to the short and readily available for visual analysis of matrix procedures.

11.7 Analysis of Redundant Frames

Now we consider application of the matrix procedures for analysis of the simple frame (Fig. 11.26a); $l = h$, EI is constant. The frame has two unknowns of the displacement method. They are the angular displacement at joint 1 and linear displacement of cross bar. The primary system is shown in Fig. 11.26b.

The ancillary ***J-L***, ***Z-P***, and ***S-e*** diagrams are presented in Fig. 11.26c–e, respectively.

The vector of external equivalent joint loads on the basis of the ***J-L*** and ***Z-P*** diagrams becomes $\vec{\mathbf{P}} = \lfloor 0 \ P \rfloor^{\mathrm{T}}$ Since the load P is applied at joint then the fixed-end

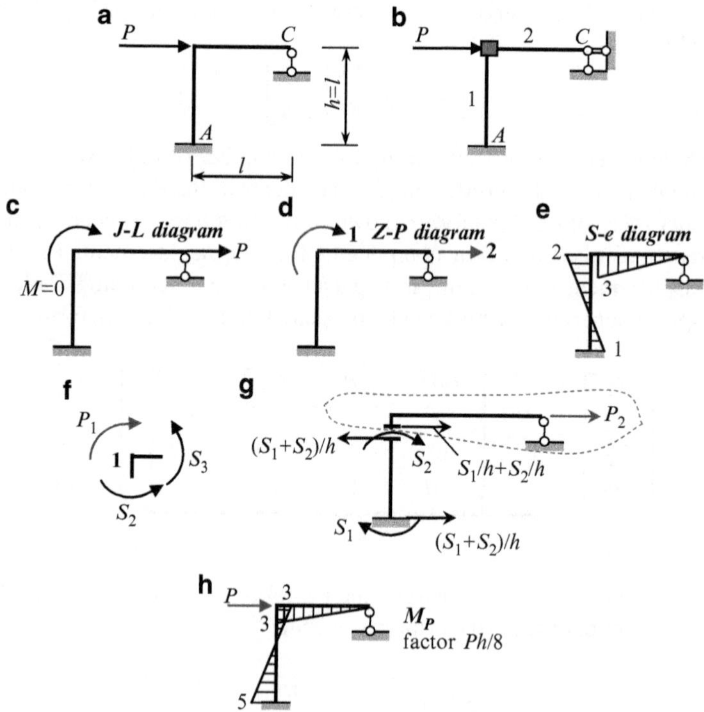

Fig. 11.26 (**a, b**) Design diagram of the frame, primary system and finite elements; (**c–e**) Ancillary diagrams; (**f**) Construction of static matrix; (**h**) Final bending moment diagram

moments are zero. Therefore, the vector of fixed-end moments (vector of internal forces of the first state) at sections 1–3 on the basis of the M_P^0 and ***S-e*** diagrams becomes $\vec{\mathbf{S}}_1 = \lfloor 0\ 0\ 0 \rfloor^{\mathrm{T}}$.

The static matrix is constructed on the basis of the ***Z-P*** and ***S-e*** diagrams. Figure 11.26f shows the free body diagram for joint 1 subjected to possible moment P_1 and two unknown internal forces in vicinity of joint 1 (bending moments S_2 and S_3). Equilibrium condition is $P_1 = S_2 + S_3$.

It is obvious that

$$P_2 = -\frac{S_1}{h} - \frac{S_2}{h}.$$

Thus, the static matrix becomes

$$\mathbf{A} = \begin{bmatrix} 0 & 1 & 1 \\ -1/h & -1/h & 0 \end{bmatrix}$$

Stiffness matrix for each finite element and stiffness matrix $\tilde{\mathbf{k}}$ for all structure in local coordinates are

$$\mathbf{k}_1 = \frac{EI_1}{l_1}\begin{bmatrix} 4 & 2 \\ 2 & 4 \end{bmatrix} = \frac{EI}{h}\begin{bmatrix} 4 & 2 \\ 2 & 4 \end{bmatrix},\ \mathbf{k}_2 = \frac{EI_2}{l_2}[3] = \frac{EI}{l}[3]$$

Thus, the stiffness matrices of the structure in the local coordinates

$$\tilde{\mathbf{k}} = \begin{bmatrix} \mathbf{k}_1 & 0 \\ 0 & \mathbf{k}_2 \end{bmatrix} = \frac{EI}{h}\begin{bmatrix} 4 & 2 & 0 \\ 2 & 4 & 0 \\ 0 & 0 & 3 \end{bmatrix}$$

Matrix procedures: Intermediate matrix complex

$$\tilde{\mathbf{k}}\mathbf{A}^{\mathrm{T}} = \frac{EI}{h}\begin{bmatrix} 4 & 2 & 0 \\ 2 & 4 & 0 \\ 0 & 0 & 3 \end{bmatrix} \cdot \begin{bmatrix} 0 & -1/h \\ 1 & -1/h \\ 1 & 0 \end{bmatrix} = \frac{EI}{h}\begin{bmatrix} 2 & -6/h \\ 4 & -6/h \\ 3 & 0 \end{bmatrix}$$

Stiffness matrix for whole structure in global coordinates

$$\mathbf{K} = \mathbf{A}\cdot\tilde{\mathbf{k}}\mathbf{A}^{\mathrm{T}} = \begin{bmatrix} 0 & 1 & 1 \\ -1/h & -1/h & 0 \end{bmatrix} \underbrace{\frac{EI}{h}\begin{bmatrix} 2 & -6/h \\ 4 & -6/h \\ 3 & 0 \end{bmatrix}}_{\tilde{\mathbf{k}}\mathbf{A}^{\mathrm{T}}} = \frac{EI}{h}\begin{bmatrix} 7 & -6/h \\ -6/h^2 & 12/h^2 \end{bmatrix}$$

For 2×2 matrix, we can use the following useful relationship: if

$$R = \begin{bmatrix} a & c \\ b & d \end{bmatrix}, \text{ then } R^{-1} = \frac{1}{\det R}\begin{bmatrix} d & -c \\ -b & a \end{bmatrix}.$$

In our case the inverse matrix becomes

$$\mathbf{K}^{-1} = \frac{h}{EI}\frac{1}{48/h^2}\begin{bmatrix} 12/h^2 & 6/h \\ 6/h & 7 \end{bmatrix} = \frac{h^3}{48EI}\begin{bmatrix} 12/h^2 & 6/h \\ 6/h & 7 \end{bmatrix}$$

Resolving equations: Vector of required displacements (angular of rigid joint and linear of cross bar) are

$$\vec{\mathbf{Z}} = \begin{bmatrix} Z_1\,(\text{rad}) \\ Z_2\,(\text{m}) \end{bmatrix} = \mathbf{K}^{-1}\vec{\mathbf{P}} = \frac{h^3}{48EI}\begin{bmatrix} 12/h^2 & 6/h \\ 6/h & 7 \end{bmatrix} \cdot \begin{bmatrix} 0 \\ P \end{bmatrix} = \frac{Ph^3}{48EI}\begin{bmatrix} 6/h \\ 7 \end{bmatrix}$$

We can see that in order to calculate the vector of end displacements **Z**, in fact, we need to have three matrices: they are static matrix **A**, internal stiffness matrix $\tilde{\mathbf{k}}$, and vector of external loads $\vec{\mathbf{P}}$.

Vector of internal unknowns bending moments is $\vec{\mathbf{S}}_{\text{fin}} = \vec{\mathbf{S}}_1 + \vec{\mathbf{S}}_2$, where moments of the first state $\vec{\mathbf{S}}_1 = \lfloor 0 \quad 0 \quad 0 \rfloor^{\mathrm{T}}$ because external load is applied at joint only. Therefore, the final vector of bending moments at sections 1–3 is

$$\vec{\mathbf{S}}_{\text{fin}} = \vec{\mathbf{S}}_2 = \tilde{\mathbf{k}}\mathbf{A}^{\mathrm{T}}\vec{\mathbf{Z}} = \underbrace{\frac{EI}{h}\begin{bmatrix} 2 & -6/h \\ 4 & -6/h \\ 3 & 0 \end{bmatrix}}_{\tilde{\mathbf{k}}\mathbf{A}^{\mathrm{T}}} \underbrace{\frac{Ph^3}{48EI}\begin{bmatrix} 6/h \\ 7 \end{bmatrix}}_{\vec{\mathbf{Z}}} = \frac{Ph}{8}\begin{bmatrix} -5 \\ -3 \\ 3 \end{bmatrix}$$

Corresponding bending moment diagram is shown in Fig. 11.26h. For calculation of reactions, the following algorithm can be applied: bending moments – shear forces – axial forces – reactions.

Now we show the application of the matrix displacement method for analysis of the frame in Fig. 11.27a; the relative bending stiffnesses are presented in circle. Analysis of this frame has been performed early by the both classical method, so this frame may be treated as the etalon one. Comparing with displacement method in canonical form will allow us to understand a physical meaning of an each matrix procedure.

The frame has two unknowns of the displacement method. They are the angular displacement at joint 1 and linear displacement of cross bar 1-C. The primary system and M_P^0 diagram are shown in Fig. 11.27b.

Ancillary ***Z-P*** diagram (Fig. 11.27c) shows that the structure has one possible angular displacement of joint 1 and corresponding possible external joint moment, as well as one horizontal displacement 2 of cross bar and corresponding force. Unknown internal forces (moments S_1–S_3) and their positive directions are shown on ***S-e*** diagram (Fig. 11.27d).

The finite elements are A-1, 1-B, and 1-C. The fixed end moments at joint 1 are

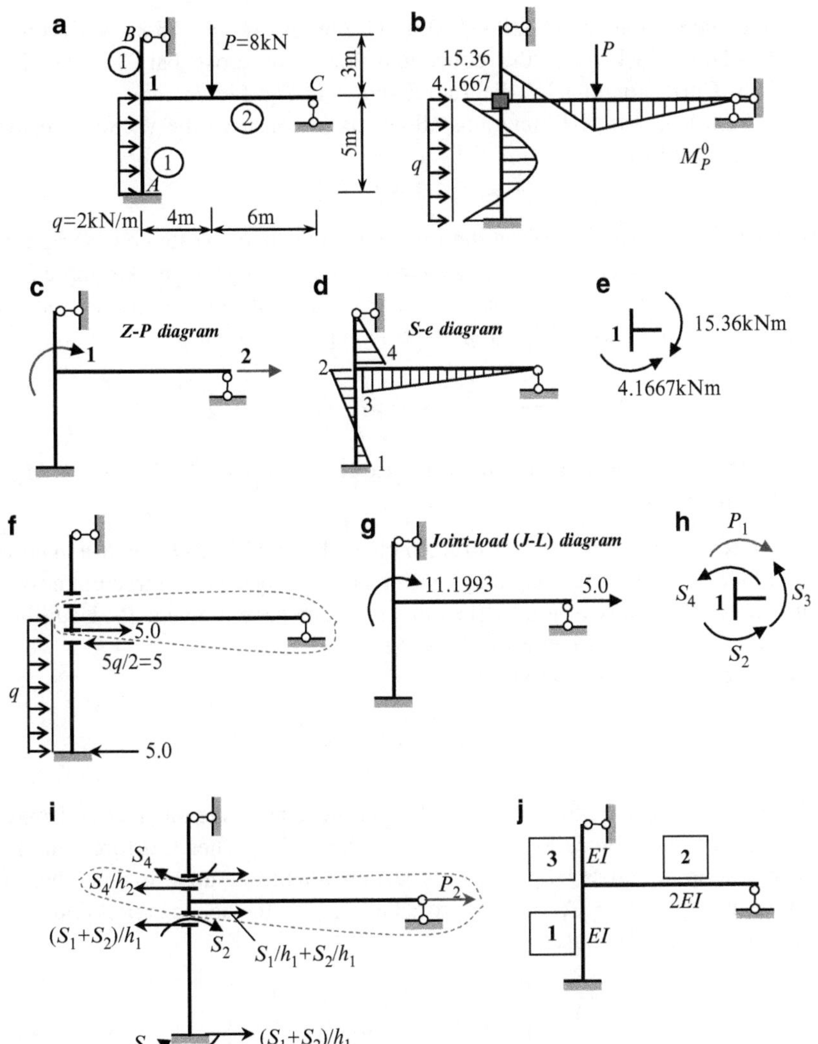

Fig. 11.27 (**a, b**) Etalon frame. Design diagram and calculation of fixed end moments; (**c, d**) ***Z-P*** and ***S-e*** diagrams; (**e–g**) Calculation of equivalent joint moment and force and joint-load diagram; (**h, i**) Construction of a static matrix; (**j**) Finite elements

$$M_{1-A} = M_{A-1} = \frac{ql_{1-A}^2}{12} = 4.1667\,\text{kNm}; \quad M_{1-B} = 0;$$

$$M_{1-C} = \frac{Pl}{2}\upsilon\left(1-\upsilon^2\right) = \frac{8\times 10}{2}\times 0.6\left(1-0.6^2\right) = 15.36\,\text{kNm}.$$

The equivalent moment at joint 1 (Fig. 11.27e) is $M = 15.36 - 4.1667 = 11.1993$ kNm (clockwise); the equivalent force for cross bar (Fig. 11.27f) is $P = 5$ kN. Corresponding ***J-L*** diagram is shown in Fig. 11.27g.

The joint-load and ***Z-P*** diagrams allows us to construct the vector of external equivalent joint loads

$$\vec{\mathbf{P}} = \lfloor 11.1993 \quad 5.0 \rfloor^{\mathrm{T}}.$$

The entries of this vector present the free terms, which are written on the right side of canonical equations of the Displacement method (compare with Example 8.2).

The vector of fixed-end moments (vector of internal forces of the first state) at sections 1–4 on the basis of the M_P^0 and ***S-e*** diagrams becomes

$$\vec{\mathbf{S}}_1 = \lfloor -4.1667 \quad 4.1667 \quad -15.36 \quad 0 \rfloor^{\mathrm{T}}.$$

This vector corresponds to the last term in expression $M = \overline{M}_1 Z_1 + \cdots + M_P^0$ for final bending moment.

Static matrix This matrix is constructed on the basis of the ***Z-P*** and ***S-e*** diagrams. Figure 11.27h shows free body diagram for joint 1 subjected to three unknown internal forces in vicinity of joint 1 (bending moments) and moment P_1. Equilibrium condition for joint 1 is $P_1 = S_2 + S_3 + S_4$.

It is not difficult to show that

$$P_2 = -\frac{S_1}{h_1} - \frac{S_2}{h_1} + \frac{S_4}{h_2}$$

Indeed, the positive moments S_1 and S_2 at the ends of the member A-1 may be equilibrated by two forces $S_1/h_1 + S_2/h_1$ (Fig. 11.27i). Then this force should be transmitted on the cross bar; similar procedure should be done for member 1-B. Equilibrium equation $\sum X = 0$ for cross bar leads to the above expression for P_2, so the static matrix becomes

$$\mathbf{A} = \begin{bmatrix} 0 & 1 & 1 & 1 \\ -1/h_1 & -1/h_1 & 0 & -1h_2 \end{bmatrix}$$

For given parameters h_1 and h_2 we get

$$\mathbf{A} = \begin{bmatrix} 0 & 1 & 1 & 1 \\ -0.2 & -0.2 & 0 & 0.333 \end{bmatrix}$$

Stiffness matrices of the elements in the local coordinates The finite elements are shown in Fig. refch11:fig11.27j; stiffness matrix for each member in local coordinates is

$$\mathbf{k}_1 = \frac{EI_1}{l_1}\begin{bmatrix} 4 & 2 \\ 2 & 4 \end{bmatrix} = \frac{EI}{5}\begin{bmatrix} 4 & 2 \\ 2 & 4 \end{bmatrix},$$

$$\mathbf{k}_2 = \frac{EI_2}{l_2}[3] = \frac{2EI}{10}[3] = \frac{EI}{5}[3],$$

$$\mathbf{k}_3 = \frac{EI_3}{l_3}[3] = \frac{EI}{3}[3] = \frac{EI}{5}[5].$$

For whole structure the stiffness matrix in local coordinates is

$$\tilde{\mathbf{k}} = \begin{bmatrix} \mathbf{k}_1 & 0 & 0 \\ 0 & \mathbf{k}_2 & 0 \\ 0 & 0 & \mathbf{k}_3 \end{bmatrix} = \frac{EI}{5}\begin{bmatrix} 4 & 2 & 0 & 0 \\ 2 & 4 & 0 & 0 \\ 0 & 0 & 3 & 0 \\ 0 & 0 & 0 & 5 \end{bmatrix}$$

Matrix procedures. For whole structure the stiffness matrix in global coordinates

$$\mathbf{K} = \mathbf{A}\tilde{\mathbf{k}}\mathbf{A}^{\mathrm{T}} = \begin{bmatrix} 0 & 1 & 1 & 1 \\ -0.2 & -0.2 & 0 & 0.333 \end{bmatrix} \frac{EI}{5}\begin{bmatrix} 4 & 2 & 0 & 0 \\ 2 & 4 & 0 & 0 \\ 0 & 0 & 3 & 0 \\ 0 & 0 & 0 & 5 \end{bmatrix} \cdot \begin{bmatrix} 0 & -0.2 \\ 1 & -0.2 \\ 1 & 0 \\ 1 & 0.333 \end{bmatrix}$$

$$= EI\begin{bmatrix} 2.4 & 0.093 \\ 0.093 & 0.207 \end{bmatrix}$$

The entries of this matrix are unit reactions of the displacement method in canonical form (Example 8.2).

The determinant of this matrix is detK = 0.48815, so the inverse matrix

$$\mathbf{K}^{-1} = \frac{1}{EI}\begin{bmatrix} 0.4241 & -0.1905 \\ -0.1905 & 4.9165 \end{bmatrix}.$$

The matrix resolving equation $\mathbf{K}\vec{\mathbf{Z}} = \vec{\mathbf{P}}$ allows us to find the vector of unknown displacements

$$\vec{\mathbf{Z}} = \begin{bmatrix} Z_1\,(\mathrm{rad}) \\ Z_2\,(\mathrm{m}) \end{bmatrix} = \mathbf{K}^{-1}\vec{\mathbf{P}} = \frac{1}{EI}\begin{bmatrix} 0.4241 & -0.1905 \\ -0.1905 & 4.9165 \end{bmatrix} \cdot \begin{bmatrix} 11.193 \\ 5 \end{bmatrix}$$

$$= \frac{1}{EI}\begin{bmatrix} 3.7944 \\ 22.452 \end{bmatrix}$$

These values present the angle of rotation of the rigid joint and linear displacement of the cross-bar; they have been obtained previously by the displacement method (Example 8.2).

Vector of internal unknowns bending moments is $\vec{\mathbf{S}}_{\text{fin}} = \vec{\mathbf{S}}_1 + \vec{\mathbf{S}}_2$, where

$$\vec{\mathbf{S}}_2 = \tilde{\mathbf{k}}\mathbf{A}^{\mathrm{T}}\vec{\mathbf{Z}} = \frac{EI}{5}\begin{bmatrix} 4 & 2 & 0 & 0 \\ 2 & 4 & 0 & 0 \\ 0 & 0 & 3 & 0 \\ 0 & 0 & 0 & 5 \end{bmatrix} \cdot \begin{bmatrix} 0 & -0.2 \\ 1 & -0.2 \\ 1 & 0 \\ 1 & 0.333 \end{bmatrix} \frac{1}{EI}\begin{bmatrix} 3.7944 \\ 22.452 \end{bmatrix} = \begin{bmatrix} -3.8707 \\ -2.3529 \\ 2.2766 \\ 11.2709 \end{bmatrix}$$

Control: $\mathbf{A}\overline{\mathbf{S}} = \vec{\mathbf{P}}$. In our case

$$\begin{bmatrix} 0 & 1 & 1 & 1 \\ -0.2 & -0.2 & 0 & 0.333 \end{bmatrix} \cdot \begin{bmatrix} -3.8707 \\ -2.3529 \\ 2.2766 \\ 11.2709 \end{bmatrix} = \begin{bmatrix} 11.1946 \\ 4.9979 \end{bmatrix}$$

Final bending moments at specified sections 1–4 are

$$\vec{\mathbf{S}}_{\text{fin}} = \vec{\mathbf{S}}_1 + \vec{\mathbf{S}}_2 = \begin{bmatrix} -4.1667 \\ 4.1667 \\ -15.36 \\ 0 \end{bmatrix} + \begin{bmatrix} -3.8707 \\ -2.3529 \\ 2.2766 \\ 11.2709 \end{bmatrix} = \begin{bmatrix} -8.0374 \\ 1.8138 \\ -13.0834 \\ 11.2709 \end{bmatrix}$$

This vector allows us to construct the bending moment diagram. All ordinates should be plotted according ***S-e*** diagram (Fig. 11.27d). For example, $M_1 = -8.037\,\text{kNm}$ should be plotted at the support A left at neutral line. Final bending moment diagram is presented in Fig. 8.2g. Note that stiffness matrix method is precise and some disagreement with data obtained previously is a result of the rounding off.

11.8 Analysis of Statically Indeterminate Trusses

Figure 11.28a presents the statically indeterminate truss; the stiffness EA for all members is equal. We need to compute the displacements of the all joints, and calculate the internal forces.

First let us construct the ***Z-P*** diagram (Fig. 11.28b). This diagram shows possible joint displacements and corresponding possible loads. After that we can construct the vector of external forces

$$\vec{P} = \begin{bmatrix} 0 & -4 & -2 & 0 & 0 \end{bmatrix}^{\mathrm{T}}.$$

Vector has five entries because the given structure allows five possible joint displacements. The first entry of this matrix (0) means that in the possible direction 1 the active force is absent.

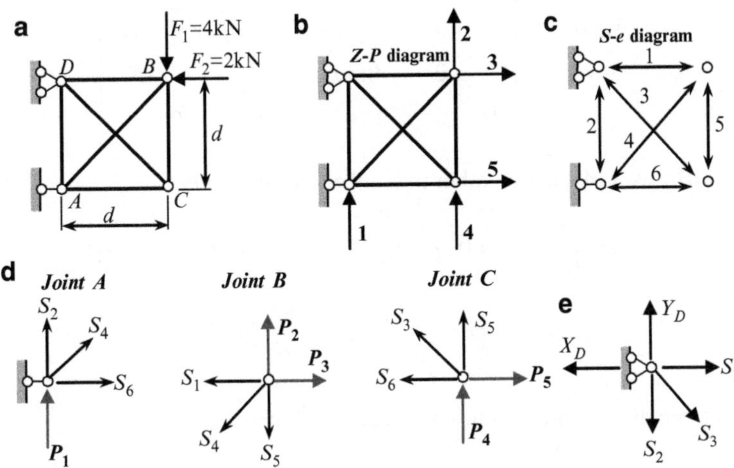

Fig. 11.28 (**a-c**) Truss and corresponding ***Z-P*** and ***S-e*** diagrams; (**d**) Free-body diagram for each joint; (**e**) Computation of reactions at0 support D

To construct the static matrix let us show the ***S-e*** diagram (Fig. 11.28c). Then we need to consider free body diagram for joints which have possible displacements (all joints except the pinned support D) and express the possible forces P_1–P_5 in terms of unknown internal forces S_1–S_6. This step is presented in Fig. 11.28d.

$$\text{Joint A}: \sum Y = 0 \to P_1 = -S_2 - 0.707S_4$$

$$\text{Joint B}: \sum Y = 0 \to P_2 = 0.707S_4 + S_5$$

$$\sum X = 0 \to P_3 = S_1 + 0.707S_4$$

$$\text{Joint C}: \sum Y = 0 \to P_4 = -0.707S_3 - S_5$$

$$\sum X = 0 \to P_5 = 0.707S_3 + S_6$$

Thus, the static matrix becomes

$$\mathbf{A}_{(5\times 6)} = \begin{bmatrix} 0 & -1 & 0 & -0.707 & 0 & 0 \\ 0 & 0 & 0 & 0.707 & 1 & 0 \\ 1 & 0 & 0 & 0.707 & 0 & 0 \\ 0 & 0 & -0.707 & 0 & -1 & 0 \\ 0 & 0 & 0.707 & 0 & 0 & 1 \end{bmatrix}$$

The stiffnesses of each member in local coordinates are

$$k_1 = \frac{EA}{l_1} = \frac{EA}{d}[1]; \quad k_2 = k_5 = k_6 = k_1 = \frac{EA}{d}[1];$$

$$k_3 = k_4 = \frac{EA}{d\sqrt{2}}[1] = \frac{EA}{d}[0.707].$$

So stiffness matrix of all structure in local coordinates is

$$\tilde{\mathbf{k}} = \frac{EA}{d}\begin{bmatrix} 1 & 0 & 0 & 0 & 0 & 0 \\ 0 & 1 & 0 & 0 & 0 & 0 \\ 0 & 0 & 0.707 & 0 & 0 & 0 \\ 0 & 0 & 0 & 0.707 & 0 & 0 \\ 0 & 0 & 0 & 0 & 1 & 0 \\ 0 & 0 & 0 & 0 & 0 & 1 \end{bmatrix}$$

Stiffness matrix of all structure in global coordinates is

$$\mathbf{K} = \mathbf{A}\tilde{\mathbf{k}}\mathbf{A}^{\mathrm{T}} = \frac{EA}{d}\begin{bmatrix} 1.3534 & -0.3534 & -0.3534 & 0 & 0 \\ -0.3534 & 1.3534 & 0.3534 & -1 & 0 \\ -0.3534 & 0.3534 & 1.3534 & 0 & 0 \\ 0 & -1 & 0 & 1.3534 & -0.3534 \\ 0 & 0 & 0 & -0.3534 & 1.3534 \end{bmatrix}$$

Inverse matrix may be calculated by computer using a standard program. This matrix is

$$\mathbf{K}^{-1} = \frac{d}{EA}\begin{bmatrix} 0.8965 & 0.5 & 0.1035 & 0.3965 & 0.1035 \\ 0.5 & 2.4149 & -0.5 & 1.9149 & 0.5 \\ 0.1035 & -0.5 & 0.89655 & -0.3965 & -0.1035 \\ 0.3965 & 1.9149 & -0.3965 & 2.313 & 0.6035 \\ 0.1035 & 0.5 & -0.1035 & 0.6035 & 0.8965 \end{bmatrix}$$

Vector displacements

$$\vec{\mathbf{Z}} = \begin{bmatrix} Z_1 \\ Z_2 \\ Z_3 \\ Z_4 \\ Z_5 \end{bmatrix} = \mathbf{K}^{-1}\vec{\mathbf{P}} = \frac{d}{EA}\begin{bmatrix} -2.2071 \\ -8.6594 \\ 0.2071 \\ -6.8665 \\ -1.7929 \end{bmatrix}$$

The negative sign for Z_1 indicates that the joint A according ***Z-P*** diagram has a negative displacement, i.e., downwards.

Vector of internal forces

$$\vec{\mathbf{S}} = \begin{bmatrix} S_1 \\ S_2 \\ S_3 \\ S_4 \\ S_5 \\ S_6 \end{bmatrix} = \tilde{\mathbf{k}}\mathbf{A}^{\mathrm{T}}\vec{\mathbf{Z}} = \begin{bmatrix} 0.2071 \\ 2.2071 \\ 2.5360 \\ -3.1217 \\ -1.7929 \\ -1.7929 \end{bmatrix}$$

The negative sign for S_4 according ***S-e*** diagram means that the diagonal member 4 is compressed. The units for all internal forces are kN.

Control: $\mathbf{A}\overline{\mathbf{S}} = \vec{\mathbf{P}}$. In our case

$$\mathbf{A}\vec{\mathbf{S}} = \begin{bmatrix} 0 & -1 & 0 & -0.707 & 0 & 0 \\ 0 & 0 & 0 & 0.707 & 1 & 0 \\ 1 & 0 & 0 & 0.707 & 0 & 0 \\ 0 & 0 & -0.707 & 0 & -1 & 0 \\ 0 & 0 & 0.707 & 0 & 0 & 1 \end{bmatrix} \cdot \begin{bmatrix} 0.2071 \\ 2.2071 \\ 2.5360 \\ -3.1217 \\ -1.7929 \\ -1.7929 \end{bmatrix} = \begin{bmatrix} 0 \\ -4 \\ -2 \\ 0 \\ 0 \end{bmatrix}$$

For calculation of reaction of supports, we need to consider equilibrium conditions for rolled and pinned supports. For example, the free-body diagram for pinned support D and corresponding equilibrium equations are shown in Fig. 11.28e.

$$\sum X = 0 \rightarrow X_D = S_1 + 0.707S_3 = 2.0\,\mathrm{kN},$$
$$\sum Y = 0 \rightarrow Y_D = S_2 + 0.707S_3 = 4.0\,\mathrm{kN}.$$

Of course, for this *externally* statically determinate structure all reactions may be determined considering the structure in whole. However, for calculation of reactions we used typical approach as for any statically indeterminate structure.

11.9 Summary

1. The MSM is modern effective method for analysis of any deformable structures subjected to arbitrary actions. Among them are external loads, change of temperature, settlements of supports. For trusses and frames with straight members, the MSM leads to the exact results. Curvilinear member should be replaced by the set of inscribed straight members. In this case, MSM leads to the approximate results.
2. Design diagram of the MSM presents the set of uniform straight members connected by hinged or fixed joints. Any action should be replaced by the equivalent loads (moments and forces), which should be applied at the joints. The primary unknowns of the MSM and their number are same as at displacement method in canonical form, i.e., the angular displacements of the fixed joints and independent linear displacement of joints. The unknown forces (in the simplest version of the MSM) are axial forces for truss members and bending moment at the fixed ends for the bending members.
3. Arbitrary external exposure (loads, change of temperature, and settlement of supports) should be transformed into equivalent joint loads and presented in the form of the ***J-L*** diagram.
4. The ***Z-P*** diagram contains information about possible displacements of the joints and type of corresponding external load. In case of truss the ***Z-P*** diagram shows linear displacement of joints and concentrated forces along these displacements. For rigid joint of a frame, the ***Z-P***diagram shows angular displacement of the joint and couple; in case of linear displacement of the joints, the ***Z-P***diagram shows independent linear displacement and force.
5. The ***S-e*** diagram contains information about location and signs of the required internal forces S. For truss the unknown forces are axial forces at the each member; for bending member the unknown forces are bending moment at the fixed joint of the primary system.
6. The static matrix **A** connects the possible external loads P and required internal forces S. For computation of members of this matrix, it is necessary to express each possible external load P in terms of unknown internal forces S. The member a_{ik} is coefficient at unknown force S_k in an equation for P_i. The number m of rows of matrix **A** equals to the number of possible external forces P; the number n of columns equals to the number of the unknown internal forces. If $m > n$ then structure is geometrically changeable, if $m = n$ then structure is statically determinate; if $m < n$ then structure is statically indeterminate. The entries a_{ik} may be positive, negative, or zero.
7. Deformation matrix **B** connects the end deformation of each finite element in the primary system of the displacement method and unit displacement of introduced constraints. The member b_{ik} is displacement in direction of unknown force S_i due to unit displacement of introduced constraint k. The number of rows of matrix **B** equals to the number of required internal forces S; the number of columns equals to the number of the introduced constraints of displacement method. The deformation and static matrices obey to equation $\mathbf{B} = \mathbf{A}^{\mathrm{T}}$.

8. The stiffness matrix **k** of finite members in local coordinate connects unknown internal force S and displacement of the end of the element. These matrices may be presented in truncated or in expanded form. In truncated form, the stiffness matrices for truss member and bending fixed-pinned member presents a scalar, while for fixed-fixed member (2×2) matrix. The stiffness matrix in local coordinate for whole structure $\tilde{\mathbf{k}}$ presents matrix which contains on the principal diagonal the stiffness matrices of separate members. This matrix is square, symmetrical, and all entries are positive.
9. Juxtapose ***Z-P*** and ***J-L*** diagrams leads to the vector of external forces **P**. The number of entries of this vector equals to the number of primary unknown of the displacement method. For truss, the number of entries of this vector equals to the number of the possible displacements of the joints. If external joint load at the direction of the possible displacement is absent then corresponding entry of the vector **P** is zero. If a structure is subjected to different groups of external loads, then **P** will present a matrix. The number of the columns equals to the number of the set loading.
10. The stiffness matrix **K** of whole structure in global coordinate is symmetrical square $(n \times n)$ matrix where n is a number of primary unknowns of displacement method. The members of this matrix are the unit reactions r_{ik} of the displacement method in canonical form; equation $\mathbf{KZ} = \mathbf{P}$ is exactly canonical equations of displacement method.
11. Juxtapose ***S-e*** and M_P^0 diagrams leads to the vector of internal forces $\mathbf{S}_1$ of the first state. The number of entries of this vector equals to the number of unknown bending moments at the rigid joints. For truss $\mathbf{S}_1 = \mathbf{0}$.
12. The final results may be presented using two matrices as follows: The vector of joint displacement is $\mathbf{Z} = \mathbf{K}^{-1}\mathbf{P}$ and the internal forces is $\mathbf{S} = \mathbf{S}_1 + \mathbf{S}_2$, where the vector of unknown internal forces of the second state is $\mathbf{S}_2 = \tilde{\mathbf{k}}\mathbf{A}^{\mathrm{T}} \cdot \mathbf{Z}$. Thus, the joint displacements and the distribution of internal forces of any structure are defined only by three matrices. They are the static matrix **A** of a structure, stiffness matrix $\tilde{\mathbf{k}}$ of a structure in local coordinates, and the vector of external forces **P**.
13. To plot the final bending moment diagram, the signs of obtained final moments should be juxtaposed with **S-e** diagram. The shear force can be calculated on the basis of the bending moment diagram; the axial forces can be calculated on the basis of the shear diagram. Reactions of supports can be calculated on the basis of the axial and shear forces and bending moment diagrams.

Problems

11.1a-c. The uniform two-span beam is subjected to fixed load as shown in Fig. P11.1. The flexural rigidity of the beam is EI. Determine the angle of rotation at support 1and the bending moments at specified points.

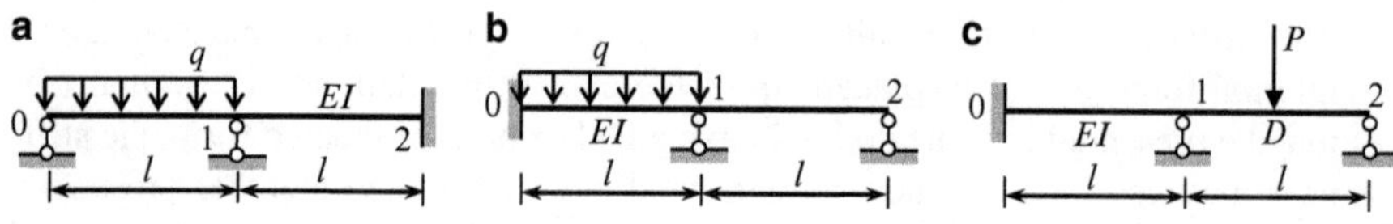

Fig. P11.1

Ans. (a) $Z_1 = -\dfrac{ql^3}{56EI}$, $M_1 = -\dfrac{ql^2}{14}\left(\dfrac{-}{+}\right)$; (b) $Z_1 = -\dfrac{ql^3}{84EI}$, $M_1 = -\dfrac{ql^2}{28}$;

(c) $Z_1 = \dfrac{3Pl^2}{112EI}$.

11.2. The uniform three-span beam with different spans is subjected to uniformly distributed load q (Fig. P11.2). The flexural rigidity of the beam is EI. Determine the angle of rotation at support 1 and the bending moments at specified points. Compare the result with data in Table A.12.

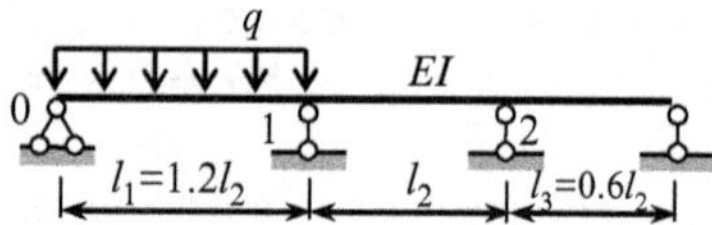

Fig. P11.2

Ans. $M_1 = -0.0734ql_1^2$

11.3. The uniform three-span beam with equal spans l is subjected to the settlement of support 1 as shown in Fig. P.11.3. The flexural rigidity of the beam is EI. Determine the angle of rotation at support 1 and the bending moments at specified points. Compare with data at the Table A18.

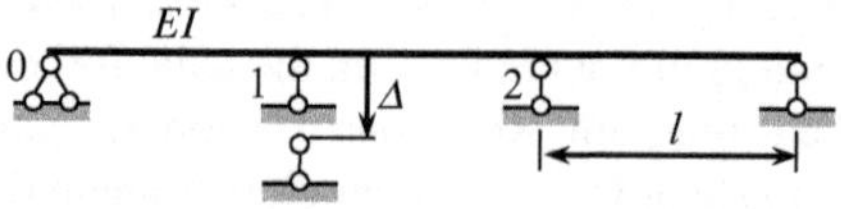

Fig. P11.3

Ans. $M_1 = 3.6\dfrac{EI}{l^2}\Delta, \quad M_2 = -2.4\dfrac{EI}{l^2}\Delta$

11.4. Design diagram of the uniform two-span continuous beam is presented in Fig. P11.4. Construct the influence line for angle of rotation and bending moment at the support B (section 6). Compare with data at the Table A.9

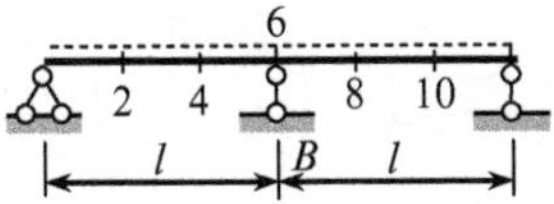

Fig. P11.4

11.5a,b. The frames in Fig. P11.5 are subjected to uniformly distributed load q. Construct the bending moment diagrams. Present your answer in terms of *EI*, l, and q. Compare distribution of bending moment within the columns and explain their difference.

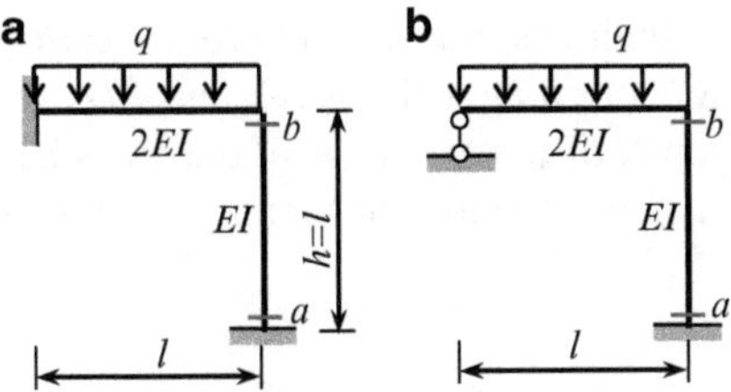

Fig. P11.5

Ans. (a) $M_a = -\frac{ql^2}{72}, \quad M_b = \frac{ql^2}{36}$; (b) $M_a = M_b = \frac{ql^2}{56}$

11.6a,b. The combined structures are subjected to load as shown in Fig. P11.6a, b. Construct the internal force diagrams. Calculate the reactions of support and provide check of results. For vertical members are used the steel wide-flange shape W150 × 22($I = 12.1 \times 10^6\,\mathrm{mm}^4$, for tie rod Ø4.3 cm; ($\frac{A}{I} = 125\,\mathrm{m}^{-2}$).

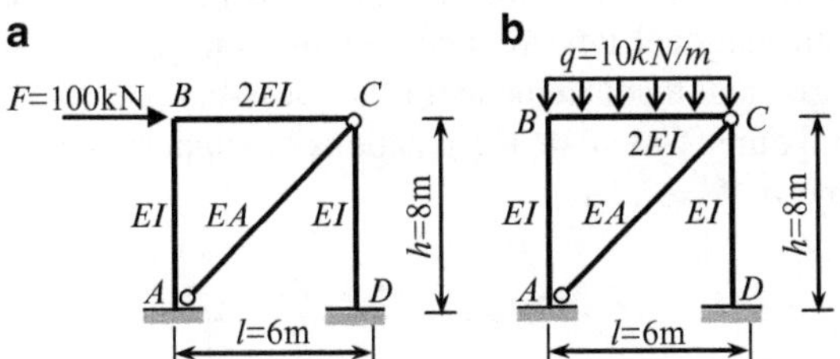

Fig. P11.6

Ans. (a) $N_{A-C} = 165.8$ (kN) ; (b) $N_{A-C} = 4.663$ (kN)

11.7a,b. The portal frames with absolutely rigid cross bar are subjected to load as shown in Fig. P11.7a, b. Calculate the horizontal displacement of the cross-bar. Construct the internal force diagrams. Calculate the reactions of support and axial force S in the cross bar. Provide your answer in terms of *EI*, h, and given load.

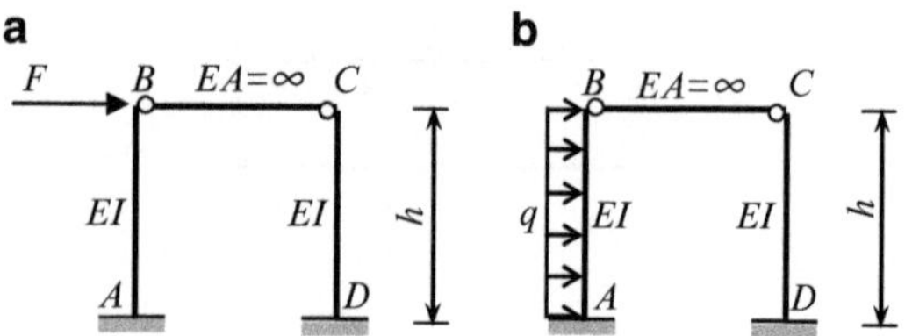

Fig. P11.7

Ans. (a) $M_A = M_D = Fh/2, \quad S = F/2;$

$$\text{(b) } M_A = \frac{5}{16}qh^2, \quad M_D = \frac{3}{16}qh^2, \quad S = \frac{3}{16}qh \text{ (compr)}$$

11.8. The portal frame with absolutely rigid cross bar BC is subjected to load F as shown in Fig. P11.8. Both diagonal ties AC and BD are not connected at point K. Calculate the horizontal displacement Z of the cross-bar. Construct the internal force diagrams. Calculate the reactions of support and axial force S in the cross bar. Assume $A/I = 60\,\text{m}^{-2}$.

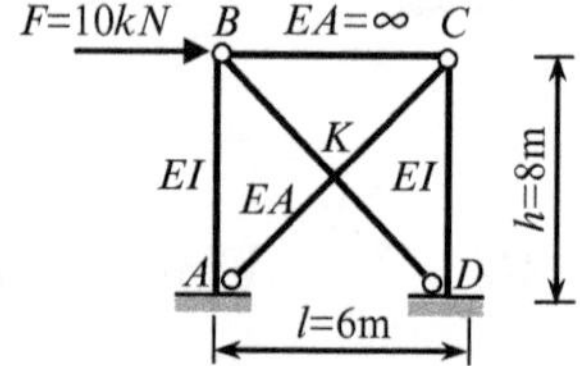

Fig. P11.8

Ans. $Z_{BC} = 2.308/EI, \ M_A = M_D = 0.108\,\text{kNm}, \ S_{AC} = -S_{BD} = 8.31\,\text{kN},$ $S_{BC} = F/2.$

11.9. The portal frame with absolutely rigid cross bar BC is subjected to load F (Fig. P11.9). Both diagonal ties AC and BD are connected by hinge at point K. Calculate the horizontal displacements of the cross-bar and joint K. Construct the internal force diagrams. Calculate the reactions of support and axial force S in the cross bar. Assume $A/I = 60\,\text{m}^{-2}$.

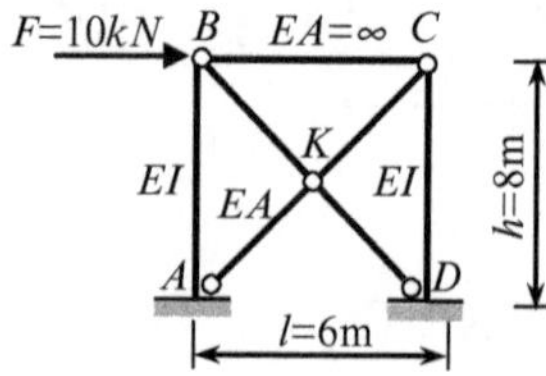

Fig. P11.9

Ans. $Z_{BC} = 2.308/EI, \ Z_K^{\text{hor}} = 1.154/EI, \ M_A = M_D = 0.108\,\text{kNm}, S_{KC} =$ $S_{AK} = 8.31\,\text{kN}$

11.10. Calculate the displacement of the joints and internal forces in all member of the truss; *EA*= constant for all elements, $d = 1$ m (Fig. P11.10).

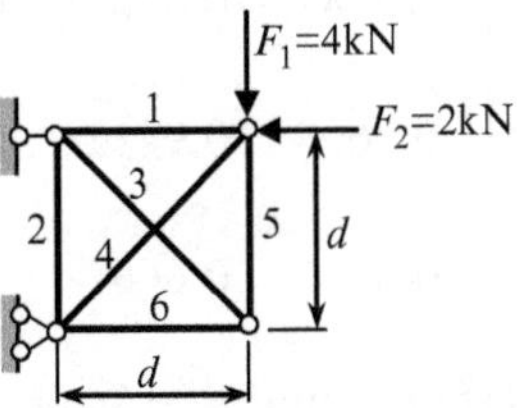

Fig. P11.10

Ans. $S_1 = 0.6212;\ S_2 = S_5 = S_6 = -1.3788;\ S_3 = 1.9503;\ S_4 = -3.7074$

11.11. The externally and internally statically indeterminate truss is subjected to two groups of external loads. The first group is $F_1 = 12$ kN and $F_2 = 5$ kN; the second group is $N_1 = 16$ kN and $N_2 = 4$ kN as shown in Fig. P11.11. For each set of load calculate internal forces in each element. Knowing internal forces compute the reactions of supports. EA = constant for all elements.

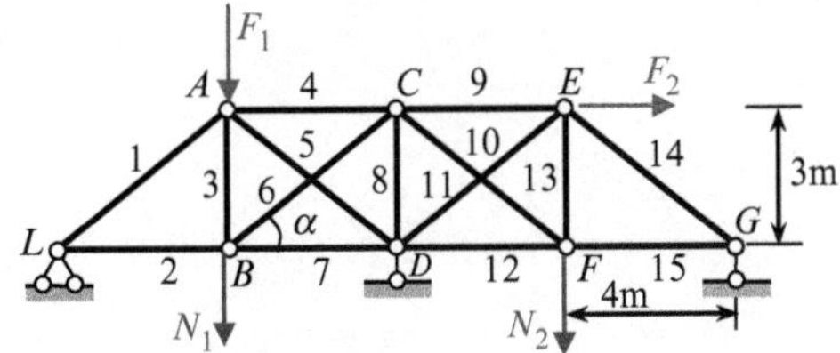

Fig. P11.11

Ans. Set F: $S_1 = -6.2910;\ S_2 = 10.0328;\ldots,\ S_{15} = -0.4672.\ R_D = 8.57606$ kN

11.12. Construct the influence lines for internal force in each member of the internally statically indeterminate truss shown in Fig. P11.12. *EA* = constant for all elements.

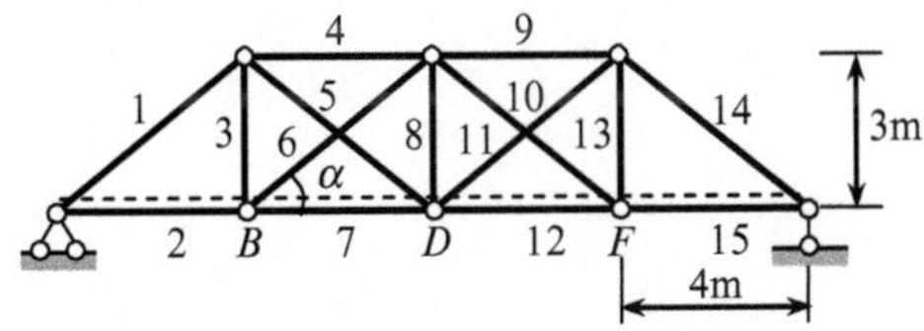

Fig. P11.12

Ans. Ordinates of influence lines

$$\mathbf{S} = \begin{matrix} S_1 \\ S_2 \\ \cdots \\ S_7 \\ \cdots \\ S_{15} \end{matrix} \begin{bmatrix} -1.250 & -0.8333 & -0.4167 \\ 1.000 & 0.6667 & 0.3333 \\ \cdots & \cdots & \cdots \\ 0.7608 & 0.9412 & 0.4941 \\ \cdots & \cdots & \cdots \\ 0.3333 & 0.6667 & 1.000 \end{bmatrix}$$

The columns 1–3 corresponds to location $P = 1$ at joint B, D and F, respectively.

Part III
Special Topics

Chapter 12
Plastic Behavior of Structures

This chapter is devoted to the analysis of a structure, taking into account the plastic properties of material. Such analysis allows the use of the reserves of strength of material, which remains unused considering the material of structure as elastic. Therefore, plastic analysis allows us to define the limit load on the structure and to design a more economical structure. Fundamental idea of the plastic analysis is discussed using the direct method. Kinematical and statical methods of calculation of the limit loads are considered. Detailed plastic analysis of the beams and frames are presented.

12.1 Idealized Stress–Strain Diagrams

In the previous chapters, we considered structures taking into account only elastic properties of materials for all members of a structure. Analysis of a structure based on elastic properties of material is called the elastic (or linear) analysis. Elastic analysis does not allow us to find out the *reserve* of strength of the structure beyond its elastic limit. Also this analysis cannot answer the question: what would happen with the structure, if the stresses in its members will be larger than the proportional limit? Therefore, a problem concerning to the actual strength of a structure cannot be solved using elastic analysis.

The typical stress–strain diagram for the specimen of structural steel is presented in Fig. 12.1a. Elastic analysis corresponds to the initial straight portion of the $\sigma-\varepsilon$ diagram. If a specimen is loaded into the proportional limit (or below) and then released, then material will unload along the loading path back to the origin (Fig. 12.1a). So, there are no residual strains. This property of unloaded specimen to return to its original dimensions is called *elasticity*, and material in this region is called the *linearly elastic*. Within the elastic region, a relationship between stress and strain obey to Hooke's law $\sigma = E\varepsilon$.

Let a specimen is loaded into the elastic limit. The stress at this point slightly exceeds the proportional limit. From this point, the material unloads along the line that is parallel to straight portion of the diagram and thus, the material has the very small *residual* strain (Fig. 12.1a).

I.A. Karnovsky and O. Lebed, *Advanced Methods of Structural Analysis*,
DOI 10.1007/978-1-4419-1047-9_12, © Springer Science+Business Media, LLC 2010

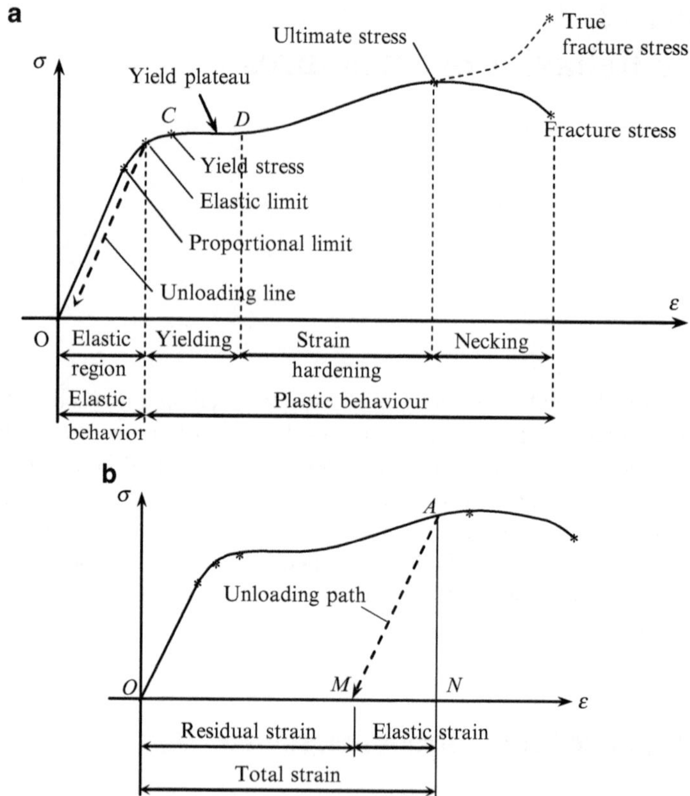

Fig. 12.1 (**a**) Typical stress–strain diagram for structural steel. (**b**) Loading-unloading diagram

Plastic behavior starts at the elastic limit. The region *CD* is referred as the *perfect plastic* zone. In this region, the specimen continues to elongate without any increase in stress. Above the yield plateau, starting from point *D*, the behavior of the specimen is described by nonlinear relationships $\sigma - \varepsilon$. If the specimen will be unloaded at point *A* (Fig. 12.1b), then unloading line will be parallel to the load straight line, so the specimen returns only partially to its original length. Total strain of the specimen is *ON*, while the strain *MN* has been recovered elastically and the strain *OM* remains as residual one.

If the material remains within the elastic region, it can be loaded, unloaded, and loaded again without significantly changing the behavior. However, when the load is reapplied in a plastic region, the internal structure of material is altered, its properties change, and the material obeys to Hook's law within the straight line *MA*; it means that the proportional limit of the material has been increased. This process is referred to as the *strain-hardening*.

For plastic analysis, we change the typical diagram by its idealized diagram. Different idealized diagrams are considered in engineering practice. Some of idealized models are presented in Table 12.1.

Table 12.1 Idealized $\sigma-\varepsilon$ diagrams for axially loaded members

Material			
Elasto-plastic	Rigid-plastic	Elasto-plastic with linear hardening	Rigid-plastic with linear hardening
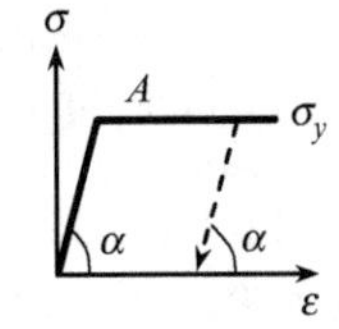 σ, A, σ_y, α, α, ε	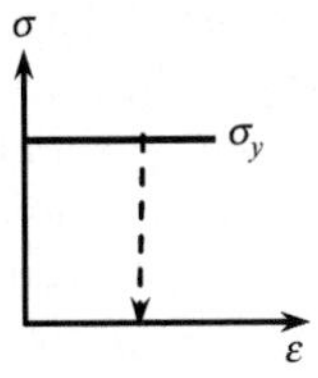 σ, σ_y, ε	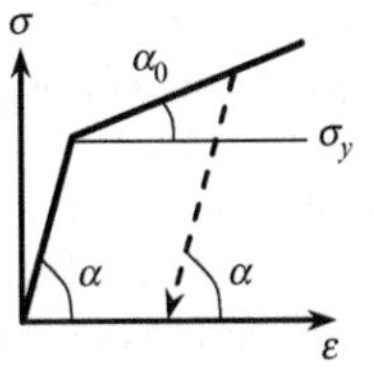 σ, α_0, σ_y, α, α, ε	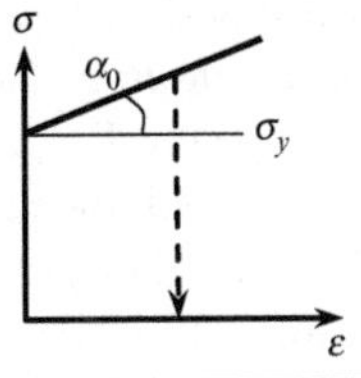 σ, α_0, σ_y, ε

For further analysis, we will consider idealized elasto-plastic material and rigid-plastic material. We start from elasto-plastic material; corresponding diagram is called Prandtl diagram. This diagram has two portions – linear "stress–strain" part and the yield plateau. Elastic properties of material hold up to yield point stress σ_y. The yield plateau shows that displacement of material can become indeterminately large under the same stress. Idealized elasto-plastic material does not have the effect of hardening. This diagram may be applicable for a structural steel and for reinforced concrete. Structural analysis on the basis of idealized diagram is referred as the *plastic analysis*. The quantitative results of plastic analysis are much closer to the actual behavior of a structure than the results obtained on the basis of elastic properties of material.

In case of statically determinate structure, yielding of any member leads to the failure of the structure as a whole. Other situation occurs in case of statically indeterminate structure. Assume that for all members of the structure, the Prandtl diagram is applicable. In the first stage, when loads are small, behavior of all members follows the first portion of the Prandtl diagram. Proportional increase of all loads leads to the yielding in the most loaded member. It means that the degree of statical indeterminacy is decreased by one. The following proportional increase of all loads leads to the following effect: the internal force in the *yielding* member remains the same, while the forces in the *other* members will be increased. This effect will be continued until the next member starts to yield. Finally, the structure becomes statically determinate and yielding of any member of this structure immediately leads to the failure of the structure, since the structure is transformed into a mechanism. In general, if the structure has n redundant constrains, then its failure occurs when the number of yielding member becomes $n + 1$. Its means that capability of a structure to carry out the increasing load has been exhausted. This condition is called *limit equilibrium condition*. In this condition, the limit loads and internal forces satisfy to equilibrium condition. The following increase of a load is impossible. In this condition, the displacement of the structure becomes undefined. While the linear portion of typical stress–strain diagram leads to linear problems of structural analysis (elastic problems), the Prandtl diagram leads to nonlinear problems of plastic behavior of structures. Indeed, the design diagram of a structure is changed upon different levels of loads. Transition from one design diagram to another happens abruptly.

Let us consider plane bending of a beam of a rectangular ($b \times h$) cross section. In the elastic region of the stress–strain diagram, the normal stresses are distributed within the height of a cross-section of the beam linearly. The maximum tensile and compressed stresses are located at the extreme fibers of the beam. The stress σ_y corresponds to yield plateau (Fig. 12.2a). Increasing of the load leads to appearance and developing of the yield zone and decreasing of the "elastic core" of the section of the beam. Diagrams in Fig. 12.2b, c correspond to partially plastic bending of a beam, which means that the middle part of the cross-section is in elastic condition, while the bottom and top parts of the beam are in plastic condition. Further increasing of load leads to complete plastic state (Fig. 12.2d), which corresponds to the limit equilibrium, i.e., we are talking about appearance of so-called *plastic hinge* (Fig. 12.2d, e). It is obvious that all sections of the beam are in different states. Defining of the location of the plastic hinge is an additional problem of plastic analysis. This problem will be considered below.

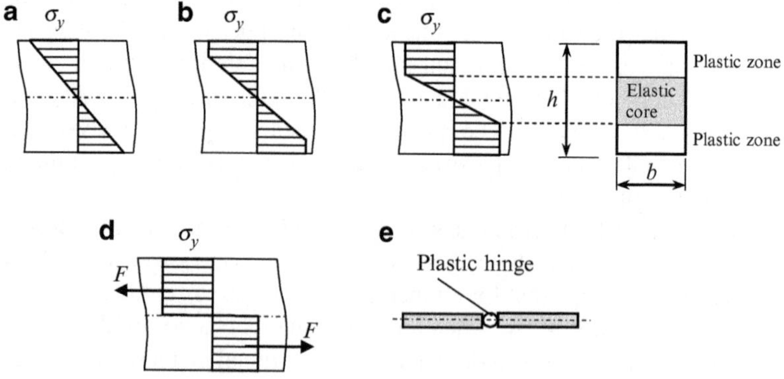

Fig. 12.2 Distribution of normal stresses within the height of a beam

What is the difference between plastic and ideal hinge? First, the plastic hinge disappears if the structure is unloaded, so the plastic hinge may be considered as fully recoverable or one-sided hinge. Second, in the ideal hinge, the bending moment equals to zero, while plastic hinge is characterized by the appearance of bending moment, which is equal to the limit (or plastic) moment of internal forces $F = \sigma_y \frac{bh}{2}$ (Fig. 12.2d). A bearing capability of a structure is characterized by the plastic moment

$$M_p = F\frac{h}{2} = \sigma_y \frac{bh^2}{4}.$$

Plastic analysis involves determination of plastic load or limit load, which structure can resist before full failure due to yielding of some elements. The limiting load does not depend on settlements of supports, errors of fabrication, prestressed tension, and temperature changes; this is a fundamental difference between plastic and elastic analysis. In the following sections, we will consider different methods of determining plastic loads.

12.2 Direct Method of Plastic Analysis

The fundamental concept of plastic analysis of a structure may be clearly presented using the direct method. Let us consider the structure shown in Fig. 12.3a subjected to load P at point K. The horizontal rod is absolutely rigid. All hangers have constant stiffness *EA*. The plastic analysis must be preceded by elastic analysis.

Elastic analysis This analysis should be performed on the basis of any appropriate method of analysis of statically indeterminate structures. Omitting this analysis, which is familiar for reader and presents no difficulties, the distribution of internal forces in members 1–4 of the structure is as follows (Fig. 12.3b):

$$N_1 = 0.4P; \quad N_2 = 0.3P; \quad N_3 = 0.2P; \quad N_4 = 0.1P$$

Plastic analysis
Step 1: Increasing of load leads to the appearance of the yield stresses. They are reached in the most highly stressed member. In our case, this member is element 1. Let N_1 become equal to limit load, i.e. $N_1 = N_y$. Since $N_1 = 0.4P$, then it occurs if external load would be equal to $P = \frac{N_y}{0.4} = 2.5N_y$. For this load P, the limit tension will be reached in the first hanger. Internal forces in another members are (Fig. 12.3c)

$$N_2 = 0.3P = 0.3 \cdot 2.5N_y = 0.75N_y; \quad N_3 = 0.5N_y; \quad N_4 = 0.25N_y$$

Step 2: If load P will be increased by value ΔP_2, then $N_1 = N_y$ remains without changes. It means that additional load will be distributed between three members 2, 3, and 4, i.e., the design diagram had been changed (Fig. 12.3d). This structure is once statically indeterminate. Elastic analysis of this structure due to load ΔP_2 leads to the following internal forces

$$\overline{N}_2 = 0.833\Delta P_2; \quad \overline{N}_3 = 0.333\Delta P_2; \quad \overline{N}_4 = -0.167\Delta P_2.$$

As always, the most highly stressed member will reach the yield stress first. Since first hanger is already in yield condition (and cannot resist any additional load), the most highly stressed member due to load ΔP_2 is the second hanger. The total limit load in this element equals

$$N_2 = 0.75N_y + 0.833\Delta P_2.$$

In this formula, the first term corresponds to initially applied load $P = 2.5N_y$ (Fig. 12.3c), while the second term corresponds to additional load ΔP_2.

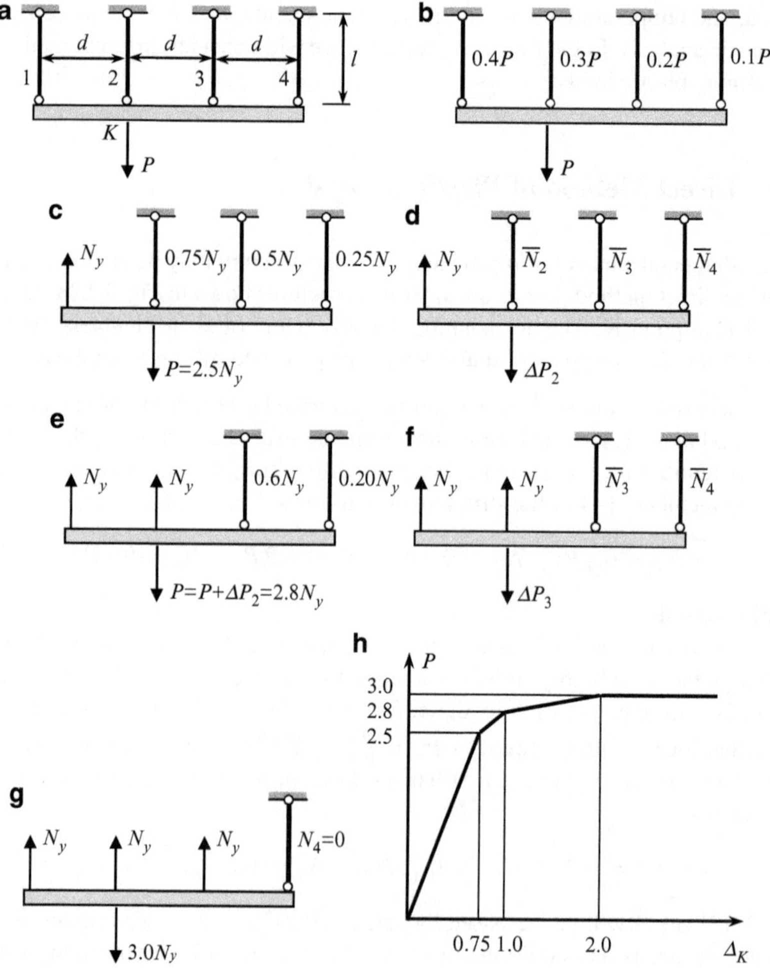

Fig. 12.3 (**a, b**) Design diagram and distribution of internal forces according to elastic analysis; (**c**) Step 1 – Plastic state in the member 1 and internal forces in the rest members; (**d**) Step 2 – Internal forces in the members 2–4 due to load ΔP_2; (**e**) Step 2 – Plastic state in the members 1 and 2; (**f**) Step 3 – Internal forces in the members 3–4 due to load ΔP_3; (**g**) Step 4 – Plastic state in the members 1, 2, and 3; (**h**) $P - \Delta_K$ diagram in plastic analysis

The limit load for the second hanger is $N_2 = N_y$. Thus equation $N_2 = 0.75N_y + 0.833\Delta P_2 = N_y$ leads to the following value for increment of load $\Delta P_2 = \frac{0.25N_y}{0.833} = 0.3N_y$.

Thus, the value $\Delta P_2 = 0.3N_y$ represents additional load, which is required so that the second hanger reaches its yielding state. Therefore, if load

$$P = 2.5N_y + 0.3N_y = 2.8N_y,$$

then both members 1 and 2 reach their limit state. As this takes place, the internal forces in hangers 3 and 4 are (Fig. 12.3e) are following

$$N_3 = 0.5N_y + 0.333 \cdot 0.3N_y = 0.6N_y; \quad N_4 = 0.25N_y - 0.167 \cdot 0.3N_y = 0.20N_y$$

Step 3: Since internal forces in hangers 1 and 2 reached the limit values, then the following increase of the load by value ΔP_3 (Fig. 12.3f) affects the members 3 and 4 only. Elastic analysis of this statically determinate structure leads to the following internal forces in members 3 and 4: $\overline{N}_3 = 2\Delta P_3$ and $\overline{N}_4 = -\Delta P_3$.
Step 4: Similarly as above, the limit state for this case occurs if internal force in hanger 3 reaches its limit value

$$N_3 = 0.6N_y + 2\Delta P_3 = N_y.$$

This equation leads to the following value for increment of the load

$$\Delta P_3 = \frac{0.4N_y}{2} = 0.2N_y$$

The total value of external force (Fig. 12.3g)

$$P = 2.5N_y + \Delta P_2 + \Delta P_3 = 2.5N_y + 0.3N_y + 0.2N_y = 3.0N_y$$

The first term in this formula corresponds to limit load in the first member; increment of the force by $0.3N_y$ leads to the limit state in the second member. The following increment of the force by $0.2N_y$ leads to the limit state in the third member. After that the load carrying capacity of the structure is exhausted. From the equilibrium equation for the entire structure, we can see that on this stage $N_4 = 0$ (Fig. 12.3g).

All forces satisfy to equilibrium condition. Plastic behavior analysis leads to the increment of the limit load by $\frac{3-2.5}{2.5}100\% = 20\%$.

Plastic displacements If some of the elements reached its limiting value and the load continues to increase, then we cannot determine displacements of the system using only elastic analysis. However, plastic analysis allows calculating displacements of a structure on the each stage of loading. Let us show the graph of displacement of the point application of force P (point K).

If load $P = 2.5N_y$, then internal force in second element equals $0.75N_y$ (see Fig. 12.3c) and vertical displacement of point K is $\Delta_K = 0.75\frac{N_y l}{EA}$.

If load $P = 2.8N_y$, then internal force in second element equals N_y (see Fig. 12.3e) and vertical displacement of point K is $\Delta_K = \frac{N_y l}{EA}$.

If load $P = 3.0N_y$, then internal force in third element equals N_y (see Fig. 12.3g) and deflection of this element equals $\frac{N_y l}{EA}$. Since internal force in fourth element equals zero, its deflection is zero and required displacement $\Delta_K = 2\frac{N_y l}{EA}$. Corresponding $P - \Delta_K$ diagram is shown in Fig. 12.3h; the factors l/EA and N_y for horizontal and vertical axis, respectively. This diagram shows that plastic analysis is nonlinear analysis.

12.3 Fundamental Methods of Plastic Analysis

Analysis of the plastic behavior of a structure may be performed also by kinematical and static methods. Both these methods are exact and much easier than the direct method. The first of these methods deals with various *forms* of failure, while the second method deals with various *distributions* of internal force satisfying equilibrium conditions. The idea of these methods is explained below.

Absolutely rigid rod is suspended by four hangers 1–4 as shown in Fig. 12.3a. The axial stiffness of all members *EA* is constant. Find the limit load considering both methods.

12.3.1 Kinematical Method

This method requires consideration of different *forms of failure* of a structure. For each form of failure, there is a corresponding well-defined value of the failure load. The actual limit load is a *minimum* load among all possible failure loads.

Let us consider all possible forms of the failure of the structure; they are shown in Fig. 12.4a. Each form 1–4 presents position of the rigid rod MK for different scheme of failures. Assume that elastic displacements of hangers are much less then plastic ones and may be ignored. Therefore, each position is obtained by rotation of the rod around the point of connection of a hanger and the rod. Internal forces, which arise in each hanger, are N_y.

Now for each scheme of failure, we need to find the load P using equilibrium condition.

Scheme 1: $P = \frac{N_y(d+2d+3d)}{d} = 6N_y$
Scheme 2: $P = \infty$, since the moment arm of the force P with respect to point of rotation is zero
Scheme 3: $P = \frac{N_y(2d+d+d)}{d} = 4N_y$
Scheme 4: $P = \frac{N_y(3d+2d+d)}{2d} = 3N_y$

Corresponding values of P is shown on the schemes 1–4. The minimum value of $P = 3N_y$. This case corresponds to rotation of the rod around the point D. This result coincides with result, which was already obtained by direct method.

12.3.2 Static Method

According to this method, it is necessary to find all possible *distributions* of internal forces in statically indeterminate structure; we assume that for each distribution, the internal forces do not exceed the limit load. For each distribution of internal force, there is a corresponding well-defined value of the external load P. Actual limit load is a maximum load among all possible limit loads.

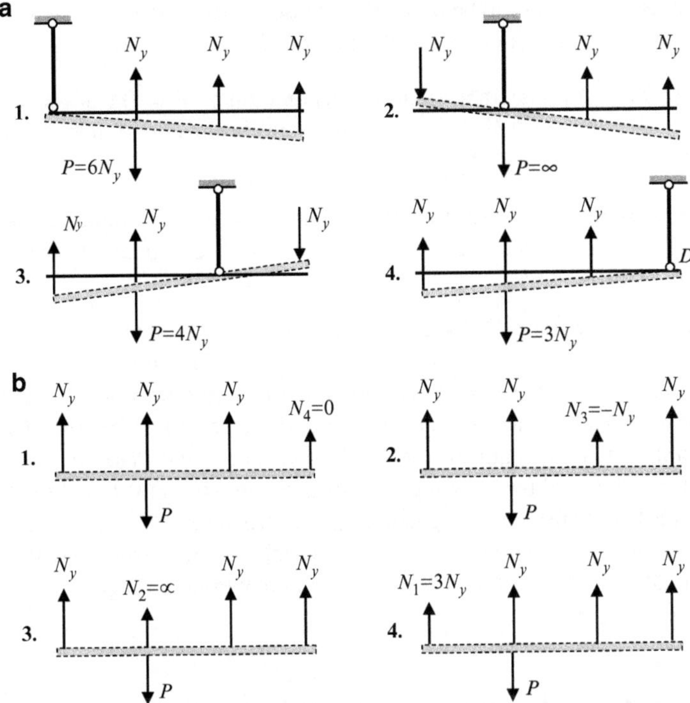

Fig. 12.4 (**a**) Kinematical method. Different forms of failure of the structure; (**b**) Static method. Different stress conditions of the structure

Let us consider various *distributions of internal forces* in the structure. For all these schemes, the following two conditions should be met:

1. Since the structure contains four hangers, then forces in any three members have to be equal to the limit load N_y, which means that these three members correspond to horizontal part of Prandtl diagram
2. The force in the remaining forth member should be less than N_y.

There is possible the following cases.

1. $N_1 = N_2 = N_3 = N_y$
2. $N_1 = N_2 = N_4 = N_y$
3. $N_1 = N_3 = N_4 = N_y$
4. $N_2 = N_3 = N_4 = N_y$

The force in the remaining forth member should be defined from equilibrium condition. The sum of the moments with respect to point of application of the force P leads to the following results:

$$1.\, N_4 = 0, \quad 2.\, N_3 = -N_y, \quad 3.\, N_4 = \infty, \quad 4.\, N_4 = 3N_y$$

These internal forces are shown in Fig. 12.4b.

The cases 3 and 4 should be omitted, because internal forces are greater than N_y. Now we need to consider cases 1 and 2 in more detail.

Case 1: Since $N_4 = 0$, then equilibrium equation $\sum Y = 0$ leads to $P = 3N_y$.
Case 2: Since $N_3 = -N_y$, then the same equilibrium equation leads to result: $P = 3N_y - N_y = 2N_y$.

The maximum load which the system can resist is $3N_y$; thus, the actual limit plastic load corresponds to case 1. This result have been obtained earlier by direct and kinematical methods.

Summary For any statically indeterminate structure, there are a number of various *forms of failure*, which are possible. The *kinematical theorem* states that the true form of collapse is that one that corresponds to the minimum value of the limit load.

For any statically indeterminate structure, there are a number of *internal force distributions* satisfying equilibrium conditions. The *static theorem* states that the distribution of internal forces, which occurs at the maximum value of the limit load, corresponds to exhausted bearing capacity of a structure.

Both the methods express two extreme properties of plastic load for statically indeterminate structure, all members of which obey Prandtl diagram.

12.4 Limit Plastic Analysis of Continuous Beams

So far we have discussed plastic analysis of a structure in case of ideal elasto-plastic material. Now we will consider the ideal rigid-plastic material (column 2, Table 12.1). When this stress–strain diagram may be adopted? Since elastic displacements of a structure are significantly less than plastic displacements, then these elastic displacements may be ignored. In this case, the material of the structure is called as idealized rigid-plastic material, and the structure is called as rigid-plastic one. This material is not real, but using this material, the procedure for plastic analysis of elasto-plastic structures may be simplified. This simplification is based on the following fact: if two structures, which are made from elasto-plastic and rigid-plastic materials, have the same limit plastic load, then the limit condition of the elasto-plastic structure asymptotically approaches the limit condition of the rigid-plastic structure.

Let a structure is subjected to different loads simultaneously. Assume that each load may increase independently of each other. In this case, the limit conditions may be approached under different values of loads. What will be the limit load in this case? Concept of "limit load" becomes unclear. Therefore, let us assume that the loading is simple. It means that if the structure is subjected to different loads P_1 and $P_2 = \lambda P_1$ acting simultaneously, then these loads are increasing together and parameter λ of the load (coefficient between loads) remains constant during entire loading process.

A plastic analysis of idealized rigid-plastic structures may be performed using two principal methods, namely static and kinematical methods. Fundamental condition for both the methods is that in the limit plastic state, the bending moments at all plastic hinges are equal to yielding moment M_y, which is a characteristic of material and cross-section of the beam. For plastic analysis by both methods, it is necessary to show a collapse mechanism first.

Let us consider application of static and kinematical methods for plastic analysis of continuous beams. Two-span beam of constant cross-section subjected to two equal forces P is shown in Fig. 12.5a. It is required to determine the limit load.

12.4.1 Static Method

First, let us consider the behavior of structure subjected to given load and the failure mechanism.

First stage In the elastic condition, the bending moment diagram is presented in Fig. 12.5b. Increasing of the loads leads to the increasing of bending moment ordinates. Since maximum bending moment occurs at the support 1 ($M_1 = 3Pl/16$), then a material of the beam begins to yield at this support. Spreading of the plastic zone at the support 1 during the increasing loading is shown in Fig. 12.2. Thus, the first plastic hinge appears at this support, and the entire two-span continuous beam is transformed into two simply supported beams. The maximum possible bending moment at the support 1 will be equal to the limit bending moment M_y. So the moment at plastic hinge will be M_y. Corresponding design diagram is presented in Fig. 12.5c.

Second stage Each of these simply supported beams is subjected to force P and plastic moment M_y at support 1. Corresponding design diagram is shown in Fig. 12.5d (in fact, Fig. 12.5c and 12.5d are equivalent). Bending moment at the point of application of force equals to

$$M\left(\frac{l}{2}\right) = \frac{Pl}{4} - \frac{M_y}{2}.$$

Again, we will increase the loads P. It is clear that the maximum moment occurs at the point of application of load P. The spreading of elastic zone of material at this point is as shown in Fig. 12.2. Finally, a new plastic hinge occurs within the span (in this simplest case, this plastic hinge will be located at the point of application of the load). As the result, three hinges will be located on the on the each span of the beam; they are – hinge at support A, hinge under the point of application of the load P, and hinge at support 1. The nature of these hinges is different. Hinge at the point A is ideal one, which represents support, while two other hinges are plastic ones, and they are the result of exhausted bearing capability of the beam. The same situation is with the second beam 1-B. Even though the hinges are of different nature, but since they are located on one line, this leads to the failure of the structure.

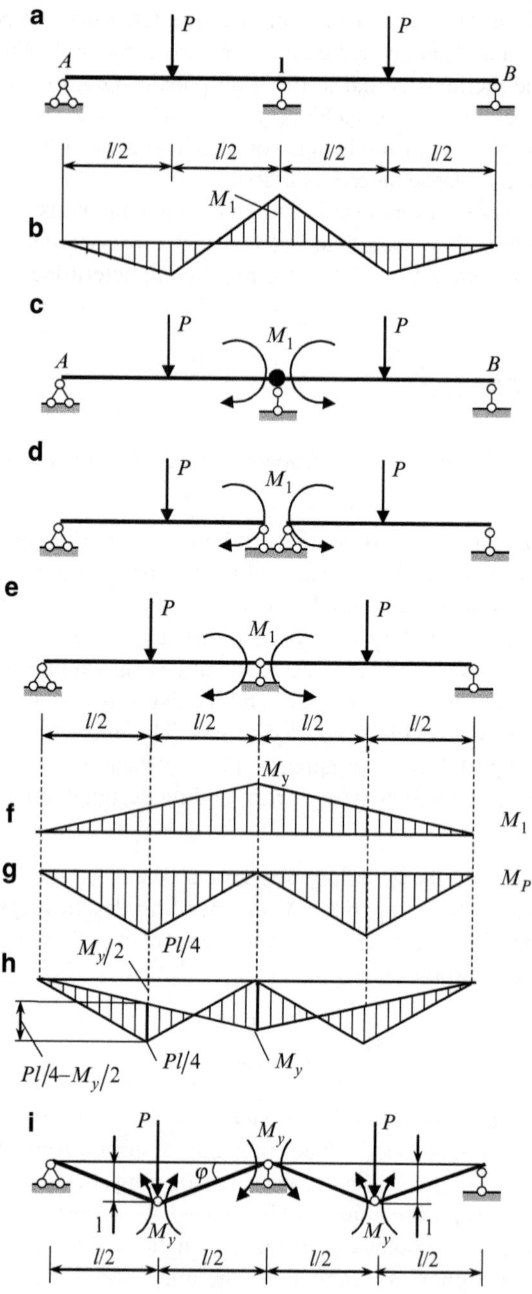

Fig. 12.5 (**a, b**) Continuous beam. Loading and bending moment diagram in elastic state; (**c, d**) Plastic hinge at the support 1 and presenting the continuous beam as two separate beams; (**e–h**) Finding of limit load by graphical method; (**i**) Calculation of limit plastic load by kinematical method

In the limit state, the bending moment at the point of application of force P should be equal to the limit moment, i.e.,

$$\frac{Pl}{4} - \frac{M_y}{2} = M_y. \tag{12.1}$$

This equation leads to the limit load

$$P_{\text{lim}} = \frac{6M_y}{l}.$$

Limit load can be also found using graphical procedure, based on (12.1) using the superposition of two bending moment diagrams; this procedure is presented in Fig. 12.5e–h.

These diagrams are caused by limit plastic moment at the plastic hinge 1 and load P. Plastic hinge at support 1 with plastic moment $M_1 = M_y$ is shown in Fig. 12.5e; corresponding bending moment diagram is shown in Fig. 12.5f. Bending moment diagram for two simply supported statically determinate beams with forces P is shown in Fig. 12.5g; the final bending moment at the force point is $\frac{Pl}{4} - \frac{M_y}{2}$. If this ordinate is less than plastic moment M_y, then the limit state at the point of application of force does not occur.

Let us find such value of P so that $\frac{Pl}{4} - \frac{M_y}{2} = M_y$. This could be found by the procedure of equalizing of the final bending moment at the point of application of force and the limit moment at support 1 as shown in Fig. 12.5h by bold lines. This procedure leads to the value of limit load $P_{\text{lim}} = \frac{6M_y}{l}$.

12.4.2 Kinematical Method

This method is based on the following idea: in the limit state, the total work done by unknown plastic loads and all plastic bending moments M_y is zero. This method consists of the following steps:

1. Identify the location of the potential plastic hinges. These hinges may be located at supports and at points of concentrated loads. In case of distributed load, plastic hinge may appear at point of zero shear. Also, plastic hinges may occur at the joints of the frame. Thus, we identify possible failure mechanisms.
2. Equilibrium equations should be written for each failure mechanism. In our case, the failure mechanism is shown in Fig. 12.5i. This failure mechanism is as follows: plastic hinge first appears at the middle support and thus, turning the original statically indeterminate beam into two statically determinate simply supported beams. With further increase of load P, plastic hinges appear at the points of application of forces P. As the result, three hinges are located on one line and thus, the system becomes instantaneously changeable, which corresponds to limit state.

Assume that in the limit state, the vertical displacement at the each force P equals unity. Then the angle of inclination of the left and right parts of each mechanism is

$$\varphi = \pm\frac{1}{l/2} = \pm\frac{2}{l}.$$

The work done by limit loads equals $W(P) = 2P_{\text{lim}} \cdot 1$. The work W_1 produced by limit moments M_y at the intermediate support equals $W_1 = -2M_y\varphi$. The work W_2 produced by limit moments at the points of load P equals $W_2 = -4M_y\varphi$ (these plastic moments are shown according to location of extended fibers in elastic stage). Since $W(P) + W_1 + W_2 = 0$, then this condition can be rewritten as follows:

$$2P_{\text{lim}} \cdot 1 = 2M_y\varphi + 4M_y\varphi = \frac{12M_y}{l}.$$

The limit load becomes

$$P_{\text{lim}} = \frac{6M_y}{l}.$$

As it was expected, results obtained by both static and kinematical methods are identical.

Some typical examples of plastic analysis of statically indeterminate beams are shown below.

Example 12.1. Design diagram of pinned-clamped beam, which is subjected to concentrated load P, is presented in Fig. 12.6a. Calculate the limit load P.

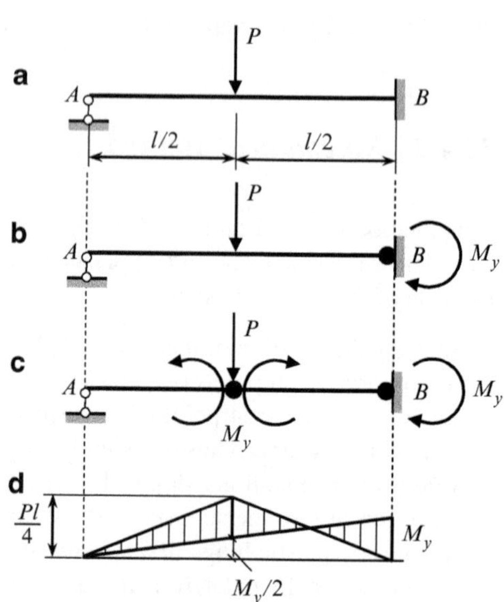

Fig. 12.6 Plastic analysis and graphical calculation of limit load for pinned-clamped beam

Solution. From the elastic analysis of the beam we know that the maximum moment occurs at the support B. Therefore, just here will be located the first plastic hinge. It is shown by solid circle; corresponding plastic moment is M_y (Fig. 12.6b). Now we have simply supported beam, i.e., the appearance of the plastic hinge do not destroy the beam but led to the changing of design diagram. Therefore, we can increase the load P until the second plastic hinge appears at the point of application of the load. Thus, we have three hinges (Fig. 12.6c), which are located on the one line, so this structure becomes instantaneously changeable.

Using superposition principle, the moment at the point of application of force is

$$M\left(P_{\text{lim}}\right)+M\left(M_y\right)=\frac{P_{\text{lim}}l}{4}-\frac{M_y}{2}.$$

In the limit condition, the moment at the point of application of force equals to plastic moment:

$$\frac{P_{\text{lim}}l}{4}-\frac{M_y}{2}=M_y. \tag{a}$$

This equation leads to the limit load P

$$P_{\text{lim}}=\frac{6M_y}{l}.$$

The procedure (a) may be presented graphically as shown in Fig. 12.6d.

Example 12.2. Design diagram of a pinned-clamped beam is presented in Fig. 12.7. Calculate the limit load q and find the location of a plastic hinges.

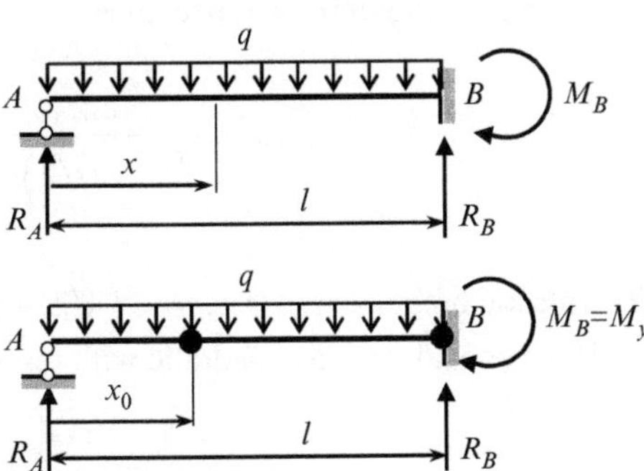

Fig. 12.7 Plastic analysis of pinned-clamped beam

Solution. It is obvious that the first plastic hinge appears at the clamped support B; corresponding plastic moment is M_y. Now we have simply supported beam AB (this design diagram is not shown), so we can increase the load q until the second plastic hinge appears. Location of this plastic hinge will coincide with position of maximum bending moment of simply supported beam subjected to plastic moment at the support B and given load q.

For this beam, the general expressions for shear and bending moment at any section x are

$$Q(x) = R_A - qx = \left(\frac{ql}{2} - \frac{M_B}{l}\right) - qx,$$

$$M(x) = R_A x - \frac{qx^2}{2} = \left(\frac{ql}{2} - \frac{M_B}{l}\right) x - \frac{qx^2}{2}.$$

In these expressions, the moment M_B at the support B equals to plastic moment M_y.

The maximum moment occurs at the point where $Q(x) = 0$. This condition leads to $x_0 = \frac{l}{2} - \frac{M_B}{ql}$.

Corresponding bending moment equals

$$M_{\max}(x_0) = \left(\frac{ql}{2} - \frac{M_B}{l}\right) \cdot \left(\frac{l}{2} - \frac{M_B}{ql}\right) - \frac{q}{2}\left(\frac{l}{2} - \frac{M_B}{ql}\right)^2.$$

The limit condition becomes when this moment and the moment at the support B will be equal to M_y, i.e. $M_{\max} = M_y$. This condition leads to the following equation

$$M_y^2 - 3M_y ql^2 + \frac{q^2 l^4}{4} = 0. \quad \text{(a)}$$

If we consider this equation as quadratic with respect to M_y, then solution of this equation is $M_y = \frac{ql^2}{2}\left(3 \pm \sqrt{8}\right)$. Minimum root is $M_y = \frac{ql^2}{2}\left(3 - 2\sqrt{2}\right) = 0.08578ql^2$. The limit load becomes

$$q_{\lim} = \frac{2M_y}{\left(3 - 2\sqrt{2}\right) l^2} = 11.657 \frac{M_y}{l^2}. \quad \text{(b)}$$

The plastic hinge occurs at $x_0 = l\left(\sqrt{2} - 1\right) = 0.4142l$.

If we consider (a) as quadratic with respect to q, then solution of this equation is

$$q_{\max} = \frac{2M_y}{l^2}\left(3 + 2\sqrt{2}\right).$$

This result coincides with (b).

Example 12.3. Two-span beam with overhang is subjected to force P and uniformly distributed load q as shown in Fig. 12.8a. The loading of the beam is simple; assume that relationship between loads is always $P = 2ql_2$. Determine the limit load, if $a = 3\,\text{m}$, $b = 4\,\text{m}$, $l_2 = 6\,\text{m}$, $l_3 = 2\,\text{m}$, and bearing capacity of all cross sections within the beam is $M_y = 60\,\text{kNm}$.

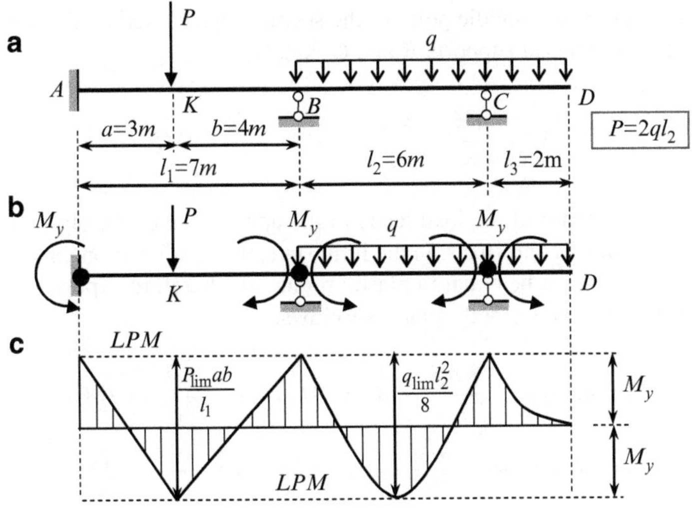

Fig. 12.8 (**a**) Design diagram; (**b**) Beam with plastic hinges at supports; (**c**) Bending moment diagram inscribed between two limit plastic moments (LPM)

Solution. The given structure has two redundant constraints. There exist different failure mechanisms. Let us consider one of them. The progressive increase of the loads leads to the appearance of the plastic hinge at one of the supports, so the structure becomes statically indeterminate of the first degree. Further increase of the loads leads to the appearance of the plastic hinge at another support. Finally, plastic hinge happens at the last support. It means that the entire continuous beam is being transformed into two simply supported beams subjected to given loads and plastic moments M_y at the supports A, B and C as shown in Fig. 12.8b. Direction of moments M_y is shown according to location of extended fibers in elastic analysis. The order of appearance of the plastic hinges on supports depends on the relationships between force P and load q as well as geometrical parameters of the beam.

This failure mechanism allows developing the theory of plastic analysis of continuous beams subjected to several loads. However, it does not mean that exactly the above sequence of formation of plastic hinges will be realized. For example, if load q is small then plastic hinges can appear first of all at the supports A, B, and at point K, and after that at support C and in the second span. Real sequence of plastic hinges may be defined only after determination of limit load as shown below.

Bending moment at the point K of the first span caused by force P as well as plastic moments at supports A and B equals

$$M_K = \frac{Pab}{l_1} - \frac{M_y b}{l_1} - \frac{M_y a}{l_1} = \frac{Pab}{l_1} - M_y. \tag{a}$$

Bending moment at the middle point of the second span caused by distributed load q and plastic moments at supports B and C equals

$$M = \frac{ql_2^2}{8} - \frac{M_y}{2} - \frac{M_y}{2} = \frac{ql_2^2}{8} - M_y. \tag{b}$$

The following increase of the load leads to the appearance of the plastic moments within the first and second spans. In the limit state, the bending moments of the first and second spans must be equal to plastic moments; therefore expressions (a) and (b) should be rewritten for both spans as follows:

$$\text{For first span} \quad \frac{Pab}{l_1} - M_y = M_y \quad \text{or} \quad \frac{Pab}{l_1} = 2M_y. \tag{c}$$

$$\text{For second span} \quad \frac{ql_2^2}{8} - M_y = M_y \quad \text{or} \quad \frac{ql_2^2}{8} = 2M_y. \tag{d}$$

Equations (c) and (d) show that in limit state, the maximum bending moment caused by external load equals twice plastic moment. This can be presented geometrically as shown in Fig. 12.8c. Limit plastic moments (*LPM*) are shown by two horizontal dotted lines. These lines show that limit moments for supports and for any section of the beam are equal, may be negative or positive; however, they cannot be more than M_y. Now we can fit a space between two *LPM* lines by bending moment diagrams caused by external load for each simply supported beam. This procedure is called equalizing of bending moments and can be effectively applied for any continuous beam.

For the first span, (c) allows calculating the limit force P

$$\frac{P_{\lim}ab}{l_1} = 2M_y \rightarrow P_{\lim} = \frac{2M_y l_1}{ab} = \frac{2 \times 60\,(\text{kNm}) \times 7\,(\text{m})}{3\,(\text{m}) \times 4\,(\text{m})} = 70\,\text{kN}. \tag{e}$$

For the second span, (d) allows calculating the limit distributed load q

$$\frac{q_{\lim}l_2^2}{8} = 2M_y \rightarrow q_{\lim} = \frac{16M_y}{l_2^2} = \frac{16 \times 60\,(\text{kNm})}{36\,(\text{m}^2)} = 26.67\ \text{kN/m}.$$

Now we need to take into account the condition of simple loading as well as limiting force $P = 70\,\text{kN}$ and load $q = 26.67\,\text{kN/m}$.

Knowing the distributed load we can calculate, according condition $P = 2ql_2$, corresponding limit force P

$$P_{\lim} = 2.0 \times 26.67\,(\text{kN/m}) \times 6\,(\text{m}) = 320\ \text{kN} > 70\ \text{kN}. \tag{f}$$

For part *CD*

$$\frac{q_{\lim} l_3^2}{2} = M_y \to q_{\lim} = \frac{2M_y}{l_3^2} = \frac{2 \times 60\,(\text{kNm})}{2^2\,(\text{m}^2)} = 30\ \text{kN/m},$$

which leads to

$$P_{\lim} = 2.0 \times 30 \times 6 = 360\ \text{kN} > 70\ \text{kN}. \tag{g}$$

Thus, limit distributed load q in both cases leads to the limit force P which can not be accepted.

The final limit load is governed by minimum value given by formulae (e), (f), and (g), so the limit load $P_{\lim} = 70\,\text{kN}$, and corresponding $q_{\lim} = \frac{P_{\lim}}{2l_2} = \frac{70}{2\times 6} = 5.83\,\text{kN/m}$.

Discussion:

On the basis of obtained numerical results, we can explain the order of appearance of the plastic hinges. The limit load $P = 70\,\text{kN}$ and corresponding load q are determined from the conditions of appearance of plastic hinges at the supports A, B and at point K of application force P. Thus, the failure of the structure in whole is defined by a failure of the span *AB* because this simply supported beam is being transformed in mechanism. In this case, with further increase of the load q, the relationship $P = 2ql_2$ is not held anymore, since load P cannot reach the greater value than $P_{\lim}$. It is obvious that span *BC* still can resist to increased load q, however, the structure in whole is differ from the original one.

The problem of determination of limit load for continuous beam with given bearing capacity has unique solution.

12.5 Limit Plastic Analysis of Frames

A frame can be failed by different ways. The different schemes of failure are presented in Fig. 12.9. They are following: beam mechanism of failure (B_1, B_2, B_3), mechanism of sidesway failure (S), joint failure (J), framed (F) and different combined mechanisms.

The type of mechanism of failure, which will occur, is not known in advance. This is the principal difficulty for plastic analysis of frames. Therefore, for each mechanism of failure and their different combinations, the equilibrium conditions should be considered and then that mechanism of failure should be adopted, which occurs at the minimum load.

Let us consider the limit load determination for different types of failure. Design diagram of the portal frame is presented in Fig. 12.10a. The loading of the frame is simple. Assume that $Q = 2P$ and the limit moments for vertical and horizontal members satisfy to condition $M_y^{\text{hor}} = 2M_y^{\text{vert}}$.

Now let us consider the different mechanisms of failure.

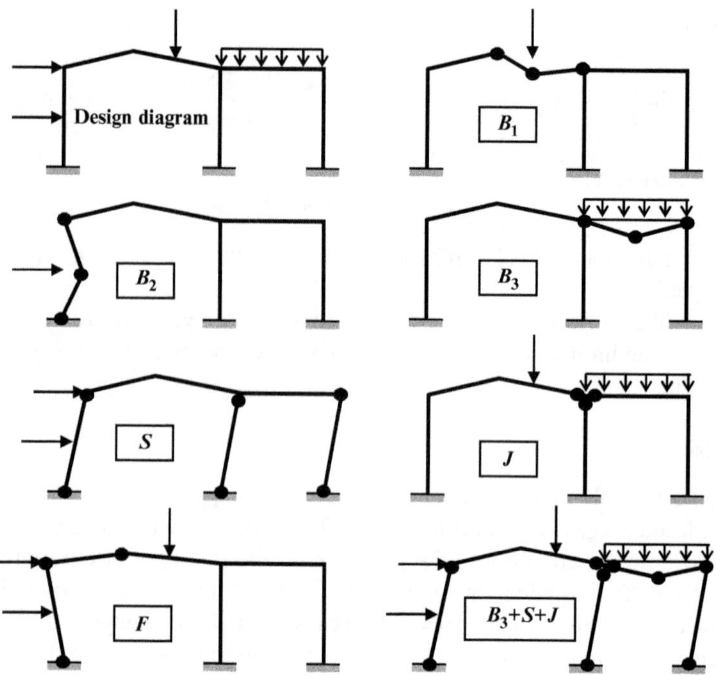

Fig. 12.9 Design diagram of the frame and possible mechanisms of failure

12.5.1 Beam Failure

This scheme is characterized by failure of its horizontal element only; the horizontal displacement of the frame is absent (Fig. 12.10b). In this case, the plastic hinges appear at the joints C (and D) as well as at the point K. Corresponding plastic moments are denoted as M_y^{vert} and M_y^{hor}. Their directions are shown according to the location of extended fibers; location of such fibers are easy to find if bending moment diagram is plotted in elastic state. The inclination of the parts CK and KD is denoted by α.

Equation of limit equilibrium for beam scheme of failure can be found using kinematical method (in the limit condition the total work done by unknown plastic loads and all plastic bending moments M_y equals to zero):

$$Q_{\lim}\frac{\alpha l}{2} - 2M_y^{\text{hor}}\alpha - 2M_y^{\text{vert}}\alpha = 0.$$

We can see that beam mechanism is realized only by load Q. Since $M_y^{\text{hor}} = 2M_y^{\text{vert}}$, then the solution of this equation leads to the following limit load $Q_{\lim} = \frac{12M_y^{\text{vert}}}{l}$. Since $Q = 2P$, then corresponding limit load P is following:

$$P_{\lim} = \frac{6M_y^{\text{vert}}}{l}. \tag{12.2}$$

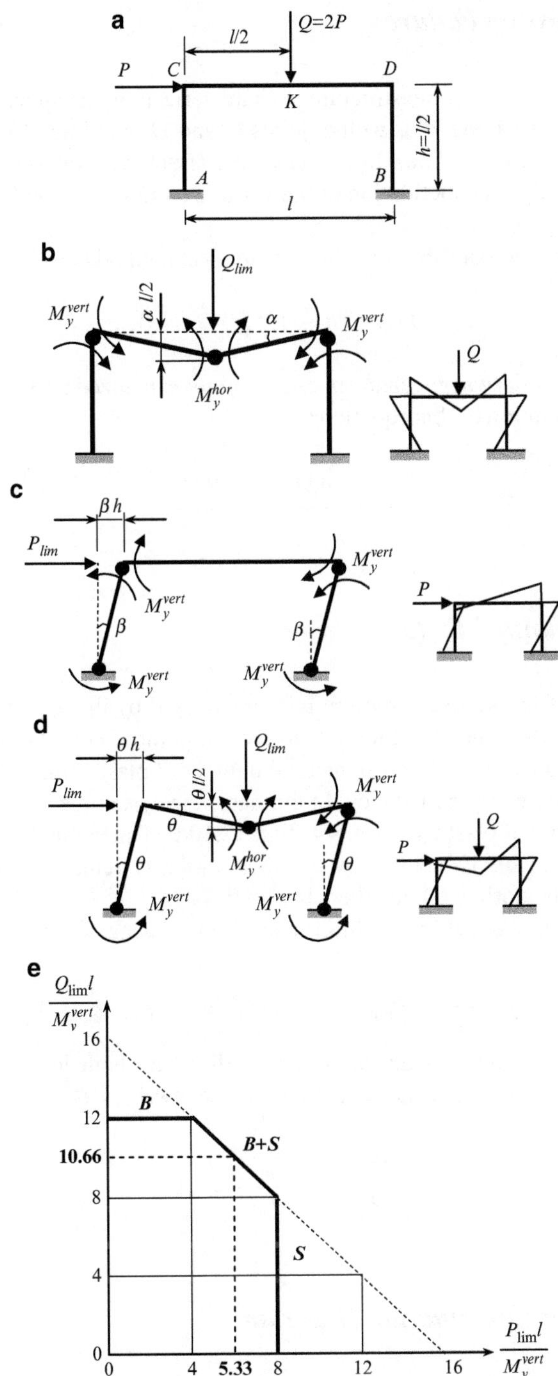

Fig. 12.10 (**a**) Design diagram of a portal frame; (**b**) Beam mechanisms of failure (B); (**c**) Sidesway mechanisms of failure (S); (**d**) Combined mechanisms of failure (B + S); (**e**) Graph of limit combination of loads

12.5.2 Sidesway Failure

This scheme of failure of the structure is characterized by the appearance of plastic hinges at supports A and B as well at joints C and D; a failure of its horizontal element is absent (Fig. 12.10c). In this case, the structure is ongoing the horizontal displacement only. The inclination of the vertical members AC and BD are denoted by β.

Equation of limit equilibrium using kinematical method is following:

$$P_{\lim} \cdot \beta h - 4M_y^{\text{vert}} \cdot \beta = 0.$$

We can see that sidesway mechanism of the failure is realized only by load P. Since $l = 2h$, then solution of this equation

$$P_{\lim} = \frac{4M_y^{\text{vert}}}{h} = \frac{8M_y^{\text{vert}}}{l}. \tag{12.3}$$

12.5.3 Combined Failure

This scheme of failure of a structure is characterized by the appearance of plastic hinges at supports A and B, at joint D as well at point K (Fig. 12.10a). In order to the system becomes a mechanism the total number of plastic hinges must be $n + 1$, where n is a degree of redundancy. In our case, the system becomes a mechanism when the number of plastic hinges is 4 (Fig. 12.10d). The inclination of the vertical members AC and BD is denoted by θ. Since a joint C remains a rigid ones, then inclination of the horizontal members is also θ.

As before, the equation of limit equilibrium using kinematical method is following:

$$Q_{\lim}\frac{\theta l}{2} + P_{\lim} \cdot \theta h - 4M_y^{\text{vert}}\theta - 2M_y^{\text{hor}}\theta = 0.$$

We can see that combined mechanism is realized by both loads Q and P. For $M_y^{\text{hor}} = 2M_y^{\text{vert}}$ and $l = 2h$; last equation may be rewritten as

$$\frac{Q_{\lim}l}{M_y^{\text{vert}}} + \frac{P_{\lim}l}{M_y^{\text{vert}}} = 16. \tag{12.4}$$

12.5.4 Limit Combination Diagram

The result of the plastic analysis may be presented by diagram shown in Fig. 12.10e. The designations of the axis are $\frac{P_{\lim}l}{M_y^{\text{vert}}}$ and $\frac{Q_{\lim}l}{M_y^{\text{vert}}}$.

For beam failure $Q_{\lim} = \frac{12M_y^{\text{vert}}}{l}$. This case is shown on the Fig. 12.10e by line B, which is parallel to horizontal axis because the beam mechanism is realized only by load Q.

For sidesway failure $P_{\lim} = \frac{8M_y^{\text{vert}}}{l}$. This case is shown in the Fig. 12.10e by line S, which is parallel to vertical axis because the sidesway mechanism is realized only by load P.

For combined failure relationships between limit loads is given formula (12.4). Corresponding line is shown in Fig. 12.10e by line $B + S$. Since $Q = 2P$, then

$$P_{\lim} = \frac{16M_y^{\text{vert}}}{3l} = \frac{5.33M_y^{\text{vert}}}{l}. \tag{12.5}$$

Formulae (12.2), (12.3), and (12.5) present limit load P, which corresponds to different mechanisms of failure. Failure is governed by minimum load

$$P_{\lim} = \frac{5.33M_y^{\text{vert}}}{l},$$

which corresponds to combined mechanism of failure; therefore, this type of failure will happens.

Note that increasing of all geometrical dimensions of the frame (l, h) by n-times leads to decreasing of the limiting loads by n-times.

Problems

12.1. A straight rod with cross sectional area A is located between two rigid supports M and N and subjected to axial load P (Fig. P12.1). The yield stress of material is σ_y. Define the limit load P. Solve this problem by direct and static methods.

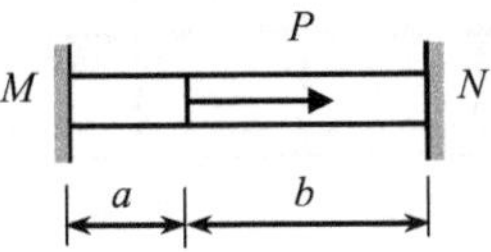

Fig. P12.1

Ans. $P_{\lim} = 2\sigma_y A$.

12.2. A straight uniform rod with cross sectional area A is attached to a rigid support at end M, while there is a gap of Δ between the right end of rod and the rigid support at N; the rod is subjected to axial load P (Fig. P12.2). The yield stress of material is σ_y. Calculate the limit load P. Compare result with problem 12.1 and explain your answer.

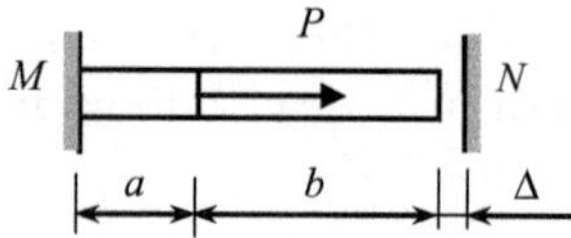

Fig. P12.2

Ans. $P_{\text{lim}} = 2\sigma_y A$.

12.3. Symmetrical structure is subjected to load P as shown in Fig. P12.3. The cross sectional area of the vertical member is A, and for rest members are kA, where k is any number. Yielding stress for all members is σ_y. Perform the elastic analysis and calculate P_{allow}. Calculate the limit load and determine ratio $P_{\text{lim}}/P_{\text{allow}}$. Use the static method.

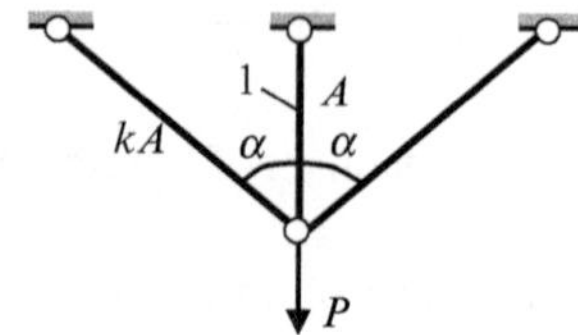

Fig. P12.3

Ans. $P_{\text{lim}} = \sigma_y A\,(1 + 2k\cos\alpha)$.

12.4. Absolutely rigid rod is suspended by vertical hangers as shown in Fig. P12.4. The axial stiffness of all members EA is constant. The limit internal force for each element is N_y. Find the limit load P for each design diagram.

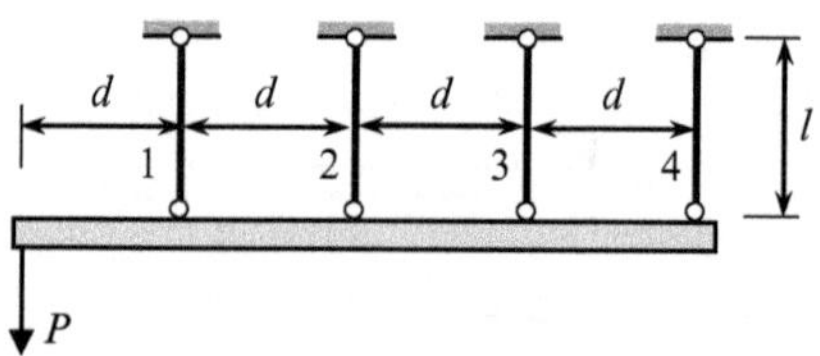

Fig. P12.4

Ans. $P = 1.33N_y$

12.5. A concentrated force P acts on the uniform pinned-clamped beam as shown in Fig. P12.5. The limit moment is M_y. Determine the limit load P_{lim}.

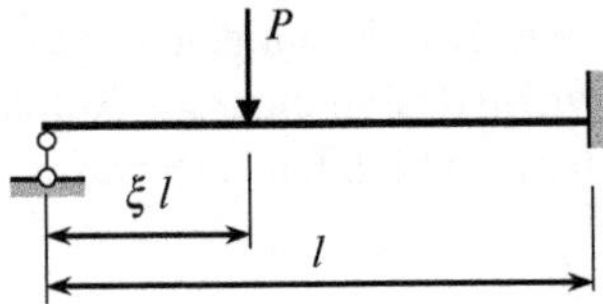

Fig. P12.5

$$\text{Ans. } P_{\lim} = \frac{(1+\xi)}{\xi \cdot (1-\xi)} \frac{M_y}{l}$$

12.6. Determine the limit concentrated load P for beam with clamped supports (Fig. P12.6). The limit moment is M_y.

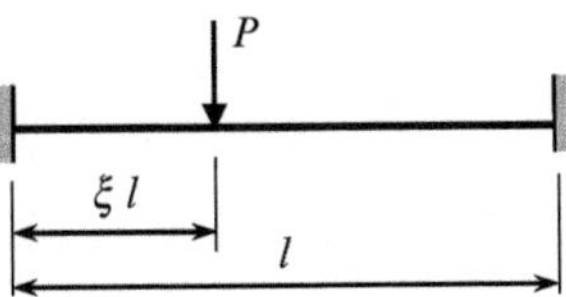

Fig. P12.6

$$\text{Ans. } P_{\lim} = \frac{2}{\xi \cdot (1-\xi)} \frac{M_y}{l}$$

12.7. Determine the limit distributed load q for beam with clamped supports.

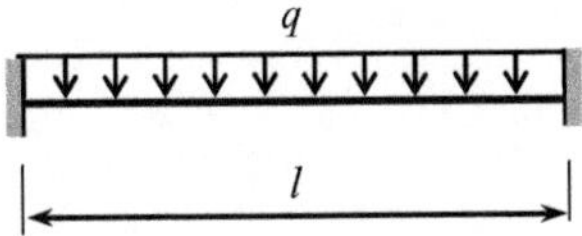

Fig. P12.7

$$\text{Ans. } q_{\lim} = \frac{16M_y}{l^2}$$

12.8. Two-span uniform beam ABC is subjected to uniformly distributed load as shown in Fig. P12.8. Find a location of the plastic hinge. Calculate the limit load.

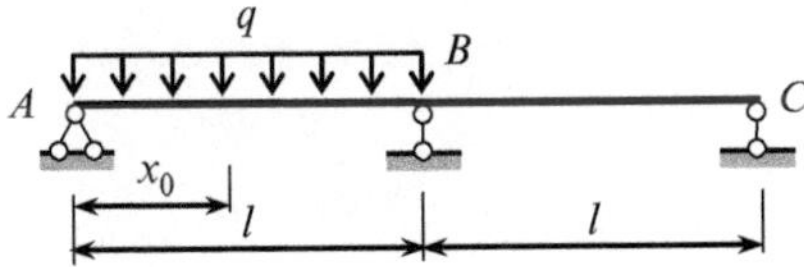

Fig. P12.8

$$\text{Ans. } x_0 = l\left(\sqrt{2}-1\right);\ q_{\lim} = \frac{M_y}{l^2}\left(6+4\sqrt{2}\right) = 11.657\frac{M_y}{l^2}$$

12.9. Two-span uniform beam ABC is subjected to uniformly distributed load and concentrated load P as shown on design diagram. Both loads are satisfy to condition $P = kql$. Find the parameter k, which leads to plastic conditions at the both spans at once.

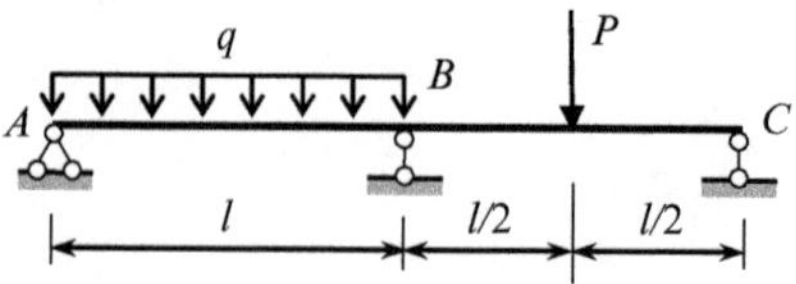

Fig. P12.9

Ans. $k = 0.5147$.

Chapter 13
Stability of Elastic Systems

Theory of structural stability is a special branch of structural analysis. This theory explores the very important phenomenon that is observed in the behavior of the structures subjected to compressed loads. This phenomenon lies in the abrupt change of initial form of equilibrium. Such phenomenon is called loss of stability. As a rule, the loss of stability of a structure leads to it collapse. Engineering practice knows a lot of examples when ignoring this feature of a structure led to its failure.

This chapter is an introduction to stability analysis of engineering structures subjected to compressed loads. Among them are structures that contain nondeformable members as well as beams, frames, and arches. Classical methods of analysis will be discussed.

13.1 Fundamental Concepts

We will differentiate two types of structures, mainly, the structures consisting of absolutely rigid bodies connected by elastic constrains and structures consisting of deformable members; it is possible to combine in one structure both types of members, i.e., absolutely rigid discs with deformable members.

Some examples of these structures under compressed loads are shown in Fig. 13.1. Structures that contain absolutely rigid members ($EI = \infty$) are shown in Fig. 13.1a, b; these design diagrams present structures with elastic joints. Elastic joint means that the angle between two adjacent members is changed upon load application. Figure 13.1c, d presents structures, which contain deformable elements; structure in Fig. 13.1e contains the absolutely rigid part AC and deformable part CB.

Structures, which are subjected to compressed loads, may be either in stable or in unstable equilibrium. Stability is a property of a structure to keep its initial position or initial deformable shape. Stable structure will regain to its original state if any disturbed factor changes the initial state and after it is removed.

A structure subjected to compressed loads may be disturbed from initial equilibrium state by, for example, a small lateral load. After removing this disturbance, a

I.A. Karnovsky and O. Lebed, *Advanced Methods of Structural Analysis*,
DOI 10.1007/978-1-4419-1047-9_13,

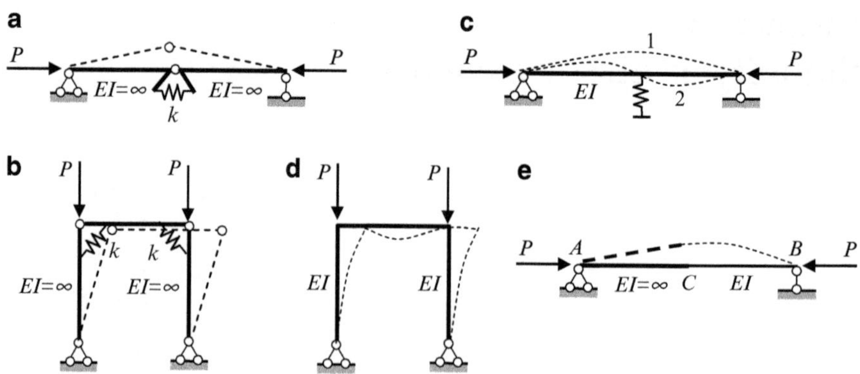

Fig. 13.1 Type of structures and form of buckling: (**a, b**) structures with absolutely rigid members; (**c, d**) structures with deformable members only; (**e**) structure with absolutely rigid and deformable members

structure can return into the initial state, or *tends* to return to the initial state, or remains in the new state, or even tends to switch into new state. Behavior of a structure after removing a disturbance depends on the value of compressed loads. For small compressed load, a structure will return to the initial state, i.e., this equilibrium state is a stable one, while for larger compressed load, a structure will not return to the initial state, i.e., this equilibrium state is unstable. However, what does "small and large" load mean? It is obvious, that the behavior of a structure after removing a disturbance load depends not only on the value of the compressed load, but also on the types of supports, the length of compressed members, and their cross sections. Any compressed load for tall column with small cross section may be treated as a large load, while for short column with large cross section the same load may be treated as a small one. The theory of elastic stability gives the precise quantitative characteristics of compressed loads for different types of structures, which leads to well-defined state of a structure, and allows us to understand the influence of parameters of a structure (boundary conditions, cross-section, and properties of material) on the value of this load and corresponding behavior of a structure.

Let us introduce the following definitions:

Stable equilibrium state means that if the structure under compressed load is disturbed from an initial equilibrium state and after all disturbing factors are removed, then the structure returns to the initial equilibrium state. This is concerning to the elastic structures. If a structure consists of plastic or elasto-plastic elements, then a complete returning to the initial state is impossible. However, equilibrium state is assumed to be stable, if a structure even tends to return to the initial equilibrium state. In case of absolutely rigid bodies, we are talking about stable *position* of a structure, while in case of deformable elements, we are talking about stable equilibrium *form* of a deformable state. In all these cases, we say that the acting compressed load is less than the critical one. Definition of the critical load will be given later.

Unstable equilibrium state means that if a structure under compressed load is disturbed from an initial equilibrium state and after all disturbing factors are removed, then the structure does not return to the initial equilibrium state. In this case, we say that the acting compressed load is larger than the critical one.

Change of configuration of a structure under the action of compressed load is called a loss of stability of the initial form of equilibrium or a buckling. If the compressed load is a static one, then this case is referred as the static loss of stability. In this chapter, we will consider absolutely rigid and absolutely elastic structures subjected to static compressed loads only. If a structure switches to other state (as a result of loss of stability) and remains in this state in equilibrium, then this new equilibrium state is called the adjacent form of equilibrium.

The static load may be of two types: conservative and nonconservative. The work done by conservative forces is determined only by the initial and final position of points of application of a force. Example of a conservative force is a force that keeps its direction (Fig. 13.2a).

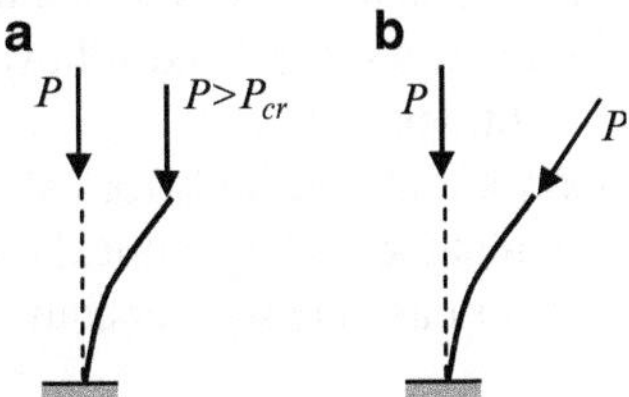

Fig. 13.2 The column under conservative and tracking force

The work done by nonconservative forces is determined by trajectory between the initial and final position of points of application of the force. Example of nonconservative force is a *tracking* force whose orientation depends on the slope of the elastic curve at the point of application of the force (Fig. 13.2b).

The corresponding systems are called conservative and nonconservative. In this chapter, only conservative systems are considered.

The *critical force* P_{cr} is the maximum force at which the structure holds its initial equilibrium form (the structure is still stable), or minimum force, at which the structure no longer returns to the initial state (the structure is already unstable) if all disturbing factors are removed.

The state of a structure that corresponds to critical load is called the critical state. The switching of a structure into new state occurs suddenly and as a rule leads to the collapse of a structure. The theory of static stability of the structures is devoted to methods of calculation of critical loads.

Degree of freedom is a fundamental concept of stability analysis. Degrees of freedom present independent parameters, which define the structure's configuration. The structures, which contain only rigid elements, have the finite number degrees of freedom. Each deformable element should be considered as a member with distributed parameters, so the structures that contain only deformable elements have infinite number degrees of freedom. Structures presented in Fig. 13.1a, b have one degree of freedom, while the structures presented in Fig. 13.1c, d have infinitely many degrees of freedom.

The difference of this concept used in different parts of the structural theory, such as the kinematical analysis, matrix stiffness method, and stability of structures, is obvious.

Generalized coordinates are independent parameters, which uniquely defines configuration of a system in new arbitrary state. A structure with n degrees of freedom has n generalized coordinates. A structure with n degrees of freedom has n critical loads. Each critical load corresponds to one specified form of equilibrium. For structure with one degree of freedom, there exists only the unique form of the loss of stability (Fig. 13.1a, b) and its corresponding unique critical load. For structure with infinitely many degrees of freedom, there exist infinitely many critical loads and its corresponding forms of loss of stability. Figure 13.1d shows only the first buckling form of a frame. In Fig. 13.1c, the numbers 1 and 2 indicate the first and second forms of the loss of stability of a beam. It is very important to define the smallest critical load, because it leads to the loss of stability accordingly first form, i.e., to the failure of the structure. The second and following forms may be realized upon the special additional conditions.

There exist precise and approximate methods for calculating critical loads. Precise methods are static, energy method and dynamical ones. These methods reflect the fact that the concepts "stable or unstable state of equilibrium" may be considered from different points of view.

Static method (or equilibrium method) is based on the consideration of equilibrium of a structure in a new configuration. The critical force is such a minimum force, which can hold the structure in equilibrium in the adjacent condition or maximum force, for which initial straight form of equilibrium is yet possible.

Energy method requires consideration of total energy of a structure in a new configuration. This energy U equals to stress energy U_0 and potential W of external forces

$$U = U_0 + W. \tag{13.1}$$

The potential W of external forces equals to work, which is produced by external forces on the displacement from *final* state into *initial* one. The stable equilibrium of the structure corresponds to a minimum of the total energy.

For system with n degrees of freedom, the critical load may be calculated from the set of equations

$$\frac{\partial U}{\partial q_1} = 0; \quad \frac{\partial U}{\partial q_2} = 0; \quad \cdots \quad \frac{\partial U}{\partial q_n} = 0, \tag{13.2}$$

where q_i are the generalized coordinates of the structure. This method is equivalent to the virtual displacement method, according to which the sum of the work done by all forces on any virtual displacements is zero. The static and energy methods are considered later.

13.2 Stability of Structures with Finite Number Degrees of Freedom

This section is devoted to the calculation of critical load for structures containing only absolutely rigid bars, with elastic constraints. Three types of elastic constraints will be considered. They are the following:

1. Elastic support, which allows an *angular* displacement. Rigidity of support is k_{rot}. The reactive moment of such support and its angular displacement θ are related as $M = k_{\text{rot}}\theta$.
2. Elastic support, which allows a *linear* displacement. Rigidity of support is k. Reaction of support and its linear displacement f are related as $R = kf$.
3. Elastic connections between absolutely rigid members (elastic hinged joint) of a structure. Rigidity of connection is k_{rot}. The moment of this joint and *mutual angular* displacement θ of two adjacent absolutely rigid bars are related as $M = k_{\text{rot}}\theta$.

From methodical point of view, it is reasonable that all structures with finite number degrees of freedom divide into two large groups. They are structures with one degree of freedom and structures with two or more degrees of freedom.

13.2.1 Structures with One Degree of Freedom

A simplest structure with one degree of freedom is shown in Fig. 13.3a. Absolutely rigid vertical weightless column is placed on rigid supporting plate. Since the foundation is flexible, the column in whole may rotate around the fixed point O. The column is subjected to axial compressed static force P. Assume that an *angular* rigidity of elastic support is k_{rot}; such type of elastic support is presented as *two* springs of equal stiffness. Within their deformation, the forces that arise in both springs create the elastic couple.

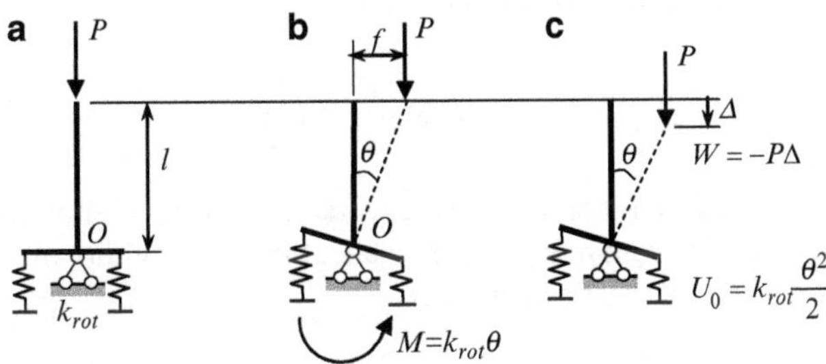

Fig. 13.3 Absolutely rigid vertical cantilever bar on the elastic support: (**a**) Design diagram; (**b**) Static method; (**c**) Energy method

To determine the critical load by static or energy method, first of all, we need to accept a generalized coordinate. Let an angular displacement θ of the supporting plate being the generalized coordinate. This parameter describes completely a perturbed configuration of the structure. The structure in the strained state is shown by dotted line.

Static method (Fig. 13.3b). The moment that is produced in the support is $M = \theta k_{\text{rot}}$. The moment due to external load N with respect to support point O is Pf. We assume that angular displacement θ is small, and therefore $f = l\sin\theta \approx l\theta$. Equilibrium equation $\sum M_0 = 0$ leads to the stability equation

$$Pf - \theta k_{\text{rot}} = Pl\theta - \theta k_{\text{rot}} = 0. \tag{13.3}$$

This equation is obtained on the basis of linearization procedure $\sin\theta \approx \theta$; therefore, the equilibrium equation (13.3) is called linearized *stability equation.*

Equation (13.3) is satisfied at two special cases:

1. The angle $\theta = 0$. It means that initial vertical form of the column remains vertical for any force P. This is a trivial solution, which corresponds to the initial form of equilibrium.
2. The angle $\theta \neq 0$. It means that the strained form of equilibrium is possible for load $P_{\text{cr}} = \dfrac{k_{\text{rot}}}{l}$. This load is critical.

If the given structure is subjected to load $P < P_{\text{cr}}$, then initial vertical position of the column is the only equilibrium position; therefore, if the structure would be disturbed, then it returns to its initial position. If the structure is subjected to load $P \geq P_{\text{cr}}$, then additional equilibrium state is possible. Pay attention that the value of the angle θ cannot be determined on the basis of linearized stability equation.

Energy method (Fig. 13.3c). Since the vertical displacement of the point of application of the force P is

$$\Delta = l\,(1 - \cos\theta) \cong \frac{l\theta^2}{2},$$

then the potential of external load N is

$$W = -P\Delta = -Pl\frac{\theta^2}{2}.$$

This expression has negative sign since the work of the force P should be calculated on the displacement Δ from *final to initial* state. Since the rigidity of support is k_{rot}, then the strain energy of elastic support is

$$U_0 = k_{\text{rot}}\frac{\theta^2}{2}.$$

The total energy of the system is

$$U = U_0 + W = k_{\text{rot}}\frac{\theta^2}{2} - Pl\frac{\theta^2}{2}.$$

Condition (13.2)

$$\frac{\partial U}{\partial \theta} = k_{\text{rot}}\theta - Pl\theta = 0$$

leads to the same critical load $P_{\text{cr}} = k_{\text{rot}}/l$.

Example 13.1. The structure contains two absolutely rigid bars ($EI = \infty$), which are connected by hinge C, and supported by elastic support at this point; the rigidity of elastic supports is k. The structure is subjected to axial compressed force P (Fig. 13.4a). Calculate the critical force by the static and energy methods.

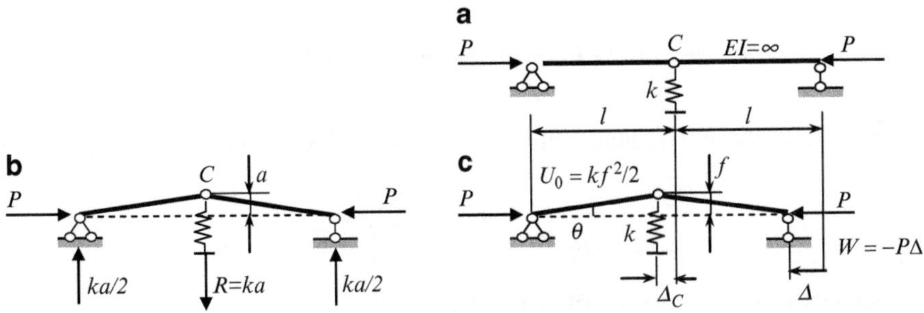

Fig. 13.4 Absolutely rigid 2-elements structure with elastic support: (**a**) Design diagram; (**b**) Static method; (**c**) Energy method

Solution. *Static method* (Fig. 13.4b). The structure has one degree of freedom. Let a vertical displacement a of the hinge C be a generalized coordinate, so the reaction of elastic support is $R = ka$ and reactions of pinned and rolled supports are $ka/2$. Bending moment at hinge C equals zero, therefore stability equation becomes

$$M_C^{\text{left}} = Pa - \frac{ka}{2}l = 0.$$

This equation immediately leads to the critical force becomes

$$P_{\text{cr}} = \frac{kl}{2}.$$

Energy method (Fig. 13.4c). Let an angular displacement θ at the support points be a generalized coordinate. The vertical displacement of hinge C equals $f = l\sin\theta \cong l\theta$, while the horizontal displacement at the same point C is $\Delta_C = l(1-\cos\theta)$, so horizontal displacement at the point of application P becomes

$$\Delta = 2\Delta_C = 2l\,(1 - \cos\theta) \cong 2l\frac{\theta^2}{2}.$$

Strain energy accumulated in elastic support

$$U_0 = R\frac{f}{2}$$

where reaction of *elastic* support is $R = kf$, and vertical displacement of hinge C in terms of generalized coordinate is $f = l\sin\theta \cong l\theta$, so

$$U_0 = k\frac{f^2}{2} = k\frac{l^2\theta^2}{2}.$$

The potential of external force is $W = -P\Delta = -Pl\theta^2$, so the total energy is

$$U = U_0 + W = k\frac{l^2\theta^2}{2} - Pl\theta^2.$$

Derivative of this expression with respect to generalized coordinate equals to zero, i.e.,

$$\frac{\partial U}{\partial\theta} = kl^2\theta - 2Pl\theta = 0,$$

which leads to the above determined critical load.

Discussion

1. The static method requires calculation of reactions of *all* supports, while energy method requires calculation of reactions only for *elastic* supports.
2. Why potential of external force equals $W = -P\Delta$, while the energy accumulated in elastic support contains coefficient 0.5? If a structure switches into the new position, a compressed force P remains the same; that is why potential of this force equals $W = -P\Delta$. The internal forces in the elastic constrain increase from zero to maximum values; that is why the expression for accumulated energy contains coefficient 0.5.

So far we have considered only two types of joints, mainly hinged and rigid joints. Let us introduce a concept of *elastic* joint. If a structure is subjected to any loading then for such joint the initial angle between members forming the joint changes. Each elastic joint is characterized by the rigidity of the elastic hinge k_{rot}. The moment which arise at elastic connection C is $M = k_{\text{rot}}\theta$, where θ is a mutual angular displacement of the members at the elastic joint.

Let us show an application of static and energy methods for determination of critical load for following structure: two absolutely rigid rods are connected by elastic hinge C, as presented in Fig. 13.5a. The rigidity of the hinge is k_{rot}; the structure is subjected to axial static force P. The structure has one degree of freedom. Take angle α as a generalized coordinate. A new form of equilibrium is shown by solid line.

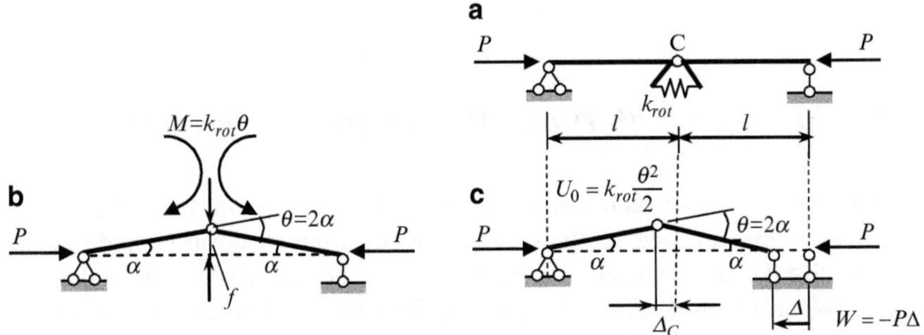

Fig. 13.5 Absolutely rigid elements with elastic joint C: (**a**) Design diagram; (**b**) Static method; (**c**) Energy method

Static method (Fig. 13.5b). The vertical displacement of the joint C is $f = l\sin\alpha \cong l\alpha$. The mutual angle of rotation of the bars is $\theta = 2\alpha$, so the moment, which arises in elastic connection at point C, equals $M = k_{\text{rot}}\theta = 2k_{\text{rot}}\alpha$. The vertical reactions of supports are zero. The bending moment at the hinge C equals zero, so the stability equation becomes

$$\sum M_C^{\text{left}} = 0: \quad Pf - k_{\text{rot}}\theta = 0 \rightarrow Pl\alpha - 2k_{\text{rot}}\alpha = 0$$

Nontrivial solution of this equation is

$$P_{\text{cr}} = \frac{2k_{\text{rot}}}{l}.$$

Energy method (Fig. 13.5c). Horizontal displacement of the point of application of the force P equals

$$\Delta = 2\Delta_C = 2l\left(1 - \cos\alpha\right) \approx 2l\frac{\alpha^2}{2},$$

so potential of the external force is $W = -P\Delta = -Pl\alpha^2$. Mutual angle of rotation at joint C is $\theta = 2\alpha$, the strain energy accumulated in elastic connection is

$$\frac{k_{\text{rot}}\left(2\alpha\right)^2}{2} = 2k_{\text{rot}}\alpha^2,$$

so the total energy is $U = -Pl\alpha^2 + 2k_{\text{rot}}\alpha^2$. Condition $\dfrac{\partial U}{\partial \alpha} = 0$ leads to the following stability equation

$$-2Pl\alpha + 4k_{\text{rot}}\alpha = 0,$$

and for critical force we get the same expression

$$P_{\text{cr}} = \frac{2k_{\text{rot}}}{l}.$$

13.2.2 Structures with Two or More Degrees of Freedom

For stability analysis of such structures, first, it is necessary to define the number of degrees of freedom, sketch the structure in any arbitrary position, and to chose the generalized coordinates. As for structures with one degree of freedom, the static method requires consideration of equilibrium conditions. Energy method requires the calculation of potential of external forces and energy accumulated in elastic constraints. The static method requires calculation of reactions of *all* supports, while energy method requires calculation of reactions only for *elastic* supports.

Static equations (for static method) and conditions (13.2) for energy method lead to n algebraic homogeneous equations with respect to unknown generalized coordinates. Trivial solution corresponds to initial (or unperturbed) state of equilibrium. To obtain nontrivial solution, it is necessary to equate the determinant of coefficients before generalized coordinates to zero. This equation serves for calculation of critical loads. Their number equals to the number degrees of freedom. Each critical load corresponds to specified shape of loss of stability.

Example 13.2. The structure contains three absolutely rigid bars ($EI = \infty$), which are connected by hinges C_1 and C_2, and supported by elastic supports at these points; the rigidity of elastic supports is k. The structure is subjected to axial compressed force N as shown in Fig. 13.6a. Calculate the critical load.

Solution. Let us consider this problem by the static and energy methods.

Static method The structure has two degrees of freedom. Displacements of the hinges C_1 and C_2 are a_1 and a_2; they are considered as generalized coordinates. The reactions of elastic supports are $R_1 = ka_1$ and $R_2 = ka_2$; reactions of the left and right supports are shown in Fig. 13.6b.

Bending moments at hinges C_1 and C_2 are equal to zero, therefore

$$M_1^{\text{left}} = Na_1 - \frac{k(2a_1 + a_2)}{3}l = 0,$$
$$M_2^{\text{right}} = Na_2 - \frac{k(a_1 + 2a_2)}{3}l = 0.$$

This system may be rewritten as homogeneous algebraic equations with respect to unknown generalized coordinates a_1 and a_2

$$\begin{aligned} a_1(3N - 2kl) - a_2kl &= 0, \\ -a_1kl + a_2(3N - 2kl) &= 0. \end{aligned} \qquad \text{(a)}$$

The trivial solution $a_1 = 0,\ a_2 = 0$ corresponds to initial unstrained configuration of the structure.

Nontrivial solution of this system occurs if determinant of the system equals zero:

$$\begin{bmatrix} 3N - 2kl & -kl \\ -kl & 3N - 2kl \end{bmatrix} \rightarrow (3N - 2kl)^2 - (kl)^2 = 0.$$

The roots of this equation present the critical loads; they are

$$N_{1\text{cr}} = \frac{kl}{3}, \quad N_{2\text{cr}} = kl. \tag{b}$$

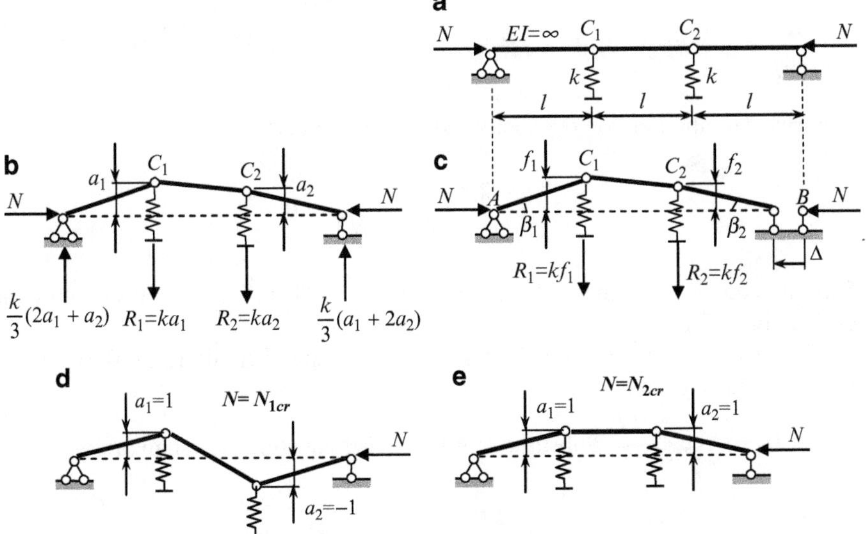

Fig. 13.6 (**a–c**) Structure with two degrees of freedom: (**a**) Design diagram; (**b**) Static method; (**c**) Energy method. (**d, e**) First and second form of the loss of stability

Energy method Let the generalized coordinates be angles β_1 and β_2 (Fig. 13.6c). The angle between portion C_1C_2 and horizontal line is $\beta_1 - \beta_2$, so horizontal displacement of the point of application force N is

$$\begin{aligned} \Delta &= l\,(1 - \cos\beta_1) + l\,(1 - \cos\beta_2) + l\,(1 - \cos(\beta_1 - \beta_2)) \\ &\cong \frac{l}{2}\left[\beta_1^2 + \beta_2^2 + (\beta_1 - \beta_2)^2\right] = l\left(\beta_1^2 + \beta_2^2 - \beta_1\beta_2\right). \end{aligned}$$

The potential of the force N is $W = -N\Delta$.

The vertical displacements of points C_1 and C_2 equal to $f_1 = l\tan\beta_1 \cong l\beta_1$ and $f_2 = l\tan\beta_2 \cong l\beta_2$, respectively. Therefore, reactions of elastic supports are

$R_1 = kf_1 = kl\beta_1$ and $R_2 = kf_2 = kl\beta_2$. The energy accumulated in elastic supports is

$$\sum \frac{R_i f_i}{2} = \frac{k}{2}(l\beta_1)^2 + \frac{k}{2}(l\beta_2)^2.$$

The total energy

$$U = -N\Delta + \sum \frac{R_i f_i}{2} = -Nl\left(\beta_1^2 + \beta_2^2 - \beta_1\beta_2\right) + \frac{k}{2}l^2\beta_1^2 + \frac{k}{2}l^2\beta_2^2.$$

Derivative of the total energy with respect to generalize coordinate leads to the following equations

$$\begin{aligned} \frac{\partial U}{\partial \beta_1} &= -2Nl\beta_1 + Nl\beta_2 + kl^2\beta_1 = 0 \\ \frac{\partial U}{\partial \beta_2} &= -2Nl\beta_2 + Nl\beta_1 + kl^2\beta_2 = 0 \end{aligned} \quad or \quad \begin{aligned} (kl - 2N)\,\beta_1 + N\beta_2 &= 0 \\ N\beta_1 + (kl - 2N)\,\beta_2 &= 0 \end{aligned} \qquad \text{(c)}$$

Nontrivial solution of homogeneous system (c) occurs if

$$\begin{vmatrix} kl - 2N & N \\ N & kl - 2N \end{vmatrix} = 0$$

Stability equation becomes $(kl - 2N)^2 - N^2 = 0$. This equation leads to the same critical loads (b).

Each critical load corresponds to a specified shape of equilibrium. Both critical loads should be considered.

1. Let $N = N_{1\text{cr}} = \dfrac{kl}{3}$. Substituting it in the *first* equation of system (a), we obtain

$$a_1\left(3\frac{kl}{3} - 2kl\right) - a_2kl = 0,$$

   and relationship between generalized coordinates is $a_1 \div a_2 = -1$, which determines the first form of a loss of stability; considering the *second* equation of system (a) we will get the same result. Corresponding equilibrium form is presented in Fig. 13.6d.
2. Let $N = N_{2\text{cr}} = kl$. Substituting it in the first equation of system (a), we obtain

$$a_1\,(3kl - 2kl) - a_2kl = 0,$$

   and the second form of the loss of stability is defined by relationship between generalized coordinates as $a_1 \div a_2 = +1$. Corresponding equilibrium form is presented in Fig. 13.6e. Note that for the each critical load, we cannot define the displacements a_1 and a_2 *separately*. However, the *shape* of the loss of stability is defined by their relationships.

13.3 Stability of Columns with Rigid and Elastic Supports

Elastic bar presents a structure with infinite number of degrees of freedom. Such structures are called the structures with distributed parameters. Their stability analysis may be effectively performed on the basis of differential equation of the elastic curve and initial parameter method. Both methods are presented below.

13.3.1 *The Double Integration Method*

Stability analysis of the uniform compressed columns is based on the moment-curvature equation

$$EI\frac{\mathrm{d}^2 y}{\mathrm{d}x^2} = M(x), \tag{13.4}$$

where x and y are the coordinate and lateral displacement of any point of the beam; EI is the flexural rigidity of the beam; $M(x)$ is the bending moment at the section x of the beam caused by given loads.

This equation allows us to find exact value of the critical load for columns with rigid and/or elastic supports. To apply this equation, we need to show a column in deflected state, then it is necessary to construct the expression for bending moment in terms of lateral displacement y of any point of the column, and to write the differential equation (13.4). As a result we get the ordinary differential equation, which could be homogeneous or nonhomogeneous. Then for each specific problem, we need to integrate this equation and find the constant of integration, using the boundary conditions. For typical supports they are the following:

Pinned support: $y = 0$ and $y'' = 0$
Clamped support: $y = 0$ and $y' = 0$
Sliding support: $y' = 0$ and $y''' = 0$
Free end: $y'' = 0$ and $y''' = 0$

For computation of unknown parameters, we get a system of the homogeneous linear algebraic equations.

Nontrivial solution of this system leads to the equation of stability. Solution of this equation leads to the expression for critical load.

13.3.1.1 Uniform Clamped-Free Column

Let the column is subjected to axial compressed force P (Fig. 13.7). Elastic curve of the column is shown by dotted line.

If the lateral displacement of the free end is f, then the bending moment is $M(x) = P(f - y)$. Corresponding differential equation of the compressed column becomes

$$EI\frac{\mathrm{d}^2 y}{\mathrm{d}x^2} = P(f - y) \quad or \quad EI\frac{\mathrm{d}^2 y}{\mathrm{d}x^2} + Py = Pf. \tag{13.5}$$

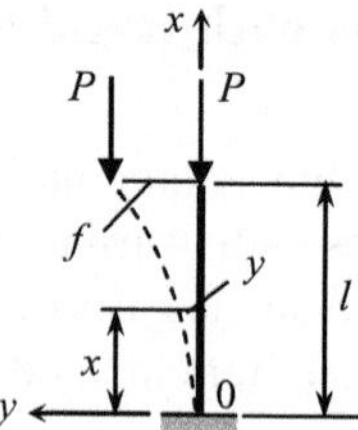

Fig. 13.7 Buckling of clamped-free column

This equation may be transformed to the form

$$\frac{d^2y}{dx^2} + n^2y = n^2f \ , n = \sqrt{\frac{P}{EI}}\left[\frac{1}{\text{length}}\right]. \tag{13.6}$$

Equation (13.6) is nonhomogeneous linear differential equation of order two in one variable x with constant coefficient n^2. Therefore, the solution of this equation should be presented in the form

$$y = A\cos nx + B\sin nx + y^*,$$

where A and B are constants of integration. The partial solution y^* we will find in the form of the right part of (13.6). Since the right part is constant (does not depend on y), then suppose $y^* = C$. Substitution of this expression into (13.6) leads to $n^2C = n^2f \rightarrow C = f$. Therefore, the general solution of (13.6) and corresponding slope are

$$\begin{aligned} y &= A\cos nx + B\sin nx + f, \\ y' &= -An\sin nx + Bn\cos nx. \end{aligned}$$

To determine unknown parameters, let us consider the following boundary conditions:

1. At $x = 0$ (fixed end) the slope $y' = 0$. Expression for slope leads to the $B = 0$
2. At $x = 0$ the displacement $y = 0$. Expression for y leads to the $A = -f$

Thus the displacement of the column becomes

$$y = f\,(1 - \cos nx)\,.$$

At $x = l$ (free end) the displacement $y = f$. Therefore, $f = f(1 - \cos nl)$, which holds if

$$\cos nl = 0.$$

This equation is called the stability equation for given column; the smallest root equals $nl = \pi/2$. The value $n = \pi/2l$ is called the critical parameter. Thus the smallest critical load for uniform clamped-free column becomes

$$P_{\text{cr}} = n_{\text{cr}}^2 EI = \frac{\pi^2 EI}{4l^2}.$$

13.3.1.2 Uniform Columns with Elastic Supports

Now let us consider the compressed column with elastic supports at the both ends (Fig. 13.8a). The flexural stiffness of the column is EI = constant; the stiffness coefficients of elastic supports are k_1 (kN/m) and k_2 (kN m/rad) for linear and angular displacements, respectively.

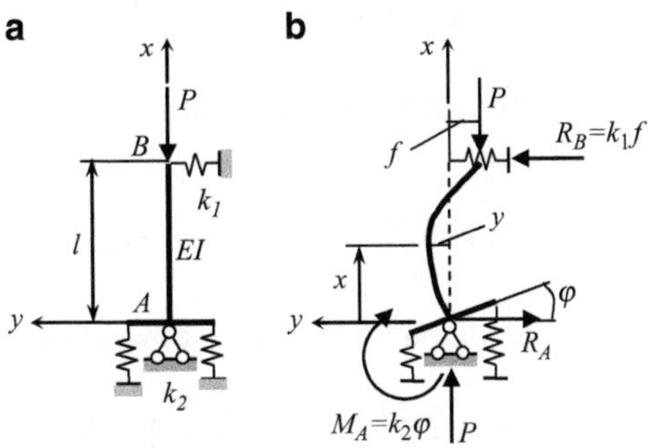

Fig. 13.8 Column with elastic supports

The cross section at support A can rotate through angle φ while the support B has a linear displacement f. Thus reactions of elastic supports are moment at the supports A and force at support B. They are equal $M_A = k_2\varphi$ and $R_B = k_1 f$. Deformable state, elastic curve, and all reactions are shown in Fig. 13.8b.

The bending moment at any section x equals

$$M(x) = -P(f + y) + k_1 f(l - x).$$

Substituting this expression into (13.4) leads to the buckling differential equation of the column

$$EI\frac{d^2y}{dx^2} + Py = f[k_1(l - x) - P]. \tag{13.7}$$

This is the second-order nonhomogeneous differential equation. In case of $k_1 = \infty$, we get homogeneous differential equation.

The partial solution of (13.7) is y^*. Substituting this constant into (13.7) leads to the following expression:

$$y^* = f\left[\frac{k_1}{P}(l - x) - 1\right].$$

The general solution of differential equation (13.7) and corresponding slope are

$$y = C_1 \cos nx + C_2 \sin nx + y^*, \quad n^2 = \frac{P}{EI},$$

$$\frac{dy}{dx} = y' = -C_1 n \sin nx + C_2 n \cos nx - \frac{f k_1}{P}.$$

Unknown parameters C_1, C_2, and f may be determined using the following boundary conditions:

$$(1)\ y\,(0) = 0; \quad (2)\ y'\,(0) = \phi; \quad (3)\ y\,(l) = -f$$

1. The first boundary condition leads to equation

$$C_1 + f\left(\frac{k_1 l}{N} - 1\right) = 0.$$

2. At point $A(x = y = 0)$ the reactive moment equals $M_A = f(k_1 l - P)$, thus the angle of rotation at A is

$$\varphi = \frac{M_A}{k_2} = \frac{f}{k_2}\,(k_1 l - P)\,,$$

so the second boundary condition leads to the following equation

$$C_2 n - f\left(\frac{k_1}{P} + \frac{k_1 l - P}{k_2}\right) = 0.$$

3. The third boundary condition leads to the following equation

$$C_1 \cos nl + C_2 \sin nl = 0.$$

Conditions 1–3 may be rewritten in the form of the homogeneous algebraic equations with respect to unknowns C_1, C_2, and f. Equation for critical load is presented as determinant from coefficients at these unknowns, i.e.,

$$D = \begin{vmatrix} 1 & 0 & \dfrac{k_1 l}{P} - 1 \\ 0 & n & -\left(\dfrac{k_1}{P} + \dfrac{k_1 l - P}{k_2}\right) \\ \cos nl & \sin nl & 0 \end{vmatrix} = 0 \quad or$$

$$D = \begin{vmatrix} 1 & 0 & \dfrac{k_1 l}{P} - 1 \\ 0 & n & -\left(\dfrac{k_1}{P} + \dfrac{k_1 l - P}{k_2}\right) \\ 1 & \tan nl & 0 \end{vmatrix} = 0$$

This equation may be rewritten as the transcendental equation with respect to parameter n

$$\tan nl = nl \frac{\dfrac{k_1 l}{n^2 EI} - 1}{\dfrac{k_1 l}{n^2 EI} + \dfrac{\left(k_1 l - n^2 EI\right) l}{k_2}}.$$

For given parameters l, EI, k_1 and k_2 of the structure, solution of this equation leads to parameter n of critical load. The critical load is $P_{\mathrm{cr}} = n^2 EI$. Table 13.1

presents the columns with specified supports and corresponding stability equations. The stability equations for cases 4–6 contain dimensionless parameter (α or β); the roots for these cases can be calculated numerically for specified α (or β).

Table 13.1 Limiting cases for columns with elastic supports (the length of column l and flexural stiffness EI)

Case	1  $k_1=0$, $k_2=\infty$	2 $k_1=\infty$, $k_2=\infty$ 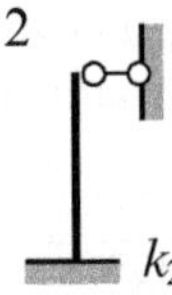	3 $k_1=\infty$, $k_2=0$ 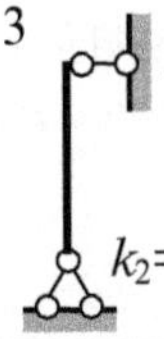
Stability equation	$\cos nl = 0$	$\tan nl = nl$	$\tan nl = 0$
Root $(nl)_{\min}$	$\pi/2$	4.493	π
Case	4  $k_1=0$, k_2	5 $k_1=\infty$, k_2 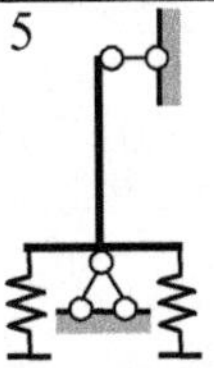	6 k_1, $k_2=\infty$ 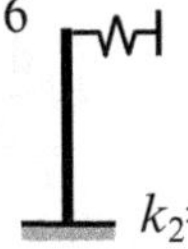
Stability equation	$\tan nl = \dfrac{1}{nl\alpha}$, $\alpha = \dfrac{EI}{k_2 l}$	$\tan nl = \dfrac{nl}{n^2l^2\alpha + 1}$, $\alpha = \dfrac{EI}{k_2 l}$	$\tan nl = nl(1 - n^2l^2\beta)$, $\beta = \dfrac{EI}{k_1 l^3}$

Limiting case. For case 4, the stability equation

$$\tan nl = \frac{1}{nl\alpha}, \quad \alpha = \frac{EI}{k_2 l}$$

may be presented as

$$\frac{\tan nl}{nl} = \frac{k_2}{Pl}.$$

For absolutely rigid body $EI = \infty$, and

$$n = \sqrt{\frac{P}{EI}} \to 0.$$

Since

$$\lim_{nl \to o} \frac{\tan nl}{nl} = 1$$

then a critical force becomes

$$P_{\text{cr}} = \frac{k_2}{l}.$$

This result has been obtained in Sect. 13.2.1.

13.3.2 Initial Parameters Method

This method may be effectively applied for stability analysis of the columns with rigid and elastic supports and columns with step-variable cross section. Moreover, this method allows us to derive the useful expressions, which will be applied for stability analysis of the frames.

Let us consider a beam of constant cross section. The beam is subjected to axial compressed force P. Differential equation of the beam is

$$EI\frac{d^2y}{dx^2} + Py = 0,$$

where y is a lateral displacement. Twice differentiation of this equation leads to fourth-order differential equation

$$EI\frac{d^4y}{dx^4} + P\frac{d^2y}{dx^2} = 0, \quad or \quad \frac{d^4y}{dx^4} + n^2\frac{d^2y}{dx^2} = 0, \tag{13.8}$$

where

$$n = \sqrt{\frac{P}{EI}}$$

Solution of (13.8) may be presented as

$$y(x) = C_1 \cos nx + C_2 \sin nx + C_3 x + C_4, \tag{13.8a}$$

where C_i are unknown coefficients. It is easy to check that this solution satisfies (13.8).

The slope, bending moment and shear are

$$\begin{aligned}
\varphi(x) &= y'(x) = -C_1 n \sin nx + C_2 n \cos nx + C_3, \\
M(x) &= -EIy''(x) = EI(C_1 n^2 \cos nx + C_2 n^2 \sin nx), \\
Q(x) &= -EIy'''(x) = -EI(C_1 n^3 \sin nx - C_2 n^3 \cos nx).
\end{aligned}$$

At $x = 0$ the initial kinematical and static parameters become

$$\begin{aligned}
y(0) &= y_0 = C_1 + C_4, \\
\varphi(0) &= \varphi_0 = C_2 n + C_3, \\
M(0) &= M_0 = C_1 n^2 EI, \\
Q(0) &= Q_0 = C_2 n^3 EI,
\end{aligned}$$

where y_0, φ_0, M_0, Q_0 are lateral displacement, angle of rotation, bending moment, and shear at the origin (Fig. 13.9, y-axis is directed downward). They arise because the rod lost the stability.

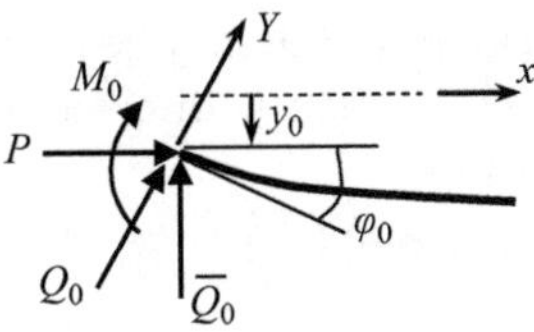

Fig. 13.9 Initial parameters of a beam

Constants C_i may be expressed in terms of kinematical initial parameter (displacement y_0 and slope $\varphi_0 = y' = \frac{dy}{dx}$) and static initial parameters (bending moment $M_0 = -EIy_0''$ and shear $Q_0 = -EIy_0'''$) as follows

$$C_1 = \frac{M_0}{n^2 EI} = \frac{M_0}{P}, \quad C_2 = \frac{Q_0}{n^3 EI} = \frac{Q_0}{nP}, \quad C_3 = \varphi_0 - \frac{Q_0}{P}, \quad C_4 = y_0 - \frac{M_0}{P}$$

Substitution of these constants C_i in (13.8a) leads to the following expressions in terms of initial parameters

$$\begin{aligned} y(x) &= y_0 + \varphi_0 x - M_0 \frac{1 - \cos nx}{P} - Q_0 \frac{nx - \sin nx}{nP}, \\ y'(x) &= \varphi_0 - M_0 \frac{n \sin nx}{P} - Q_0 \frac{1 - \cos nx}{P}, \\ M(x) &= M_0 \cos nx + Q_0 \frac{\sin nx}{n}, \\ Q(x) &= -M_0 n \sin nx + Q_0 \cos nx. \end{aligned} \tag{13.9}$$

These equations present the first form of the initial parameter method for compressed columns. It can be seen that, in spite of the external lateral load being absent, the shear $Q(x)$ is variable along the column. It happens because (13.9) is presented in terms of Q_0, which is directed as perpendicular to the tangent of an *elastic* line of the beam.

Let us calculate the critical load for uniform clamped-free column (Fig. 13.7); EI = constant. The origin is placed at the clamped support. The geometrical initial parameters are $y_0 = 0$ and $\varphi_0 = 0$. The third equation of system (13.9) becomes

$$M(x) = M_0 \cos nx + Q_0 \frac{\sin nx}{n}.$$

Since the bending moment for free end of the column ($x = l$) is zero, then

$$M_l = M_0 \cos nl + Q_0 \frac{\sin nl}{n} = 0.$$

It is obvious that $Q_0 = 0$ and $M_0 \neq 0$, therefore the stability equation becomes $\cos nl = 0$. This result had been obtained using integration of the differential equation (13.4).

The initial parameters equations (13.9) may be presented in equivalent form (second form), i.e., in terms of $\overline{Q}_0$, which is directed as perpendicular to the *initial* straight line of the rod (Fig. 13.9).

According to equilibrium equation $\sum Y = 0$ (axis Y is directed along Q_0) we have

$$Q_0 = \overline{Q}_0 \cos\varphi_0 + P\sin\varphi_0 \cong \overline{Q}_0 + P\varphi_0.$$

Then, substitution of this expression in (13.9) gives the initial parameters equations in terms of $\overline{Q}_0$

$$\begin{aligned} y(x) &= y_0 + \varphi_0 \frac{\sin nx}{n} - M_0 \frac{1-\cos nx}{P} - \overline{Q}_0 \frac{nx - \sin nx}{nP}, \\ y'(x) &= \varphi_0 \cos nx - M_0 \frac{n \sin nx}{P} - \overline{Q}_0 \frac{1-\cos nx}{P}, \\ M(x) &= \varphi_0 nEI \sin nx + M_0 \cos nx + \overline{Q}_0 \frac{\sin nx}{n}, \\ \overline{Q}(x) &= Q_0. \end{aligned} \tag{13.10}$$

In second form (13.1), the shear along the beam is constant, since an external lateral load is absent. Both forms (13.9) and (13.10) are equivalent.

Now we illustrate the application of (13.10) for stability analysis of the stepped fixed-free column shown in Fig. 13.10; the upper and lower portions of the column be l_1 and l_2, respectively. The bending stiffness for both portions are EI_1 and EI_2. The column is loaded by two compressed axial forces $P_1 = P$ and $P_2 = \beta P$, where β is any positive number. It means that growth of all loads up to critical condition of a structure occurs in such way that relationships between all loads remain constant (simple loading).

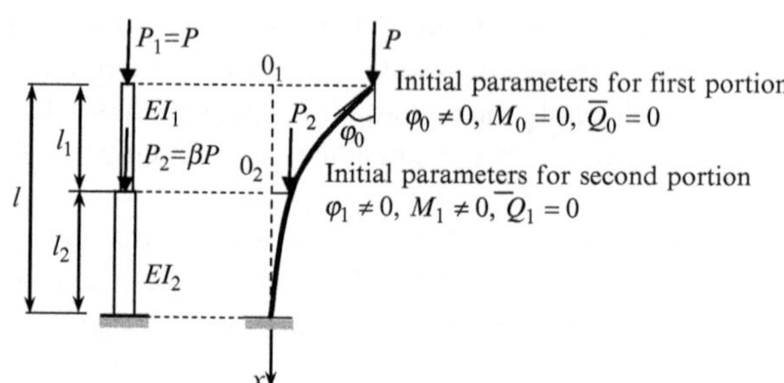

Fig. 13.10 Stepped clamped-free column

Let the origin 0_1 is located at the free end (Fig. 13.10). Initial parameters (at free end) for upper portion are $\varphi_0 = y_0' \neq 0,\ M_0 = 0;\ \overline{Q}_0 = 0$. The slope and bending moment at the end of the first portion (at the $x = l_1$) according to (13.10) are

$$\begin{aligned} \varphi_1 (x = l_1) &= \varphi_0 \cos n_1 l_1, \\ M_1 (x = l_1) &= \varphi_0 n_1 EI_1 \sin n_1 l_1, \\ n_1 &= \sqrt{\frac{P_1}{EI_1}}. \end{aligned} \tag{a}$$

For the second portion of the column, the origin 0_2 is placed at the point where force P_2 is applied. Initial parameters for this portion coincide with corresponding parameters at the end of the first portion (at $x = l_1$); they are $\varphi_1 = y_1' \neq 0$, $M_1 \neq 0$; $\overline{Q}_1 = 0$. The slope at the end of the second portion (at the $x = l_2$) according second equation (13.10) can be presented as

$$\varphi_2\,(x = l_2) = \varphi_1 \cos n_2 l_2 - M_1 \frac{n_2 \sin n_2 l_2}{P_1 + P_2}, \; n_2 = \sqrt{\frac{P_1 + P_2}{EI_2}} \tag{b}$$

In this equations φ_1 and M_1 are initial parameters for portion 2. Substitution of (a) in (b) yields

$$\varphi_2\,(x = l_2) = \varphi_0 \left(\cos n_1 l_1 \cdot \cos n_2 l_2 - n_1 EI_1 \sin n_1 l_1 \frac{n_2 \sin n_2 l_2}{P_1 + P_2} \right).$$

For a clamped support the slope $\varphi_2(x = l_2) = 0$. Since $\varphi_0 \neq 0$, then stability equation becomes

$$\cos n_1 l_1 \cdot \cos n_2 l_2 - n_1 EI_1 \sin n_1 l_1 \frac{n_2 \sin n_2 l_2}{P_1 + P_2} = 0.$$

After a simple rearrengement, this equation may be presented as

$$\tan n_1 l_1 \cdot \tan n_2 l_2 - \frac{n_1}{n_2}(1 + \beta) = 0.$$

This equation may be presented in other form. Let the total length of column $l_1 + l_2 = l$ and $l_2 = \alpha l$, where α is any positive number. In this case $l_1 = (1 - \alpha) l$ and stability equations becomes

$$\tan\left[n_1 (1 - \alpha)\, l\right] \cdot \tan n_2 \alpha\, l - \frac{n_1}{n_2}(1 + \beta) = 0. \tag{c}$$

Limiting cases

1. Let $\alpha = 0$. This case corresponds to uniform column of length l, stiffness EI_1 and loaded by P. Stability equation becomes $\tan n_1 l = \infty$. The root of equation is $(n_1 l)_{\min} = \pi/2$, so the critical load $P_{\text{cr1}} = \dfrac{\pi^2 EI_1}{4l^2}$.
2. Let $\alpha = 1$. This case corresponds to the uniform column of length l, stiffness EI_2 and loaded by force P. Stability equation becomes $\tan n_2 l = \infty$, so the critical load $P_{\text{cr2}} = \dfrac{\pi^2 EI_2}{4l^2}$.

In general case, the equation (c) should be solved numerically. Let $\alpha = 0.5$, $EI_2 = 2EI_1$, and $\beta = 3$. For these parameters, the stability equation becomes $\tan\varphi \cdot \tan\sqrt{2}\varphi = \sqrt{2}$, where $\varphi = 0.5 n_1 l$. The root of this equation is 0.719, thus the critical load

$$P_{1\text{cr}} = \frac{2.0678EI_1}{l^2}.$$

Summary. Initial parameters method may be applied for stability analysis of the stepped columns subjected to several forces. In this case, the origin should be shifted for each following portion. Initial parameters for each following portion coincide with parameters at the end of the previous portion.

Example 13.3. Design diagram of the beam with elastic support is presented in Fig. 13.11. The flexural stiffness of the beam is EI. Stiffness coefficient of elastic support is k [kN m/rad]. Derive the stability equation.

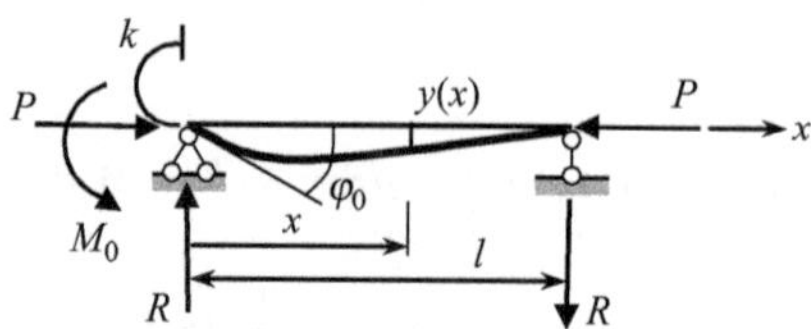

Fig. 13.11 Beam with elastic support

Solution. Suppose that the section at the left end of the beam is rotated clockwise. Corresponding reactions M_0 (which arise in elastic support) and R of the beam are shown in Fig. 13.11. The lateral displacement of the beam according to the first equation of the system (13.10)

$$y(x) = \varphi_0 \frac{\sin nx}{n} - M_0 \frac{1 - \cos nx}{P} - \overline{Q}_0 \frac{nx - \sin nx}{nP}. \tag{a}$$

Initial parameters are $M_0 = -k\varphi_0$, and shear

$$\overline{Q}_0 = R = \frac{k\varphi_0}{l}.$$

Their signs are accepted according to Fig. 13.9. Thus, (a) may be rewritten as

$$y(x) = \varphi_0 \frac{\sin nx}{n} + k\varphi_0 \frac{1 - \cos nx}{P} - \frac{k\varphi_0}{l} \frac{nx - \sin nx}{nP}. \tag{b}$$

Boundary condition: at $x = l \quad y = 0$. Therefore

$$y(l) = \varphi_0 \left[\frac{\sin nl}{n} + k \frac{1 - \cos nl}{P} - \frac{k}{l} \frac{nl - \sin nl}{nP} \right] = 0.$$

Since $\varphi_0 \neq 0$, then

$$\frac{\sin nl}{n} + k \frac{1 - \cos nl}{P} - \frac{k}{l} \frac{nl - \sin nl}{nP} = 0.$$

This expression leads to the following stability equation

$$\tan nl = \frac{nl}{n^2 l^2 \alpha + 1}, \tag{c}$$

where dimensionless parameter

$$\alpha = \frac{EI}{kl}.$$

Equation (c) had been derived earlier (Table 13.1, case 5).

Discussion:

1. It may appear that we used only one boundary condition. However, it is not true. There are two boundary conditions used: the second boundary condition is $M(l) = 0$ and this condition allows write expression for $R = \frac{k\varphi_0}{l}$.
2. If $k = \infty(\alpha = 0)$ then the stability equation becomes $\tan nl = nl$. This case corresponds to clamped-pinned beam (Table 13.1, case 2). If $k = 0$ (simply-supported beam) then the stability equation becomes $\tan nl = 0$ (Table 13.1, case 3).
3. The initial parameter method may be effectively applied for beams with over-hang, intermediate hinges, sliding, and elastic supports.

13.4 Stability of Continuous Beams and Frames

This section is devoted to the stability analysis of compressed continuous beams and frames. We assume that beams are subjected to axial forces only. Frames are subjected to external compressed loads, which are applied in the joints of a frame. If several different compressed loads P_i acts on a structure, we assume that all loads can be presented in terms of *one* load P. It means that growth of all loads up to critical condition of a structure occurs in such way, so that relationships between all loads remain constant, i.e., the loading is simple.

The both classical methods can be applied for stability analysis of continuous beams and frames. However, for such structures, the displacement method often occurs more effective than the force method. The primary system of the displacement method must be constructed as usual, i.e., by introducing additional constraints, which prevent angular and linear displacements of the joints. The primary unknowns are linear and angular displacements of the joints. However, calculation of the unit reactions has specific features.

13.4.1 *Unit Reactions of the Beam-Columns*

A primary system of the displacement method contains the one-span standard (pinned-clamped, clamped-clamped, etc.) members. These members are subjected not only to the settlements (angular or linear) of the introduced constraints, but also to the axial compressed load P. For stability analysis, this is a fundamental feature.

As earlier, we need to have the reactions of standard elements, which are subjected to unit settlements of constraints and axial compressed load. Calculation of reactions for typical member is shown below.

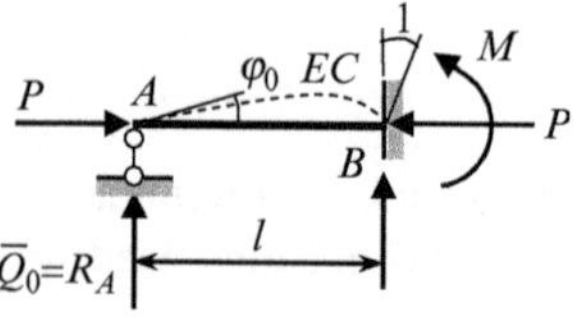

Fig. 13.12 Reaction of compressed beam subjected to unit angular displacement at support B

The pinned-clamped beam is subjected to unit angular displacement of support B and axial compressed force P. The length of the beam is l, $EI = \text{const}$. Figure 13.12 presents the elastic curve (EC) and positive unknown reactions R and M. For their determination we can use the initial parameters method.

The origin is placed at support A. Since $y_0 = 0$ and $M_0 = 0$, then (13.10) becomes

$$y(x) = \varphi_0 \frac{\sin nx}{n} - \overline{Q}_0 \frac{nx - \sin nx}{nP},$$

$$y'(x) = \varphi_0 \cos nx - \overline{Q}_0 \frac{1 - \cos nx}{P}.$$

For calculation of two unknown initial parameters φ_0 and $\overline{Q}_0 = R$ we have two boundary conditions: $y(l) = 0$ and $y'(l) = 1$, therefore

$$y(l) = \varphi_0 \frac{\sin nl}{n} - \overline{Q}_0 \frac{nl - \sin nl}{nP} = 0,$$

$$y'(l) = \varphi_0 \cos nl - \overline{Q}_0 \frac{1 - \cos nl}{P} = 1.$$

Solution of these equations is

$$\overline{Q}_0 = -\frac{3i}{l} \frac{\upsilon^2 \tan \upsilon}{3(\tan \upsilon - \upsilon)},$$

where $i = EI/l$; the dimensionless parameter of critical force is

$$\upsilon = nl = l\sqrt{\frac{P}{EI}}.$$

Since $\overline{Q}_0$ is negative, then reaction R_A is directed downward. Reactive moment at support B is

$$M = R_A l = -3i \frac{\upsilon^2 \tan \upsilon}{3(\tan \upsilon - \upsilon)}.$$

The negative sign means that the reactive moment acts clockwise; indeed direction of moment coincides with direction of angular displacement of clamped support.

The expressions for unit reactions for standard members subjected to compressed force P and special types of displacements can be derived similarly; for typical uniform beams they are presented in Table A.22. In all cases, the length of the beam is l, the flexural stiffness is EI, bending stiffness per unit length is $i = EI/l$. Corresponding elastic curve is shown by dotted line; the graphs present the real direction of reactions; the bending moment diagrams are plotted on the tensile fibers.

The row 3 of the Table A.22 presents the bending moment diagrams when a clamped support rotates and switches in transversal direction simultaneously. While the angle of rotation is fixed as $Z = 1$, the displacement Δ does not require indication of its value. These cases may be used for analysis of compressed beams with elastic supports. Thus, in case of frames with sidesway, the primary system is obtained by introducing constraints, which prevents only angular displacements, and bending moment diagrams should be traced as for member with elastic supports. It is recommended to show elastic curves and remember that bending moment has one-sign ordinates.

Expressions of special functions in two forms are presented in Table A.24; the more preferable is form 2. Also this table contains approximate presentation of these special functions in the form of Maclaurin series. Numerical values of these functions in terms of dimensionless parameter υ are presented in Table A.25.

13.4.2 Displacement Method

Canonical equations of the displacement method for structure with n unknowns Z_j, $(j = 1, 2, \ldots, n)$ are

$$\begin{aligned} r_{11}Z_1 + r_{12}Z_2 + \cdots + r_{1n}Z_n &= 0, \\ r_{21}Z_1 + r_{22}Z_2 + \cdots + r_{2n}Z_n &= 0, \\ \cdot \quad \cdot \quad \cdot \quad \cdot \quad \cdot \quad \cdot \quad \cdot & \quad \cdot \quad \cdot \\ r_{n1}Z_1 + r_{n2}Z_2 + \cdots + r_{nn}Z_n &= 0. \end{aligned} \tag{13.11}$$

Features of (13.11):

1. Since the forces P_i are applied only at the joints, then the canonical equations are homogeneous ones.
2. Bending moment diagram caused by unit displacements of introduced constrains within the *compressed* members are *curvilinear*. Reactions of constraints depend on axial forces in the members of the frame, i.e., contain parameter υ of critical load. If a frame is subjected to different forces P_i, then critical parameters should be formulated for each compressed member $\upsilon_i^2 = \frac{P_i l_i^2}{(EI)_i}$ and after that all of these parameters should be expressed in terms of parameter υ for specified basic member. Thus, the unit reactions are functions of parameter, i.e., $r_{ik}(\upsilon)$.

The trivial solution ($Z_i = 0$) of (13.11) corresponds to the initial nondeformable design diagram. Nontrivial solution ($Z_i \neq 0$) corresponds to the new form of equilibrium. This occurs if the determinant, which is consisting of coefficients of unknowns, equals zero, i.e.,

$$\det \begin{bmatrix} r_{11}(\upsilon) & r_{12}(\upsilon) & \cdots & r_{1n}(\upsilon) \\ r_{21}(\upsilon) & r_{22}(\upsilon) & \cdots & r_{2n}(\upsilon) \\ \cdots & \cdots & \cdots & \cdots \\ r_{n1}(\upsilon) & r_{n2}(\upsilon) & \cdots & r_{nn}(\upsilon) \end{bmatrix} = 0 \tag{13.12}$$

Condition (13.12) is called the stability equation of a structure in the form of displacement method. For practical engineering, it is necessary to calculate the minimum root of the above equation. This root defines the smallest parameter υ of critical force or smallest critical force.

It is obvious that condition (13.12) leads to transcendental equation with respect to parameter υ. Since the functions $\varphi(\upsilon)$ and $\eta(\upsilon)$ are tabulated (Tables A.24 and A.25), then a solution of stability equation may be obtained by graphical method. Since the determinant is very sensitive with respect to parameter υ, it is recommended to solve the equation (13.12) using a graphing calculator or computer.

The displacement method is effective for stability analysis of stepped continuous beams on rigid supports with several axial compressed forces along the beam and for frames with/without sidesway.

Let us derive the stability equation and determine the critical load for frame shown in Fig. 13.13a. This frame has one unknown of the displacement method. The primary unknown is the angle of rotation of rigid joint. Figure 13.13b shows the primary system, elastic curve, and bending moment diagram caused by unit rotation of introduced constrain. The bending moments diagram for compressed vertical member of the frame is curvilinear. The ordinate for this member is taken from Table A.22, row 1.

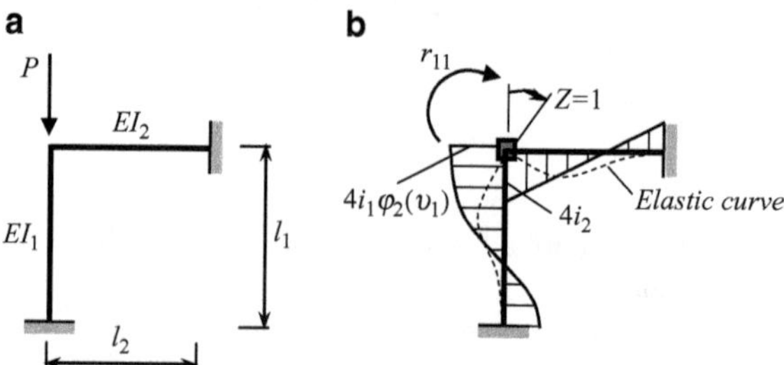

Fig. 13.13 (**a**) Design diagram; (**b**) Primary system of the displacement method and unit bending moment diagram

The bending moment diagram yields $r_{11} = 4i_1\varphi_2(\upsilon_1) + 4i_2$, where parameter of critical load

$$\upsilon_1 = l_1\sqrt{\frac{P}{EI_1}}.$$

Note that subscript 2 at function φ is related to the clamped–clamped member subjected to angular displacement of the one support (Table A.22), while the subscript 1 at the parameter υ is related to the compressed-bent member 1. Canonical equation of the displacement method is $r_{11}(\upsilon_1)Z = 0$. Nontrivial solution of this equation leads to equation of stability $r_{11} = 0$ or in expanded form

$$r_{11} = \frac{4EI_1}{l_1}\varphi_2(\upsilon_1) + \frac{4EI_2}{l_2} = 0.$$

Special cases:

1. Assume that $l_2 \to 0$. In this case, the second term $\frac{4EI_2}{l_2} \to \infty$, rigid joint is transformed to clamped support and the initial frame in whole is transformed into the vertical clamped–clamped column. Stability equation becomes $\varphi_2(\upsilon_1) = -\infty$. Root of this equation (Table A.25) is $\upsilon_1 = 2\pi$ and critical force becomes

$$P_{cr} = \frac{\upsilon_1^2 EI}{l_1^2} = \frac{4\pi^2 EI}{l_1^2} = \frac{\pi^2 EI}{(0.5l_1)^2},$$

   where $\mu = 0.5$ is effective-length factor for clamped–clamped column.
2. Assume $EI_2 \to 0$. In this case, the rigid joint is transformed to hinge and the initial frame is transformed into the vertical clamped-pinned column. Stability equation becomes $\varphi_2(\upsilon_1) = 0$. Root of this equation is $\upsilon_1 = 4.488$ (Table A.25) and critical force $P_{cr} = \frac{\upsilon_1^2 EI}{l_1^2} = \frac{4.488^2 EI}{l_1^2} = \frac{\pi^2 EI}{(0.7l_1)^2}$, $\mu = 0.7$
3. If $l_1 = l_2$, $EI_1 = EI_2$, then stability equation becomes $\varphi_2(\upsilon_1) + 1 = 0$.

The root of this equation is $\upsilon_1 = 5.3269$ and critical load equals

$$P_{cr} = \frac{\upsilon_1^2 EI}{l_1^2} = \frac{28.397EI}{l_1^2}.$$

Now let us consider a nonuniform two-span continuous beam shown in Fig. 13.14a. We need to derive a stability equation and determine the critical load.

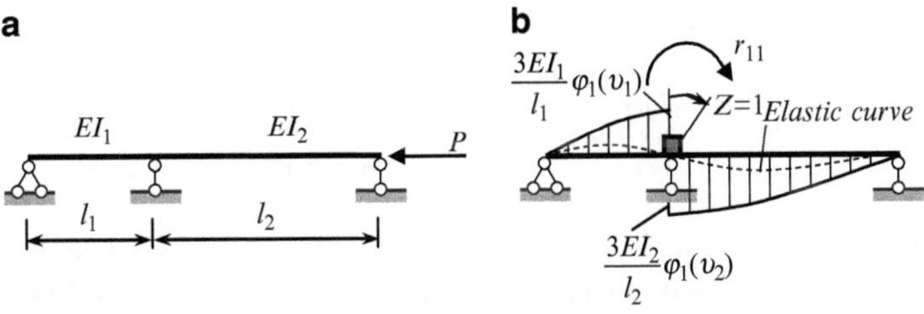

Fig. 13.14 Continuous compressed beam: (**a**) Design diagram; (**b**) Primary system and unit bending moment diagram

The axial compressed forces in both spans are equal, so the dimensionless parameters υ_1 and υ_2 for both spans are

$$\upsilon_1 = l_1\sqrt{\frac{P}{EI_1}}; \quad \upsilon_2 = l_2\sqrt{\frac{P}{EI_2}}.$$

Let the left span is considered as the basic member, so $\upsilon_1 = \upsilon$. In this case, the parameter υ_2 can be presented in term of the basic parameter υ as follows

$$\upsilon_2 = \upsilon_1\frac{l_2}{l_1}\sqrt{\frac{EI_1}{EI_2}} = \upsilon\alpha, \quad \alpha = \frac{l_2}{l_1}\sqrt{\frac{EI_1}{EI_2}}.$$

The primary unknown of the displacement method is the angle of rotation at the intermediate support. The primary system and bending moment diagram due to unit rotation of the introduced constrain are presented in Fig. 13.14b. Since both spans are compressed, then the bending moment diagrams are curvilinear. According to Table A.22, the moment for clamped-pinned beam in case of angular displacement has the multiple $\varphi_1(\upsilon)$. Unit reaction, which arises in introduced constraint, is

$$r_{11}(\upsilon) = \frac{3EI_1}{l_1}\varphi_1(\upsilon_1) + \frac{3EI_2}{l_2}\varphi_1(\upsilon_2) = \frac{3EI_1}{l_1}\varphi_1(\upsilon) + \frac{3EI_2}{l_2}\varphi_1(\alpha\upsilon).$$

As before, the subscript 1 at the function $\varphi_1(\upsilon)$ reflects the *type of the beam* and *type of displacement* (Table A.22), while subscripts 1 and 2 at the parameter υ denote the number of the span.

Canonical equation of the displacement method is $r_{11}(\upsilon)Z = 0$. Nontrivial solution of this equation leads to equation of a critical force

$$r_{11}(\upsilon) = \frac{3EI_1}{l_1}\varphi_1(\upsilon) + \frac{3EI_2}{l_2}\varphi_1(\alpha\upsilon) = 0.$$

Special cases:

1. Assume $l_2 \to 0$. In this case, intermediate pinned support is transformed into *clamped* support and initial beam becomes one-span pinned-clamped beam length l_1. Stability equation becomes $\varphi_1(\upsilon_1) = -\infty$. Root of this equation is $\upsilon_1 = 4.488$ and critical force

$$P_{cr} = \frac{\upsilon_1^2 EI}{l_1^2} = \frac{4.488^2 EI}{l_1^2} = \frac{\pi^2 EI}{(0.7l_1)^2}, \quad \mu = 0.7$$

2. Assume $l_1 = l_2 = l$ and $EI_1 = EI_2 = EI$. In this case, parameter $\alpha = 1$, the stability equation becomes $\varphi_1(\upsilon) = 0$, and parameter of critical load $\upsilon = \pi$. So the critical load $P_{cr} = \dfrac{\pi^2 EI}{l^2}$. This critical load corresponds to column with pinned-rolled supports.

3. Assume $l_2 = 2l_1$ and $EI_1 = EI_2 = EI$. In this case parameter $\alpha = \frac{l_2}{l_1}\sqrt{\frac{EI_1}{EI_2}} = 2$ and critical load equation becomes $\varphi_1(\upsilon) + 0.5\varphi_1(2\upsilon) = 0$; parameter of critical load $\upsilon = 1.967$. So the critical load $P_{\text{cr}} = \frac{3.869EI}{l_1^2}$. It can be seen that increasing of the one span by two times has profound effect on the critical load (coefficient 3.869 instead of π^2 for two-span beam with equal spans).

Let us illustrate the displacement method in canonical form for stability analysis of the multistory frame.

Example 13.4. The two-story frame in Fig. 13.15a is subjected to compressed axial forces P and αP. Geometrical parameters of the frame are h and $l = \beta h$, the bending stiffness of the members are EI for members of the second level and kEI for column of the first level. Derive the stability equation in terms of arbitrary positive numbers α, β, and k.

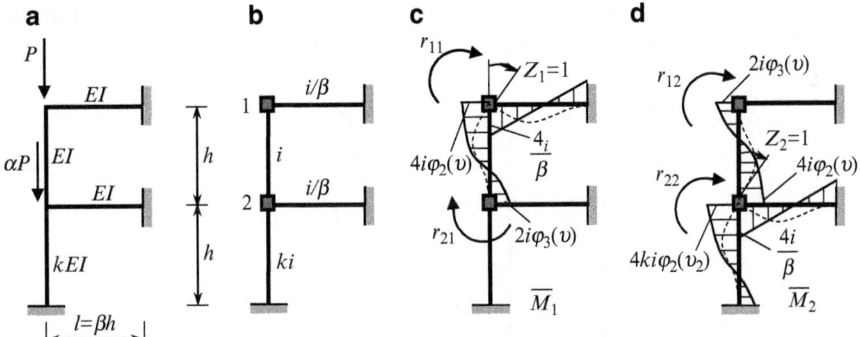

Fig. 13.15 Two-story frame

Solution. The primary system is presented in Fig. 13.15b. Let us assign member 1–2 as basic one; its flexural stiffness per unit length equals $i = EI/h$. The flexural stiffness per unit length for each member are shown in Fig. 13.15b. Parameters of critical force for member 1–2 and 2-A are determined as follows:

$$\upsilon_1 = h\sqrt{\frac{P}{EI}} = \upsilon, \quad \upsilon_2 = h\sqrt{\frac{P+\alpha P}{kEI}} = \upsilon\sqrt{\frac{1+\alpha}{k}}.$$

Bending moment diagrams caused by $Z_1=1$ and $Z_2=1$ are shown in Fig. 13.15c, d, respectively.

Unit reactions are

$$r_{11} = 4i\varphi_2(\upsilon) + \frac{4i}{\beta}; \quad r_{12} = r_{21} = 2i\varphi_3(\upsilon);$$

$$r_{22} = 4i\varphi_2(\upsilon) + 4ki\varphi_2(\upsilon_2) + \frac{4i}{\beta}.$$

Stability equation in general and expanded forms are

$$\begin{vmatrix} r_{11} & r_{12} \\ r_{21} & r_{22} \end{vmatrix} = 0$$

$$4\left[\varphi_2(\upsilon) + \frac{1}{\beta}\right] \cdot \left[\varphi_2(\upsilon) + k\varphi_2(\upsilon_2) + \frac{1}{\beta}\right] - \varphi_3^2(\upsilon) = 0.$$

Let $\alpha = 3,\ \beta = 1,\ k = 4$. In this case, $\upsilon_2 = \upsilon$ and stability equation becomes

$$4\left[\varphi_2(\upsilon) + 1\right] \cdot \left[5\varphi_2(\upsilon) + 1\right] - \varphi_3^2(\upsilon) = 0.$$

The root of this equation is $\upsilon = 4.5307$. The critical load is

$$P_{\text{cr}} = \frac{\upsilon^2 EI}{h^2} = \frac{4.5307^2 EI}{h^2}.$$

Example 13.5. The frame in Fig. 13.16a is loaded by two forces at the joints. Derive the stability equation and find the critical load P.

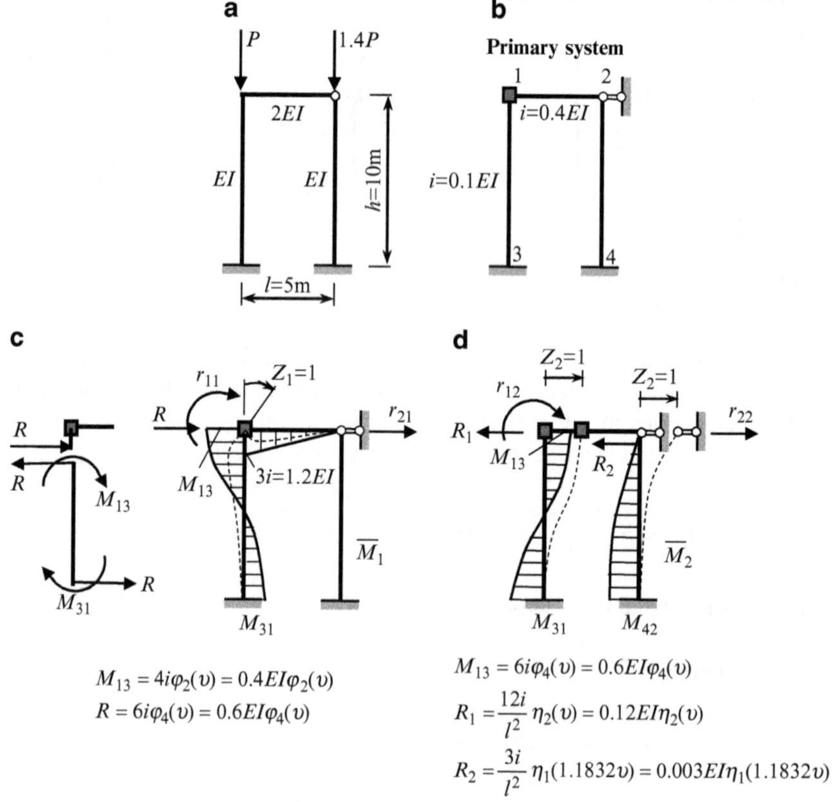

Fig. 13.16 (**a, b**) Design diagram of the frame and primary system. (**c, d**) Unit bending moment diagrams

Solution. The frame has two unknowns of the displacement method. They are the angle of rotation Z_1 of rigid joint and horizontal displacement Z_2 of the cross bar. The primary system is presented in Fig. 13.16b.

Bending moment diagrams caused by unit displacement of the introduced constraints are presented in Fig. 13.16c, d; elastic curves are shown by dotted line. The diagrams within members 1–3 and 2–4 are curvilinear. Direction of R for $\overline{M}_1$ diagram is explained for the left column (Fig. 13.16c). Similarly may be explained directions for R_1 and $R_2(\overline{M}_2$ diagram).

Parameters of a critical load are $\upsilon_{13} = \upsilon = h\sqrt{\frac{P}{EI}}$; $\upsilon_{24} = h\sqrt{\frac{1.4P}{EI}} = 1.1832\upsilon$.

Canonical equations and unit reactions are

$$\begin{aligned} r_{11}Z_1 + r_{12}Z_2 &= 0, \\ r_{21}Z_1 + r_{22}Z_2 &= 0, \end{aligned} \tag{a}$$

where

$$\begin{aligned} r_{11} &= [0.4\varphi_2(\upsilon) + 1.2]EI; r_{21} = r_{12} = -0.6EI\varphi_4(\upsilon); \\ r_{22} &= [0.012\eta_2(\upsilon) + 0.003\eta_1(1.1832\upsilon)]EI \end{aligned}$$

Stability equation becomes:

$$\begin{vmatrix} 0.4\varphi_2(\upsilon) + 1.2 & -0.6\varphi_4(\upsilon) \\ -0.6\varphi_4(\upsilon) & 0.012\eta_2(\upsilon) + 0.003\eta_1(1.1832\upsilon) \end{vmatrix} = 0$$

Solution of this equation leads to parameter of critical load $\upsilon = 2.12$. The critical load is

$$P_{\text{cr}} = \frac{\upsilon^2 EI}{h^2} = \frac{4.49EI}{h^2}.$$

Thus the frame becomes unstable if it will be loaded by two forces P and $1.4P$ simultaneously.

Example 13.6. Design diagram of the multispan frame is presented in Fig. 13.17a. Derive the stability equation and calculate the critical force.

Solution. The primary system is shown in Fig. 13.17b. The introduced constraint 1 prevents linear displacement. Canonical equation is $r_{11}Z_1 = 0$, so stability equation is $r_{11} = 0$. Bending moment diagram caused by unit displacement of the introduced support is presented in Fig. 13.17b. Within the second and third columns, the bending moment diagrams are curvilinear.

Free body diagram of a cross-bar is shown in Fig. 13.17c. Shear forces for compressed-bent members are presented in Table A.22. Parameter of critical load $\upsilon = h\sqrt{\frac{P}{EI}}$. Unit reaction equals

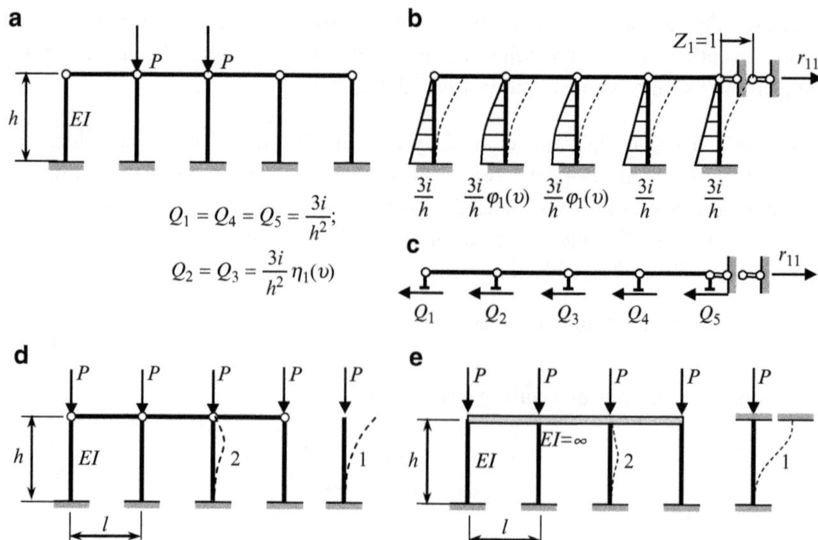

Fig. 13.17 **(a–c)** Multispan frame. Design diagram, primary system and free-body diagram. **(d, e)** Multispan regular frames; a number of columns is k

$$r_{11} = 3 \cdot \frac{3i}{h^2} + 2 \cdot \frac{3i}{h^2}\eta_1(\upsilon). \tag{a}$$

The stability equation becomes $3 + 2\eta_1(\upsilon) = 0$. The minimum positive root of this equation, i.e., the parameter of lowest critical load is $\upsilon = 2.4521$. Critical load

$$P_{\text{cr}} = \frac{\upsilon^2 EI}{h^2} = \frac{6.0128EI}{h^2}.$$

Discussion. Analysis of this frame allows easy considering of two important typical cases. Both cases are related to the regular multispan frame with k columns loaded by equal forces P at the each joint. Figure 13.17d presents regular frame with hinged joints, while the Fig. 13.17e shows the frame with absolutely rigid cross-bar and fixed joints. In both cases, the frame has one unknown of the displacement method. Introduced constraint prevents linear displacement; this constraint is not shown.

For both cases, the loss of stability is possible according two different forms. The first form occurs with horizontal displacement of the cross bar (dotted line 1) and the second form without of such displacement (dotted line 2). For both cases, the lowest critical load corresponds to the first form.

For case of hinged joints (Fig. 13.17d), the horizontal reaction of introduced constraint is

$$r_{11} = k \cdot \frac{3i}{h^2}\eta_1(\upsilon).$$

Stability equation becomes $\eta_1(\upsilon) = 0$ and the lowest root is $\upsilon = \pi/2$. The critical load is equal to

$$P_{\text{cr}} = \frac{\pi^2 EI}{4h^2};$$

This case corresponds to single fixed-free column. For second form of buckling we get $P_{\text{cr}} = \pi^2 EI/(0.7h)^2$ as for simple fixed-pinned column.

For case of absolutely rigid cross bar, the lowest critical load also corresponds to the first form. This form is characterized by horizontal displacement of cross bar, while the fixed joint have not angular displacements because for cross-bar $EI = \infty$. In this case the unit horizontal reaction equals

$$r_{11} = k \cdot \frac{12i}{h^2}\eta_2(\upsilon).$$

Stability equation becomes $\eta_2(\upsilon) = 0$ and the lowest root is $\upsilon = \pi$. The critical load is equal to

$$P_{\text{cr}} = \frac{\pi^2 EI}{h^2};$$

This case corresponds to single column with fixed ends while the one fixed support permit the horizontal displacement. For second form of buckling, we get $P_{\text{cr}} = 4\pi^2 EI/h^2$ as for fixed-fixed column.

We can see that for these regular frames, a critical load P does not depend of the number of columns k and length l of each span.

13.4.3 Modified Approach of the Displacement Method

In general case, the displacement method requires introducing constraints, which prevent to angular displacement of rigid joints and independent linear displacements of joints. However, in stability problems of a frame with sidesway, it is possible some modification of the classical displacement method. Using modified approach, we can introduce a new type of constraint, mainly the constraint, which prevents to angular displacement, but *simultaneously* has a linear displacement. This type of constraint is presented for pinned-clamped and clamped-clamped members in Table A.22, line 3.

Example below presents analysis of the frame with sidesway (Fig. 13.18a) using two approaches. The first approach corresponds to classical primary system of the displacement method; the primary system contains *two* introduced constraints, one of which prevents angular and another prevents linear displacements. The second approach corresponds to modified primary system of the displacement method; the primary system contains *one* introduced constraint, which prevents angular displacement and allows linear displacement Δ. We will derive the stability equation using both approaches and calculate the critical load.

First approach. The primary unknowns are angle of rotation of rigid joint and linear displacement of a cross-bar. Figure 13.18b shows the primary system, elastic curve,

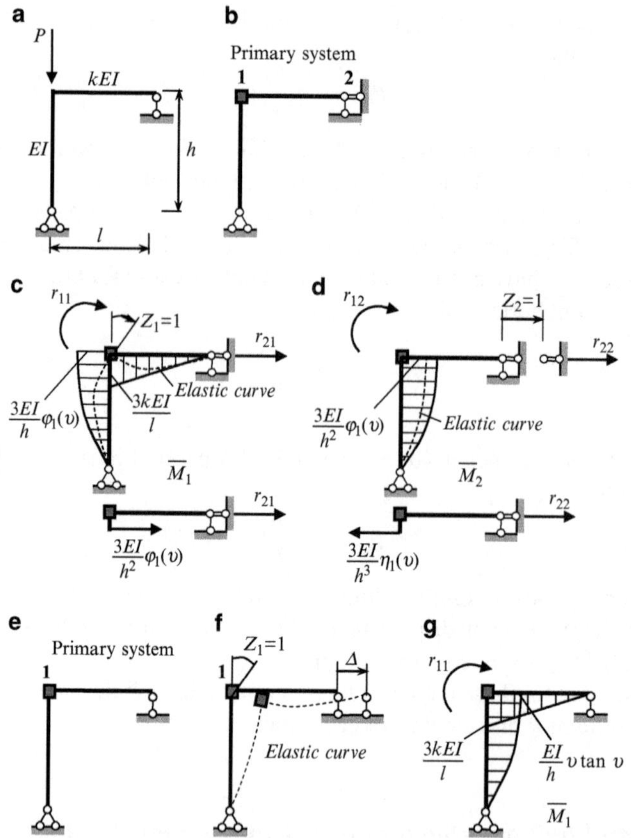

Fig. 13.18 (**a**) Design diagram of the frame. (**b–d**) First approach – primary system and corresponding bending moment diagrams in the unit conditions. (**e–g**) Second approach – primary system and corresponding bending moment diagrams in the unit conditions

and bending moment diagrams caused by the unit rotation and linear displacement of induced constrains 1 and 2.

Bending moments diagram is curvilinear for compressed vertical member of the frame. The ordinates are found in accordance with the Table A.22. The bending moment diagrams yield

$$r_{11} = \frac{3EI}{h}\varphi_1(\upsilon) + \frac{3kEI}{l}; \quad r_{12} = -\frac{3EI}{h^2}\varphi_1(\upsilon)$$
$$r_{21} = -\frac{3EI}{h^2}\varphi_1(\upsilon); \quad r_{22} = -\frac{3EI}{h^3}\eta_1(\upsilon)$$

where parameter of stability $\upsilon = h\sqrt{\frac{P}{EI}}$. Again, the subscript 1 at functions φ and η is concerning to the pinned-clamped member subjected to angular and linear displacements of the clamped support.

Nontrivial solution of canonical equations of the displacements method leads to the following stability equation

$$\begin{vmatrix} r_{11} & r_{12} \\ r_{21} & r_{22} \end{vmatrix} = 0$$

If we assume that $l = h$, and take into account the expressions for functions $\varphi_1(\upsilon)$ and $\eta_1(\upsilon)$, then stability equation after rearrangements becomes

$$\upsilon^3\,(3k - \upsilon\tan\upsilon) = 0.$$

The root of this equation is $\upsilon = 0$ and corresponds to initial condition of the frame.

Condition

$$3k - \upsilon\tan\upsilon = 0$$

allows to calculate the critical parameter υ for any value of k. Some results are presented in Table 13.2.

Table 13.2 Critical load in terms of parameter k

Parameter k	Root of equation υ	Critical load P_{cr} (factor EI/h^2)
1	1.193	1.423
10	1.521	2.313
∞	1.57	2.465

Second approach. In this case only constraint 1 is introduced (Fig. 13.18e). However, this constrain *allows* the linear displacement Δ. This case is presented in Table A.22, row 3. Elastic curve caused by unit angular displacements (if linear displacement Δ occurs) and corresponding bending moment diagram are shown in Fig. 13.18f, g.

Unit reaction

$$r_{11} = -\frac{EI}{h}\upsilon\tan\upsilon + \frac{3kEI}{l}$$

so the stability equation becomes

$$-\upsilon\tan\upsilon + 3k\frac{h}{l} = 0.$$

If $l = h$, then this stability equation is the same as was obtained above. The second approach is more effective than the first one.

13.5 Stability of Arches

Stability analysis of the different types of arches is based on a solution of a differential equation. Precise analytical solution may be obtained only for specific arches and their loading. In this section, we will consider a plane uniform arch with constant radius R of curvature (circular arch), which is subjected to uniformly distributed pressure normal to the axis of the arch. In this case, only axial compressed forces arise before the buckling arch. Thus this problem, as all previous problems in this chapter, has a general feature – a structure before buckling is subjected to only compressed load.

Figure 13.19a, b presents statically indeterminate arches with pinned and fixed supports as well as three-hinged arch. The loss of stability of an arch may occur in two simplest forms. They are symmetrical form (a), when elastic curve is symmetrical with respect to vertical axis of symmetry and, otherwise, antisymmetrical form (b).

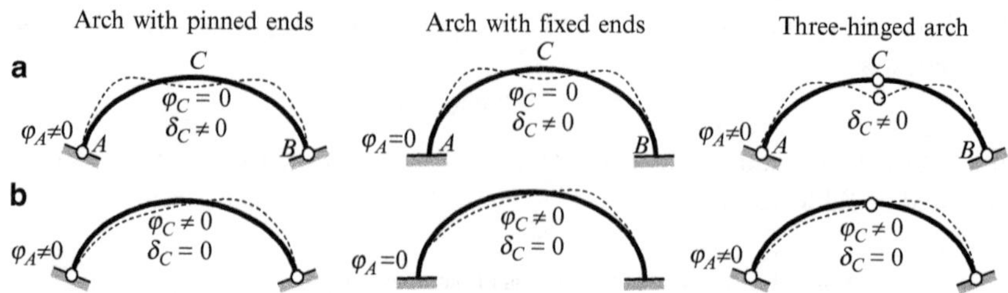

Fig. 13.19 Arches with different boundary conditions. (**a**) symmetrical and (**b**) antisymmetrical buckling forms

The symmetrical and antisymmetrical forms mean that both supports rotate in the opposite directions and the same directions, respectively. As it shown by analytical analysis and experiments, the smallest critical load for hingeless and two-hinged arches corresponds to antisymmetrical buckling form.

13.5.1 Circular Arches Under Hydrostatic Load

Assume that symmetrical circular arch of radius R has the elastic-fixed supports; their rotational stiffness coefficient is k [kN m/rad]. The central angle of the arch is 2α; the flexural rigidity is EI, and the intensity of the radial uniformly distributed load is q (Fig. 13.20a).

For stability analysis of the arch, we will use the following differential equation

$$\frac{\mathrm{d}^2 w}{\mathrm{d}s^2} + \frac{w}{R^2} = -\frac{M}{EI}, \tag{13.13}$$

where w is a displacement point of the arch in radial direction (Fig. 13.20b), and M is bending moment which is produced in the cross sections of the arch when it loss a stability.

Let ds presents the arc, which corresponds to central angle dθ. Since

$$\frac{\mathrm{d}w}{\mathrm{d}s} = \frac{\mathrm{d}w}{\mathrm{d}\theta}\frac{\mathrm{d}\theta}{\mathrm{d}s} = \frac{1}{R}\frac{\mathrm{d}w}{\mathrm{d}\theta} \quad \text{and} \quad \frac{\mathrm{d}^2 w}{\mathrm{d}s^2} = \frac{1}{R^2}\frac{\mathrm{d}^2 w}{\mathrm{d}\theta^2},$$

then (13.13) may be presented in polar coordinates in the following form

$$\frac{\mathrm{d}^2 w}{\mathrm{d}\theta^2} + w = -\frac{M}{EI} R^2, \tag{13.13a}$$

which is called the Boussinesq's equation (1883).

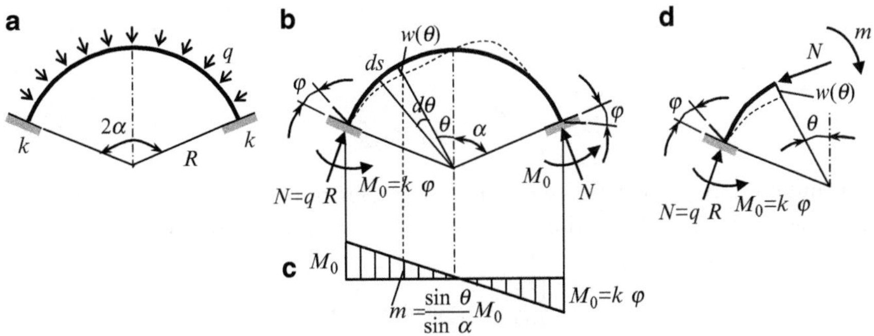

Fig. 13.20 (**a**) Design diagram of the circular arch with elastic supports; (**b**) Reactions and antisymmetrical buckling forms; (**c**) Distribution of bending moments caused by two reactive moments M_0; (**d**) Free-body diagram for portion of the arch (load q is not shown)

It is easy to show that the axial compressive force of the arch caused by uniformly distributed hydrostatic load q is $N = qR$. Indeed, the total load within the portion ds (Fig. 13.20b) equals $q\mathrm{d}s = qR\mathrm{d}\theta$, and all load which acts on the left half-arch is perceived by the left support, so the horizontal and vertical components of reaction N are

$$H = -\int_0^{\alpha} qR\sin\theta\,\mathrm{d}\theta = qR\cos\alpha \quad \text{and} \quad V = \int_0^{\alpha} qR\cos\theta\,\mathrm{d}\theta = qR\sin\alpha,$$

so

$$N = \sqrt{H^2 + V^2} = qR.$$

The slope φ at the elastic support and corresponding reactive moment M_0 are related as $M_0 = k\varphi$. Distribution of the bending moments caused by two antisymmetrical angular displacements φ of elastic supports (or reactive moments M_0) is presented in Fig. 13.20c. Bending moment at section with central angle θ caused by only reactive moments M_0 equals m. The total bending moment at any section, which is characterized by displacement w (free-body diagram is shown in Fig. 13.20d), equals

$$M = qR\,w - \frac{\sin\theta}{\sin\alpha}k\varphi. \tag{13.14}$$

For two-hinged arch, the second term, which takes into account moment due to elastic supports, should be omitted.

Thus, differential equation (13.13a) becomes

$$\frac{d^2w}{d\theta^2} + \left(1 + \frac{qR^3}{EI}\right) w = \frac{k\varphi}{EI} \frac{R^2}{\sin\alpha} \sin\theta. \tag{13.15}$$

Denote

$$n^2 = 1 + \frac{qR^3}{EI}. \tag{13.15a}$$

$$C = \frac{k\varphi R^2}{EI \sin\alpha}. \tag{13.15b}$$

Pay attention that C is unknown, since the angles of rotation φ of the supports are unknown. Differential equation (13.15) may be rewritten as follows

$$\frac{d^2w}{d\theta^2} + n^2 w = C \sin\theta. \tag{13.16}$$

Solution of this equation is

$$w = A\cos n\theta + B\sin n\theta + w^*. \tag{13.17}$$

The partial solution w^* should be presented in the form of the right part of (13.16), mainly $w^* = C_0 \sin\theta$, where C_0 is a new unknown coefficient. Substituting of this expression into (13.16) leads to equation

$$-C_0 \sin\theta + n^2 C_0 \sin\theta = C \sin\theta,$$

so

$$C_0 = \frac{C}{n^2 - 1}.$$

Thus the solution of equation (13.15) becomes

$$w = A\cos n\theta + B\sin n\theta + \frac{C}{n^2 - 1}\sin\theta. \tag{13.17a}$$

Unknown coefficients A, B, and C may be obtained from the following boundary conditions:

1. For point of the arch on the axis of symmetry ($\theta = 0$), the radial displacement $w = 0$ (because the antisymmetrical form of the loss of stability); this condition leads to $A = 0$;
2. For point of the arch at the support ($\theta = \alpha$) the radial displacement is $w = 0$, so

$$B\sin n\alpha + \frac{C}{n^2 - 1}\sin\alpha = 0. \tag{13.18}$$

3. Using expression (13.17a), the slope is

$$\frac{\mathrm{d}w}{\mathrm{d}\theta} = Bn\cos n\theta + \frac{C\cos\theta}{n^2-1}. \tag{13.19}$$

The slope at the support is

$$\frac{\mathrm{d}w}{\mathrm{d}s} = -\varphi,$$

the negative sign means the reactive moment M_0 and angle φ have the opposite directions. On the other hand

$$\frac{\mathrm{d}w}{\mathrm{d}s} = \frac{\mathrm{d}w}{R\mathrm{d}\theta},$$

so

$$\frac{\mathrm{d}w}{\mathrm{d}\theta} = -R\varphi.$$

According to (13.15b) we get

$$\varphi = C\frac{EI\sin\alpha}{kR^2}, \text{ so } \frac{\mathrm{d}w}{\mathrm{d}\theta} = -C\frac{EI\sin\alpha}{kR}.$$

If $\theta = \alpha$ then the expression (13.19) becomes

$$Bn\cos n\alpha + \frac{C\cos\alpha}{n^2-1} = -C\frac{EI\sin\alpha}{kR}.$$

After rearrangements this expression may be presented in form

$$Bn\cos n\alpha + C\left(\frac{\cos\alpha}{n^2-1} + \frac{EI\sin\alpha}{kR}\right) = 0. \tag{13.20}$$

Equations (13.18) and (13.20) are homogeneous linear algebraic equations with respect to unknown parameters B and C. The trivial solution $B = C = 0$ corresponds to state of the arch before the loss of stability. Nontrivial solution occurs if the following determinant is zero:

$$D = \begin{vmatrix} \sin n\alpha & \dfrac{\sin\alpha}{n^2-1} \\ n\cos n\alpha; & \dfrac{\cos\alpha}{n^2-1} + \dfrac{EI\sin\alpha}{kR} \end{vmatrix} = 0 \tag{13.21}$$

The stability equation (13.21) may be presented as follows:

$$\tan n\alpha = \frac{nk_0}{k_0\cot\alpha + (n^2-1)}, \quad k_0 = \frac{kR}{EI}. \tag{13.22}$$

Solution of this transcendental equation for given α and dimensionless parameter k_0 is the critical parameter n. According to (13.15a) the critical load

$$q_{\mathrm{cr}} = \left(n^2 - 1\right) \frac{EI}{R^3}. \tag{13.23}$$

If the central angle $2\alpha = 60°$, then the roots of equation (13.22) for different k_0 are presented the Table 13.3:

Table 13.3 Critical parameter n for circular arch with elastic clamped supports, $2\alpha = 60°$

k_0	0.0	1.0	10	100	1000	10^5
n	6.0000	6.2955	7.5294	8.4628	8.6051	8.621

Limiting cases. The general stability equation (13.22) allows us to consider some specific arches.

1. *Two-hinged arch.* In this case the stiffness $k = 0$ and stability equation (13.22) is $\tan n\alpha = 0$. The minimum root of this equation is $n\alpha = \pi$, so $n = \dfrac{\pi}{\alpha}$ and corresponding critical load equals

$$q_{\mathrm{cr\ min}} = \left(\frac{\pi^2}{\alpha^2} - 1\right) \frac{EI}{R^3}. \tag{13.24}$$

Critical load for $\alpha = \pi/2$ (half-circular arch with pinned supports) equals

$$q_{\mathrm{cr\,min}} = 3\frac{EI}{R^3}.$$

2. *Arch with fixed supports.* In this case the stiffness $k = \infty$ and stability equation (13.22) becomes

$$\tan n\alpha = n \cdot \tan \alpha \tag{13.25}$$

In case $\alpha = \pi/2$ (half-circular arch) this equation can be presented in the form

$$\cot \frac{n\pi}{2} = 0, \text{ so} \frac{n\pi}{2} = \frac{\pi}{2}, \frac{3\pi}{2}, \ldots$$

Solution $n = 1$ is trivial because this solution, according to expression (13.15a), corresponds to $q = 0$. Thus, minimum root is $n = 3$. Thus for a half-circular arch with clamped supports the critical load equals

$$q_{\mathrm{cr\ min}} = 8\frac{EI}{R^3}.$$

Roots n of stability equation (13.25) for different angle α are presented in the Table 13.4.

Table 13.4 Critical parameter n for circular arch with fixed supports

α	30°	45°	60°	90°
N	8.621	5.782	4.375	3.000

It is worth to present the critical load for three-hinged symmetrical arch under hydrostatic load. The critical load for antisymmetric buckling form coincides with critical load for two-hinged arch. In case of symmetrical buckling form, the critical load should be calculated by using the formula

$$q_{\mathrm{cr}} = \left(\frac{4u^2}{\alpha^2} - 1\right)\frac{EI}{R^3}, \tag{13.26}$$

where parameter u is a root of transcendental equation

$$\frac{\tan u - u}{u^3} = 4\frac{(\tan\alpha - \alpha)}{\alpha^3}. \tag{13.27}$$

Roots of this equation are presented in the Table 13.5.

Table 13.5 Circular three-hinged arch. Critical parameter u for symmetrical buckling form

α	30°	45°	60°	90°
U	1.3872	1.4172	1.4584	$\pi/2 = 1.5708$

For all above cases, the critical load may be calculated by the formula

$$q_{\mathrm{cr}} = K\frac{EI}{R^3},$$

where parameter K is presented in the Table 13.6.

Table 13.6 Parameter K for critical hydrostatic load of circular arches with different boundary conditions

Types of arch	$\alpha = 15°$	30°	45°	60°	75°	90°
Hingeless	294	73.3	32.4	18.1	11.5	8
Two-hinged	143	35	15	8	4.76	3
Three-hinged (symmetrical form)	108	27.6	12	6.75	4.32	3

Table 13.6 indicates that for three-hinged arch the lowest critical hydrostatic load corresponds to loss of stability in symmetrical form. For arches with elastic supports, factor $\mathcal{K}$ satisfies condition

$$K_1 \le K \le K_2,$$

where K_1 and K_2 are related to hingeless and two-hinged arches.

13.5.2 Complex Arched Structure: Arch with Elastic Supports

In practical engineering, the stiffness coefficient k of the elastic supports is not given; however, in special cases it can be determined from an analysis of adjacent parts of the arch. Let us consider a structure shown in Fig. 13.21a. The central part of the structure presents the circular arch; supports of the arch are rigid joints of the frames. The arch is subjected to uniformly distributed hydrostatic load q. Assume that $R = 20\,\text{m}$ and the central angle $2\alpha = 60°$. The stiffness of all members of the structure is EI.

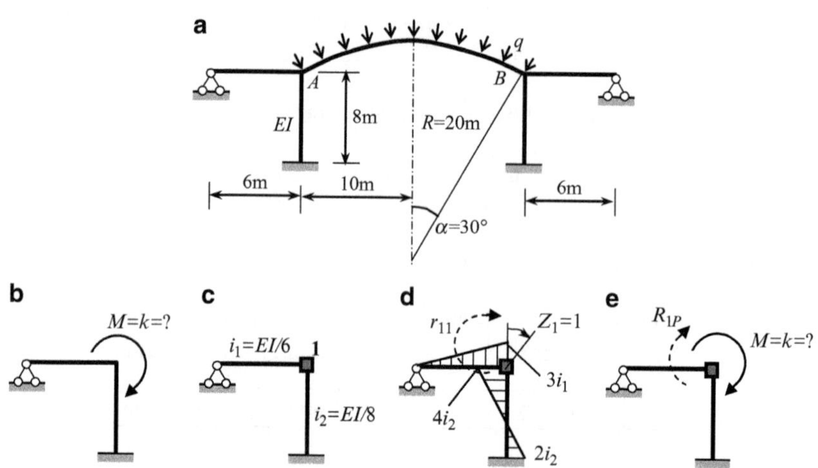

Fig. 13.21 (**a**) Design diagram of the structure. (**b–e**) Calculation of the stiffness k of the elastic supports of the arch

Since the left and right frames are deformable structures, then the each joint A and B has some angle of rotation, so the arch AB should be considered as arch with elastic supports with rotational stiffness k. For this case of circular arch with given type of load, the stability equation according (13.22) becomes:

$$\tan n\alpha = \frac{n}{\cot\alpha + \dfrac{\left(n^2 - 1\right)EI}{kR}}. \tag{13.28}$$

Rotational stiffness coefficient k is a couple M, which arises at elastic support of the arch if this support rotates through the angle $\varphi = 1$. Since the joints A and B are rigid, so the angle of rotation for frame and arch are same. Therefore, for calculation of the stiffness k we have to calculate the couple M, which should be applied at the rigid joint A of the frame in order to rotate this joint by angle $\varphi = 1$.

The frame subjected to unknown moment $M = k$ is shown in Fig. 13.21b. For solving of this problem, we can use the displacement method.

Primary system of the displacement method is shown in Fig. 13.21c. Canonical equation is

$$r_{11}Z_1 + R_{1P} = 0.$$

Displacement $Z_1 = 1$ and corresponding bending moment diagram is shown in Fig. 13.21d. The unit reaction

$$r_{11} = 3i_1 + 4i_2 = 0.5EI + 0.5EI = 1.0EI.$$

The primary system subjected to external unknown couple M is presented in Fig. 13.21e, so $R_{1P} = -M$. The canonical equation becomes $1EI \cdot Z_1 - M = 0$. If the angle of rotation $Z_1 = 1$, then $M = k = 1.0EI$. For given parameters R and α the stability equation (13.28) of the structure becomes

$$\tan n\frac{\pi}{6} = \frac{n}{\cot\dfrac{\pi}{6} + \dfrac{(n^2-1)}{20}} \quad \text{or} \quad \tan(0.5236n) = \frac{20n}{33.64 + n^2}.$$

The root of this equation $n = 7.955$. The critical load is

$$q_{\text{cr}} = \left(7.955^2 - 1\right)\frac{EI}{R^3} = 62.28\frac{EI}{R^3}. \tag{13.29}$$

According to Table 13.6, the critical load for arch with fixed supports and for two-hinged arch (the central angle in both cases is $2\alpha = 60°$) are $q_{\text{cr}} = 73.3\frac{EI}{R^3}$ and $q_{\text{cr}} = 35\frac{EI}{R^3}$, respectively. Above calculated critical load (13.29) is located between two limiting cases.

13.6 Compressed Rods with Lateral Loading

In the previous chapters, the bending structure had been analyzed on the basis of a nondeformable scheme. However, the axial compressed force P creates additional moment on the displacements δ due to lateral load. Even if a small lateral load leads to small lateral displacements, the compressed load on these small displacements leads to additional moments and displacements. Influence of the compressed axial force becomes especially significant for tall structures.

A straight member that is simultaneously subjected to axial compression and lateral bending is called a beam-column. Often analysis for such structures is refers as P-delta analysis. Other title of such analysis is analysis of a structure on the basis of deformable scheme. This analysis is nonlinear one and it allows finding a more real distribution of internal forces and deflections, while the analysis on the basis of the nondeformable design diagram leads to theirs underestimating.

Two different methods for beam-columns analysis are presented below. They are the double integration and initial parameters methods.

13.6.1 Double Integration Method

Figure 13.22 shows a simply supported beam subjected to lateral force F and compressed force P. We need to derive the expressions for deflection and internal forces.

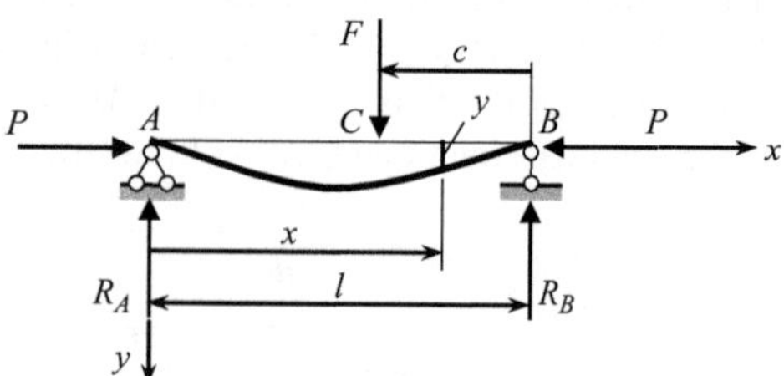

Fig. 13.22 Simply supported beam subjected to compressive axial load P and lateral load F

Differential equation of elastic curve of the beam for left and right parts (portions 1 and 2, respectively) may be written as

$$EI\frac{\mathrm{d}^2 y_1}{\mathrm{d}x^2} = -Py_1 - R_A x = -Py - \frac{Fc}{l}x, \quad x \le l - c,$$
$$EI\frac{\mathrm{d}^2 y_2}{\mathrm{d}x^2} = -Py_2 - R_B(l-x) = -Py - \frac{F(l-c)}{l}(l-x), \quad x > l - c. \tag{13.30}$$

General solution of these equations is

$$y_1 = C_1 \cos nx + D_1 \sin nx - \frac{Fc}{Pl}x. \tag{13.31}$$

$$y_2 = C_2 \cos nx + D_2 \sin nx - \frac{F}{Pl}(l-c)(l-x), \tag{13.32}$$

where

$$n = \sqrt{\frac{P}{EI}}$$

is parameter of compressed load P.

At the points $A(x = 0)$ and $B(x = l)$ the displacement y is zero. Equations (13.31) and (13.32) lead to $C_1 = 0$ and $C_2 = -D_2 \tan nl$. Therefore, expressions for displacements within the left and right portions are

$$y_1 = D_1 \sin nx - \frac{Fc}{Pl}x,$$
$$y_2 = -D_2 \tan nl \cos nx + D_2 \sin nx - \frac{F}{Pl}(l-c)(l-x). \tag{13.33}$$

For calculation of unknown coefficients D_1 and D_2 we can use the following conditions at the point C:

$$y_1 = y_2 \quad \text{and} \quad \frac{\mathrm{d}y_1}{\mathrm{d}x} = \frac{\mathrm{d}y_2}{\mathrm{d}x}.$$

In expanded form these conditions are

$$D_1 \sin n(l-c) = D_2 [\sin n(l-c) - \tan nl \cos n(l-c)],$$
$$D_1 n \cos n(l-c) = D_2 n [\cos n(l-c) + \tan nl \sin n(l-c)] + \frac{F}{P}. \quad (13.34)$$

Solution of these equations is

$$D_1 = \frac{F \sin nc}{Pn \sin nl}, \quad D_2 = -\frac{F \sin n(l-c)}{Pn \tan nl}.$$

Displacements at any point of the left part of the beam

$$y_1 = \frac{F \sin nc}{Pn \sin nl} \sin nx - \frac{Fc}{Pl} x. \quad (13.35)$$

Knowing (13.35) we may construct the equations for slope and internal forces. In particularly, the bending moment for left part of the beam

$$M(x) = -EI \frac{d^2 y}{dx^2} = EI \frac{Fn \sin nc}{P \sin nl} \sin nx. \quad (13.35a)$$

Limiting case

If a force F is placed at the middle span ($c = 0.5l$), then for section $x = 0.5l$ from (13.35) and (13.35a), we get the maximum deflection and bending moment

$$y\left(\frac{l}{2}\right) = \frac{Fl^3}{48EI} \frac{3(\tan\vartheta - \vartheta)}{\vartheta^3}, \quad \vartheta = \frac{l}{2}\sqrt{\frac{P}{EI}} = \frac{nl}{2},$$
$$M\left(\frac{l}{2}\right) = \frac{Fl}{4} \frac{\tan\vartheta}{\vartheta}.$$

If a compressive force $P \to 0$, then

$$\frac{3(\tan\vartheta - \vartheta)}{\vartheta^3} \to 1, \quad \frac{\tan\vartheta}{\vartheta} \to 1, \text{ so } y\left(\frac{l}{2}\right) \to \frac{Fl^3}{48EI} \text{ and } M\left(\frac{l}{2}\right) \to \frac{Fl}{4}.$$

Let us evaluate numerically effect of the compressive force P. Since a critical force for simply supported column is

$$P_{\text{cr}} = \frac{\pi^2 EI}{l^2},$$

then the dimensionless parameter ϑ may be rewritten as

$$\vartheta = \frac{\pi}{2}\sqrt{\frac{P}{P_{\text{crit}}}}.$$

For $P = 0.36P_{\text{crit}}$ we get

$$\vartheta = 0.94248 \text{ and } \frac{3(\tan\vartheta - \vartheta)}{\vartheta^3} = 1.555,$$

i.e., a small compressive load increases the deflection at force F by 55%. In this case

$$\frac{\tan\vartheta}{\vartheta} = 1.4604,$$

so the maximum bending moment increases by 46%. Thus, compressive force has unfavorable effect on the state of the beam-column and therefore, P-delta analysis should not be ignored.

Notes:

1. The displacement and *lateral force* F according to (13.35) are related by linear law. Since the axial force P appear in parameter $n = \sqrt{\frac{P}{EI}}$, then the displacement and *axial compressed* force P in the same equation are related according to nonlinear law. It means that superposition principle is applicable only for lateral loads.
2. Equation (13.35a) may be treated as the expression for influence line for bending moment of a simple-supported compressed beam. For this purpose, we need to consider the section x as a fixed one, while a location of the unit force F is defined by a variable parameter c.

13.6.2 Initial Parameters Method

This method is effective for P-delta analysis in case of general case of beam-column loading. A straight element is subjected to axial compressed force P as well as lateral loads F_i and uniformly distributed load q (Fig. 13.23a); dotted line shows the initial nondeformable position (*INDP*) of the element; the initial parameters are y_0, θ_0, M_0, and $\overline{Q}_0$. The shear $\overline{Q}_0$ is directed to perpendicular to nondeformed axis of the beam.

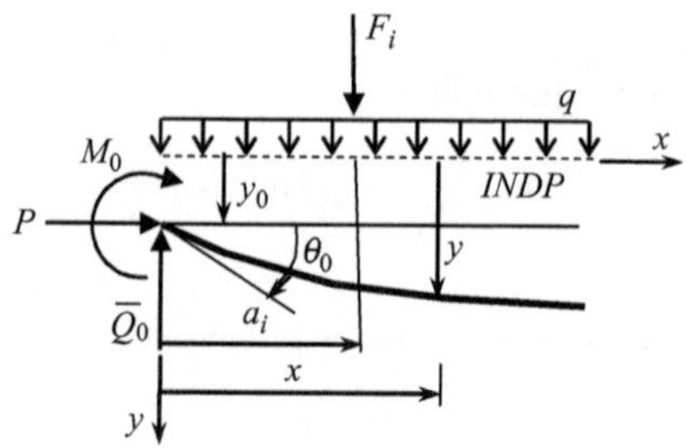

Fig. 13.23 Loading of the beam-column and positive initial parameters

Differential equation of elastic curve is

$$EI\frac{\mathrm{d}^2y}{\mathrm{d}x^2} = -M(x),$$

where the bending moment at any section x is

$$M(x) = P(y - y_0) + M_0 + \overline{Q}_0 x - \sum F_i(x - a_i) - \frac{qx^2}{2}.$$

Differential equation becomes

$$\frac{\mathrm{d}^2y}{\mathrm{d}x^2} + n^2y = -\frac{1}{EI}\left[M_0 + \overline{Q}_0 x - Py_0 - \sum F_i(x - a_i) - \frac{qx^2}{2}\right], \; n = \sqrt{\frac{P}{EI}}. \tag{13.36}$$

Solution of this equation is

$$\begin{aligned} y = {} & C_1\cos nx + C_2\sin nx + \\ & + y_0 - \frac{1}{n^2EI}\left(M_0 + \overline{Q}_0 x\right) + \frac{1}{n^3EI}\sum F_i[n(x - a_i) - \sin n(x - a_i)] \\ & - \frac{q}{n^4EI}\left[1 - \frac{n^2x^2}{2} - \cos nx\right]. \end{aligned} \tag{13.36a}$$

The first and second terms of this expression are solution of homogeneous equation (13.36), while other terms are partial solution of nonhomogeneous equation. For calculation of unknowns C_1 and C_2, we can use the following boundary conditions: at $x = 0$ initial displacement and slope are $y = y_0$ and $y' = \theta_0$. These conditions lead to

$$C_1 = \frac{M_0}{n^2EI} \text{ and } C_2 = \frac{1}{n}\left(\theta_0 + \frac{\overline{Q}_0}{n^2EI}\right).$$

Substitution of these constants into expression for y and differentiation with respect to x leads to the following formulas for displacement, angle of rotation, bending moment and shear:

$$\begin{aligned} y(x) &= y_0 + \theta_0\cdot\frac{\sin nx}{n} - \frac{M_0}{EI}\cdot\frac{1 - \cos nx}{n^2} - \frac{\overline{Q}_0}{EI}\cdot\frac{nx - \sin nx}{n^3} + y^*; \\ \theta(x) &= \frac{\mathrm{d}y}{\mathrm{d}x} = \theta_0\cdot\cos nx - \frac{M_0}{EI}\cdot\frac{\sin nx}{n} - \frac{\overline{Q}_0}{EI}\cdot\frac{1 - \cos nx}{n^2} + \theta^*. \\ M(x) &= -EI\frac{\mathrm{d}^2y}{\mathrm{d}x^2} = \theta_0 EI\cdot n\sin nx + M_0\cos nx + \overline{Q}_0\cdot\frac{\sin nx}{n} + M^*; \\ Q(x) &= \frac{\mathrm{d}M}{\mathrm{d}x} = \theta_0 EI\cdot n^2\cos nx - M_0 n\sin nx + \overline{Q}_0\cdot\cos nx + Q^*. \end{aligned} \tag{13.37}$$

Each formula of system (13.37) contains of two parts. The first part takes into account the initial parameters of the structure. The second part depends on the lateral load; these terms are denoted by symbol (*) and they are presented in Table 13.7.

Table 13.7 Additional terms of (13.36); $n=\sqrt{\frac{P}{EI}}$

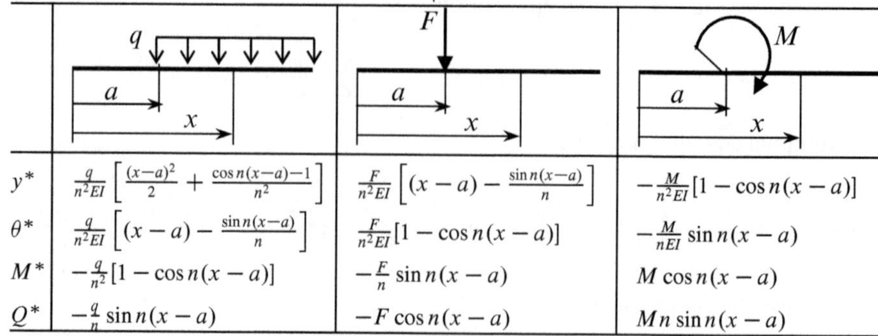

y^*	$\frac{q}{n^2EI}\left[\frac{(x-a)^2}{2}+\frac{\cos n(x-a)-1}{n^2}\right]$	$\frac{F}{n^2EI}\left[(x-a)-\frac{\sin n(x-a)}{n}\right]$	$-\frac{M}{n^2EI}[1-\cos n(x-a)]$
θ^*	$\frac{q}{n^2EI}\left[(x-a)-\frac{\sin n(x-a)}{n}\right]$	$\frac{F}{n^2EI}[1-\cos n(x-a)]$	$-\frac{M}{nEI}\sin n(x-a)$
M^*	$-\frac{q}{n^2}[1-\cos n(x-a)]$	$-\frac{F}{n}\sin n(x-a)$	$M\cos n(x-a)$
Q^*	$-\frac{q}{n}\sin n(x-a)$	$-F\cos n(x-a)$	$Mn\sin n(x-a)$

Superposition principle for *different lateral* loads is applicable for the *same compressed* load. For example, in case of several concentrated forces F_i, which are applied at $x=a_i$, and uniformly distributed load q within the whole span, the additional terms are

$$y^*=\frac{1}{n^3EI}\sum F_i\left[n\left(x-a_i\right)-\sin n\left(x-a_i\right)\right]+\frac{q}{n^4EI}\left(\frac{n^2x^2}{2}+\cos nx-1\right).$$

Note, that for given section x, only loads located to the left of section x, should be considered.

Let us illustrate the initial parameters method for free-clamped beam subjected to axial compressed force P and lateral force F (Fig. 13.24). It is necessary to compute the vertical and angular displacements at the free end and bending moment at the fixed support and to provide the numerical P-delta analysis.

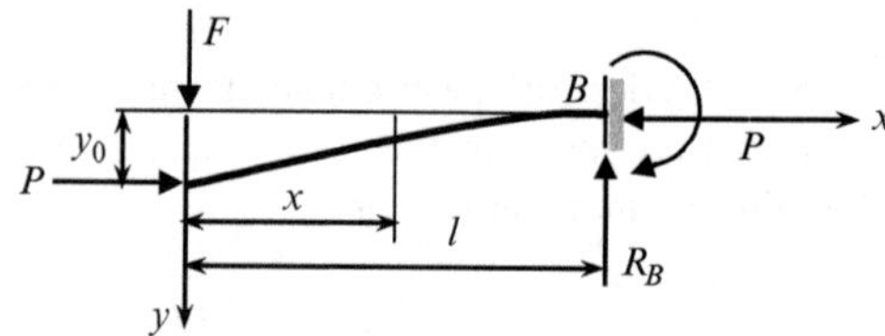

Fig. 13.24 Free-clamped compressed-bending beam

According to Table 13.7 the terms, which take into account a lateral load are

$$\begin{aligned}
y^* &= \frac{F}{n^2EI}\left[(x-a)-\frac{\sin n\left(x-a\right)}{n}\right]=\frac{F}{n^2EI}\left[x-\frac{\sin nx}{n}\right]\\
&= \frac{F}{n^3EI}\left(nx-\sin nx\right),\\
\theta^* &= \frac{F}{n^2EI}\left[1-\cos n\left(x-a\right)\right]=\frac{F}{n^2EI}\left(1-\cos nx\right),\ a=0,\\
M^* &= -\frac{F}{n}\sin n\left(x-a\right).
\end{aligned}\tag{13.38}$$

Since $Q_0 = 0, \quad M_0 = 0$, then (13.37) become

$$y(x) = y_0 + \theta_0 \cdot \frac{\sin nx}{n} + \frac{F}{n^3 EI}(nx - \sin nx);$$
$$\theta(x) = \theta_0 \cdot \cos nx + \frac{F}{n^2 EI}(1 - \cos nx). \tag{13.39}$$

These equations contain two unknown parameters. They are θ_0 and y_0. Boundary conditions are:

1. At $x = l$ (support B) the slope of elastic curve $\theta = 0$, so

$$\theta(l) = \theta_0 \cdot \cos nl + \frac{F}{n^2 EI}(1 - \cos nl) = 0,$$

which leads immediately to the slope at the free end

$$\theta_0 = -\frac{F}{n^2 EI}\frac{1 - \cos nl}{\cos nl} = -\frac{Fl^2}{2EI}\frac{2(1 - \cos \upsilon)}{\upsilon^2 \cos \upsilon}, \upsilon = nl = l\sqrt{\frac{P}{EI}}. \tag{13.40}$$

2. At $x = l$ the vertical displacement $y = 0$, so

$$y(l) = y_0 + \theta_0 \cdot \frac{\sin nl}{n} + \frac{F}{n^3 EI}(nl - \sin nl) = 0.$$

Taking into account (13.40), the vertical displacement at the free end becomes

$$\begin{aligned} y_0 &= \frac{Fl^2}{\upsilon^2 EI}\frac{1 - \cos \upsilon}{\cos \upsilon} \cdot \frac{\sin \upsilon}{n} - \frac{F}{n^3 EI}(\upsilon - \sin \upsilon) \\ &= \frac{Fl^3}{\upsilon^3 EI}\left[\frac{1 - \cos \upsilon}{\cos \upsilon} \cdot \sin \upsilon - (\upsilon - \sin \upsilon)\right] = \frac{Fl^3}{3EI}\frac{3(\tan \upsilon - \upsilon)}{\upsilon^3} = \frac{Fl^3}{3EI}\varphi_y. \end{aligned} \tag{13.41}$$

If a beam is subjected to lateral force F only, then a transversal displacement at the free end equals $Fl^3/3EI$. However, if additional axial force P acts then the factor

$$\varphi_y = \frac{3(\tan \upsilon - \upsilon)}{\upsilon^3}$$

must be included.

3. The moment at clamped support equals

$$\begin{aligned} M(l) = \theta_0 EI \cdot n \sin nl + M^* = -\frac{Fl^2}{\upsilon^2 EI}\frac{1 - \cos \upsilon}{\cos \upsilon} \cdot EI \cdot n \sin \upsilon - \\ \frac{F}{n}\sin \upsilon = -Fl\frac{\tan \upsilon}{\upsilon} = -Fl\varphi_M. \end{aligned} \tag{13.42}$$

Let's the compressive force P and critical force P_{cr} are related as follows:

$$P = kP_{\text{cr}} = k\frac{\pi^2 EI}{4l^2},$$

where k is any positive number, $k \le 1$. In this case the dimensionless parameter is

$$\upsilon = l\sqrt{\frac{P}{EI}} = \frac{\pi}{2}\sqrt{k}.$$

Coefficients φ_y and φ_M for different parameters k are presented in Table 13.8.

Table 13.8 Coefficients φ_y and φ_M for different parameters $k = P/P_{\text{cr}}$

	Parameter k					
	0.0	0.2	0.4	0.6	0.8	1.0
φ_y	**1.0**	1.2467	1.6576	2.4792	4.9434	∞
φ_M	**1.0**	1.2051	1.5453	2.2234	4.2562	∞

Bolded data corresponds to case $P = 0$. Even if $k = P/P_{\text{cr}} = 0.2$, the transversal displacement at free end and bending moment at the fixed support are 24.6% and 20.5% higher than for same beam without compressive load.

Some important cases in the expanded form are presented in Table A.23; note that this table contains formulas in terms of parameter

$$\upsilon = nl = l\sqrt{\frac{P}{EI}}.$$

This table allows performing the analysis for limiting cases $(P = 0)$. For example, for pinned-clamped beam subjected to lateral concentrated load P only we get

$$\lim_{\upsilon \to 0} M(l) = -Fl \cdot \lim_{\upsilon \to 0} \frac{\sin\frac{\upsilon}{2}\left(1 - \cos\frac{\upsilon}{2}\right)}{\sin\upsilon - \upsilon\cos\upsilon} = -\frac{3}{16}Fl.$$

This result is presented in Table A.3, pos. 3 for $u = 0.5$.

Summary. Beam-column analysis allows determining the real deflections and internal forces, which are much higher than those, which would be determined without taking into account compressive load. This is an essence of the P-delta analysis. Neglecting of the compressive forces may lead to failure of a structure; therefore, for high and multileveled structures, analysis on the basis of deformable design diagram (P-delta analysis) is necessary.

For P-delta analysis we can use two different approaches. In the first approach, we need to complete and solve the second-order differential equation of the beam for each specified loading. In the second approach (initial parameters method), we use the once derived formulas for displacements, slope, bending moments, and shear. These formulas can be used for any loads and procedure integration do not required.

13.6.3 P-Delta Analysis of the Frames

Analysis of a structure on the basis of deformable design diagram (P-delta analysis) contains two steps. The first step presents the classical analysis of structure on the basis of nondeformable scheme. If a frame is subjected to arbitrary set of external forces, including compressive forces at the joints, then on the first stage of analysis these compressive forces should be omitted. This classical analysis is performed by any appropriate methods, which were discussed previously. For each member of the frame, the axial forces must be calculated. The second step begins from calculation of axial forces with taking into account external axial compressive forces. For each member, dimensionless parameter of compressive load

$$\upsilon = l\sqrt{\frac{P}{EI}}$$

should be calculated. After that, the internal force diagrams of the first step should be reconstructed taking into account compressive load. The displacement method on this stage is more preferable. Unit reactions depend on parameter υ; the unit reactions are presented in Table A.22. The free terms of canonical equations depends on lateral loads. The following example demonstrates application of a beam-column theory for detailed analysis of two-span sidesway frame.

Example 13.7. The frame in Fig. 13.25a is subjected to horizontal force F at the level of the cross bar and compressed forces P. The cross bar of the frame is connected with vertical members by means of the hinges. The stiffnesses of the columns and cross bar are EI and $2EI$. Construct the internal force diagrams. Cross section for vertical members is W150 × 37($I = 22.2 \times 10^6\,\text{mm}^4$), the modulus of elasticity $E = 2 \times 10^5(\text{N/mm}^2)$.

Solution. The primary system of the displacement method is shown in Fig. 13.25b. The introduced constraints are 1 and 2. The primary unknowns are angular displacement Z_1 of constraint 1 and linear displacement Z_2 of constraint 2. The bending stiffness per unit length for members are $i_{\text{vert}} = EI/5 = 0.2EI$ and $i_{\text{hor}} = 2EI/6 = 0.333EI$.

Stage 1. Analysis of the frame on the basis of nondeformable design diagram. In this step, we need to provide analysis *without* the axial compressive forces N. This analysis is presented in Chap. 8, Example 8.1. The internal forces are the following: the bending moments at the fixed supports are

$$M_A = M_B = M_C = 50\ \text{kNm},$$

the axial forces at the columns are $N_A = N_B = N_C = 0$ and axial forces at the left and right cross bar are

$$N_{\text{left}} = -20\ \text{kN}, \quad N_{\text{right}} = -10\ \text{kN}$$

Stage 2. Analysis of the frame on the basis of deformable design diagram. This step should be performed taking into account axial forces N. Axial forces that arise in each member are $P_1 = N_0 + P$, where N_0 is axial force in specified member as a result of the stage 1 and P is external compressive load. Since for all columns $N_0 = 0$ then for left and right columns, the axial force equals to given load P.

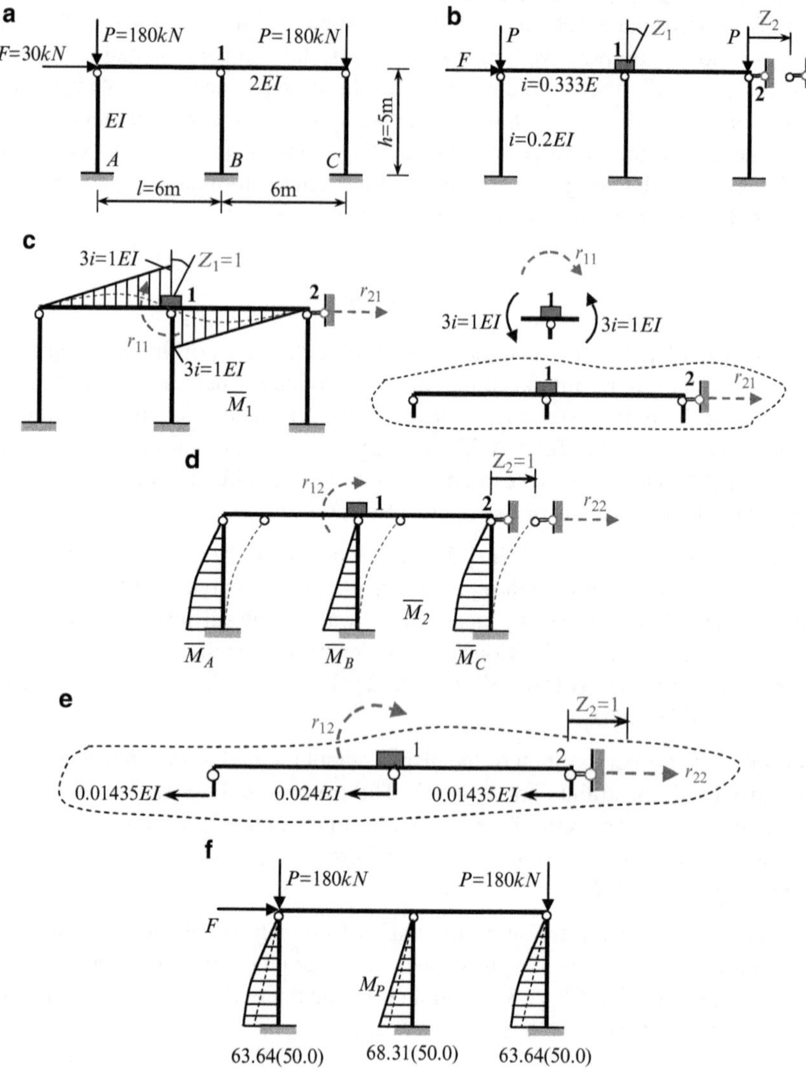

Fig. 13.25 (**a, b**) Design diagram of the frame and primary system. (**c**) Unit bending moment diagram due to $Z_1 = 1$ and calculation of unit reactions. (**d**) Unit bending moment diagram due to $Z_2 = 1$ and unit reactions. (**e**) Free body diagram and computation of the unit reactions. (**f**) Final bending moment diagram

Parameters of compressive load

$$\upsilon = l\sqrt{\frac{P}{EI}}$$

for columns are

$$\upsilon_A = \upsilon_C = 5{,}000\,(\text{mm})\cdot\sqrt{\frac{18\times10^4\,(N)}{2\times10^5\left(\dfrac{N}{\text{mm}^2}\right)\times 22.2\times10^6(\text{mm}^4)}} = 1.007;\ \upsilon_B = 0.$$

Parameters of compressive load for cross bar are

$$\upsilon_{\text{left}} = 6{,}000\,(mm)\cdot\sqrt{\frac{2\times10^3\,(N)}{2\times10^5\left(\dfrac{N}{\text{mm}^2}\right)\times 2\times 22.2\times10^6\,(\text{mm}^4)}} = 0.0028;$$

$$\upsilon_{\text{right}} = 0.0020.$$

According Table A.25, we can assume $\upsilon_A = \upsilon_C = 1.0$ and $\upsilon_{\text{left}} = \upsilon_{\text{right}} = 0.0$.

Canonical equations of the displacement method are:

$$\begin{aligned} r_{11}Z_1 + r_{12}Z_2 + R_{1P} &= 0,\\ r_{21}Z_1 + r_{22}Z_2 + R_{2P} &= 0. \end{aligned} \tag{a}$$

The bending moment diagram $\overline{M}_1$ in the primary system due to the rotation of induced constraint 1 is shown in Fig. 13.25c. Since parameters υ for cross bars are zero, then this member may be considered without effect of compressive load, so the bending moment diagram is bounded by straight lines. It is obvious that

$$r_{11} = 2EI\left(\frac{\text{kN m}}{\text{rad}}\right) \text{ and } r_{21} = 0.$$

Figure 13.25d presents the bending moment diagram $\overline{M}_2$ in the primary system due to the linear displacement of induced constraint 2. For columns A and C, we need to take into account parameter υ because these members are subjected to axial forces. Therefore, bending moment diagrams along these members are curvilinear.

The corrected functions according Table A.25 are $\varphi_1(\upsilon) = \varphi_1(1.0) = 0.9313$, $\eta_1(\upsilon) = \eta_1(1.0) = 0.5980$.

Specified ordinates of the bending moment diagram are

$$\overline{M}_A = \overline{M}_C = \frac{3i}{l}\varphi_1(\upsilon) = \frac{3\cdot 0.2EI}{5}\cdot 0.9313 = 0.1118EI,$$

$$\overline{M}_B = \frac{3i}{l} = \frac{3\cdot 0.2EI}{5} = 0.12EI.$$

Shear forces at specified sections are

$$\overline{Q}_A = \overline{Q}_C = \frac{3i}{l^2}\eta_1(\upsilon) = \frac{3\cdot 0.2EI}{5^2}\cdot 0.5980 = 0.01435EI,$$

$$\overline{Q}_B = \frac{3i}{l^2} = \frac{3\cdot 0.2EI}{5^2} = 0.024EI.$$

It is obvious that $r_{12} = 0$. The free-body diagram for the cross bar is presented in Fig. 13.25e.

The equilibrium equation $\sum X = 0$ for the cross bar becomes

$$r_{22} = 2 \cdot 0.01435EI + 0.024EI = 0.0527EI \,(\text{kN/m})$$

It is obvious that the free terms of canonical equations are

$$R_{1P} = 0, \text{ and } R_{2P} = -F = -30\,\text{kN}.$$

The canonical equations become

$$\begin{aligned} 2EI \cdot Z_1 + 0 \cdot Z_2 &= 0, \\ 0 \cdot Z_1 + 0.0527EI \cdot Z_2 - F &= 0. \end{aligned} \qquad \text{(b)}$$

The roots of these equations are

$$Z_1 = 0, \qquad Z_2 = \frac{F}{0.0527EI} = \frac{569.25}{EI}.$$

The bending moment diagram can be constructed using the principle of superposition:

$$M_P = \overline{M}_1 \cdot Z_1 + \overline{M}_2 \cdot Z_2 + M_P^0 \qquad \text{(c)}$$

Since $Z_1 = 0$ and acting load P does not cause bending of the members, then formula (c) becomes $M_P = \overline{M}_2 \cdot Z_2$. The bending moments at the clamped supports are

$$\begin{aligned} M_A = M_C &= 0.1118EI \cdot \frac{569.25}{EI} = 63.64 \quad (\text{kNm}), \\ M_B &= 0.12EI \cdot \frac{569.25}{EI} = 68.31 \quad (\text{kNm}). \end{aligned}$$

The resulting bending moment diagram is presented in Fig. 13.25f. The number in parenthesis is bending moments calculated on the basis of the nondeformable design diagram (Example 8.1, $F = 30\,\text{kN}$).

We can see that P-delta effect is significant. Increasing of the horizontal displacement Z_2 is 36.6%.

13.6.4 Graph Multiplication Method for Beam-Columns

Vereshchagin rule for computation of displacements may be modified for case of the uniform members subjected to any transversal load and compressive force P. The unit state should be created as usual. The bending moment diagrams in the actual and unit states are plotted without compressive load. However, axial load is taking into account by factors depending of axial compressive force. Two important cases are presented below.

13.6.4.1 Simply Supported Beam-Column

The bending moment diagrams in unit and actual states are shown in Fig. 13.26a. The bending moment diagram M_P due to transversal load (this load is not shown) is curvilinear.

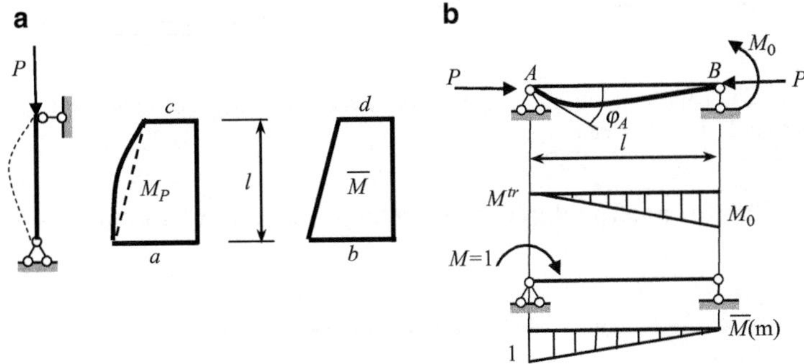

Fig. 13.26 (**a**) Graph multiplication method for simply supported beam-column; (**b**) Calculation of the slope φ_A

Graph multiplication method leads to the following result

$$\Delta = \frac{l}{6EI}\left[(2ab + 2cd)\,\alpha(\upsilon) + (ad + bc)\,\beta(\upsilon)\right],\ \upsilon = nl = l\sqrt{\frac{P}{EI}},$$

$$\alpha(\upsilon) = \frac{3}{\upsilon}\left(\frac{1}{\upsilon} - \frac{1}{\tan\upsilon}\right),\quad \beta(\upsilon) = \frac{6}{\upsilon}\left(\frac{1}{\sin\upsilon} - \frac{1}{\upsilon}\right).$$

In case of $P = 0$ this formula and trapezoid rule (6.22) coincide.

Let a simply supported beam AB is loaded by couple M_0 at support B and compressed force P (Fig. 13.26b). For computation of the slope at the support A we need to construct the bending moment diagram (M^{tr}) for transversal load *without* compressive load and bending moment diagram for unit state $\overline{M}$. Graph multiplication method leads to the following result

$$\begin{aligned}\varphi_A &= \frac{M^{\text{tr}} \times \overline{M}}{EI}\\ &= \frac{l}{6EI}\left[(2\cdot 0\cdot 1 + 2\cdot M_0\cdot 0)\,\alpha(\upsilon) + (0\cdot 0 + M_0\cdot 1)\,\beta(\upsilon)\right] = \frac{lM_0}{6EI}\beta(\upsilon).\end{aligned}$$

13.6.4.2 Fixed-Free Beam-Column

The beam of length l is loaded by any transversal load and compressed by force P (Fig. 13.27a). Notation of the bending moment ordinates at the free end and clamped support caused by transversal loads and unit load are shown in Fig. 13.27a.

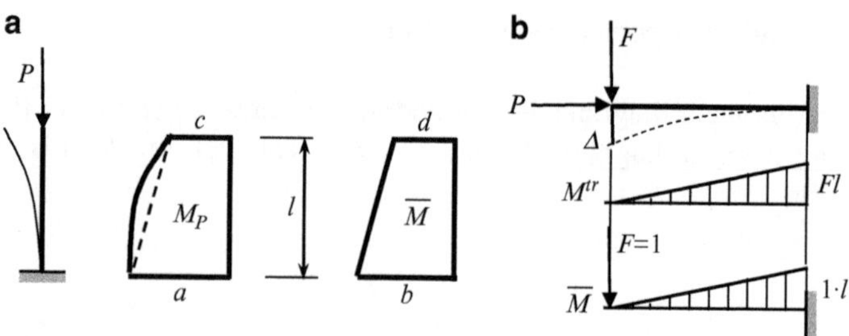

Fig. 13.27 (**a**) Graph multiplication method for clamped-free beam-column; (**b**) Calculation of the deflection Δ at the free end

According to graph multiplication method, the general formula for displacement is

$$\Delta = \frac{l}{6EI}\left[2ab\,\theta_1(\upsilon) + 2cd\,\theta_2(\upsilon) + (ad + bc)\,\theta_3(\upsilon)\right],\ \upsilon = nl = l\sqrt{\frac{P}{EI}}$$

$$\theta_1(\upsilon) = \frac{\tan\upsilon}{\upsilon}\alpha(\upsilon);\ \theta_2(\upsilon) = \alpha(\upsilon) + \frac{\upsilon\tan\upsilon}{12}\beta^2(\upsilon);\ \theta_3(\upsilon) = \frac{\tan\upsilon}{\upsilon}\beta(\upsilon)$$

For beam in Fig. 13.27b for vertical displacement at the free end we get:

$$\begin{aligned}\Delta &= \frac{l}{6EI}\left[2Fl\cdot 1\cdot l\cdot\theta_1(\upsilon) + 2\cdot 0\cdot 0\cdot\theta_2(\upsilon) + (0\cdot 1\cdot l + 0\cdot F\cdot l)\,\theta_3(\upsilon)\right]\\ &= \frac{Fl^3}{3EI}\theta_1(\upsilon)\end{aligned}$$

This result had been obtained previously by the initial parameter method.

Problems

13.1. Absolutely rigid column is loaded by two forces $P_1 = P$ and $P_2 = \alpha P$ as shown in Fig. P13.1; α is any positive number. Determine a critical force.

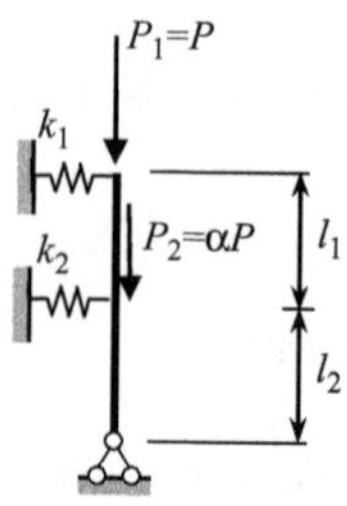

Fig. P13.1

$$\text{Ans. } P_{\text{cr}} = \frac{k_1 (l_1 + l_2)^2 + k_2 l_2^2}{l_1 + l_2 (1 + \alpha)}.$$

13.2. Two absolutely rigid bodies ($EI = \infty$) connected by hinge at point C (Fig. P13.2). Stiffness coefficient of each elastic support is k. Derive the stability equation, find critical forces and corresponding shapes.

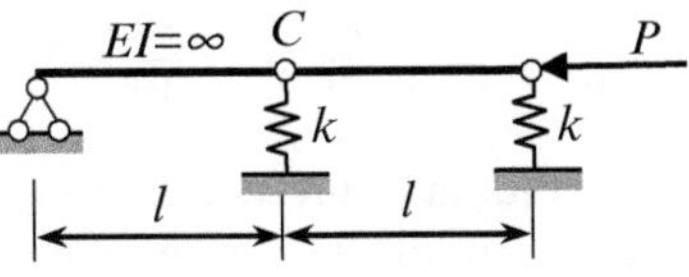

Fig. P13.2

$$\text{Ans. } P_{1\text{cr}} = \frac{3 - \sqrt{5}}{2} kl = 0.3819kl; \quad P_{2\text{cr}} = \frac{3 + \sqrt{5}}{2} kl = 2.6180kl.$$

13.3. Design diagram of two-story frame with elastic constraints between adjacent absolutely rigid members is shown in Fig. P13.3. Calculate the critical load using energy method.

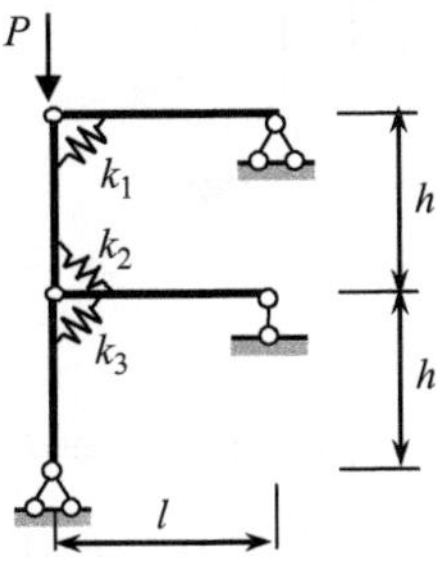

Fig. P13.3

$$\text{Ans. } P_{\text{cr}} = \frac{k_1 + k_2 + k_3}{2h}.$$

13.4. Pinned-pinned beam with elastic support A is subjected to axial compressive force P (Fig. P13.4). The restoring moment in rotational spring is $M = k\varphi 0$. Derive the stability equation. Apply the double integration and initial parameter methods.

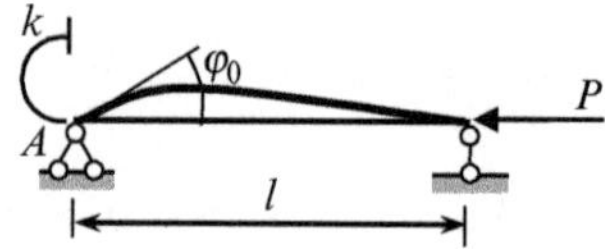

Fig. P13.4

$$\text{Ans. } \tan\lambda l = \frac{\lambda l}{\lambda^2 l^2 \alpha + 1}, \quad \lambda = \sqrt{\frac{P}{EI}}, \alpha = \frac{EI}{kl}$$

13.5. Design diagram of the column is presented in Fig. P13.5. The total length of the column is l, while the length of the bottom part is αl. The column is subjected to forces N_1 and N_2. Relationship between these forces remains constant, i.e., $N_2 = \beta N_1$. Parameters α and β are the given fixed numbers. Stiffness rigidity of the portions are EI_1 and EI_2. Derive the stability equation. Consider three possible buckling forms. Apply double integration method.

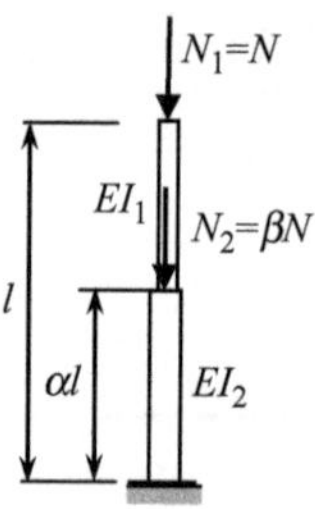

Fig. P13.5

Ans. (1) $\cos(1-\alpha)n_1 l = 0$; (2) $\cos n_2 \alpha l = 0$; (3) For smallest critical force $\tan(1-\alpha)n_1 l \cdot \tan n_2 \alpha l - \frac{n_1}{n_2}(1+\beta) = 0, \quad n_1 = \sqrt{\frac{N_1}{EI_1}}, \quad n_2 = \sqrt{\frac{(1+\beta)N}{EI_2}}$

13.6. Derive the stability equation for uniform columns subjected to axial compressive force. Consider the following supports: (a) clamped-free; (b) clamped-pinned; (c) pinned-pinned. Apply the initial parameter method. Compare the results with those are presented in Table 13.1.

13.7. The clamped beam with elastic support is subjected to axial force N (Fig. P13.7). Derive the stability equation. (Hint: Reaction at the right support is ky_1, where y_1 is vertical displacement at the right support. The moment at the left support should be calculated taking into account the lateral displacement y_1 of axial force N, i.e., $M_0 = ky_1 l - Ny_1$).

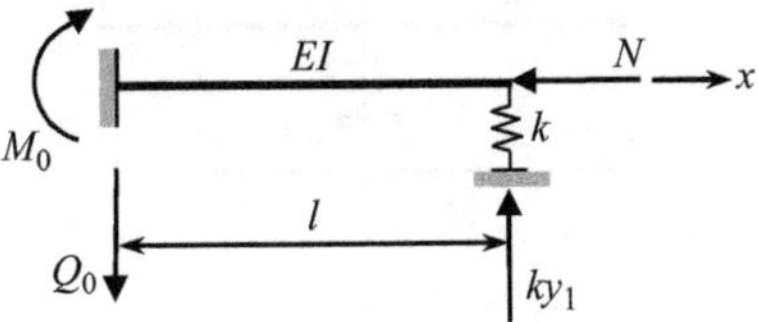

Fig. P13.7

Ans. $\tan nl = nl\left(1 - \frac{n^2 l^2 EI}{kl^3}\right), \quad n = \sqrt{\frac{N}{EI}}$

13.8. The uniform beam with overhang is subjected to axial compressive force P (Fig. P13.8). Derive the stability equation. Use the initial parameter method.

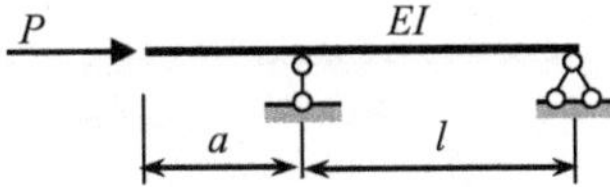

Fig. P13.8

Ans. $nl\,(\tan nl + \tan na) = \tan nl \cdot \tan na,\ n = \sqrt{\frac{P}{EI}}$

13.9. The uniform pinned-clamped beam with intermediate hinge C is subjected to axial compressive force P (Fig. P13.9). Derive the stability equation by initial parameter method. Calculate the critical load and trace elastic curve; (a) Consider the special cases ($a = 0, a = l$).

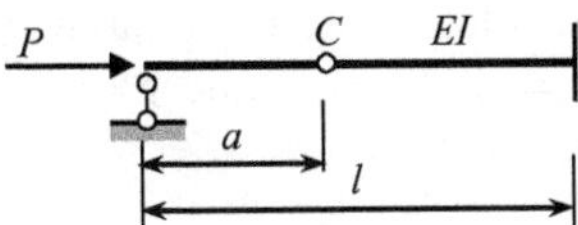

Fig. P13.9

Ans. $\sin na\,[nl \cos n\,(l - a) - \sin n\,(l - a)] = 0$

13.10. The uniform pinned-pinned beam with elastic support is subjected to axial compressed force P (Fig. P13.10); stiffness coefficient of elastic support is k. The total length of the beam is $l = l_1 + l_2$. Derive the stability equation. Use the initial parameter method. Consider two special cases; (a) $k = 0$; (b) $l_1 = l_2$ and $k = \infty$.

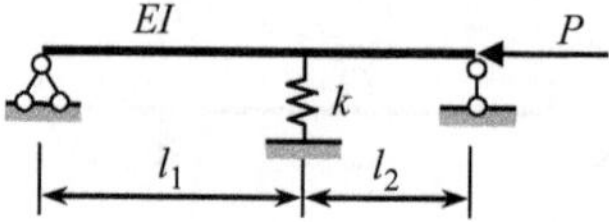

Fig. P13.10

Ans. $\sin nl_1 \sin nl_2 = nl \sin nl \cdot \left(\frac{l_1 l_2}{l^2} - \frac{P}{kl}\right), \quad n = \sqrt{\frac{P}{EI}}$

For problems 13.11 through 13.15, it is recommended to apply the Displacement method. All stability functions $\phi_1, \phi_2, \ldots$ are presented in Table A25.

13.11. Design diagram of the continuous beam is presented in Fig. P13.11. Derive the equation for critical load in term of parameter α. Consider a special case for $\alpha = 0.5$.

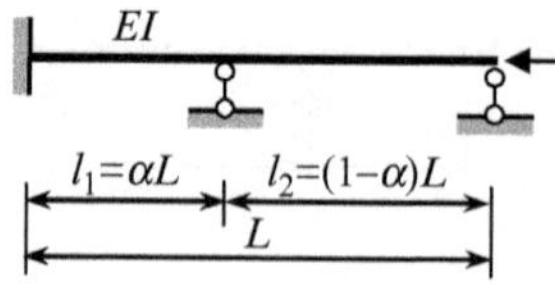

Fig. P13.11

Ans. $\frac{4}{\alpha}\varphi_2(\alpha \upsilon_0) + \frac{3}{1-\alpha}\varphi_1((1-\alpha)\upsilon_0) = 0, \ \upsilon_1 = l_1\sqrt{\frac{P}{EI}} = \alpha L\sqrt{\frac{P}{EI}} = \alpha \upsilon_0,$ $\upsilon_2 = (1-\alpha)L\sqrt{\frac{P}{EI}} = \upsilon_0(1-\alpha)$

13.12. Two-span beam of spans l_1 and $l_2 = \beta l_1$ is subjected to axial forces P and αP (Fig. P13.12). The flexural rigidity for the left and right spans are *EI* and *kEI*. Derive the equation for critical load in term of parameters α, β and k. Consider the special cases: a) $\alpha = 3, \beta = 1, k = 4$; and b) $\alpha = 0, k = 1, \beta = 1$. Explain obtained results.

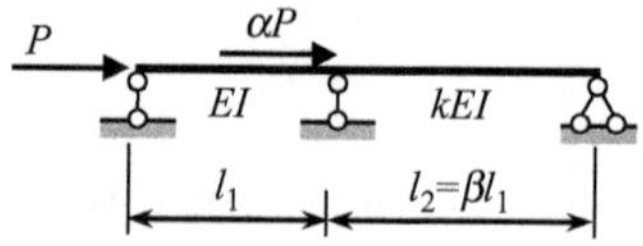

Fig. P13.12

Ans. $\varphi_1(\upsilon_1) + \frac{k}{\beta}\varphi_1(\upsilon_2) = 0, \ \upsilon_1 = l_1\sqrt{\frac{P}{EI}}, \quad \upsilon_2 = l_2\sqrt{\frac{P+\alpha P}{kEI}} = \upsilon_1\beta\sqrt{\frac{1+\alpha}{k}}$

13.13. Design diagram of the frames with deformable cross bar are presented in Fig. P13.13. Derive the equation for critical load.

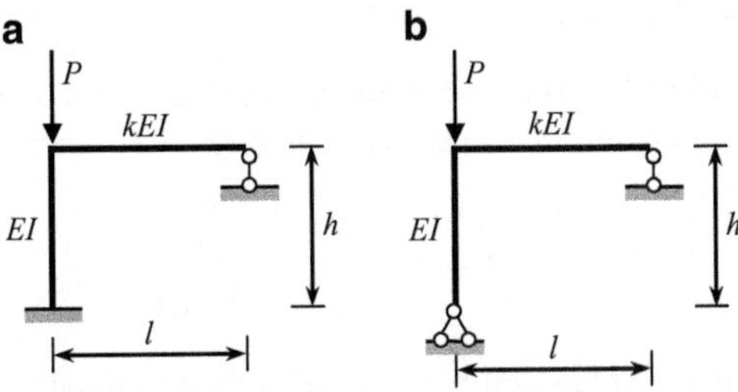

Fig. P13.13

Ans. (a) $3k\frac{h}{l}+\frac{\upsilon}{\tan\upsilon}=0,\quad \upsilon=h\sqrt{\frac{P}{EI}}$; (b) $\upsilon\tan\upsilon=3k\frac{h}{l}$

13.14. The frame with absolutely rigid cross bar is presented in Fig. P13.14. Derive the stability equation and calculate the critical load.

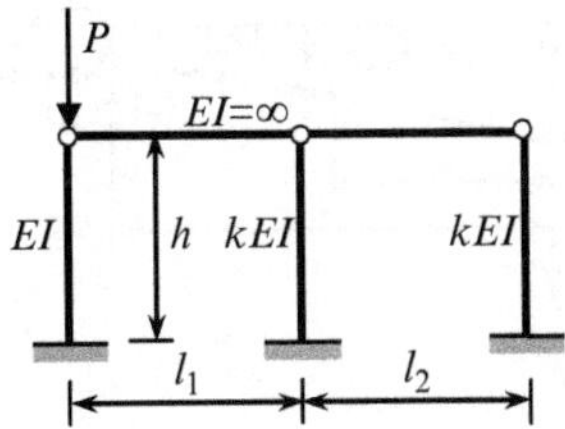

Fig. P13.14

Ans. $\eta_1(\upsilon)+2k=0,\quad \upsilon=h\sqrt{\frac{P}{EI}}$. If $k=1$, then $\upsilon=2.67.P_{\text{cr}}=\frac{7.13EI}{h^2}$.

13.15. Design diagram of the frame is presented in Fig. P13.15. Relationship between two forces is constant. Derive the stability equation and find critical load.

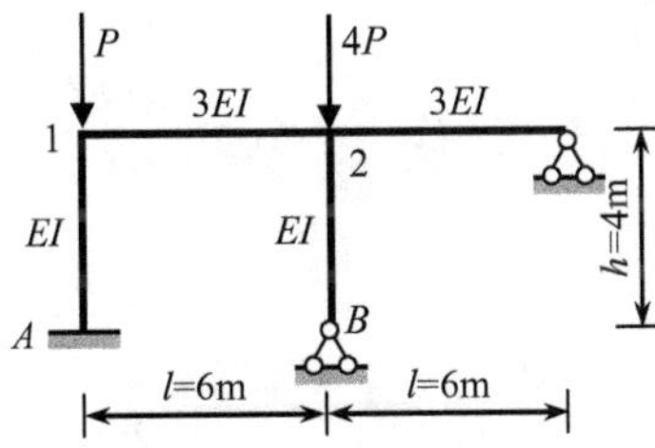

Fig. P13.15

Ans. $\begin{vmatrix} \varphi_2(\upsilon)+2 & 1.0 \\ 1.0 & 0.75\varphi_1(2\upsilon)+3.5 \end{vmatrix} = 0$, $\upsilon_{1A} = \upsilon = h\sqrt{\dfrac{P}{EI}}$; $\upsilon_{2B} = h\sqrt{\dfrac{4P}{EI}} = 2\upsilon$, $\upsilon_{\min} = 2.097$.

In problems 13.16–13.19 it is necessary to analyze the structure on the basis of deformable design diagram. In all cases parameter

$$\upsilon = nl = l\sqrt{\frac{P}{EI}} = l\sqrt{\frac{kP_{\text{cr}}}{EI}}.$$

13.16. The uniform free-clamped beam is subjected to axial compressive force P and lateral distributed load q (Fig. P13.16). Calculate the vertical and angular displacement at free end, and the bending moment at fixed support. Estimate effect of axial load for different parameter K,

$$P = kP_{\text{cr}}, \quad P_{\text{cr}} = \frac{\pi^2 EI}{4l^2}.$$

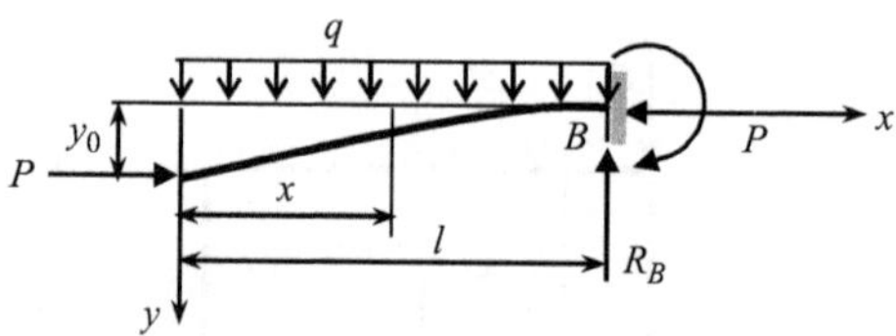

Fig. P13.16

Ans. $\theta_0 = -\dfrac{ql^3}{6EI}\cdot\dfrac{6(\upsilon - \sin\upsilon)}{\upsilon^3\cos\upsilon}$; $y_0 = -\dfrac{ql^4}{8EI}\cdot\dfrac{8}{\upsilon^4}\left(\dfrac{\upsilon\sin\upsilon + \cos\upsilon - 1}{\cos\upsilon} - \dfrac{\upsilon^2}{2}\right)$; $M_B = -\dfrac{ql^2}{2}\cdot\dfrac{2(\cos\upsilon + \upsilon\sin\upsilon - 1)}{\upsilon^2\cos\upsilon}$

13.17. The uniform beam AB is subjected to axial compressive force P and lateral uniform distributed load q (Fig. P13.17). Calculate slope at the support, the deflection and bending moment at the middle of the beam. Estimate effect of axial load for different parameters k,

$$P = kP_{\text{cr}}, \quad P_{\text{cr}} = \frac{\pi^2 EI}{l^2}.$$

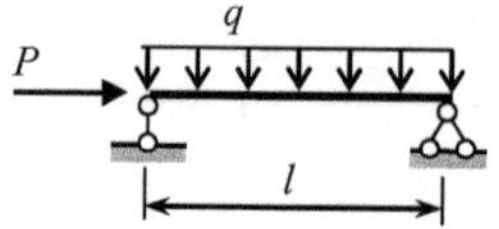

Fig. P13.17

$$\text{Ans. } \theta_0 = \frac{ql^3}{24EI}\cdot\frac{12}{\upsilon^3}\left(2\tan\frac{\upsilon}{2}-\upsilon\right);\; y\left(\frac{l}{2}\right) = \frac{5ql^4}{384EI}\cdot\frac{384}{5\upsilon^4}\left(\sec\frac{\upsilon}{2}-1-\frac{\upsilon^2}{8}\right);$$

$$M\left(\frac{l}{2}\right) = \frac{ql^2}{8}\cdot\frac{8}{\upsilon^2}\left(\sec\frac{\upsilon}{2}-1\right)$$

13.18. The uniform beam AB is subjected to axial compressive force P and external moment M_0 at the support A (Fig. P13.18). Calculate the slope at the support A and reactions of supports A and B. Estimate effect of axial load for different parameters k,

$$P = kP_{\rm cr}, \quad P_{\rm cr} = \frac{\pi^2 EI}{(0.7l)^2}.$$

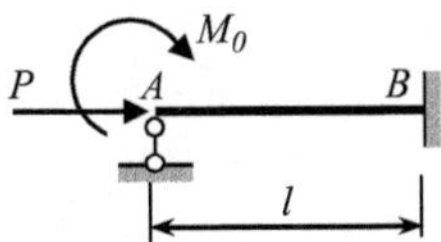

Fig. P13.18

$$\text{Ans. } Q_0 = \frac{M_0}{l}\cdot\frac{\upsilon(\cos\upsilon-1)}{\sin\upsilon-\upsilon\cos\upsilon},\; M_B = M_0\frac{\sin\upsilon-\upsilon}{\sin\upsilon-\upsilon\cos\upsilon}.$$

13.19. The uniform beam AB is subjected to axial compressive force P and transversal force F at the middle point of the beam. Calculate slope at he support A, and linear displacement and bending moment at the point of force F.

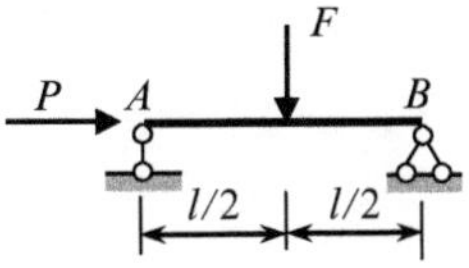

Fig. P13.19

$$\text{Ans. } \theta_0 = \frac{Fl^2}{16EI}\cdot\frac{8}{\upsilon^2}\left(\sec\frac{\upsilon}{2}-1\right),\; y\left(\frac{l}{2}\right) = \frac{Fl^3}{48EI}\cdot\frac{24}{\upsilon^3}\left(\tan\frac{\upsilon}{2}-\frac{\upsilon}{2}\right);$$

$$M\left(\frac{l}{2}\right) = \frac{Fl}{4}\cdot\frac{2}{\upsilon}\tan\frac{\upsilon}{2},\; \upsilon = l\sqrt{\frac{P}{EI}}$$

Chapter 14
Dynamics of Elastic Systems

Structural dynamics is a special branch of structural analysis, which studies the behavior of structures subjected to dynamical loads. Such loads develop the dynamical reactions, dynamical internal forces, and dynamical displacements of a structure. They all change with time, and maximum values often exceed static ones. Dynamical analysis of structure is based on the free vibration analysis.

This chapter is devoted to linear free vibration analysis of elastic structures with lumped and distributed parameters. The fundamental methods of structural analysis (force and displacement methods) are applied for calculation of frequencies of the free vibration and corresponding mode shape of vibration. They are inherent to the structure itself and are called as the eigenvalues and eigenfunctions.

14.1 Fundamental Concepts

14.1.1 Kinematics of Vibrating Processes

The simplest periodic motion can be written as

$$y(t) = A\sin(\omega t + \phi_0),$$

where A is the amplitude of vibration, φ_0 is the initial phase of vibration, t is time. This case is presented in Fig. 14.1a. The initial displacement $y_0 = A\sin\varphi_0$ is measured from the static equilibrium position. The number of cycles of oscillation during 2π seconds is referred to as circular (angular or natural) frequency of vibration $\omega = 2\pi/T$ (radians per second or s^{-1}), T(s) is the period of vibration. Figure 14.1b, c presents the damped and increased vibration with constant period.

14.1.2 Forces Which Arise at Vibrations

During vibration a structure is subjected to different forces. These forces have a different nature and exert a different influence on the vibrating process. All forces

I.A. Karnovsky and O. Lebed, *Advanced Methods of Structural Analysis*,
DOI 10.1007/978-1-4419-1047-9_14, © Springer Science+Business Media, LLC 2010

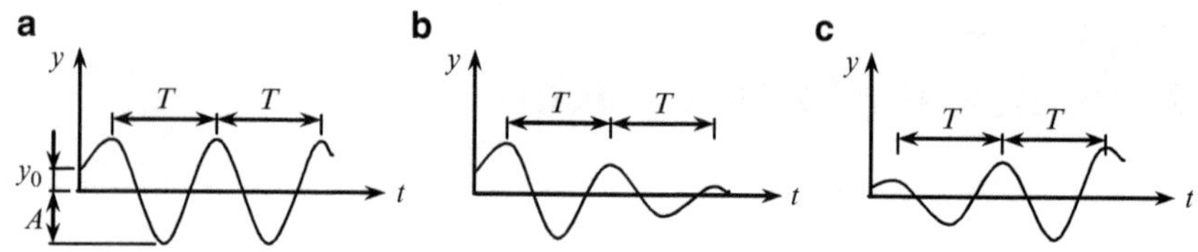

Fig. 14.1 Types of oscillatory motions

may be divided into the following groups: disturbing forces, positional (restoring) forces, resisting forces, and forces of the mixed character.

1. *Disturbing forces* may be of the following types:
   (a) Immovable periodical loads are produced by stationary units and mechanisms with moving parts. These loads have a periodical, but not necessary a harmonic character and generally do not depend on the elastic properties of the structure.
   (b) Impact (impulsive) loads are produced by falling weights or collision of bodies. Impulsive loads are characterized by very short duration of their action and depend on the elastic properties of the structure, which is subjected to such loads.
   (c) Moving loads act on the structures through wheels of a moving train or truck. The availability of the rail joins on the railway bridge or irregularities of the deck on the car bridge lead to appearance of inertial forces. These type of loads should be distinguished from moving one, which has been studied in the sections "Influence lines" because unit moving load $P = 1$ had been considered without dynamical effects.
   (d) Seismic loads arise due to earthquakes. The reason of the seismic load on the structure is the acceleration of the supports caused by acceleration of the ground. This type of disturbance is called kinematical. The acceleration of supports leads to the acceleration of the separate parts of the structure, and as a result inertial forces act on these parts. Seismic forces, which arise in the members of the structure, depend on the type and the amount of ground acceleration, distribution of the mass within the members of the structure and their elastic properties.

2. *Restoring forces* depend on the displacement of the structure, arise due to deviation of system from a static equilibrium position, and tend to return the system to its initial position. Restoring properties of a system are described by its elastic characteristic $P = P(y)$, where P is a static force, which is applied to the structure. Characteristic $P - y$ may be linear or nonlinear. Some types of characteristics $P - y$ are presented in Table 14.1; in all cases y is the displacement at the point of P.
3. *Resisting forces.* The forces of inelastic resistance (friction or damping forces) depends on the velocity υ of motion, $R = R(\upsilon)$, and always act in the opposite direction of velocity. These forces are a result of internal friction in the material of a structure and/or in the connections of a system.

Table 14.1 Types of elastic members and their characteristics

Design diagram	Characteristic P-y	Design diagram	Characteristic P-y
 P, y	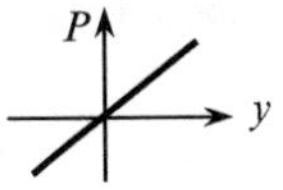 P, y	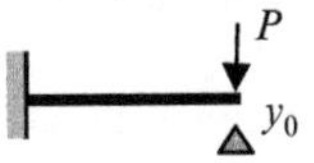 P, y_0	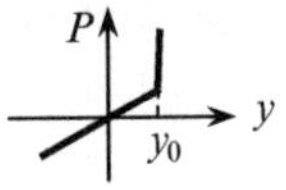 P, y, y_0
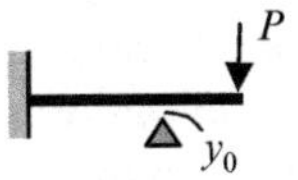 P, y_0	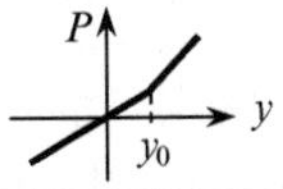 P, y, y_0	 P	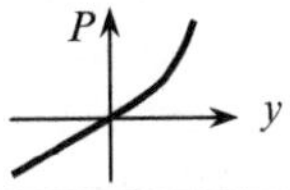 P, y

Different types of forces acting on a structure lead to different types of vibration. Among them are two general classes – they are free and forced vibration.

Vibrations of a system in which disturbing forces are absent are called free vibrations. At free vibration, the system is subjected to forces inherent to the system itself, i.e., the restoring and resisting forces.

To impose free vibrations, nonzero initial conditions should be created, which means some initial displacement and initial velocity. Free vibration may be linear or nonlinear depending on the characteristics of restoring and resisting forces. Absence of resisting forces leads to the *free undamped* vibrations; in this case, the system is subjected only to a restoring force.

Vibration of a system caused by any disturbing forces is called a forced vibration. Absence of resisting forces leads to the *forced undamped* vibration. Just as the free vibration, the forced vibration may be linear and nonlinear.

14.1.3 Degrees of Freedom

The fundamental difference of the concept of "degrees of freedom" in static and structural dynamics in spite of the same definition (a number of independent parameters, which uniquely defines the positions of all points of a structure) is as follows: In statics, the number of degrees of freedom is related to a structure consisting of *absolutely rigid* discs. If the degree of freedom is greater than or equal to one, then the system is geometrically changeable and cannot be assumed as an engineering structure; when the degree of freedom equals to zero, it means that a system is geometrically unchangeable and statically determined. In structural dynamics, the number of degrees of freedom is determined by just taking into account the *deformation* of the members. If the degree of freedom equals to zero, then a system presents an absolutely rigid body and all displacements in space are absent.

All structures may be divided into two principal classes according to their degrees of freedom. They are the structure with concentrated and distributed parameters.

Members with concentrated parameters assume that the distributed mass of the member itself may be neglected in comparison with the lumped mass, which is located on the member. The structure with distributed parameters is characterized by uniform or nonuniform distribution of mass within its parts.

From mathematical point of view, the difference between the two types of systems is the following: the systems of the first class are described by ordinary differential equations, while the systems of the second class are described by partial differential equations.

Figure 14.2a, b shows a massless statically determinate and statically indeterminate beam with one lumped mass. These structures have one degree of freedom, since transversal displacement of the lumped mass defines position of all points of the beam. Note that these structures, from point of view of their *static analysis*, have the number degrees of freedom $W = 0$ for statically determinate beam (a) and $W = 3D - 2H - S_0 = 3 \cdot 1 - 2 \cdot 0 - 5 = -2$ for statically indeterminate beam (b) with two redundant constraints. It is obvious that a massless beam in Fig. 14.2c has three degrees of freedom. It can be seen that introducing of additional constraints on the structure increases the stiffness of the structure, i.e., increase the degrees of static indeterminacy, while introducing additional masses increase the degrees of freedom.

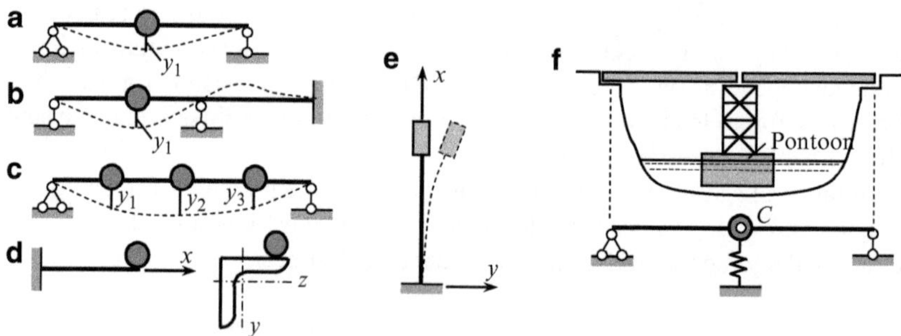

Fig. 14.2 (a–f) Design diagrams of structures

Figure 14.2d presents a cantilevered massless beam carrying one lumped mass. However, this case is not a plane bending, but bending combined with torsion, because mass is not applied at the shear center. That is why this structure has two degrees of freedom, such as the vertical displacement and angle of rotation in y-z plane with respect to the x-axis. A structure in Fig. 14.2e presents a massless beam with an absolutely rigid *body*. The structure has two degrees of freedom, such as the lateral displacement y of the body and angle of rotation of the body in y-x plane. Figure 14.2f presents a bridge, which contains two absolutely rigid bodies. These bodies are supported by a pontoon. Corresponding design diagram shows two absolutely rigid bodies connected by hinge C with elastic support. So this structure has one degree of freedom.

The plane and spatial bars structures and plane truss are presented in Fig. 14.3. In all cases, we assume that all members of a structure do not have distributed masses. The lumped mass (Fig. 14.3a) can move in vertical and horizontal directions; therefore, this structure has two degrees of freedom. Similarly, the statically indeterminate structure shown in Fig. 14.3b has two degrees of freedom. However, if we assume that horizontal member is absolutely rigid in axial direction (axial stiffness $EA = \infty$), then the mass can move only in vertical direction and the structure has one degree of freedom. Structure 14.3c has three degrees of freedom; they are displacement of the lumped mass in x-y-z directions. Introducing additional deformable but massless members does not change the number of degrees of freedom.

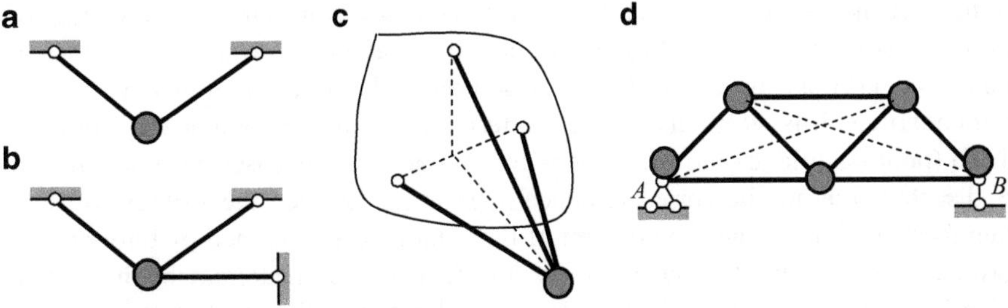

Fig. 14.3 (**a–d**) Design diagrams of bar structures and truss

The truss (Fig. 14.3d) contains five concentrated masses. The mass at the joint A is fixed, the mass at the joint B can move only in horizontal direction, and the rest masses can move in horizontal and vertical directions. So, this structure has seven degrees of freedom. If we assume that horizontal displacements of the joints may be negligible in comparison with vertical displacements, then this truss may be considered as a statically determinate structure with three degrees of freedom. If additional members will be introduced in the truss (shown by dotted lines), then this truss should be considered as two times statically indeterminate structure with three degrees of freedom.

Figure 14.4 presents plane frames and arches. In all cases, we assume that all members of a structure do not have distributed masses. The lumped mass M in Fig. 14.4a, b can move in vertical and horizontal directions, so these structures have two degrees of freedom. Figure 14.4c shows the two-story frame containing absolutely rigid cross bars (the total mass of each cross bar is M). This frame may be presented as shown in Fig. 14.4d.

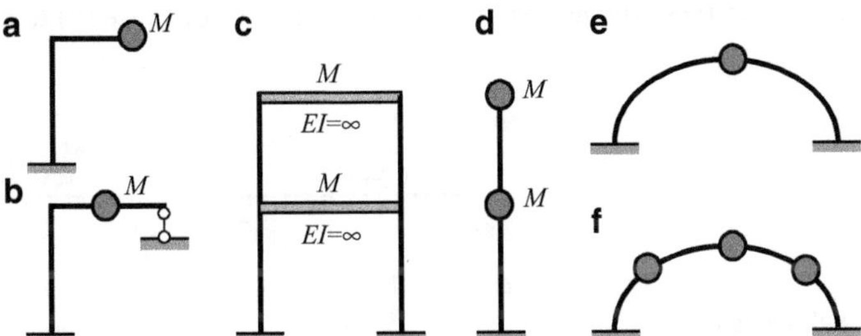

Fig. 14.4 (**a–f**) Design diagrams of frames and arches

Arches with one and three lumped masses are shown in Fig. 14.4e, f. Taking into account their vertical and horizontal displacements, the number of their degrees of freedom will be 2 and 6, respectively. For *gently sloping* arches, the horizontal displacements of the masses may be neglected; in this case, the arches should be considered as structures having one and three degrees of freedom in the *vertical* direction.

All cases shown in Figs. 14.2–14.4 present the design diagrams for systems with lumped (concentrated) parameters. Since masses are concentrated, the configuration of a structure is defined by displacement of each mass as a function of time, i.e., $y = y(t)$ and behavior of such structures is described by *ordinary* differential equations. It is worth discussing the term "concentrated parameters" for cases 14.2f (pontoon bridge) and 14.4c (two-story frame). In both cases, the mass, in fact, is distributed along the correspondence members. However, the stiffness of these members is infinite, and the position of these members is defined by only *one* coordinate. For the structure given in Fig. 14.2f, such coordinate may be the vertical displacement of the pontoon or the angle of inclination of the span structure, and for the two-story frame (Fig. 14.4c), it may be the horizontal displacements of the each cross bar.

The structures with distributed parameters are more difficult for analysis. The simplest structure is a beam with a distributed mass m. In this case, the configuration of a system is determined by displacement of each elementary mass as a function of time. However, since masses are distributed, then the displacement of any point is a function of a time t and location x of the point, i.e., $y = y(x,t)$, so behavior of the structures is described by *partial* differential equations.

Figure 14.5 presents the four possible design diagrams for dynamical analysis of the beam subjected to a moving concentrated load. Parameters that are taken into account (mass of moving load M and distributed mass of a beam m) are shown by bold and thick solid lines. The scheme (a) does not take into account the mass of the beam and mass of the load; therefore, the inertial forces are absent. This case corresponds to static loading, and parameter υ only means that force P may be located at any point. Just this case of loading is assumed for construction of influence lines. Case (b) takes into account only the mass of the moving load. Case (c) corresponds to the motion of the massless load along the beam with distributed mass m. Case (d) takes into account the mass of the load and mass of the beam. The difficulty in solving these dynamical problems increases from case (b) to case (d).

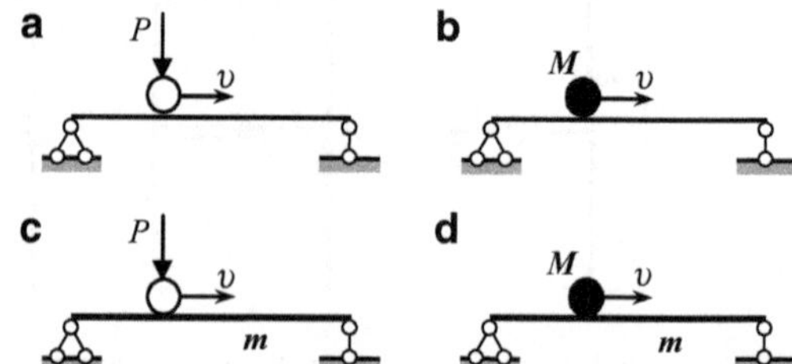

Fig. 14.5 (**a–d**) Design diagrams for beam subjected to moving load

It is possible to obtain a combination of the members with concentrated and distributed parameters. Figure 14.6 shows a frame with a massless strut BF ($m = 0$), members AB and BC with distributed masses m and absolutely rigid member CD ($EI = \infty$). The simplest form of vibration is shown by the dotted line.

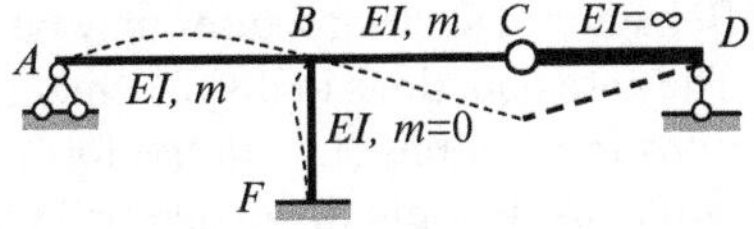

Fig. 14.6 Frame with distributed and concentrated parameters

14.1.4 *Purpose of Structural Dynamics*

The two fundamental problems of dynamical analysis are the following:

1. Determine the internal forces and deflections of a structure caused by dynamical loads
2. Determine the dynamical displacements, velocities, and accelerations. These quantities are transferred onto the equipment operators and other types of special equipment, which are located on the structure, and must not exceed allowable quantities

The solution of these problems is based on determining the very important characteristic of a structure – its frequencies and shapes of free vibration.

Free vibration of a structure occurs with some frequencies. These frequencies depend on only the parameters of the structure (boundary conditions, distribution of masses and stiffnesses within the members, etc.) and does not depend on the reason of vibration. Therefore, these frequencies are often called as eigenfrequencies, because these frequencies are inherent to the given structure. The number of frequency vibrations coincides with the number of degrees of freedom. The structure with distributed parameters has infinity number of degrees of freedom. The set of frequency vibration presents the frequencies spectrum of a structure. Each mode shape of vibration shows the form of elastic curve, which corresponds to specific frequency.

This chapter contains only free vibration analysis.

14.1.5 *Assumptions*

Free vibration analysis presented in this textbook is based on the following assumptions:

1. Only linear vibrations are considered
2. Damping effects are ignored
3. Stiffness and inertial effects of the structure are time independent

14.2 Free Vibrations of Systems with Finite Number Degrees of Freedom: Force Method

Behavior of such structures may be described by two types of differential equations. They are equations in displacement (i.e., in the form of the force method) and equations in reactions (i.e., in the form of the displacement method). In both cases, we will consider only the undamped vibration.

14.2.1 *Differential Equations of Free Vibration in Displacements*

In essence, the first approach consists of expressing the forces of inertia as function of *unit displacements*. For the derivation of differential equations, let us consider a structure with concentrated masses (Fig. 14.7).

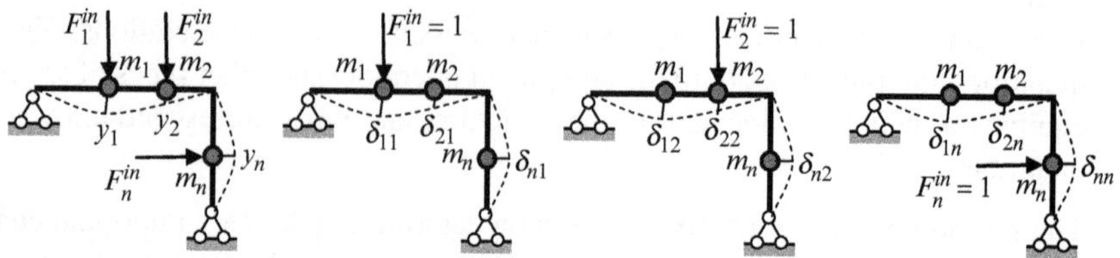

Fig. 14.7 Design diagram and unit states

In case of free vibration, each mass is subjected to forces of inertia only. Displacement of each mass may be presented as

$$
\begin{aligned}
y_1 &= \delta_{11} F_1^{\text{in}} + \delta_{12} F_2^{\text{in}} + \ldots + \delta_{1n} F_n^{\text{in}}, \\
y_2 &= \delta_{21} F_1^{\text{in}} + \delta_{22} F_2^{\text{in}} + \ldots + \delta_{2n} F_n^{\text{in}}, \\
&\cdots\cdots\cdots\cdots \\
y_n &= \delta_{n1} F_1^{\text{in}} + \delta_{n2} F_2^{\text{in}} + \ldots + \delta_{nn} F_n^{\text{in}},
\end{aligned}
\tag{14.1}
$$

where δ_{ik} is displacement in i-th direction caused by unit force acting in the k-th direction.

Since the force of inertia of mass m_i is $F_i^{\text{in}} = -m_i \ddot{y}_i$, then the equations (14.1) become

$$
\begin{aligned}
&\delta_{11} m_1 \ddot{y}_1 + \delta_{12} m_2 \ddot{y}_2 + \ldots + \delta_{1n} m_n \ddot{y}_n + y_1 = 0, \\
&\cdots\cdots\cdots\cdots \\
&\delta_{n1} m_1 \ddot{y}_1 + \delta_{n2} m_2 \ddot{y}_2 + \ldots + \delta_{nn} m_n \ddot{y}_n + y_n = 0.
\end{aligned}
\tag{14.2}
$$

Each equation of (14.2) presents the compatibility condition. The differential equations of motion are coupled dynamically because the second derivative of all coordinates appears in each equation.

We can see that the idea of force method has not been used above. Alternatively, these same equations may be obtained by force method. In this case, unknown inertial forces should be considered as primary unknowns of the force method. Therefore, hereafter (14.2) will be called the differential equations of free undamped vibration in *displacements* or canonical equations in form of the *force method*.

In matrix form this system may be presented as

$$\mathbf{F}\mathbf{M}\ddot{\mathbf{Y}} + \mathbf{Y} = \mathbf{0}, \tag{14.2a}$$

where **F** is the flexibility matrix (or matrix of unit displacements), **M** is the diagonal mass matrix and **Y** represents the vector displacements

$$\mathbf{F} = \begin{bmatrix} \delta_{11} & \delta_{12} & \dots & \delta_{1n} \\ \delta_{21} & \delta_{22} & \dots & \delta_{2n} \\ \dots & \dots & \dots & \dots \\ \delta_{n1} & \delta_{n2} & \dots & \delta_{nn} \end{bmatrix}, \quad \mathbf{M} = \begin{bmatrix} m_1 & 0 & \dots & 0 \\ 0 & m_2 & \dots & 0 \\ \dots & \dots & \dots & \dots \\ 0 & 0 & \dots & m_n \end{bmatrix}, \quad \mathbf{Y} = \begin{bmatrix} y_1 \\ y_2 \\ \dots \\ y_n \end{bmatrix}. \tag{14.2b}$$

14.2.2 Frequency Equation

Solution of system of differential equations (14.2) is

$$y_1 = A_1 \sin(\omega t + \varphi_0), \quad y_2 = A_2 \sin(\omega t + \varphi_0), \quad y_3 = A_3 \sin(\omega t + \varphi_0), \tag{14.3}$$

where A_i are the amplitudes of the corresponding masses m_i and φ_0 is the initial phase of vibration

The second derivatives of these displacements over time are

$$\begin{aligned} \ddot{y}_1 &= -A_1\omega^2 \sin(\omega t + \varphi_0), \quad \ddot{y}_2 = -A_2\omega^2 \sin(\omega t + \varphi_0), \\ \ddot{y}_n &= -A_n\omega^2 \sin(\omega t + \varphi_0). \end{aligned} \tag{14.3a}$$

By substituting (14.3) and (14.3a) into (14.2) and reducing by $\omega^2 \sin(\omega t + \varphi_0)$ we get

$$\begin{aligned} &\left(m_1\delta_{11}\omega^2 - 1\right) A_1 + m_2\delta_{12}\omega^2 A_2 + \ldots + m_n\delta_{1n}\omega^2 A_n = 0, \\ &m_1\delta_{21}\omega^2 A_1 + \left(m_2\delta_{22}\omega^2 - 1\right) A_2 + \ldots + m_n\delta_{2n}\omega^2 A_n = 0, \\ &\cdot \quad \cdot \quad \cdot \quad \cdot \quad \cdot \quad \cdot \quad \cdot \quad \cdot \quad \cdot \quad \cdot \quad \cdot \quad \cdot \quad \cdot \\ &m_1\delta_{31}\omega^2 A_1 + m_2\delta_{32}\omega^2 A_2 + \ldots + \left(m_n\delta_{nn}\omega^2 - 1\right) A_n = 0. \end{aligned} \tag{14.4}$$

The equations (14.4) are homogeneous algebraic equations with respect to unknown amplitudes A. Trivial solution $A_i = 0$ corresponds to the system at rest. Nontrivial solution (nonzero amplitudes A_i) is possible, if the determinant of the coefficients of amplitude is zero.

$$D = \begin{bmatrix} m_1\delta_{11}\omega^2 - 1 & m_2\delta_{12}\omega^2 & \ldots & m_n\delta_{1n}\omega^2 \\ m_1\delta_{21}\omega^2 & m_2\delta_{22}\omega^2 - 1 & \ldots & m_n\delta_{2n}\omega^2 \\ \ldots & \ldots & \ldots & \ldots \\ m_1\delta_{n1}\omega^2 & m_2\delta_{n2}\omega^2 & \ldots & m_n\delta_{nn}\omega^2 - 1 \end{bmatrix} = 0. \tag{14.5}$$

This equation is called the frequency equation in terms of displacements. Solution of this equation $\omega_1, \omega_2, \ldots, \omega_n$ presents the eigenfrequencies of a structure. The number of the frequencies of free vibration equals to the number of degrees of freedom.

14.2.3 Mode Shapes Vibration and Modal Matrix

The set of equations (14.4) are homogeneous algebraic equations with respect to unknown amplitudes A. This system does not allow us to find these amplitudes. However, we can find the ratios between different amplitudes. If a structure has two degrees of freedom, then the system (14.4) becomes

$$\begin{aligned} \left(m_1\delta_{11}\omega^2 - 1\right) A_1 + m_2\delta_{12}\omega^2 A_2 = 0, \\ m_1\delta_{21}\omega^2 A_1 + \left(m_2\delta_{22}\omega^2 - 1\right) A_2 = 0. \end{aligned} \tag{14.4a}$$

From these equations, we can find the following ratios

$$\frac{A_2}{A_1} = -\frac{m_1\delta_{11}\omega^2 - 1}{m_2\delta_{12}\omega^2} \text{ or } \frac{A_2}{A_1} = -\frac{m_1\delta_{21}\omega^2}{m_2\delta_{22}\omega^2 - 1}. \tag{14.6}$$

If we substitute the first frequency of vibration ω_1 into any of the two equations (14.6), then we can find $(A_2/A_1)_{\omega_1}$. Then we can assume that $A_1 = 1$ and calculate the corresponding A_2 (or vice versa). The numbers $A_1 = 1$ and A_2 defines the distribution of amplitudes at the first frequency of vibration ω_1; such distribution is referred as the first mode shape of vibration. This distribution is presented in the form of vector-column φ_1, whose elements are $A_1 = 1$ and the calculated A_2; this column vector is called a first eigenvector φ_1. Thus the set of equations (14.4a) for ω_1 define the first eigenvector to within an arbitrary constant.

Second mode shape of vibration or second eigenvector, which corresponds to the second frequency vibration ω_2, can be found in a similar manner. After that we can construct a modal matrix $\mathbf{\Phi} = \lfloor \varphi_1 \; \varphi_2 \rfloor$.

If a structure has n degrees of freedom, then the modal matrix $\mathbf{\Phi} = \lfloor \varphi_1\ \varphi_2\ \dots\ \varphi_n \rfloor$.

Example 14.1. Design diagram of the frame is shown in Fig. 14.8a. Find eigenfrequincies and mode shape vibration.

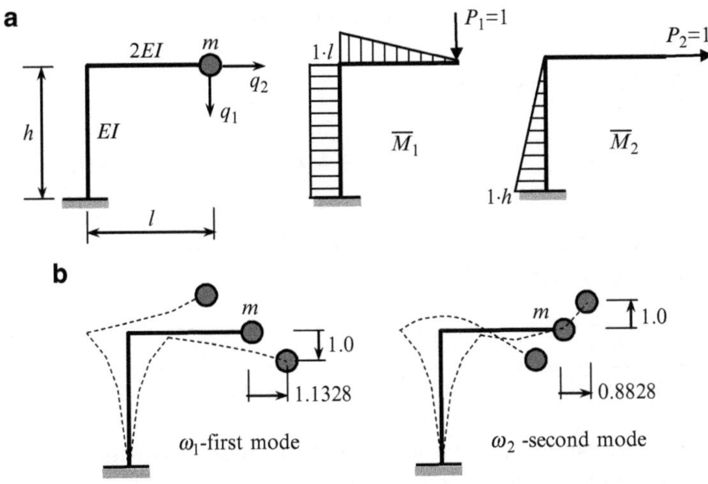

Fig. 14.8 (**a**) Design diagram of the frame and unit states; (**b**) Mode shapes of vibration

Solution. The system has two degrees of freedom. Generalized coordinate are q_1 and q_2. We need to apply unit forces in direction of q_1 and q_2, and construct the bending momens deagram. Unit displacements are

$$\delta_{11} = \frac{\overline{M}_1 \times \overline{M}_1}{EI} = \frac{1}{2EI} \cdot \frac{1}{2} \cdot 1 \cdot l \cdot l \cdot \frac{2}{3} \cdot 1 \cdot l + \frac{1}{EI} \cdot 1 \cdot l \cdot h \cdot 1 \cdot l = \frac{l^3}{6EI} + \frac{l^2 h}{EI};$$

$$\delta_{22} = \frac{\overline{M}_2 \times \overline{M}_2}{EI} = \frac{1}{EI} \cdot \frac{1}{2} \cdot 1 \cdot h \cdot h \cdot \frac{2}{3} 1 \cdot h = \frac{h^3}{3EI};$$

$$\delta_{12} = \delta_{21} = \frac{\overline{M}_1 \times \overline{M}_2}{EI} = \frac{1}{EI} \cdot \frac{1}{2} \cdot 1 \cdot h \cdot h \cdot 1 \cdot l = \frac{h^2 l}{2EI}$$

Let $h = 2l$ and $\delta_0 = l^3/6EI$. In this case $\delta_{11} = 13\delta_0; \delta_{22} = 16\delta_0;$ $\delta_{12} = \delta_{21} = 12\delta_0$.

Equation for calculation of amplitudes (14.4)

$$\begin{aligned} \left(13\delta_0 m\omega^2 - 1\right) A_1 + 12\delta_0 m\omega^2 A_2 &= 0, \\ 12\delta_0 m\omega^2 A_1 + \left(16\delta_0 m\omega^2 - 1\right) A_2 &= 0. \end{aligned} \tag{a}$$

Let

$$\lambda = \frac{1}{\delta_0 m\omega^2} = \frac{6EI}{m\omega^2 l^3}.$$

In this case equation (a) may be rewritten

$$\begin{aligned} (13-\lambda)\,A_1 + 12A_2 &= 0, \\ 12A_1 + (16-\lambda)\,A_2 &= 0. \end{aligned} \qquad \text{(b)}$$

Frequency equation becomes

$$D = \begin{bmatrix} 13-\lambda & 12 \\ 12 & 16-\lambda \end{bmatrix} = (13-\lambda)(16-\lambda) - 144 = 0.$$

Roots in *descending order* are $\lambda_1 = 26.593$; $\lambda_2 = 2.4066$
Eigenfrequencies *in increasing order* are

$$\omega_1 = \sqrt{\frac{6EI}{\lambda_1 m l^3}} = 0.4750\sqrt{\frac{EI}{ml^3}}, \quad \omega_2 = \sqrt{\frac{6EI}{\lambda_2 m l^3}} = 1.5789\sqrt{\frac{EI}{ml^3}}.$$

Mode shape vibration may be determined on the basis of equations (b).
For first mode ($\lambda_1 = 26.593$) ratio of amplitudes are

$$\frac{A_2}{A_1} = -\frac{13-\lambda}{12} = -\frac{13-26.593}{12} = 1.1328,$$
$$\frac{A_2}{A_1} = -\frac{12}{16-\lambda} = -\frac{12}{16-26.593} = 1.1328.$$

Assume that $A_1 = 1$, so the first eigenvector φ becomes

$$\varphi = \lfloor \varphi_{11}\ \varphi_{21} \rfloor^{\mathrm{T}} = \lfloor 1 \quad 1.1328 \rfloor^{\mathrm{T}}$$

For second mode ($\lambda_2 = 2.4066$) ratio of amplitudes are

$$\frac{A_2}{A_1} = -\frac{13-\lambda}{12} = -\frac{13-2.4066}{12} = -0.8828$$
$$\frac{A_2}{A_1} = -\frac{12}{16-\lambda} = -\frac{12}{16-2.4066} = -0.8828$$

The modal matrix $\mathbf{\Phi}$ is then defined as

$$\Phi = \begin{bmatrix} 1 & 1 \\ 1.1328 & -0.8828 \end{bmatrix}.$$

Corresponding mode shapes of vibration are shown in Fig. 14.8b.

Example 14.2. Design diagram of structure containing two hinged-end members is shown in Fig. 14.9a. Modulus of elasticity E and area of cross section A are constant for both members; l and $l\sqrt{3}$ are length of the members, $\alpha = 60°, \beta = 30°$. Find eigenfrequencies, modal matrix and present the mode shapes.

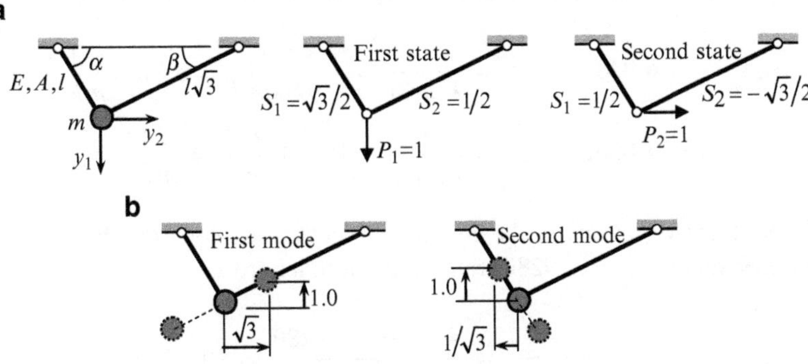

Fig. 14.9 (**a**) Design diagram of the structure and unit states; (**b**) Mode shapes vibrations

Solution. The structure has two degrees of freedom. The first and second unit states and corresponding internal forces for each member are shown in Fig. 14.9a.

Equations (14.4) for unknown amplitudes

$$\begin{aligned}(m\delta_{11}\omega^2 - 1)\,A_1 + m\delta_{12}\omega^2 A_2 = 0,\\ m\delta_{21}\omega^2 A_1 + (m\delta_{22}\omega^2 - 1)\,A_2 = 0.\end{aligned} \tag{a}$$

Unit displacements

$$\delta_{11} = \sum \int \tfrac{S_1 \cdot S_1}{EA} \mathrm{d}s = \tfrac{1}{EA}\left(\tfrac{\sqrt{3}}{2}\cdot\tfrac{\sqrt{3}}{2}\cdot l + \tfrac{1}{2}\cdot\tfrac{1}{2}\cdot l\sqrt{3}\right) = \tfrac{l}{4EA}\left(3+\sqrt{3}\right),$$

$$\delta_{22} = \sum \int \tfrac{S_2 \cdot S_2}{EA} \mathrm{d}s = \tfrac{1}{EA}\left(\tfrac{1}{2}\cdot\tfrac{1}{2}\cdot l + \tfrac{\sqrt{3}}{2}\cdot\tfrac{\sqrt{3}}{2}\cdot l\sqrt{3}\right) = \tfrac{l}{4EA}\left(1+3\sqrt{3}\right),$$

$$\delta_{12} = \delta_{21} = \sum \int \tfrac{S_1 \cdot S_2}{EA} \mathrm{d}s = \tfrac{1}{EA}\left(\tfrac{\sqrt{3}}{2}\cdot\tfrac{1}{2}\cdot l - \tfrac{1}{2}\cdot\tfrac{\sqrt{3}}{2}\cdot l\sqrt{3}\right) = \tfrac{l}{4EA}\left(\sqrt{3}-3\right).$$

Let us denote $\delta_0 = l/4EA$, $\lambda = 1/m\delta_0\omega^2$, then $\delta_{11} = \delta_0(3+\sqrt{3})$, $\delta_{22} = \delta_0(1+3\sqrt{3})$, $\delta_{12} = \delta_{21} = \delta_0(\sqrt{3}-3)$ and equation (a) becomes

$$\begin{aligned}\left(3+\sqrt{3}-\lambda\right)A_1 + \left(\sqrt{3}-3\right)A_2 = 0\\ \left(\sqrt{3}-3\right)A_1 + \left(1+3\sqrt{3}-\lambda\right)A_2 = 0\end{aligned} \quad \text{or} \quad \begin{aligned}(4.7320-\lambda)\,A_1 - 1.2679A_2 = 0\\ -1.2679A_1 + 6.1961A_2 = 0.\end{aligned} \tag{b}$$

Frequency equation

$$D = \begin{bmatrix} 4.7320 - \lambda & -1.2679 \\ -1.2679 & 6.1961 - \lambda \end{bmatrix} = 0$$

Roots of frequency equation and corresponding eigenfrequencies are

$$\lambda_1 = 6.9280 \to \omega_1^2 = \frac{1}{\lambda_1 m \delta_0} = \frac{4EA}{6.9280ml} = 0.5774\frac{EA}{ml},$$

$$\lambda_2 = 4.0002 \to \omega_2^2 = \frac{1}{\lambda_2 m \delta_0} = 0.9999\frac{EA}{ml} \cong \frac{EA}{ml}.$$

Mode shape vibration may be determined on the base equation (b).
For first mode ($\lambda_1 = 6.9280$) ratio of amplitudes are

$$\frac{A_2}{A_1} - \frac{3+\sqrt{3}-\lambda_1}{\sqrt{3}-3} = \frac{4.7320 - 6.9280}{1.2679} = -1.73 = -\sqrt{3},$$

$$\frac{A_2}{A_1} - \frac{\sqrt{3}-3}{1+3\sqrt{3}-\lambda_1} = \frac{1.2679}{6.1961 - 6.9280} = -\sqrt{3}.$$

Assume that $A_1 = 1$, so the first eigenvector φ becomes

$$\varphi = \lfloor \varphi_{11}\ \varphi_{21} \rfloor^{\mathrm{T}} = \lfloor 1 \quad -\sqrt{3} \rfloor^{\mathrm{T}}$$

For second mode ($\lambda_2 = 4.0002$) ratio of amplitudes are

$$\frac{A_2}{A_1} = \frac{4.7320 - 4.0002}{1.2679} = 0.577 = \frac{1}{\sqrt{3}},$$

$$\frac{A_2}{A_1} = \frac{1.2679}{6.1961 - 4.0002} = 0.577 = \frac{1}{\sqrt{3}}.$$

The modal matrix $\mathbf{\Phi}$ is then defined as

$$\mathbf{\Phi} = \begin{bmatrix} 1 & 1 \\ -\sqrt{3} & 1/\sqrt{3} \end{bmatrix}.$$

Corresponding mode shapes of vibration are shown in Fig. 14.9b.

Example 14.3. The beam in Fig. 14.10a carries three equal concentrated masses m_i. The length of the beam is $l = 4a$, and flexural stiffness beam *EI*. The mass of the beam is neglected. It is necessary to find eigenvalues and mode shape vibrations.

Solution. The beam has three degrees of freedom. The bending moment diagrams caused by unit inertial forces are shown in Fig. 14.10b.

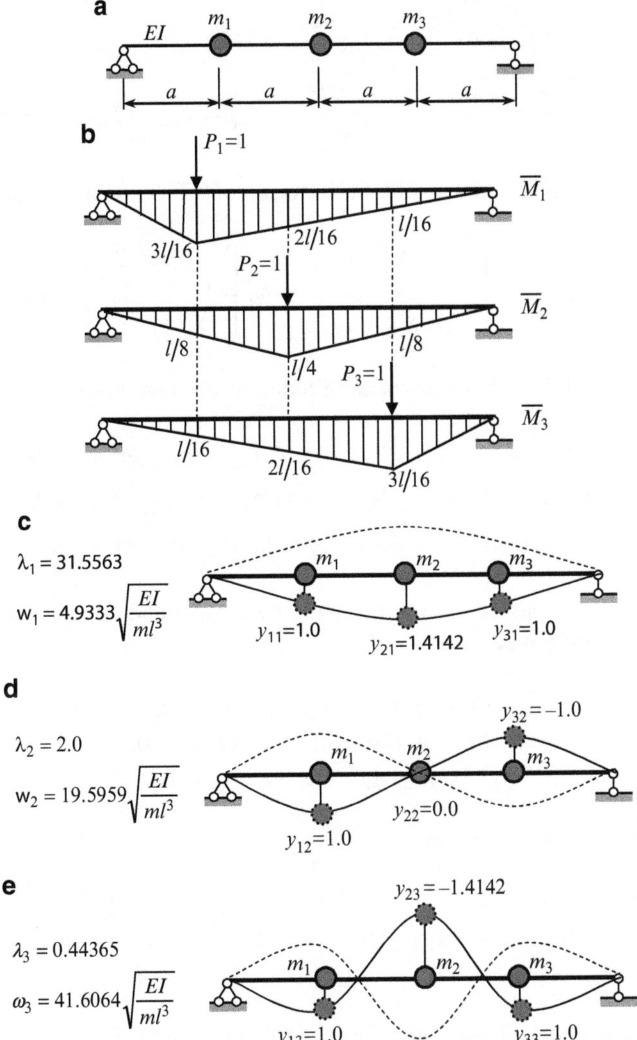

Fig. 14.10 (**a**) Design diagram of the beam; (**b**) Unit bending moment diagrams; (**c**) Mode shape vibration which corresponds to fundamental (lowest) frequency; (**d**) Second mode of vibration; (**e**) Third mode of vibration

Multiplication of corresponding bending moment diagrams leads to the following results for unit displacements

$$\delta_{11} = \int \frac{\overline{M}_1 \overline{M}_1}{EI} \mathrm{d}x = \frac{1}{EI}\left(\frac{1}{2}\frac{l}{4}\frac{3l}{16}\frac{2}{3}\frac{3l}{16} + \frac{1}{2}\frac{3l}{4}\frac{3l}{16}\frac{2}{3}\frac{3l}{16}\right) = \frac{9}{768}\frac{l^3}{EI},$$

$$\delta_{22} = \int \frac{\overline{M}_2 \overline{M}_2}{EI} \mathrm{d}x = \frac{16}{768}\frac{l^3}{EI}, \quad \delta_{33} = \int \frac{\overline{M}_3 \overline{M}_3}{EI} \mathrm{d}x = \frac{9}{768}\frac{l^3}{EI},$$

$$\delta_{12} = \delta_{21} = \int \frac{\overline{M}_1 \overline{M}_2}{EI} \mathrm{d}x = \frac{11}{768} \frac{l^3}{EI},$$

$$\delta_{13} = \delta_{31} = \int \frac{\overline{M}_1 \overline{M}_3}{EI} \mathrm{d}x = \frac{7}{768} \frac{l^3}{EI}, \quad \delta_{23} = \delta_{32} = \delta_{12} = \delta_{21} = \frac{11}{768} \frac{l^3}{EI}.$$

Let $\delta_0 = l^3/768EI$. Matrix of unit displacements **F** (Flexibility matrix) is

$$\mathbf{F} = [\delta_{ik}] = \begin{bmatrix} \delta_{11} & \delta_{12} & \delta_{13} \\ \delta_{21} & \delta_{22} & \delta_{23} \\ \delta_{31} & \delta_{32} & \delta_{33} \end{bmatrix} = \delta_0 \begin{bmatrix} 9 & 11 & 7 \\ 11 & 16 & 11 \\ 7 & 11 & 9 \end{bmatrix} = \delta_0 \mathbf{F}_0.$$

Equations (14.4) with unknown amplitudes A_i of mass m_i are

$$\begin{aligned} &\left(m_1\delta_{11}\omega^2 - 1\right) A_1 + m_2\delta_{12}\omega^2 A_2 + m_3\delta_{13}\omega^2 A_3 = 0, \\ &m_1\delta_{21}\omega^2 A_1 + \left(m_2\delta_{22}\omega^2 - 1\right) A_2 + m_3\delta_{23}\omega^2 A_3 = 0, \\ &m_1\delta_{31}\omega^2 A_1 + m_2\delta_{32}\omega^2 A_2 + \left(m_3\delta_{33}\omega^2 - 1\right) A_3 = 0. \end{aligned} \tag{a}$$

In our case all masses $m_i = m$. Divide by $m\delta_0\omega^2$ and denote $\lambda = 1/m\delta_0\omega^2$. Equations for amplitudes A_i

$$\begin{aligned} &(9 - \lambda) A_1 + 11A_2 + 7A_3 = 0, \\ &11A_1 + (16 - \lambda) A_2 + 11A_3 = 0, \\ &7A_1 + 11A_2 + (9 - \lambda) A_3 = 0. \end{aligned} \tag{b}$$

Frequency equation becomes

$$\det [\, \mathbf{F}_0 - \lambda \, \mathbf{I}] = \det \begin{bmatrix} 9 - \lambda & 11 & 7 \\ 11 & 16 - \lambda & 11 \\ 7 & 11 & 9 - \lambda \end{bmatrix} = 0,$$

where **I** is unit matrix. Eigenvalues are in descending order

$$\begin{aligned} &\lambda_1 = 31.5563, \\ &\lambda_2 = 2.0, \\ &\lambda_3 = 0.44365. \end{aligned} \tag{c}$$

Verification:

1. The sum of the eigenvalues is $\lambda_1 + \lambda_2 + \lambda_3 = 31.5563 + 2.0 + 0.44365 = 34$; On the other hand, the trace of the matrix $Tr(\mathbf{F}_0) = 9 + 16 + 9 = 34$.
2. The multiplication of the eigenvalues is $\lambda_1 \cdot \lambda_2 \cdot \lambda_3 = 31.5563 \cdot 2.0 \cdot 0.44365 = 28$; On the other hand, $\det \mathbf{F}_0 = 28$. (Note, that determinant of unit displacement matrix is strictly positive, i.e. $\det \mathbf{F} > 0$).

Frequencies of the free vibration in increasing order

$$\omega_1^2 = \frac{1}{\lambda_1 m\delta_0} = \frac{768}{31.5563}\frac{EI}{ml^3} = 24.337\frac{EI}{ml^3} \rightarrow \omega_1 = 4.9333\sqrt{\frac{EI}{ml^3}},$$
$$\omega_2^2 = \frac{1}{\lambda_2 m\delta_0} = \frac{768}{2.0}\frac{EI}{ml^3} = 384\frac{EI}{ml^3} \rightarrow \omega_2 = 19.5959\sqrt{\frac{EI}{ml^3}}, \quad \text{(d)}$$
$$\omega_3^2 = \frac{1}{\lambda_3 m\delta_0} = \frac{768}{0.44365}\frac{EI}{ml^3} = 1731.09\frac{EI}{ml^3} \rightarrow \omega_3 = 41.6064\sqrt{\frac{EI}{ml^3}}.$$

For each i-th eigenvalue, the set of equation (b) for calculation of amplitudes is

$$\begin{aligned} &(9-\lambda_i)A_1 + 11A_2 + 7A_3 = 0, \\ &11A_1 + (16-\lambda_i)A_2 + 11A_3 = 0, \\ &7A_1 + 11A_2 + (9-\lambda_i)A_3 = 0. \end{aligned} \quad \text{(e)}$$

Equations (e) divide by A_1. Let $\rho_2 = A_2/A_1,\ \rho_3 = A_3/A_1$.
Equations for modes become

$$\begin{aligned} &(9-\lambda_i) + 11\rho_{2i} + 7\rho_{3i} = 0, \\ &11 + (16-\lambda_i)\rho_{2i} + 11\rho_{3i} = 0, \\ &7 + 11\rho_{2i} + (9-\lambda_i)\rho_{3i} = 0. \end{aligned} \quad \text{(f)}$$

Assuming $A_1 = 1$ we can calculate ρ_2 and ρ_3 for each calculated eigenvalue. For their calculation we can consider set of *any* two equations.

1. Eigenvalue $\lambda_1 = 31.5563$

$$\begin{aligned} &(9-31.5563) + 11\rho_2 + 7\rho_3 = 0, \\ &11 + (16-31.5563)\rho_2 + 11\rho_3 = 0. \end{aligned}$$

Solution of these equations is $\rho_2 = 1.4142,\ \rho_3 = 1.0$. Therefore, the first (principal) mode is defined as $y_{11},\ y_{21} = \sqrt{2}y_{11},\ y_{31} = y_{11}$. If we assume that $y_{11} = 1$, then the eigenvector $\boldsymbol{\varphi}_1$ which corresponds to the frequency ω_1 is $\boldsymbol{\varphi}_1 = \lfloor 1.0\ \sqrt{2}\ 1 \rfloor^{\mathrm{T}}$.
Corresponding mode shape vibration is shown in Fig. 14.10c.

$$\lambda_1 = 31.5563$$
$$\omega_1 = 4.9333\sqrt{\frac{EI}{ml^3}}$$

Note, that substitution of ρ_2 and ρ_3 into third equation (f) leads to the identity.

2. Eigenvalue $\lambda_2 = 2.0$. In this case

$$\begin{aligned} &(9-2.0) + 11\rho_2 + 7\rho_3 = 0, \\ &11 + (16-2.0)\rho_2 + 11\rho_3 = 0. \end{aligned}$$

Solution is $\rho_2 = 0.0$, $\rho_3 = -1.0$, and second eigenvector becomes $\boldsymbol{\varphi}_2 = \lfloor 1.0 \ 0.0 \ -1 \rfloor^{\mathrm{T}}$; this mode shape vibration is shown in Fig. 14.10d.

$$\lambda_2 = 2.0,$$
$$\omega_2 = 19.5959\sqrt{\frac{EI}{ml^3}}.$$

3. Eigenvalue $\lambda_3 = 0.44365$. In this case

$$(9 - 0.44365) + 11\rho_2 + 7\rho_3 = 0,$$
$$11 + (16 - 0.44365)\,\rho_2 + 11\rho_3 = 0.$$

Solution $\rho_2 = -1.4142$, $\rho_3 = 1.0$, and Eigenvector $\boldsymbol{\varphi}_3 = \lfloor 1.0 \ -\sqrt{2} \ 1 \rfloor^{\mathrm{T}}$. Third mode shape vibration is shown in Fig. 14.10e.

$$\lambda_3 = 0.44365,$$
$$\omega_3 = 41.6064\sqrt{\frac{EI}{ml^3}}.$$

We can see that a number of the nodal points of the mode of shape vibration one less than the number of the mode.

The modal matrix is defined as

$$\Phi = \lfloor \boldsymbol{\varphi}_1 \ \ \boldsymbol{\varphi}_2 \ \ \boldsymbol{\varphi}_3 \rfloor = \begin{bmatrix} \varphi_{11} & \varphi_{12} & \varphi_{13} \\ \varphi_{21} & \varphi_{22} & \varphi_{23} \\ \varphi_{31} & \varphi_{32} & \varphi_{33} \end{bmatrix} = \begin{bmatrix} 1 & 1 & 1 \\ \sqrt{2} & 0.0 & -\sqrt{2} \\ 1 & -1 & 1 \end{bmatrix}, \quad \text{(g)}$$

where the i-th and k-th indexes at φ mean the number of mass and number of frequency, respectively.

14.3 Free Vibrations of Systems with Finite Number Degrees of Freedom: Displacement Method

Now we will consider a dynamical analysis of the structures with finite number degrees of freedom using the concept of unit *reactions*. For several types of structures, displacement method is more preferable than the force method.

14.3.1 Differential Equations of Free Vibration in Reactions

In essence, this method consists in expressing the forces of inertia as function of *unit reactions*. According to the displacement method, we need to introduce additional constraints which prevent each displacement of the mass. Thus, the total number of constraints equal to the number of degrees of freedom. Let us consider a structure with concentrated masses m_i, $i = 1,\ldots,n$ (Fig. 14.11a). This structure has n degrees of freedom. They are lateral displacements of the frame points at the each mass. Primary system of the displacement method is shown in Fig. 14.11b.

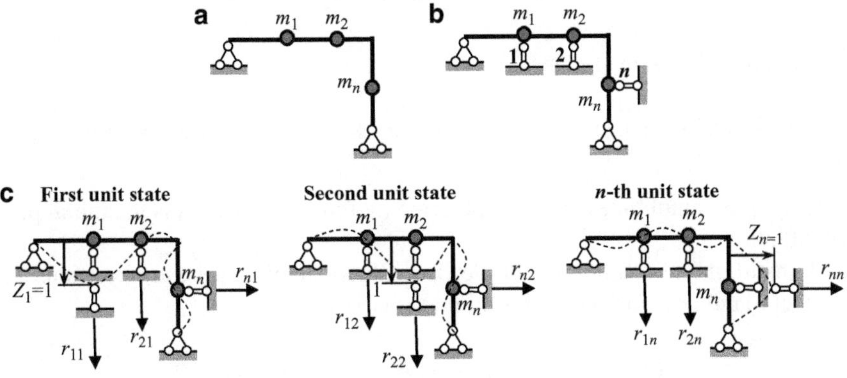

Fig. 14.11 (**a**) Design diagram; (**b**) Primary system; (**c**) Unit states

Displacement of each lumped mass (or displacement of each introduced constraints) are y_i. Inertial forces of each mass may be presented in terms of unit reactions r_{ik} as follows

$$\begin{aligned} -m_1\ddot{y}_1 &= r_{11}y_1 + r_{12}y_2 + \ldots + r_{1n}y_n, \\ &\cdots \\ -m_n\ddot{y}_n &= r_{n1}y_1 + r_{n2}y_2 + \ldots + r_{nn}y_n. \end{aligned} \tag{14.7}$$

The coefficient r_{ik} presents the reaction in i-th introduced constraint caused by unit displacement of k-th introduced constraint. The term $r_{ik}y_k$ means reaction in i-th introduced constraint caused by real displacement of k-th introduced constraint. Each equation of (14.7) describes the equilibrium condition.

Equations (14.7) lead to the following differential equations of undamped free vibration of the multi-degree of freedom system

$$\begin{aligned} m_1\ddot{y}_1 + r_{11}y_1 + r_{12}y_2 + \ldots + r_{1n}y_n &= 0, \\ &\cdots \\ m_n\ddot{y}_n + r_{n1}y_1 + r_{n2}y_2 + \ldots + r_{nn}y_n &= 0. \end{aligned} \tag{14.7a}$$

These equations are coupled statically, because the generalized coordinates appears in each equation.

In matrix form, the system (14.7a) may be written as

$$\mathbf{M}\ddot{\mathbf{Y}} + \mathbf{SY} = \mathbf{0}, \tag{14.7b}$$

where the mass and stiffness matrices as well as displacement vector are

$$\mathbf{M}=\begin{bmatrix} m_1 & 0 & \dots & 0 \\ 0 & m_2 & \dots & 0 \\ \dots & \dots & \dots & \dots \\ 0 & 0 & \dots & m_n \end{bmatrix}, \quad \mathbf{S}=\begin{bmatrix} r_{11} & r_{12} & \dots & r_{1n} \\ r_{21} & r_{22} & \dots & r_{2n} \\ \dots & \dots & \dots & \dots \\ r_{n1} & r_{n2} & \dots & r_{nn} \end{bmatrix}, \quad \mathbf{Y}=\begin{bmatrix} y_1 \\ y_2 \\ \dots \\ y_n \end{bmatrix},$$

Solution of system (14.7)

$$y_1 = A_1 \sin(\omega t + \varphi_0), \quad y_2 = A_2 \sin(\omega t + \varphi_0), \ \dots \ y_3 = A_3 \sin(\omega t + \varphi_0), \tag{14.8}$$

where A_i are amplitudes of the displacement of mass m_i and φ_0 is the initial phase of vibration.

Substituting (14.8) into (14.7a) leads to algebraic homogeneous equations with respect to unknown amplitudes of lumped masses

$$\begin{aligned} &\left(r_{11} - m_1\omega^2\right) A_1 + r_{12}A_2 + \dots + r_{1n}A_n = 0, \\ &r_{21}A_1 + \left(r_{22} - m_2\omega^2\right) A_2 + \dots + r_{2n}A_n = 0, \\ &\cdot \quad \cdot \quad \cdot \quad \cdot \quad \cdot \quad \cdot \quad \cdot \quad \cdot \quad \cdot \quad \cdot \quad \cdot \\ &r_{n1}A_1 + r_{n2}A_2 + \dots + \left(r_{nn} - m_n\omega^2\right) A_n = 0. \end{aligned} \tag{14.9}$$

14.3.2 Frequency Equation

Nontrivial solution (nonzero amplitudes A_i) is possible, if the determinant of the coefficients to amplitude is zero.

$$D=\begin{bmatrix} r_{11} - m_1\omega^2 & r_{12} & \dots & r_{1n} \\ r_{21} & r_{22} - m_2\omega^2 & \dots & r_{2n} \\ \dots & \dots & \dots & \dots \\ r_{n1} & r_{n2} & \dots & r_{nn} - m_n\omega^2 \end{bmatrix} = 0. \tag{14.10}$$

This equation is called the frequency equation in form of the displacement method. Solution of this equation presents the eigenfrequencies of a structure. The number of the frequencies of free vibration is equal to the number of degrees of freedom.

14.3.3 Mode Shape Vibrations and Modal Matrix

Equations (14.9) are homogeneous algebraic equations with respect to unknown amplitudes A. This system does not allow us to find these amplitudes. However, we can find the ratios between different amplitudes. If a structure has two degrees of freedom, then system (14.9) becomes

$$\begin{aligned}\left(r_{11}-m_1\omega^2\right)A_1+r_{12}A_2&=0,\\ r_{21}A_1+\left(r_{22}-m_2\omega^2\right)A_2&=0.\end{aligned}$$

From these equations, we can find following ratios

$$\frac{A_2}{A_1}=-\frac{r_{11}-m_1\omega^2}{r_{12}} \quad \text{or} \quad \frac{A_2}{A_1}=-\frac{r_{21}}{r_{22}-m_2\omega^2}. \tag{14.11}$$

If we assume $A_1=1$, then entries $\lfloor 1 \;\; A_2 \rfloor^{\mathrm{T}}$ defines for each eigenfrequency, the corresponding column φ of the modal matrix $\mathbf{\Phi}$. The formulas (14.11) and (14.6) lead to the same result.

Let us show application of the displacement method for free vibration analysis of a beam with three equal lumped masses (Fig. 14.12a); previously this structure had been analyzed by force method (Example 14.3). It is necessary to find eigenvalues and modal matrix.

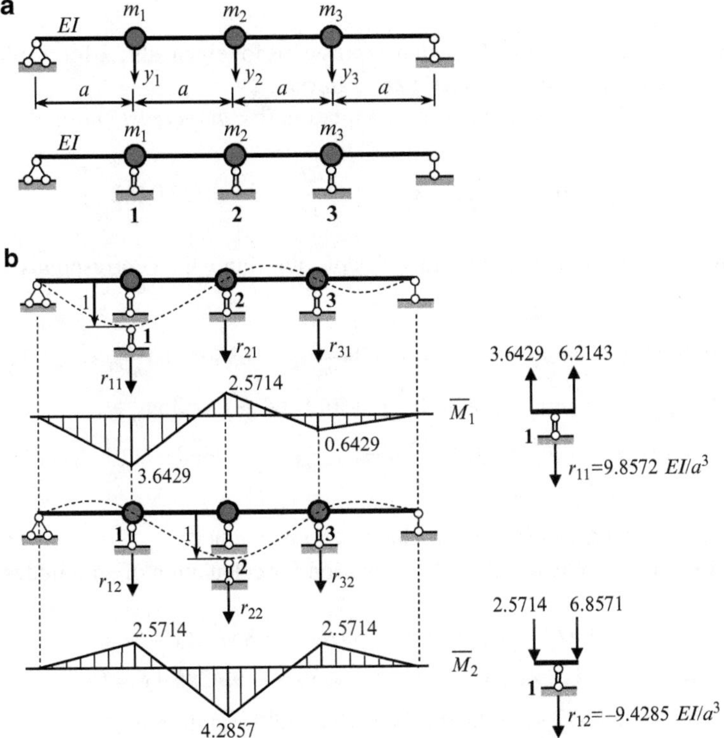

Fig. 14.12 (**a**) Design diagram of the beam and primary system; (**b**) Unit displacements and corresponding bending moment diagrams (factor EI/a^2); calculation of r_{11} and r_{12}

The introduced constraints 1, 2, 3 which prevent to displacement y_i are shown in Fig. 14.12a.

For calculation of unit reactions, we need to construct bending moment diagram due to unit displacements of each introduced constraint. These diagrams are presented in Fig. 14.12b; they are constructed using Table A.18. Since a structure is symmetrical, then bending moment diagram caused by unit displacement constraint 3 is not shown.

Calculation of unit reactions has no difficulties. All shear and unit reactions have multiplier EI/a^3.

Since $a = l/4$, then the stiffness matrix

$$\mathbf{S} = \frac{64EI}{l^3}\begin{bmatrix} 9.8572 & -9.4285 & 3.8572 \\ -9.4285 & 13.7142 & -9.4285 \\ 3.8572 & -9.4285 & 9.8572 \end{bmatrix}. \tag{a}$$

Since all masses are equal, then (14.10) may be rewritten as

$$\begin{bmatrix} 9.8572-\lambda & -9.4285 & 3.8572 \\ -9.4285 & 13.7142-\lambda & -9.4285 \\ 3.8572 & -9.4285 & 9.8572-\lambda \end{bmatrix} = 0, \tag{b}$$

where parameter $\lambda = m\omega^2$. Note that expressions for eigenvalue λ for displacement and force methods $\left(\lambda = 1/m\delta_0\omega^2\right)$ are different.

The eigenvalues present the roots of equation (b); *in increasing order* they are

$$\lambda_1 = 0.3804\frac{EI}{a^3}, \quad \lambda_2 = 6.0\frac{EI}{a^3}, \quad \lambda_3 = 27.0482\frac{EI}{a^3}. \tag{c}$$

Now we can calculate the frequencies of vibration which corresponds to eigenvalues.

$$\begin{aligned} &\lambda_1 = 0.3804\tfrac{EI}{a^3} = m\omega_1^2 \to \omega_1^2 = 0.3804\tfrac{EI}{ma^3} = 0.3804\cdot 64\tfrac{EI}{ml^3} = 24.345\tfrac{EI}{ml^3},\\ &\lambda_2 = 6.0\tfrac{EI}{a^3} = m\omega_2^2 \to \omega_2^2 = 6.0\tfrac{EI}{ma^3} = 6.0\cdot 64\tfrac{EI}{ml^3} = 384\tfrac{EI}{ml^3},\\ &\lambda_3 = 27.0482\tfrac{EI}{a^3} = m\omega_3^2 \to \omega_3^2 = 27.0482\tfrac{EI}{ma^3} = 27.0482\cdot 64\tfrac{EI}{ml^3} = 1731.08\tfrac{EI}{ml^3}. \end{aligned} \tag{d}$$

Same frequencies have been obtained by force method.

For each i-th eigenvalue the set of equation for calculation of amplitudes is

$$\begin{aligned} &(9.8572-\lambda_i)\,A_1 - 9.4285A_2 + 3.8572A_3 = 0,\\ &-9.4285A_1 + (13.7142-\lambda_i)\,A_2 - 9.4285A_3 = 0,\\ &3.8572A_1 - 9.4285A_2 + (9.8572-\lambda_i)\,A_3 = 0. \end{aligned} \tag{e}$$

Equations (e) divide by A_1. Let $\rho_2 = A_2/A_1$, $\rho_3 = A_3/A_1$. Equations for modes become

$$\begin{aligned}(9.8572-\lambda_i)-9.4285\rho_{2i}+3.8572\rho_{3i}&=0,\\ -9.4285+(13.7142-\lambda_i)\,\rho_{2i}-9.4285\rho_{3i}&=0,\\ 3.8572-9.4285\rho_{2i}+(9.8572-\lambda_i)\,\rho_{3i}&=0.\end{aligned} \tag{f}$$

1. **Eigenvalue** $\lambda_1 = 0.3804$

$$\begin{aligned}(9.8572-0.3804)-9.4285\rho_{2i}+3.8572\rho_{3i}&=0,\\ -9.4285+(13.7142-0.3804)\,\rho_{2i}-9.4285\rho_{3i}&=0.\end{aligned}$$

Solution of this equation $\rho_2 = 1.4142, \quad \rho_3 = 1.0.$
The same procedure should be repeated for $\lambda_2 = 6.0$ and $\lambda_3 = 27.0482$.
The modal matrix $\mathbf{\Phi}$ is defined as

$$\Phi = \begin{bmatrix} 1 & 1 & 1 \\ 1.4142 & 0.0 & -1.4142 \\ 1 & -1 & 1 \end{bmatrix} \tag{g}$$

Same mode shape coefficients have been obtained by force method.

Example 14.4. Design diagram of multistore frame is presented in Fig. 14.13a. The cross bars are absolutely rigid bodies; theirs masses are shown in design diagram. Flexural sriffness of the vertical members are *EI* and masses of the struts are ignored. Calculate the frequencies of vibrations and find the corresponding mode shapes.

Solution. The primary system is shown in Fig. 14.13b. For computation of unit reactions, we need to construct the bending moment diagrams due to unit diplacements of the introduced constraints and then to consider the equilibrium condition for each cross-bar.

Bending moment diagram caused by unit displacement of the constraint 1 is shown in Fig. 14.13c. Elastic curve is shown by dotted line. Since cross bars are absolutely rigid members, then joints cannot be rotated and each vertical member should be considered as fixed-fixed member. In this case, specified ordinates are $6i/h$. Bending moment diagram is shown on the extended fibers. Now we need to show free-body diagram for each vertical member. The sections are passes infinitely close to the bottom and lower joints. Bending moments are $6i/h$. Both moments may be equilibrate by two forces $12i/h^2$. These forces should be transmitted on both cross-bars. Positive unit reactions r_{11}, r_{21}, and r_{31} are shown by dotted arrows.

Equilibrium condition for each cross-bar leads to the follofing unit reactions

$$r_{11} = 2\frac{12i}{h^2} = 24\frac{i}{h^2}, \quad r_{21} = -24\frac{i}{h^2}, \quad r_{31} = 0, \quad i = \frac{EI}{h}.$$

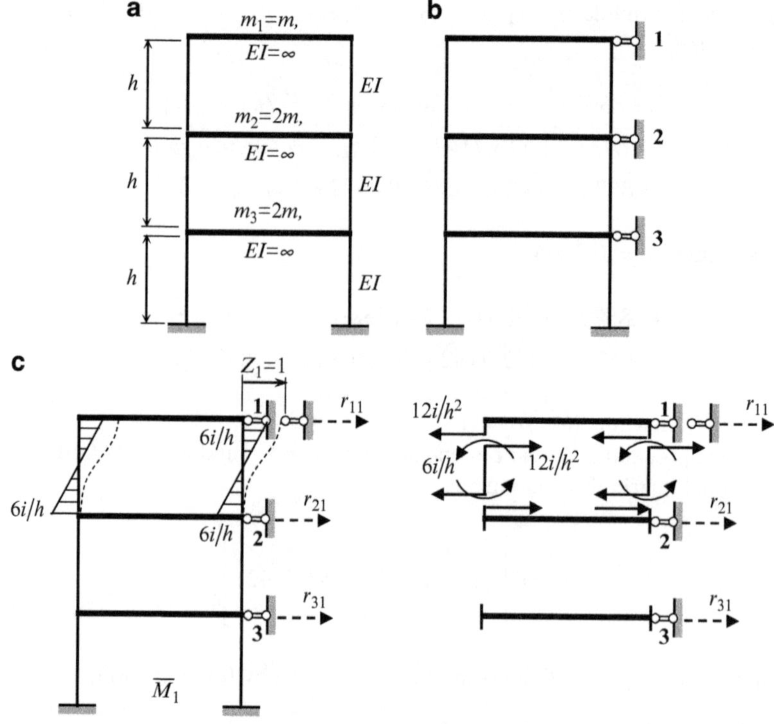

Fig. 14.13 (**a**) Design diagram of the frame; (**b**) Primary system; (**c**) Bending moment diagram caused by unit displacement of the constraint 1 and calculation of unit reactions r_{11}, r_{21}, and r_{31}

Similarly, considering the second and third unit displacements, we get

$$r_{12} = -24\frac{i}{h^2}, \quad r_{22} = 48\frac{i}{h^2}, \quad r_{32} = -24\frac{i}{h^2},$$
$$r_{13} = 0, \quad r_{23} = -24\frac{i}{h^2}, \quad r_{33} = 48\frac{i}{h^2}$$

Let $r_0 = 24\frac{i}{h^2}$. Equations (14.9) becomes

$$\begin{aligned} &\left(r_0 - m\omega^2\right) A_1 - r_0 A_2 + 0.A_3 = 0, \\ &r_0 A_1 + 2\left(r_0 - m\omega^2\right) A_2 - r_0 A_3 = 0, \\ &0.A_1 - r_0 A_2 + 2\left(r_0 - m\omega^2\right) A_3 = 0. \end{aligned} \tag{a}$$

The frequency equation is

$$D = \begin{bmatrix} r_0 - m\omega^2 & -r_0 & 0 \\ -r_0 & 2r_0 - 2m\omega^2 & -r_0 \\ 0 & -r_0 & 2r_0 - 2m\omega^2 \end{bmatrix} = 0$$

If eigenvalue is denoted as

$$\lambda = \frac{m\omega^2}{r_0} = \frac{m\omega^2 h^2}{24i},$$

then the system (a) may be rewritten as

$$\begin{aligned} &(1-\lambda) A_1 - A_2 = 0, \\ &-A_1 + 2(1-\lambda) A_2 - A_3 = 0, \\ &-A_2 + 2(1-\lambda) A_3 = 0. \end{aligned} \tag{b}$$

Eigenvalue equation is

$$D = \begin{bmatrix} 1-\lambda & -1 & 0 \\ -1 & 2(1-\lambda) & -1 \\ 0 & -1 & 2(1-\lambda) \end{bmatrix} = 0$$

$$4(1-\lambda)^3 - 3(1-\lambda) = 0 \text{ or } (1-\lambda)\left[4(1-\lambda)^2 - 1\right] = 0$$

The eigenvalues in increasing order are

$$\lambda_1 = 1 - \frac{\sqrt{3}}{2}, \quad \lambda_2 = 1, \quad \lambda_3 = 1 + \frac{\sqrt{3}}{2}.$$

Corresponding frequencies of free vibration (eigenfrequencies)

$$\omega_1^2 = 24\left(1 - \frac{\sqrt{3}}{2}\right)\frac{EI}{mh^3}, \quad \omega_2^2 = 24\frac{EI}{mh^3}, \quad \omega_3^2 = 24\left(1 + \frac{\sqrt{3}}{2}\right)\frac{EI}{mh^3}.$$

Mode shapes vibration. Now we need to consider the system (b) for each calculated eigenvalue. If denote $\rho_2 = A_2/A_1$ and $\rho_3 = A_3/A_1$, then system (b) may be rewritten as

$$\begin{aligned} &(1-\lambda) - \rho_2 = 0, \\ &-1 + 2(1-\lambda)\rho_2 - \rho_3 = 0, \\ &-\rho_2 + 2(1-\lambda)\rho_3 = 0. \end{aligned} \tag{c}$$

This system should be solved with respect to ρ_2 and ρ_3 for each eigenvalue.
First mode

$$\omega_1 \quad \left(\lambda_1 = 1 - \frac{\sqrt{3}}{2}\right).$$

Equations (c) becomes

$$\begin{aligned}&(1-\lambda_1)-\rho_2=0,\\&-1+2\,(1-\lambda_1)\,\rho_2-\rho_3=0,\\&-\rho_2+2\,(1-\lambda_1)\,\rho_3=0.\end{aligned}$$

Solution of these equation are $\rho_2=\sqrt{3}/2, \quad \rho_3=1/2$.
The same procedure should be repeated for $\lambda_2=1.0$ and $\lambda_3=1+\frac{\sqrt{3}}{2}$
The modal matrix $\mathbf{\Phi}$ is defined as

$$\Phi=\begin{bmatrix}1 & 1 & 1\\ \sqrt{3}/2 & 0 & -\sqrt{3}/2\\ 1/2 & -1 & 1/2\end{bmatrix}$$

14.3.4 Comparison of the Force and Displacement Methods

Some fundamental data about application of two fundamental methods for free vibration analysis of the structures with finite number degrees of freedom is presented in Table 14.2.

Generally, for nonsymmetrical beams, the force method is more effective than the displacement method. However, for frames especially with absolutely rigid crossbar, the displacement method is beyond the competition.

14.4 Free Vibrations of One-Span Beams with Uniformly Distributed Mass

The more precise dynamical analysis of engineering structure is based on the assumption that a structure has distributed masses. In this case, the structure has infinite number degrees of freedom and mathematical model presents a partial differential equation. Additional assumptions allow construction of the different mathematical models of transversal vibration of the beam. The simplest mathematical models consider a plane vibration of uniform beam with, taking into account only, bending moments; shear and inertia of rotation of the cross sections are neglected. The beam upon these assumptions is called as Bernoulli-Euler beam.

Table 14.2 Comparison of the force and displacement methods for free vibration analysis

	Force method (analysis in terms of displacements)	Displacement method (analysis in terms of reactions)
Coupled differential equations 1. Canonical form	$\delta_{11}m_1\ddot{y}_1+\delta_{12}m_2\ddot{y}_2+\ldots+\delta_{1n}m_n\ddot{y}_n+y_1=0$ $\cdots\cdots\cdots\cdots$ $\delta_{n1}m_1\ddot{y}_1+\delta_{n2}m_2\ddot{y}_2+\ldots+\delta_{nn}m_n\ddot{y}_n+y_n=0$	$m_1\ddot{y}_1+r_{11}y_1+r_{12}y_2+\ldots+r_{1n}y_n=0$ $\cdots\cdots\cdots\cdots$ $m_n\ddot{y}_n+r_{n1}y_1+r_{n2}y_2+\ldots+r_{nn}y_n=0$
2. Matrix form	$\mathbf{FM}\ddot{\mathbf{Y}}+\mathbf{Y}=\mathbf{0}$	$\mathbf{M}\ddot{\mathbf{Y}}+\mathbf{SY}=\mathbf{0}$
Type of coupling	Dynamical	Static
Matrices	Flexibility matrix $\mathbf{F}=\begin{bmatrix}\delta_{11} & \delta_{12} & \ldots & \delta_{1n}\\ \delta_{21} & \delta_{22} & \ldots & \delta_{2n}\\ \ldots & \ldots & \ldots & \ldots\\ \delta_{n1} & \delta_{n2} & \ldots & \delta_{nn}\end{bmatrix},\quad \mathbf{Y}=\begin{bmatrix}y_1\\ y_2\\ \ldots\\ y_n\end{bmatrix}$ **M** is diagonal mass matrix	Stiffness matrix $\mathbf{S}=\begin{bmatrix}r_{11} & r_{12} & \ldots & r_{1n}\\ r_{21} & r_{22} & \ldots & r_{2n}\\ \ldots & \ldots & \ldots & \ldots\\ r_{n1} & r_{n2} & \ldots & r_{nn}\end{bmatrix},\quad \mathbf{Y}=\begin{bmatrix}y_1\\ y_2\\ \ldots\\ y_n\end{bmatrix}$ **M** is diagonal mass matrix
Solution	$\mathbf{Y}=\mathbf{A}\sin(\omega t+\varphi_0)$	$\mathbf{Y}=\mathrm{A}\sin(\omega t+\varphi_0)$
Equations for unknowns amplitudes A_i 1. Canonical form	$(m_1\delta_{11}-1/\omega^2)A_1+m_2\delta_{12}A_2+\ldots+m_n\delta_{1n}A_n=0$ $m_1\delta_{21}A_1+(m_2\delta_{22}-1/\omega^2)A_2+\ldots+m_n\delta_{2n}A_n=0$ $\cdots\cdots\cdots\cdots$ $m_1\delta_{n1}A_1+m_2\delta_{n2}A_2+\ldots+(m_n\delta_{nn}-1/\omega^2)A_n=0$	$(r_{11}-m_1\omega^2)A_1+r_{12}A_2+\ldots+r_{1n}A_n=0$ $r_{21}A_1+(r_{22}-m_2\omega^2)A_2+\ldots+r_{2n}A_n=0$ $\cdots\cdots\cdots\cdots$ $r_{n1}A_1+r_{n2}A_2+\ldots+(r_{nn}-m_n\omega^2)A_n=0$
2. Matrix form	$\left[\mathbf{FM}-\frac{1}{\omega^2}\mathbf{I}\right]\mathbf{A}=\mathbf{0}$	$[\mathbf{S}-\omega^2\mathbf{M}]\mathbf{A}=0$
Frequency equation 1. Canonical form	$\begin{bmatrix}m_1\delta_{11}-1/\omega^2 & m_2\delta_{12} & \ldots & m_n\delta_{1n}\\ m_1\delta_{21} & m_2\delta_{22}-1/\omega^2 & \ldots & m_n\delta_{2n}\\ \ldots & \ldots & \ldots & \ldots\\ m_1\delta_{n1} & m_2\delta_{n2} & \ldots & m_n\delta_{nn}-1/\omega^2\end{bmatrix}=0$	$\begin{bmatrix}r_{11}-m_1\omega^2 & r_{12} & \ldots & r_{1n}\\ r_{21} & r_{22}-m_2\omega^2 & \ldots & r_{2n}\\ \ldots & \ldots & \ldots & \ldots\\ r_{n1} & r_{n2} & \ldots & r_{nn}-m_n\omega^2\end{bmatrix}=0$
2. Matrix form	$\mathbf{FM}-\frac{1}{\omega^2}\mathbf{I}=0$	$\mathbf{S}-\omega^2\mathbf{M}=\mathbf{0}$

14.4.1 Differential Equation of Transversal Vibration of the Beam

Differential equation of the uniform beam is

$$EI\frac{d^4 y}{dx^4} = q, \tag{14.12}$$

where y is the transverse displacement of a beam, E the modulus of elasticity, I the moment of inertia of the cross section about the neutral axis, and q the transverse load per unit length of the beam.

In case of free vibration, the load per unit length is

$$q = -\rho A\frac{d^2 y}{dt^2}, \tag{14.13}$$

where ρ is the mass density and A is the cross-sectional area.

Equations (14.12) and (14.13) lead to following differential equation of the transverse vibration of the uniform Bernoulli-Euler beam

$$EI\frac{\partial^4 y}{\partial x^4} + \rho A\frac{\partial^2 y}{\partial t^2} = 0. \tag{14.14}$$

If a beam is subjected to forced load $f(x, t)$, then the mathematical model is

$$EI\frac{\partial^4 y}{\partial x^4} + \rho A\frac{\partial^2 y}{\partial t^2} = f(x,t). \tag{14.14a}$$

Thus the transverse displacement of a beam depends on the axial coordinate x and time t, i.e., $y = y(x, t)$.

Boundary and initial conditions: The classical boundary condition takes into account only the shape of the beam deflection curve at the boundaries. The nonclassical boundary conditions take into account the additional mass, the damper, as well as the translational and rotational springs at the boundaries. The classical boundary conditions for the transversal vibration of a beam are presented in Table 14.3.

Notation: y and θ are transversal deflection and slope; M and Q are bending moment and shear force.

Initial conditions present the initial distribution of the displacement and the initial distribution of the velocities of each point of a beam at $t = 0$

$$y(x,0) = u(x); \quad \frac{dy}{dt}(x,0) = \dot{y}(x,0) = \upsilon(x).$$

Table 14.3 Classical boundary conditions

Clamped end ($y = 0,\ \theta = 0$)	Free end ($Q = 0,\ M = 0$)
$y = 0;\quad \frac{\partial y}{\partial x} = 0$	$\frac{\partial}{\partial x}\left(EI\frac{\partial^2 y}{\partial x^2}\right) = 0;\quad EI\frac{\partial^2 y}{\partial x^2} = 0$
Pinned end ($y = 0,\ M = 0$)	**Sliding end** ($Q = 0,\ \theta = 0$)
$y = 0;\quad EI\frac{\partial^2 y}{\partial x^2} = 0$	$\frac{\partial}{\partial x}\left(EI\frac{\partial^2 y}{\partial x^2}\right) = 0;\quad \frac{\partial y}{\partial x} = 0$

14.4.2 Fourier Method

A solution of differential equation (14.14) may be presented in the form

$$y(x,t) = X(x)\,T(t), \tag{14.15}$$

where $X(x)$ is the space-dependent function (shape function, mode shape function, eigenfunction); $T(t)$ is the time-dependent function.

The shape function $X(x)$ and time-dependent function $T(t)$ depends on the boundary conditions and initial conditions, respectively. Plugging the form (14.15) into the (14.14), we get

$$\frac{EIX^{IV}}{\rho AX} + \frac{\ddot{T}}{T} = 0. \tag{14.16}$$

It means that both terms are equals but have opposite signs. Let $\ddot{T}/T = -\omega^2$; then for functions $T(t)$ and $X(x)$ may be written the following differential equations

$$\ddot{T} + \omega^2 T = 0. \tag{14.17}$$

$$X^{IV}(x) - k^4 X(x) = 0, \tag{14.18}$$

where $k = \sqrt[4]{\frac{m\omega^2}{\mathrm{EI}}}$ and $m = \rho A$ is mass per unit length of the beam. Thus, instead of (14.14) containing two independent parameters (time t and coordinate x), we obtained two *uncoupled* ordinary differential equations with respect to unknown functions $X(x)$ and $T(t)$. This procedure is called the separation of variables method.

The solution of (14.17) is $T(t) = A_1 \sin\omega t + B_1 \cos\omega t$, where ω is frequency of vibration. This equation shows that displacement of vibrating beam obey to harmonic law; coefficients A_1 and B_1 should be determined from *initial* conditions.

The general solution of (14.18) is

$$X(x) = A\cosh kx + B\sinh kx + C\cos kx + D\sin kx, \tag{14.19}$$

where A, B, C, and D may be calculated using the *boundary* conditions.

The natural frequency ω of a beam is defined by equation

$$\omega = k^2\sqrt{\frac{EI}{m}} = \frac{\lambda^2}{l^2}\sqrt{\frac{EI}{m}}, \quad \text{where } \lambda = kl. \tag{14.20}$$

To obtain frequency equation using general solution (14.19), the following algorithm is recommended:

Step 1. Represent the mode shape in the general form (14.19), which contains four unknown constants.

Step 2. Determine constants using the boundary condition at $x = 0$ and $x = l$. Thus, the system of four homogeneous algebraic equations is obtained.

Step 3. The nontrivial solution of this system represents a frequency equation.

Example 14.5. Calculate the frequencies of free vibration and find the corresponding mode shapes for pinned-pinned beam. The beam has length l, mass per unit length m, modulus of elasticity E, and moment of inertia of cross-sectional area I.

Solution. The shape of vibration may be presented in form (14.19). For pinned-pinned beam displacement and bending moment at $x = 0$ and at $x = l$ equal zero. Expression for bending moment is

$$X''(x) = k^2\left(A\cosh kx + B\sinh kx - C\cos kx - D\sin kx\right).$$

Conditions $X(0) = 0$ and $X''(0) = 0$ leads to the equations

$$\begin{aligned} A + C &= 0 \\ A - C &= 0 \end{aligned}$$

Thus $A = C = 0$.

Conditions $X(l) = 0$ and $X''(l) = 0$ leads to the equations

$$\begin{aligned} B\sinh kl + D\sin kl &= 0 \\ B\sinh kl - D\sin kl &= 0 \end{aligned}$$

Thus $B = 0$ and $D\sin kl = 0$. Non-trivial solution occurs, if $\sin kl = 0$. This is frequency equation. Solution of this equation is $kl = \pi, \quad 2\pi, \ldots$ Thus, the frequencies of vibration are

$$\omega = k^2\sqrt{\frac{EI}{m}}, \quad \omega_1 = \frac{3.1416^2}{l^2}\sqrt{\frac{EI}{m}}, \quad \omega_2 = \frac{6.2832^2}{l^2}\sqrt{\frac{EI}{m}}$$

The mode shape of vibration is

$$X_i(x) = D \sin k_i x = D \sin \frac{i\pi}{l} x, \quad i = 1, 2, 3, \ldots$$

14.4.3 Krylov–Duncan Method

A general solution of differential equation (14.18) may be presented in the form

$$X(kx) = C_1 S(kx) + C_2 T(kx) + C_3 U(kx) + C_4 V(kx), \tag{14.21}$$

where $X(kx)$ is the general expression for mode shape; $S(kx)$, $T(kx)$, $U(kx)$, $V(kx)$ are the Krylov–Duncan functions (Krylov, 1936; Duncan, 1943). They are present the combination of trigonometric and hyperbolic functions.

$$\begin{aligned} S(kx) &= \frac{1}{2}(\cosh kx + \cos kx) \\ T(kx) &= \frac{1}{2}(\sinh kx + \sin kx) \\ U(kx) &= \frac{1}{2}(\cosh kx - \cos kx) \\ V(kx) &= \frac{1}{2}(\sinh kx - \sin kx) \end{aligned} \tag{14.22}$$

The constants C_i may be expressed in terms of initial parameters as follows

$$C_1 = X(0),\ C_2 = \frac{1}{k}X'(0),\ C_3 = \frac{1}{k^2}X''(0),\ C_3 = \frac{1}{k^3}X'''(0) \tag{14.23}$$

Each combination (14.22) satisfies to equations of the free vibration of a uniform Bernoulli-Euler beam. The functions (14.22) have the following important properties:

1. Krylov–Duncan functions and their derivatives result in the unit matrix at $x = 0$.

$$\begin{array}{llll} S(0) = 1 & S'(0) = 0 & S''(0) = 0 & S'''(0) = 0 \\ T(0) = 0 & T'(0) = 1 & T''(0) = 0 & T'''(0) = 0 \\ U(0) = 0 & U'(0) = 0 & U''(0) = 1 & U'''(0) = 0 \\ V(0) = 0 & V'(0) = 0 & V''(0) = 0 & V'''(0) = 1 \end{array} \tag{14.24}$$

2. Krylov–Duncan functions and their derivatives satisfy to circular permutations (Fig. 14.14, Table 14.4).

These properties of the functions (14.22) may be effectively used for deriving of the frequency equation and mode shape of free vibration.

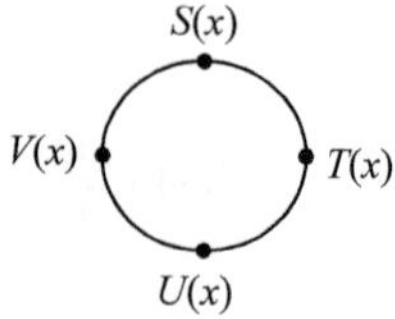

Fig. 14.14 Krylov-Duncan functions circular permutations

Table 14.4 Krylov-Duncan functions and their derivatives

Function	First derivative	Second derivative	Third derivative	Fourth derivative
$S(x)$	$kV(x)$	$k^2\,U(x)$	$k^3\,T(x)$	$k^4\,S(x)$
$T(x)$	$kS(x)$	$k^2\,V(x)$	$k^3\,U(x)$	$k^4\,T(x)$
$U(x)$	$kT(x)$	$k^2\,S(x)$	$k^3\,V(x)$	$k^4\,U(x)$
$V(x)$	$kU(x)$	$k^2\,T(x)$	$k^3\,S(x)$	$k^4\,V(x)$

To obtain frequency equation using Krylov–Duncan functions, the following algorithm is recommended:

Step 1. Represent the mode shape in the form that satisfies boundary conditions at $x = 0$. This expression will have only two Krylov–Duncan functions and, respectively, two constants. The decision of what Krylov–Duncan functions to use is based on (14.24) and the boundary condition at $x = 0$.

Step 2. Determine constants using the boundary condition at $x = l$ and Table 14.4. Thus, the system of two homogeneous algebraic equations is obtained.

Step 3. The nontrivial solution of this system represents a frequency equation.

Example 14.6. The beam has length l, mass per unit length m, modulus of elasticity E, and moment of inertia of cross-sectional area I (Fig. 14.15). Calculate the frequency of vibration and find the mode of vibration.

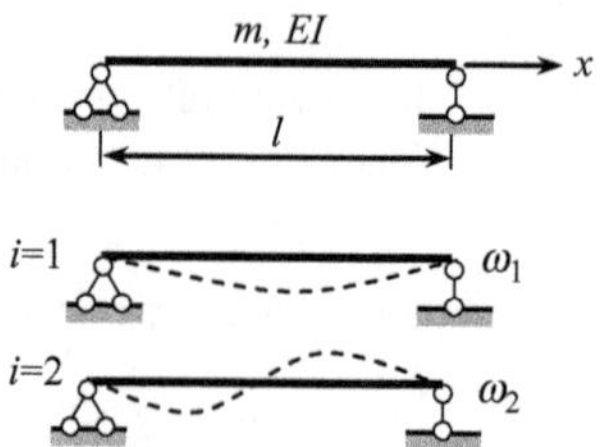

Fig. 14.15 Design diagram for simply supported beam and mode shapes vibration for $i = 1$ and $i = 2$

Solution. At the left end ($x = 0$) deflection and the bending moment are zero:

$$1.\ X(0) = 0,$$
$$2.\ X''(0) = 0.$$

At $x = 0$ the Krylov–Duncan functions and their second derivatives equal zero. According to properties (14.24), only $T(kx)$ and $V(kx)$ functions satisfy these conditions. So, the expression for the mode shape is

$$X(x) = C_2 T(kox) + C_4 V(kx).$$

Constants C_2 and C_4 are calculated from boundary conditions at $x = l$: $X(l) = 0;\ X''(l) = 0$

$$\begin{aligned} X(l) &= C_2 T(kl) + C_4 V(kl) = 0, \\ X''(l) &= k^2 [C_2 V(kl) + C_4 T(kl)] = 0. \end{aligned} \tag{a}$$

A nontrivial solution of the above system is the frequency equation

$$\begin{vmatrix} T(kl) & V(kl) \\ V(kl) & T(kl) \end{vmatrix} = 0 \rightarrow T^2(kl) - V^2(kl) = 0.$$

According (14.22), the last formula may be presented as $\sin kl = 0$. The roots of the equation are

$$\lambda = kl = \pi,\ 2\pi, \ldots$$

So the frequencies of the free vibration are

$$\omega_i = \frac{\lambda_i^2}{l^2}\sqrt{\frac{EI}{m}}, \quad \omega_1 = \frac{\pi^2}{l^2}\sqrt{\frac{EI}{m}}, \qquad \omega_2 = \frac{4\pi^2}{l^2}\sqrt{\frac{EI}{m}}, \ldots$$

The mode shape of vibration is

$$X(x) = C_2 T(kx) + C_4 V(kx) = C_2 \left[T(k_i x) + \frac{C_4}{C_2} V(k_i x) \right]$$

Since the ratio C_4/C_2 from first and second equations (a) are

$$\frac{C_4}{C_2} = -\frac{T(k_i l)}{V(k_i l)} = -\frac{V(k_i l)}{T(k_i l)},$$

then i-th mode shape (eigenfunction) corresponding to i-th frequency of vibration (eigenvalue) is

$$\begin{aligned} X(x) &= C_2 T(kx) + C_4 V(kx) = C \left[T(k_i x) - \frac{T(k_i l)}{V(k_i l)} V(k_i x) \right] \\ &= C \left[T(k_i x) - \frac{V(k_i l)}{T(k_i l)} V(k_i x) \right], \end{aligned} \tag{b}$$

Since the Krylov–Duncan functions $T(\pi) = V(\pi),\ T(2\pi) = V(2\pi), \ldots$ so the mode shapes are

$$X_i(x) = C[T(k_i x) - V(k_i x)] = C \sin k_i x = C \sin \frac{i\pi}{l} x, \quad i = 1, 2, \ldots \tag{c}$$

The first and second modes are shown in Fig. 14.15.

Fundamental data for one-span uniform beams with classical boundary conditions are presented in Table A.26. This table contains the frequency equation, and the first, second and third eigenvalues. For the nodal points of the mode shapes of a free vibration, the origin is placed on the left end of the beam.

Problems

14.1. Statically determinate beam carried one lumped mass. Determine the frequency of free vibration.

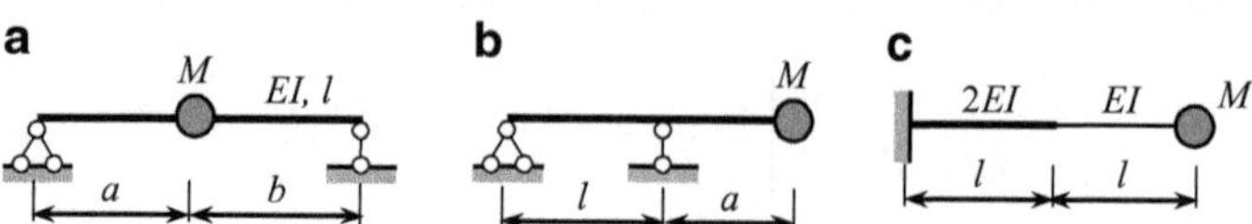

Fig. P14.1

$$\text{Ans. (a) } \omega = \sqrt{\frac{3lEI}{a^2b^2M}}; \text{ (c) } \omega = \sqrt{\frac{2EI}{3l^3M}}.$$

14.2. Statically indeterminate beam carried one lumped mass (Fig. P14.2). Determine the frequency of free vibration.

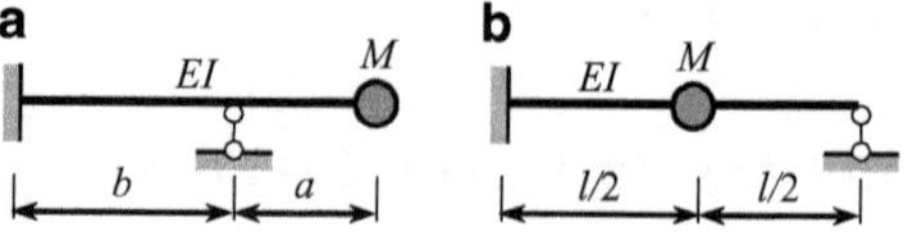

Fig. P14.2

$$\text{Ans. (a) } \omega = \sqrt{\frac{12EI}{a^2\,(3b+4a)\,M}}; \text{ (b) } \omega = \sqrt{\frac{768EI}{7l^3M}}.$$

14.3. Symmetrical frame with absolutely rigid cross bar of total mass M is shown in Fig. P14.3. Find the frequency of free horizontal vibration.

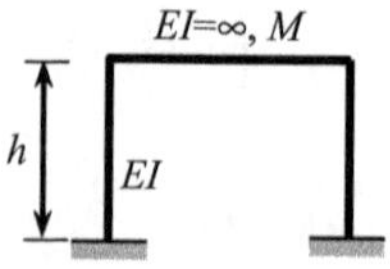

Fig. P14.3

$$\text{Ans. } \omega = \sqrt{\frac{24EI}{h^3M}}$$

14.4. Symmetrical statically indeterminate frame carried one lumped mass. Determine the frequency of free horizontal and vertical vibrations.

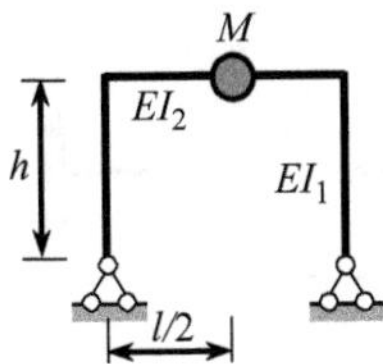

Fig. P14.4

Ans. $\omega_{\text{hor}} = \sqrt{\dfrac{12EI_2}{h^3M\,(2\beta + k)}}, \; \beta = \dfrac{I_2}{I_1}, \; k = \dfrac{l}{h}; \quad \omega_{\text{vert}} = \sqrt{\dfrac{48EI_2}{l^3M} \cdot \dfrac{12k + 8\beta}{3k + 8\beta}}$

14.5. Calculate the frequency of free vertical vibration of symmetrical truss presented in Fig. P14.5. Axial rigidity for diagonals and vertical elements *EA*, for lower and top chords is 2*EA*. A lumped mass M is placed at joint 6.

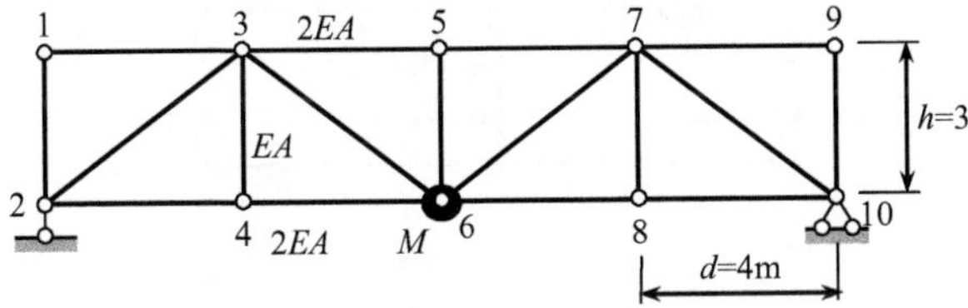

Fig. P14.5

Ans. $\omega = \sqrt{\dfrac{1}{m\delta_{11}}} = \sqrt{\dfrac{EA}{24.555M}} \; \left(s^{-1}\right).$

14.6. A massless beam carried two lumped mass (Fig. P14.6). Calculate the frequency of free vibration, find the modal matrix and show the shapes of vibrations.

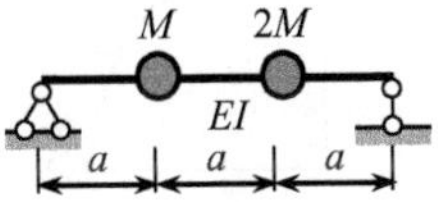

Fig. P14.6

Ans. $\omega_1 = 0.8909\sqrt{\dfrac{EI}{a^3M}}, \; \omega_2 = 3.6887\sqrt{\dfrac{EI}{a^3M}}, \; \Phi = \begin{bmatrix} 1 & 1 \\ 1.0483 & -0.4769 \end{bmatrix}$

14.7. A massless beam carried body of mass M and moment of inertia J_x with respect to x-axis (Fig. P14.7); radius of gyration is $\rho = J_x/Ml^2$. Calculate the frequencies of the free vibration and find the modal matrix.

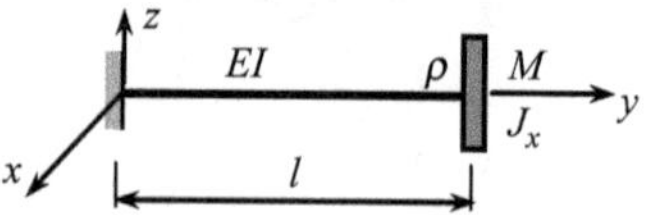

Fig. P14.7

Ans. $\omega_1^2 = \dfrac{3EI}{\lambda_1 Ml^3}, \omega_2^2 = \dfrac{3EI}{\lambda_2 Ml^3}, \lambda_{1,2} = \dfrac{1 + 3\rho \pm \sqrt{1 + 3\rho + 9\rho^2}}{2}$

14.8. A massless beam carried two lumped mass (Fig. P14.8). Calculate the frequencies of free vibrations and find the modal matrix. (Hint: For construction of bending moment diagram in unit states use the influence lines).

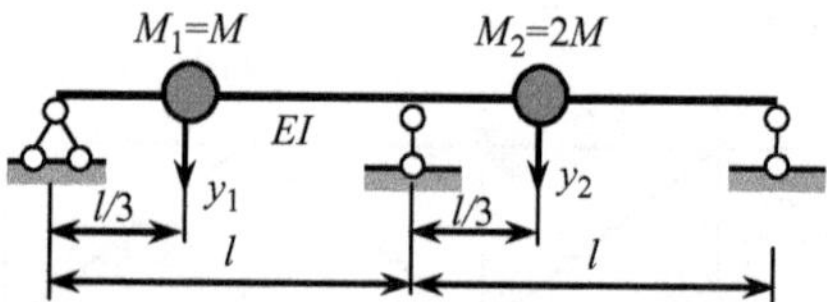

Fig. P14.8

Ans. $\omega_1 = 6.379\sqrt{\dfrac{EI}{l^3 M}}, \omega_2 = 11.047\sqrt{\dfrac{EI}{l^3 M}}, \Phi = \begin{bmatrix} 1 & 1 \\ -1.4142 & 0.3462 \end{bmatrix}$

14.9. A uniform circular rod is clamped at point $B$and carrying the lumped mass M at free end as shown in Fig. P14.9. Calculate the frequencies of free vibrations, and find the modal matrix and show the shapes of vibrations.

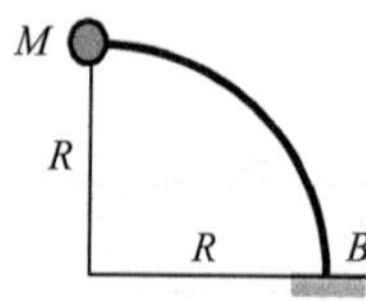

Fig. P14.9

$$\text{Ans. } \omega_i^2 = \frac{EI}{\lambda_i \cdot 0.7854MR^3},$$

$$\lambda_1 = 1.4195, \quad \lambda_2 = 0.03395 \quad \Phi = \begin{bmatrix} 1 & 1 \\ 0.659 & -1.517 \end{bmatrix}$$

14.10. Symmetrical frame with absolutely rigid cross bars of total mass M and $2M$ and massless columns is shown in Fig. P14.10. Calculate the frequencies of free vibrations, and find the modal matrix and show the shapes of vibrations.

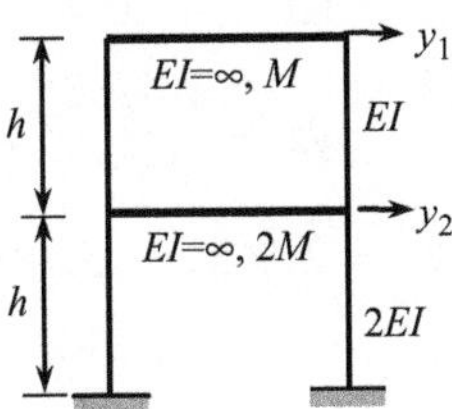

Fig. P14.10

$$\text{Ans. } \omega_1 = 3.4641\sqrt{\frac{EI}{h^3M}}, \omega_2 = 6.9282\sqrt{\frac{EI}{h^3M}}, \Phi = \begin{bmatrix} 1 & 1 \\ 0.5 & -1 \end{bmatrix}$$

14.11. Symmetrical frame with absolutely rigid cross bars of total mass M and $2M$ and massless columns is shown in Fig. P14.11. Find the frequencies of free vibrations.

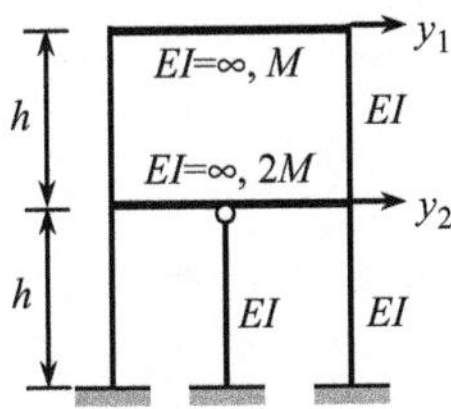

Fig. P14.11

$$\text{Ans. } \omega_1^2 = 7.7628\frac{EI}{h^3M}, \omega_2^2 = 41{,}7366\frac{EI}{h^3M}, \Phi = \begin{bmatrix} 1 & 1 \\ 0.6765 & -0.7390 \end{bmatrix}$$

14.12. Derive the frequency equation, calculate the frequencies of vibration and find the mode shape of vibrations for beam (length l, flexural stiffness EI, mass per unit length m) with following classical boundary conditions: (a) Fixed-fixed beam; (b) Clamped-pinned beam; (c) Guided-clamped beam

$$\omega_i = \frac{\lambda_i^2}{l^2}\sqrt{\frac{EI}{m}}.$$

Ans.

(a) $U^2(kl) - T(kl)V(kl) = 0 \to \cosh kl \cos kl = 1 \to \lambda = kl = 4.7300; \ldots$

$$\omega_i = \frac{\lambda_i^2}{l^2}\sqrt{\frac{EI}{m}}, \omega_1 = \frac{4.73^2}{l^2}\sqrt{\frac{EI}{m}}, \ldots,$$

$$X_i(x) = C\left[U(k_i x) - \frac{U(k_i l)}{V(k_i l)} V(k_i x)\right] = C \cdot \left[U(k_i x) - \frac{T(k_i l)}{U(k_i l)} V(k_i x)\right]$$

(b) $\tan kl = \tanh kl \to kl = 3.92266; \quad 7.0768; \quad 10.2102, \ldots$

$$\omega_1 = \frac{3.9266^2}{l^2}\sqrt{\frac{EI}{m}}, \quad \omega_2 = \frac{7.0768^2}{l^2}\sqrt{\frac{EI}{m}}, \quad \ldots$$

(c) $\tan kl + \tanh kl = 0 \to kl = 2.3650; \quad 5.4987, ..$

Appendix

Table A.1 Typical graphs, their area and centroid position

No	Shape of the graph	Area Ω	Position of the centroid	
			z_1	z_2
1	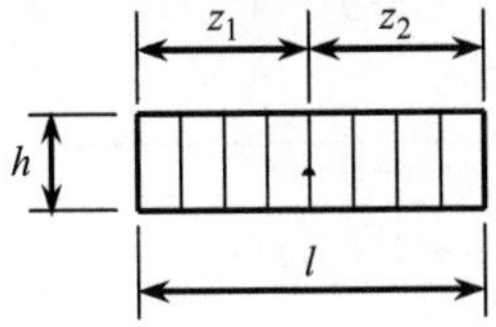 Rectangle (h, l)	hl	$\frac{l}{2}$	$\frac{l}{2}$
2	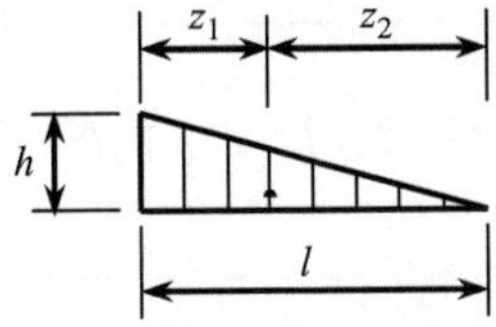 Triangle (h, l)	$\frac{hl}{2}$	$\frac{l}{3}$	$\frac{2l}{3}$
3	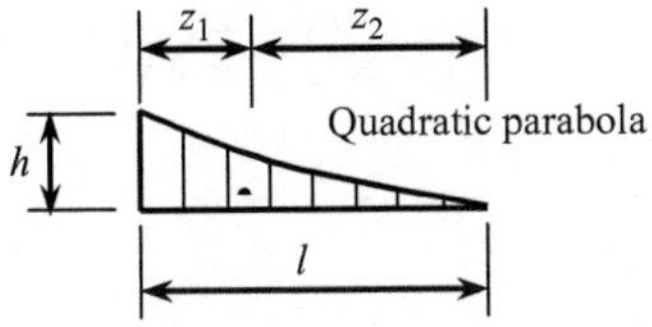 Quadratic parabola	$\frac{hl}{3}$	$\frac{l}{4}$	$\frac{3l}{4}$
4	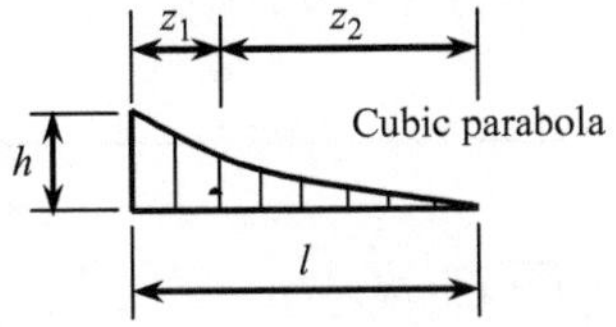 Cubic parabola	$\frac{hl}{4}$	$\frac{l}{5}$	$\frac{4l}{5}$
5	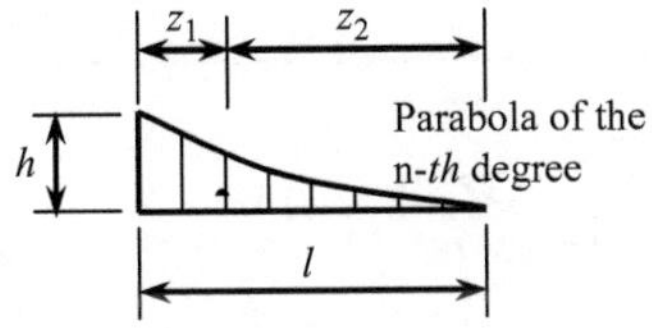 Parabola of the n-*th* degree	$\frac{hl}{n+1}$	$\frac{l}{n+2}$	$\frac{(n+1)\,l}{n+2}$
6	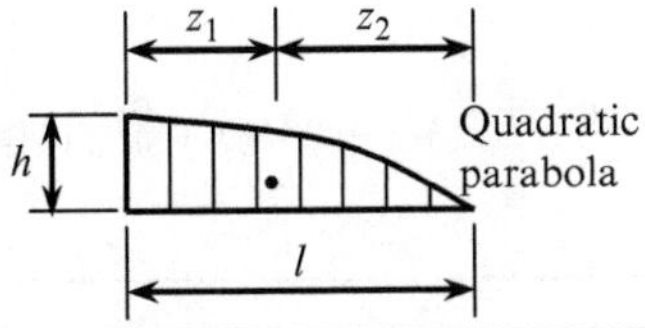 Quadratic parabola	$\frac{2hl}{3}$	$\frac{3l}{8}$	$\frac{5l}{8}$

Table A.2 Multiplication of two bending moment diagrams ($EI = \text{const}$)

No.	Diagram	$\int_0^l \overline{M} M \, dx$
	$\overline{M}_A$, $\overline{M}_B$, l	$\int_0^l \overline{M} M \, dx$
1	M_A, M_B, l	$\frac{l}{6}\left(\overline{M}_A M_B + \overline{M}_B M_A + 2\overline{M}_A M_A + 2\overline{M}_B M_B\right)$
2	M_A, l	$\frac{l}{2}\left(\overline{M}_A + \overline{M}_B\right) M_A$
3	M_A, l	$\frac{l}{6}\left(\overline{M}_B + 2\overline{M}_A\right) M_A$
4	M_A, a, l	$\frac{a}{6}\left[\overline{M}_A\left(3 - \frac{a}{l}\right) + \overline{M}_B \frac{a}{l}\right] M_A$
5	M_C, a, l	$\frac{l}{6}\left[\overline{M}_A\left(2 - \frac{a}{l}\right) + \overline{M}_B\left(1 + \frac{a}{l}\right)\right] M_C$
6	M_C, M_C, a, a, l	$\frac{l-a}{2}\left[\overline{M}_A + \overline{M}_B\right] M_C$
7	M_C, Quadratic parabola, $l/2$, l	$\frac{l}{3}\left[\overline{M}_A + \overline{M}_B\right] M_C$
8	M_A, Quadratic parabola, l	$\frac{l}{12}\left[3\overline{M}_A + \overline{M}_B\right] M_A$
9	M_C, Quadratic parabola, M_A, M_B, $l/2$, $l/2$	$\frac{l}{6}\left[\overline{M}_A (M_A + 2M_C) + \overline{M}_B (M_B + 2M_C)\right]$

Tabulated data for standard uniform beams:

Table A.3 and A.4 contain the reactions at supports for uniform fixed–pinned (rolled) and fixed–fixed beams subjected to different exposures. Table A.5 and A.6 present the reactions at supports for beams with sliding and elastic supports, respectively. These tables show the actual direction for reactive forces and moments. The bending moment diagrams are traced on the side of the extended fibers. Parameter $i = EI/l$ is the bending stiffness per unit length. Parameter u is related to the fixed support, h denotes the depth of the cross section and α is the coefficient of thermal expansion.

The tables for the reactions of single-span beams with stepped flexural stiffness are presented Table A.7 and A.8

Table A.3 Reactions of fixed-pinned beams

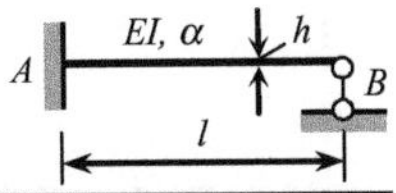

No	Loading conditions	Reactions and bending moment diagrams	Expressions for bending moments and reactions
1	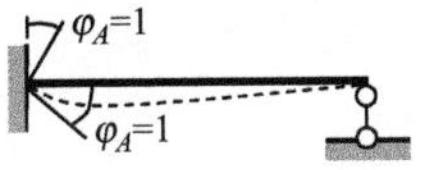 $\varphi_A=1$; $\varphi_A=1$	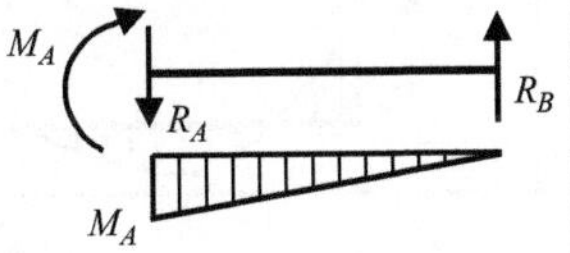 M_A, R_A, R_B, M_A	$M_A = \frac{3EI}{l} = 3i,\quad i = \frac{EI}{l}$ $R_A = R_B = \frac{3EI}{l^2} = \frac{3i}{l}$
2	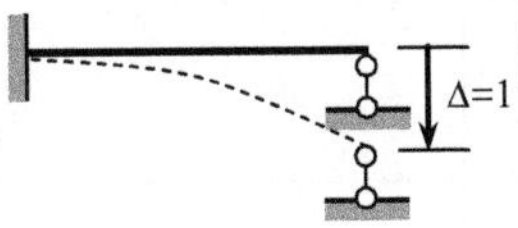 $\Delta=1$	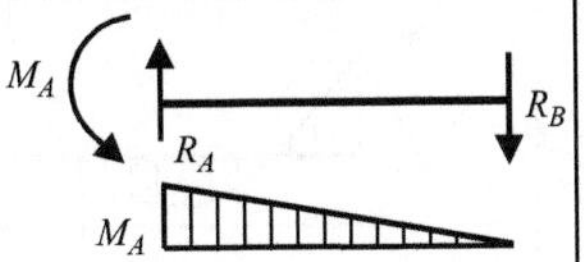 M_A, R_A, R_B, M_A	$M_A = \frac{3EI}{l^2} = \frac{3i}{l}$ $R_A = R_B = \frac{3EI}{l^3} = \frac{3i}{l^2}$
3	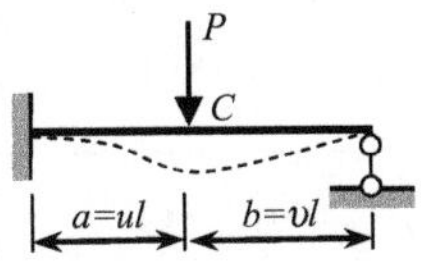 P, C, $a=ul$, $b=\upsilon l$, $(u+\upsilon=1)$	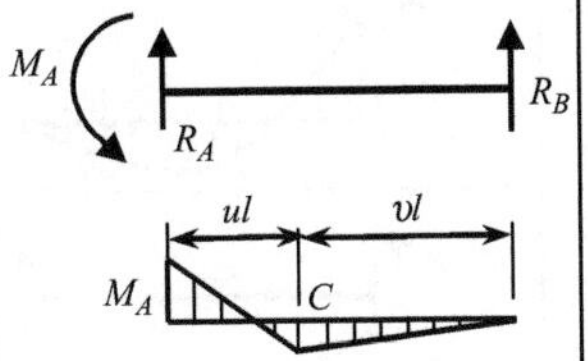 M_A, R_A, R_B, ul, υl, M_A, C	$M_A = \frac{Pl}{2}\upsilon(1-\upsilon^2)$ $M_C = \frac{Pl}{2}u^2\upsilon(3-u)$ $R_A = \frac{P\upsilon}{2}(3-\upsilon^2)$ $R_B = \frac{Pu^2}{2}(3-u)$ $u = \upsilon = 0.5$ $R_A = \frac{11}{16}P;\ R_B = \frac{5}{16}P$ $M_A = \frac{3}{16}Pl;\ M_C = \frac{5}{32}Pl$

(continued)

Table A.3 (continued)

No	Loading conditions	Reactions and bending moment diagrams	Expressions for bending moments and reactions
4	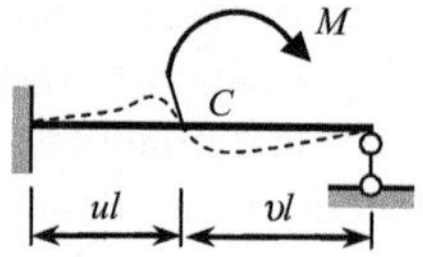 M, C, ul, υl, $(u+\upsilon=1)$	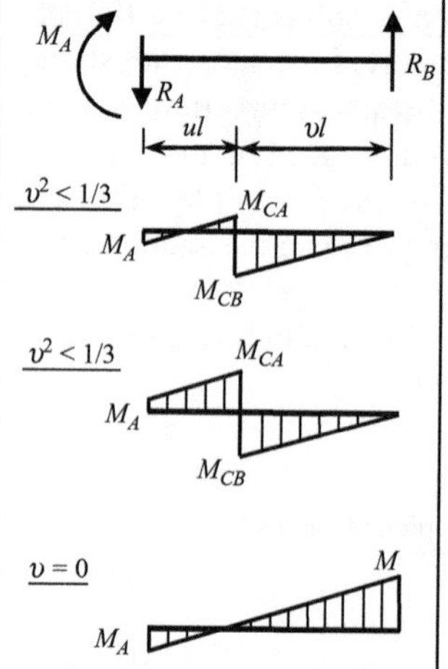 M_A, R_A, R_B, ul, υl; $\upsilon^2 < 1/3$: M_A, M_{CA}, M_{CB}; $\upsilon^2 < 1/3$: M_A, M_{CA}, M_{CB}; $\upsilon = 0$: M_A, M	$M_A = \frac{M}{2}(1-3\upsilon^2)$ $M_{CB} = \frac{3M}{2}\upsilon(1-\upsilon^2)$ $R_A = R_B = \frac{3M}{2l}(1-\upsilon^2)$ $u = \upsilon = 0.5:$ $M_A = \frac{M}{8}$ $R_A = R_A = \frac{9}{8}\frac{M}{l}$ $M_{CA} = \frac{7}{16}M$; $M_{CB} = \frac{9}{16}M$ $\upsilon = 0:$ $M_A = M/2$ $R_A = R_B = \frac{3M}{2l}$
5	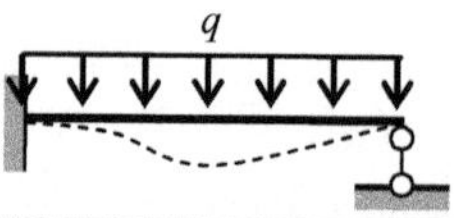 q	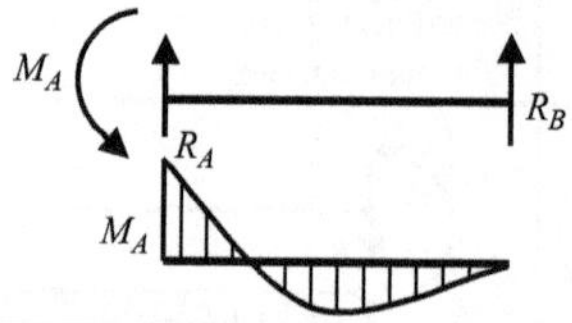 M_A, R_A, R_B, M_A	$M_A = \frac{ql^2}{8}$, $M\left(\frac{l}{2}\right) = \frac{ql^2}{16}$ $R_A = \frac{5}{8}ql$; $R_B = \frac{3}{8}ql$
6	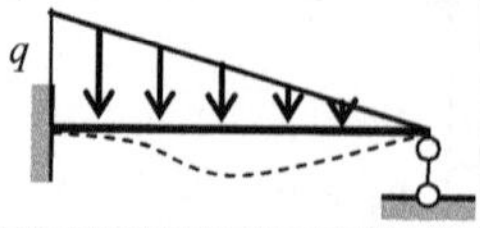 q	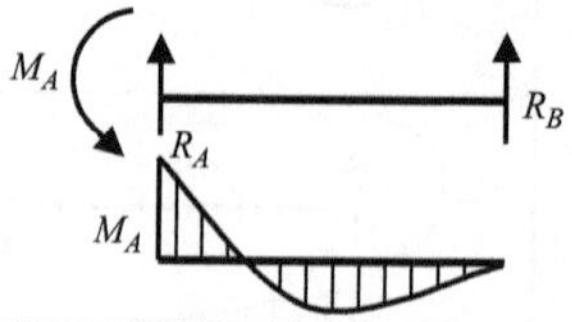 M_A, R_A, R_B, M_A	$M_A = \frac{ql^2}{15}$, $R_A = \frac{2}{5}ql$, $R_B = \frac{1}{10}ql$
7	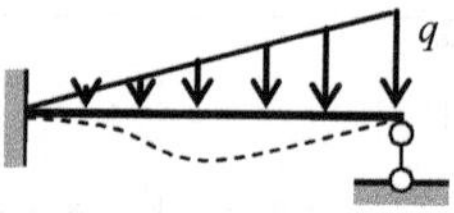 q	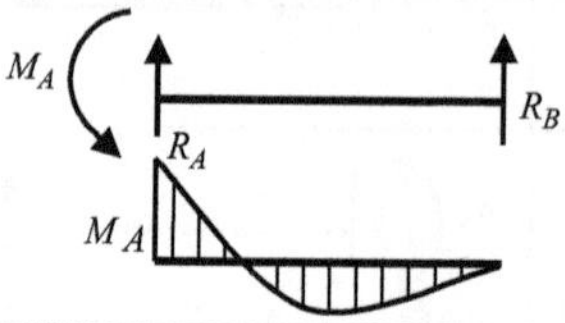 M_A, R_A, R_B, M_A	$M_A = \frac{7ql^2}{120}$, $R_A = \frac{9}{40}ql$, $R_B = \frac{11}{40}ql$
8	Temperature gradient 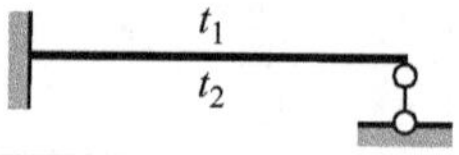 t_1 t_2	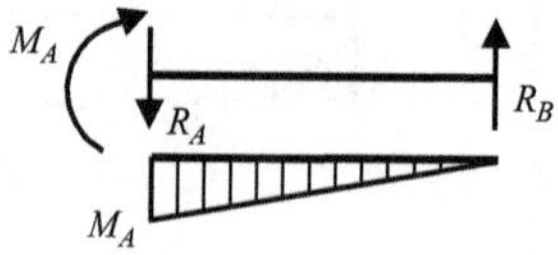 M_A, R_A, R_B, M_A	$M_A = \frac{3EI\alpha(t_1-t_2)}{2h}$ $R_A = R_B = \frac{3EI\alpha(t_1-t_2)}{2hl}$
9	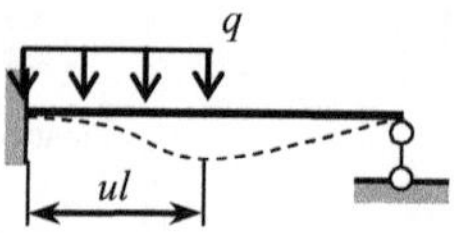 q, ul	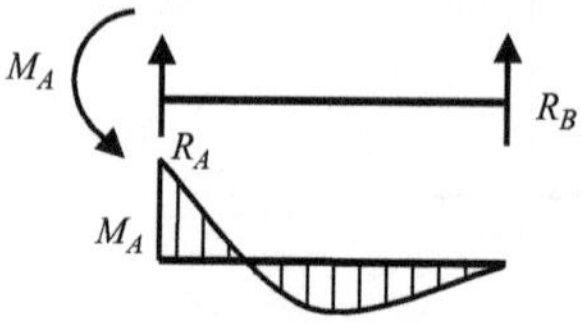 M_A, R_A, R_B, M_A	$M_A = \frac{ql^2}{8}u^2(2-u)^2 = ql^2k_1$, $R_B = \frac{qa}{2}u - qlk_1, R_A = qa - R_B$

u	k_1	u	k_1
0.2	0.0162	0.6	0.0882
0.4	0.0512	0.8	0.1152
0.5	0.0703	1.0	**0.1250**

Table A.4 Reactions of fixed–fixed beams

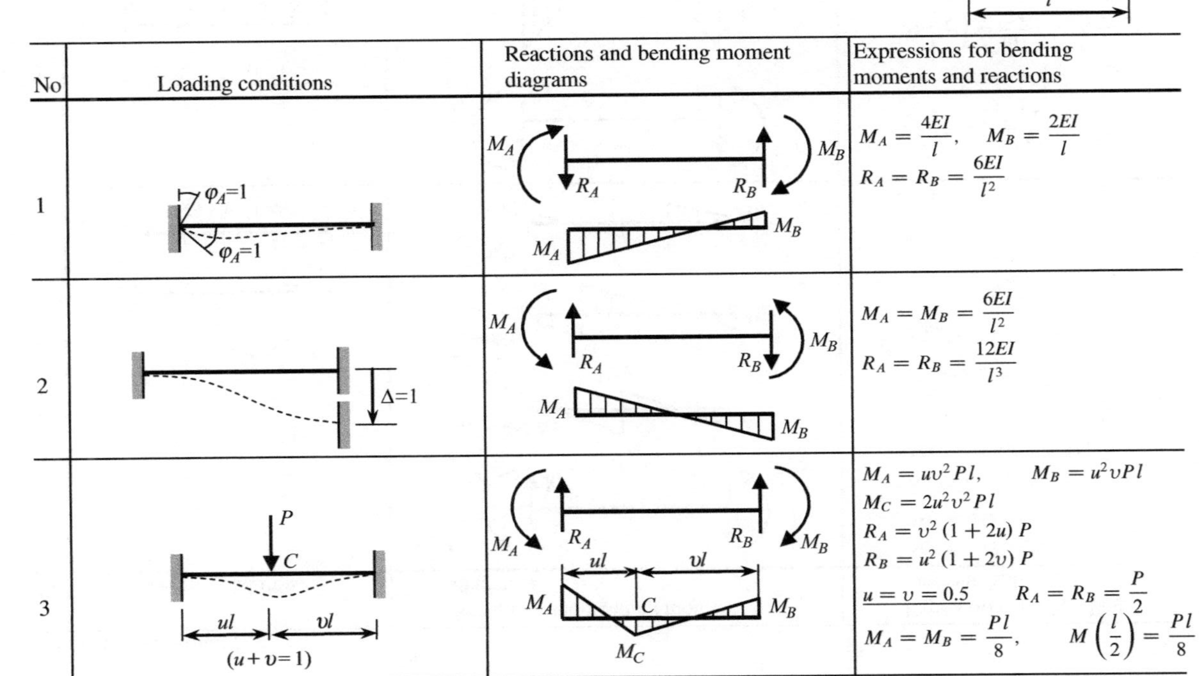

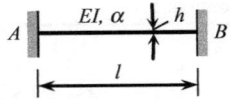

No	Loading conditions	Reactions and bending moment diagrams	Expressions for bending moments and reactions
1			$M_A = \frac{4EI}{l}$, $M_B = \frac{2EI}{l}$ $R_A = R_B = \frac{6EI}{l^2}$
2			$M_A = M_B = \frac{6EI}{l^2}$ $R_A = R_B = \frac{12EI}{l^3}$
3			$M_A = u\upsilon^2 Pl$, $M_B = u^2\upsilon Pl$ $M_C = 2u^2\upsilon^2 Pl$ $R_A = \upsilon^2 (1 + 2u) P$ $R_B = u^2 (1 + 2\upsilon) P$ $\underline{u = \upsilon = 0.5}$ $R_A = R_B = \frac{P}{2}$ $M_A = M_B = \frac{Pl}{8}$, $M\left(\frac{l}{2}\right) = \frac{Pl}{8}$

Table A.4 (continued)

No	Loading conditions	Reactions and bending moment diagrams	Expressions for bending moments and reactions
4	M C ul υl $(u+\upsilon=1)$	M_A R_A R_B M_A $1/3 < u < 2/3$ M_{CA} M_A M_{CB} M_B $\upsilon > 2/3$ M_A M_B $u > 2/3$ M_A M_B	$M_A = \upsilon\,(2-3\upsilon)\,M$ $M_B = u\,(2-3u)\,M$ $R_A = R_B = 6u\upsilon\dfrac{M}{l}$ $M_{CA} = -R_A ul + M_A$ <u>$u = 0.5$:</u> $M_A = M_B = \dfrac{M}{4}$ $R_A = R_B = \dfrac{3}{2}\dfrac{M}{l}$ $M_{CA} = M_{CB} = \dfrac{M}{2}$
5	q	M_A R_A R_B M_B M_A M_B	$M_A = M_B = \dfrac{ql^2}{12}$ $R_A = R_B = \dfrac{1}{2}ql$ $M\left(\dfrac{l}{2}\right) = \dfrac{ql^2}{24}$

(continued)

Table A.4 (continued)

6	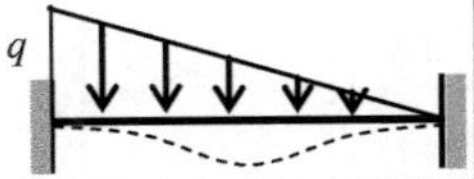 q	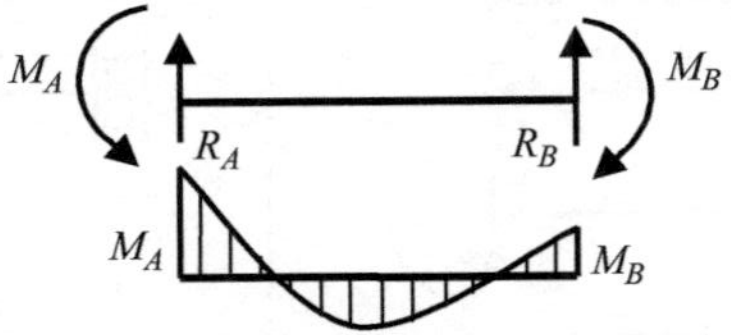 M_A, M_B, R_A, R_B; M_A, M_B	$M_A = \frac{ql^2}{20}, \qquad M_B = \frac{ql^2}{30}$ $R_A = \frac{7}{20}ql, \qquad R_B = \frac{3}{20}ql$
7	Temperature gradient 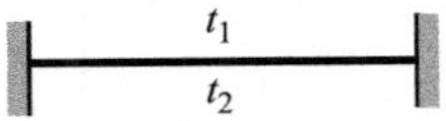 t_1 t_2	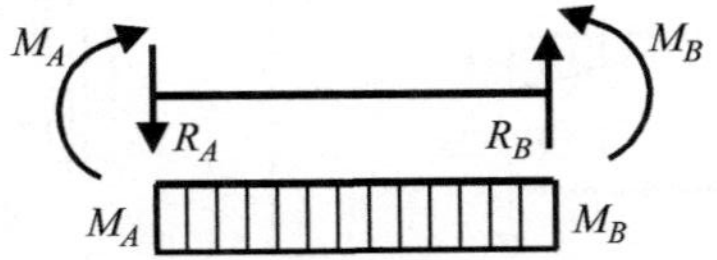 M_A, M_B, R_A, R_B; M_A, M_B	$M_A = \frac{EI\,\alpha\,(t_1 - t_2)}{h}$ $R_A = R_B = 0$
8	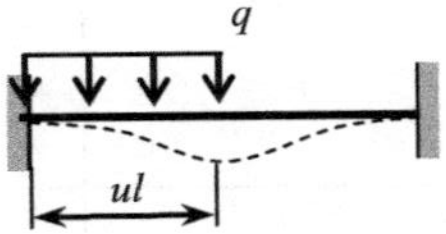 q ul	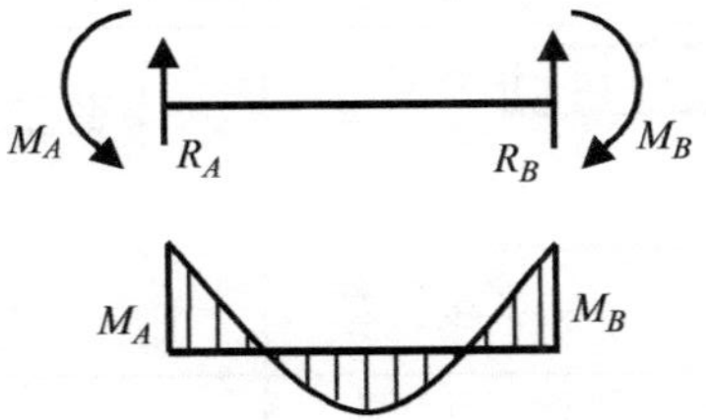 M_A, M_B, R_A, R_B; M_A, M_B	$M_A = \frac{ql^2}{6}u^2\left(3 - 4u + \frac{3}{2}u^2\right) = ql^2k_1$ $M_B = \frac{ql^2}{3}u^2\left(u - \frac{3}{4}u^2\right) = ql^2k_2$

u	k_1	k_1
0.2	0.0151	0.0023
0.4	0.0437	0.0149
0.5	0.0573	0.0260
0.6	0.0684	0.0396
0.8	0.0811	0.0683
1.0	**0.0833**	**0.0833**

Table A.5 Reactions of fixed-sliding beams

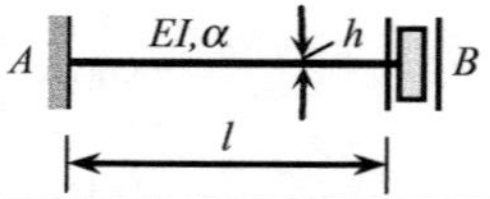

A — EI, α — h — B; l

No	Loading conditions	Reactions and bending moment diagrams	Expressions for bending moments and reactions
1	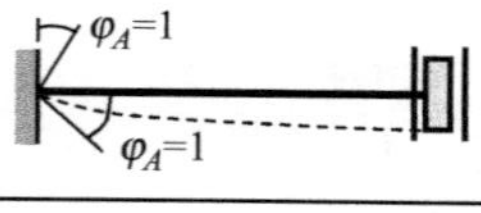 $\varphi_A=1$, $\varphi_A=1$	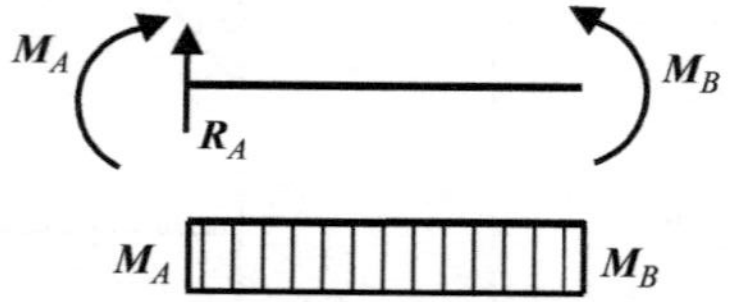 M_A, R_A, M_B; M_A, M_B	$M_A = M_B = \frac{EI}{l}$ $R_A = 0$
2[a]	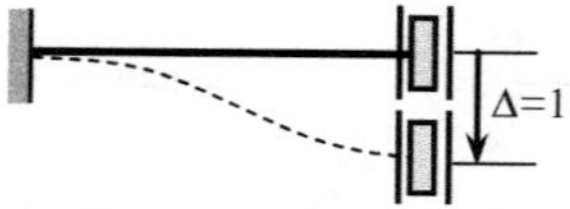 $\Delta=1$	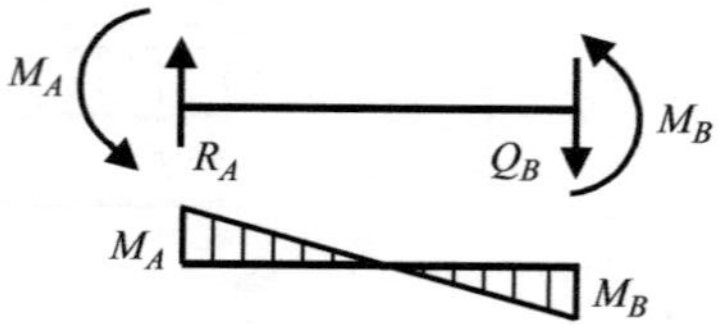 M_A, R_A, Q_B, M_B; M_A, M_B	$M_A = M_B = \frac{6EI}{l^2}$ $R_A = Q_B = \frac{12EI}{l^3}$
3	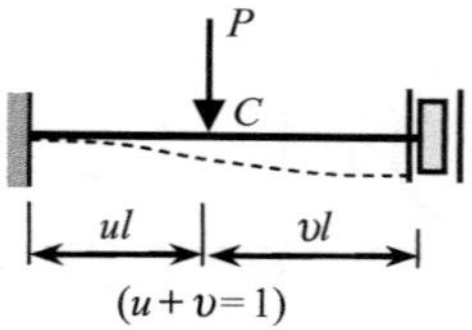 P, C; ul, υl; $(u+\upsilon=1)$	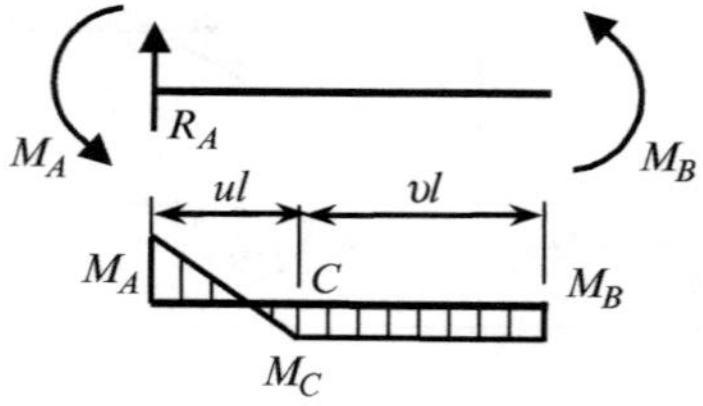 M_A, R_A, M_B; ul, υl, M_A, C, M_B, M_C	$M_A = \frac{P\,l\,u\,(2-u)}{2}$ $M_B = \frac{P\,l\,u^2}{2}$ $M_C = M_B$ $R_A = P$

(continued)

Table A.5 (continued)

No	Loading conditions	Reactions and bending moment diagrams	Expressions for bending moments and reactions
4	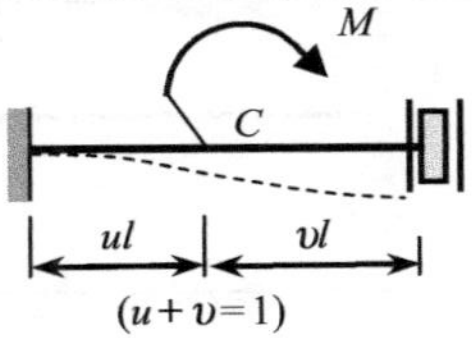 M, C, ul, vl, $(u+v=1)$	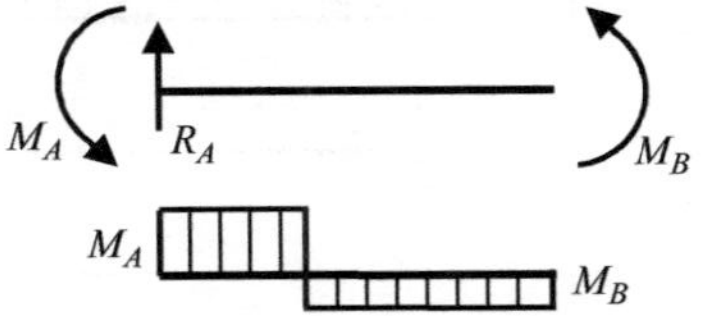 M_A, R_A, M_B	$M_A = vM$ $M_B = uM$ $R_A = 0$
5	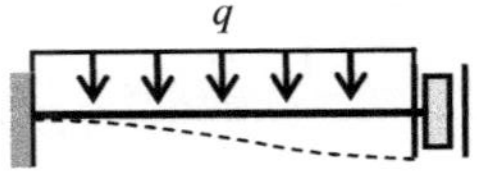 q	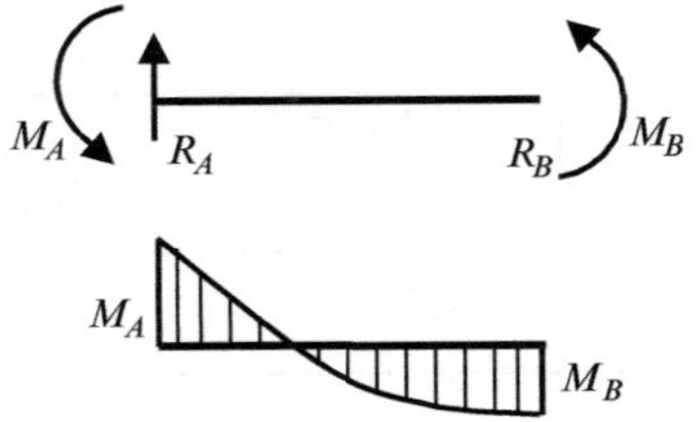 M_A, R_A, R_B, M_B	$M_A = \frac{ql^2}{3}$ $M_B = \frac{ql^2}{6}$ $R_A = ql$ $M\left(\frac{l}{2}\right) = \frac{ql^2}{24}$
6	Temperature gradient 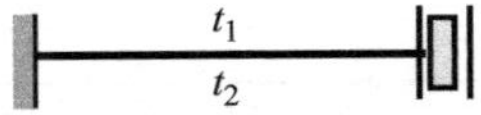 t_1 t_2	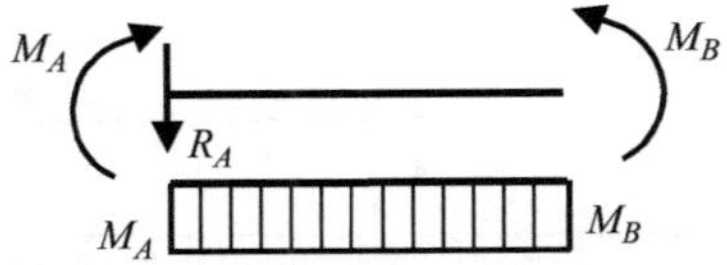 M_A, R_A, M_B	$M_A = M_B = \frac{EI\,\alpha\,(t_1 - t_2)}{h}$ $R_A = 0$

[a] Q_B is the force that should be applied to obtain vertical displacement $\Delta = 1$

Table A.6 Reactions of uniform beams with elastic supports (k and k_{rot} are the stiffnesses of the supports)

No	Loading conditions	Reactions and bending moment diagrams	Expressions for bending moments and reactions
1	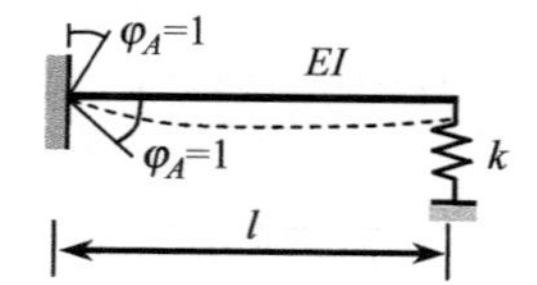 $\varphi_A=1$, EI, $\varphi_A=1$, k, l	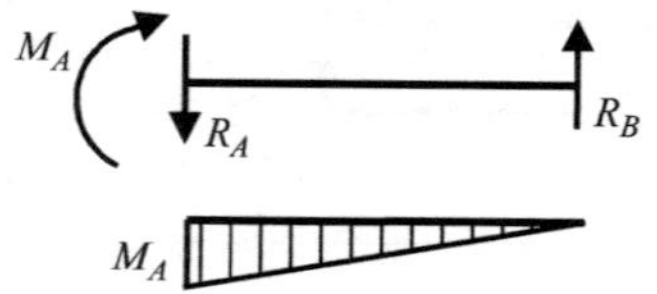 M_A, R_A, R_B, M_A	$M_A = \frac{3EI}{l}\frac{1}{1+\frac{3EI}{kl^3}}$ $R_A = R_B = \frac{3EI}{l^2}\frac{1}{1+\frac{3EI}{kl^3}}$
2	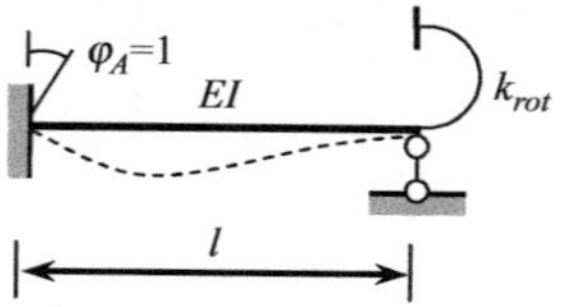 $\varphi_A=1$, EI, k_{rot}, l	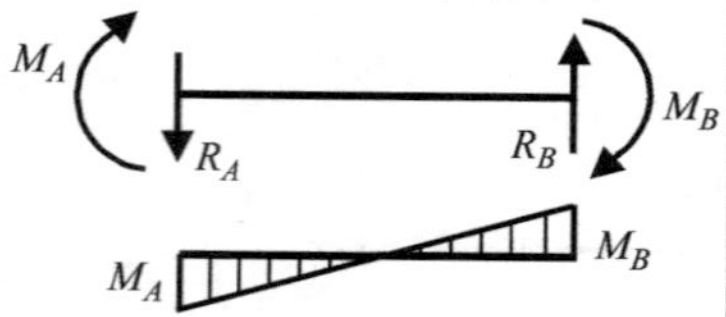 M_A, R_A, R_B, M_B, M_A, M_B	$M_A = \frac{3EI}{l}\frac{1+\frac{k_{\text{rot}}l}{3EI}}{1+\frac{k_{\text{rot}}l}{4EI}}$ $R_A = \frac{6EI}{l^2}\left(1-2\frac{3+\frac{k_{\text{rot}}l}{EI}}{4+\frac{k_{\text{rot}}l}{EI}}\right)$
3	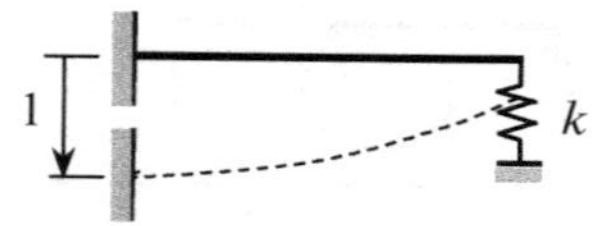 1, k	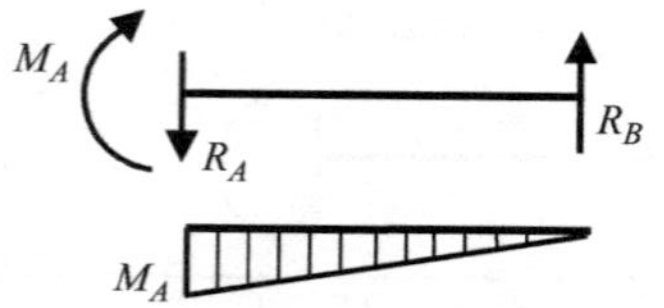 M_A, R_A, R_B, M_A	$M_A = \frac{3EI}{l^2}\frac{1}{1+\frac{3EI}{kl^3}}$ $R_A = R_B = \frac{3EI}{l^3}\frac{1}{1+\frac{3EI}{kl^3}}$

(continued)

Table A.6 (continued)

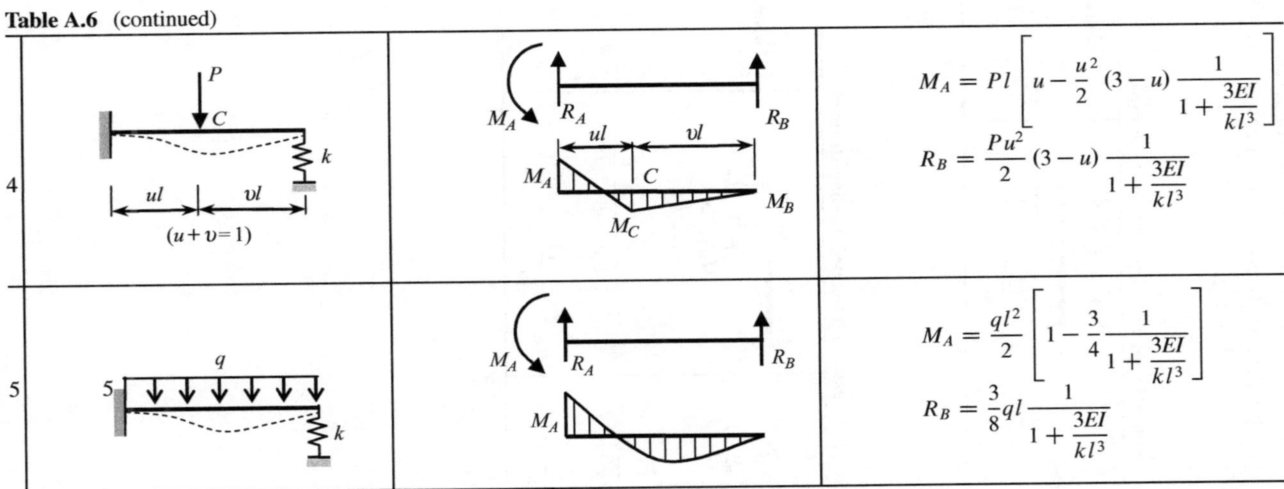

4			$M_A = Pl\left[u - \frac{u^2}{2}(3-u)\frac{1}{1+\frac{3EI}{kl^3}}\right]$ $R_B = \frac{Pu^2}{2}(3-u)\frac{1}{1+\frac{3EI}{kl^3}}$
5			$M_A = \frac{ql^2}{2}\left[1 - \frac{3}{4}\frac{1}{1+\frac{3EI}{kl^3}}\right]$ $R_B = \frac{3}{8}ql\frac{1}{1+\frac{3EI}{kl^3}}$

Unit reactions of nonuniform standard members:

Table A.7 and A.8 present the reactive moments for fixed–pinned and fixed–fixed elements with stepped bending stiffness. For both cases $\mu = (EI_1/EI_2) - 1,\ a = 1 + \varepsilon\mu,\ b = 1 + \varepsilon^2\mu,\ c = 1 + \varepsilon^3\mu,\ f = 1 + \varepsilon^4\mu$

Table A.7 Fixed–pinned beam with stepped bending stiffness

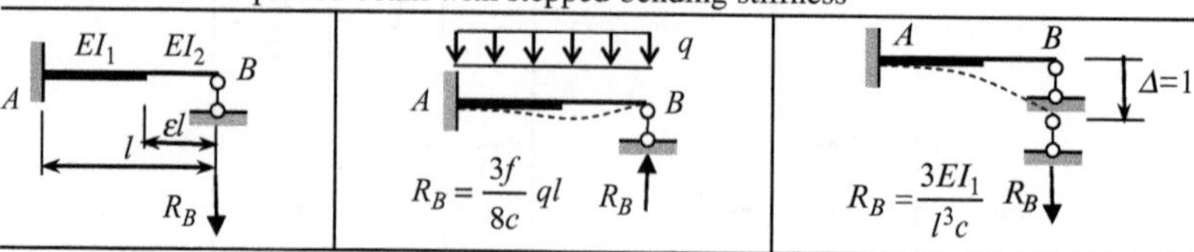

Table A.8 Fixed–fixed beam with stepped bending stiffness

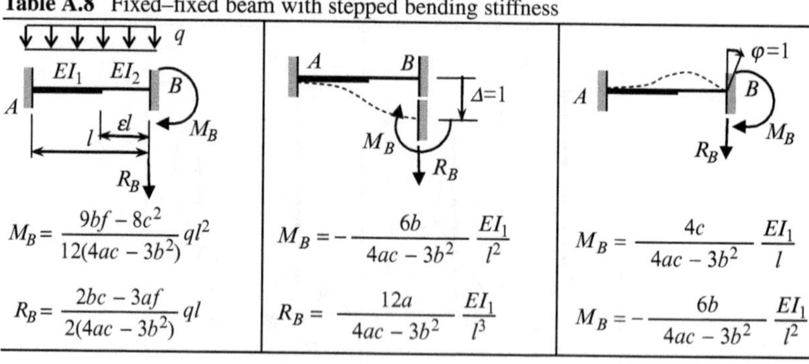

q
EI_1
EI_2
A
B
εl
l
M_B
R_B
$M_B = \frac{9bf - 8c^2}{12(4ac - 3b^2)} ql^2$
$R_B = \frac{2bc - 3af}{2(4ac - 3b^2)} ql$
Δ=1
$M_B = -\frac{6b}{4ac - 3b^2} \frac{EI_1}{l^2}$
$R_B = \frac{12a}{4ac - 3b^2} \frac{EI_1}{l^3}$
φ=1
$M_B = \frac{4c}{4ac - 3b^2} \frac{EI_1}{l}$
$M_B = -\frac{6b}{4ac - 3b^2} \frac{EI_1}{l^2}$

Table A.9 One-span beams. Influence lines for boundary effects

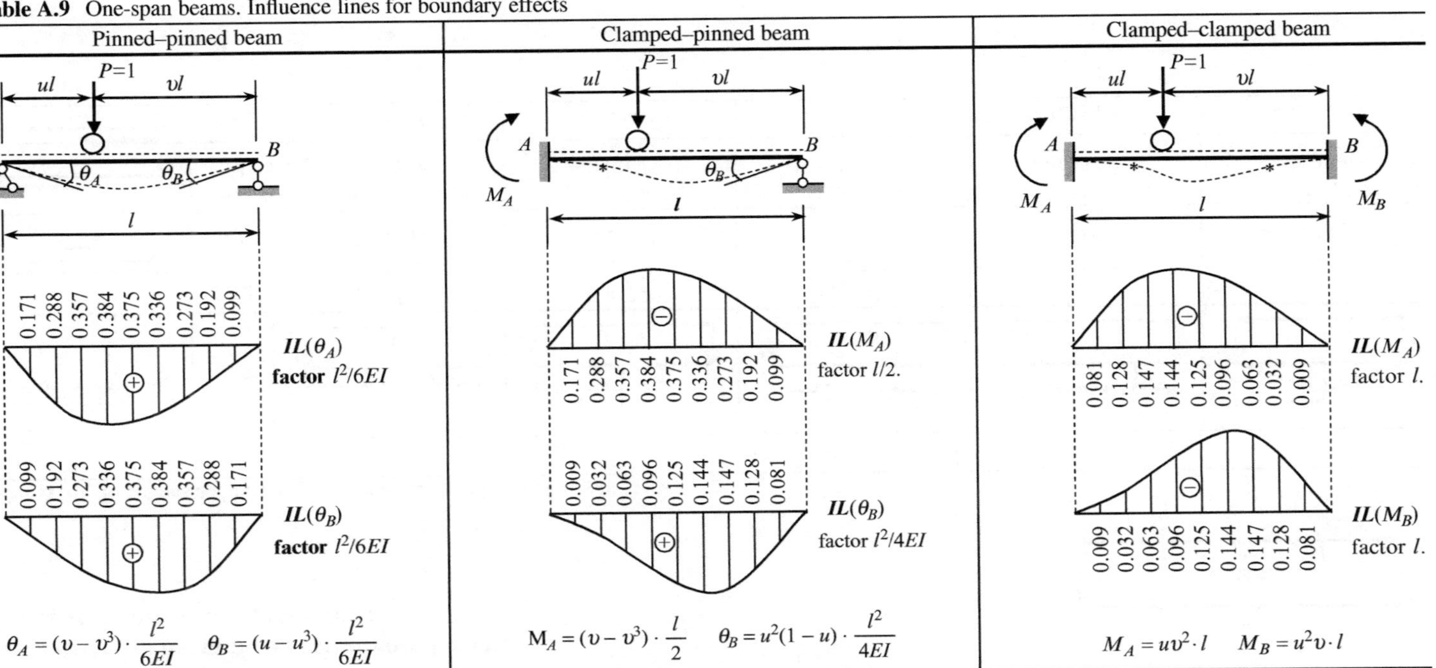

1. Positive sign for angle of rotation: for left support clockwise, for right support counterclockwise.
2. Sign * means the inflection points of elastic curve.
3. Each beam is divided by 10 equal segments.

Tabulated data for uniform continuous beams.
Two span beams with equal spans:

A B C
l l

Table A.10 Reactions and bending moments due to different types of loads

Location of the load		Type of load in the loaded span				
		q, l	a, P, $l=2a$	b, P, b, P, b, $l=3b$	c, P, c, c, P, c, $l=4c$	$l/2$, q, l
	R_A R_B M_B	$0.375ql$ $1.250ql$ $-0.125ql^2$	$0.313P$ $1.375P$ $-0.188Pl$	$0.667P$ $2.667P$ $-0.333Pl$	$1.031P$ $3.938P$ $-0.469Pl$	$0.172ql$ $0.656ql$ $-0.078ql^2$
	R_A M_B	$0.438ql$ $-0.063ql^2$	$0.406P$ $-0.094Pl$	$0.833P$ $-0.167Pl$	$1.266P$ $-0.234Pl$	$0.211ql$ $-0.039ql^2$
	R_A	$-0.063ql$	$-0.094P$	$-0.167P$	$-0.234P$	$-0.039ql$

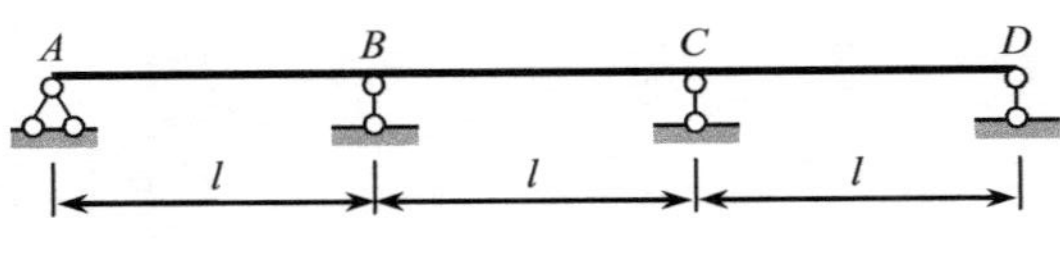

Three span beams with equal spans:

Table A.11 Reactions and bending moments due to different types of loads

Location of the load		Type of load in the loaded span				
		q, l	a, P, $l=2a$	P, P, b, b, b, $l=3b$	P, P, P, c, c, c, c, $l=4c$	$l/2$, q, l
	R_A R_B M_B	$0.400ql$ $1.100ql$ $-0.100ql^2$	$0.350P$ $1.150P$ $-0.150Pl$	$0.733P$ $2.267P$ $-0.267Pl$	$1.125P$ $3.375P$ $-0.375Pl$	$0.188ql$ $0.563ql$ $-0.063ql^2$
	R_A M_B	$0.450ql$ $-0.050ql^2$	$0.425P$ $-0.075Pl$	$0.867P$ $-0.133Pl$	$1.313P$ $-0.188Pl$	$0.219ql$ $-0.032ql^2$
	R_A M_B	$-0.050ql$ $-0.050ql^2$	$-0.075P$ $-0.075Pl$	$-0.133P$ $-0.133Pl$	$-0.188P$ $-0.188Pl$	$-0.032ql$ $-0.032ql^2$
	R_B M_B M_C	$1.200ql$ $-0.117ql^2$ $-0.033ql^2$	$1.300P$ $-0.175Pl$ $-0.050Pl$	$2.533P$ $-0.311Pl$ $-0.089Pl$	$3.750P$ $-0.438Pl$ $-0.125Pl$	$0.626ql$ $-0.073ql^2$ $-0.022ql^2$
	M_B M_C	$0.017ql^2$ $-0.067l^2$	$0.025Pl$ $-0.100Pl$	$0.044Pl$ $-0.178Pl$	$0.063Pl$ $-0.250Pl$	$0.011ql^2$ $-0.042ql^2$

Example. For beam shown below to find the bending moments at supports B and C.

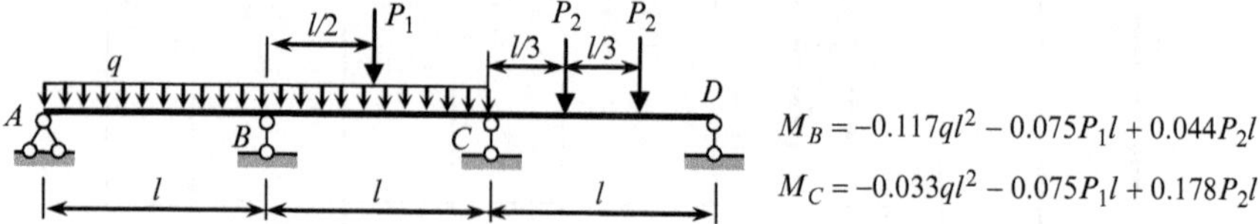

$M_B = -0.117ql^2 - 0.075P_1l + 0.044P_2l$
$M_C = -0.033ql^2 - 0.075P_1l + 0.178P_2l$

Two span beams with different spans:

The bending moment at support B and maximum bending moment at the first and second spans may be calculated by formula $M = kql^2$. Parameters k is presented in Table A.12.

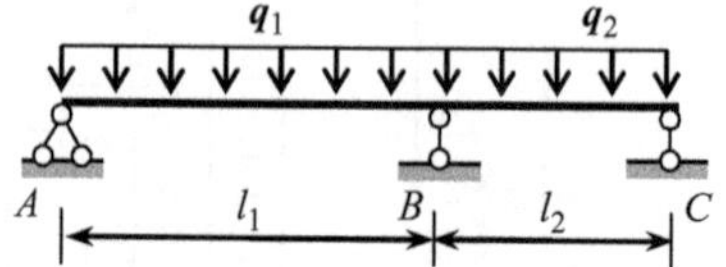

Table A.12 Bending moments due to uniformly distributed load

	Load q_1 is applied in the first span only		Load q_2 is applied in the second span only		Load $q_1 = q_2 = q$ is applied both in the first and second span		
$l_1:l_2$	M_B	$M_{1\max}$	M_B	$M_{2\max}$	M_B	$M_{1\max}$	$M_{2\max}$
1.0	−0.063	0.095	−0.063	0.095	−0.125	0.070	0.070
1.1	−0.079	0.114	−0.060	0.096	−0.139	0.090	0.065
1.2	−0.098	0.134	−0.057	0.097	−0.153	0.111	0.059
1.3	−0.119	0.155	−0.054	0.098	−0.174	0.133	0.053
1.4	−0.143	0.178	−0.052	0.099	−0.195	0.157	0.047
1.5	−0.169	0.203	−0.050	0.100	−0.219	0.183	0.040
1.6	−0.197	0.228	−0.048	0.101	−0.245	0.209	0.033
1.7	−0.227	0.256	−0.046	0.102	−0.274	0.237	0.026
1.8	−0.260	0.285	−0.045	0.103	−0.305	0.267	0.019
1.9	−0.296	0.315	−0.043	0.103	−0.339	0.298	0.013
2.0	−0.333	0.347	−0.042	0.104	−0.375	0.330	0.008
2.2	−0.416	0.415	−0.039	0.106	−0.455	0.398	0.001
2.4	−0.508	0.488	−0.037	0.107	−0.545	0.473	[a]
2.6	−0.610	0.570	−0.035	0.108	−0.645	0.553	[a]
2.8	−0.722	0.655	−0.033	0.109	−0.755	0.639	[a]
3.0	−0.844	0.743	−0.031	0.110	−0.875	0.730	[a]
Factor	$q_1l_2^2$	$q_1l_2^2$	$q_2l_2^2$	$q_2l_2^2$	ql_2^2	ql_2^2	ql_2^2

[a] Within the second span the bending moments are negative

Two span beam with equal spans:
(Odd sections are not shown)

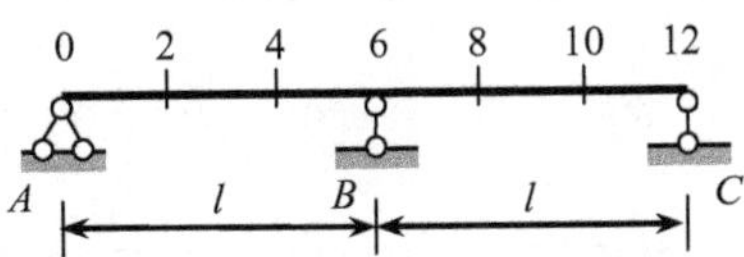

Table A.13 Influence lines for bending moments and shear forces

Position of the load $P = 1$	Ordinates of influence lines of bending moments at sections (factor l) 1	2	3	4	5	6	Ordinates of influence line Q_0
0	0.0	0.0	0.0	0.0	0.0	0.0	1.0000
1	0.1323	0.0976	0.0632	0.0285	−0.0060	−0.0405	0.7928
2	0.0988	0.1976	0.1298	0.0619	−0.0061	−0.0740	0.5927
3	0.0677	0.1354	0.2031	0.1041	+0.0051	−0.0938	0.4062
4	0.0402	0.0803	0.1205	**0.1606**	+0.0340	**−0.0926**	0.2407
5	0.0172	0.0343	0.0516	0.0687	+0.0860	−0.0636	0.1031
6	0.0	0.0	0.0	0.0	0.0	0.0	0.0
7	−0.0106	−0.0212	−0.0318	−0.0424	−0.0530	−0.0636	−0.0636
8	−0.0154	−0.0309	−0.0463	−0.0617	−0.0772	**−0.0926**	**−0.0926**
9	−0.0156	−0.0313	−0.0469	−0.0626	−0.0782	−0.0938	−0.0938
10	−0.0123	−0.0247	−0.0370	−0.0494	−0.0617	−0.0740	−0.0740
11	−0.0068	−0.0135	−0.0203	−0.0270	−0.0338	−0.0405	−0.0405
12	0.0	0.0	0.0	0.0	0.0	0.0	0.0

Example. Force P is located at section 8. Calculate the bending moment at specified points and construct the bending moment diagram.

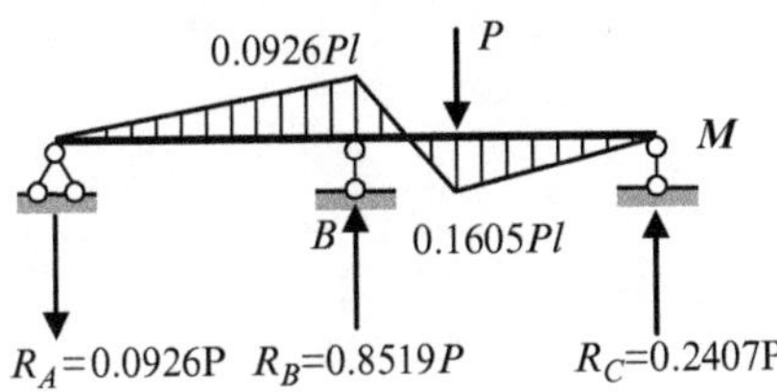

Solution. 1. Bending moment at the section 6 (support B) is $M_6 = -0.0926Pl$.
2. Reaction of support A is $R_A = 0.0926P$ kN and directed downward.
3. Reaction at support C

$$R_C \rightarrow \sum M_B = 0: \; R_C l - P\frac{l}{3} + 0.0926Pl = 0$$

$$\rightarrow R_C = \frac{P}{3} - 0.0926P = 0.2407P \text{ kN}$$

4. Reaction at support B:

$$R_B \to \sum Y = 0 : -P - 0.0926P + 0.2407P + R_B = 0 \to R_B = 0.8519P \text{ kN}$$

5. Bending moments at section 8 is

$$M_8 = R_C \frac{2}{3} l = 0.2407P \frac{2}{3} l = 0.1605Pl.$$

Since the structure is symmetrical and points 4 and 8 are symmetrically located, then bending moments M_8 (if load P is located at point 8) andM_4(if load P is located at point 4) are equal. Last bending moment may be taken immediately from Table A.13. The same result ($M_8 = 0.1605Pl$) may be obtained for point 4, if load P is located at the same point 4 (the points 4 and 8 are symmetrically located).

Beams with three different spans.
The load q is applied at the first span:

The bending moments at supports B and C are

$$M_B = -k_1 q l_1^2, \quad M_C = k_2 q l_1^2.$$

Parameters k_1 and k_2 are presented in Table A.14.

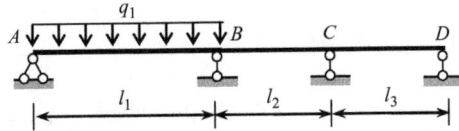

Table A.14 Bending moments at supports B and C (factor ql_1^2)

$\frac{l_1}{l_2}$		$l_3 : l_2$									
		0.3	0.4	0.6	0.8	1.0	1.2	1.4	1.6	1.8	2.0
0.3	k_1	0.034	0.033	0.033	0.032	0.032	0.032	0.031	0.031	0.031	0.031
	k_2	0.013	0.012	0.010	0.009	0.008	0.007	0.007	0.006	0.006	0.005
0.4	k_1	0.041	0.041	0.040	0.040	0.039	0.039	0.038	0.038	0.038	0.038
	k_2	0.016	0.015	0.012	0.011	0.010	0.008	0.008	0.007	0.007	0.006
0.6	k_1	0.053	0.053	0.052	0.051	0.051	0.050	0.050	0.050	0.050	0.049
	k_2	0.021	0.019	0.016	0.014	0.013	0.011	0.010	0.010	0.009	0.008
0.8	k_1	0.062	0.062	**0.061**	0.060	0.060	0.059	0.059	0.059	0.058	0.058
	k_2	0.024	0.022	**0.019**	0.017	0.015	0.013	0.012	0.011	0.010	0.010
1.0	k_1	0.069	0.069	0.068	0.067	0.067	0.066	0.066	0.066	0.065	0.065
	k_2	0.027	0.024	0.021	0.019	0.017	0.015	0.014	0.013	0.012	0.011
1.2	k_1	0.075	0.074	0.074	0.073	0.072	0.072	0.072	0.071	0.071	0.071
	k_2	0.029	0.027	0.023	0.020	0.018	0.016	0.015	0.014	0.013	0.012
1.4	k_1	0.079	0.079	0.078	0.078	0.077	0.077	0.076	0.076	0.076	0.076
	k_2	0.031	0.028	0.024	0.021	0.019	0.017	0.016	0.015	0.013	0.013
1.6	k_1	0.083	0.083	0.082	0.081	0.081	0.080	0.080	0.080	0.080	0.079
	k_2	0.032	0.029	0.026	0.023	0.020	0.018	0.017	0.015	0.014	0.013
1.8	k_1	0.086	0.086	0.085	0.085	0.084	0.084	0.084	0.083	0.083	0.083
	k_2	0.033	0.031	0.027	0.023	0.021	0.019	0.017	0.016	0.015	0.014
2.0	k_1	0.089	0.089	0.088	0.087	0.087	0.087	0.086	0.086	0.086	0.086
	k_2	0.034	0.032	0.028	0.024	0.022	0.020	0.018	0.017	0.015	0.014

Example. Calculate the bending moment at the supports B and C if $l_1 = 8\,\text{m}$, $l_2 = 10\,\text{m}$, $l_3 = 6\,\text{m}$ and uniformly distributed load $q = 2kN/m$ is applied at the first span.

Solution. Relationships $l_1/l_2 = 0.8$, $l_3/l_2 = 0.6$. For this case $k_1 = 0.061$, $k_2 = 0.019$.

Bending moments at supports B and C are:

$$M_B = -0.061 \times 64 \times 2 = -7.808\,\text{kN m}, \quad M_C = 0.019 \times 64 \times 2 = 2.432\,\text{kNm}.$$

Beam with three different spans.
The load q is applied at the second span:

The bending moments at supports B and C are

$$M_B = -k_1 q l_2^2, \quad M_C = -k_2 q l_2^2.$$

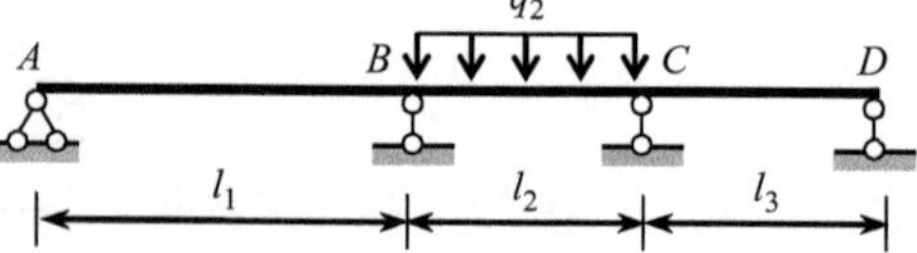

Parameters k_1 and k_2 are presented in Table A.15.

Table A.15 Bending moments at supports B and C (factor ql_2^2)

$\frac{l_1}{l_2}$		$L_3 : l_2$ 0.3	0.4	0.6	0.8	1.0	1.2	1.4	1.6	1.8	2.0
0.3	k_1	0.070	0.072	0.075	0.078	0.080	0.081	0.083	0.083	0.085	0.086
	k_2	0.070	0.064	0.055	0.048	0.043	0.038	0.035	0.031	0.030	0.027
0.4	k_1	0.064	0.066	0.069	0.072	0.073	0.075	0.076	0.077	0.079	0.079
	k_2	0.072	0.066	0.056	0.050	0.044	0.040	0.036	0.033	0.031	0.028
0.6	k_1	0.055	0.057	0.060	0.062	0.063	0.064	0.066	0.067	0.068	0.069
	k_2	0.077	0.069	0.060	0.052	0.047	0.042	0.038	0.035	0.033	0.030
0.8	k_1	0.048	0.050	0.052	0.054	0.056	0.057	0.058	0.059	0.060	0.061
	k_2	0.078	0.071	0.062	0.054	0.049	0.044	0.040	0.037	0.034	0.032
1.0	k_1	0.043	0.044	0.047	0.048	0.050	0.051	0.052	0.053	0.054	0.054
	k_2	0.080	0.074	0.064	0.056	0.050	0.045	0.041	0.038	0.035	0.033
1.2	k_1	0.038	0.040	0.042	0.044	0.045	0.046	0.047	0.048	0.049	0.049
	k_2	0.082	0.075	0.065	0.057	0.051	0.046	0.042	0.039	0.036	0.033
1.4	k_1	0.035	0.036	0.038	0.040	0.041	0.042	0.043	0.044	0.044	0.045
	k_2	0.082	0.077	0.066	0.058	0.052	0.047	0.043	0.040	0.037	0.034
1.6	k_1	0.032	0.033	0.035	0.037	0.038	0.038	0.039	0.040	0.041	0.041
	k_2	0.084	0.078	0.068	0.059	0.053	0.048	0.044	0.040	0.037	0.035
1.8	k_1	0.030	0.031	0.033	0.034	0.035	0.036	0.036	0.037	0.038	0.038
	k_2	0.085	0.078	0.067	0.060	0.054	0.049	0.044	0.041	0.038	0.035
2.0	k_1	0.027	0.029	0.030	0.032	0.033	0.034	0.034	0.035	0.035	0.036
	k_2	0.086	0.079	0.069	0.061	0.055	0.049	0.045	0.041	0.038	0.036

Beam with three different spans.
The load q is applied at the third span:

The bending moments at supports B and C are

$$M_B = k_1 q l_3^2, \quad M_C = -k_2 q l_3^2.$$

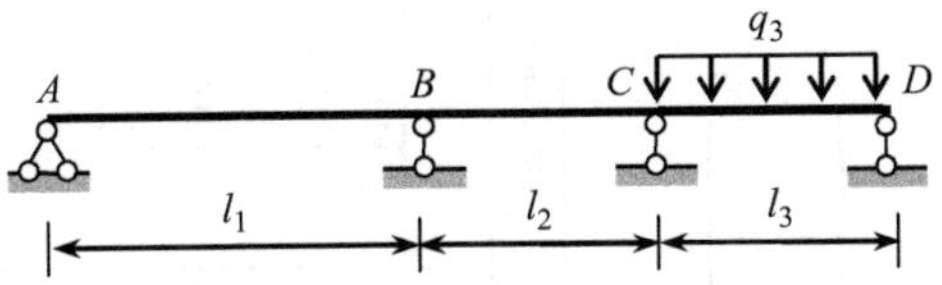

Parameters k_1 and k_2 are presented in Table A.16.

Table A.16 Bending moments at supports B and C (factor ql_3^2)

$\frac{l_1}{l_2}$		$l_3 : l_2$									
		0.3	0.4	0.6	0.8	1.0	1.2	1.4	1.6	1.8	2.0
0.3	k_1	0.013	0.016	0.021	0.024	0.027	0.029	0.031	0.032	0.033	0.034
	k_2	0.034	0.041	0.053	0.062	0.069	0.075	0.079	0.083	0.086	0.089
0.4	k_1	0.012	0.015	0.019	0.022	0.024	0.027	0.028	0.029	0.031	0.032
	k_2	0.033	0.041	0.053	0.062	0.069	0.074	0.079	0.083	0.086	0.089
0.6	k_1	0.010	0.012	0.016	0.019	0.021	0.023	0.024	0.026	0.027	0.028
	k_2	0.033	0.040	0.052	0.061	0.068	0.074	0.078	0.082	0.085	0.088
0.8	k_1	0.009	0.011	0.014	0.017	0.019	0.020	0.021	0.023	0.023	0.024
	k_2	0.032	0.040	0.051	0.060	0.067	0.073	0.078	0.081	0.085	0.087
1.0	k_1	0.008	0.010	0.013	0.015	0.017	0.018	0.019	0.020	0.021	0.022
	k_2	0.032	0.039	0.051	0.060	0.067	0.072	0.077	0.081	0.084	0.087
1.2	k_1	0.008	0.008	0.011	0.013	0.015	0.016	0.017	0.018	0.019	0.020
	k_2	0.032	0.039	0.050	0.059	0.066	0.072	0.077	0.080	0.084	0.087
1.4	k_1	0.007	0.008	0.010	0.012	0.014	0.015	0.016	0.017	0.017	0.018
	k_2	0.031	0.038	0.050	0.059	0.066	0.072	0.076	0.080	0.084	0.086
1.6	k_1	0.006	0.007	0.010	0.011	0.013	0.014	0.015	0.015	0.016	0.017
	k_2	0.031	0.038	0.050	0.059	0.066	0.071	0.076	0.080	0.083	0.086
1.8	k_1	0.006	0.007	0.009	0.010	0.012	0.013	0.013	0.014	0.015	0.015
	k_2	0.031	0.038	0.050	0.058	0.065	0.071	0.076	0.080	0.083	0.086
2.0	k_1	0.005	0.006	0.008	0.010	0.011	0.012	0.013	0.013	0.014	0.014
	k_2	0.031	0.038	0.049	0.058	0.065	0.071	0.076	0.079	0.083	0.086

Beam with three equal spans

(Odd sections are not shown)

Table A.17 Influence lines for bending moments and shear forces

Position of the load	Ordinates of influence lines of bending moments at sections (factor l)									Ordinates of Inf. Lines	
$P = 1$	1	2	3	4	5	6	7	8	9	Q_0	Q_6^{right}
0	0.0	0.0	0.0	0.0	0.0	0.0	0.0	0.0	0.0	1.0000	0.0
1	0.1318	0.0967	0.0618	0.0267	−0.0083	−0.0432	−0.0342	−0.0252	−0.0162	0.7901	0.0540
2	0.0980	0.1960	0.1273	0.0585	−0.0102	−0.0790	−0.0625	−0.0461	−0.0296	0.5877	0.0987
3	0.0667	0.1333	0.2000	0.1000	0.0	−0.1000	−0.0792	−0.0583	−0.0375	0.4000	0.1250
4	0.0391	0.0782	0.1174	0.1565	0.0289	−0.0987	−0.0782	−0.0576	−0.0370	0.2346	0.1234
5	0.0165	0.0329	0.0495	0.0659	0.0826	−0.0677	−0.0536	−0.0395	−0.0254	0.0990	0.0846
6	0.0	0.0	0.0	0.0	0.0	0.0	0.0	0.0	0.0	0.0	**0.00/1.00***
7	−0.0095	−0.0190	−0.0285	−0.0379	−0.0474	−0.0569	0.0872	0.0644	0.0418	−0.0569	0.8639
8	−0.0132	−0.0263	−0.0395	−0.0526	−0.0658	**−0.0789**	0.0364	**0.1516**	0.1002	**−0.0789**	**0.6913**
9	−0.0125	−0.0250	−0.0375	−0.0500	−0.0625	−0.0750	0.0083	0.0917	0.1750	−0.0750	0.5000
10	−0.0090	−0.0181	−0.0271	−0.0362	−0.0452	**−0.0543**	−0.0028	0.0487	0.1002	−0.0543	0.3087
11	−0.0044	−0.0088	−0.0131	−0.0175	−0.0219	−0.0263	−0.0036	0.0191	0.0418	−0.0263	0.1361
12	0.0	0.0	0.0	0.0	0.0	0.0	0.0	0.0	0.0	0.0	0.0
13	0.0028	0.0057	0.0085	0.0113	0.0141	0.0169	0.0028	−0.0113	−0.0254	0.0169	−0.0846
14	0.0041	0.0082	0.0123	0.0165	0.0206	0.0247	0.0041	−0.0165	−0.0370	0.0247	−0.1234
15	0.0042	0.0083	0.0125	0.0167	0.0208	0.0250	0.0042	−0.0167	−0.0375	0.0250	−0.1250
16	0.0033	0.0066	0.0099	0.0132	0.0165	0.0197	0.0033	−0.0132	−0.0296	0.0197	−0.0987
17	0.0018	0.0036	0.0054	0.0072	0.0090	0.0108	0.0018	−0.0072	−0.0162	0.0108	−0.0540
18	0.0	0.0	0.00	0.0	0.0	0.0	0.0	0.0	0.0	0.0	0.0

Influence lines for shear force at section 6 are shown below.

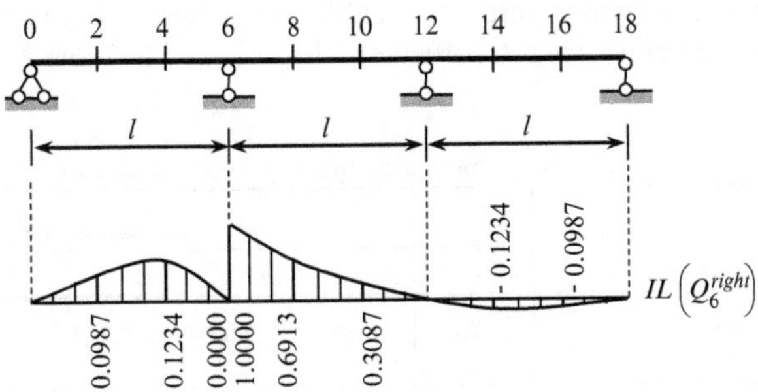

Example. Force P is applied at point 8. Construct the bending moment diagram.

Solution.

1. Bending moment at point 6 (support B) is $M_6 = -0.0789Pl$.
2. Ordinate of influence line $Q_0 = -0.0789$, so reaction of support A is $R_A = 0.0789P$ and directed downward.
3. Since $Q_6^{\text{right}} = -R_A + R_B = 0.6913P$, then reaction of support B is

$$R_B = R_A + 0.6913P = 0.0789P + 0.6913P = 0.7702P.$$

4. Reaction of support D: $R_D \to \sum M_C = -R_D l + R_A 2l - R_B l + P\frac{2}{3}l = 0 \to R_D = 0.0543P$
5. Bending moment at point 12 (support C) is $M_{12} = -0.0543Pl$. The same result may be taken immediately from Table A.15 for section 6, if load P is located at section 10.

Final bending moment diagram is presented below.

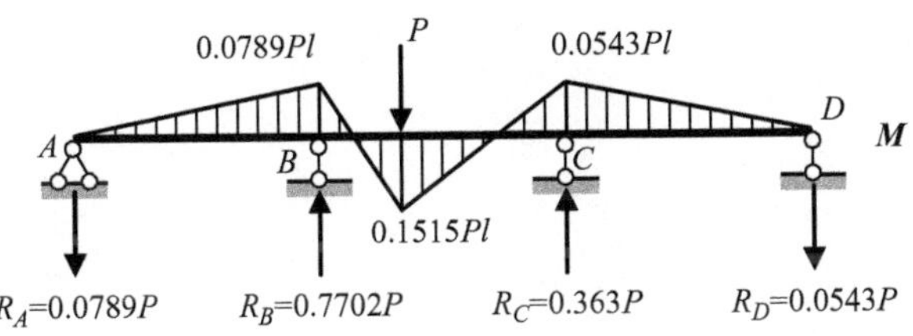

Beams with equal spans. Settlements of supports

Bending moments at supports are $M = k(EI/l^2)\Delta$, where Δ is a vertical settlement of support directed downward. Coefficient k is presented in Table A.18.

Table A.18 Bending moments due to the settlements of supports (factor $EI\Delta/l^2$)

Design diagram of the beam	Supports moments	Settlement of support A	B	C	D	E
A B C	M_B	−1.500	3.000	−1.500	–	–
A B C D	M_B	−1.600	3.600	−2.400	0.400	–
	M_C	0.400	−2.400	3.600	−1.600	–
A B C D E (l, l, l, l)	M_B	−1.6072	3.6429	−2.5714	0.6429	−0.1072
	M_C	0.4286	−2.5714	4.2857	−2.5714	0.4286
	M_D	−0.1072	0.6429	−2.5714	3.6429	−1.6072

Example. Three-span uniform beam *ABCD* with equal spans has settlement $\Delta_B(\downarrow)$ of support B. Calculate the bending moment at all supports.

Solution.

1. Bending moment at support B is $M_B = 3.6(EI/l^2)\Delta$ (extended fibers below of the neutral line).
2. Bending moment at support C is $M_C = -2.4(EI/l^2)\Delta$ (extended fibers above of the neutral line).

A. Foci points:

Each span of a beam contains two foci points. They are left and right points. If *loaded* spans are located to the *right* of the span l_n, then the bending moment diagram in all left spans is passing through *left* foci points only. These foci points are indicated F_n^L, F_{n-1}^L.

If loaded spans are located to the *left* of the span l_n, then the bending moment diagram in all right spans is passing through *right* foci points only, which indicated as F_n^R, F_{n+1}^R.

The left and right foci quotients connect consecutive support moments as follows:

$$k_{n-1}^L = -\frac{M_{n-1}}{M_{n-2}}, \quad k_n^L = -\frac{M_n}{M_{n-1}}$$
$$k_{n+1}^R = -\frac{M_n}{M_{n+1}}, \quad k_n^R = -\frac{M_{n-1}}{M_n}$$

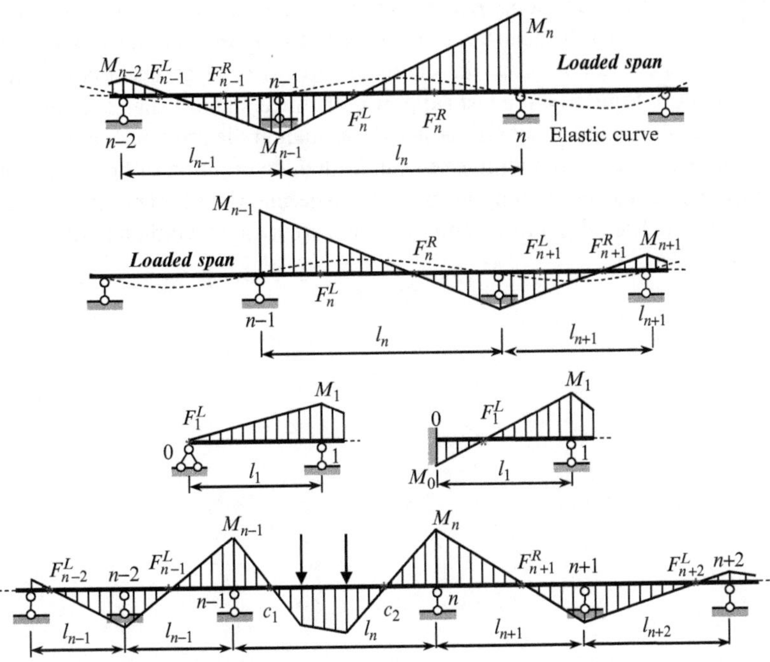

Fig. A.1 (**a**) Explanation of the left foci points. (**b**) Explanation of the right foci points. (**c**) Focus relationships for pinned and fixed supports. (**d**) Continuous beam. Notation, bending moment diagram in the given structure due to applied load, left and right foci points

The left recursive relationship for *next* $(n+1)$ span in terms of the *previous left* foci quotient k_n^L is as follows

$$k_{n+1}^L = 2 + \frac{l_n}{l_{n+1}}\left(2 - \frac{1}{k_n^L}\right).$$

The right recursive relationship for *previous* $(n-1)$-th span in terms of the *next right* foci quotient k_n^R is

$$k_{n-1}^R = 2 + \frac{l_n}{l_{n-1}}\left(2 - \frac{1}{k_n^R}\right).$$

Case a. If the very *left* support 0 is pinned then the left focus for the *first* span coincides with support 0 and the *left* focus quotient becomes $k_1^L = -(M_1/M_0) = \infty$.
Similarly, if the very *right* support is pinned then for the *last* span, the *right* focus quotient is $k_{\text{last}}^R = \infty$.

Case b. If the very left support 0 is clamped, then the left focus quotient is for first span equals two. It means that if a very left support is clamped and first span is unloaded then a moment at the clamped support is twice less than in the next pinned support. Similarly, if a last support is clamped and last span is unloaded then a moment at the clamped support is twice less than in the previous pinned support.

If load is applied only within one span l_n then distribution of bending moments is shown below. The left foci points on right spans and right foci points on left spans are not shown. The nil points c_1 and c_2, which are located within a loaded span, are not foci.

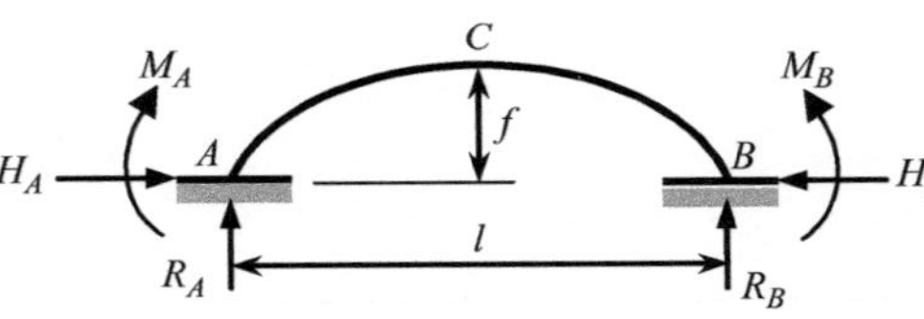

Hingeless parabolic arches $(f/l) \leq (1/4)$.

Cross section of the arch uniform or slightly changing through the span:

Table A.19 Reactions and bending moments due to different types of loads

	Loading	R_A	H	M_A	M_B	M_C
1	P, $l/2$	$\frac{P}{2}$	$\frac{15}{64}\frac{Pl}{f}$	$\frac{Pl}{32}$	$\frac{Pl}{32}$	$\frac{3}{64}Pl$
2	P, ul	$P(1-u)^2(1+2u)$	$\frac{15}{4}\frac{Pl}{f}u^2(1-u)^2$	$-\frac{Pl}{2}u(1-u)^2(2-5u)$	$\frac{Pl}{2}u^2(1-u)(3-5u)$	$u \leq 0.5$ $-\frac{Pl}{4}u^2(3-10u+5u^2)$
3	q, $l/2$	$\frac{13}{32}ql$	$\frac{ql^2}{16f}$	$-\frac{ql^2}{64}$	$\frac{ql^2}{64}$	0

(continued)

Table A.19 (continued)

	Loading	R_A	H	M_A	M_B	M_C
4	q, ul	$\frac{ql}{2}u\,(2-2u^2+u^3)$	$\frac{ql^2}{8f}u^3\,(10-15u+6u^2)$	$-\frac{ql^2}{2}u^2\,(1-3u+3u^2-u^3)$	$\frac{ql^2}{2}u^3\,(1-2u+u^2)$	$u \le 0.5$ $-\frac{ql^2}{8}u^3\,(2-5u+2u^2)$
5	$\varphi=1$	$-\frac{6EI}{l^2}$	$\frac{15}{2}\frac{EI}{lf}$	$\frac{9EI}{l}$	$\frac{3EI}{l}$	$-\frac{3}{2}\frac{EI}{l}$
6	$\Delta=1$	0	$\frac{45}{4}\frac{EI}{lf^2}$	$\frac{15}{2}\frac{EI}{lf}$	$\frac{15}{2}\frac{EI}{lf}$	$-\frac{15}{4}\frac{EI}{lf}$
7[a]	$t°$, $t°$	0	$\frac{45}{4}\frac{EI}{f^2}\alpha_t t$	$\frac{15}{2}\frac{EI}{f}\alpha_t t$	$\frac{15}{2}\frac{EI}{f}\alpha_t t$	$-\frac{15}{4}\frac{EI}{f}\alpha_t t$

[a] α_t is a thermal expansion coefficient

Hingeless non-uniform parabolic arches

$$I_x = \frac{I_C}{\cos\alpha}, \quad \nu = \frac{45}{4}\frac{I_C}{A_C f^2}, \quad k = \frac{1}{1+\nu}a$$

Table A.20 Reactions and bending moments due to different types of loads

	Loading	R_A	H	M_A	M_B	M_C
1	P; l/2	$\frac{P}{2}$	$\frac{15}{64}\frac{Pl}{f}k$	$\frac{Pl}{8}\left(\frac{5}{4}k-1\right)$	$\frac{Pl}{8}\left(\frac{5}{4}k-1\right)$	$\frac{Pl}{8}\left(1-\frac{5}{8}k\right)$
2	P; ul	$P\upsilon^2(1+2u)$ $\upsilon = 1-u$	$\frac{15}{4}\frac{Pl}{f}u^2\upsilon^2 k$	$Plu\cdot\upsilon^2\left(\frac{5k}{2}u-1\right)$	$Plu^2\cdot\upsilon\left(\frac{5k}{2}\upsilon-1\right)$	$u \le 0.5$ $\frac{Pl}{2}u^2\left(1-\frac{5k}{2}\upsilon^2\right)$
3	q; l/2	$\frac{13}{32}ql$	$\frac{ql^2}{16f}k$	$-\frac{ql^2}{192}(11-8k)$	$\frac{ql^2}{192}(8k-5)$	$\frac{ql^2}{48}(1-k)$
4	$\varphi=1$	$-\frac{6EI_C}{l^2}$	$\frac{15}{2}\frac{EI_C}{lf}$	$\frac{9EI_C}{l}$	$\frac{3EI_C}{l}$	$-\frac{3}{2}\frac{EI_C}{l}$
5	$\Delta=1$	0	$\frac{45}{4}\frac{EI_C}{lf^2}$	$\frac{15}{2}\frac{EI_C}{lf}$	$\frac{15}{2}\frac{EI_C}{lf}$	$-\frac{15}{4}\frac{EI_C}{lf}$
6	$t°$; $t°$; Increase of temperature	0	$\frac{45}{4}\frac{EI_C}{f^2}\alpha_t t$	$\frac{15}{2}\frac{EI_C}{f}\alpha_t tk$	$\frac{15}{2}\frac{EI_C}{f}\alpha_t tk$	$-\frac{15}{4}\frac{EI_C}{f}\alpha_t tk$

[a] I_C, A_C are moment of inertia and area of cross section at the crown C. If axial forces are neglected, then $\upsilon = 0, k = 1$

Table A.21 Load characteristic $D = \int_0^l Q^2(x)\mathrm{d}x$ for analysis of gentile cables

	Loading	Load characteristic D
1	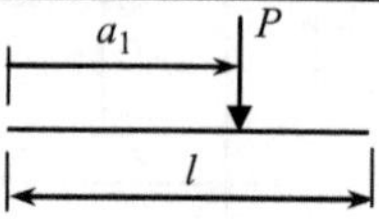	$P^2 l \xi_1 (1-\xi_1), \quad \xi_1 = \frac{a_1}{l}$ $\frac{P^2 l}{4}$, for $a_1 = \frac{l}{2}$
2	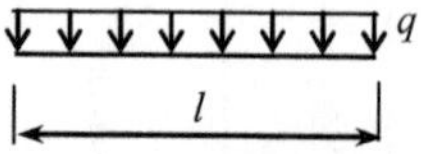	$\frac{q^2 l^3}{12}$
3	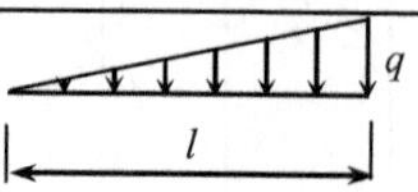	$\frac{q^2 l^3}{45}$
4	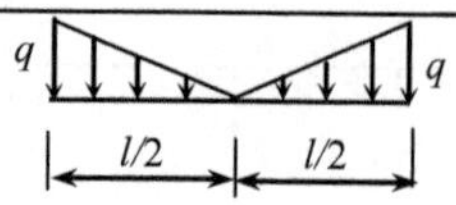	$\frac{q^2 l^3}{80}$
5	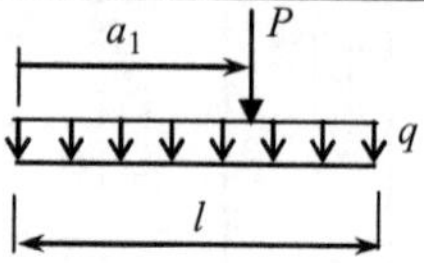	$\frac{q^2 l^3}{12}\left[1 + 12\xi_1\gamma_1(1-\xi_1)(1+\gamma_1)\right], \quad \xi_1 = \frac{a_1}{l}, \quad \gamma_1 = \frac{P}{ql}$ $\frac{q^2 l^3}{12}\left[1 + 3\gamma_1 + 3\gamma_1^2\right]$ for $a_1 = \frac{l}{2}$
6	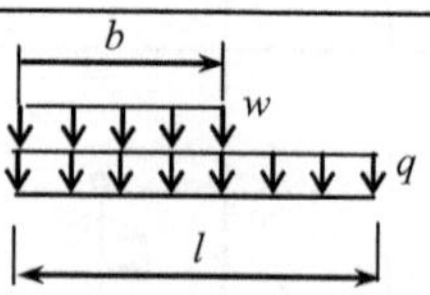	$\frac{q^2 l^3}{12}\left[1 + (4-3\beta)\beta^3\gamma^2 + (6-4\beta)\beta^2\gamma\right], \quad \beta = \frac{b}{l}, \quad \gamma = \frac{w}{q}$ $\frac{q^2 l^3}{12}\left[1 + \gamma + \frac{5}{16}\gamma^2\right]$ for $b = \frac{l}{2}$; $\frac{q^2 l^3}{12} \cdot 4$ for $b = l$, $w = q$
7	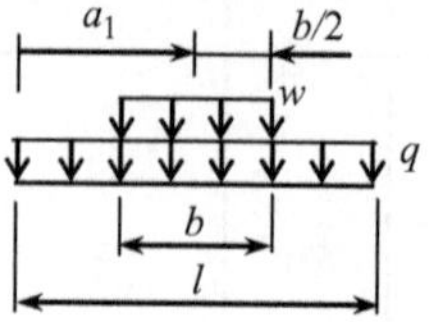	$\frac{q^2 l^3}{12}\left[1 + \left(12\xi_1 - 12\xi_1^2 - 2\beta\right)\beta^2\gamma^2 + \left(12\xi_1 - 12\xi_1^2 - \beta^2\right)\beta\gamma\right]$, $\xi_1 = \frac{a_1}{l}, \quad \beta = \frac{b}{l}, \quad \gamma = \frac{w}{q}$ $\frac{q^2 l^3}{12} 4$ for $b = l, w = q$
8	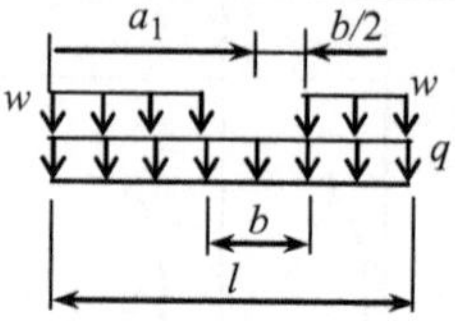	$\frac{(q+w)^2 l^3}{12}\left[1 + \left(12\xi_1 - 12\xi_1^2 - 2\beta\right)\beta^2\gamma_2^2 - \left(12\xi_1 - 12\xi_1^2 + \beta^2\right)\beta\gamma_2\right]$, $\xi_1 = \frac{a_1}{l}, \quad \beta = \frac{b}{l}, \quad \gamma_2 = \frac{w}{q+w}$ $\frac{q^2 l^3}{12} 4$ for $b = 0$, $w = q$

Table A.22 Reactions of beams subjected to compressed load and unit settlement of support, $\upsilon = l\sqrt{\dfrac{P}{EI}}$

	Pinned-clamped beam	Clamped-clamped beam
1	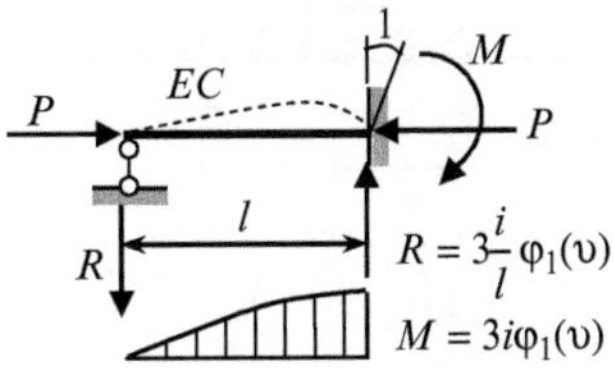 P, EC, 1, M, P, l, R; $R = 3\frac{i}{l}\varphi_1(\upsilon)$; $M = 3i\varphi_1(\upsilon)$	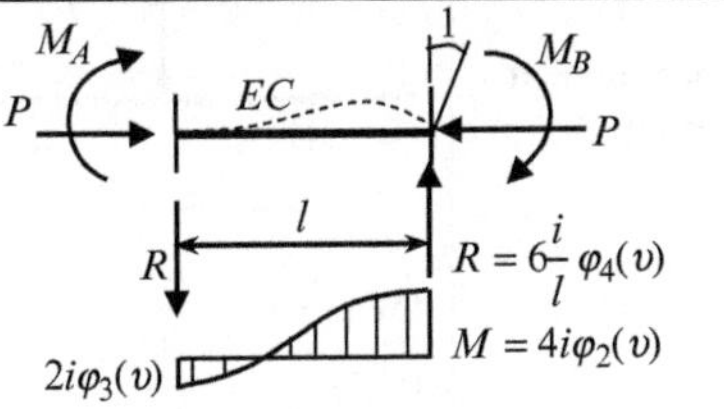 M_A, P, EC, 1, M_B, P, l, R; $R = 6\frac{i}{l}\varphi_4(\upsilon)$; $2i\varphi_3(\upsilon)$; $M = 4i\varphi_2(\upsilon)$
2	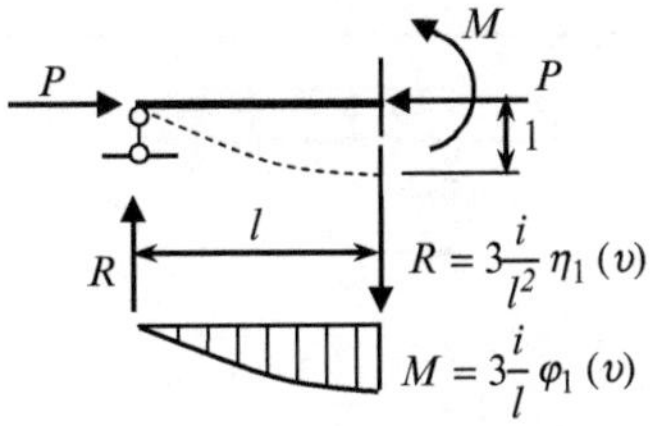 M, P, P, 1, l, R; $R = 3\frac{i}{l^2}\eta_1(\upsilon)$; $M = 3\frac{i}{l}\varphi_1(\upsilon)$	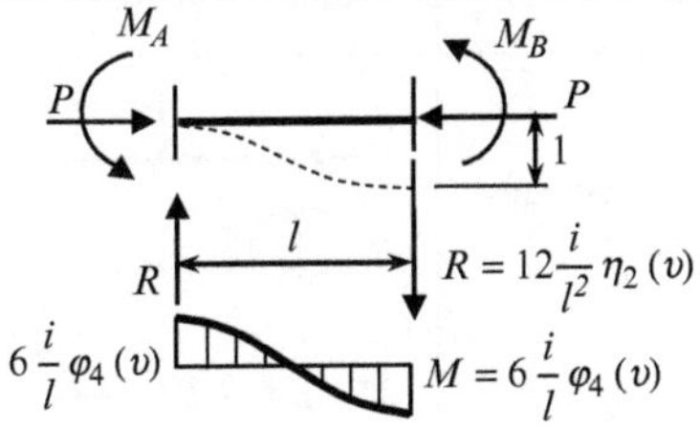 M_A, M_B, P, P, 1, l, R; $R = 12\frac{i}{l^2}\eta_2(\upsilon)$; $6\frac{i}{l}\varphi_4(\upsilon)$; $M = 6\frac{i}{l}\varphi_4(\upsilon)$
3	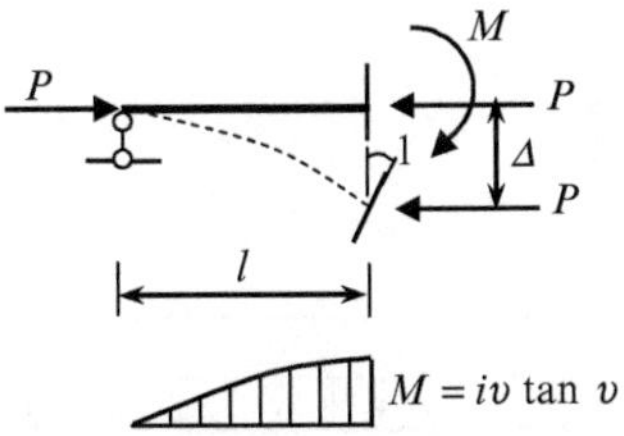 M, P, P, 1, Δ, P, l; $M = i\upsilon \tan \upsilon$	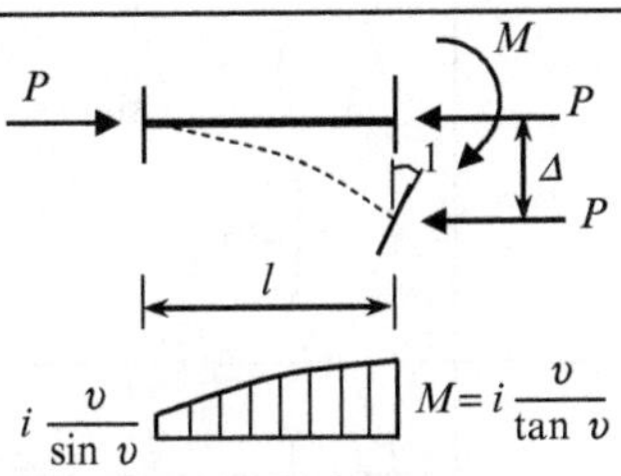 M, P, P, 1, Δ, P, l; $i\frac{\upsilon}{\sin \upsilon}$; $M = i\frac{\upsilon}{\tan \upsilon}$
4	Pinned-pinned beam 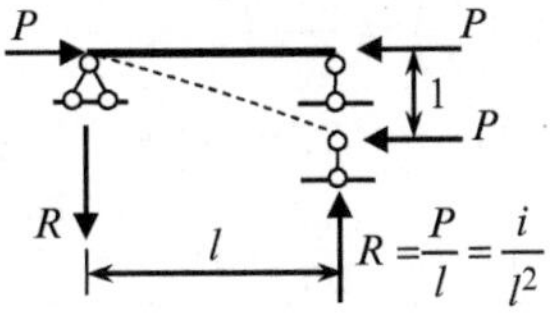 P, P, 1, P, R, l; $R = \frac{P}{l} = \frac{i}{l^2}$	Clamped-free beam 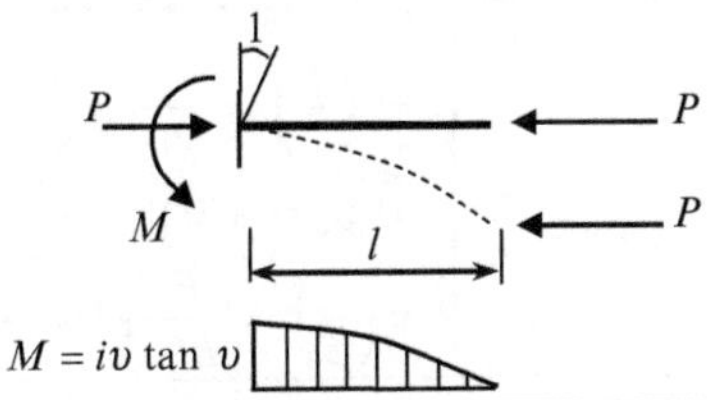 1, P, M, P, P, l; $M = i\upsilon \tan \upsilon$

Table A.23 Reactions of compressed-bent beams; $\upsilon = l\sqrt{\frac{P}{EI}}$

	Beam under concentrated load F and axial compressive force P	Beam under uniformly distributed load q and axial compressive force P
Pinned–pinned beam	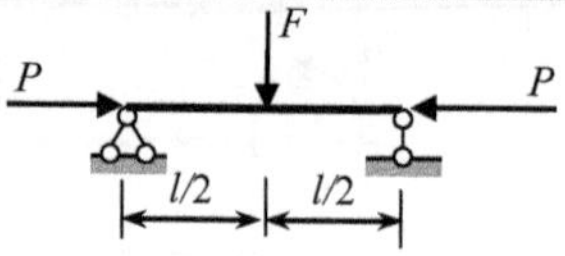 $M\left(\frac{l}{2}\right) = \frac{Fl}{2\upsilon}\tan\frac{\upsilon}{2}$ $y\left(\frac{l}{2}\right) = \frac{Fl^3}{2EI\upsilon^3}\left(\tan\frac{\upsilon}{2} - \frac{\upsilon}{2}\right)$	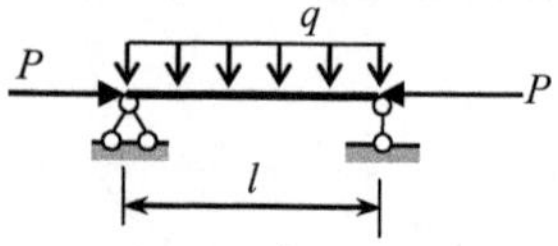 $M\left(\frac{l}{2}\right) = \frac{ql^2}{\upsilon^2}\left(\sec\frac{\upsilon}{2} - 1\right)$ $y\left(\frac{l}{2}\right) = \frac{ql^4}{EI\,\upsilon^4}\left(\sec\frac{\upsilon}{2} - 1 - \frac{\upsilon^2}{8}\right)$
Pinned-clamped beam	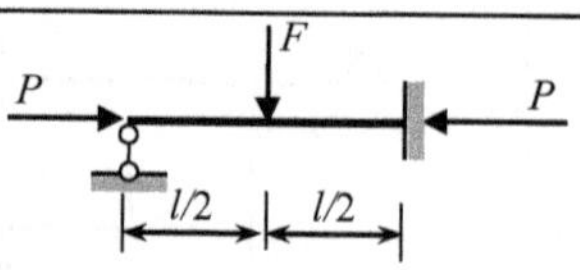 $M(l) = -Fl\frac{\sin\frac{\upsilon}{2}\left(1 - \cos\frac{\upsilon}{2}\right)}{\sin\upsilon - \upsilon\cos\upsilon}$ $M\left(\frac{l}{2}\right) = Fl\frac{\sin\frac{\upsilon}{2}}{\sin\upsilon - \upsilon\cos\upsilon}\cdot\left(\cos\frac{\upsilon}{2} - \frac{\sin\frac{\upsilon}{2}}{\upsilon} - \frac{1}{2}\right)$	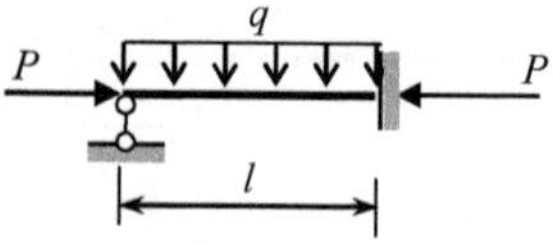 $M(l) = -\frac{ql^2}{2\upsilon}\frac{2 - \upsilon\sin\upsilon - 2\cos\upsilon}{\sin\upsilon - \upsilon\cos\upsilon}$
Clamped-clamped beam	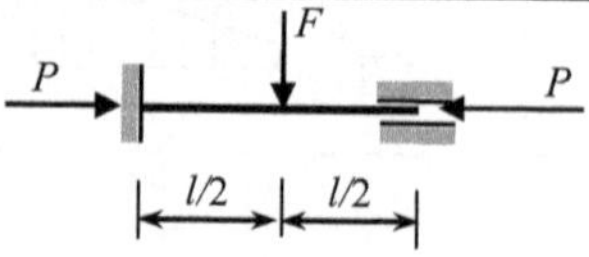 $M\left(\frac{l}{2}\right) = \frac{Fl}{2\upsilon}\tan\frac{\upsilon}{4}$ $M(0) = M(l) = -\frac{Fl}{2\upsilon}\tan\frac{\upsilon}{4}$ $y\left(\frac{l}{2}\right) = \frac{Fl^3}{EI\,\upsilon^3}\left(\tan\frac{\upsilon}{4} - \frac{\upsilon}{4}\right)$	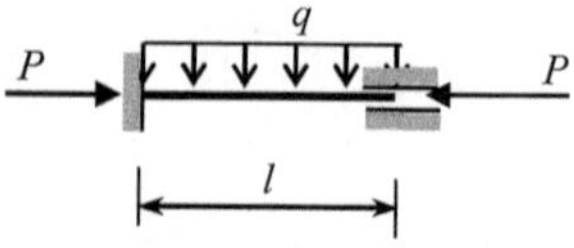 $M\left(\frac{l}{2}\right) = \frac{ql^2}{2\upsilon^2}\left(\upsilon\csc\frac{\upsilon}{2} - 2\right)$ $M(0) = M(l) = -\frac{ql^2}{\upsilon^2}\left(1 - \frac{\upsilon}{2}\cot\frac{\upsilon}{2}\right)$ $y\left(\frac{l}{2}\right) = \frac{ql^4}{2EI\,\upsilon^3}\left(\tan\frac{\upsilon}{4} - \frac{\upsilon}{4}\right)$

Table A.24 Special functions for stability analysis

Functions	Form 1	Form 2	Maclaurin series
$\varphi_1(\upsilon)$	$\dfrac{\upsilon^2\tan\upsilon}{3(\tan\upsilon-\upsilon)}$	$\dfrac{1}{3}\dfrac{\upsilon^2\sin\upsilon}{\sin\upsilon-\upsilon\cos\upsilon}$	$1-\dfrac{\upsilon^2}{15}-\dfrac{\upsilon^4}{525}+\cdots$
$\varphi_2(\upsilon)$	$\dfrac{\upsilon(\tan\upsilon-\upsilon)}{8\tan\upsilon\left(\tan\dfrac{\upsilon}{2}-\dfrac{\upsilon}{2}\right)}$	$\dfrac{1}{4}\dfrac{\upsilon\sin\upsilon-\upsilon^2\cos\upsilon}{2-2\cos\upsilon-\upsilon\sin\upsilon}$	$1-\dfrac{\upsilon^2}{30}-\dfrac{11\upsilon^4}{25200}+\cdots$
$\varphi_3(\upsilon)$	$\dfrac{\upsilon(\upsilon-\sin\upsilon)}{4\sin\upsilon\left(\tan\dfrac{\upsilon}{2}-\dfrac{\upsilon}{2}\right)}$	$\dfrac{1}{2}\dfrac{\upsilon(\upsilon-\sin\upsilon)}{2-2\cos\upsilon-\upsilon\sin\upsilon}$	$1+\dfrac{\upsilon^2}{60}+\dfrac{13\upsilon^4}{25200}+\cdots$
$\varphi_4(\upsilon)$	$\varphi_1\left(\dfrac{\upsilon}{2}\right)$	$\dfrac{1}{6}\dfrac{\upsilon^2\sin\upsilon}{2\sin\upsilon-\upsilon-\upsilon\cos\upsilon}$	$1-\dfrac{\upsilon^2}{60}-\dfrac{\upsilon^4}{84000}+\cdots$
$\eta_1(\upsilon)$	$\dfrac{\upsilon^3}{3(\tan\upsilon-\upsilon)}$	$\dfrac{1}{3}\dfrac{\upsilon^3\cos\upsilon}{\sin\upsilon-\upsilon\cos\upsilon}$	$1-\dfrac{2\upsilon^2}{5}-\dfrac{\upsilon^4}{525}+\cdots$
$\eta_2(\upsilon)$	$\eta_1\left(\dfrac{\upsilon}{2}\right)$	$\dfrac{1}{12}\dfrac{\upsilon^3(1+\cos\upsilon)}{2\sin\upsilon-\upsilon-\upsilon\cos\upsilon}$	$1-\dfrac{\upsilon^2}{10}-\dfrac{\upsilon^4}{8400}+\cdots$
$\dfrac{\upsilon}{\sin\upsilon}$	$\dfrac{\upsilon}{\sin\upsilon}$	$\dfrac{\upsilon}{\sin\upsilon}$	$1+\dfrac{\upsilon^2}{6}+\dfrac{7\upsilon^4}{360}+\cdots$
$\dfrac{\upsilon}{\tan\upsilon}$	$\dfrac{\upsilon}{\tan\upsilon}$	$\dfrac{\upsilon\cos}{\sin\upsilon}$	$1-\dfrac{\upsilon^2}{3}-\dfrac{\upsilon^4}{45}+\cdots$
$\upsilon\tan\upsilon$	$\upsilon\tan\upsilon$	$\dfrac{\upsilon\sin\upsilon}{\cos\upsilon}$	$0+\upsilon^2+\dfrac{\upsilon^4}{3}+\cdots$

Numerical values of these functions in terms of dimensionless parameter υ are presented in Table A.25

Table A.25 Special functions for stability analysis by Displacement method

υ	$\varphi_1(\upsilon)$	$\varphi_2(\upsilon)$	$\varphi_3(\upsilon)$	$\varphi_4(\upsilon)$	$\eta_1(\upsilon)$	$\eta_2(\upsilon)$
0.0	1.0000	1.0000	1.0000	1.0000	1.0000	1.0000
0.2	0.9973	0.9980	1.0009	0.9992	0.9840	0.9959
0.4	0.9895	0.9945	1.0026	0.9973	0.9362	0.9840
0.6	0.9756	0.9881	1.0061	0.9941	0.8557	0.9641
0.8	0.9566	0.9787	1.0111	0.9895	0.7432	0.9362
1.0	0.9313	0.9662	1.0172	0.9832	0.5980	0.8999
1.1	0.9164	0.9590	1.0209	0.9798	0.5131	0.8789
1.2	0.8998	0.9511	1.0251	0.9757	0.4198	0.8557
1.3	0.8814	0.9424	1.0298	0.9715	0.3181	0.8307
1.4	0.8613	0.9329	1.0348	0.9669	0.2080	0.8035
1.5	0.8393	0.9226	1.0403	0.9619	0.0893	0.7743
π/2	0.8225	0.9149	1.0445	0.9620	**0.0000**	0.7525
1.6	0.8153	0.9116	1.0463	0.9566	−0.0380	0.7432
1.7	0.7891	0.8998	1.0529	0.9509	−0.1742	0.7100
1.8	0.7609	0.8871	1.0600	0.9448	−0.3191	0.6747
1.9	0.7297	0.8735	1.0676	0.9382	−0.4736	0.6374
2.0	0.6961	0.8590	1.0760	0.9313	−0.6372	0.5980
2.1	0.6597	0.8437	1.0850	0.9240	−0.8103	0.5565
2.2	0.6202	0.8273	1.0946	0.9164	−0.9931	0.5131

(continued)

Table A.25 (continued)

υ	$\varphi_1(\upsilon)$	$\varphi_2(\upsilon)$	$\varphi_3(\upsilon)$	$\varphi_4(\upsilon)$	$\eta_1(\upsilon)$	$\eta_2(\upsilon)$
2.3	0.5772	0.8099	1.1050	0.9083	−1.1861	0.4675
2.4	0.5304	0.7915	1.1164	0.8998	−1.3895	0.4198
2.5	0.4793	0.7720	1.1286	0.8909	−1.6040	0.3701
2.6	0.4234	0.7513	1.1417	0.8814	−1.8299	0.3181
2.7	0.3621	0.7294	1.1559	0.8716	−2.0679	0.2641
2.8	0.2944	0.7064	1.1712	0.8613	−2.3189	0.2080
2.9	0.2195	0.6819	1.1878	0.8506	−2.5838	0.1498
3.0	0.1361	0.6560	1.2057	0.8393	−2.8639	0.0893
3.1	0.0424	0.6287	1.2252	0.8275	−3.1609	0.0207
π	**0.0000**	0.6168	1.2336	0.8225	−3.2898	**0.0000**
3.2	−0.0635	0.5997	1.2463	0.8153	−3.4768	−0.0380
3.3	−0.1847	0.5691	1.2691	0.8024	−3.8147	−0.1051
3.4	−0.3248	0.5366	1.2940	0.7891	−4.1781	−0.1742
3.5	−0.4894	0.5021	1.3212	0.7751	−4.5727	−0.2457
3.6	−0.6862	0.4656	1.3508	0.7609	−5.0062	−0.3191
3.7	−0.9270	0.4265	1.3834	0.7457	−5.4903	−0.3951
3.8	−1.2303	0.3850	1.4191	0.7297	−6.0436	−0.4736
3.9	−1.6268	0.3407	1.4584	0.7133	−6.6968	−0.5542
4.0	−2.1726	0.2933	1.5018	0.6961	−7.5058	−0.6372
4.1	−2.9806	0.2424	1.5501	0.6783	−8.5836	−0.7225
4.2	−4.3155	0.1877	1.6036	0.6597	−10.196	−0.8103
4.3	−6.9949	0.1288	1.6637	0.6404	−13.158	−0.9004
4.4	−15.330	0.0648	1.7310	0.6202	−21.780	−0.9931
4.5	227.80	−0.0048	1.8070	0.5991	+221.05	−1.0884
4.6	14.669	−0.0808	1.8933	0.5772	7.6160	−1.1861
4.7	7.8185	−0.1646	1.9919	0.5543	0.4553	−1.2865
4.8	5.4020	−0.2572	2.1056	0.5304	−2.2777	−1.3895
4.9	4.1463	−0.3612	2.2377	0.5054	−3.8570	−1.4954
5.0	3.3615	−0.4772	2.3924	0.4793	−4.9718	−1.6040
5.2	2.3986	−0.7630	2.7961	0.4234	−6.6147	−1.8299
5.4	1.7884	−1.1563	3.3989	0.3621	−7.9316	−2.0679
5.6	1.3265	−1.7481	4.3794	0.2944	−9.1268	−2.3189
5.8	0.9302	−2.7777	6.2140	0.2195	−10.283	−2.5939
6.0	0.5551	−5.1589	10.727	0.1361	−11.445	−2.8639
6.2	0.1700	−18.591	37.308	0.0424	−12.643	−3.1609
2π	**0.0000**	**−∞**	**+∞**	0.0000	−13.033	−3.2898

Table A.26 One span beams with classical boundary conditions. Frequency equation and eigenvalues

#	Type of beam	Frequency equation	n	Eigen value λ_n	Nodal points $\xi = x/l$ of mode shape X
1	Pinned–pinned	$\sin k_n l = 0$	1	3.14159265	0; 1.0
			2	6.28318531	0; 0.5; 1.0
			3	9.42477796	0; 0.333; 0.667; 1.0
2	Clamped–clamped	$\cos k_n l \cosh k_n l = 1$	1	4.73004074	0; 1.0
			2	7.85320462	0; 0.5; 1.0
			3	10.9956079	0; 0.359; 0.641; 1.0
3	Pinned–clamped	$\tan k_n l - \tanh k_n l = 0$	1	3.92660231	0; 1.0
			2	7.06858275	0; 0.440; 1.0
			3	10.21017612	0; 0.308; 0.616; 1.0
4	Clamped–free	$\cos k_n l \cosh k_n l = -1$	1	1.87510407	0
			2	4.69409113	0; 0.774
			3	7.85475744	0; 0.5001; 0.868
5	Free–free	$\cos k_n l \cosh k_n l = 1$	1	0	Rigid-body mode
			2	4.73004074	0.224; 0.776
			3	7.85320462	0.132; 0.500; 0.868
6	Pinned–free	$\tan k_n l - \tanh k_n l = 0$	1	0	Rigid-body mode
			2	3.92660231	0; 0.736
			3	7.06858275	0; 0.446; 0.853

Bibliography

Textbooks

Clough RW, Penzien J (1975) Dynamics of structures. McCraw Hill, New York

Cook RD, Malkus DS, and Plesha ME (1989) Concepts and applications of finite element analysis, 3rd edn. Wiley, New York

Darkov, A (ed) (1989) Structural mechanics. Mir Publishers, Moscow

Fertis DG (1996) Advanced mechanics of structures. Marcel Dekker, New York

Ghali A, Neville AM, Cheung YK (1997) Structural analysis. A unified classical and matrix approach, 4th edn. E & FN SPON, London

Kiselev VA (1969) Structural mechanics. Special course. ILSA, Moscow (in Russian)

Leet KM, Uang CM, Gilbert AM (2008) Fundamentals of structural analysis, 3rd edn. McGraw-Hill, New York

Nelson JK, McCormac JC (2003) Structural analysis: Using classical and matrix methods, 3rd edn. Wiley, New York

Smirnov AF et al (1984) Structural mechanics. Dynamics and stability of structures. Stroizdat, Moscow (in Russian)

Timoshenko SP, Young DH (1965) Theory of structures, 2nd edn. McGraw-Hill, New York

Manuals

Anokhin NN (2000) Structural mechanics in examples and solutions, Part 2. Statically indeterminate systems. ACB, Moscow (in Russian)

Klein GK (ed) (1980) Manual for structural mechanics. Statics of bar systems. Vusshaya Shkola, Moscow (in Russian)

Tuma JJ, Munshi RK (1971) Theory and problems of advanced structural analysis, Schaum's outline series. McGraw-Hill, New York

Handbooks

Blevins RD (1979) Formulas for natural frequency and mode shape. Van Nostrand Reinhold, New York

Gaylord EH, Gaylord ChN, Stallmeyer JE (1997) Structural engineering. Handbook, 4th edn. McGraw-Hill, New York, Chapters 1–32

Karnovsky IA, Lebed OI (2003) Free vibrations of beams and frame. Eigenvalues and eigenfunctions. McGraw-Hill, New York
Roark RJ (1975) Formulas for stress and strain, 5th edn. McGraw-Hill, New York
Umansky AA (ed) (1972) Handbook of designer, vol. 1,2. Stroizdat, Moscow (in Russian)

History

Benvenuto E (1991) An introduction to the history of structural mechanics. Springer, York
Bernshtein SA (1957) Essays in structural mechanics history. GILSA, Moscow (in Russian)
Oravas GA, McLean L (1966) Historical development of energetical principles in elastomechanics. Appl Mech Rev Part 1, 19(8); Part 2, 19(11)
Timoshenko SP (1953) History of strength of materials. McGraw Hill, New York
Todhunter I, Pearson K (1960) A history of the theory of elasticity and of the strength of materials, vols. I and II. Dover, New York (originally published by the Cambridge University Press in 1886 and 1893)
Westergaard HM (1930) One hundred fifty years advance in structural analysis. Trans Am Soc Civil Eng 94:226–246

Index

A

Ancillary diagram
- displacement-load (Z-P), 372, 376–379
- internal forces-deformation (S-e), 372, 377–378
- joint-load (J-L), 372–376

Arches
- askew, 97–100, 106
- Boussinesq equation, 485
- differential equation, 484, 486
- geometry parameters, 79–80
- hingeless, 237, 325, 331–339, 484, 489, 577
- influence lines, 86–93
- nil points, 94, 102
- three-hinge, 77–107
- two-hinged, 77, 237–239, 243, 267, 366, 484, 485, 488, 489

B

Beam-columns
- deflection, 504
- differential equation, 540–541

Beams
- continuous, 432–441, 471–483, 508, 564, 574
- with elastic supports, 560
- free vibration equation, 520–521, 530–532
- Gerber–Semikolenov, 39–42, 45, 46, 212
- universal equation, 148, 150–152, 155, 194

Bendixen, 271

Bendixen Bernoulli-Euler beam, 538, 540, 543

Betti theorem, 189–191

Boundary condition, 117, 150, 152, 154, 160, 194, 309, 360, 450, 461, 464, 470–472, 484, 489, 519, 540–542, 544, 545, 549, 585

Boussinesq equation, 485

Buckling, xxii, 450–452, 462, 463, 481, 484, 485, 488, 489, 506

C

Cable
- arbitrary load, 122–125
- catenary, 110, 125–128, 131, 133–137
- differential equation, 119
- direct problem, 114–117, 131
- inverse problem, 110–111, 115–121, 123, 125

Castigliano theorem, 147, 181, 195

Cauchy–Clebsch condition, 149

Change of temperature, 40, 73, 145, 159, 165–170, 195, 228, 251–253, 257, 258, 293, 303, 369, 374, 379, 414

Chebushev formula, 11–13, 78

Clebsch, 271

Combined method, 302, 303, 305, 312

Comparison of methods, 291–294, 355–358, 538

Compressed force
- conservative, 451
- critical, 450–456
- nonconservative, 451

Conjugate beam method, 181, 185, 189, 194, 195

Connecting line, 34, 46, 53, 54, 61, 67, 94, 96

Constraints
- required, 6, 8
- redundant, 6, 12, 211, 212, 214, 216, 218–220, 234, 244, 271, 439, 516
- replacing, 6, 212, 218

Continuous beams
- change of temperature, 228, 238, 251–269
- foci method, 575–576
- influence lines, 326–331
- plastic analysis, 432–441
- settlement of supports, 246–251
- stability equation, 471–483

Coordinates
global, 369–371, 377, 390–392, 395, 398, 401, 405, 409, 412, 415
local, 369–371, 377, 381, 387–392, 395, 397–399, 405, 408–409, 412, 415
Critical load (Load unfavorable position), xxii, 4, 31, 33, 450–456, 458–462, 464, 467, 469, 474–481, 483, 484, 488, 489, 491, 505, 507–509
Critical load (stability)
critical load parameter, 476, 479

D
Deflection computation
beam-columns, 185
Castigliano theorem, 145
conjugate beam method, 181
double integration method, 148, 181, 194
elastic load method, 185–189, 195, 206
graph multiplication method, 176–185, 195, 202
initial parameters method, 147–158, 181, 189, 194
Maxwell–Morh integral, 147, 159–170, 176, 177, 181, 188, 195, 198, 201, 206
statically indeterminate structures, 158, 165, 189, 193
work–energy method, 195
Deflection types
angular, 146, 147, 149, 150, 159, 161, 162, 167, 170, 171, 176, 191, 192, 194, 204
linear, 145–147, 157, 161, 162, 167, 170, 171, 175, 191, 192, 194, 204
mutual angular, 162, 165, 167, 204, 205
mutual linear, 162, 167, 204
Degree of indeterminacy
kinematical, 272–274
mixed, 313–314
static, 222–244
Degrees of freedom
Chebushev formula, 11–13
for dynamical analysis, 518
for stability analysis, 458
for static analysis, 516
Design diagrams of structures
deformable, 498, 499, 510
non-deformable, 491, 499, 502
Dimensions of
ordinates of influence lines, 31
unit displacements, 280
unit reactions, 279–280
Displacement method
canonical equation, 269, 276–290
conception, 275–276
influence lines, 271, 290
kinematical indeterminacy, 272–274
matrix form, 292
primary system, 274–275
primary unknown, 274–275
settlements of support, 296–299
stability analysis, 271
Dummy load method
actual state, 162
unit state, 162

E
Elastic curve, 146–148, 150–159, 181, 183, 185, 193, 194, 196–198, 207, 216, 226, 227, 229, 231, 261–263, 265, 272, 275, 277, 278, 281, 282, 288, 298, 301, 303, 307, 309, 310, 312, 358, 359, 361–363, 367, 372, 451, 461, 463, 472–474, 479, 481, 483, 484, 492, 495, 497, 507, 519, 535, 563
Elastic load method, 147, 164, 185–189, 195, 206, 341, 343, 361
Elastic supports, 154, 159, 198, 260, 262, 268, 275, 309, 453–456, 458, 459, 461–471, 473, 485, 490–491, 504–507, 516, 553, 560
Errors of fabrication, 145, 170–176, 211, 238, 271, 292, 296, 300–302, 426

F
Failure, 40, 425, 426, 430–433, 435, 439, 441–445, 449, 452, 498
Fictitious beam, 186, 188, 189, 194, 341, 342
Finite element method, 370, 407
Flexibility matrix, 521, 528
Foci points, 363, 364, 575–576
Focus, 363, 575, 576
Force method
canonical equation, 217–222
change of temperature, 251–259, 372
concept, 371
conception, 217–219
degree of redundancy, 211–214
errors of fabrication, 238
fixed loads, 393
influence lines, 399
matrix form, 217
primary system, 211–214
primary unknowns, 211–214

resolving equations, 387
settlement of supports, 246–251, 379
static indeterminacy, 222–244
symmetrical unknowns, 333
Fourier method, 541–543
Frames
plastic analysis, 441–445
with sidesway, 226, 473, 481
stability, 471–483
statically determinate, 147, 161, 165, 167, 168, 170, 171, 176, 183, 189, 195, 201, 203
statically indeterminate, 157, 158, 212–214, 224–233, 273, 294, 296, 313, 353, 547
symmetrical, 302, 303, 320, 546, 549
three-hinged, 205, 213
Frequency of free vibration
displacement method, 530–538
force method, 520–530
Krylov–Duncan method, 543–546
separation variable method, 541

G

Generalized coordinates, 452, 458–460, 531
Generalized forces, 166, 167, 170, 171, 175, 195
Graph multiplication method, for beam-columns, 502–504
Gvozdev theorem, 193

H

Helmholtz, 189
Henneberg method, 69
Hinge
simple, 3, 11–13, 41
multiple, 3, 12

I

Indirect load application, 33–36, 46, 52, 88, 94, 96, 103, 364
Influence lines method
analytical construction, 15–26
application, 27–33
connecting line, 34
indirect load application, 33–35
models, 30, 31, 33
Initial parameters method, 147–158, 466–472, 491, 494–498, 504
Interaction scheme, 40–42

K

Kinematical analysis
degrees of freedom, 11–13
geometrically changeable structures, 3
geometrically unchangeable structures, 3, 5–6
infinitesimally changeable structure, 5
infinitesimally rigid structure, 5
redundant constrains, 6
Kinematical method for influence lines, 358–363
Krylov–Duncan functions, 543–550

L

Leites, 271
Leites Load
conservative (nonconservative), 451
critical, xxii, 4, 31, 33, 450–456, 458–462, 464, 467, 469, 471, 474–481, 483, 484, 488, 489, 491, 505, 507–508
dynamical, xxii, 519
fixed (static), xxiii, xxiv, 15, 16, 27–30, 35–38, 49, 67, 70, 83, 92, 97, 111, 159–165, 176, 242, 246, 306, 324, 330, 331, 338, 342, 344, 345, 353, 364, 367, 368, 393, 394, 415, 451, 518
moving, xxii, xxiv, 15, 16, 27–38, 42, 50, 53, 58, 66–68, 70, 72, 76, 87, 100, 323, 324, 326, 339, 344, 345, 358, 361, 363, 364, 379, 400, 401, 403, 514, 518
seismic, xxii, 514
tracking, 451
unfavorable, xxiii, 15, 30–33, 358
Load path (interaction diagram), 41–42
Load unfavorable position, 15, 30–33, 358
Loss of stability (buckling), 451, 452, 458–460, 480, 484, 486, 487, 489

M

Manderla, 271
Manderla Matrices
ancillary, 372–378
deformation, 386–387
flexibility, 521, 528
modal, 522–530, 532–538, 547–549
static, 381–385
stiffness, 387–390
Matrix stiffness method, 369–419
change of temperature, 372
concept, 371

fixed loads, 393
influence lines, 399
resolving equations, 387
settlement of supports, 379
Maxwell theorem, 190–191
Maxwell–Mohr integral, 159–170
Method
displacement, 271–312
dummy load (Maxwell–Mohr integral), 147, 160–162, 195
elastic load, 147, 185–189, 195, 206, 341–343, 358, 361
force, 211–269
graph multiplication, 176–185, 195, 202, 219, 228, 238, 326, 502–504
influence line, 323–368
initial parameters, 147–158, 466–471, 494–498
mixed, 313–321
strain energy, 195
Mixed method, 313–321
Modal matrix, 522–530, 532–538
Mode shape vibration, 523, 524, 526, 529, 530, 532–538
Müller–Breslau's principle, 358, 361, 363
Multispan redundant beams (continuous beams), 40, 42, 155, 222–224, 248, 262, 280, 301, 325–331, 346–353, 355, 358, 359, 361, 362, 364, 371, 372, 374, 375, 378, 382, 383, 386, 393–404, 416, 432–441, 471–483, 508, 564, 575
Multispan statically determinate beams
influence lines, 42–45
structural presentation, 43

N
Nil points method, 94–97, 99, 102, 103, 576
Nonlinear analysis, 429

O
Ostenfeld, 271

P
P-Δ effect, xxii, 491, 494, 496, 498–502
Plastic analysis
direct method, 427–430
kinematical theorem, 430–442, 444
schemes of failure, 425, 426, 430–433, 435, 439, 441–445,
static theorem, 430–436
Plastic hinge, 426, 433–435, 437–439, 441, 442, 444, 447
Prandtl diagram, 425, 431, 432
Primary system
displacement method, 274–275
force method, 211–214
mixed method, 314–315
Primary unknown
displacement method, 274–275
force method, 211–214
mixed method, 313–314
Principle
Müller–Breslau, 358, 361, 363
superposition, 27, 28, 69, 214–217, 220, 225, 226, 243, 245, 250, 257, 259, 276, 290, 299, 305, 435, 437, 494, 496
virtual displacements, 171, 173–175, 452

R
Rayleigh theorem, 192–193, 246
Reciprocal theorems
displacements, 190–191
displacements and reactions, 193
reactions, 192–193
works, 189–192
Redundancy degree, 211–214, 224, 227–234, 247, 252, 325, 339, 444

S
Sag, 112, 114–118, 122–125, 128–130, 135–143
Separation variable method, 541
Settlement of supports, 40, 73, 104, 145–147, 157, 158, 170–176, 211, 228, 246–251, 292, 296–299, 303, 307, 369, 375, 379, 396, 397, 414, 426, 574, 581
Settlements of supports, 40, 171–174, 228, 246–251, 268, 291, 296–299, 303, 307, 308, 372, 375, 395, 396, 413, 426, 448, 573
Simpson rule, 178–181, 227, 245
Stability methods
double integration, 461–465, 491–494, 505, 506
energy, 452–460, 505
initial parameter, 461, 466–472, 491, 494–502, 504–507, 543
static, 451–459, 466, 467
Statically determinate
arches, 97–103
beams, 26, 39–46, 147, 203, 212, 216, 435, 516, 546
frames, 146, 147, 161, 167, 168, 170, 171, 176, 182–185, 195, 201, 203–205
trusses, 47–49, 56

Statically indeterminate
arches, 237–243, 293, 333, 339, 484
beams, 151, 215, 222–224, 280, 294, 331, 358, 369, 393, 435, 436, 516, 546
displacement computation, 146, 147, 159, 164, 170, 171, 175, 176, 179–181, 185, 189, 193, 195, 236–238, 240, 241, 243, 361, 502
frames, 157, 158, 212–214, 224–233, 273, 280, 294, 296, 302, 313, 353, 547
trusses, 72, 233–237, 257, 258, 325, 339–343, 358, 410–412, 418
Stiffness matrix
global coordinate, 369–371, 390–392, 395, 398, 401, 405, 409, 412, 415
local coordinate, 369–371, 381, 387–390, 395, 398, 399, 404, 409, 412, 414, 415
Strain energy method, 189, 190, 195, 454, 456, 457
Structure
geometrically changeable, 3, 5, 8–13, 47, 59, 69, 212, 213, 273, 382, 414, 515
geometrically unchangeable, 3, 5–6, 9–13, 39, 41–43, 56, 58, 62, 65, 69, 77, 78, 97, 211, 212, 233, 272, 273, 515
instantaneously changeable, 5–11, 435, 437
instantaneously rigid, 5
statically determinate, ix, 15, 28, 36, 39, 45, 62, 97, 165, 170, 211, 216, 220, 244, 325, 339, 413, 425, 429, 517
statically indeterminate, ix, 16, 28, 36, 165, 189, 191–193, 211, 212, 214, 216, 217, 219, 222–244, 246, 251, 257, 271, 272, 275, 323–325, 344, 345, 355, 358, 364, 382, 384, 413, 414, 427, 430, 432, 448, 516, 517
Substitute bar method, 69–71
Superposition principle, xxiii, 27, 28, 69, 214–217, 220, 225, 226, 243–245, 250, 257, 259, 276, 290, 299, 305, 435, 437, 494, 496
Symmetrical frames, 302, 303, 312, 546, 549

T
Temperature changes, 40, 73, 145, 159, 165–170, 195, 211, 219, 228, 251–269, 271, 293, 303, 369, 372, 374, 375, 414, 427
Thrust, 62–64, 66–68, 72, 76, 78, 79, 81, 82, 84, 86–88, 90–93, 97–106, 109–119, 122–126, 128–130, 133–143, 175, 176, 242, 335, 338
Trapezoid rule, 178, 179, 182, 503
Trusses analysis
classical methods, 68
elastic load method, 31
force method, 233–237
influence lines, 50–53
Trusses general
fundamental conception, 72
kinematical analysis, 56–57
statically indeterminacy, 234, 235
Trusses generation
complex, 47, 49
compound, 47–49
simple, 47
with subdivided panels, 47, 54–61
Trusses specified types
Baltimore truss, 57–61
K-truss, 75
Pratt truss, 54, 57
three-hinged, 61–64, 68
Warren truss, 57–61
Wichert truss, 49, 68–70, 72

U
Unit displacements
computations, 237, 238, 240–242
dimensions, 160, 171, 280, 386
Unit reactions
computations, 277–285, 288, 294, 296, 298, 300, 345, 353, 500, 534, 536
dimensions, 280
Unit state
displacement method, 278, 279, 295, 301
force method, 224, 225, 227, 228, 240, 241, 243–245, 253–256
mixed method, 316–318
Unknown
antisymmetrical, 303, 305, 335, 336
displacement method, 280, 390, 391
force method, 211–214
mixed method, 313–314
symmetrical, 303, 305, 333, 336

V
Vereshchagin's rule
Simpson rule, 178–181, 241, 245
Trapezoid rule, 178, 179, 182
Vibration methods
displacement method, 520–521, 530–538
Krylov–Duncan, 543–550
separation variable, 541

W
Winkler, 271
Winkler Work
external forces, 452
internal forces, 162, 163, 169, 248

CPI Antony Rowe
Chippenham, UK
2018-10-18 10:42

Advanced Methods of Structural Analysis